Plant Cell Walls

FROM CHEMISTRY TO BIOLOGY

Plant Cell Walls
FROM CHEMISTRY TO BIOLOGY

PETER ALBERSHEIM

ALAN DARVILL

KEITH ROBERTS

RON SEDEROFF

ANDREW STAEHELIN

Garland Science

Garland Science
Vice President: Denise Schanck
Project Editor: Sigrid Masson
Illustration: Nigel Orme
Production Editor, Design, and Layout: EJ Publishing Services
Deputy Production Manager: Simon Hill
Copyeditor: Richard K. Mickey
Indexer: Liza Furnival
Permissions Coordinator: Mary Dispenza

Front Cover
Fluorescence micrograph of water-storage cells in mesophyll of *Sansevieria deserti* leaf, free-hand section in calcofluor white. (Courtesy of Alan L. Koller, PhD, JD)

Back Cover
Photograph courtesy of Stefan Eberhard.

Peter Albersheim is Emeritus Director of the Complex Carbohydrate Research Center at the University of Georgia. He received his PhD from the California Institute of Technology. He and Alan Darvill founded the CCRC in September 1985.
Alan Darvill is Director of the CCRC at the University of Georgia, Director of the Department of Energy (DOE)-funded Center for Plant and Microbial Complex Carbohydrates, and is UGA Lead in the DOE-funded BioEnergy Science Center. He received his PhD from the University College of Wales. **Keith Roberts** is Emeritus Fellow at the John Innes Centre, Norwich. He received his PhD from the University of Cambridge. **Ron Sederoff** is Professor of Forestry and Co-Director of the Forest Biotechnology Group at North Carolina State University. He received his PhD from the University of California, Los Angeles. **Andrew Staehelin** is Emeritus Professor at the University of Colorado at Boulder. He received his PhD from the Swiss Federal Institute of Technology.

Library of Congress Cataloging-in-Publication Data
Plant cell walls / by Peter Albersheim ... [et al.].
    p. cm.
  Includes index.
  ISBN 978-0-8153-1996-2
  1. Plant cell walls. 2. Plant anatomy. 3. Plant molecular biology. I. Albersheim, Peter.
  QK725.P4725 2010
  571.6'82--dc22

                    2010004911

Published by Garland Science, Taylor & Francis Group, LLC, an informa business, 270 Madison Avenue, New York NY 10016, USA, and 2 Park Square, Milton Park, Abingdon, OX14 4RN, UK.

Printed in the United States of America
15 14 13 12 11 10 9 8 7 6 5 4 3 2 1

Taylor & Francis Group, an informa business

Visit our website at http://www.garlandscience.com

Dedication

This book is dedicated, with great affection, to our colleague
Joseph E. Varner
October 7, 1921 — July 4, 1995

Preface

It is impossible to imagine a world without plants. They are the dominant species in most of our ecosystems, and the adaptation of plants to the land depended on the evolution and diversity of their cell walls. All plants evolved from a leafless, rootless, freshwater alga about 400 million years ago. To survive on the land over this time, plants were able to transport water, often to great heights, resist strong mechanical forces, withstand transient changes in drought, heat, cold, attacks by pests and pathogens, and adapt to long-term climate change.

Our human existence and our survival on this Earth are absolutely dependent on plants and their cell walls. With their unique ability to capture carbon, plants are at the core of our human food chain. They gave rise to much of our fossil fuels and their cell walls provide our wood, pulp and paper, our cotton and linen, gums and thickeners, and some chemical feedstocks. Crucial aspects of the plant's own biology are also underpinned by their cell walls, from their growth and development, breeding potential and agricultural utility to their responses to disease and stress. Cell walls affect almost all areas of plant biology and it is hard to underestimate their importance. Two factors make it particularly timely to survey this important area of science in its broader context.

First is the leap in our understanding of cell wall biology in the last couple of decades. New imaging techniques, advances in carbohydrate chemistry, plant genome sequencing and the power of molecular genetics, have all combined to enrich our appreciation of the beauty and complexity of walls and their functions. Second is our increasing awareness of the importance of cell walls, as a renewable resource and as carbon sinks in the global carbon cycle. With the potential for conversion of plant cell walls to liquid biofuels, interest has turned to the utility of plant walls as readily renewable lignocellulosic biomass, in contrast to our non-renewable sources of energy, coal and petroleum.

The book you hold is designed to provide a cell wall-oriented view of much of contemporary plant biology. It covers in some detail, plant anatomy, plant biochemistry and metabolism, plant cell biology, cell wall biosynthesis, plant growth and development, plant-microbe interactions and our use of walls as a renewable material resource. Much of basic plant science is inextricably wrapped up with the science of cell walls; that curious extracellular matrix that co-evolved with the plant's sessile and photosynthetic lifestyle. This complex structure lies at the heart of modern molecular plant biology.

Our book is aimed generally at university students, including both upper-level undergraduates, and graduate students, who want to understand plants better, and in their broader context. It is also a sufficiently up-to-date and detailed account to serve as a professional-level reference book.

The book is written and organized in a "concept" format, using short, didactically helpful concept heads; a device we feel helps both the reader's perspective and understanding of the material. The book's internal cross referencing system is designed so that the reader is directed to the relevant concept, as follows: Concept 6B11 refers to Chapter 6, section B, heading 11. The text is visually enriched with an integrated illustration program of over 400 figures, panels and micrographs. Additionally, the images and tables from the book are available in JPEG and can be accessed at www.garlandscience.com/pcw. At the end of each chapter are sets of key terms to help review learning, and a set of literature references for each concept heading, which provide an entry to the relevant research literature. Panels throughout the book bring together important reference material that is useful beyond the confines of its immediate context. A good example is Panel 6.2, *Plant cell wall architecture: models*. This panel condenses a lot of structural data about which molecules are where within cell walls, and presents a series of provisional structural models that help visualize cell wall architecture. It is reprinted as a frontispiece, for easy reference, because it is relevant to all chapters in the book.

This book has taken a long time to complete. When we started to write the book, many factors, such as biomass for energy, had much less significance, but we are in fact pleased that the book appears now, when the science is so much better understood and the environmental and economic relevance is that much higher. The book had its genesis in several meetings, convened initially by Peter Albersheim and with encouragement and financial support provided by Bob Rabson of the Department of Energy, where we argued fervently about what should and should not be in it, concept by concept. These long and energetic meetings were held in the wonderful remoteness of the English Manor Inns in Clayton, Georgia, close to the place chosen to film *Deliverance*! At that point Joe Varner, an inspirational plant biochemist, was an active, engaged and vocal team member, up until, indeed, his sad death in 1995. The book was reenergized when Ron Sederoff joined the team, and more regular writing sessions were held in the more salubrious facilities at the CCRC labs in Athens, Georgia. The final push to update and finish came partly through the increasing pace of remarkable discoveries in the field, all demanding space in the book, but also through the patient, tireless, energy and enthusiasm of our editor at Garland Science, Denise Schanck, without whom we would never have made it.

Denise was not alone in support of our endeavor, and we have been helped by a large number of talented individuals along the way. They include Rosemary Nuri and Karen Howard at the CCRC and Sigrid Masson, Emma Jeffcock and, in the early stages, Miranda Robertson, at Garland Science and our colleagues who generously commented on chapters. Nigel Orme transformed our labored hand-drawn figures into graphically pleasing output. And our friends, colleagues, lab members and families provided that essential, loyal but critical bedrock without which books like this would never emerge.

The whole area of plant science is in a state of flux as genomes become sequenced, bioinformatic and genetic tools become more sophisticated, structural and chemical methods advance and commercial interest in cell walls and their uses becomes more obvious. While celebrating the insights we now have into the cell walls of plants, we also want to emphasize that all is certainly not known. Huge areas of real ignorance remain, yet the potential impact that cell wall research could have for enhanced crop protection, agricultural yield, biomass use, and renewable chemical feedstock has never been more obvious. For all those new cell wall researchers, who will solve many of these problems, this book is written.

Peter Albersheim, Alan Darvill, Keith Roberts, Ron Sederoff, Andrew Staehelin

March 2010

Acknowledgments

While writing this book we benefited not only from the constructive input of our immediate lab colleagues, but also from a generous set of expert reviewers who read one or more chapters and provided much valued advice and criticism. We hope we have responded adequately to these comments; certainly the text is the better for their input:

Tony Bacic (The University of Melbourne), Dianna Bowles (University of York), Nicholas Carpita (Purdue University), Pamela Diggle (University of Colorado at Boulder), Brian Gunning (Australian National University), Michael Hahn (University of Georgia), Martha Hawes (University of Arizona), Michael Jarvis (University of Glasgow), Jonathan Jones (John Innes Centre), Serge Kauffmann (University of Strasbourg), John Labavitch (University of California, Davis), Simcha Lev-Yadun (University of Haifa-Oranim), Clive Lloyd (John Innes Centre), Debra Mohnen (University of Georgia), Jocelyn Rose (Cornell University), Bruce Stone (La Trobe University, Australia), Chris Somerville (Stanford University), Bjorn Sundberg (The Swedish University of Agricultural Sciences), Jonathan Walton (Michigan State University).

Composition and Architecture of Primary Plant Cell Walls

THE PRINCIPLE OF THE MODEL

Although they display considerable heterogeneity, all cell walls are built using the same structural or architectural principles. They are fiber-reinforced composite materials. The cellulose-hemicellulose network resists tension, while the coextensive, intermeshed pectin network resists compression and shearing forces. The middle lamella is an adhesive layer sticking adjacent cells together.

In this schematic diagram the cellulose microfibrils are shown aligned at right angles to the direction of cell elongation (*double arrow*).

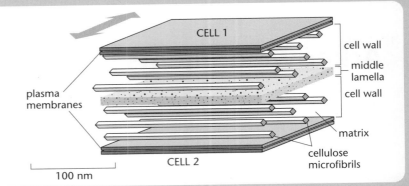

THE PRIMARY CELL WALL

The cellulose-hemicellulose network

This is a schematic section through a wall similar to that shown above. Each lamella of cellulose microfibrils is represented by a single microfibril, and all are oriented in the same direction. They are interconnected within each wall by cross-linking hemicellulose molecules that are hydrogen-bonded to their surfaces. The resultant network is resistant to tensions in the wall. Each lamella can be oriented differently with respect to its neighbors depending on the developmental stage and fate of the cell. In each wall shown here, there are only three lamellae, whose total thickness is about the minimum needed to maintain a primary cell wall. In this diagram all the various components are drawn approximately to scale.

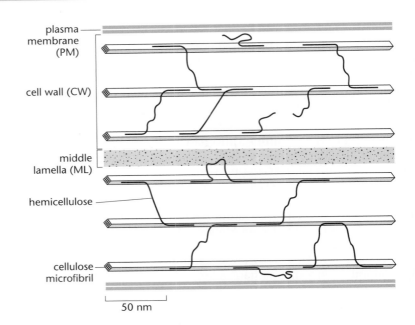

The pectic polysaccharide network

The pectic polysaccharides assemble into a coextensive network with the cellulose-hemicellulose network, intermeshing with it to create a water-retentive matrix that resists the forces of compression and shear and controls wall permeability. The pectin is probably anchored to the plasma membrane in some way, through such molecules as Wak proteins. At the middle lamella they form an adhesive mesh involving calcium cross-bridges between homogalacturonans. Further cross-links include the borate diester bridges between RG-II molecules. It is entirely possible that the pectin molecules are far more ordered than is schematically shown here.

THE PRIMARY CELL WALL: STRUCTURAL WALL PROTEINS

Arabinogalactan protein (AGP) distribution

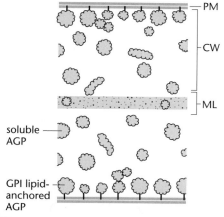

soluble AGP

GPI lipid-anchored AGP

PM
CW
ML

There is an enormous heterogeneity in AGPs, both in kind and in which cells they are expressed. They form a hydrated interface between plasma membrane and cell wall.

Extensin 3 networks (cell plate/early cell wall)

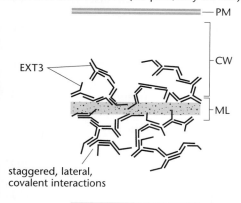

EXT3

staggered, lateral, covalent interactions

PM
CW
ML

Extensin is a hydroxyproline-rich glycoprotein (HRGP).

Wound-extensin network

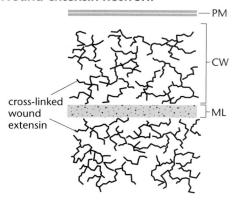

cross-linked wound extensin

PM
CW
ML

Extensins are not found in all cell types, but can be induced in walls by wounding or stress to provide mechanical protection.

THE OUTER EPIDERMAL WALL

All walls are different, some subtly but others more dramatically. While respecting the same overall architectural principles, they vary in the proportions and nature of the individual components to produce considerable heterogeneity such as the lignified walls of xylem vessels. One example of wall specialization is shown here, the primary wall of the outer shoot epidermis, in which this wall, unlike the basal and side walls, is thickened and encrusted with cutin and waxes.

The cellulose-hemicellulose distribution

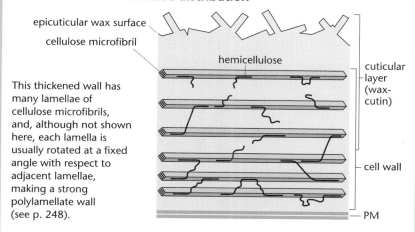

epicuticular wax surface
cellulose microfibril
hemicellulose
cuticular layer (wax-cutin)
cell wall
PM

This thickened wall has many lamellae of cellulose microfibrils, and, although not shown here, each lamella is usually rotated at a fixed angle with respect to adjacent lamellae, making a strong polylamellate wall (see p. 248).

The distribution of AGPs and pectic polysaccharides

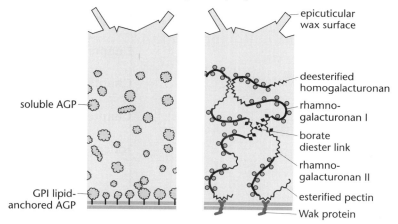

soluble AGP

GPI lipid-anchored AGP

epicuticular wax surface
deesterified homogalacturonan
rhamno-galacturonan I
borate diester link
rhamno-galacturonan II
esterified pectin
Wak protein

The deposition of cutin and waxes

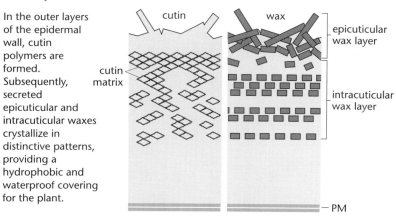

cutin
wax
epicuticular wax layer
cutin matrix
intracuticular wax layer
PM

In the outer layers of the epidermal wall, cutin polymers are formed. Subsequently, secreted epicuticular and intracuticular waxes crystallize in distinctive patterns, providing a hydrophobic and waterproof covering for the plant.

Contents

Chapter 1   Cell Walls and Plant Anatomy   1

Chapter 2   The Structural Polysaccharides of the Cell Wall
and How They Are Studied   43

Chapter 3   Biochemistry of the Cell Wall Molecules   67

Chapter 4   Membrane Systems Involved in Cell Wall Assembly   119

Chapter 5   Biosynthesis of Cell Wall Polymers   161

Chapter 6   Principles of Cell Wall Architecture and Assembly   227

Chapter 7   The Cell Wall in Growth and Development   273

Chapter 8   Cell Walls and Plant–Microbe Interactions   319

Chapter 9   Plant Cell Walls: A Renewable Material Resource   365

Index   411

Detailed Contents

Chapter 1 Cell Walls and Plant Anatomy   1

A. The Derivation of Cells and Their Walls   1
1. Cells arise in specialized regions of the plant called meristems.   1
2. Walls originate in dividing cells.   3
3. Plant organ development depends on precise control of the plane of cell division and of cell expansion.   4
4. Cytoskeletal elements predict the position of the new cross-wall before the cell divides.   8
5. Actin filaments help to position new cross-walls.   9
6. The new cross-wall must join and fuse with the mother cell wall.   10
7. A plant is constructed from two compartments: the apoplast and the symplast.   10

B. Walls in Cell Growth and Differentiation   12
1. Cells become organized at an early stage into three major tissue systems.   12
2. When they have stopped dividing, cells usually continue to grow in size.   16
3. Depending on its position in the plant, a cell differentiates into a specific cell type.   17
4. A distinction is often made between primary and secondary walls.   18
5. Differentiated cell types are often characterized by functionally specialized cell walls.   20

C. Plant Cell Types and Their Walls   21
1. Shoot epidermal cells produce a waterproof cuticle that helps provide protection.   21
2. Stomatal guard cells have asymmetrically thickened walls that allow them to reversibly change shape.   22
3. Cells in ground tissues fill the space between epidermis and vascular tissue.   24
4. The endodermis regulates the flow of solutes through the apoplast of the root.   25
5. The pericycle encircles the vascular tissue and can give rise to lateral roots.   26
6. Xylem vessel elements are pipes that conduct water at negative pressure.   27
7. Phloem sieve tube elements are high-pressure tubes for conducting sugar solutions.   29
8. Transfer cells load solutes into and out of conducting elements.   31
9. Plasmodesmata allow adjacent cells to communicate with each other across their intervening walls.   32
10. Collenchyma and sclerenchyma provide support in young tissues.   34
11. The shaping of mesophyll cells generates large air spaces that facilitate gaseous diffusion.   34

12. Pollen grains have a unique and sculptured protective wall.   37
13. Some plant cells have no walls.   37
Key Terms   39
References   39

Chapter 2 The Structural Polysaccharides of the Cell Wall and How They Are Studied   43

A. Molecules of the Wall   43
1. Primary walls isolated in a variety of ways contain the same structural polymers.   43
2. Cellulose is a β-1,4-linked glucan with a disaccharide repeating unit.   44
3. Pectins and hemicelluloses are the noncellulosic polysaccharides of primary walls.   45
4. Wall polysaccharides are composed of 13 different sugars.   45
5. Homogalacturonan, rhamnogalacturonan I, and rhamnogalacturonan II are three pectic polysaccharides in primary walls.   45
6. Xyloglucans and arabinoxylans are the two major hemicellulosic polysaccharides in primary walls.   47
7. The structural characterization of polysaccharides is difficult.   47
8. Cell walls contain proteins, glycoproteins, and proteoglycans that perform enzymatic, structural, defensive, and other functions.   48
9. Primary cell walls contain an aqueous phase.   49
10. Secondary walls are composed of primary walls plus additional layers of polymers.   49

B. Methods for Characterizing the Structural Polysaccharides of the Cell Wall   52
1. Obtaining pure polysaccharides is a prerequisite for primary structure determination.   52
2. Extracellular polysaccharides secreted by cultured cells are excellent models for wall polysaccharides.   53
3. Enzymatic and chemical cleavage of polysaccharides has made structural studies possible.   53
4. Several methods are used for glycosyl composition analysis.   54
5. Glycosyl linkage compositions are determined by methylation analysis.   57
6. Glycosyl sequencing of oligo- and polysaccharides is not routine.   57
7. Nuclear magnetic resonance (NMR) spectroscopy and mass spectrometry (MS) are important tools for determining the structures of polysaccharides.   58
8. The conformations of cell wall oligo- and polysaccharides can be investigated using NMR and computational methods.   62

9. The structure of cellulose has been determined by X-ray diffraction techniques. 63
Key Terms 64
References 64

Chapter 3 Biochemistry of the Cell Wall Molecules 67

A. Cellulose 67
1. β-Glucan chains hydrogen-bond together to form cellulose microfibrils. 67
2. The glucan chains of cellulose microfibrils all have the same orientation. 67

B. Hemicelluloses 68
1. Xyloglucan is a principal hemicellulose of cell walls. 68
2. The structural features of xyloglucan have been defined by analysis of endoglucanase-released oligosaccharides. 69
3. Side chain substitutions on the backbone of xyloglucan affect endoglucanase cleavage of the backbone. 71
4. Arabinoxylan is the predominant hemicellulose of grasses. 72
5. The elucidation of arabinoxylan structures has been facilitated by characterizing oligosaccharides generated by endoxylanases. 73
6. Secondary walls contain xylans and glucomannans. 74
7. A galacturonic acid–containing oligosaccharide is found at the reducing end of glucuronoxylans. 74

C. Pectic Polysaccharides 75
1. Homogalacturonans are gel-forming polysaccharides. 75
2. Rhamnogalacturonan I (RG-I) is a family of large polysaccharides with a backbone of repeating disaccharide units. 75
3. Rhamnogalacturonan I (RG-I) has many arabinosyl and galactosyl residue–containing side chains. 77
4. Rhamnogalacturonan II (RG-II) is the most complex polysaccharide known. 78
5. Rhamnogalacturonan II (RG-II) has an oligogalacturonide backbone. 80
6. Rhamnogalacturonan II (RG-II) has side chains with unique structures. 80

D. Other Wall Polysaccharides 81
1. β-1,3/1,4-(mixed-linked) glucans are found in grasses and horsetail. 81
2. Enzyme-generated oligosaccharides have been used to structurally characterize mixed-linked glucans. 82
3. Callose is a β-1,3-D-glucan. 82
4. Substituted galacturonans are found in some cell walls. 83
5. Galactoglucomannans are found in Solanaceae cell walls and in the growth medium of Solanaceae cultured cells. 83
6. Some seed cell walls contain storage polysaccharides. 84
7. Gums are secreted in response to stress and are not normally considered to be wall components. 84

E. Proteins and Glycoproteins 85
1. Plant cell walls contain structural, enzymatic, lipid transfer, signaling, and defense proteins. 85
2. The repeated sequence motifs of structural proteins define their classes and their physical properties. 86
3. Sequence motifs define proline hydroxylation, glycosylation, and intra- and intermolecular cross-linking sites. 87
4. HRGP and PRP expression is controlled by developmental programs and triggered by external stimuli. 88
5. Extensin-type HRGPs are rod-shaped molecules that become insolubilized by a peroxide-mediated cross-linking process. 89
6. Arabinogalactan proteins form a family of complex proteoglycans located at the cell-surface, and in the cell wall and intercellular space. 90
7. Classical arabinogalactan proteins are GPI-anchored proteoglycans. 92
8. GRPs are structural cell wall proteins that participate in the assembly of vascular bundles and are present in root cap mucilage. 94

9. AGPs, GRPs, HRGPs, and PRPs may have a common evolutionary origin. 95

F. Lignin 95
1. Lignin is a hydrophobic polymer of secondary cell walls. 95
2. Our knowledge of lignin composition is based on the chemical hydrolysis of wood. 96
3. Lignins in plant tissues, cells, and cell walls may be detected by histochemical staining or immunochemical localization. 99
4. The complexity of lignin is increased by the diversity of intermolecular linkages. 100
5. Models for higher-order structure of lignin. 102
6. Lignin is cross-linked to polysaccharide components in lignin-carbohydrate complexes. 103
7. Lignin composition and content vary greatly among the major groups of higher plants. 104
8. Lignin content and composition vary in wood formed during increased mechanical stress. 105

G. Suberin, Cutin, Waxes, and Silica 106
1. Suberin is a complex hydrophobic material that forms physical and biological barriers essential for plants. 106
2. Cutin is an aliphatic polyester that is part of the plant cuticle. 110
3. Epicuticular waxes form the primary interface between the outside of the plant and the aerial environment. 111
4. Silica deposits are formed on the surface of epidermal cells as well as within the cell walls of internal tissues of many herbaceous and woody plants. 111
Key Terms 113
References 113

Chapter 4 Membrane Systems Involved in Cell Wall Assembly 119

A. Sites of Cell Wall Polymer Assembly in Interphase Cells 120
1. Different polymers are synthesized at different locations, in the ER and the Golgi apparatus, at the plasma membrane, or in the cell wall. 120
2. The endoplasmic reticulum consists of a three-dimensional membrane network that extends throughout the cytoplasm and is differentiated into specialized domains. 121
3. Golgi stacks consist of sets of flattened cisternae that are structurally distinct and exhibit a polar architecture. 122
4. Plant Golgi stacks are dispersed throughout the cytoplasm, travel along actin filament tracks with myosin motors, and stop at ER export sites. 124
5. COPII vesicles bud from the ER with an external scaffold that is transferred with the vesicles to the cis side of the Golgi stacks, giving rise to the Golgi scaffold (matrix). 125
6. Intra-Golgi transport of membrane and cargo molecules is best explained by the cisternal progression/maturation model. 127
7. The trans Golgi network cisternae of plants are transient organelles that sort and package Golgi products into secretory and clathrin-coated vesicles. 128
8. Maintenance of an appropriate pH in different Golgi compartments is essential for their functions. 131
9. Golgi stacks are structurally and functionally differentiated in a tissue- and developmental stage–specific manner. 132
10. The process of exocytosis delivers matrix molecules to the cell wall and membrane molecules to the plasma membrane. 133
11. Turgor pressure has profound effects on vesicle-mediated secretion, membrane recycling from the plasma membrane, and the movement of secreted molecules through cell walls. 134
12. Cellulose microfibrils, callose, and mixed-linked glucans are made by enzyme complexes in the plasma membrane. 136
13. The enzymes that synthesize cellulose microfibrils in higher plants are organized into complexes that visually look like rosettes. 137

14. Cellulose synthase complexes are pushed forward in the plasma membrane by the growing cellulose microfibrils that they extrude into the cell wall.   137
15. Rosette complexes are concentrated in domains of the plasma membrane underlying sites of rapid cellulose synthesis.   139
16. The half-life of rosette complexes determines the length of cellulose microfibrils.   141
17. The cellulose microfibrils of primary cell walls are shorter than those of secondary cell walls.   142
18. Callose appears transiently during growth and development and is also formed in response to biotic and abiotic stresses.   142
19. Callose is a transient cell wall component that is critically important for many growth and developmental processes.   143
20. Stress-induced callose synthesis serves to protect cells from potentially lethal biotic and abiotic insults.   144

B. **Membrane Systems Involved in *De Novo* Cell Wall Assembly during Cytokinesis**   **145**
1. Plant Golgi stacks multiply by division.   145
2. During mitosis and cytokinesis cytoplasmic streaming stops and the Golgi stacks redistribute to specific locations.   145
3. Somatic-type cell plate formation can be divided into four distinct phases.   147
4. Dynamin-GTPases create dumbbell-shaped, cell plate–forming vesicles containing dehydrated and possibly gelled cell wall–forming molecules.   150
5. The association of cisternae of the endoplasmic reticulum with forming cell plates increases over time, but the functional importance of this spatial relationship has yet to be fully explored.   152
6. Formation of cell walls in the syncytial endosperm and in meiocytes involves a special kind of cell plate, the syncytial-type cell plate.   153
Key Terms   157
References   157

Chapter 5 Biosynthesis of Cell Wall Polymers   **161**
A. **General Mechanisms of Polymer Assembly**   **161**
1. Polymers are assembled from building blocks.   161
2. Nucleotide sugars are the source of the glycosyl residue building blocks used in the synthesis of cell wall polysaccharides and glycoproteins.   162
3. Transport of nucleotide sugars into ER and Golgi cisternae is mediated by specific NDP-sugar/NMP antiporters.   166
4. Wall polysaccharides are made by membrane-bound polysaccharide synthases and glycosyltransferases.   167
5. Glycosyltransferases are generally nonabundant proteins.   167
6. All glycosidases and glycosyltransferases involved in the modification of glycoproteins and polysaccharides have a common topology and common Golgi retention mechanisms.   168
7. The synthesis of polysaccharide backbones involves initiation, elongation, and termination reactions.   168
8. Polysaccharide synthases can be identified biochemically by substrate and activator binding activities, by product entrapment techniques, and by catalytic activities detected in nondenaturing gels.   172
9. The three-dimensional structure of the backbone of cell wall polysaccharides suggests synthesis by glycosyltransferases with two active sites.   172
10. Polysaccharide synthesis is carefully controlled, but the control points are still poorly understood.   175

B. **Assembly and Processing of Glycoproteins and Proteoglycans**   **175**
1. *N*-linked glycans of glycoproteins are assembled on dolichol lipids.   175
2. Newly synthesized *N*-linked glycans are transferred *en bloc* to nascent polypeptides and immediately subjected to processing by two glucosidases.   177

3. Processing of *N*-linked glycans in the plant Golgi apparatus is similar but not identical to the processing in other organisms.   178
4. The sites of *O*-linked glycosylation are defined in part by rules of proline hydroxylation.   179
5. The GPI anchors of arabinogalactan proteins are both synthesized and attached to the protein backbone in the ER.   181

C. **Assembly of Polysaccharides in the Endomembrane System**   **182**
1. The backbones of some complex polysaccharides are synthesized by Golgi-located enzymes encoded by genes of the cellulose synthase-like (CSL) gene family.   182
2. Synthesis of the glucan backbone of xyloglucan in *trans* Golgi cisternae is enhanced by the cooperative assembly of glucosyl and xylosyl residues.   184
3. The XG-fucosyl and XG-galactosyltransferases can add fucosyl and galactosyl residues during or after synthesis of the XG backbone.   185
4. The assembly of xyloglucan occurs in *trans* Golgi and early *trans* Golgi network cisternae.   186
5. The large number of enzymes needed to make pectic polysaccharides makes understanding their synthesis a challenge.   186
6. Synthesis of pectic polysaccharides involves enzymes localized to late *cis*, medial, and *trans* Golgi cisternae.   189
7. Only a few steps of the pectic polysaccharide synthesis pathway have been studied biochemically *in vitro*.   189
8. The biosynthesis of galactomannans has parallels to the biosynthesis of xyloglucans.   191
9. Several glycosyltransferases involved in xylan biosynthesis have been identified, but the mechanism of xylan biosynthesis remains obscure.   192

D. **Assembly of Polysaccharides at the Plasma Membrane**   **193**
1. Studies of cellulose synthesis by *Gluconacetobacter xylinus* (*Acetobacter xylinum*) provided many of the paradigms for similar studies in higher plants.   193
2. Hydrophobic cluster analysis played a critical role in the identification of the plant *CESA* genes.   193
3. Plant *CESA* genes appear to be derived from cyanobacterial precursors.   194
4. The cellulose synthase (CESA) proteins appear to correspond to the catalytic subunits of cellulose synthase complexes.   195
5. The CESA protein is an integral protein with several domains characteristic of processive glycosyltransferases.   196
6. The Zn-binding domains of CESA and some CSLD proteins appear to mediate dimerization of the catalytic subunits of cellulose synthases.   197
7. The cellulose-synthesizing rosette complexes are composed of three different types of CESA proteins.   198
8. The involvement of KORRIGAN, a β-1,4-glucanase, in cellulose synthesis has yet to be proven conclusively.   198
9. UDP-glucose, the substrate for the cellulose and callose synthase systems, may be produced by a membrane-bound form of sucrose synthase.   199
10. There are two types of callose synthase systems, a Ca^{2+}-dependent and a Ca^{2+}-independent type.   200
11. Callose synthase is a large protein that differs in many ways from cellulose synthase.   201
12. The ratio of tri- and tetrasaccharide units in mixed-linked glucans made *in vitro* can be altered experimentally.   202

E. **Polymer Assembly in the Wall: Biosynthesis of Lignins, Waxes, Cutins, and Suberins**   **203**
1. The secondary cell wall provides a unique environment for the polymerization of lignins from cinnamyl alcohols.   203
2. Cinnamyl alcohols are the predominant precursors for lignin biosynthesis.   203
3. Monolignols are formed by successive enzymatic modifications of phenylalanine.   206

4. Hydroxylation and *O*-methylation at the 3 and 5 positions on the aromatic ring are unlikely to take place at the level of cinnamic acids.   207

5. Coniferaldehyde is the branch point for the biosynthesis of coniferyl alcohol and sinapyl alcohol.   208

6. Monolignols are glucosylated, stored, transported, and deglycosylated before polymerization.   208

7. Lignin polymers are formed through enzymatic oxidation of monolignols.   210

8. Lignin polymerization is primarily due to the addition of a monolignol to a lignin polymer.   212

9. Oxidative carriers or radical mediators may be involved in the polymerization process.   213

10. A nonenzymatic model for lignin polymerization has been proposed involving "dirigent" or guide proteins.   214

11. Nontraditional monomers are readily incorporated into lignin, indicating a high level of metabolic plasticity.   215

12. The deposition of lignin is both temporally and spatially controlled.   216

13. The composition of lignin differs between cell wall domains and cell types.   217

14. The biosynthesis of cutins, suberins, and waxes involves enzymes in chloroplasts, the endoplasmic reticulum, and cell walls.   218

Key Terms   221

References   222

Chapter 6 Principles of Cell Wall Architecture and Assembly   227

A. Cross-Links between Wall Polymers   227

1. Wall polymers are cross-linked by covalent bonds as well as by noncovalent bonds and interactions.   227

2. Hemicelluloses bind strongly to cellulose microfibrils by noncovalent bonds.   228

3. Hemicelluloses cross-link cellulose microfibrils.   229

4. Pectic polysaccharides are probably covalently inter-connected by glycosidic bonds.   229

5. Homogalacturonans and partially methylesterified homogalacturonans form gels.   230

6. Borate esters cross-link RG-II dimers in the wall.   231

7. Diferulic acids probably cross-link wall polysaccharides.   232

8. Transesterification may produce other possible cross-links.   233

9. Some wall proteins become insolubilized by covalent cross-links.   233

10. Lignin is covalently attached to hemicelluloses, and their interaction adds strength to the secondary cell wall.   234

B. Architectural Principles: Putting the Polymers Together   235

1. Two coextensive polysaccharide networks underlie the structure of the primary cell wall.   235

2. Walls are constructed of lamellae that are one cellulose microfibril thick.   236

3. The primary wall is usually composed of only a small number of lamellae.   236

4. The spacing of cellulose microfibrils is determined by matrix polysaccharides.   237

5. The walls of cells in which cellulose synthesis has been inhibited are composed largely of a pectin network.   240

6. The pectin network limits the porosity of the wall.   241

7. The cell controls the thickness of its wall.   242

8. How proteins are integrated into wall architecture is unclear.   243

9. The orientation of newly synthesized microfibrils is determined by the cell.   244

10. The orientation of microfibrils within a wall may change during growth.   246

11. The self-ordering properties of polylamellate walls can lead to high degrees of structural order.   248

12. Cell wall assembly is a hierarchical process.   249

C. Architectural Variations: The Mosaic Wall   250

1. Polymers are not evenly distributed within a wall.   250

2. The middle lamella forms an adhesive boundary between adjacent cells.   252

3. Three-way junctions are rich in protein, pectin, and phenolics.   256

4. The middle lamella is involved in cell separation.   257

5. The intercellular spaces form a continuum.   258

6. The expression of arabinogalactan proteins is developmentally regulated.   260

7. A cell wall is the product of both a cell's internal developmental program and its environmental history.   261

8. Wall composition and architecture, together with anatomy, contribute to the mechanical properties of plants and their products.   262

9. Models of the cell wall have helped us to think more clearly about their construction and function.   264

Key Terms   268

References   268

Chapter 7 The Cell Wall in Growth and Development   273

A. Interactions Between the Cytoskeleton and the Wall   273

1. Cortical microtubule reorientation changes the orientation of cellulose microfibril deposition during growth.   273

2. Intracellular factors can alter the orientation of cortical microtubules.   276

3. Bundles of microtubules can define distinct domains of the wall that will thicken during cell development.   277

4. The arrangement of cytoskeletal elements in one cell often relates to that in a neighboring cell.   278

5. The wall is attached to receptor-like proteins in the plasma membrane.   281

B. Cell Expansion   282

1. A key driver of plant growth is postmitotic cell expansion.   282

2. Cell expansion is usually accompanied by the deposition of new wall material.   284

3. Wall architecture underpins anisotropic cell expansion.   287

4. Dynamic remodeling of wall architecture facilitates cell expansion.   290

5. Proteins that enhance wall expansion have been identified.   292

6. Acidification of the wall, enhanced by auxin, may promote cell expansion.   294

C. Turnover, Remodeling, and Breakdown of the Wall   294

1. Although plant cell walls are relatively stable compartments, many matrix polysaccharides and proteoglycans do turn over.   294

2. The *de novo* insertion of plasmodesmata across established walls requires wall remodeling.   296

3. Local removal of wall material is used to create conducting elements from files of cells.   298

4. Abscission of leaves and fruit is an active process that involves controlled cell separation.   299

5. Fruit softening depends upon the expression and activity of cell wall–modifying proteins.   300

6. In some seeds, wall polysaccharides can form a food reserve to be used during germination.   302

7. Lignin and suberin provide physical barriers to the turnover and degradation of secondary walls.   303

8. Gravity sensing and mechanical stress lead to compensatory changes in cell wall synthesis and architecture.   304

D. Cell Wall–Derived Signals in Growth and Development   306

1. Many signals combine to regulate plant growth.   306

2. Oligogalacturonides can modulate development in tobacco explants.   308

3. Xyloglucan-derived oligosaccharins can affect the rate of elongation growth.   308

4. Lipo-oligosaccharides synthesized by rhizobia regulate nodule development in host plants.   310

5. Chitinases may function during normal plant development. 311
Key Terms 313
References 313

Chapter 8 Cell Walls and Plant–Microbe Interactions 319

Introduction 319

A. How Plants Detect and Respond to Microbes 320
1. Most plants are immune to most pathogens. 320
2. Signals from both pathogen and host can elicit a common defense response. 321
3. Cell surface receptors recognize common molecular patterns. 322
4. The plant's responses to danger involve a common sequence of events. 324
5. Some specialist pathogens deliver effectors to suppress the basal defense response. 326
6. Some defenses are preformed. 328
7. Defense can also be mounted at a distance. 329

B. The Wall as a Battleground 331
1. Pathogens enter either by force, through wounds, or through natural openings. 331
2. Plasmodesmata provide a route for pathogen spread. 333
3. Callose deposition is a local response to both pathogens and wounding. 335
4. Wall strengthening helps contain the pathogen. 336
5. Cell polarization is a common response to a pathogen. 338
6. Localized cell death restricts pathogen spread. 340
7. Cell wall fragments act as danger signals. 341
8. Plants and pathogens battle to control the release of oligogalacturonides. 343
9. PR proteins attack the walls of fungi and bacteria. 344
10. Carbohydrate-binding modules (CBMs) help enzymes attach to the wall. 345
11. Cell wall integrity is sensed by the host. 346
12. Responses to wounding and pathogens overlap. 348
13. Wall-degrading enzymes and their inhibitors coevolve. 349

C. Recycling of Cell Walls 351
1. Removing pectin is a key early step in dismantling the wall. 352
2. Lyases are important pectic enzymes for both necrotrophic pathogens and saprophytic microbes. 353
3. To fully digest hemicelluloses, enzymes act in concert. 356
4. Cellulose is tough, and its disassembly usually requires special machinery. 357
Key Terms 360
References 360

Chapter 9 Plant Cell Walls: A Renewable Material Resource 365

A. Effects of Cell Walls on the Nutritional Quality and Texture of Foods and Forage Crops 366
1. Digestibility of forage crops depends upon the properties of plant cell walls and is mainly restricted by the lignin content. 366
2. The texture of our fruits and vegetables depends to a large extent on the properties of their cell walls. 368
3. Gums, water-soluble cell wall–associated polymers, are used to stabilize emulsions and to modify the texture of processed foods and other industrial products. 369
4. Pectins are used as thickeners, texturizers, stabilizers, and emulsifiers in the food and pharmaceutical industries. 370

B. Medicinal and Physiological Properties of Cell Wall Molecules 371
1. Dietary fiber in the human diet is a residual component of plant cell walls. 371
2. The consumption of dietary fiber has been associated with benefits in human health. 372

3. Gum arabic and pectins soothe irritated or inflamed mucosal tissues. 374
4. Pectic polysaccharides from medicinal plants stimulate the immune system via Peyer's patch cells in the intestine. 374
5. Bioactive pectins from some medicinal herbs promote the proliferation of immune system cells and their activities. 375
6. Bioactive pectins inhibit tumor growth by inducing cancer cell apoptosis. 376
7. Coating of medical devices with cell wall polysaccharides alters their bioactive properties. 377

C. Cell Wall Fibers Used in Textiles 378
1. Commercially important plant fibers are derived from many types of plants and possess physical properties that are exploited in textile products. 378
2. Cotton is the most widely used plant textile fiber. 378
3. Cotton fibers are epidermal trichomes that develop on the surface of ovules. 380
4. Coconut-derived coir fibers are used extensively in tropical countries. 381
5. The physical properties of different bast fibers make them suitable for use in specialized types of textiles. 381
6. Leaf fibers are used for ropes, tea bags, currency paper, and filter-tipped cigarettes. 383

D. Wood: an Essential Product for Construction and Paper Production 384
1. Wood is one of the world's most abundant industrial raw materials. 384
2. Wood and paper properties are derived from the composition and morphology of the xylem cell walls. 385
3. The physical strength of wood depends on the multilayered structure of the wood cell walls and cell morphology. 387
4. Cell wall–degrading enzymes are increasingly used for pulp and paper processing for environmental and economic reasons. 389
5. Bark is a source of cork, tannins, waxes, and drugs. 390

E. Plant Cell Walls: the Most Important Renewable Source of Biofuels and Chemical Feedstocks 391
1. Humankind's survival may depend on the wise use of renewable natural resources. 391
2. Plant cell walls and their derivatives comprise a major fraction of the terrestrial biomass and carbon content. 394
3. Photosynthesis-driven biomass production by plants constitutes a major and sustainable energy source. 395
4. Half of the wood harvested around the world is burnt as fuel. 396
5. Producing bioethanol from lignocellulosic biomass at competitive prices will require significant investments in research. 396
6. Overcoming the lignin barrier is essential to the success of the lignocellulosic biomass–based liquid biofuel industry. 398
7. Major improvements in fermenting enzyme systems are needed to improve the efficiency of ethanol production. 400
8. New types of biorefineries are needed to produce ethanol from lignocellulosic biomass at increased efficiencies and lower cost. 400
9. Plant cell walls are a renewable source of industrial chemicals. 402
10. Large-scale biofuel and chemical feedstock production is likely to affect food prices and biodiversity. 404
11. Cell walls offer unique challenges and opportunities for genetic engineering. 405
Key Terms 407
References 407

Cell Walls and Plant Anatomy

1

Examine a thin section of any higher plant organ with the light microscope, and you will immediately notice two features (**Figure 1.1A**). First, the plant is built from more than one type of cell, and each type can be identified by its size, its location, and the thickness, organization, and structure of the wall that surrounds it. Second, the different cell types are neatly cemented to their neighbors by their walls (Figure 1.1B) in beautiful and reproducible patterns. The structure and function of cells and the developmental patterns they form are the subject matter of plant anatomy. In this chapter we introduce the role of the cell wall in the development of the various cell types that are used to build higher plants. We discuss how and where new cells and therefore new walls arise, emphasizing the involvement of intracellular structures. We then examine the way in which new walls allow cells to expand and how walls are modified when cells differentiate. Last, we discuss the basic cell types of the plant body and how the structural adaptations of their walls suit their particular function.

A. The Derivation of Cells and Their Walls

1. Cells arise in specialized regions of the plant called meristems.[ref1]

Plants, like ourselves, start off with the fusion of an egg and a sperm to form a zygote. However, the plant zygote's subsequent development differs in at least two major respects from our own. First, the cell movements and migrations that characterize animal embryo development are not possible within a plant, whose cells are stuck to each other by their surrounding cell walls. The second is that plant development, that is, the elaboration of the adult plant from the fertilized egg, is not a continuous process. Instead, a programmed series of cell divisions gives rise to an embryo that contains

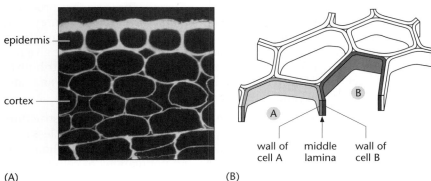

epidermis

cortex

(A)

wall of cell A   middle lamina   wall of cell B

(B)

Figure 1.1 Cell walls are cemented to each other. (A) Section of *Arabidopsis* stem stained with calcofluor, a dye that binds to the cellulose in walls, making them fluorescent. (B) It is important to realize that in sections, what we generally refer to as the *cell wall* can also be viewed as a composite structure formed from two walls, one contributed by each adjoining cell, firmly stuck together at a region called the middle lamella.

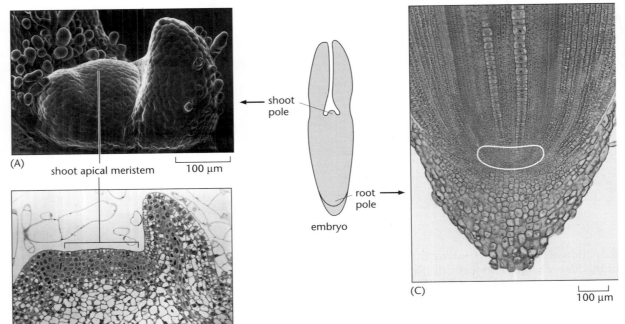

Figure 1.2 Apical meristems derive from the root and shoot poles of the embryo. (A) A scanning electron micrograph shows the shoot apex with two sequentially emerging leaf primordia, seen here as lateral swellings on either side of the domed apical meristem. (B) A thin section of a similar apex shows that the youngest leaf primordium arises from a small group of cells (about 100) in the outer four or five layers of cells. (C) The root meristem and root cap of a corn root, showing the orderly files of cells produced. (A and B, from R.S. Poethig and I.M. Sussex, *Planta* 165:158–169, 1985. C, from P.H. Raven, R.F. Evert, and S.E. Eichhorn, Biology of Plants, 4th ed. New York: Worth, 1986.)

a shoot pole, a root pole, and either one or more seed leaves; and at this point the embryo's development is commonly arrested in a seed. Within the maternal ovule tissue of the seed, the embryo will remain dormant until seed dispersal and suitable environmental conditions permit it to germinate and grow.

The young plant seedling will grow by a coordinated combination of cell enlargement and the production of new cells. However, cell proliferation does not occur evenly throughout the plant body. Following germination, two specialized groups of cells at opposite ends of the seedling axis, the shoot and root poles, begin to proliferate actively. Although cell divisions can occur elsewhere in the plant, these two groups of cells, now called the *shoot apical meristem* and the *root meristem*, respectively, will become a major source of new cells in the plant. They both perpetuate themselves and also give rise to new meristems in an iterative process that produces the characteristic structure of the mature plant, including structures such as leaves and flowers (**Figure 1.2**). Other groups of dividing cells can arise in established tissues, where they are called *lateral meristems*. These give rise to lateral roots and, through the cambium and cork cambium contribute to growth in thickness of the plant. A simple overview of higher plant anatomy is discussed later (see Panel 1.1).

Meristems themselves have a relatively defined, species-specific structure, but all are composed of small, densely cytoplasmic cells, generally less than 10 μm in diameter and having thin cell walls (**Figure 1.3**). Both shoot apical and root meristems contain two distinct populations of cells. A small subset of cells, called initials, or *stem cells*, divide surely but slowly. When

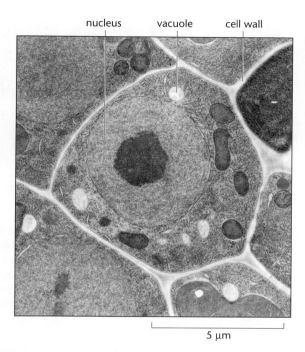

nucleus   vacuole   cell wall

5 μm

Figure 1.3 Meristematic cells. Typical meristematic plant cells are small, with very small vacuoles and thin cell walls. This example is from a shoot apical meristem. (Courtesy of Ichirou Karahara.)

they divide, they produce one daughter cell that remains a stem cell and another daughter cell that enters the cell population of the rest of the meristem. These cells may divide further, but eventually they leave the meristem. It is largely within these populations of dividing cells that new primary walls are produced.

2.   Walls originate in dividing cells.[ref2]

If we consider a single cell within an apical meristem, it is tempting to think that its surrounding wall arose in a single process of wall deposition around the living protoplast. But that would be to ignore the course of development. In fact, every facet of its polyhedral wall was laid down at a different time, namely, when that cell was cut off from one of its immediate neighbors by a cross wall during cell division. In some cases this might be a recent event, but in others the wall might derive, by growth, from one laid down during a cell division much earlier, for example, in the embryo. Each wall is thus a patchwork coat, with each patch reflecting events that happened at different times in the cell's development. Each cell, consequently, carries within its wall a historical relationship with all of its neighbors (**Figure 1.4**).

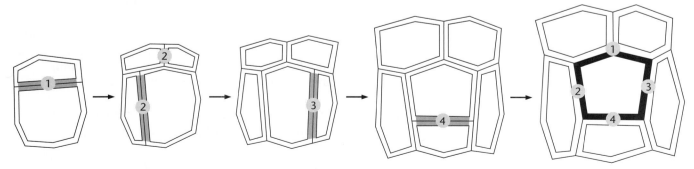

Figure 1.4 Sequential cell divisions create "walls" with different histories. Schematic diagram, showing how a sequence of cell divisions creates a series of intersecting cell walls of different ages (*shaded gray*). The resultant final cell pattern is shown on the right with the wall of the central cell shown in black. This is usually interpreted as a single cell wall, but developmentally it has been made by the consecutive addition of different cell plates that eventually intersect with each other. A particular wall (for example, wall 1) has a history in common with the walls of adjacent cells with which it shares ancestry, rather than necessarily with wall 3, for example, which has only recently been cut and pasted into it. This example emphasizes the problem of whether a "wall" stops at the middle lamella or is a single partition that includes the contribution from two neighboring cells.

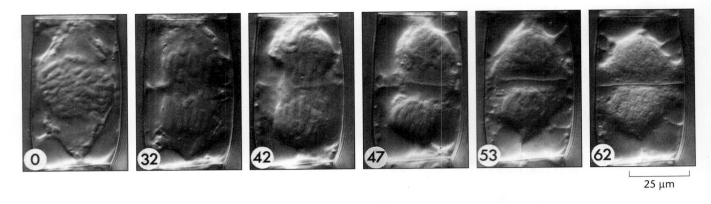

25 µm

Figure 1.5 The origin of a new cell wall. Sequential light micrographs of a dividing stamen hair cell. The elapsed time in minutes is shown at the bottom left corner of each photograph. By 42 minutes the nucleus has undergone mitosis, cytokinesis is under way, and the early cell plate can be seen between the two daughter nuclei. This extends rapidly outward until it reaches and fuses with the mother cell wall. (Courtesy of Peter Hepler.)

With very few exceptions, new cross-walls arise only when cells divide and, following nuclear division or mitosis, they partition the protoplast in a process called *cytokinesis*. Plant cytokinesis is an inside-outward event that starts with the formation of a small, disklike wall, or *cell plate*, inside the cell, between the two daughter nuclei. This cell plate extends radially outward until it finally reaches and fuses with the mother cell wall, thus cutting the cell in two (**Figure 1.5**). The structure that assembles this cell plate is called the *phragmoplast*; it originates most commonly from the microtubules of the two half spindles of the mitotic apparatus and has a disklike structure, with the plus ends of the microtubules on either side ending in a plane at right angles to the spindle axis. Associated with this structure are actin filaments, closely aligned with the microtubules. The phragmoplast is surrounded by numerous Golgi stacks, which produce vesicles that are transported along the microtubules to the center of the phragmoplast. Here, the vesicles fuse to form a membrane-enclosed disk, the cell plate (**Figure 1.6**). The vesicles carry pectic polysaccharides, xyloglucan, and proteins to the cell plate; this subject is discussed in greater detail in Chapter 4. The polysaccharide callose, a β1-3 glucan, can be detected in the plate at an early stage, but the cellulose, characteristic of the final wall, is not deposited until the latest stages of cell plate formation. Intact phragmoplasts that can continue to make wall material for the cell plate can be isolated from synchronized cells in suspension culture (**Figure 1.7**).

For complete partition, even in a small meristematic cell, the cell plate must be extended at its edge until it finally makes contact with the mother cell wall to produce a cross-wall. The phragmoplast achieves this by continuously disassembling the microtubules at its center and reassembling them at the edge of the growing cell plate, where they function in guiding new vesicles to fuse with and extend its margin (**Figure 1.8**).

At about the time the cell plate finally reaches the mother cell wall and separates the two daughter cells, callose is removed and cellulose deposition begins on each face of the plate, allowing us to finally distinguish two separate "walls" within the partition, one being made by and belonging to one daughter cell and one being made by and belonging to the other.

3. Plant organ development depends on precise control of the plane of cell division and of cell expansion.[ref3]

A single isolated plant cell grown in culture can successfully divide and expand or elongate. If its cell wall is removed, however, the remaining protoplast, although living, can neither divide nor elongate unless it regenerates its wall. Correspondingly, the walled cell cannot divide or elongate unless its cytoskeleton is both intact and functional. The proliferation and growth of plant cells depends on a functional interaction between elements inside the plasma membrane—in particular the cytoskeleton—and the cell wall outside the plasma membrane. However, during tissue development, these

cell A in telophase | cell B in cytokinesis

new cell plate

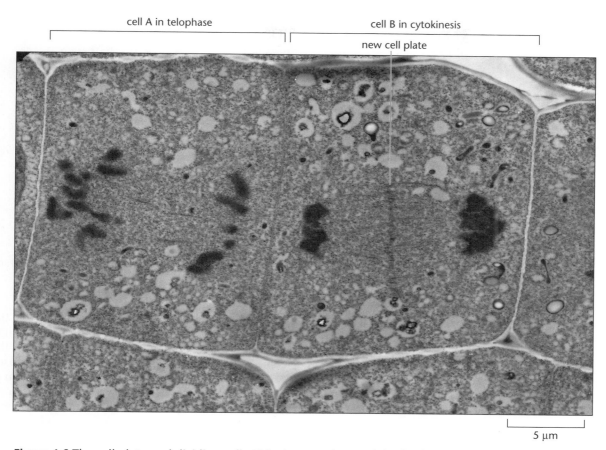

5 μm

Figure 1.6 The cell plate and dividing cells. This electron micrograph is of a thin longitudinal section through a maize root meristem. The long axis of the root runs from left to right. The large central cell has already divided once to form two new cells, A and B, with a thin young primary wall separating them. The two daughter cells are each now dividing again. Cell A is in late telophase, and the two separated sets of chromosomes are visible. Cell B has reached the next stage of cell division, and between the two re-forming nuclei an early stage of vesicle accumulation and cell plate formation can be seen. (Courtesy of Adrian Turner.)

cell-autonomous activities are tightly coupled with those of neighboring cells to generate the precise and reproducibly structured organs we find in the mature plant.

There are additional constraints on the generation of plant form. We have already mentioned that the cells of plants are immobile and that, with very few exceptions, adjacent cells are firmly stuck together and are thus prevented from expanding at different rates and "slipping" relative to one another. Thus, local cell-cell interactions are central to the coordination and control of plant growth.

We are now in a position to appreciate that, in many cases, it is the accurate placing of new cross-walls during cell division, followed by the subsequent controlled expansion and/or elongation of the resultant cells, that has

Figure 1.7 Polysaccharide synthesis in isolated phragmoplasts. Radioactive precursors are incorporated into wall polysaccharides by phragmoplasts isolated from tobacco cells in culture. These retain all the cytoplasmic organelles usually associated with the phragmoplast, including Golgi stacks. (A) When the phragmoplasts are fed with radioactive UDP-xylose, this marker is incorporated into matrix polysaccharides in the Golgi stacks surrounding the cell plate (*arrows*) and can be revealed by autoradiography as black silver grains. (Golgi stacks cannot transport the product to the cell plate within the isolated phragmoplast.) (B) When phragmoplasts are fed with labeled UDP-glucose, the label is incorporated into callose by enzymes located in the membrane of the cell plate. Here the silver grains are located directly over the cell plate (*arrows*). (From T. Kakimoto and H. Shibaoka, *Plant Cell Physiol.* 33:353–361, 1992.)

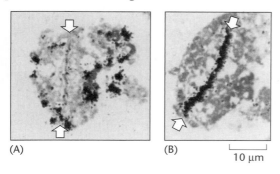

(A)   (B)   10 μm

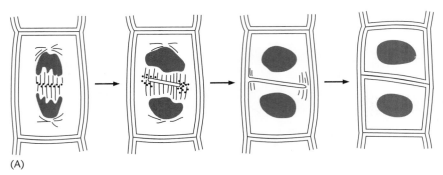

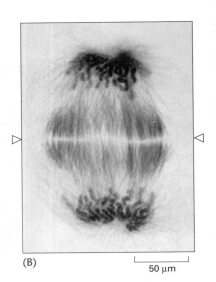

Figure 1.8 The phragmoplast and cell plate. (A) Sequence of events in phragmoplast formation. (B) A light micrograph of a plant cell entering cytokinesis. The cell has been stained to show both the microtubules and the two sets of daughter chromosomes. The clear region where the new cell plate is being assembled is indicated by the arrowheads. (B, courtesy of Andrew Bajer.)

50 μm

a major influence on plant organ development. For example, in the first early divisions of the zygote of the model plant *Arabidopsis* (**Figure 1.9**), the precise sequential placing of the first cross-walls that produces the precise pattern of cells in the embryo appears to be essential for subsequent development. Mutations that affect these early embryonic division planes produce severely deformed or nonviable embryos. Later in development, common patterns of division and expansion often emerge. Thus, meristems often contain cells that reiterate transverse divisions to produce extended files of cells (**Figure 1.10**).

The continuous cell proliferation activity of meristems creates a population of cells that will divide less frequently, will expand, and will eventually differentiate into mature cell types. Both the spatial and temporal controls over the plant cell cycle and the spatial and temporal controls over cell expansion have to be tight and, more important, coordinated with neighboring cells. Although such controls are likely to include mechanical constraints, environmental and physiological cues, and the action of conventional plant growth factors, the detailed molecular machinery through which they work has not been fully elucidated.

Despite our emphasis so far on the pattern of sequential divisions, it should be noted that plant structure is characterized by flexibility and plasticity, and it is the interaction between the separate processes of cell proliferation and cell expansion that generates plant form. In some cases, properties of the whole organ can override or entrain the contributions of division and expansion at the cellular level, as in the maize mutant *tangled*. In a normal maize plant, the leaf epidermis has a highly ordered pattern of longitudinal and transverse divisions that contribute to the width and length, respectively, of the leaf. By contrast, in the *tangled* mutant, the leaf epidermis has large numbers of cells with aberrant division planes. Despite the rather

Figure 1.9 Early sequence of divisions in the *Arabidopsis* zygote. The fertilized egg (A) divides transversely to form an upper cell (*gray*) that will form the embryo proper, and a lower cell (*white*) that will form the suspensor and part of the root meristem (B). The embryo proper goes through a series of stereotyped divisions (C–F) that include, for example, some periclinal divisions (E) that distinguish an outer layer of cells that will form the entire shoot epidermis (*dark gray*). In some other species, early embryonic divisions are much less organized.

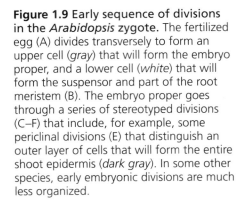

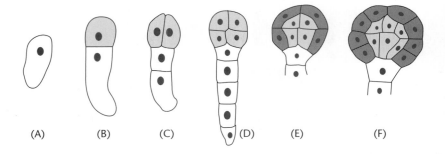

(A)   (B)   (C)   (D)   (E)   (F)

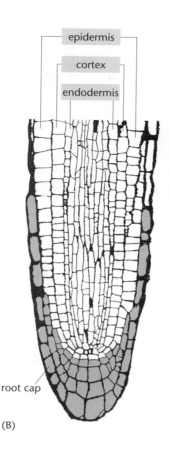

(A)

50 µm

(B)

Figure 1.10 Root meristems produce organized files of cells. (A) A scanning electron micrograph of an *Arabidopsis* root tip showing the longitudinal files of cells produced by transverse divisions of the epidermal cells. The diameter of this root is exactly comparable to a human hair. (B) A longitudinal section of a similar root shows that the files of cells can be traced back to their origins within the meristem. The epidermis in this section is largely covered by cells of the root cap. (A, courtesy of Paul Linstead.)

muddled pattern of resulting cell walls, the overall shape of the mutant leaf is remarkably similar to that of the wild type (**Figure 1.11**). The *TANGLED* gene is required for the proper positioning of the cytoskeletal arrays involved in the formation and placement of new cell plates, which we discuss next. The way in which larger-scale structures and their growth patterns can affect cell division and expansion patterns remains obscure.

(A)   (B)

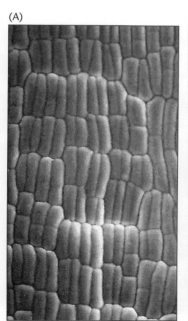

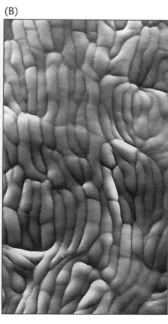

Figure 1.11 Division plane control. Scanning electron micrographs of the surface of maize leaf primordia. (A) Wild type and (B) the mutant *tangled*. Despite the disorganized division planes, the mature leaf that finally develops looks remarkably normal in shape. (Courtesy of Laurie Smith.)

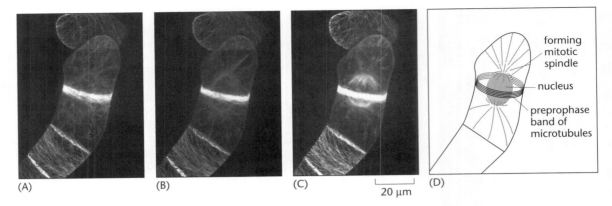

(A)   (B)   (C)   (D)

20 µm

forming
mitotic
spindle

nucleus

preprophase
band of
microtubules

Figure 1.12 The preprophase band. In these living tobacco cells in culture, the microtubules are stained because the cell is expressing a green fluorescent protein fused to a microtubule-binding protein. The microtubules in the dividing cell gradually accumulate in the preprophase band (A–C) and also begin to form the mitotic spindle (C). Cortical microtubules can be seen in the nondividing cells. (D) The preprophase band encircles the cell, defining the future division plane. (A–C, courtesy of Henrik Buschmann.)

4. Cytoskeletal elements predict the position of the new cross-wall before the cell divides.[ref4]

Thin sections of healthy plant tissue show clear, reproducible, and species-specific patterns of cells. This consistent anatomy strongly indicates that, during the cell divisions that give rise to the tissue, the positioning of new cross-walls is an extremely accurate process. What mechanisms exist within plant tissues to ensure such precision in cell partitioning? In almost every case, shortly after DNA replication has finished, cytoskeletal elements rearrange themselves within the cell into a structure just inside the plasma membrane that predicts the site where the new cell plate will later meet the mother cell wall. The most obvious feature of this structure is a tight bundle of microtubules lying just beneath the plasma membrane that has been called the *preprophase band* (**Figure 1.12**). The ability of this band to prefigure the exact future division plane is true for both symmetric and asymmetric cell divisions (**Figure 1.13**). The nucleus at the early prophase stage of mitosis, with its condensing chromosomes, typically lies at the center of the plane defined by the preprophase band and appears to be held there by cytoplasmic strands that also contain both microtubules and actin filaments. One might think of the whole structure as rather like a bicycle wheel in which the rim represents the preprophase band, the hub represents the nucleus, and the spokes represent the microtubules and actin filaments radiating from the nucleus to the preprophase band (**Figure 1.14** and **Figure 1.15**).

Figure 1.13 A preprophase band predicts a future asymmetric cell division. In the epidermis of a grass leaf (rye) a series of divisions gives rise to a mature stoma flanked by two guard cells and two subsidiary cells. (A) The subsidiary cells arise by an asymmetric division in the subsidiary cell mother cell (B) that is predicted by a curved preprophase band. (C) The immunofluorescence image shows the microtubules of the preprophase band revealed by an antibody to tubulin. (C, from S.-O. Cho and S.M. Wick, *J. Cell Sci.* 92:581–594, 1989.)

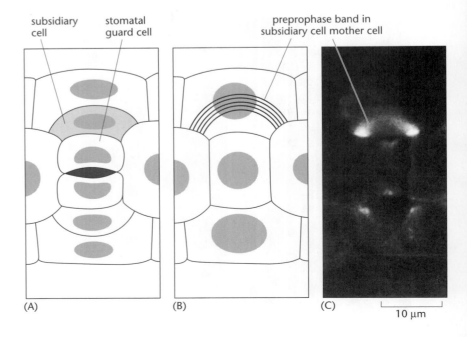

subsidiary
cell

stomatal
guard cell

preprophase band in
subsidiary cell mother cell

(A)   (B)   (C)

10 µm

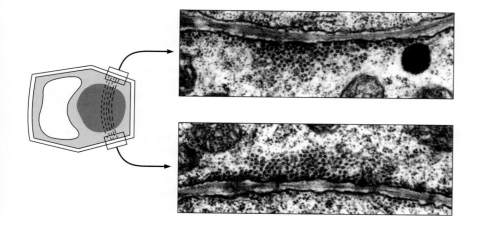

Figure 1.14 Microtubules in the preprophase band. These electron micrographs of a thin section of a leaf epidermal cell from sugar cane show the preprophase band of microtubules just beneath the plasma membrane. The cell in this case is part of the stomatal complex and is dividing asymmetrically to produce a smaller daughter cell that will form the stomatal subsidiary cell. (Courtesy of C.H. Busby.)

It may seem paradoxical that the microtubules of the preprophase band, which appear to predict so accurately the position of the future cross-wall, actually disappear as the cell enters mitosis. The nature of the "molecular memory" that remains at the cell surface appears to be produced by remodeling of the plasma membrane in the region of the preprophase band by endocytic vesicles that remove selected membrane proteins. The significance of this is explored in the next section.

The signals or cues that are used by a cell to position the preprophase band are largely unknown. However, one signal used by many cells to determine the division plane is mechanical stress. Artificially applied pressure on living plant tissue commonly triggers new cell divisions in which the new cross-walls are oriented at right angles to the applied pressure. To what extent such cues operate to determine the position of preprophase bands during normal development is not known.

5.   Actin filaments help to position new cross-walls.[ref5]

We have seen how placement of the new cross-wall during cell division is faithfully predicted by the position of the preprophase band of microtubules. Both environmental cues—for example, physical pressure—and local factors can influence cross-wall positioning. In particular, the shape of the cell and the local cellular geometry of neighboring cells influence the decision about where the preprophase band, and hence the future cross-wall, will be located. For example, long cells tend to divide with the new cross-wall at right angles to the long axis of the cell. A cross-wall in one cell also tends to inhibit the placing of a new cross-wall directly opposite in the neighboring cell. In other words, *three-way junctions* (places at which three cells contact each other) are strongly favored over four-way junctions (**Figure 1.16**). In groups of cells in which the final pattern of divisions is highly repeatable, such as the stomatal divisions in Figure 1.13, the extending cell plate meets the mother cell wall at the predetermined site with an accuracy of less than 1 μm, and we must consider now how this relates to the earlier positioning of the preprophase band.

Following the disappearance of the microtubules from the preprophase band, the actin filaments that remain appear to have several major roles in subsequent events. First, an actin-free zone persists in place of the preprophase band, probably originating by local endocytosis removing membrane-associated actin-binding proteins. This in some way marks that cortical region of the cell with which the new cross-wall will fuse. Groups of actin filaments, extending from the surface of the nucleus to a region at each end of the cell, create a force on the developing spindle that helps align it perpendicular to the plane of the preprophase band (**Figure 1.17A–C**). This ensures that, following chromosome separation, the developing phragmo-

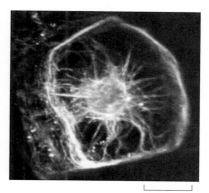

20 μm

Figure 1.15 A preprophase band in a large dividing vacuolate epidermal cell. The cell is fluorescently labeled with an antibody to tubulin and the original image is three-dimensional, made by combining a stack of separate optical sections. The nucleus, in the center of the cell, is clearly connected to the cortical preprophase band of microtubules by radial bundles of microtubules. (From C.W. Lloyd, ed., Cytoskeletal Basis of Plant Growth and Form. London: Academic Press, pp. 245–257, 1991.)

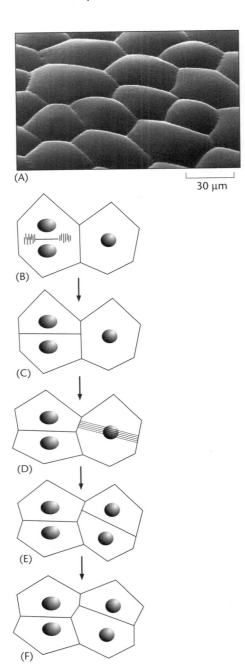

(A)

30 µm

(B)

(C)

(D)

(E)

(F)

Figure 1.16 Cell division planes and the preference for three-way junctions. (A) A scanning electron micrograph of the surface of an oilseed rape embryo reveals that junctions are almost invariably between three cells rather than four or more. (B–F) These preferred Y-shaped, three-way junctions arise because new cell walls that are inserted during cell division generally avoid a preexisting junction as is shown below schematically. (B, C) A new division wall is inserted transversely in the left-hand cell of two young expanding cells. (D) As the cells expand it is thought that the new wall lags behind and the common wall buckles to produce a Y-shaped three-way junction. When the right-hand cell divides, the preprophase band avoids this vertex, and when the new division wall is complete (E), it is staggered with respect to the earlier wall. (F) This junction, too, will eventually become Y-shaped. Similar, and well understood, geometries occur in rafts of soap bubbles, but the molecular basis for selecting the preprophase band site, although influenced in part by geometry and mechanical forces, is by no means completely understood. (A, courtesy of Lloyd Peto and Kim Findlay.)

plast, and therefore the developing cell plate, is in the plane defined by the preprophase band. Second, the groups of radial actin filaments persist and become attached to the edge of the growing cell plate. These filaments exert tension and guide the extending plate accurately to its final cortical position where the actin-free zone was established. The accuracy of this guidance system is particularly obvious in the highly vacuolate cells of the cambium, where the growing edge of the plate can be guided over a considerable distance to its final predetermined site (Figure 1.17D–F).

6. The new cross-wall must join and fuse with the mother cell wall.[ref6]

When the extending cell plate finally meets its appointed destination on the mother cell wall, further events must occur before the formation of two new daughter cells is completed. The membrane that surrounds the cell plate is topologically equivalent to the plasma membrane of the mother cell, and when these meet they fuse, initially at local sites, and later completely to effect the separation of two daughter protoplasts. At this stage, the new cross-wall, as it has now become, has a relatively homogeneous interior, and at this stage or shortly before, cellulose begins to be deposited. Since the cellulose is made at the plasma membrane (see Chapters 4 and 5), production of this structural polysaccharide takes place simultaneously at both faces of the new cross-wall. Cellulose fibers from one face do not intermingle with those from the other, and therefore, an "exclusion zone," rich in pectic polysaccharides, is created in the center of the wall. This is called the *middle lamella*. At the junction between the mother cell and the new cross-wall, further events take place. In this region, local dissolution of the mother cell wall takes place, allowing the newly created middle lamella to extend and fuse with the middle lamella of the surrounding mother cell wall. It is as though the old wall were cut and the new walls were pasted in (**Figure 1.18**). The two new cross-walls, as they now are, become successfully integrated into the structure of the mother cell wall and finally create two daughter cells, each with its own new wall. The new three-way junctions of middle lamella that are created are important sites for future events (such as air space formation), which are discussed in Chapter 6 (Concept 6C5, which corresponds to Chapter 6, Section C, Heading 5 on page 258).

7. A plant is constructed from two compartments: the apoplast and the symplast.[ref7]

During the formation of new cross-walls, tubular elements of endoplasmic reticulum become trapped in the cell plate at the stage of vesicle fusion.

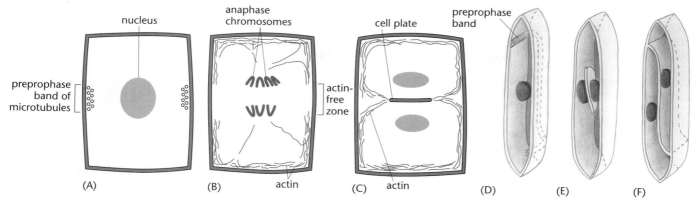

Figure 1.17 The role of the cytoskeleton in division plane alignment. (A) As a vacuolated dividing cell enters prophase, the preprophase band of microtubules at the cell cortex marks the site of the future division plane. Microtubules and actin also extend through the cytoplasm and connect with the nucleus (see Figure 1.15). (B) As the chromosomes align at metaphase, the microtubules in the preprophase band have disassembled, leaving in place an actin-free zone at the cortex throughout mitosis. (C) Actin persists elsewhere in the cell and helps guide the extending cell plate during cytokinesis to the correct site at the mother cell wall. On completion of cytokinesis, the cortical array of microtubules is reestablished. (D–F) Cutaway diagrams showing the division of an elongated and vacuolate cambium cell. These show the position of the preprophase band of microtubules (D), the extension of the cell plate following nuclear division (E), and the completed division (F) in which the cell plate has fused with the mother cell wall at the predicted site to create two new cambial cells.

The resulting channels are structurally elaborated into more permanent pores called *plasmodesmata* (singular, plasmodesma). Each pore is lined by plasma membrane that is continuous from cell to cell and usually contains a rodlike structure derived from the endoplasmic reticulum that is called the *desmotubule*. While many plasmodesmata are formed during cell plate formation, it is important to realize that they may also be formed *de novo* in more established cell walls. Although their diameter is on the order of 50 nm or so, the aqueous channel between the plasma membrane and the desmotubule that extends from cell to cell is rather tightly regulated and in general will allow the free passage of only small molecules. (Plasmodesmata are discussed in more detail in Concept 1C9.)

The presence of abundant plasmodesmata means that all of the cells in a higher plant, with very few exceptions (stomatal guard cells and germ cells), share, topologically speaking, a single plasma membrane and thus the cells are cytoplasmically coupled to each other. This cytosolic continuity within the plant creates a single, extended topological space, bounded by shared plasma membrane, that is called the *symplast* (**Figure 1.19**). The remainder of the plant, which includes everything outside the symplast, also forms a continuous space, occupied by the cell walls, the intercellular spaces, and the contents of such dead cells as xylem vessels that have lost their plasma membrane. This continuum is called the *apoplast* (Figure 1.19) and is broadly equivalent to what we know as the *extracellular matrix*.

Figure 1.18 Steps in completing a new cross wall and generating two new daughter cells. After the membrane around the cell plate has completely fused with the plasma membrane of the mother cell (A), cellulose begins to be deposited within the plate and a new middle lamella region is created in the central region (B). (C) Wall hydrolytic enzymes aid the controlled dissolution of a region of the mother cell wall that allows the cutting and pasting of the new wall into the old. (D) The middle lamellae become continuous, the two daughter cells acquire their own complete walls, and a new three-way junction is created.

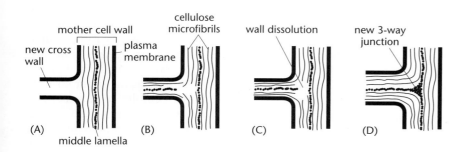

Figure 1.19 Apoplast and symplast.
Most but not all cells in the plant
are connected to their neighbors by
plasma membrane–lined cytoplasmic
channels called plasmodesmata. These
connections create two discrete topological
compartments within the plant: the
apoplast, consisting of everything outside
the cell's plasma membrane, including the
dead xylem vessels and their contents; and
the symplast, which consists of everything
inside the plasma membrane—the plant's
collective cytoplasm. As shown in the inset,
the cytoplasm of cell A is continuous with
that of cell B via the (deliberately simplified)
plasmodesma.

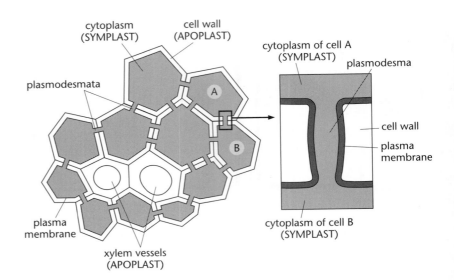

Both apoplast and symplast can provide routes for the local transport of
water and solutes. Water, ions, and small signaling molecules, for example,
may be transported through the apoplast. The same molecules, together
with sugars, amino acids, and many other small organic molecules, may be
transported from cell to cell within the symplast. The plasma membrane,
which forms the boundary between apoplast and symplast throughout
the plant, has an important function in regulating the transport of solutes
between the two compartments. In some cases the movement of materials
may be restricted to either the apoplast or the symplast. Symplastic restriction is exemplified by stomatal guard cells (see Concept 1C2). These become
symplastically isolated from their neighbors early in their differentiation,
when all of the plasmodesmata connecting them to adjacent cells are
removed. Apoplastic restriction can be seen in the waterproof, suberinized
wall layer, known as the casparian band, which acts as a barrier between the
endodermal cells of the root to the movement of water and solutes through
the wall (see Concept 1C4). The idea of apoplast and symplast helps focus
our attention on the three-dimensional organized continuity of both cells
and their extracellular matrix.

B. Walls in Cell Growth and Differentiation

1. Cells become organized at an early stage into three major tissue systems.[ref8]

In the previous section we discussed how and where new cells and new
walls arise. In this section we examine how cells grow and differentiate into
a variety of cell types and how these cell types are assembled into the tissues, tissue systems, and organs of the mature plant.

It is convenient, at this stage, to have a descriptive framework to make some
sense of the complex variety of cell patterns we find in sections of different plant parts. A simple hierarchical system of anatomy is widely used in
which plants are first broken down into *organs* (for example, root, stem, or
leaf), each of which in turn is built from different arrangements of three
tissue systems, the dermal system, the ground system, and the vascular
system (**Panel 1.1**). The *dermal tissue system* comprises the outer covering
of the plant, that is, the epidermis and its more complex replacements in
older organs such as bark. The *ground tissue system* consists of supportive
tissues (for example, parenchyma, collenchyma, and sclerenchyma, which
help support and protect the vascular tissue system) together with storage

and photosynthetic tissues. The *vascular tissue system* contains the two important tissues, phloem and xylem, that are used for moving water and dissolved solutes around the plant.

These three systems, dermal, ground, and vascular (which are discussed in more detail later in the chapter), are established early in plant development and can be seen in the early heart-shaped embryo (**Figure 1.20**). Our classification of tissue systems is not an artificial one but has biological relevance, since in some cases (for example, roots) it is known that each system is perpetuated by the proliferative activity of separate groups of progenitor stem cells known as initials. Each tissue system can, in turn, be subdivided into separate *tissues*. For example, the vascular tissue system will usually contain at least phloem, xylem, and parenchyma tissues as well as cambium cells. In turn, each tissue may be made up of more than one *cell type*. Phloem, for example, is a tissue that contains an organized arrangement of cell types, including sieve tube elements, companion cells, parenchyma cells, and phloem fibers. Compared with animals, plants have relatively few clearly defined cell types, although subtle distinctions between apparently similar cells but in different positions within the plant are being identified. Although some features of the protoplast, for example, plastids, are also important, individual cell types can generally be distinguished visually by a combination of three criteria: the shape of the cell, its position in the organ with respect to its neighbors, and the nature of its cell wall. The walls of different cell types may differ in the relative amounts of the main structural polysaccharides, their thickness, and the presence or absence of particular proteins. The above categories of organs, tissue systems, and tissues should not be interpreted as having rigid boundaries. Rather, we should be aware

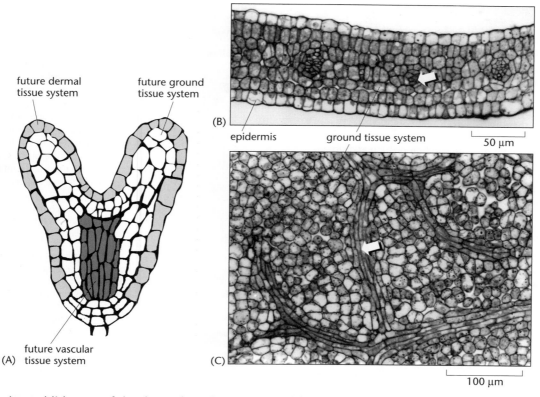

future dermal tissue system

future ground tissue system

(B)

epidermis   ground tissue system   50 μm

future vascular
(A) tissue system

(C)

100 μm

Figure 1.20 Early establishment of the three plant tissue systems. (A) A longitudinal section through a late heart-shaped embryo of *Arabidopsis thaliana* already displays the groups of cells that will later form the three tissue systems of the mature plant: dermal, ground, and vascular. During plant development new organs are similarly divided into the same three tissue systems. (B) A cross section of a developing leaf, and (C) a section at right angles to this. In this very young leaf of *Arabidopsis,* the dermal tissue system (epidermis), the ground tissue system, and the more elongated cells that will form the vascular tissue system (minor veins) can be clearly seen (*arrowheads*). (B and C, from P.M. Donnelly et al., *Plant Cell* 9:1121–1135, 1999.)

14

Panel 1.1 The cell types and tissues from which higher plants are constructed.

THE THREE TISSUE SYSTEMS

Cell division, growth, and differentiation give rise to tissue systems with specialized functions.

DERMAL TISSUE (▭): This is the plant's protective outer covering in contact with the environment. It facilitates water and ion uptake in roots and regulates gas exchange in leaves and stems.

VASCULAR TISSUE: Together the phloem (▬) and the xylem (▬) form a continuous vascular system throughout the plant. This tissue conducts water and solutes between organs and also provides mechanical support.

GROUND TISSUE (▭): This packing and supportive tissue accounts for much of the bulk of the young plant. It also functions in food manufacture and storage.

The young flowering plant shown on the *right* is constructed from three main types of organs: leaves, stems, and roots. Each plant organ in turn is made from three tissue systems: ground (▭), dermal (▬), and vascular (▬).

All three tissue systems derive ultimately from the cell proliferative activity of the shoot or root apical meristems, and each contains a relatively small number of specialized cell types. These three common tissue systems, and the cells that comprise them, are described in this panel.

THE PLANT

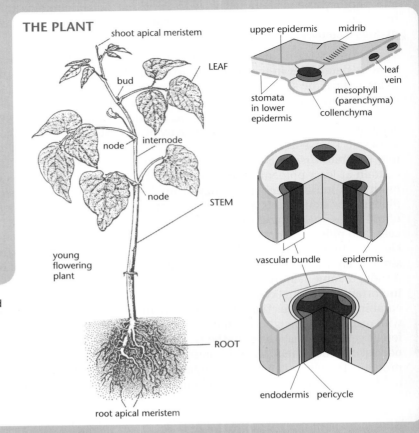

GROUND TISSUE

The ground tissue system contains three main cell types called parenchyma, collenchyma, and sclerenchyma.

Parenchyma cells are found in all tissue systems. They are living cells, generally capable of further division, and have a thin primary cell wall. These cells have a variety of functions. The apical and lateral meristematic cells of shoots and roots provide the new cells required for growth. Food production and storage occur in the photosynthetic cells of the leaf and stem (called mesophyll cells); storage parenchyma cells form the bulk of most fruits and vegetables. Because of their proliferative capacity, parenchyma cells also serve as stem cells for wound healing and regeneration.

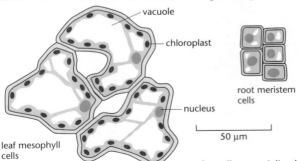

A transfer cell, a specialized form of the parenchyma cell, is readily identified by elaborate ingrowths of the primary cell wall. The increase in the area of the plasma membrane beneath these walls facilitates the rapid transport of solutes to and from cells of the vascular system.

Collenchyma cells are living cells similar to parenchyma cells except that they have much thicker cell walls and are usually elongated and packed into long ropelike fibers. They are capable of stretching and provide mechanical support in the ground tissue system of the elongating regions of the plant. Collenchyma cells are especially common in subepidermal regions of stems.

typical locations of supporting groups of cells in a stem

sclerenchyma fibers
vascular bundle
collenchyma

Sclerenchyma cells, like collenchyma cells, have strengthening and supporting functions. However, they are usually dead cells with thick, lignified secondary cell walls that prevent them from stretching as the plant grows. Two common types are fibers, which often form long bundles, and sclereids, which are shorter branched cells found in seed coats and fruit.

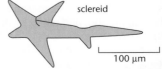

DERMAL TISSUE

The epidermis is the primary outer protective covering of the plant body. Cells of the epidermis are also modified to form stomata and hairs of various kinds.

Epidermis

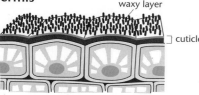

The epidermis (usually one layer of cells deep) covers the entire stem, leaf, and root of the young plant. The cells are living, have thick primary cell walls, and are covered on their outer surface by a special cuticle with an outer waxy layer. The cells are tightly interlocked in different patterns.

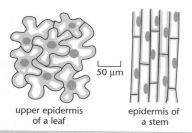

upper epidermis of a leaf — 50 μm — epidermis of a stem

Stomata

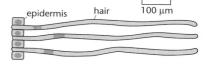

guard cells — air space — 5 μm

Stomata are openings in the epidermis, mainly on the lower surface of the leaf, that regulate gas exchange in the plant. They are formed by two specialized epidermal cells called guard cells, which regulate the diameter of the pore. Stomata are distributed in a distinct species-specific pattern within each epidermis.

Hairs (or trichomes) are appendages derived from epidermal cells. They exist in a variety of forms and are commonly found in all plant parts. Hairs function in protection, absorption, and secretion; for example,

epidermis — hair — 100 μm

young, single-celled hairs in the epidermis of the cotton seed. When these grow, the walls will be secondarily thickened with cellulose to form cotton fibers.

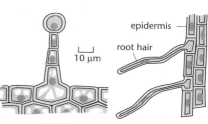

epidermis — root hair — 10 μm

a multicellular secretory hair from a geranium leaf | single-celled root hairs have an important function in water and ion uptake.

VASCULAR BUNDLES

Roots usually have a single vascular bundle, but stems have several bundles. These are arranged with strict radial symmetry in dicots, but they are more irregularly dispersed in monocots.

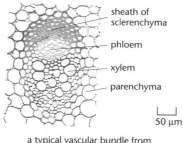

sheath of sclerenchyma — phloem — xylem — parenchyma — 50 μm

a typical vascular bundle from the young stem of a buttercup

VASCULAR TISSUE

The phloem and the xylem together form a continuous vascular system throughout the plant. In young plants they are usually associated with a variety of other cell types in vascular bundles. Both phloem and xylem are complex tissues. Their conducting elements are associated with parenchyma cells that maintain the elements and exchange materials with them. In addition, groups of collenchyma and sclerenchyma cells provide mechanical support.

Phloem

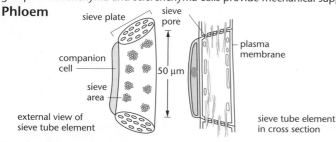

sieve plate — sieve pore — plasma membrane — companion cell — 50 μm — sieve area — external view of sieve tube element — sieve tube element in cross section

Phloem is involved in the transport of organic solutes in the plant. The main conducting cells (elements) are aligned to form tubes called sieve tubes. The sieve tube elements at maturity are living cells, interconnected by perforations in their end walls formed from enlarged and modified plasmodesmata (sieve plates). These cells retain their plasma membrane, but they have lost their nuclei and much of their cytoplasm; they therefore rely on associated companion cells for their maintenance. These companion cells have the additional function of actively transporting soluble food molecules into and out of sieve tube elements through porous sieve areas in the wall.

Xylem

Xylem carries water and dissolved ions in the plant. The main conducting cells are the vessel elements shown here, which are dead cells at maturity that lack a plasma membrane. The cell wall has been secondarily thickened and heavily lignified. As shown below, its end wall is largely removed, enabling very long, continuous tubes to be formed.

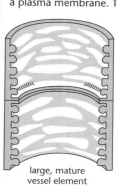

small vessel element in root tip — large, mature vessel element

The vessel elements are closely associated with xylem parenchyma cells, which actively transport selected solutes into and out of the elements across the parenchyma cell plasma membrane.

xylem parenchyma cells — vessel element

TABLE 1.1 THE VOLUMES OF VARIOUS MATURE CELLS RELATIVE TO THE VOLUME OF A TYPICAL MERISTEMATIC CELL TAKEN AS A 10-μm-SIDED CUBE	
CELL TYPE	**RELATIVE VOLUME**
Phloem parenchyma cell in *Arabidopsis* root	0.01
Shoot apical meristem cell (10-μm cube)	1
Parenchyma cell, dicot stem	50
Parenchyma cell of potato or apple	5000
Large xylem vessel in tree wood	50,000

that there are subtle gradations between categories of cell and tissue that all function in harmony in the plant. A brief overview of plant anatomy is shown in Panel 1.1.

2. When they have stopped dividing, cells usually continue to grow in size.[ref9]

All of the cells of the plant body derive ultimately from the proliferative activity of cells in the root and shoot apical meristems. As the cells produced by this activity steadily exit from the cell cycle, they almost invariably continue to grow. It is important to recognize that the dividing cells in the meristem also have to grow to double their size before dividing; if they did not, the daughter cells would get progressively smaller! In many cases cells that are no longer dividing expand to hundreds of even thousands of times their original volume. Meristematic cells are commonly 5 to 10 μm in diameter; cells in a mature structure, such as a potato, may be 100 to 200 μm in diameter. It is difficult to overemphasize the importance of cell expansion in plant development. Without it, trees would be no bigger than small shrubs, and apples would be smaller than a pea. Nutrient supply, environmental signals, and endogenous growth are all well-characterized factors that influence cell expansion, but the molecular details of how these factors control growth are less well understood (**Table 1.1**).

Cell expansion does not go hand in hand with the production of more cytoplasm. Instead, the bulk of the increase in cell volume is accounted for by the increase in size of the vacuole, a sensible economy in the use of materials (**Figure 1.21**). In some cases the ploidy of the nucleus increases. In all cases, however, cell expansion is matched by the uniform addition of new material to the expanding wall. Indeed, it is the properties of the wall and the production of new wall material that appear crucial to determine when and how a cell will expand, and also when it will stop. This is discussed in

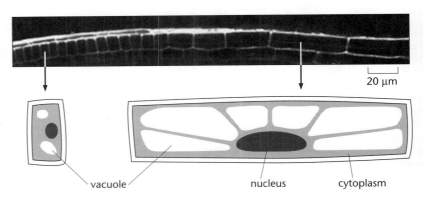

Figure 1.21 Cell expansion is accompanied by vacuole enlargement. Cells produced in the root meristem, shown here on the left of the micrograph, rapidly elongate (with relatively little increase in width). The net increase in volume is largely accounted for by growth of the vacuole, as shown in the drawings. (Courtesy of Takashi Ishii.)

20 μm

vacuole   nucleus   cytoplasm

more detail in Chapter 7. Since cells in tissues are firmly cemented to one another, it also follows that the rate of cell expansion must be integrated with that of adjacent cells and cannot be a cell-autonomous property. Each cell is entrained by its neighbors, and the growth of tissues and organs will finally depend on the coordinated expansion of all their cells.

3. Depending on its position in the plant, a cell differentiates into a specific cell type.[ref10]

In addition to growing, cells produced as a result of meristem activity ultimately become progressively committed to a particular fate and differentiate into an appropriate mature cell type. The process of *differentiation* involves a series of steps in which the pattern of gene expression and the cell's complement of organelles, proteins, polysaccharides, and other components becomes progressively geared to the particular function of the mature cell type. This can be seen clearly by following cells in a root, as they are left further and further behind the advancing root meristem. Here the cells divide predominantly transversely, giving rise to long files of cells (see Figure 1.21). By following cells up such a file we can see directly the structural changes that occur as they differentiate into, for example, phloem sieve elements or protoxylem elements (**Figure 1.22**). In this case differentiation commences before the cell has fully finished expanding: the two processes of growth and differentiation overlap.

It would be tempting, looking at the files of cells in the root, to think that cell fate is in some way a function of cell lineage; in other words, a cell's ancestors determine what it will become. In fact, a large body of evidence suggests

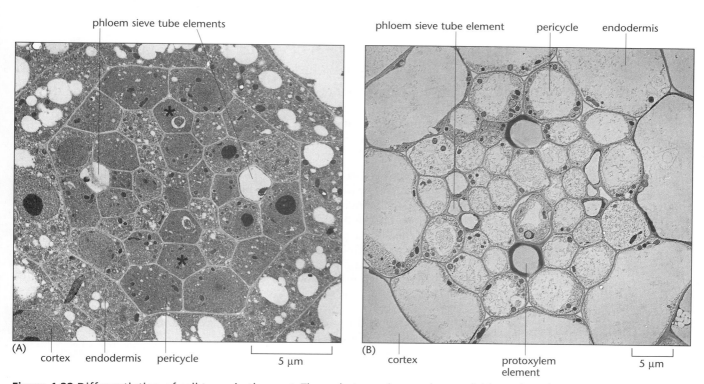

Figure 1.22 Differentiation of cell types in the root. These electron micrographs are of thin sections through the tip of a young *Arabidopsis* root. The vascular tissue system at the center of the root is shown, and cells in particular positions within this simple and regular structure are beginning to differentiate. (A) Clearly visible in section near the tip of the root are the two young phloem sieve tube elements that have already differentiated. The positions of the two future protoxylem elements are indicated by an asterisk. Other differentiating cell types are also labeled. (B) A few millimeters back from the tip of the root, the two protoxylem elements have differentiated. Cell elongation has occurred at this level, and the dramatic increase in the amount of vacuole can be seen, particularly in the endodermis and pericycle. (A, courtesy of Paul Linstead; B, courtesy of Glenn Freshour.)

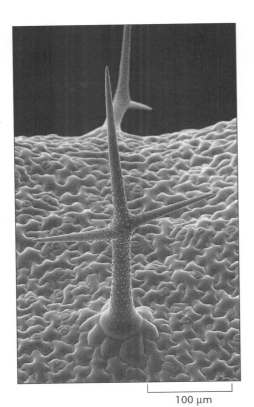

Figure 1.23 The leaf hair, or trichome, of *Arabidopsis.* This scanning electron micrograph shows the large, three-pointed single-cell trichome rising from the young leaf surface. Trichomes probably have a role in insect resistance, but their precise shape has made them a valuable genetic model for cellular morphogenesis. (Courtesy of Paul Linstead.)

that this is not the case. Instead it seems that, as in animals, cell fate is largely determined by local signals from neighboring cells, in other words, from a cell's position. Of course, early patterned divisions in the embryo help to establish the relative positions of the initials, and in roots at least, files of related cells arising from these initials help establish the basic cellular pattern of the root. All the evidence, however, suggests that the fine-grained pattern of final cell types depends on early signaling between neighbors, and that the cell's fate is established by this *positional information*. The molecular machinery used by cells to register positional information is beginning to be understood, and it is likely to involve interactions between a variety of plant growth factors and their corresponding intracellular signal transduction pathways, mobile transcription factors, and direct signaling between molecules expressed at the surfaces of adjacent cells.

A key part of the differentiation process is the set of downstream changes that take place in the wall. We explore in the next section some of the specific wall-related features of a selection of differentiated cell types, but the point to emphasize here is that each cell type has a characteristic size, shape, and cell wall structure and composition and that this alone is often sufficient for us to identify it. A good example is the leaf *trichome*. In *Arabidopsis* this sharp three-pointed cell on the leaf helps protect the leaf against insect attack (**Figure 1.23**). Each trichome starts as an epidermal cell and then grows and shapes itself, modifying its cytoskeleton, selecting new growth sites, orienting its spines with respect to the leaf axis, and depositing oriented and strengthening cellulose in its wall. The complex pathway of *cell morphogenesis* followed by this particular cell type is being dissected genetically by characterizing a large number of mutations that affect the process and identifying the genes that act in the pathway. The analysis of these genes is likely to throw light on how cells shape and construct their characteristic walls.

4. A distinction is often made between primary and secondary walls.[ref11]

It is common in the plant literature to make a formal distinction between primary and secondary cell walls. Although the distinction is useful, the definitions we adopt may differ from their usage elsewhere. We have seen that new cells arise by division, they grow, and they mature. What happens to their walls?

It is simplest to describe each newly established wall as the *primary cell wall.* Almost without exception this wall is thin, between 50 and 200 nm; is separated from its neighbors by a middle lamella; and has a relatively constant proportion of the four main classes of wall macromolecules: protein and the polysaccharides cellulose, hemicellulose, and pectin (**Table 1.2**). An important defining feature of this primary wall is that it is capable of sustaining cell expansion, growing in area itself in the process. Depending on its fate, a cell may follow one of two options. It may simply retain its thin primary cell wall, with little elaboration, as it adopts its final size and fate.

TABLE 1.2 COMPOSITION OF A TYPICAL PRIMARY CELL WALL, %		
MACROMOLECULE	**DICOTS AND MONOCOTS (EXCLUDING GRASSES)**	**GRASSES**
Cellulose	25–30	20–30
Pectin	30	10–15
Hemicellulose	20-30	40–50
Protein	1–10	1–5

This is true, for example, in many parenchyma cells such as leaf mesophyll cells or potato storage cells. Alternatively, at a point in its differentiation pathway when a cell has finished expanding, it may begin to add successive new layers to its wall, always on the side of the wall next to the plasma membrane, to create what is designated as the *secondary cell wall*. In some cases these new layers may be similar in composition to the remaining primary wall (for example, in collenchyma, in the outer epidermal wall, or in phloem sieve elements), although it is usual for the proportion of pectic polysaccharide in the wall to drop and of cellulose in the wall to rise. In other cases the layers may have very different compositions; for example, the thickened wall of cotton fiber cells is almost pure cellulose. Last, these additional layers may become impregnated with other materials, most commonly lignin, as in the case of xylem vessels, fibers, and sclereids (**Figure 1.24**). In extreme cases the secondary wall may occupy the bulk of the cell's volume, as in lignified fiber formation.

In general, secondary walls are deposited in cells that have finished expanding, but this is by no means universal. Even young *protoxylem elements*, which have lignin deposited in secondary thickenings, are structured in such a way that the inextensible lignin is arranged in rings or helices that still allow the primary cell wall between to grow and the cell to elongate (see Figure 1.40). The addition of large amounts of secondary wall may prevent a cell from ever dividing and growing again, clearly the case for dead xylem vessels! Plants, however, have evolved strategies that in many cases of stress or trauma do allow for the reprogramming of mature cell types. Simple examples include the production of adventitious roots from groups of leaf or stem cells and the repair of vascular continuity around a cut in a stem in which parenchyma cells redifferentiate into xylem tracheary elements (**Figure 1.25**). This remarkable reprogramming can even be achieved *in vitro*, using isolated leaf cells in culture (Figure 1.25C). Many apparently determined cell types do have considerable plasticity, either to redifferen-

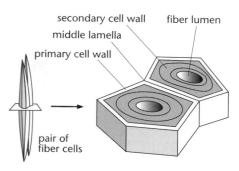

Figure 1.24 Secondary cell walls. This schematic drawing of a section of two adjacent fiber cells shows how successive tough secondary cell wall layers are laid down within the thin preexisting primary cell wall. In the case of fibers the cell that existed in the reduced space of the lumen usually dies and is removed, creating an empty cavity, and the thick secondary cell wall is impregnated with lignin (see also Concept 1C10).

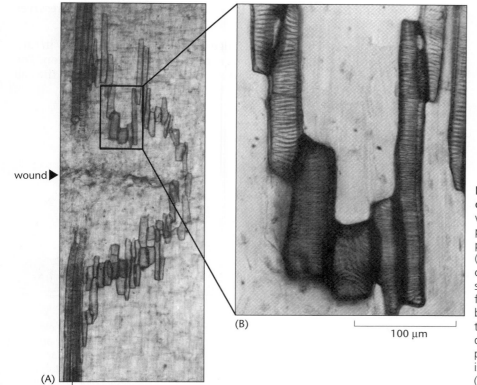

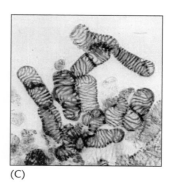

Figure 1.25 The repair of vascular continuity in response to a wound. A small vascular bundle in a stem of *Coleus,* a house plant, has been cut with a scalpel and the plant left for a week to repair the damage. (A) The micrograph shows that ground tissue cells have redifferentiated into a connected set of vascular elements that reestablish the flow of liquid through the original vascular bundle. (B) The magnified portion shows the wall thickenings typical of normal differentiated xylem vessels. (C) The same process can be achieved in cell culture, using isolated leaf cells from the plant *Zinnia.* (A and B, from E.G. Cutter, Plant Anatomy, 2nd ed. London: Edward Arnold, 1978. C, courtesy of Zheng-Hua Ye.)

tiate into a different cell type or even to re-enter the cell cycle. Cells in an excised plant organ—a stem, for example—can be induced to proliferate to form a mass of cells called a *callus*. Under suitable conditions, whole new plants can be regenerated from this callus material, an important procedure both in plant research and in commercial plant propagation. All of these events necessarily involve a wall that is not irrevocably locked into a final state, and all of the cells that can adapt in this way must retain the ability to expand or remodel their primary walls.

5. Differentiated cell types are often characterized by functionally specialized cell walls.[ref12]

In our discussion of the origin and diversity of primary and secondary walls, it is important to emphasize that this diversity has functional significance. Cell walls, with their remarkable chemical, physical, and mechanical properties, provide exceptional building materials for the plant (and also indeed for humans in the form of wood and its by-products; see Chapter 9). Their ability to withstand both tension and compression provides adaptive strength for both young plants and old trees. Epidermal cells and the protective layers of cells or periderm that later replace them (for example, bark) both have resistant waterproof walls that help to protect the stems of plants against water loss and pathogens. The specialized architecture of different secondary walls allows the construction of both high-pressure and low-pressure tubes to transport water and solutes around the plant, and yet in leaves the walls are transparent enough to allow through the light required for photosynthesis. In some cells these transparent properties are used to advantage. In the petals of many flowers, for example, the epidermal cell outer wall is shaped into a conelike structure that acts like a prism, intensifying the reflected color of the pigments inside the cell that, together with the surface texture that the cones provide, are responsible for attracting pollinating insects (**Figure 1.26**). The ability of cells to lay down extra wall layers is used by fibers, for example, to construct tough, stretch-resistant cables to reinforce soft tissues, and by cells in some seeds to store the food reserves that will be needed for the early growth of the young embryo.

All of these examples highlight the range of wall properties and how in turn these have been used and exploited by plants during their evolution. The next section looks in more detail at the functional specialization of the cell wall in a variety of common cell types found in present-day plants.

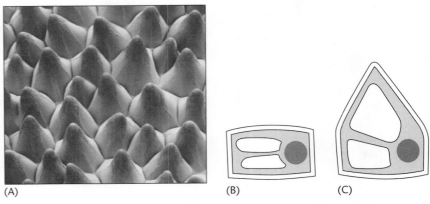

(A)   (B)   (C)

Figure 1.26 Cone cells on the inner petal of snapdragon. (A) Scanning electron micrograph of the outer surface of the petal epidermis showing the conical apex of each cell. The production of this cone requires the presence of a particular transcription factor (a Myb family member), and when the gene encoding this is mutated, the cone fails to form. Sections through a mutant petal epidermal cell (B) and a wild-type epidermal cell (C) show the local wall elaboration required to construct the cone. (Courtesy of Cathie Martin and Beverley Glover.)

C. Plant Cell Types and Their Walls

1. Shoot epidermal cells produce a waterproof cuticle that helps provide protection.[ref13]

The outermost cell layer of the plant, the *epidermis,* is strategically placed between the plant and its environment. In the aerial parts of the plant, the cells of the epidermis have become functionally modified to serve two purposes. The first is to avoid water loss, and the second is to provide physical protection against predators, pathogens, and mechanical insults (see Chapter 8). To this end, the epidermal cells covering the aerial parts of the plant (leaf and stem) develop an unusually thickened and waterproofed wall on their outer face. In leaves these cells adhere to each other particularly strongly, often with cells in an interlocking jigsaw-like pattern that forms a coherent sheet with increased tensile strength (**Figure 1.27**). Epidermal cells typically lack chloroplasts, but those in petals often accumulate pigments such as anthocyanins in their vacuoles.

Almost invariably, the thickened outer wall is covered by a layer containing *cutin* (whose chemical structure is discussed in Chapter 3) that is relatively impermeable to water. This in turn often has a layer of wax deposited on its outer surface. The thickness of the cutin layer, or *cuticle* (**Figure 1.28**), varies considerably from species to species, but it is also to some extent under environmental control. Thus, in plants suffering water stress, the cuticle may be thickened. Somewhat paradoxically, neither the thickness of the cuticle nor the amount of wax correlates well with how effective the cuticle is as a water barrier. Instead, water transport depends on the microarchitecture of the cuticular layers, on the chain-length distribution of the aliphatic wax constituents, and on the physical arrangement of the waxes within the cuticle.

Cutin is a tough polymer and is not readily broken down, which accounts for its persistence in plant fossils. In surface view the cuticle forms species-specific patterns that relate to the shapes of the underlying epidermal cells. The cuticle of some leaves and fruit may be further modified by the deposi-

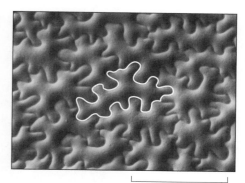

Figure 1.27 The leaf epidermis. This scanning electron micrograph of the surface of a tomato leaf demonstrates the jigsaw-like strategy adopted that results in a tough, coherent sheet of cells to cover the leaf. For clarity, a single epidermal cell has been outlined in white. (Courtesy of Kim Findlay.)

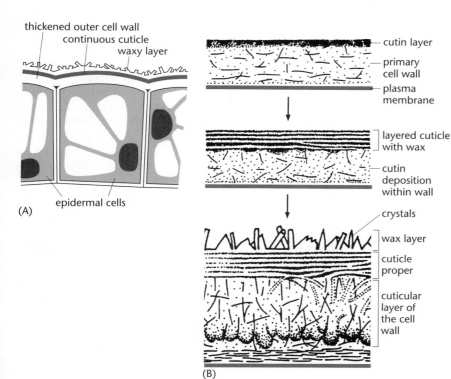

Figure 1.28 The plant cuticle. (A) The outer epidermal cell wall, for example, in a leaf, is usually thickened and overlaid by a continuous layer, the cuticle. In some cases this has a layer of waxes on its outer face that often crystallize into characteristic shapes. (B) The development of the cuticle involves the gradual deposition of cutin at the surface, as well as the impregnation of the wall itself, followed last by wax layer deposition. (B, adapted from C.E. Jeffree, in G. Kersteins, ed., Plant Cuticles: An Integrated Functional Approach, pp. 33–82. Oxford: BIOS Scientific, 1996.)

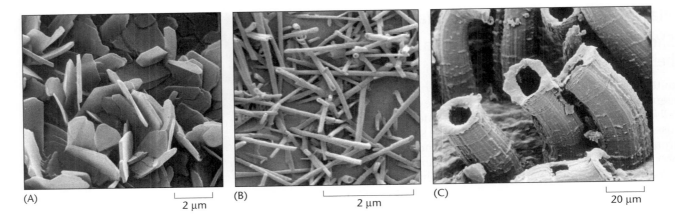

(A)   2 µm   (B)   2 µm   (C)   20 µm

Figure 1.29 The wax layer. Scanning electron micrographs showing the variety of crystalline wax deposits on the cuticle of different plants. (A) Wax plates on a leaf of *Lecythis chartacea*. (B) Tubular wax crystals on the surface of a wheat leaf. (C) Wax chimneys around stomata on a leaf of *Colletia cruciata*. (A and C, courtesy of Kerstin Koch; B, courtesy of Kim Findlay.)

tion of waterproof waxes on its outer face. This is often visible as a "bloom" such as that seen on grapes and on some leaves (**Figure 1.29**). Wax deposition is again to some extent under environmental control. In rare cases sufficient wax is deposited for it to be commercially useful. Carnauba wax, used in polishes, is derived from the surface of the leaves of the carnauba palm. The precursors for both cutin and wax layers of the cuticle are transported through the outer epidermal cell wall, probably in association with specialized lipid transport proteins that are expressed only in epidermal cells, where they are targeted to the outer wall.

2. Stomatal guard cells have asymmetrically thickened walls that allow them to reversibly change shape.[ref14]

In the epidermis of leaves, and of many stems, specialized groups of cells arise that create small pores, called *stomata* (singular, *stoma*), that regulate the exchange of carbon dioxide, oxygen, and water vapor between the underlying leaf cells and the atmosphere. Each stoma is located over a corresponding air space created between the underlying photosynthetic cells (**Figure 1.30**). The opening and closing of stomata each confer both costs and benefits to the plant. Stomatal opening allows the plant access to carbon dioxide, but equally facilitates water loss through the open pores. Stomata can balance these costs and benefits by their exquisitely sensitive opening and closing responses to water availability and carbon requirements.

The cells that create the stoma consist of two *stomatal guard cells* that create the pore and control its opening and closing, together with a variable number of subsidiary cells depending on the species. The whole cell group is commonly called the *stomatal complex*, and it arises during development

Figure 1.30 Sections of the stomatal complex. (A) A vertical section through the lower epidermis of a young tobacco leaf viewed in the electron microscope. A closed stomatal complex with its two stomatal guard cells can be seen over a large underlying air space within the leaf. (B) A transverse section of a stoma in a sugar beet leaf. The common wall has separated along part of its length to create the pore. Unlike other leaf epidermal cells, the guard cells have numerous chloroplasts. (A, courtesy of Cathie Martin. B, from K. Esau, Anatomy of Seed Plants, 2nd ed. New York: John Wiley & Sons, 1977.)

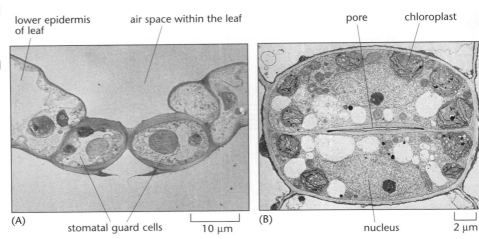

lower epidermis of leaf   air space within the leaf   pore   chloroplast

(A)   stomatal guard cells   10 µm   (B)   nucleus   2 µm

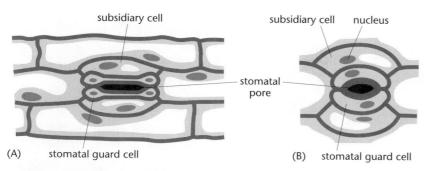

(A) subsidiary cell

stomatal guard cell

(B) subsidiary cell   nucleus

stomatal pore

stomatal guard cell

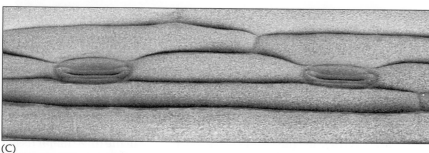

(C)

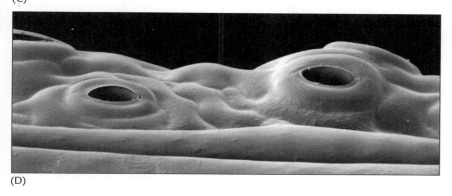

(D)

by a very precise sequence of epidermal cell divisions whose pattern is species-specific (**Figure 1.31**; see also Figure 1.13).

The two stomatal guard cells are the daughters of a common mother cell, and as they differentiate they deposit a secondary wall that is rich in cellulose and that is thickened in the region that faces the future pore. The guard cells have a cuticle. The middle lamella between the two guard cells, which are commonly kidney-shaped, is removed in a defined central region to create a space or pore between them (the stoma). This region of wall is rich in callose that has been deposited there during the division of the guard cell mother cell, and this may play a part in pore formation. Because each guard cell has unequal wall thickenings on either side, changes in turgor pressure that result indirectly from rapid effluxes and influxes of potassium ions can cause a small but predictable distortion of cell shape, enough to affect the opening and closing of the pore between the two guard cells (**Figure 1.32**). One can think of each guard cell as a partially inflated sausage-shaped balloon, with strong sticky tape stuck along one side to represent the reinforced side wall. If the balloon is further inflated, unequal expansion of the two sides will cause it to distort into a kidney shape. The generation of large turgor pressure fluctuations in the guard cells, without affecting their neighbors, is possible only because plasmodesmatal connections between them are broken during development and the guard cells become symplastically isolated. Environmental control of stomatal opening and closing is mediated by a complex signal transduction pathway involving multiple ion channels regulated in part by the plant signaling molecule abscisic acid.

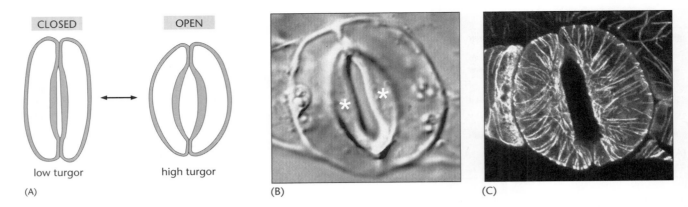

Figure 1.32 Stomatal opening and closing. Stomatal opening is a turgor-driven process that is dependent on the water status of the leaf (A) and is thought to be mediated by the arrangement of cellulose microfibrils in the wall of the guard cells and the thickened wall (*asterisks*) adjacent to the pore (B). (C) The microtubules in the young stomatal guard cells mirror the orientation of the cellulose microfibrils in the wall. (B and C, courtesy of Geoff Wasteneys.)

Stomatal complexes in the leaf are also important entry sites for pathogenic bacteria and fungi, and these are discussed later in Chapter 8.

3. Cells in ground tissues fill the space between epidermis and vascular tissue.[ref15]

As we discussed earlier, the ground tissue forms one of the three major tissue systems of the plant (see also Panel 1.1). The most common cell type found in ground tissue is *parenchyma*. The cortex and pith of stems and roots, the mesophyll of leaves, and the flesh of fruit, bulbs, and tubers are all largely composed of parenchyma. As well as having structural roles, parenchyma cells may function in photosynthesis, respiration, and storage and are generally large, highly vacuolate cells with relatively unmodified primary cell walls. Although they are differentiated cells, the absence of a thickened secondary wall means that they are capable not only of further growth, but also in some cases of re-entering the cell cycle and proliferating. Indeed, in many cases they are the source of cells for renewed growth or repair after wounding. Root nodules, cell cultures derived from explants, and roots and shoots regenerated from cuttings commonly arise from parenchyma cells that are triggered to divide.

A consequence of the thin walls of parenchyma cells is that their shape may be modified by external forces. Many cortex and pith cells, for example, are nearly isodiametric polyhedra, whose shapes are largely determined by pressures exerted by the epidermis, like little rubber balls squeezed tightly into a jar (**Figure 1.33**). Thin-walled ground-tissue parenchyma cells can generate mechanical support and force through their internal turgor pres-

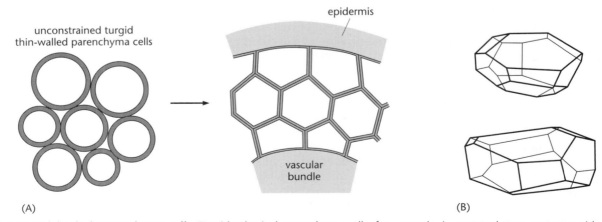

Figure 1.33 Polyhedral parenchyma cells. Turgid spherical parenchyma cells, for example, in root and stem cortex or pith, are often under pressure from the constraining dermal tissue system. (A) This constraint forces the spheres into typical polyhedral cell shapes, rather like soap bubbles. (B) In three dimensions geometrical considerations suggest that such cells commonly have on average 14 sides or facets. (B, adapted from R.L. Hulbary, *Am. J. Bot.* 31:561–580, 1944.)

intercellular spaces

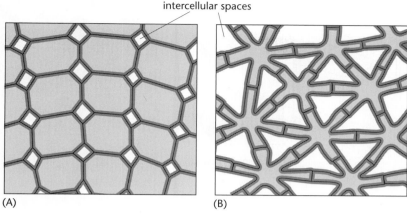

(A)   (B)

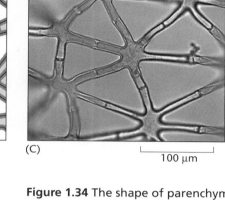

(C)

100 µm

sure. The tissue forces generated by turgid parenchyma, when constrained by a strong epidermal layer, can be used to provide mechanical support to organs and the force, for example, to drive root tip growth in soil. In some cases, more highly organized patterns of parenchyma cells are seen, as in the cortex of some roots, while in other cases parenchyma cells can undergo controlled cell separation at particular sites on the cell leading to the formation of gaps or intercellular spaces (**Figure 1.34A**). The degree of wall separation at the three-way junctions between parenchyma cells can vary widely, and this is discussed further in Chapter 6. In some tissues there are no intercellular spaces to speak of, while in more extreme cases such as the cortex of waterlogged roots or the pith cells of some rushes, enormous extended air spaces are created between cells (Figure 1.34B and C). Plant tissues, it seems, can respond to conditions of low oxygen tension and yet at the same time balance the resulting increased air spaces with the mechanical constraints imposed by increased cell separation.

Figure 1.34 The shape of parenchyma cells. (A) The cortex of some roots has regular files of parenchyma cells with well-developed air spaces at the junctions (B). Extreme development of functional air spaces arises when localized cell-cell attachments are maintained while local cell elongation events occur, as in the branched, or stellate, parenchyma found in the pith of rushes and other waterlogged plants (B). (C) A light micrograph of stellate parenchyma in a rush stem. (C, courtesy of Ales Kladnik.)

4.   The endodermis regulates the flow of solutes through the apoplast of the root.[ref16]

In roots of higher plants, the innermost cell layer of the cortex is called the *endodermis* (**Figure 1.35**). This single-cell layer is thought to act as a barrier to prevent the free apoplastic flow of water and ions between the cells of the cortex outside the endodermis and the vascular tissue inside the endodermis. This barrier function resides in a remarkable structural modification of a local region of the wall of the endodermal cells. As the cylinder of endodermal cells is first established in the root meristem, a region within their radial walls becomes identifiable by the lack of pectic polysaccharides deposited there. Further back in the root, as the endodermal cells differentiate fully, usually at about the level in the root where root hairs appear, this wall region becomes thoroughly impregnated with a dense, cross-linked set of hydrophobic polymers called *suberin,* whose chemistry is discussed later (see Concept 3G2). The ring of waterproof material in each wall is called the *casparian band,* and each ring is continuous with that in the wall of every other endodermal cell in such a way that an impermeable apoplastic cylinder is formed around the vascular tissue (Figure 1.35). For ions and water to enter the vascular system, they must leave the apoplast, cross the plasma membrane of the endodermal cell, and enter the symplast. The seal is made more effective by the very tight anchoring of the plasma membrane to the hydrophobic casparian band. This anchor remains tight, even if the rest of the protoplast is pulled away from the wall by severe plasmolysis (**Figure 1.36**).

When mixtures of wall-degrading enzymes are added to young roots to solubilize their walls and the cytoplasmic contents are also removed, what

Figure 1.35 The endodermis and the casparian band. (A) Schematic diagram showing the ring of endodermal cells in the root that separates the cortex outside from the vascular tissue system inside. The casparian band encircling each endodermal cell is neatly connected to that of its neighbors (above, below, and on either side) to create an apoplastic barrier to the movement of water and solutes into the vascular tissue. (B) Electron micrograph showing a casparian band from the endodermis of an *Arabidopsis* root. The cells have been immunogold-labeled (*black dots*) with an antibody against arabinogalactans associated with the plasma membrane and cell wall interface. Label is excluded from the region of the casparian band, emphasizing the special nature of the casparian band–plasma membrane interaction. (B, courtesy of Glenn Freshour.)

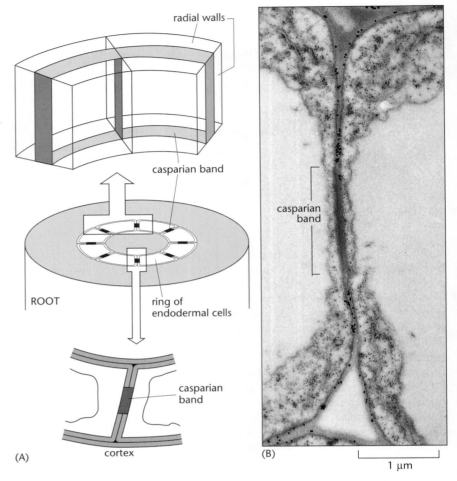

remains behind is a netlike cylinder of enzyme-resistant and interconnected casparian bands (**Figure 1.37**). This elegant purification method should allow the molecules involved in anchoring the plasma membrane, as well as more detailed chemistry of the band itself, to be elucidated.

5. The pericycle encircles the vascular tissue and can give rise to lateral roots.[ref17]

We have seen earlier how cells left behind by the apical meristem enlarge and differentiate, many of them by elaborating thickened secondary cell walls. In roots, however, the cells produced by the root meristem include a population that retains the ability to divide and enlarge and that can later produce lateral roots. This population of cells forms a cylinder of small,

Figure 1.36 Plasma membrane attachment to the casparian band. (A) A section of a severely plasmolyzed pea root is shown here at low power in the electron microscope to reveal the sites in the endodermal cells where the plasma membrane remains attached to the cell wall. These sites (*arrows*) correspond to the casparian bands, one of which is shown at higher magnification (B). (From I. Karahara and H. Shibaoka, *Plant Cell Physiol.* 33:555–561, 1992.)

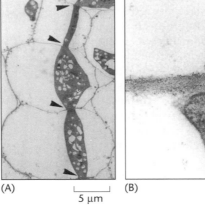

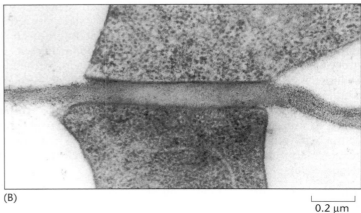

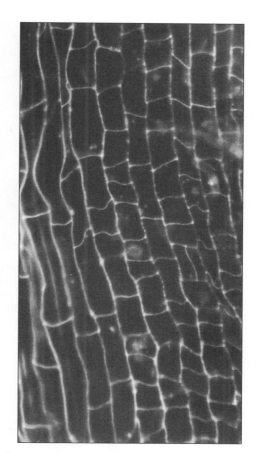

Figure 1.37 The endodermal "net" of casparian bands. When the cell contents and primary cell walls of a pea root are completely digested away, the interconnected cylindrical "net" of endodermal casparian bands remains and can be seen in the light microscope, where it fluoresces in ultraviolet light. (From I. Karahara and H. Shibaoka, *Plant Cell Physiol.* 33:555–561, 1992.)

thin-walled cells, one cell thick, lying between the endodermis and the vascular tissue, called the *pericycle* (**Figure 1.38**; see also Figures 1.10 and 1.43). A stimulus, thought to include the plant growth factor auxin, triggers specific groups of pericycle cells (whose position usually bears a defined relationship to the pattern of xylem vessels in the vascular tissue) to divide, first transversely and then periclinally, to produce a mound of cells that will push through the overlying root tissues, develop within it a new root meristem, and emerge as a lateral root (**Figure 1.39**).

Small, thin-walled cells, of which the pericycle is but one example, present within the body of the plant are an important source of potential stem cells for the plant. In cases where renewed cell proliferation is necessary, for example, the production of callus to protect a wounded stem, or the regeneration of shoots or roots, the plant can call on these reservoirs of cells that are still competent to re-enter the cell cycle and thereby initiate the formation of a new organ.

6. Xylem vessel elements are pipes that conduct water at negative pressure.[ref18]

In addition to anchoring the plant in the soil, the root system supplies the rest of the plant with water and dissolved ions. The vascular tissue responsible for transporting this water throughout the plant is the *xylem*. Although other cell types are present in the xylem, it is the water-conducting *vessels* that are the most conspicuous. Each vessel is constructed from a series of tubelike cells called *vessel elements,* placed end to end to form a long continuous pipe. Since water is pulled up through the vessels by suction, largely generated by water evaporating from the leaves, the vessels have to withstand the compressive forces that result from the negative pressures within. A series of thickened ridges on the inner face of their cell walls prevents the vessels from collapse, in much the same way as flexible vacuum cleaner tubing is strengthened by rings or helices of wire inserted in the casing. In the protoxylem of young growing tissues, the pattern of wall thickenings is similar (**Figure 1.40**), typically with rings or helices of thickened cell wall

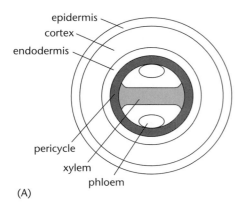

10 µm

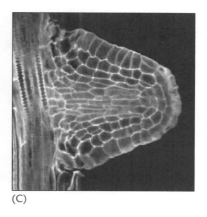

Figure 1.38 The pericycle. (A) Schematic drawing of the cells in the root of *Arabidopsis,* showing the ring of pericycle cells. (B) Longitudinal section of a young root showing early stages of lateral root formation. (C) A later stage of lateral root emergence is shown in a confocal section in which the cell walls have been labeled with a fluorescent stain. (B, from J.E. Malamy and P.N. Benfey, *Development* 124:33–44, 1997. C, from E. Truernit et al., *Plant Cell* 20:1494–1503, 2008.)

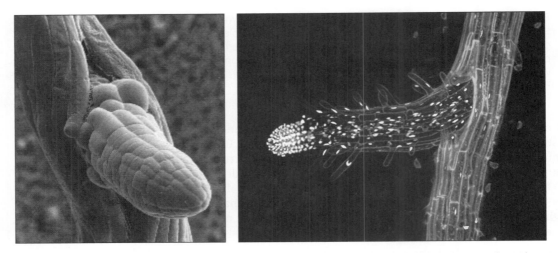

Figure 1.39 Lateral root emergence. (A) This scanning electron micrograph shows a young lateral root of *Arabidopsis* that has arisen by divisions of a group of pericycle cells within the primary root. It is seen emerging through the disrupted endodermis, cortex, and epidermis cell layers of the primary root. (B) A three-dimensional confocal image of a similar lateral root emerging from the primary root. The nuclei are fluorescing and clearly mark out the new lateral root meristem. (A, courtesy of Paul Linstead; B, courtesy of Jim Haseloff.)

characterized by the deposition of *lignin*, a tough waterproof phenolic polymer whose structure will be described later (see Concept 3F1). The thickenings are elaborated early in cell differentiation and initially are largely composed of cellulose and xylan. In addition, a large number of proteins are

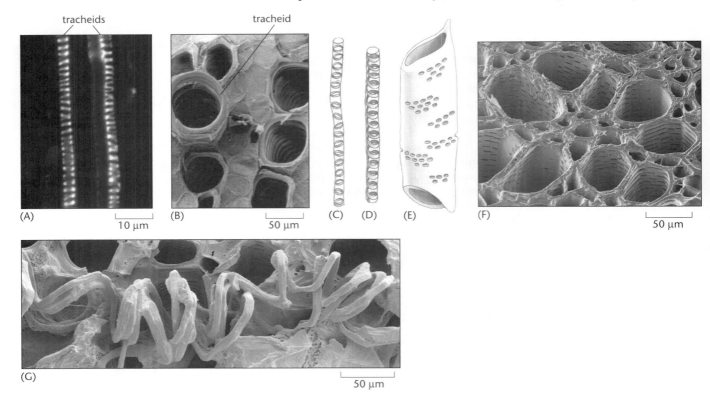

Figure 1.40 Wall thickening in primary xylem. In the young seedling root of *Arabidopsis* two primary xylem strands develop to carry water to the shoot. (A) The young xylem cells, or tracheids, have regular lignified annular thickenings that fluoresce in ultraviolet light. (B) In the small vascular bundles of a stem, the primary xylem elements often have a helical arrangement of lignified thickenings, as seen in this scanning electron micrograph of a young *Zinnia* stem. Primary xylem vessel elements and tracheids, therefore, can have annular, ringlike thickenings (C) or helical thickenings (D), while more mature vessel elements, for example, in secondary xylem or wood, may be much more extensively thickened with only small pitted regions and the end walls open and unlignified (E, F). (G) In a scanning electron micrograph of a cut *Zinnia* stem, the lignified spiral thickenings, which have been pulled clear of the tracheids, can be seen. (B, courtesy of Kim Findlay. F and G, courtesy of Preeti Dahiya.)

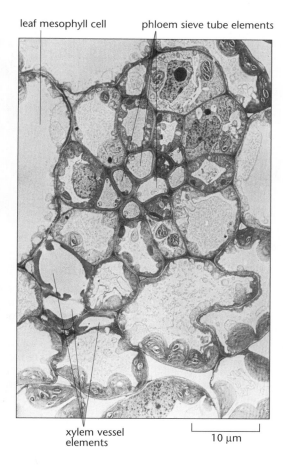

leaf mesophyll cell   phloem sieve tube elements

xylem vessel elements

10 μm

Figure 1.41 Vascular bundles. Phloem tissue and xylem tissue, together with other cell types, are organized into vascular bundles that transport water and solutes around the plant. The vascular tissue system can range in size and complexity from the major conducting tissues of a large tree trunk to the minor veins in a leaf, similar in fact to the differences between our arteries and capillaries. Shown here is a section of a small vein in a tobacco leaf. A group of hollow xylem vessel elements and phloem sieve tube elements are seen in this small vascular bundle, associated with a number of parenchyma cells. (Courtesy of Cathie Martin.)

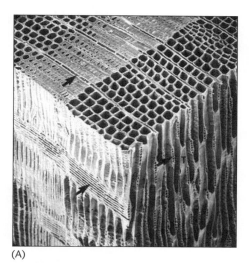

(A)

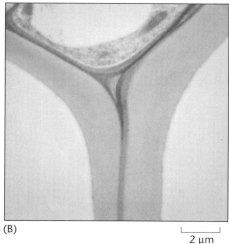

(B)   2 μm

Figure 1.42 Secondary xylem. (A) Scanning electron micrograph of pine wood showing secondary xylem. The arrows point to files of living ray parenchyma cells that cross the dead tubes of the secondary xylem. (B) The thick lignified secondary cell walls of two xylem vessels are seen in this section of root from *Arabidopsis*. (A, courtesy of W.A. Coté. B, from S. Persson et al., *Plant Cell* 19:237–255, 2007.)

known to be specifically associated with the wall thickenings during vessel formation. These include a glycine-rich protein associated with the non-thickened region of the wall and arabinogalactan proteins associated with the wall thickenings (see Concept 3E). Lignin is then laid down within the matrix of the thickened regions and can account for more than 25% of the dry weight of the mature wall. The final event in vessel element differentiation is the controlled dissolution of the cell contents and the selective removal of the end walls of the cells, leaving as a result a completely hollow, toughened tube made of dead cell walls.

In the stems, roots, and leaves of young plants, xylem is usually found associated with the conducting elements of the phloem (another conducting tissue, discussed next), in organized vascular bundles (**Figure 1.41**). As the plant grows and develops, secondary xylem can be produced from cells that arise by cell divisions in the *cambium*, a lateral meristem. This secondary thickening becomes most significant in trees and other woody plants. Wood contains large numbers of secondary xylem vessels and tracheids (**Figure 1.42**). In these the surface area of the wall that is occupied by lignified thickenings increases, sometimes leaving only small pits in the wall (Figure 1.40). The genetic and molecular controls of xylem differentiation are not well understood, but this is an active area of research because of commercial interest in modifying the properties of wood.

7. Phloem sieve tube elements are high-pressure tubes for conducting sugar solutions.[ref19]

In leaves photosynthesis fixes carbon in the form of sucrose, which is then translocated to all the other living cells in the plant. The vascular tissue responsible for the transport of sucrose and other necessary metabolites

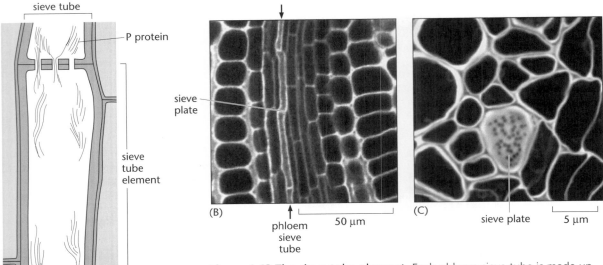

sieve tube

P protein

sieve tube element

sieve plate

companion cell

(A)

sieve plate

phloem sieve tube

50 μm

(C)

sieve plate

5 μm

(B)

Figure 1.43 The sieve tube element. Each phloem sieve tube is made up from numerous thick-walled cells, sieve tube elements (A) connected end to end to form a continuous pipe or tube for conducting sucrose around the plant (B). (C) The end walls are perforated to form a sieve plate. When phloem tissue is damaged, callose is rapidly deposited at the sieve plate to close the pores, and P-protein blocks the remaining gaps to reduce the flow of liquid through the tube. B and C are confocal microscope images of *Arabidopsis* hypocotyl and stem, respectively. (B and C, from E. Truernit et al., *Plant Cell* 20:1494–1503, 2008.)

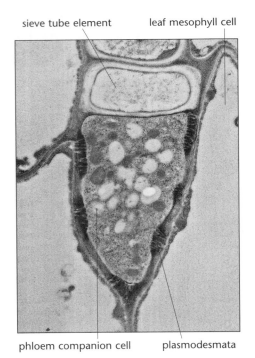

sieve tube element

leaf mesophyll cell

phloem companion cell

plasmodesmata

Figure 1.44 Phloem cells. This cross section of a part of a minor leaf vein shows a hollow, thick-walled phloem sieve tube element and its associated densely cytoplasmic companion cell. The latter, in this case, is connected to its neighboring leaf mesophyll cells by numerous branched plasmodesmata. (Courtesy of Adrian Turner.)

is the *phloem* (see Figure 1.41). In flowering plants, the cells in the phloem that actually carry the sucrose solution are the *sieve elements.* The sieve elements are placed end to end to create long sieve tubes, whose walls are specialized for translocation (**Figure 1.43**; see also Figure 1.22).

During its differentiation, a sieve element loses most of its cellular contents, including the nucleus, but unlike a xylem element it retains its plasma membrane. This membrane remains in continuity, through numerous plasmodesmata, with that of a neighboring sister cell called the *companion cell,* which provides essential materials to its sieve element during the course of its life and in addition helps load and unload sucrose at the appropriate time and place (**Figure 1.44**; see also Figure 1.22).

Active sucrose loading occurs through a sucrose/proton co-transporter in the plasma membrane of the sieve element and/or its companion cell, driven by a proton gradient generated at the expense of ATP. The high concentrations of sucrose (0.3 to 0.9 M) in the sieve elements of the leaf or source tissue cause water supplied by the xylem/apoplast to be taken up by osmosis, generating high pressures within the sieve tubes. An osmotically generated pressure gradient is formed relative to sink tissues, such as the root, where sucrose is unloaded, and this pressure gradient is thought to drive the bulk flow of sucrose solution at an average rate of about 1 m/h through the sieve tubes. The individual sieve element cells need, therefore, to withstand this high-pressure translocation. They do this by depositing evenly thickened and reinforced secondary cell walls. These have high levels of circumferentially arranged cellulose microfibrils that resist tensions in the wall. The sieve element's internal pressure is also partially relieved by the pressure exerted by the surrounding parenchyma tissue.

The end walls of each sieve element are modified to create a continuous tube that at the same time offers enough resistance to the bulk flow of liquid to help maintain the hydrostatic pressure gradient. A series of small holes, up to 1 μm or so in diameter, are created by localized wall dissolution forming a porous end wall called a *sieve plate* (Figure 1.43). The sieve plate is also elegantly adapted to cope with insults. If the sieve element is damaged, two events happen that together quickly seal off the tube and prevent the leakage of sucrose solution. Very rapid synthesis of callose (see Concept 4A18) around the pores in the sieve plate helps to close them, and loose fibrous proteins, called P-proteins, in the sieve elements are pushed onto the sieve plate as the wound releases the internal pressure. The fibrous P-proteins appear to function by physically blocking the sieve plate pores (Figure 1.43). The ability to close off sieve elements by callose deposition, presumably in a reversible manner, is also used to slow down translocation in the plant during periods of dormancy.

8. Transfer cells load solutes into and out of conducting elements.[ref20]

Phloem companion cells actively concentrate the sucrose supplied by surrounding photosynthetic cells in the leaf and deliver it to the sieve element. In most plants the cell wall plays an important part in sucrose translocation. Sucrose is exported from the source cells at some point into the wall, from where the companion cell then actively takes it up across its plasma membrane. The local high transfer rate of sucrose and other solutes across the plasma membrane between apoplast and symplast often induces the formation of elaborate wall ingrowths that correspondingly increase the surface area of plasma membrane available for transport. Cells adapted in this way are called *transfer cells*. It is important to realize that transfer cells are not a special cell type in their own right. Indeed any cell involved in high local rates of solute transfer between apoplast and symplast may develop characteristic wall ingrowths and thus qualify as a transfer cell. Although the wall ingrowths of transfer cells are not absolutely required, the larger surface area they create increases the cell's capacity to transfer materials.

Transfer cells have been found in a wide variety of locations in plants, each associated with high local rates of solute transfer. They include cells associated with the movement of water and solutes out of the xylem in minor veins and nodes (**Figure 1.45**), with secretory cells like nectaries and some hair cells, and with the cells at the base of cereal seeds that are involved in transferring reserves into the developing endosperm. Support for the idea that a high local flux of solute can itself act as the signal for a cell to develop as a transfer cell comes from experiments in which tissues grown on high sugar concentrations develop transfer cell-like wall ingrowths.

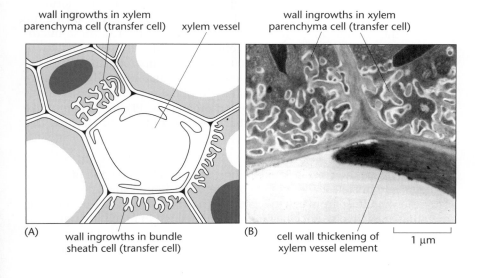

wall ingrowths in xylem parenchyma cell (transfer cell)   xylem vessel

wall ingrowths in xylem parenchyma cell (transfer cell)

(A)   wall ingrowths in bundle sheath cell (transfer cell)

(B)   cell wall thickening of xylem vessel element

1 μm

Figure 1.45 Transfer cells. Cells involved in high local rates of solute transport often develop elaborate wall ingrowths and are then called transfer cells. (A) Shown schematically are two kinds of cells that border a xylem vessel in a leaf vein, both of which have developed wall ingrowths adjacent to the xylem vessel and therefore qualify as transfer cells. (B) An electron micrograph of the wall ingrowths in a xylem parenchyma cell (transfer cell) from the stem vascular bundle entering a leaf of *Zinnia*. Here the wall ingrowths facilitate the unloading of water and salts from the xylem to the leaf. (B, courtesy of P. Dahiya and B. Wells.)

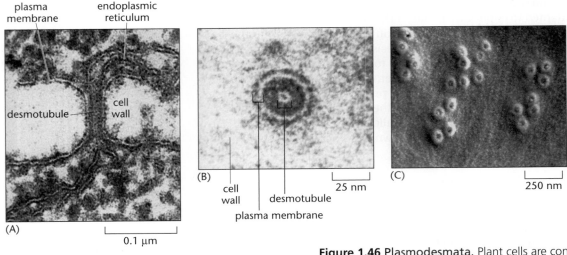

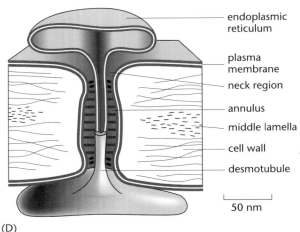

Figure 1.46 Plasmodesmata. Plant cells are connected to their neighbors by plasma membrane–lined tubes, called plasmodesmata, that traverse the intervening cell walls. (A) Section illustrating that the plasmodesma contains a rod, or desmotubule, that is connected to endoplasmic reticulum at either end. (B) The desmotubule can be seen more clearly in this cross section of a plasmodesma. Other proteins, structural and regulatory, occupy the annulus and restrict the free flow of molecules through the channel. They can be seen clearly as electron-dense material in the cross section (B), but the precise arrangement of proteins within plasmodesmata is still unclear. (C) In many older cells, plasmodesmata occur in groups, called pit fields, as seen in this shadowed replica of some isolated maize cell walls. (D) Schematic diagram showing some of the structural features of a simple plasmodesma. (A and B, from R.L. Overall, J. Wolfe and B.E.S. Gunning, *Protoplasma* 111:134–150, 1982. C, courtesy of Adrian Turner.)

9.   Plasmodesmata allow adjacent cells to communicate with each other across their intervening walls.[ref21]

Local, regulated cell-to-cell communication is central to the development and function of multicellular organisms. As discussed earlier, plant cells are in cytoplasmic communication with each other by plasmodesmata, plasma membrane-lined tubes (outside diameter approximately 50 nm) that cross the cell wall and contain a central cylinder of modified endoplasmic reticulum, the desmotubule (**Figure 1.46**). Proteins connect the desmotubule to the surrounding specialized plasma membrane and thereby divide the cytoplasmic annulus into "microchannels." The diameter of these microchannels (about 2.5 nm) appears to define the basal size-exclusion limit of plasmodesmata, which is close to 1000 D. This is comparable to the size-exclusion limit of animal gap junctions and suggests that the need to exchange ions, metabolites, and possibly small, hydrophilic regulatory molecules is common to all multicellular eukaryotic organisms. Plasmodesmata, like gap junctions, are gated channels that close in response to elevated free Ca^{2+} levels in the cytosol.

Plasmodesmata, however, differ from gap junctions by being able to transiently increase their size-exclusion limit and thereby allow the passage of proteins and nucleic acid–protein complexes. Although small proteins can often slowly diffuse from cell to cell through plasmodesmata, this opening, or gating, may involve specific targeting sequences, in much the same way that transport of proteins through nuclear pores is regulated. Plants exploit targeted protein trafficking through plasmodesmata to dispatch regulatory

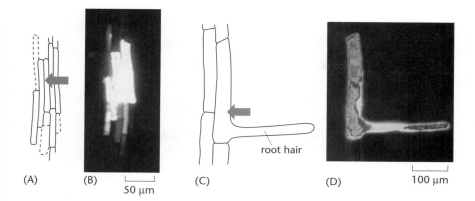

(A) (B) 50 µm (C) root hair (D) 100 µm

Figure 1.47 Symplastic domains. The developmental gating of plasmodesmata is demonstrated by the following experiment. If a small fluorescent dye is carefully injected into a living cell, its movement can be followed over time. Injections into young epidermal cells close to the root meristem (A) result in the spread of the dye away from the injected cell through plasmodesmata (B). If the dye is injected into a fully differentiated root hair cell (C), however, the dye does not move (D). (Adapted from C.M. Duckett et al., *Development* 120:3247–3255, 1994.)

proteins such as transcription factors that help coordinate development in adjacent cell layers. The spread of viruses in susceptible host plants also occurs in part through plasmodesmata and involves the participation of specific, virally encoded movement proteins.

During the early stages of embryo formation, all cells are interconnected by plasmodesmata. This implies that the cells are integrated into a single symplast. Later in development, subsets of plasmodesmata are closed or removed, leading to the symplastic isolation of single cells or groups of cells and to the formation of distinct subcompartments within the plant (**Figure 1.47**). A plant's ability to increase or decrease the number of plasmodesmata, and their gating properties, between different sets of cells is important both for controlling development and for regulating the transport of metabolites between source and sink compartments.

Plasmodesmata are constructed from a limited number of different proteins. Although the actual channel-forming molecules have yet to be identified, a family of membrane receptor-like proteins have been identified that not only localize to plasmodesmata but appear to function in regulating cell-cell trafficking (**Figure 1.48**). The proteins have a single transmembrane domain, which is sufficient to target the proteins to plasmodesmata, and can also be used to target heterologous proteins to the same location. Among other proteins localized to plasmodesmata by immunocytochemistry, several hint at intriguing mechanistic properties. For example, the presence of a protein kinase, actin, and a form of myosin suggests that gating of the channels could involve the phosphorylation of channel proteins and a mechanical displacement of the desmotubule with respect to the plasma membrane–lined tube. Calreticulin, a multifunctional, Ca^{2+}-binding protein also found in plasmodesmata, may serve as a regulator of callose synthase, found in the plasma membrane around the plasmodesmatal openings. Callose is usually present in the form of annuli around plasmodesmata, and it has been implicated in the reversible closing of plasmodesmata in response to stress or wounding, possibly mediated by a rise in intracellular calcium concentration (see Concept 8B2).

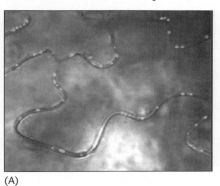
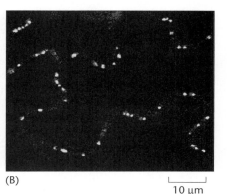

(A) (B) 10 µm

Figure 1.48 Protein targeting to the plasmodesma. A membrane receptor-like protein, PDLP1, is localized to plasmodesmata by its single transmembrane domain. Fused to GFP, and expressed in *Arabidopsis* leaf cells, it reveals the punctate distribution of plasmodesmata in living cells. (Adapted from C.L. Thomas et al., *PLoS Biol.* 6(1): e7. doi:10.1371/journal.pbio.0060007.)

Figure 1.49 Collenchyma. Collenchyma is a protective and supportive tissue found in young growing tissues. Composed of living cells with thickened primary walls, it is arranged in young stems and petioles either as discrete strands (A), as in a stick of celery; or as a continuous supportive cylinder (B).

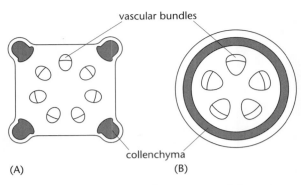

(A)   (B)

10. Collenchyma and sclerenchyma provide support in young tissues.[ref22]

Young herbaceous plants need both some physical protection and assistance to stay upright. They require strong but flexible support for the conducting vessels in the vascular bundles and mechanical protection against the alternate bending and stretching forces delivered by the wind. Two cell types provide this protection, and each has a specialized cell wall.

Collenchyma is a supportive tissue present in most stems and leaves. Groups of collenchyma cells commonly form strands or cylinders that help support and protect the stem, but near the surface of stems they can also form tough cables that resist tensile forces (**Figure 1.49**). A common everyday example is the fibers found in the ridges in celery stems. The strength of collenchyma cells derives from their dramatically thickened cellulose-rich cell walls. Since they develop in growing tissues, collenchyma cells have to elongate and to thicken their walls. Because they are still growing, strictly speaking they have very thick primary walls rather than secondary walls, and collenchyma cells are alive. The cellulosic thickenings of collenchyma cells are usually laid down asymmetrically within the wall, two common examples being shown in **Figure 1.50**. The distribution of wall thickenings in these cells and the plywoodlike arrangement of successive cellulosic layers in the thickenings confer great strength, allowing the cells to resist both compressive and tensile forces (see Concept 6B11). An important feature of collenchyma cell development is its responsiveness to environmental control. Herbaceous plants subjected to regular mechanical shaking commonly develop more collenchyma tissue as a response to the stress than the corresponding unstressed plants.

Sclerenchyma, like collenchyma, is a tissue that provides resistance to bending stresses and protection for stems. In this case, however, the cells have secondary cell walls. The most common cell type in sclerenchyma is the fiber, an elongated cell, often pointed at each end, with a very thick and usually lignified cell wall. Groups of sclerenchyma fiber cells, which are usually dead at maturity, are often tightly cemented to each other to form a fiber bundle. These strong bundles of cells have commercial significance when they occur in plants such as jute and flax (see Concept 9C5). Sclerenchyma cells are also commonly associated with the vascular bundles in the stems and leaves of monocot plants (**Figure 1.51**). Other specialized sclerenchyma cells exist, such as the large pitted *sclereids* that make the flesh of pears seem gritty in the mouth and the giant star-shaped sclereids that provide support in the leaves of some water plants (Figure 1.51).

11. The shaping of mesophyll cells generates large air spaces that facilitate gaseous diffusion.[ref23]

Leaves, the powerhouses of the plant, are beautifully specialized to capture light for photosynthesis and to control the exchange of gases between the atmosphere and cells within the leaf. The internal cells of the leaf that

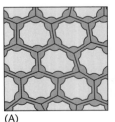 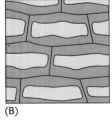

(A)   (B)

Figure 1.50 Collenchyma wall thickenings. The thickening of the walls in growing collenchyma cells is often localized. (A) In bundles of collenchyma cells wall thickening often occurs preferentially around three-way cell junctions. (B) In cylinders of collenchyma tissue, thickening may take place in the tangential, or periclinal, walls rather than in the radial walls.

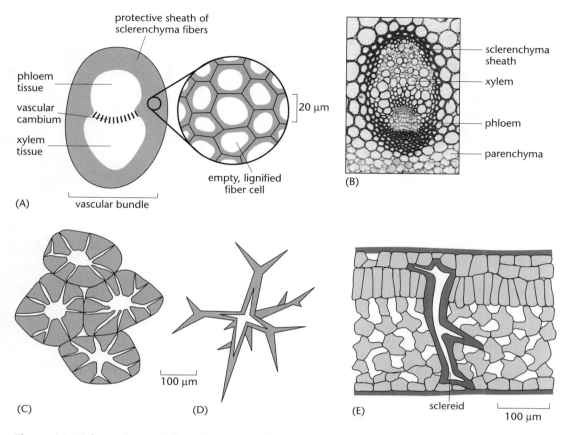

Figure 1.51 Sclerenchyma. Sclerenchyma provides mechanical support in a variety of plant organs and is made of dead cells with thickened and lignified walls. (A) Groups of long, tightly adhering sclerenchyma fiber cells can form a protective cylinder around a vascular bundle, as shown here schematically in transverse section in the stem of a buttercup. (B) A thin section of a vascular bundle from a buttercup stem. Surrounding the xylem and phloem tissues is a sheath of sclerenchyma fiber cells. Sclereids are large sclerenchyma cells that occur singly or in small groups. Shown here are the large isodiametric pitted sclereids that give pear flesh its characteristic gritty texture (C) and a large star-shaped sclereid that can be found spanning and supporting the mesophyll tissue of some leaves (D). (E) Some leaf sclereids are large enough to form a structural support between the top and bottom of a leaf. (B, courtesy of Alison Roberts).

carry out photosynthesis are called *mesophyll* cells. These are thin-walled parenchyma cells whose shape creates a large interconnected set of air spaces throughout the leaf and a corresponding increased cell surface area for gaseous diffusion. Between the upper and lower leaf epidermis, most leaves of eudicots contain two clearly distinguishable kinds of mesophyll cell, both of which contain abundant chloroplasts. The upper layer or layers of cells within the leaf are roughly cylindrical or columnar in shape and are called *palisade cells*. Below them are cells that are more irregular in shape, with increased air spaces between them; these are called *spongy mesophyll cells* (**Figure 1.52**). The young, growing spongy mesophyll cells locally strengthen and loosen different regions of their cell wall causing the cells to expand irregularly. Corresponding localized cell separation results in a loose, spongelike network of cells whose walls are connected by their middle lamella only at regions where the cells meet their neighbors (**Figure 1.53**). The mechanism underlying this differential cell expansion is not fully understood, but is thought to involve complex local rearrangements of the cytoskeleton (see Figure 7A3.3). The result is a network of cells surrounded by a system of air spaces that allow for the efficient exchange of oxygen, carbon dioxide, and water vapor between the mesophyll cells and the atmosphere.

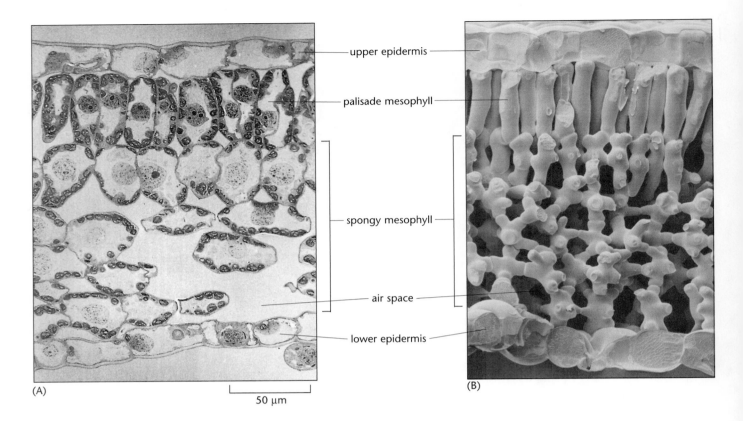

(A)

(B)

50 μm

Figure 1.52 Mesophyll cells in leaf architecture. (A) A vertical section through a young tobacco leaf shown in the electron microscope. Between the upper and lower epidermis can be seen the palisade mesophyll cells and the spongy mesophyll cells. These photosynthetic cells contain abundant chloroplasts that are preferentially located adjacent to the numerous air spaces. (B) Scanning electron micrograph showing a different view of the leaf interior. In this image of a *Zinnia elegans* leaf, the air spaces can be clearly seen as can the increased mesophyll cell surface available for gaseous exchange. (Courtesy of Kim Findlay.)

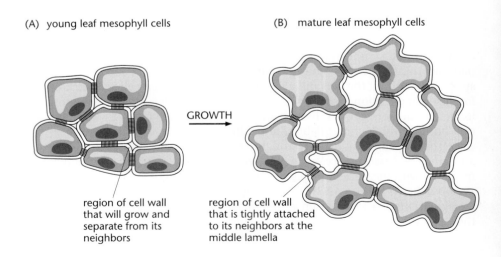

(A) young leaf mesophyll cells

(B) mature leaf mesophyll cells

GROWTH

region of cell wall that will grow and separate from its neighbors

region of cell wall that is tightly attached to its neighbors at the middle lamella

Figure 1.53 Air space creation between leaf mesophyll cells. (A) In the very young leaf the mesophyll cells in the interior are relatively tightly packed with only small air spaces between them. (B) As the leaf grows, wall expansion in regions of cell separation and localized regions of tight cell-cell adhesion combine to produce irregular cells with large three-dimensional air spaces surrounding them, thus facilitating gas exchange between the mesophyll cells and the atmosphere.

12. Pollen grains have a unique and sculptured protective wall.[ref24]

Pollen grains have some of the most elaborate walls found in nature. Each grain is surrounded by a complex wall whose outer layer is sculpted into a characteristic pattern that is uniquely diagnostic of a particular plant species (**Figure 1.54**). The exine, or outer layer of the wall, is laid down around the young pollen grain. Its initial composition is rather typical of a primary wall, but as the pollen grain grows, a new polymer, *sporopollenin*, is deposited to build the complex and distinctively patterned coat with its columns, rods, cavities, and pores (**Figure 1.55**). There are large pores present in the pollen grain wall to allow for the eventual emergence of the pollen tube. Sporopollenin is an extremely resistant material whose chemistry is not well understood. It is believed to be largely an oxidative polymer of carotenoids and/or carotenoid esters, but other compounds are also thought to be present. The carotenoids account for the orange or yellow color of pollen. The polymer remains untouched by strong alkali and acid (even sulfuric and hydrofluoric acids). Sporopollenin's extreme resistance to degradation accounts for the richness of pollen remains in fossil deposits and ancient peat bogs, where they act as useful markers for plant species that died thousands or even millions of years ago.

Although the pollen grain initially makes its own cellulosic primary wall, or intine, the precursors required for the elaboration of the exine are not provided by the pollen grain itself. Instead, they are supplied by the tapetal cells that line the pollen sacs in the anther, where the pollen grains develop (see Figure 1.56). The cavities in the mature exine layer of the pollen wall house a variety of important soluble polymers, among which are the glycoproteins that in some plants are involved in the recognition event between the pollen and the stigma it lands on, as well as the pollen antigens responsible for hay fever.

13. Some plant cells have no walls.[ref25]

Although the majority of cell types in plants have cellulose-containing walls, there are two exceptions. First, there are a few cell types that have no requirement for mechanical strength. An example is the tapetum, a layer of cells lining the cavity in the anther where pollen grains are produced. The major function of tapetal cells is to secrete materials needed by developing pollen grains. Although they arise as walled cells, mature tapetal cells (**Figure 1.56**) have a very thin and insubstantial extracellular matrix that contains no detectable primary wall polysaccharides such as cellulose and pectic polysaccharides. Indeed, these cells more closely resemble an animal

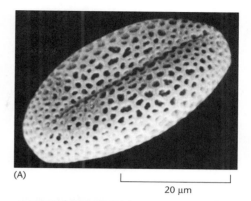
(A)
20 μm

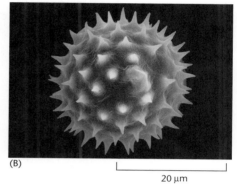
(B)
20 μm

Figure 1.54 Pollen grains. (A) and (B) Pollen grains from two different plant species have distinct sculptured patterns in their outer cell wall layer, or exine. (A, courtesy of Kim Findlay. B, courtesy of Colin MacFarlane and Chris Jeffree.)

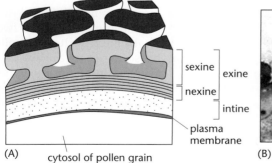
sexine
exine
nexine
intine
plasma membrane
(A)   cytosol of pollen grain

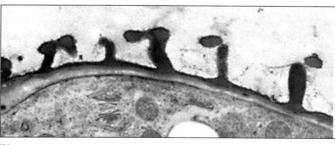
(B)

Figure 1.55 Cell wall layers of a pollen grain. (A) Schematic diagram showing the two major layers of a pollen grain wall, the exine and the intine. While the intine is a more typical cellulose fibril–containing wall, the outer layer, or exine, is often elaborately sculpted into characteristic patterns that contain cavities commonly filled with proteins. (B) Electron micrograph showing a cross section of the wall of an *Arabidopsis* pollen grain. (B, courtesy of José-Maria Segui-Simarro.)

Figure 1.56 Tapetal cells. (A) Transverse section of an immature anther showing the four pollen sacs, each containing young pollen mother cells and each surrounded by a layer of cells called the tapetum. The tapetal cells supply a variety of materials required by the developing pollen grains, and at maturity have no cell walls. (B) A section of an oilseed rape anther shows, by autofluorescence, the position of the tapetum and the pollen grains. (C) When the tapetal cells are stained with a fluorescent cell wall probe, it is clear that they have no wall. (A, from E.G. Cutter, Plant Anatomy, 2nd ed. London: Edward Arnold, 1978. B and C, from R.I. Pennell et al., *Plant Cell* 3:1317–1326, 1991.)

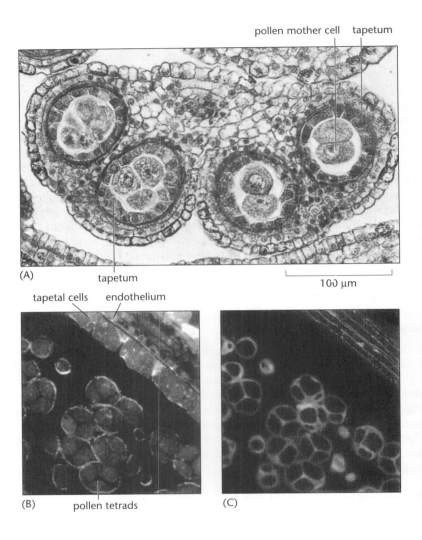

(A)

100 μm

(B)   pollen tetrads   (C)

epithelial cell sheet, with an apical surface active in secretion. As soon as pollen development is complete, the tapetal cells are no longer required, and they undergo programmed dissolution.

Second, there are two cells, the egg and the sperm, whose very function demands that they be naked. The two gametes, whose plasma membranes have to contact each other directly and fuse to effect zygote formation, both arise during complex cell division patterns (including meiosis) in the ovary and anther. The result is an egg cell within the embryo sac and two sperm cells within each germinated pollen grain (**Figure 1.57**). The two naked sperm cells are delivered to the embryo sac by the growth of the pollen tube, which, on penetrating the embryo sac, bursts to release the two sperm, one of which fuses with and fertilizes the naked egg cell to produce the zygote. One of the first things the zygote does is to assemble a brand-new primary cell wall around itself so that it can begin to divide and embark on embryogenesis. This newly formed wall is the start of the surface covering of the next generation!

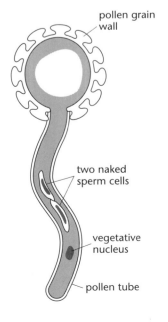

Figure 1.57 Sperm cells. When a pollen grain germinates, the growing pollen tube contains, in addition to the pollen grains' vegetative nucleus, two naked sperm cells, usually loosely associated with each other. When delivered to the embryo sac, one of the naked sperm cells fuses with the naked egg cell to form the zygote.

Key Terms

apoplast
callus
cambium
casparian band
cell morphogenesis
cell plate
cell type
collenchyma
companion cell
cuticle
cutin
cytokinesis
dermal tissue system
desmotubule
differentiation
endodermis
epidermis
ground tissue system
lateral meristem

lignin
mesophyll
middle lamella
organ
palisade cell
parenchyma
pericycle
phloem
phragmoplast
plasmodesma
positional information
preprophase band
primary cell wall
protoxylem element
root meristem
sclereid
sclerenchyma
secondary cell wall
shoot apical meristem

sieve element
sieve plate
spongy mesophyll cell
sporopollenin
stem cell
stoma
stomatal complex
stomatal guard cell
suberin
symplast
three-way junction
tissue system
tissue
transfer cell
trichome
vascular tissue system
vessel element
vessel
xylem

References

[1] Cells arise in specialized regions of the plant called meristems.

Huala E & Sussex IM (1993) Determination and cell interactions in reproductive meristems. *Plant Cell* 5, 1157–1165.

Lenhard M & Laux T (1999) Shoot meristem formation and maintenance. *Curr. Opin. Plant Biol.* 2, 44–50.

Lyndon RF (1990) Plant Development: The Cellular Basis. London: Unwin Hyman.

Park S & Harada JJ (2008) *Arabidopsis* embryogenesis. *Methods Mol. Biol.* 427, 3–16.

Smith A, Coupland G, Dolan L et al. (2010) Plant Biology. New York: Garland Science.

Steeves TA & Sussex IM (1989) Patterns in Plant Development, 2nd ed. Cambridge, UK: Cambridge University Press.

Turner S (2001) Lateral Meristems. Encyclopedia of Life Sciences. John Wiley & Sons. doi: 10.1038/npg.els.0002051.

Wang Y & Li J (2008) Molecular basis of plant architecture. *Annu. Rev. Plant Biol.* 59, 253–279.

[2] Walls originate in dividing cells.

Gunning BES (1982) The cytokinetic apparatus: Its development and spatial regulation. In The Cytoskeleton in Plant Growth and Development (CW Lloyd ed), pp 229–292. London: Academic Press.

Heese M, Mayer U & Jurgens G (1998) Cytokinesis in flowering plants: cellular process and developmental integration. *Curr. Opin. Plant Biol.* 1, 986–999.

Hepler PK & Jackson WT (1968) Microtubules and early stages of cell plate formation in the endosperm of *Haemanthus katherinea* Baker. *J. Cell Biol.* 38, 437–446.

Jürgens G (2005) Cytokinesis in higher plants. *Annu. Rev. Plant Biol.* 56, 281–299.

Kakimoto T & Shibaoka H (1992) Synthesis of polysaccharides in phragmoplasts isolated from tobacco BY-2 cells. *Plant Cell Physiol.* 33, 353–361.

Van Damme D, Vanstraelen M & Geelen D (2007) Cortical division zone establishment in plant cells. *Trends Plant Sci.* 12, 458–464.

[3] Plant organ development depends on precise control of the plane of cell division and of cell expansion.

Jürgens G (2005) Cytokinesis in higher plants. *Annu. Rev. Plant Biol.* 56, 281–299.

Lloyd CW (ed) (1991) The Cytoskeletal Basis of Plant Growth and Form. London: Academic Press.

Otegui M & Staehelin LA (2000) Cytokinesis in flowering plants: more than one way to divide a cell. *Curr. Opin. Plant Biol.* 3, 493–502.

Sinnott EW & Bloch R (1941) Division in vacuolated plant cells. *Am. J. Bot.* 28, 225–232.

Smith LG, Hake S & Sylvester AW (1996) The *tangled-1* mutation alters cell division orientations throughout maize leaf development without altering leaf shape. *Development* 122, 481–489.

[4] Cytoskeletal elements predict the position of the new cross-wall before the cell divides.

Gunning BES & Wick SM (1985) Preprophase bands, phragmoplasts, and spatial control of cytokinesis *J. Cell Sci. Suppl.* 2, 157–179.

Pickett Heaps JD & Northcote DH (1966) Organization of microtubules and endoplasmic reticulum during mitosis and cytokinesis in wheat meristems. *J. Cell Sci.* 1, 109–120.

Wick SM (1991) The preprophase band. In The Cytoskeletal Basis of Plant Growth and Form (CW Lloyd ed), pp 231–244. London: Academic Press.

van Damme D (2009) Division plane determination during plant somatic cytokinesis. *Curr. Opin. Plant Biol.* 12, 745–751.

[5]Actin filaments help to position new cross-walls.

Hoshino H, Yoneda A, Kumagai F & Hasezawa S (2003) Roles of actin-depleted zone and preprophase band in determining the division site of higher-plant cells, a tobacco BY-2 cell line expressing GFP-tubulin. *Protoplasma* 222, 157–165.

Hussey PJ, Ketelaa T & Deeks MJ (2006) Control of the actin cytoskeleton in plant cell growth. *Annu. Rev. Plant Biol.* 57, 109–125.

Lloyd CW (1991) Cytoskeletal elements of the phragmosome establish the division plane in vacuolated higher plant cells. In The Cytoskeletal Basis of Plant Growth and Form (Lloyd CW ed), pp 245–257. London: Academic Press.

Smith LG (2001) Plant cell division: building walls in the right places. *Nat. Rev. Mol. Cell Biol.* 2, 33–39.

[6]The new cross-wall must join and fuse with the mother cell wall.

Heese M, Mayer U & Jürgens G (1998) Cytokinesis in flowering plants: cellular process and developmental integration. *Curr. Opin. Plant Biol.* 1, 986–999.

Hepler PK & Newcomb EH (1967) Fine structure of cell plate formation in the apical meristem of *Phaseolus* roots. *J. Ultrastruct. Res.* 19, 499–513.

Samuels AL, Giddings TH & Staehelin LA (1995) Cytokinesis in tobacco BY-2 and root tip cells: a new model of cell plate formation in higher plants. *J. Cell Biol.* 130, 1345–1357.

Smith LG (1999) Divide and conquer: cytokinesis in plant cells. *Curr. Opin. Plant Biol.* 2:447–453.

[7]A plant is constructed from two compartments: the apoplast and the symplast.

Hepler PK (1982) Endoplasmic reticulum in the formation of the cell plate and plasmodesmata. *Protoplasma* 111, 121–133.

Staehelin L & Hepler PK (1996) Cytokinesis in higher plants. *Cell* 84:821–824.

[8]Cells become organized at an early stage into three major tissue systems.

Dickison WC (2000) Integrative Plant Anatomy. San Diego: Harcourt Academic Press.

Esau K (1977) Anatomy of Seed Plants, 2nd ed. New York: John Wiley & Sons.

Leyser O & Day S (2003) Mechanisms in Plant Development. Oxford: Blackwell Science.

[9]When they have stopped dividing, cells usually continue to grow in size.

Inzé D & De Veylder L (2006) Cell cycle regulation in plant development. *Annu. Rev. Genet.* 40, 77–105.

Lyndon RF (1990) Plant Development: The Cellular Basis. London: Unwin Hyman.

Steeves TA & Sussex IM (1989) Patterns in Plant Development, 2nd ed. Cambridge, UK: Cambridge University Press.

[10]Depending on its position in the plant, a cell differentiates into a specific cell type.

Dolan L, Janmaat K, Willemsen V et al. (1993) Cellular organisation of the *Arabidopsis* root. *Development* 119, 71–84.

Hülskamp M, Miséra S & Jürgens G (1994) Genetic dissection of trichome cell development in *Arabidopsis. Cell* 76, 555–566.

Schellmann S & Hülskamp M (2005) Epidermal differentiation: trichomes in *Arabidopsis* as a model system. *Int. J. Dev. Biol.* 49, 579–584.

Tucker MR & Laux T (2007) Connecting the paths in plant stem cell regulation. *Trends Cell Biol.* 17, 403–410.

Van den Berg C, Willemsen V, Hage W et al. (1995) Cell fate in the *Arabidopsis* root meristem determined by directional signaling. *Nature* 378, 62–65.

[11]A distinction is often made between primary and secondary walls.

Cutter EG (1978) Plant Anatomy, pt 1: Cells and Tissues, 2nd ed. London: Edward Arnold.

Fukuda H (2004) Signals that control plant vascular cell differentiation. *Nat. Rev. Mol. Cell Biol.* 5, 379–391.

Lyndon RF (1990) Plant Development: The Cellular Basis. London: Unwin Hyman.

Sachs T (1991) Cell polarity and tissue patterning in plants. *Development Suppl.* 1, 83–93.

[12]Differentiated cell types are often characterized by functionally specialized cell walls.

Cutter EG (1978) Plant Anatomy, pt 1: Cells and Tissues, 2nd ed. London: Edward Arnold.

Irish VF (2008) The *Arabidopsis* petal: a model for plant organogenesis. *Trends Plant Sci.* 13, 430–436.

Martin C, Bhatt K, Baumann K et al. (2002) The mechanics of cell fate determination in petals. *Philos. Trans. R. Soc. Lond. B. Biol. Sci.* 357, 809–813.

Noda K-I, Glover BJ, Linstead P & Martin C (1994) Flower colour intensity depends on specialised cell shape controlled by a Myb-related transcription factor. *Nature* 369, 661–664.

[13]Shoot epidermal cells produce a waterproof cuticle that helps provide protection.

Cutler DF, Alvin KL & Price CE (eds) (1982) The Plant Cuticle. London: Academic Press.

Hamilton RJ (ed) (1995) Waxes: Chemistry, Molecular Biology and Functions. Dundee, Oily Press.

Kersteins G (ed) (1996) Plant Cuticles: An Integrated Functional Approach. Oxford, BIOS Scientific.

Kolattakudy PE, Espelie KE & Soliday CL (1981) Hydrophobic layers attached to cell walls: cutin, suberin and associated waxes. In Encyclopedia of Plant Physiology NS, vol 13B, pp 225–254. Berlin: Springer.

Riederer M & Muller C (eds) (2006) Biology of the Plant Cuticle. Oxford: Blackwell.

Samuels L, Kunst L & Jetter R (2008) Sealing plant surfaces: cuticular wax formation by epidermal cells. *Annu. Rev. Plant Biol.* 59, 683–707.

[14]Stomatal guard cells have asymmetrically thickened walls that allow them to reversibly change shape.

Bergmann DC & Sack FD (2007) Stomatal development. *Annu. Rev. Plant Biol.* 58, 163–181.

Esau K (1965) Plant Anatomy, 2nd ed. New York: John Wiley & Sons.

Majewska-Sawka A, Münster A & Rodríguez-García MI (2002) Guard cell wall: immunocytochemical detection of polysaccharide components. *J. Exp. Bot.* 53, 1067–1079.

Martin C & Glover BJ (1998) Cellular differentiation in the shoot epidermis. *Curr. Opin. Plant Biol.* 1, 511–519.

Palevitz BA & Hepler PK (1976) Cellulose microfibril orientation and cell shaping in developing guard cells of *Allium*: the role of microtubules and ion accumulation. *Planta* 132, 71–93.

Singh AP & Srivastava LM (1973) The fine structure of pea stomata. *Protoplasma* 76, 61–82.

Willner C & Fricker M (1996) Stomata. London: Chapman & Hall.

[15]Cells in ground tissues fill the space between epidermis and vascular tissue.

Cutter EG (1977) Plant Anatomy, pt 1: Cells and Tissues, 2nd ed. New York: Edward Arnold.

[16]The endodermis regulates the flow of solutes through the apoplast of the root.

Bonnet HT (1968) The root endodermis: fine structure and function. *J. Cell Biol.* 37, 199–205.

Esau K (1965) Plant Anatomy, 2nd ed. New York: John Wiley & Sons.

Karahara I & Shibaoka H (1992) Isolation of Casparian strips from pea roots. *Plant Cell Physiol.* 33, 555–561.

Steudle E & Peterson CA (1998) How does water get through roots? *J. Exp. Bot.* 49, 775–788.

[17]The pericycle encircles the vascular tissue and can give rise to lateral roots.

De Smet I, Vanneste S, Inzé D & Beeckman T (2006) Lateral root initiation or the birth of a new meristem. *Plant Mol. Biol.* 60, 871–887.

Dolan L, Janmaat K, Willemsen V et al. (1993) Cellular organisation of the *Arabidopsis thaliana* root. *Development* 119, 71–84.

Dubrovsky JG, Doerner PW, Colón-Carmona A & Rost TL (2000) Pericycle cell proliferation and lateral root initiation in *Arabidopsis*. *Plant Physiol.* 124, 1648–1657.

Malamy JE & Benfey PN (1997) Organization and cell differentiation in lateral roots of *Arabidopsis thaliana*. *Development* 124, 33–44.

[18]Xylem vessel elements are pipes that conduct water at negative pressure.

Chaffey N (ed) (2002) Wood Formation in Trees: Cell and Molecular Biology Techniques. London, Taylor & Francis.

Esau K (1965) Plant Anatomy, 2nd ed. New York: John Wiley & Sons.

Fukuda H (1996) Xylogenesis: initiation, progression and cell death. *Annu. Rev. Plant Physiol. Plant Mol. Biol.* 47, 299–325.

Mellerowicz EJ & Sundberg B (2008) Wood cell walls: biosynthesis, developmental dynamics and their implications for wood properties. *Curr. Opin. Plant Biol.* 11, 293–300.

Turner S, Gallois P & Brown D (2007) Tracheary element differentiation. *Annu. Rev. Plant Biol.* 58, 407–433.

[19]Phloem sieve tube elements are high-pressure tubes for conducting sugar solutions.

Cronshaw J, Lucas WJ & Giaquinta RT (1986) Phloem transport (Plant Biology Vol.1). New York: Alan R. Liss.

Eleftheriou EP (1996) Developmental features of protophloem sieve elements in roots of wheat (*Triticum aestivum* L.). *Protoplasma* 193, 204–212.

Evert RF (1977) Phloem structure and histochemistry. *Annu. Rev. Plant Physiol.* 28, 199–222.

Lough TJ & Lucas WJ (2006) Integrative plant biology: role of phloem long-distance macromolecular trafficking. *Annu. Rev. Plant Biol.* 57, 203–232.

Sjolund RD (1997) The phloem sieve element: a river runs through it. *Plant Cell* 9, 1137–1146.

[20]Transfer cells load solutes into and out of conducting elements.

Gunning BES & Pate JS (1974) Transfer cells. In Dynamic Aspects of Plant Ultrastructure (AW Robards ed), pp 441–480. London: McGraw-Hill.

Henry Y & Steer MW (1980) A reexamination of the induction of phloem transfer cell development in pea leaves (*Pisum sativum*). *Plant Cell Environ.* 3, 377–380.

Pate JS & Gunning BES (1972) Transfer cells. *Annu. Rev. Plant Physiol.* 23, 173–196.

Van Bel AJE (1993) Strategies of phloem loading. *Annu. Rev. Plant Physiol. Plant Mol. Biol.* 44, 253–281.

[21]Plasmodesmata allow adjacent cells to communicate with each other across their intervening walls.

Cilia ML & Jackson D (2004) Plasmodesmata form and function. *Curr. Opin. Cell Biol.* 16, 500–506.

Lucas WJ & Lee JY (2004) Plasmodesmata as a supracellular control network in plants. *Nat. Rev. Mol. Cell Biol.* 5, 712–726.

Oparka KJ (2004) Getting the message across: how do plant cells exchange macromolecular complexes? *Trends Plant Sci.* 9, 33–41.

Zambryski P (2004) Cell-to-cell transport of proteins and fluorescent tracers via plasmodesmata during plant development. *J. Cell Biol.* 164, 165–168.

[22]Collenchyma and sclerenchyma provide support in young tissues.

Cutter EG (1978) Plant Anatomy, pt 1: Cells and Tissues, 2nd ed. London: Edward Arnold.

Esau K (1965) Plant Anatomy, 2nd ed. New York: John Wiley & Sons.

Jarvis MC (2001) Collenchyma. In Encyclopedia of Life Sciences. John Wiley & Sons. doi: 10.1038/npg.els.0002084.

Jarvis MC & His I (2002) Sclerenchyma. In Encyclopedia of Life Sciences. John Wiley & Sons. doi: 10.1038/npg.els.0002082.

[23]The shaping of mesophyll cells generates large air spaces that facilitate gaseous diffusion.

Evans JR & vonCaemmerer S (1996) Carbon dioxide diffusion inside leaves. *Plant Physiol.* 110, 339–346.

Jeffree CE, Dale JE & Fry SC (1986) The genesis of intercellular spaces in developing leaves of *Phaseolus vulgaris* L. *Protoplasma* 132, 90–98.

Raven JA (1996) Into the voids: the distribution, function, development and maintenance of gas spaces in plants. *Ann. Bot.* 78, 137–142.

[24]Pollen grains have a unique and sculptured protective wall.

Blackmore S, Wortley AH, Skvarla JJ & Rowley JR (2007) Pollen wall development in flowering plants. *New Phytol.* 174, 483–498.

Brooks J, Grant PG, Muir M, van Gijzel P et al. (eds) (1971) Sporopollenin. New York: Academic Press.

Wierman R & Gubatz S (1992) Pollen wall and sporopollenin. *Int. Rev. Cytol.* 140, 35–72.

[25]Some plant cells have no walls.

Goldberg RB, Beals TP & Sanders PM (1993) Anther development: basic principles and practical applications. *Plant Cell* 5, 1217–1229.

Hepler PK, Vidali L & Cheung AY (2001) Polarized cell growth in higher plants. *Annu. Rev. Cell Dev. Biol.* 17, 159–187.

McCormick S (1993) Male gametophyte development. *Plant Cell* 5, 1265–1275.

Zhang W, Sun Y, Timofejeva L et al. (2006) Regulation of *Arabidopsis tapetum* development and function by DYSFUNCTIONAL TAPETUM1 (DYT1) encoding a putative bHLH transcription factor. *Development* 133, 3085–3095.

The Structural Polysaccharides of the Cell Wall and How They Are Studied

2

The cell wall that surrounds plant cells is an insoluble entity that includes an aqueous environment that permeates the wall which contains many soluble proteins and ions. The walls surrounding growing cells (primary walls) are made predominantly of polysaccharides and often some structural proteins. Secondary walls surround cells that have stopped growing and have differentiated and often contain, polysaccharides and proteins, together with lignin and extra amounts of cellulose. Specialized cell walls can also contain other additional polymers.

A combination of biochemical, chemical, and physical methods are used to study the structure of the molecules of the cell wall and how they interact with each other. Structural studies require that the molecules of the insoluble wall be solubilized from the wall. Both chemical and enzymatic treatments are used for this purpose. The solubilized molecules can then be structurally characterized using methods that include mass spectrometry and nuclear magnetic resonance spectroscopy. Alternatively, physical methods such as solid-state nuclear magnetic resonance spectroscopy or Fourier transform infrared spectroscopy can be used to reveal details of the complete wall. In addition, specific localizing reagents (for example, stains, lectins, or antibodies) together with bright-field, fluorescence, or electron microscopy are used to localize specific epitopes within the cell wall structure.

A. Molecules of the Wall

1. Primary walls isolated in a variety of ways contain the same structural polymers.[ref1]

Characterization of the polymers of the primary cell wall necessitates, first, isolating the cell walls and, second, removing and purifying specific molecules for structural study. It is important for these structural studies that the molecules analyzed be truly representative of those found *in situ*, and considerable effort has been invested to ensure that the methods for isolating cell walls maintain the integrity of the various wall components.

In general, primary cell walls are obtained from young, growing plant tissues that have a high content of primary walls—such as young leaves, root tips, or shoots (**Sidebar 2.1**). The tissue is mechanically disrupted to remove its cytoplasmic content, and the insoluble walls are treated with a combination of aqueous buffers, detergents, and/or organic solvents to remove soluble contaminants and lipids. The chemical extractions are often performed at low temperature (around 4°C) to reduce the fragmentation and solubilization of cell wall components by endogenous wall enzymes. Often wall preparations are then treated with α-amylase to remove contaminating

Sidebar 2.1 Examples of tissues rich in primary cell walls that have been used for detailed chemical analysis.

Mung bean hypocotyl
Pea epicotyls
Arabidopsis leaves
Asparagus tips
Maize coleoptiles
Carrot roots
Tomato fruit
Apple fruit
Kiwi fruit
Cells in tissue culture

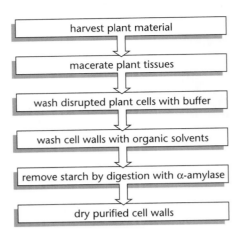

Figure 2.1 Example of wall preparation method. One possible method of preparing plant cell walls from a plant tissue such as pea epicotyls. Many variations on this general theme are described in the literature.

starch that copurifies with the cell wall material. The literature describing the preparation of primary cell walls contains many variations on this general theme, but the insoluble residue that remains at the end is, by definition, the primary cell wall (**Figure 2.1**). Although such cell wall preparations are probably depleted of some non-covalently bound cell wall components, the bulk of the cell wall macromolecules are retained, and contamination by cytoplasmic components is usually minimal.

Whole plant tissues invariably contain small but significant amounts of differentiated cells with secondary cell walls, and these will inevitably contaminate most preparations of primary cell walls. It is not surprising, therefore, during primary cell wall analysis to find small amounts of secondary cell wall components that, under ideal conditions of cell fractionation, would not be present.

Cell walls prepared by various extraction methods have been used as the starting material for the subsequent solubilization and structural characterization of cell wall structural polymers. These polymers, or fragments of them, can be released in soluble form from the insoluble cell wall either by chemical or by enzymatic methods and their structure analyzed. One of the major conclusions to be drawn from the analysis of a wide range of primary cell wall preparations is that they all contain the same three classes of structural polysaccharides—cellulose, hemicelluloses, and pectins—and many also contain structural proteins (often glycoproteins). It is the structural polymers that are consistently found in primary cell walls that will be described in the following sections.

2. Cellulose is a β-1,4-linked glucan with a disaccharide repeating unit.[ref2]

Cellulose is the most stable of all of the cell wall polysaccharides and is the major structural, or load-bearing, component of the cell wall. Yet in terms of glycosyl residue and glycosyl linkage composition, cellulose is an extremely simple polysaccharide in that it is an unbranched polysaccharide of contiguous β-1,4-linked glucopyranosyl residues. The β-1,4 linkage of the glucosyl residues requires that alternate glucosyl residues in the cellulose backbone be positioned at 180° relative to each other. Thus, the backbone of cellulose consists of repeating units of cellobiose.

On average, primary cell walls contain between 20 and 30% cellulose, while secondary cell walls can contain up to 50% cellulose. It is interesting to observe that specialized cell walls, such as those of cotton fibers, can contain greater than 95% cellulose. In chemical terms, cellulose is very stable and extremely insoluble. Its insolubility has made it extremely difficult to extract from cell walls for structural characterization. Strong chemical treatments, including use of chelating agents and base, are required to remove other cell wall components to yield cellulose preparations, and thus, as it is being analyzed, the cellulose sample may already have been modified by the harsh chemical extraction treatments. Often, minor amounts of glycosyl residues other than glucosyl residues are present due to minor contamination with other cell wall matrix polysaccharides, or possibly they are integral minor components of the cellulose molecules. Determining the molecular weight of cellulose chains has been challenging, although methods such as osmometry, ultracentrifugation, light scattering, and viscosity measurements have been used to give accurate determinations of the degrees of polymerizatrion (DPs) of cellulose. Cellulose from secondary walls has DPs between 10,000 and 15,000, whereas the cellulose of primary cell walls has been reported to be in two DP fractions, namely, 2500–4000 and 250–500.

3. Pectins and hemicelluloses are the noncellulosic polysaccharides of primary walls.[ref3]

Primary walls consist of cellulose microfibrils embedded in a matrix of polysaccharides and often, but not always, glycoproteins. The matrix polysaccharides can be divided into two major types: the pectic polysaccharides (pectins) and the hemicelluloses. Both have been traditionally defined as molecules specifically released from cell walls by the solvents that solubilized them. Classically, pectic polysaccharides have been defined as those polysaccharides that can be solubilized from walls by sequential treatment with hot water, weak acid, weak alkali, and ammonium oxalate, or by solutions of chelating agents such as EDTA or CDTA. Also traditionally, hemicelluloses have been defined as those polysaccharides solubilized from wall preparations by treatment with aqueous solutions of alkali after removal of the pectic polysaccharides. These operational definitions have proved to be inadequate, however, in light of the great heterogeneity of these fractions that more recent improved carbohydrate analyses have revealed. In this text, therefore, *pectic polysaccharides* are defined as those polysaccharides containing relatively high proportions (>20 mol %) of D-galactosyluronic acid, and *hemicelluloses* are defined as those polysaccharides that are capable of hydrogen-bonding to cellulose.

4. Wall polysaccharides are composed of 13 different sugars.[ref4]

Analysis of the glycosyl composition of primary cell wall preparations from different sources consistently indicates the presence of 13 different monosaccharides. They are depicted schematically, with the sugar ring compressed into a single plane, in **Panel 2.1**. These monosaccharides include three hexoses, two deoxyhexoses, three pentoses, two uronic acids, two ketoses (a heptulosonic and an octulosonic acid), and an acidic pentose. These 13 monosaccharides account for the monomer subunits of the matrix polysaccharides, of cellulose, and of the glyco- units of the cell wall structural glycoproteins.

Many of these monosaccharides may be structurally modified by the addition or attachment of various substituents. For example, the carboxyl groups (COOH) of some of the monosaccharides, particularly those of galacturonic acid, are often found methyl esterified. Also, the hydroxyl groups of some of the sugars are substituted with O-acetyl (CH_3CO) groups. The methyl esters and O-acetyl substituents are important structural modifications, since they can profoundly alter the physical and chemical properties of the polymers containing the modified monosaccharides, for example, by affecting the availability of polymers to enzymes. Some sugars in wall polysaccharides are also found esterified with phenolic acids, such as ferulic and coumaric acids.

5. Homogalacturonan, rhamnogalacturonan I, and rhamnogalacturonan II are three pectic polysaccharides in primary walls.[ref5]

Over the past several decades, researchers using a wide range of plant material, and different extraction techniques, have isolated many different wall fractions that contain pectic polysaccharides or fragments derived from them. It is likely that all of these arise from three pectic polysaccharides found in the primary cell walls of both angiosperms and gymnosperms. The structures of these three polysaccharides, which are known as homogalac-

Panel 2.1 The 13 monosaccharide components of primary cell walls.

HEXOSES

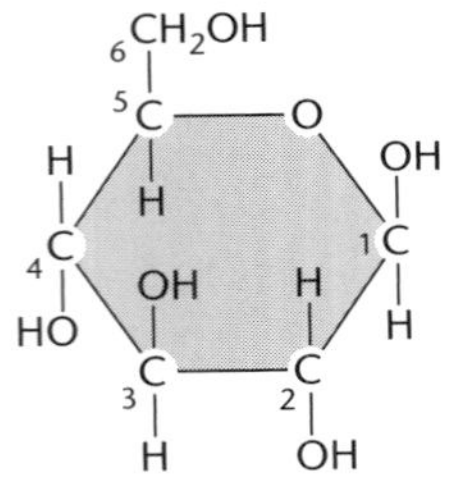

D-glucose
(Glc)

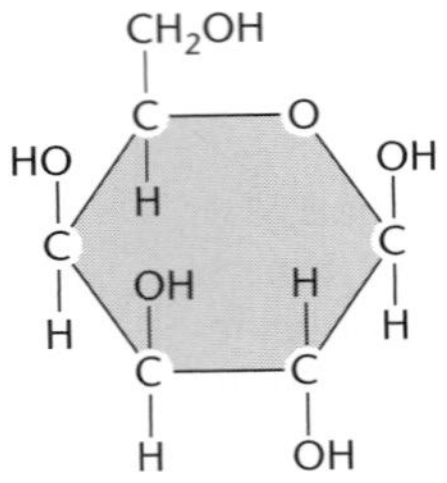

D-galactose
(Gal)

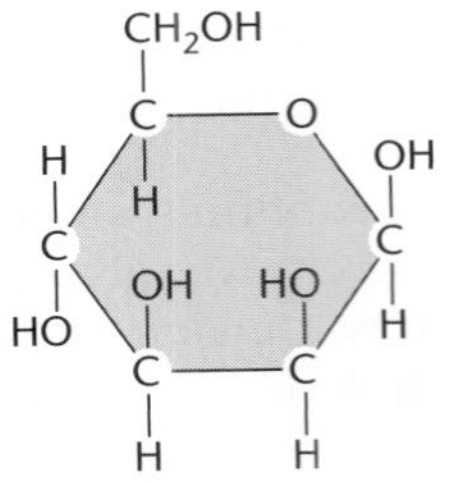

D-mannose
(Man)

The structures are shown with the sugar ring compressed into a single plane. The hexoses (six-carbon molecules) are stereoisomers, differing in the arrangement of groups attached to their carbon atoms.

DEOXYHEXOSES

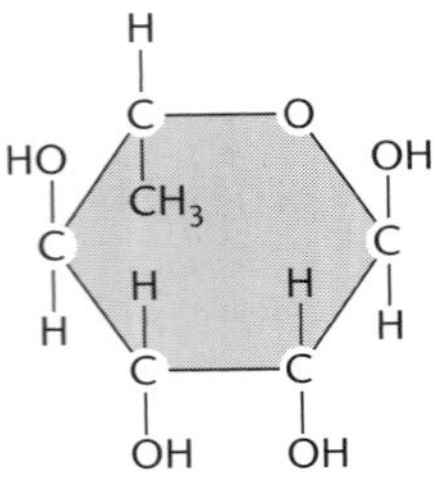

L-rhamnose
(Rha)

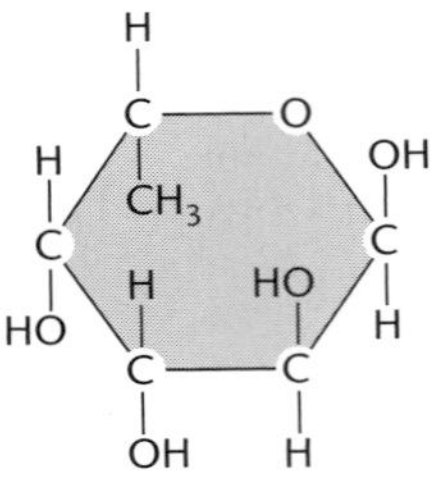

L-fucose
(Fuc)

The deoxyhexoses lack a hydroxyl (OH) group at carbon 6.

PENTOSES

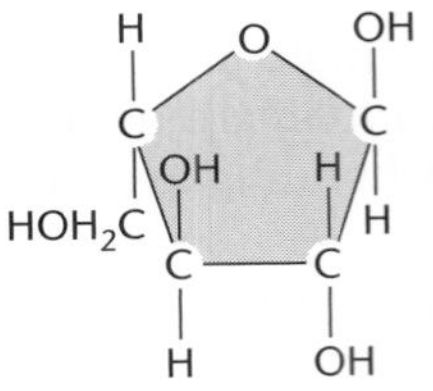

L-arabinose
(Ara)

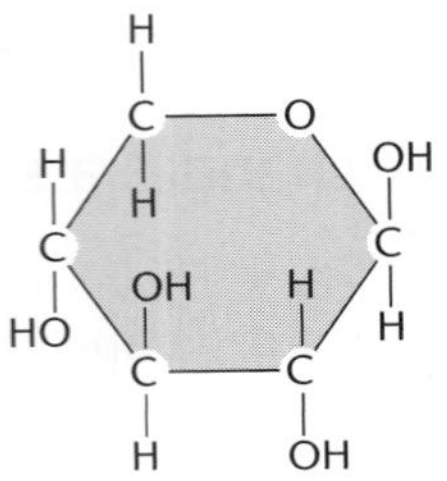

D-xylose
(Xyl)

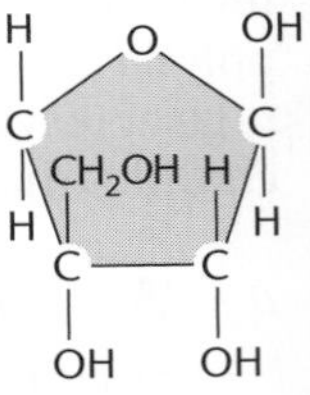

D-apiose
(Api)

The pentoses have five carbon atoms.

ACIDIC SUGARS

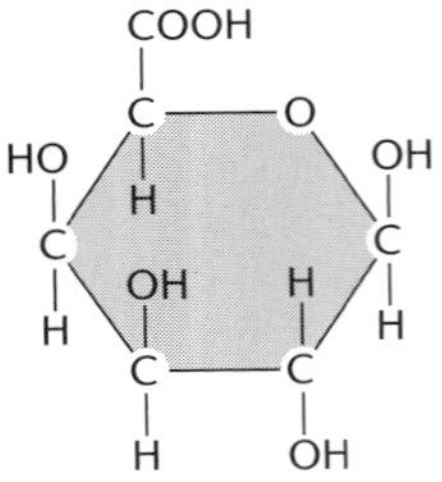

D-galacturonic acid
(GalA)

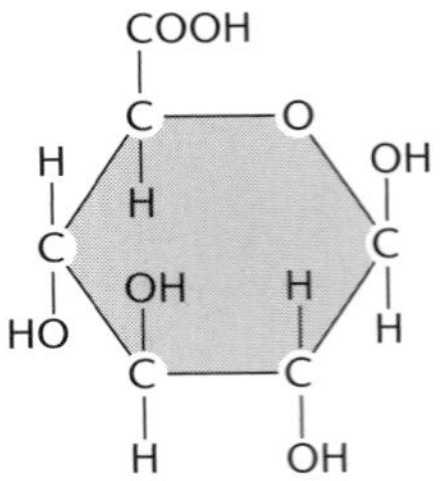

D-glucuronic acid
(GlcA)

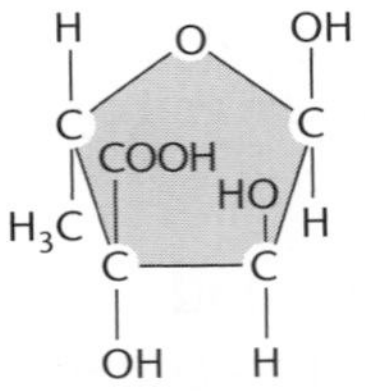

L-aceric acid
(AceA)

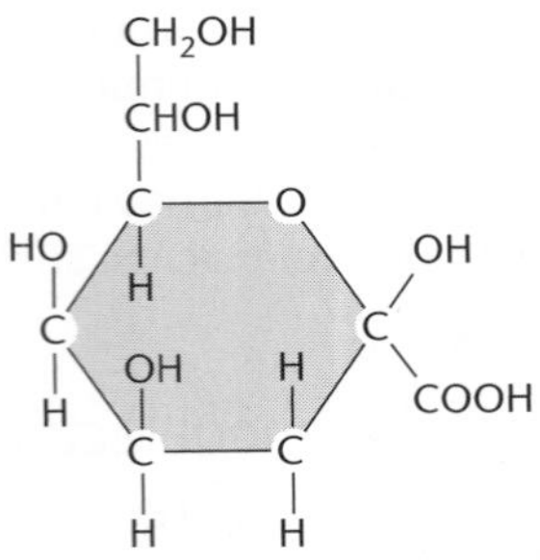

3-deoxy-D-manno-
octulosonic acid
(KDO)

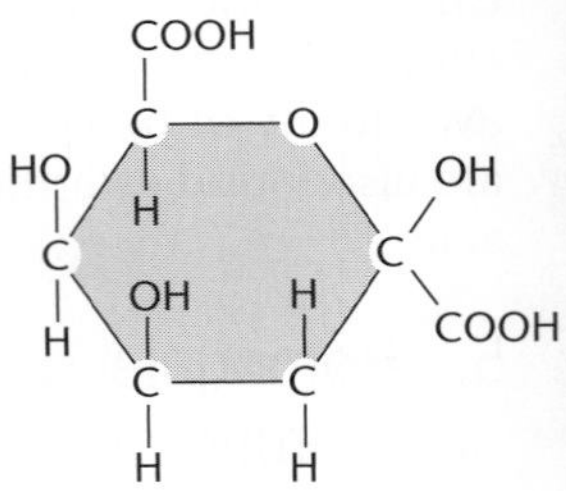

3-deoxy-D-lyxo-
2 heptulosaric acid
(DHA)

D-galacturonic acid, D-glucuronic acid, L-aceric acid, KDO, and DHA contain carboxyl groups (COOH) and thus are acidic sugars.

turonan, rhamnogalacturonan I (RG-I), and rhamnogalacturonan II (RG-II), have been extensively characterized (**Sidebar 2.2**).

Homogalacturonan is a homopolysaccharide (that is, it contains only one kind of monosaccharide) that consists exclusively of 4-linked α-D-galacto-syluronic acid residues. Within the cell wall, many of the carboxyl groups of homogalacturonan (up to 70%) are methyl esterified, and in this form homogalacturonan is often referred to as *pectin*. *RG-I* represents a closely related family of polysaccharides that contain a backbone of repeats of the disaccharide →4)-α-D-GalpA-(1→2)-α-L-Rhap-(1→, where approximately 50% of the rhamnosyl residues are substituted at C-4 with side chains containing L-arabinosyl-, D-galactosyl-, and small amounts of L-fucosyl- and D-glucosyluronic acid residues. The third pectic polysaccharide, *RG-II*, has a 4-linked α-D-galactosyluronic acid backbone with aldo- and keto-sugar-containing oligosaccharide side chains attached to C-2 or C-3.

Approximately 50% of the homogalacturonan (in the form of oligogalacturonides), 33% of the RG-I, and 100% of the RG-II are solubilized from the primary cell walls of dicots by treatment with the enzyme α-1,4-*endo*-polygalacturonase purified from fungi (Sidebar 2.2 and **Figure 2.2**). The oligogalacturonides are released by the fragmentation of the homogalacturonan backbone that probably interconnects the RG-I and RG-II molecules. It has been observed that these latter molecules are released as intact polysaccharides with tails of homogalacturonan of varying length attached to the reducing and nonreducing ends of their backbones, and this suggests that the three pectic polysaccharides are covalently attached to each other in the wall through α-1,4-linked galactosyluronic acid residues.

6. Xyloglucans and arabinoxylans are the two major hemicellulosic polysaccharides in primary walls.[ref6]

Two major hemicellulosic polysaccharides have been identified in the primary cell walls of angiosperms and gymnosperms. These are xyloglucan and arabinoxylan. Xyloglucans are isolated from primary cell walls by treatment with enzymes and with aqueous alkali. Cleavage of xyloglucan by the enzyme β-1,4-*endo*-glucanase releases oligosaccharides whose structure indicates that *xyloglucans* have a cellulose-like backbone of β-1,4-glucosyl residues with short side chains that contain various combinations of xylosyl, galactosyl, fucosyl, and arabinosyl residues distributed along the backbone. Conformational studies of xyloglucans show that the β-glucan backbone of xyloglucans is normally structured to allow it to hydrogen-bond with cellulose (Concept 6B2).

Xyloglucan is the predominant hemicellulose in dicotyledons, nongraminaceous monocotyledons, and gymnosperms (accounting for up to 35% of the cell wall), but in the graminaceous monocots the predominant hemicellulose is arabinoxylan (accounting for up to 40% of the cell wall). *Arabinoxylans* also have a cellulose-like backbone that consists of β-1,4-xylosyl rather than β-1,4-glucosyl residues, and this can also hydrogen-bond to cellulose microfibrils (Concept 6B2). The backbone residues are frequently substituted with side chains containing various combinations of xylosyl, arabinosyl glucosyluronic acid, and 4-*O*-methyl glucosyluronic acid residues.

7. The structural characterization of polysaccharides is difficult.[ref7]

The task of determining the primary structure of a polysaccharide can best be appreciated by comparing the number of potentially different structures that can be obtained for a tetrapeptide containing four different amino acids in any sequence with the number of potentially different tetrasac-

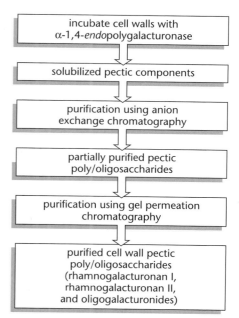

incubate cell walls with
α-1,4-*endo*polygalacturonase

↓

solubilized pectic components

↓

purification using anion
exchange chromatography

↓

partially purified pectic
poly/oligosaccharides

↓

purification using gel permeation
chromatography

↓

purified cell wall pectic
poly/oligosaccharides
(rhamnogalacturonan I,
rhamnogalacturonan II,
and oligogalacturonides)

Figure 2.2 A method for the solubilization and purification of the primary cell wall pectic components.

charides containing four different glycosyl residues linked together in any sequence. Twenty-four different tetrapeptides can be formed. Over 10,000 tetrasaccharides can result from the attachment of four different glycosyl residues, and this calculation does not include the possibility of noncarbohydrate substituents (for example, methyl esters) adding to the structural complexity. Thus, when determining primary structures of polysaccharides, the investigator has to utilize techniques that will distinguish among all possible saccharide structures. Some of the methods used for this enormous challenge are described in Section 2B.

The matrix polysaccharides of primary cell walls (that is, the pectic polysaccharides and the hemicelluloses) are the subject of intensive structural studies. Although the ultimate goal for the structural characterization of a polysaccharide is to relate the primary structure of the polysaccharide to its three-dimensional structures, to its interactions with other cell wall polysaccharides, to its physical properties, and to its biological functions, most of the effort to date has centered on purifying the matrix polysaccharides to homogeneity and determining their primary structures. Determining the primary structures of macromolecules as complex as those of the wall matrix polysaccharides is a considerable challenge. The features of polysaccharides that make their structural elucidation so difficult are highlighted in **Panel 2.2**. Determining the *primary structure* of a polysaccharide requires the following information:

1. The glycosyl residue composition
2. The absolute configuration, D or L, of each glycosyl residue
3. The glycosyl linkage composition, that is, which carbons of a sugar are involved in covalent attachment to its neighbors
4. The furanose or pyranose ring form of each glycosyl residue
5. The sequence of glycosyl residues in the polysaccharide, that is, how the various glycosyl residues are attached to each other
6. The anomeric configuration, α or β, of each glycosidic bond
7. The points of attachment of the noncarbohydrate substituents, such as methyl esters and *O*-acetyl groups

8. Cell walls contain proteins, glycoproteins, and proteoglycans that perform enzymatic, structural, defensive, and other functions.[ref8]

Isolated cell walls contain variable quantities of proteins, glycoproteins, and proteoglycans, usually in the range of 1 to 10% of the wall dry weight. These proteins can be divided into two broad groups, the insoluble structural proteins, which include the hydroxyproline-rich glycoproteins (HRGPs), the glycine-rich proteins (GRPs), and the proline-rich proteins (PRPs); and the soluble proteins, which include enzymes, transport proteins, defense proteins, lectins, and arabinogalactan proteins (AGPs). The so-called structural proteins are widespread in the walls of flowering plants, and with their characteristic amino acid repeat motifs, they are thought to play an important architectural role in the wall. The expression of genes encoding some of these proteins is regulated both during development and in response to wounding and stress, suggesting they may also have important functions in these processes. However, our knowledge of the molecular functions of this group of proteins is still very sketchy. Characteristically, they are rich in the amino acids from which they derive their names, and the HRGPs, and to a lesser extent the PRPs, are glycoproteins. The HRGPs contain between 30 and 50% glycosyl residues. They are discussed further in Section 3E.

In addition to the structural proteins, the wall contains hundreds of soluble proteins, including enzymes, and these will be discussed at more length elsewhere in the book. Here we take a quick look at some of them. Many of the enzymes are cell wall–metabolizing enzymes (for example, α-galactosidases, β-galactosidases, β-xylosidases, and β-1,4-glucanases).

Some enzymes, such as peroxidases and laccases, are likely involved in the assembly of lignins and possibly in the defense of plants against pathogens, while others (such as β-1,3-linked *endo*-glucanases and β-1,4-linked *endo*-chitinases) are also involved in the defense of plants against potential invaders (see Chapter 8). Two other defense-related but nonenzymatic proteins found in the wall are thionins and polygalacturonase-inhibiting proteins. The lipid transfer proteins represent another interesting group of nonenzymatic cell wall proteins, which appear to function in the transport of lipid molecules used in cutin biosynthesis and surface wax formation to their sites of assembly.

Arabinogalactan proteins (AGPs) are an important class of cell wall proteins that, because of their high carbohydrate content (>90%), are defined as proteoglycans. Many AGPs are soluble proteins, while others possess lipid tails that serve to anchor them to the outer surface of the plasma membrane. Many but not all AGPs interact *in vitro* with sugar-containing Yariv reagents, and their appearance is developmentally regulated, but their exact role(s) in cell walls remain to be elucidated.

9. Primary cell walls contain an aqueous phase.[ref9]

Cellulose microfibrils embedded in a matrix of pectic and hemicellulosic polysaccharides, and in some cases (glyco)proteins, together form an insoluble primary cell wall. What is often overlooked when considering the function of the wall is the aqueous (hydrophilic) environment that permeates the wall's structure. More than 70% of the fresh weight of a primary cell wall is accounted for by water. It is this aqueous phase where soluble proteins (enzymes, carrier proteins, defense proteins), inorganic cations and anions (for example, H^+, Ca^{2+}, Sr^{2+}, Na^+, Cl^-, K^+, PO_4^-, SO_4^-), soluble oligosaccharides, gums, AGPs, and newly synthesized wall components destined to become part of the insoluble wall are all found. These ions and molecules can have a considerable influence locally on the structure and function of the wall where they occur. In many cells, during the formation of secondary walls, the aqueous phase becomes reduced with a corresponding accumulation of lignin, additional cellulose, and matrix polysaccharides in the wall.

10. Secondary walls are composed of primary walls plus additional layers of polymers.[ref10]

When a cell stops growing, it differentiates into a specific cell type that is often characterized by a structurally and functionally specialized wall (see Chapter 1). Walls deposited in fully grown cells are called, by definition, secondary cell walls. When analyzing the components of the cell wall, it is therefore appropriate to consider those macromolecules that are added during secondary cell wall deposition. The data shown in **Table 2.1** indicate the bulk differences in the polymer composition found in the primary and secondary cell walls of a broad range of angiosperms (dicotyledons) and gymnosperms, and contain data from many different secondary cell wall types.

The data in Table 2.1 highlight those polymers that are added to primary walls during the transition to secondary wall deposition. The secondary walls contain higher amounts of cellulose and often large amounts of lignin, a polymer absent from primary cell walls. Together the cellulose and lignin account for well over 60% and, in some cases, up to 80% of the secondary cell walls. Secondary walls also contain additional hemicelluloses and relatively less pectin than primary cell walls. The pectin found may well be the pectin remaining in the underlying primary cell walls.

The quantitatively dominant hemicelluloses found in secondary cell walls are different from the hemicelluloses found in primary walls (see Table 2.1). In those angiosperms studied, the major hemicelluloses found in the

MONOSACCHARIDES

Monosaccharides usually have the general formula $(CH_2O)_n$ where n can be 3, 4, 5, 6, 7, or 8 and have two or more hydroxyl groups. They either contain an aldehyde group ($-C{<}^O_H$) and are called aldoses, or a ketone group ($\rangle C{=}O$) and are called ketoses.

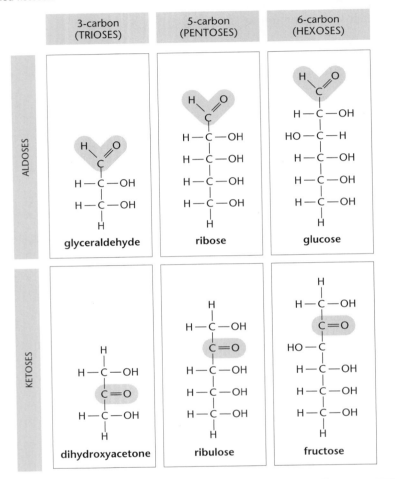

	3-carbon (TRIOSES)	5-carbon (PENTOSES)	6-carbon (HEXOSES)
ALDOSES	glyceraldehyde	ribose	glucose
KETOSES	dihydroxyacetone	ribulose	fructose

RING FORMATION

In aqueous solution, the aldehyde or ketone group of a sugar molecule tends to react with a hydroxyl group of the same molecule, thereby closing the molecule into a ring.

glucose

ribose

Note that each carbon atom has a number: glucose is in a pyranose (6-membered) ring, ribose is in a furanose (5-membered) ring.

ISOMERS

Many monosaccharides differ only in the spatial arrangement of atoms—that is, they are *isomers*. For example, glucose, galactose, and mannose have the same formula ($C_6H_{12}O_6$) but differ in the arrangement of groups around one or two carbon atoms.

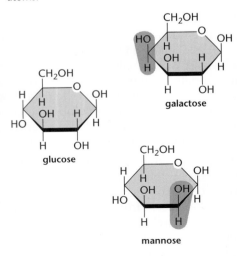

galactose

glucose

mannose

These small differences make only minor changes in the chemical properties of the sugars. But they are recognized by enzymes and other proteins and therefore can have important biological effects.

α AND β CONFIGURATION

The hydroxyl group on the carbon that carries the aldehyde or ketone can rapidly change from one position to the other. These two positions are called α and β.

β hydroxyl

α hydroxyl

As soon as one sugar is linked to another, the α or β form is frozen.

SUGAR DERIVATIVES

The hydroxyl groups of a simple monosaccharide can be replaced by other groups, for example, a carboxyl group.

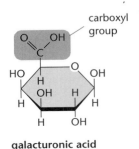

carboxyl group

galacturonic acid

D AND L FORMS

Two isomers that are mirror images of each other have the same chemistry and therefore are given the same name and distinguished by the prefix D or L.

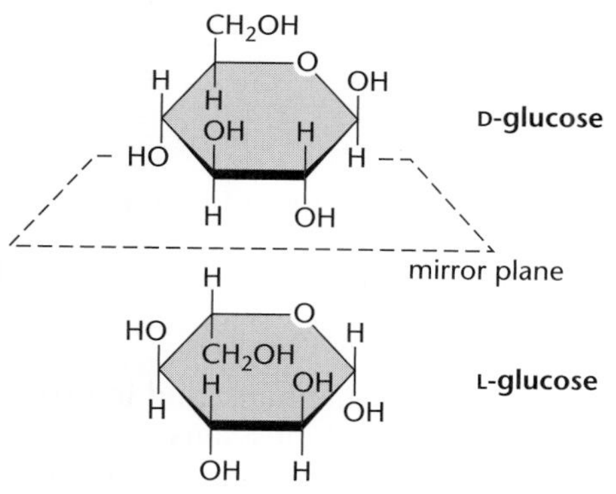

D-glucose

mirror plane

L-glucose

DISACCHARIDES

The carbon that carries the aldehyde or the ketone can react with any hydroxyl group on a second sugar molecule to form a *disaccharide*. The reaction forming sucrose is shown here.

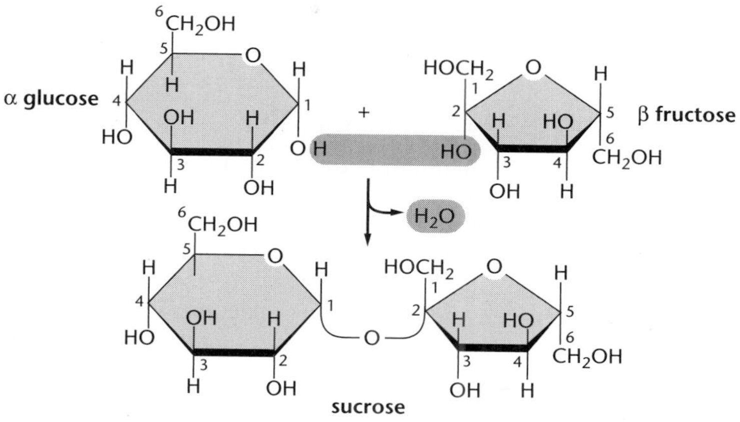

α glucose

β fructose

+

H_2O

sucrose

OLIGOSACCHARIDES AND POLYSACCHARIDES

Large linear and branched molecules can be made by linking monosaccharides together via specific carbons on each monosaccharide. Each connection is called a *glycosidic linkage* and is defined by the carbons involved and the α or β configuration. Short chains are called *oligosaccharides*, while long chains are called

polysaccharides. Xyloglucan, for example, is a plant cell wall polysaccharide that has a backbone made entirely of glucose units with branches attached that contain xylose and sometimes galactose and fucose.

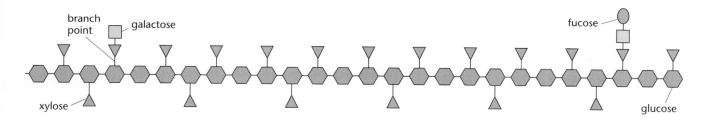

branch point — galactose

fucose

xylose

glucose

OLIGOSACCHARIDES AND POLYSACCHARIDES HAVE A REDUCING END AND A NONREDUCING END

Oligosaccharides and polysaccharides, except on very rare occasions, have a polarity to their overall structure. At one end of the molecule the aldehyde or ketone group of the terminating sugar has not been fixed into a ring structure and this sugar can freely change from a ring to a chain form. This is called the

reducing end of the molecule. The *nonreducing end* of the molecule is where the terminating sugar is fixed into a ring form via the formation of a glycosidic bond to the next neighboring sugar. The reducing and nonreducing ends of a short chain of β 1,4-linked glucose (a cellulose oligosaccharide) are shown.

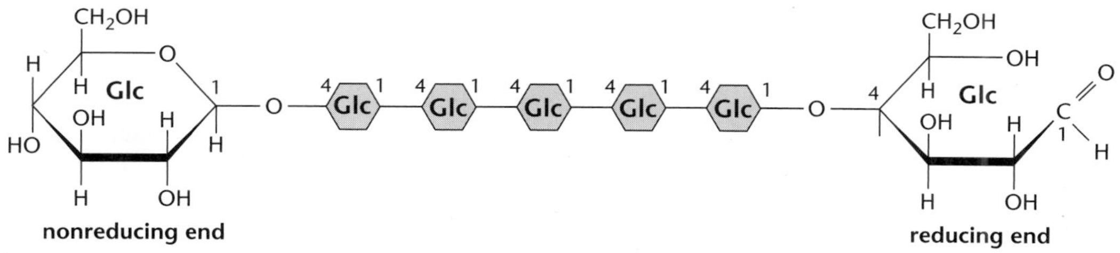

nonreducing end

reducing end

TABLE 2.1 POLYMER COMPOSITION OF PRIMARY AND SECONDARY CELL WALLS

POLYMER	CELL WALL (wt %)[a]		
	PRIMARY (dicotyledon)	SECONDARY[b]	
		ANGIOSPERM (dicotyledon)	GYMNOSPERM
Cellulose	20–30	37–57	38–52
Lignin	0	17–30	26–36
Pectin	30–35	<10	<10
Hemicellulose	25–30	20–37	16–27

[a]Values are an average from multiple species and tissue samples. Protein values are not included, because of their high variability dependent on the plant tissue analyzed.
[b]Secondary cell wall preparations will contain small amounts of primary cell wall materials. The harsh chemical extraction procedures required to fractionate secondary cell walls may contribute to some selective polymer loss.

secondary cell walls are glucuronoxylans and glucomannans; and in gymnosperms, galactoglucomannans and glucuronoarabinoxylans. Because secondary cell walls are both very tough and very insoluble, the solubilization of secondary cell wall hemicellulose polysaccharides requires harsh chemical extraction methods. The difficulty in extracting these hemicelluloses in an intact and pure form has inhibited their detailed structural analysis.

B. Methods for Characterizing the Structural Polysaccharides of the Cell Wall

1. Obtaining pure polysaccharides is a prerequisite for primary structure determination.[ref11]

The study of the primary structures of cell wall polysaccharides, like the study of all polymers, is greatly facilitated if a homogeneous polymer is available. Unfortunately for the study of plant wall polysaccharides, obtaining homogeneous polysaccharides is extremely difficult. Unlike proteins, whose structure is precisely defined in relation to a template, most cell wall polysaccharides are more variable, particularly in length. This is because the process of solubilizing polysaccharides from the insoluble wall results in some random cleavage of the polysaccharide as well as in some inherent variability in its backbone length.

To be considered pure, a preparation of a wall polysaccharide must minimally contain molecules of only one class of polysaccharide. Purification methods are available that will produce fractions of polysaccharides of this purity, containing, for example, only RG-II or only xyloglucan. However, the fractions containing these wall polysaccharides do not necessarily contain molecules of exactly the same structure. For example, a combination of ion-exchange and gel-permeation chromatography will separate the RG-I, RG-II, and oligogalacturonides released from cell walls by treatment with α-1,4-*endo*-polygalacturonase (see Figure 2.2), this results in fractions containing only RG-I, only RG-II, and only oligogalacturonides. But those fractions that contain only RG-I, for example, contain RG-I's of different sizes and thus presumably a family of closely related polysaccharides. The approximate molecular weight of RG-I is 200 kD. However, when RG-I is purified, a range of molecular weights, between 160 and 240 kD, is obtained. These molecular weight differences are due to differences in backbone length and/or differences in the number and size of side chains attached to the backbone. RG-II probably has the most consistent and homogeneous structure of any of the wall polysaccharides and can be obtained as either a pure monomer (~4.5 kD) or dimer (~9.0 kD) (see Concept 3C4). Even this polysaccharide, however, is often obtained with a variable length of backbone that reflects the method chosen to solubilize the polysaccharide from the cell wall and/or with variable side chain terminal glycosyl residues due to the source of the original plant material.

The hemicellulosic polysaccharides xyloglucan and glucuronoarabinoxylan can also be effectively purified by ion-exchange and gel-permeation chromatographies into fractions containing a single type of polysaccharide. However, as with RG-I, these polysaccharide preparations have a rather broad range of molecular weights. This is caused by the size heterogeneity of the polysaccharide within the cell wall and also by the random cleavage of backbone glycosidic bonds by the method used to solubilize the polysaccharide.

The ability to purify all of the five cell wall matrix polysaccharides (homogalacturonan, RG-I, RG-II, xyloglucan, and glucuronoarabinoxylan) into separate fractions has led to the structural characterization of these molecules described in Sections 3B and 3C.

2. Extracellular polysaccharides secreted by cultured cells are excellent models for wall polysaccharides.[ref12]

The matrix polysaccharides of cell walls are made within the cell and then secreted into the cell wall. When plant cells are grown in liquid culture, some of these polysaccharides pass into the growth medium, from which they can be recovered for structural analysis. These "extracellular polysaccharides" have been used as models for the study of primary wall polysaccharide structure, as they can readily be obtained in soluble form from the growth medium in large amounts and without using the chemical or enzymatic procedures normally used to extract polysaccharides from cell walls. Cell cultures from a variety of plant species have been used as a source of extracellular polysaccharides for structural studies (**Figure 2.3**).

For dicot plants, all of the matrix polysaccharides found in the native wall have also been identified as components of soluble extracellular polysaccharides from cell cultures, although the relative amounts of these polysaccharides differ from those in the cell wall. The use of polysaccharides from culture medium as models for cell wall polysaccharides has been most effective in the structural analysis of xyloglucans. Many of the structural characteristics of wall xyloglucans were first identified in extracellular polysaccharide xyloglucans. The subsequent identification of identical structures in xyloglucan isolated from native walls is confirmation of the validity of using extracellular polysaccharides as models for wall polymers.

3. Enzymatic and chemical cleavage of polysaccharides has made structural studies possible.[ref13]

The use of enzymes and chemical methods to cleave, with some degree of specificity, selected glycosidic bonds in cell wall polysaccharides has greatly facilitated structural studies. These treatments are effective in two ways. Initially, enzymes or chemical methods are used to release polysaccharides or fragments of the polysaccharides from the insoluble cell wall in which they are held. Ideally, enzymes are good choices for this application, because the glycosidic bonds they cleave are well known, although enzyme solubilization can produce minor size heterogeneity in the solubilized material due to the inherently random choice by the enzyme of which of several identical glycosidic bonds is actually cleaved. Chemical methods, on the other hand, have a more dramatic effect, since most of the methods utilized—such as mild acid or base treatment—can break several different covalent bonds at random, producing *major* size heterogeneity in polysaccharides or fragments thereof in the mixture. However, with careful optimization, both enzyme treatments (for example, *endo*-polygalacturo-

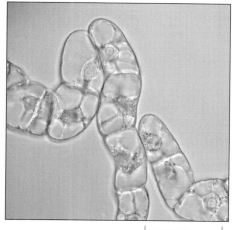

50 µm

Figure 2.3 Cultured tobacco cells. The extracellular polysaccharides secreted by the cultured cells of a variety of plant species have been used as structural models of plant cell wall polysaccharides. Single-cell or small multiple-cell groups are grown in liquid culture in "shaker flasks" on gyratory shakers to aerate the growth media. (Courtesy of Gethin Roberts.)

Sidebar 2.3 Degree of polymerization.
The *degree of polymerization* (DP) of a polysaccharide or oligosaccharide refers to the number of covalently linked monosaccharide residues in the polymer. For example, small oligogalacturonans may have a DP of, say, between 8 and 15, the backbone of RG-I may have a DP of up to 400, and a single cellulose molecule may have a DP of 15,000.

nase for pectic polysaccharides and *endo*-glucanases and *endo*-xylanases for hemicelluloses) and chemical methods (for example, strong alkali for hemicelluloses) have allowed specific cell wall polysaccharides, or fragments of polysaccharides, to be solubilized in quantities sufficient for structural characterization.

The structure of a polysaccharide or its fragments can be determined only once they are solubilized from the wall and separated into fractions that contain only one polysaccharide or fragment type. A major goal of carbohydrate structural analysis is to determine the sequence of glycosyl residues in the pure polysaccharide or fragment. The glycosyl composition and linkage data alone (Concepts 2B4 and 2B5), although very valuable, do not directly lead to full sequence data. Although oligosaccharides with a degree of polymerization on the order of 25 can be sequenced using methods currently available (see Concept 2B6), it is impossible at present to directly sequence polysaccharides. Therefore, in cases in which polysaccharides have been solubilized from cell walls, they must be broken down into oligosaccharides of sizes suitable for sequence analysis, and here again both enzymatic and chemical cleavage methods have proved valuable (**Sidebar 2.3**). *Endo*-glucanases for the study of xyloglucans and *endo*-xylanases for the study of xylans are particularly effective in this way. On the other hand, chemical treatments are effective in fragmenting the pectic polysaccharides RG-I and RG-II into oligosaccharides for structural studies.

4. Several methods are used for glycosyl composition analysis.[ref14]

The determination of the structure of a poly- or oligosaccharide first requires knowledge of which sugars are present and in what amounts, referred to as the *glycosyl residue composition*. This is achieved by fragmentation of the poly- or oligosaccharide into its monomeric building blocks, the monosaccharides, followed by their identification and quantitation. The cleavage of the poly- or oligosaccharide is generally achieved by the use of aqueous acid. Often, cell wall poly- or oligosaccharides contain several different monosaccharides whose glycosidic linkages can differ in their susceptibility to acid hydrolysis. Additionally, some monosaccharides, such as xylose, are more susceptible to acid degradation than others. Therefore, most glycosyl composition analyses use acid hydrolysis conditions that are a compromise between cleavage of the greatest number of glycosidic bonds and degradation of the least number of free monosaccharides.

Once released, the monosaccharides are separated, identified, and quantitated using a variety of approaches. An outline of two commonly used methods is shown in **Panel 2.3**. These methods involve the formation of alditol acetate derivatives or the formation of trimethylsilyl ethers of methyl glycosides and methyl esters (when analyzing uronic acids). Both methods result in the formation of volatile derivatives of the monosaccharides that can be separated by gas chromatography. Alternatively, monosaccharides can be separated, identified, and quantitated using high-performance liquid chromatography (HPLC) followed by detection using, for example, a refractive index or pulsed amperometric detector. Another detection technique that can be used with HPLC is ultraviolet (UV) absorbance, in which UV-absorbing "tags" are added to the monosaccharides to allow detection.

All of the methods of composition analysis require the use of pure sugar standards to determine accurate retention times and assist in the identification of the monosaccharides or monosaccharide derivatives as they emerge from either the gas chromatograph or the liquid chromatograph. Standard monosaccharides are also used to calibrate the detector response of the gas chromatograph and liquid chromatograph to allow accurate quantitation of each monosaccharide or monosaccharide derivative as it is analyzed. Approximately 5 to 10 μg of each cell wall matrix polysaccharide is

The two methods used for glycosyl residue composition analysis. Two methods are routinely used for glycosyl residue composition analysis. These are via the formation of alditol acetates or TMS methyl glycosides from sugar components. Both methods result in the formation of volatile derivatives that can be quantitated and identified using gas-liquid

chromatography (GLC) and/or gas-liquid chromatography–mass spectrometry (GLC-MS). When the presence of uronic acids is suspected, the TMS method is preferred, as this will result in uronic acid derivatives that can be analyzed. This is not the case with the alditol acetate method as described.

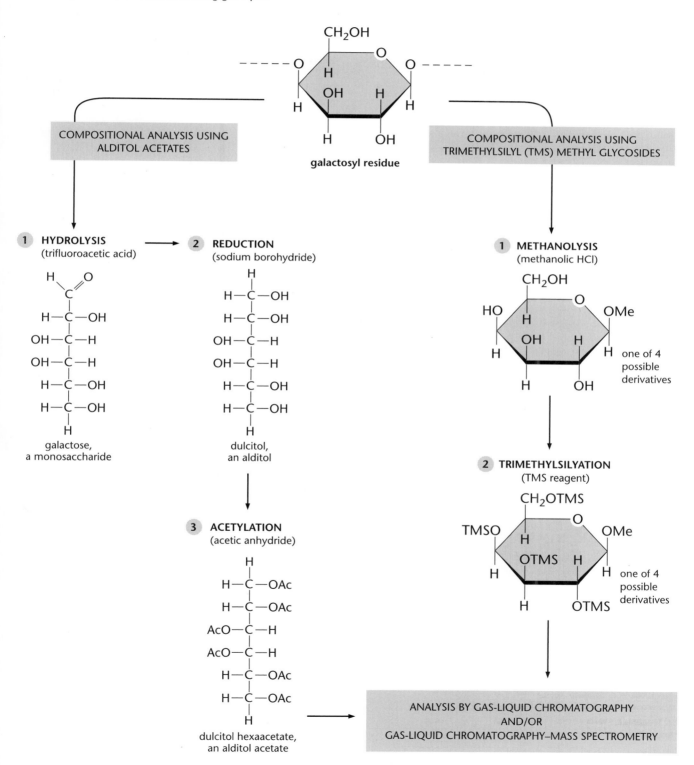

Panel 2.4 Methylation analysis.

Glycosyl linkage composition analysis allows the identification of those carbons in a glycosyl residue that are attached to other glycosyl residues in the oligo- or polysaccharide (*arrows*). This is most often determined by methylation analysis. The example shown here is of three glycosyl residues that are part of a xyloglucan polysaccharide. Several steps are required to complete a methylation analysis. These are (1) methylation, (2) hydrolysis, (3) reduction (here often sodium borodeuteride is used to uniquely label C-1 in each sugar derivative with a deuterium atom that aids in the later GLC-MS analysis), (4) acetylation, and (5) analysis by GLC and GLC-MS. In this process all free hydroxyls in the original polysaccharide are O-methylated with a

CH₃ group. After hydrolysis and reduction, all exposed hydroxyls are O-acetylated with a CH₃CO group. O-methylation is shown as OMe, and O-acetylation as OAc.

The order of the OMe and OAc groups on the backbone of each sugar derivative relative to C-1 that is labeled with the deuterium atom is determined by GLC-MS and defines the linkage of the original sugar. GLC retention time is also used to distinguish sugar isomers, that is, 4-linked glucose from 4-linked galactose. The size of the GLC peak for each sugar derivative is used for quantitation. Glycosyl residue backbone carbon atoms (1 through 6) are numbered for 4-linked glucose for help in interpretation.

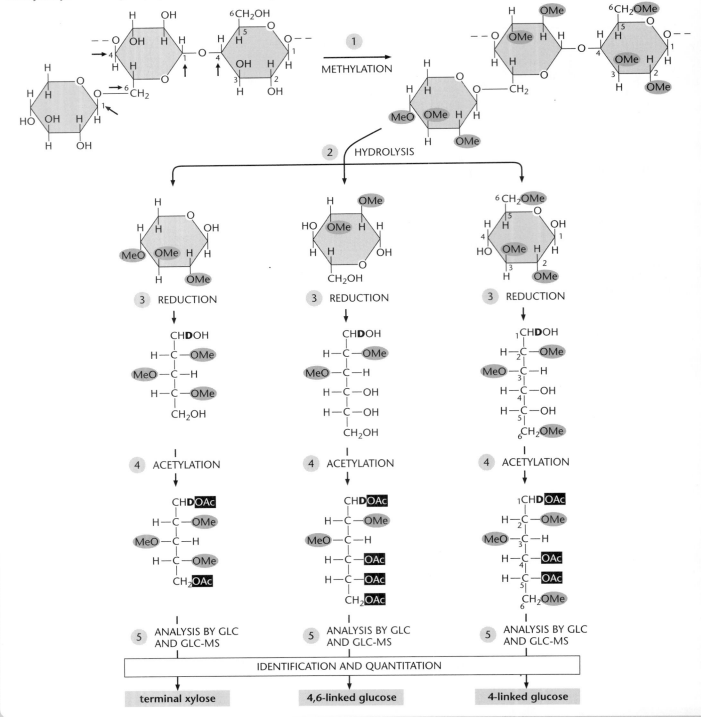

sufficient to allow the accurate determination of its glycosyl residue composition using any of the described methods. The whole procedure is relatively efficient, to the extent that a few small seedlings would provide enough wall material to do a sugar composition of all the major wall polysaccharides.

5. Glycosyl linkage compositions are determined by methylation analysis.[ref15]

To obtain a more complete structural analysis of an oligo- or polysaccharide, the next major step is to determine the glycosyl linkages present. *Glycosyl linkage composition analysis* identifies those carbons in a glycosyl residue that are attached to other glycosyl residues in the oligo- or polysaccharide. The standard method for determining the glycosyl linkage composition is called *methylation analysis*. In this procedure, all free hydroxyl groups of the oligo- or polysaccharide are first methylated, the glycosidic linkages are hydrolyzed, and the products, the partially methylated glycosyl residues, are characterized as volatile methylated alditol acetate derivatives by gas-liquid chromatography (GLC) and gas-liquid chromatography–mass spectrometry (GLC-MS). The various chemical procedures used in a methylation analysis are shown in **Panel 2.4**. The methylation of the free hydroxyl groups can be achieved using several different chemical methods. The method most routinely used is that developed by Hakomori which uses sodium methylsulfanyl anion and methyl iodide in dimethylsulfoxide. Another often used method for the derivatization procedure is sodium hydroxide and methyl iodide. Often the aldehyde at the reducing end of the oligo- or polysaccharide is reduced to the corresponding alditol prior to methylation to prevent degradation during the methylation procedure. It is also important to note that many cell wall oligo- or polysaccharides contain glycosyluronic acid residues that would be degraded during a methylation analysis. These glycosyluronic acid residues must be reduced to their corresponding glycosyl residue (often a 6,6-dideuterio glycosyl residue, following introduction of deuteride label) at some point during the methylation procedure to allow for their identification and quantitation and to prevent degradation of the glycosyluronic acid residues.

The methylated alditol acetates formed during methylation analysis are separated and quantitated by GLC. The methylated alditol acetate derivatives are identified by a combination of their retention time on a particular GLC column and by their characteristic fragmentation patterns obtained during GLC-MS. The fragmentation of all similarly linked hexoses (that is, 4-linked glucose and 4-linked galactose) will give similar fragmentation patterns, and thus the identification of a specific hexose must be determined by the retention time. The same identification procedure must be used, for example, for pentoses and deoxyhexoses. Approximately 25 to 50 µg of starting material is required for a standard methylation procedure, although microscale analyses can be performed that allow as little as 1 µg of material to be analyzed.

In the final analysis, the methylation procedure will give a determination of all the variously linked sugars in a sample. For example, this analysis will determine how many terminal, linear, and branched sugars are present and will identify the types of sugars (for example, glucose, galactose) present.

6. Glycosyl sequencing of oligo- and polysaccharides is not routine.[ref16]

The *glycosyl sequencing* of polysaccharides, and to a lesser extent oligosaccharides, is much more time-consuming and technically challenging than the sequencing of proteins or nucleic acids. The presence of different glycosyl residues in various glycosidic linkages that can be in either an α- or a β-anomeric configuration, together with the potential for branching, all contributes to the challenge involved in sequencing poly- and oligosaccharides. For these reasons the plant cell wall matrix polysaccharides RG-I and RG-II, xyloglucan, and glucuronoarabinoxylan all present glycosyl sequencing problems. However, two cell wall polysaccharides, cellulose and homogalacturonan, are relatively easy to sequence, since each contains only one type of glycosyl residue connected by a single glycosidic linkage, β-1,4- and α-1,4-, respectively. Naturally occurring polysaccharides, such as bacterial exopolysaccharides that are made up of repeating-unit oligosaccharides, are also relatively easy to sequence, as once the structure of one repeating unit is known, the structure of the entire polysaccharide is characterized.

Structural studies on the cell wall matrix polysaccharides have shown that they consist of repeating-unit backbones with a diversity of structurally different oligosaccharide side chains attached to the backbone. Thus, two problems are apparent when attempting to glycosyl-sequence these polysaccharides. The first is to sequence the glycosyl residues of both the side chains and the backbone, and the second is to determine the exact points on the polysaccharide backbone to which the specific oligosaccharide side chains are attached. To date, the complete glycosyl sequence of a cell wall polysaccharide containing more than one type of glycosyl residue has yet to be achieved. Considerable progress has been made, however, in determining the glycosyl sequences of various oligosaccharide fragments, and these give a clear overview of the parent polysaccharide structures (see Concepts 3B1 to 6 and 3C1 to 6).

Once isolated and purified to homogeneity, the structural characterization of oligosaccharide fragments can be achieved using glycosyl residue and glycosyl linkage composition analyses together with mass spectrometric (MS) and nuclear magnetic resonance (NMR) spectroscopic analyses (see Concept 2B7). The resulting glycosyl residue sequences of the oligosaccharides can be analyzed and incorporated with other oligosaccharide structures into a composite structure of the original polysaccharides. Using such methods, much of the structure of RG-I, RG-II, xyloglucan, and glucuronoarabinoxylan has been characterized, although obtaining the complete glycosyl sequence of any of these polysaccharides remains a distant goal.

It has also become apparent that considerable microheterogeneity exists in some polysaccharide structures, for example, xyloglucans. This could result from the presence of different families of xyloglucan within a single cell type, or it could arise from closely related polysaccharides isolated from different cell types from a single plant tissue. Either way, this complexity will contribute to the problems involved in obtaining complete glycosyl sequence data.

7. Nuclear magnetic resonance (NMR) spectroscopy and mass spectrometry (MS) are important tools for determining the structures of polysaccharides.[ref17]

We have seen that the structural characterization of cell wall polysaccharides is greatly facilitated by the determination of the glycosyl sequence of oligosaccharide fragments of the polysaccharides (see Concept 2B6). The structural characterization of these oligosaccharides is obtained using glycosyl composition and glycosyl linkage analysis together with MS and NMR analyses.

Mass spectrometry has proven to be invaluable in several aspects of structural characterization. Coupled to a gas chromatograph (GLC-MS), it is of considerable help in identifying the structures of methylated alditol acetates, thus allowing for the determination of glycosyl linkage (see Concept 2B5). Two types of ionization are used in this analysis. The first is electron impact ionization (EI), which leads to fragmentation of the methylated aldi-

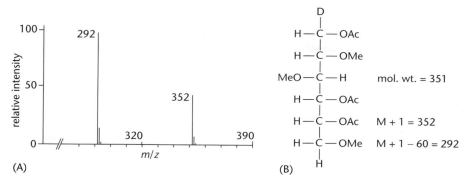

Figure 2.4 A portion of the chemical ionization mass spectrum of a methylated hexitol acetate derivative obtained after glycosyl linkage analysis (see Concept 2B5) of a 4-linked glucosyl residue. The molecular weight of this derivative is 351. The CI mass spectrum shows an ion of the protonated molecule [M + 1] and this ion minus the loss of acetic acid [M + 1 − 60]. OMe, *O*-methylation; OAc, *O*-acetylation; D, deuterium atom; *m/z*, mass/charge. (From M. McNeil and P. Albersheim, *Carbohydr. Res.* 56, 239–248, 1977.)

tol acetates, the resulting fragment ions being diagnostic of the derivative analyzed. Chemical ionization (CI) GLC-MS allows the molecular weight of the methylated alditol acetate to be determined (**Figure 2.4**). GLC-MS in both the EI and CI forms is also important for the analysis of methylated oligosaccharide alditols. Di-, tri-, and tetrasaccharides can be routinely analyzed in this way, providing both sequence information (EI analysis) and molecular weight information (CI analysis) on the derivatized oligosaccharide.

Other forms of mass spectrometry, for example, fast atom bombardment mass spectrometry (FAB-MS), matrix-assisted laser desorption/ionization time-of-flight (MALDI-TOF) mass spectrometry, and electrospray ionization (ESI) mass spectrometry, allow the molecular weight of oligosaccharides containing 2 to 16 glycosyl residues to be routinely determined. Depending on the oligosaccharide to be analyzed, these analyses are performed either on the nonderivatized oligosaccharide or on the derivatized form (for example, *O*-acetylated or *O*-methylated). Often one type of MS analysis is better for a particular type of oligosaccharide; for example, ESI-MS is often best for analyzing oligosaccharides containing acidic components. Some MS structural characteristics have been greatly facilitated by including specific exoglycosidase treatments to remove terminal glycosyl residues. MS analysis before and after such treatments provides additional structural information relative to a stand-alone MS analysis. Together with glycosyl residue composition analysis, all these MS methods allow for the identity and number of glycosyl residues within the oligosaccharide to be determined.

Nuclear magnetic resonance (NMR) spectroscopy has also proven invaluable in the structural characterization of plant cell wall polysaccharides (see **Panel 2.5**). ¹H-NMR routinely leads to the identification of the anomeric configuration of the glycosyl residues in an oligosaccharide fragment. In addition, both one- and two-dimensional (1-D and 2-D) NMR experiments can lead to the determination of the glycosyl sequence of oligosaccharides. These analyses often rely heavily on additional data obtained from glycosyl linkage and mass spectrometric analyses. Taken together, all of these methods can result in the complete structural characterization of an oligosaccharide.

It is important to note that once the one-dimensional ¹H-NMR spectrum of a structurally characterized oligosaccharide is known, it is then a much more simple task to analyze quickly and extremely accurately similar molecules or, indeed, the same molecule isolated from a different source. This is an important advance for the study of cell wall polysaccharides that are made up of several structurally similar repeating-unit oligosaccharides. Analysis of the NMR spectra of these oligosaccharides has led to refined rules for the assignments of the chemical shifts of the components of the glycosyl residues in the NMR spectra. The assignment of such chemical shift rules in NMR spectra has allowed the rapid and accurate identification of the structures of oligosaccharide subunits in both xyloglucan and xylan hemicelluloses (see Concepts 3B3 and 3B5).

Nuclear magnetic resonance (NMR) spectroscopy is one of the most versatile tools used to characterize plant cell walls. Although this technique is most often used to determine the primary structures of polysaccharides and other components of the cell wall, it can provide many other types of information. For example, NMR can be used to characterize the conformational and dynamic properties of a cell wall polysaccharide, and to investigate the geometric details of its interaction with other molecules within the wall.

THE NMR PHENOMENON

An atomic nucleus has several characteristics that may be described by analogy to an electrically charged, spinning top. For example, a spinning top has angular momentum, which leads to a characteristic type of motion, called *precession*, that occurs when the top interacts with any external force that tends to reorient its axis of rotation. Thus, although the force of gravity tends to make the spinning top lie down, interaction of this force with the top's angular momentum causes the top to precess (i.e., wobble) about the vertical line coinciding with the gravitational force. Imagine an electrically charged top spinning in deep space, where gravity is not a significant factor. If our rotating top is electrically charged, it behaves like an electromagnet, as the circulation of electric charge around its spin axis produces a magnetic field. Thus, our top would interact with any magnetic field emanating from an external source. The magnetic forces generated by this interaction would reorient the magnetic top's axis of rotation, tending to make it line up with the external magnetic field. Our cosmic top would respond to the external magnetic field by precessing about the line coinciding with the direction of the magnetic field.

characteristic *resonance frequency* that depends on the strength of the magnetic field and the physical properties of the nucleus. Each nucleus is a minute magnet, and although it would be extremely difficult to detect the magnetic field produced by a single nucleus, the magnetic field generated by the entire collection of nuclei in a sample is large enough to detect and measure. However, nuclei do not normally precess in a way that permits their resonance frequencies to be measured. At any given instant, they do not precisely line up, as they are all at different points in the precession cycle. (*See Figure, left.*) As a result, the magnetic field oscillation generated by the precession of each nucleus is canceled out by oscillations generated by other nuclei, and the *total* magnetic field generated by the collection of nuclei does not oscillate. Nevertheless, by applying a brief magnetic radio-frequency (RF) pulse that itself oscillates at the resonance frequency of the nuclei, it is possible to influence nuclei so that they precess in unison. That is, the RF pulse induces *coherent* precession of the nuclei, as illustrated on the right side of the Figure, *left*. Coherent precession leads to an oscillation of the total magnetic field, which induces an alternating current in a detector coil surrounding the sample. The nuclear resonance frequencies of the sample are represented in this current, which is amplified and digitally recorded. A computer is then used to mathematically transform the digitized data to a representation (i.e., a *frequency domain* NMR spectrum) that is more easily interpreted by the human eye.

Not all atomic nuclei are magnetically active, so not all nuclei can be observed by NMR. The most frequently observed nucleus is the proton (^{1}H), which is the nucleus of virtually all hydrogen atoms in *natural-abundance* samples (i.e., samples that have not been enriched with rare isotopes, such as ^{2}H or ^{3}H). Only 1.1% of the carbon atoms in a natural-abundance sample have a magnetically active ^{13}C nucleus, the remaining 98.9% being the magnetically inactive ^{12}C isotope. Therefore, ^{13}C-NMR is much less sensitive than ^{1}H-NMR. Oxygen consists almost entirely of the magnetically inactive ^{16}O isotope.

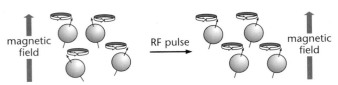

magnetic field   RF pulse →   magnetic field

Magnetic nuclei precess in a magnetic field. Coherent precession of the nuclei can be induced by a brief magnetic radio-frequency (RF) pulse.

As illustrated on the left side of the above Figure, magnetic nuclei behave just like our imaginary top. That is, they precess under the influence of an external magnetic field. This precession has a

CHEMICAL SHIFT

The resonance frequency (in Hz) of a nucleus is directly proportional to the intensity of the magnetic field *at the nucleus*. Typically, the various nuclei in a molecule have slightly different resonance frequencies, as they are exposed to slightly different magnetic field intensities. Much of this variation is due to the electrons that surround each nucleus and shield it from the external magnetic field. As the electron density varies from one location to another within a molecule, electronic shielding causes the resonance frequencies of different nuclei to vary by a few parts per million (ppm). This slight change in frequency is called the chemical shift, as it provides information regarding the chemical environment of the nucleus, encoded as a shift in the position of its NMR signal. The Figure, *right*, which depicts the ^{13}C-NMR spectrum of β-methyl cellobioside, illustrates several basic relationships between chemical shifts and molecular structure. For example, a ^{13}C nucleus is less shielded (i.e., shifted to higher frequency) when the carbon atom is directly attached to an electron-withdrawing oxygen atom. Many of the carbon atoms in β-methyl cellobioside (i.e., C-2 to C-6 and C-2' to C-6') have one directly attached oxygen atom, so their ^{13}C resonances are shifted to frequencies that are 55-90 PPM higher than that of a "standard" carbon (which has no attached oxygens). Anomeric carbon atoms, such as C-1 and C-1' of β-methyl cellobioside, are directly bound to *two* electron-withdrawing oxygen atoms. Anomeric ^{13}C resonances

are thus displaced even further toward the high-frequency end of the spectrum, typically exhibiting chemical shifts of 90–110 ppm. Electrons are also withdrawn from any non-anomeric carbon atoms that are involved in a glycosidic linkage. For example, one of the residues of β-methyl cellobioside is glycosidically linked to C-4' of the other residue, so the C-4' resonance is shifted to a frequency that is ~9 ppm higher than that of C-4, which is not involved in a glycosidic linkage. Thus, analysis of ^{13}C chemical shifts can provide considerable information regarding the molecular structure of complex carbohydrates such as the polysaccharides of the plant cell wall.

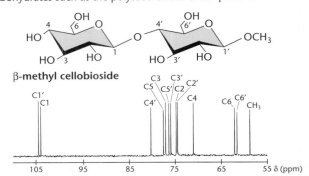

β-methyl cellobioside

DIPOLAR AND SCALAR COUPLING

The resonance frequency of each nucleus also depends on the location and spin state of other, nearby nuclei. When a pair of nuclei affect each other's resonance frequency, they are said to be *coupled*. One type of coupling, called dipolar coupling, is a *direct* effect of the minute magnetic field associated with each magnetic nucleus. This nuclear magnetism augments or reduces the total magnetic field strength at nearby nuclei, thereby altering their resonance frequencies. The magnitude of this dipolar coupling decreases rapidly as the distance between the two nuclei increases. It also depends on the orientation of the line connecting the two nuclei with respect to the external magnetic field. Thus, the effects of dipolar coupling are often clearly visible in the NMR spectra of solids, where molecular motion is limited. In contrast, the orientation of a molecule continuously and rapidly changes in solution, causing the dipolar coupling to fluctuate between positive and negative values. Thus, under most circumstances, dipolar coupling has an average value of zero in solution, and has no effect on the resonance frequency of the nuclei. However, the rapid fluctuation of dipolar coupling in solution does affect the rate at which a collection of nuclei returns to its original state after being perturbed by an RF pulse. This leads to the well-known nuclear Overhauser effect (NOE), which is often exploited to obtain information about the average distance between nuclei in biomolecules. Furthermore, new methods are being developed to extract information about molecular geometry by measuring "residual dipolar couplings," which can be observed under conditions that cause molecules to become partially aligned in solution.

Electrons provide another mechanism by which two nuclei can influence each other's resonance frequencies. This indirect, electron-mediated effect, called scalar coupling, does not depend on the overall molecular orientation, and its consequences are clearly visible in the NMR spectra of dissolved molecules. In these spectra, scalar coupling splits the NMR signals into multiplets. The magnitude (J) of the scalar coupling corresponds to the distance (in Hz) between the individual lines of the multiplet (*see the Figure left*), making it trivial to measure in many cases. Although scalar coupling is not affected by the orientation of the molecule with respect to the external magnetic field, it does depend on the relative orientation of covalent bonds within the molecule. Thus, scalar coupling provides important information regarding the internal geometry of the molecule. This is illustrated in the Figure (*left*), which depicts a small region in the ^{1}H-NMR spectrum of an oligoglycosyl alditol derived from tobacco xyloglucan. The intense resonance at 4.75 ppm is due to residual protons in the deuterated solvent. The three doublets to the left of the solvent resonance correspond to the anomeric (H-1) protons of an α-arabinofuranosyl residue (5.172 ppm, $J_{1,2}$ = 1.8 Hz) and two α-xylopyranosyl residues (5.088 ppm, $J_{1,2}$ = 3.7 Hz, and 4.938 ppm, $J_{1,2}$ = 3.7 Hz). The two doublets to the right of the solvent resonance correspond to the anomeric protons of two β-glucopyranosyl residues (4.617 ppm, $J_{1,2}$ = 8.0 Hz, and 4.550 ppm, $J_{1,2}$ = 8.0 Hz). In each case, the splitting of these signals is due to scalar coupling ($J_{1,2}$) between H-1 and H-2. The magnitude of $J_{1,2}$ depends critically on the orientation of the H-1–C-1 bond relative to the H-2–C-2 bond, which is different in α-linked residues than it is in β-linked residues. Thus, $J_{1,2}$ is less than 5 Hz for a typical α-pyranosyl residue, and more than 7 Hz for a typical β-pyranosyl residue. (Of course, there are important exceptions to this trend.) Due to the inherent flexibility of the 5-membered furanose ring, it is difficult to predict the magnitude of $J_{1,2}$ for α- or β-furanosyl residues.

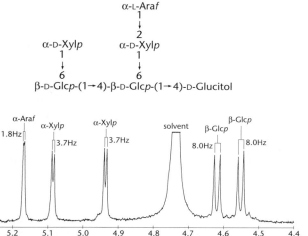

SOLID-STATE AND SOLUTION NMR

Both solid-state and solution NMR are widely used to characterize plant cell wall polysaccharides. Solution NMR is most often used to determine the primary structures of these molecules, and is also a rich source of information regarding their conformational and dynamic properties. Solid-state NMR is rarely used to determine primary structures of complex polysaccharides. However, the plant cell wall is an insoluble aggregate, and solid-state NMR can provide important information regarding the chemical environment, molecular mobility, conformational state, and orientation of its constituent polymers. Much of this information is derived from the fact that the strength of the dipolar coupling between nuclei depends on their relative orientation in the magnetic field. Thus, many of the spectroscopic effects of dipolar coupling are modulated by molecular motion. This makes it possible, for example, to distinguish nuclei in mobile components of the cell wall (e.g., xyloglucan crosslinks) from nuclei in more rigid components (e.g., cellulose microfibrils). Analysis of the plant cell wall is challenging due to its chemical and topological complexity. Nevertheless, new approaches, such as combining isotopic enrichment and sophisticated spectroscopic techniques, are providing deeper insight into the molecular processes leading to the assembly and proper function of this important organelle.

800-MHz NMR spectrometer   *Courtesy of Stefan Eberhard, Complex Carbohydrate Research Center, University of Georgia*

8. The conformations of cell wall oligo- and polysaccharides can be investigated using NMR and computational methods.[ref18]

An understanding of the structure and function of the plant cell wall will eventually require knowledge of the *conformations* (three-dimensional shapes) of the polysaccharide components of the wall. In addition, advances in the field of oligosaccharin research (Chapter 8) will also require knowledge of the conformations of these biologically active oligosaccharides as they interact with their putative receptors. The conformation of cellulose has been well described; with its unique ability to form a crystalline lattice, cellulose is suitable for X-ray analysis (see Concept 2B9). This is not the case for the other cell wall matrix polysaccharides that are soluble in solution or for oligosaccharides that originate as components of these polysaccharides. Most polysaccharides and oligosaccharides can adopt a large number of shapes in solution, and determining their favored conformation, or, in the case of oligosaccharins, the conformation that interacts with a putative receptor, will help resolve the biological functions of these molecules. The polysaccharide hemicelluloses xyloglucan and arabinoxylan are known to hydrogen-bond to cellulose within the cell wall, and thus any conformational models of these must take this interaction with cellulose into account.

The research fields of NMR spectroscopy and theoretical computational analysis do allow scientists to begin to probe the conformational structures of cell wall oligo- and polysaccharides. Scientists use molecular simulation calculations in which theoretical estimates for glycosyl residue stabilities and abilities to interact with closely associated glycosyl residues are taken into account to predict the energetically most favorable conformation of a molecule in solution. The numbers of calculations for these predictions are enormous and even today place considerable demands on the fastest and most powerful computers. Predictions that homogalacturonan or partially methylesterified homogalacturonans are potentially flat ribbons in solution have indicated how different molecular chains may interact to account for the formation of pectin gels; it is possible that the homogalacturonan molecules within the cell wall interact in a similar manner (Concept 3C1). Such predictions were possible because the polysaccharide contained only α-1,4-linked galactosyluronic acid residues. Other cell wall polysaccharides with multiple glycosyl residues linked in many different ways cannot yet be modeled, and conformational studies of these rely on predictions based on the conformational analysis of component oligosaccharides. Conformational studies with xyloglucan oligosaccharide subunits XXFG and XXXG (see xyloglucan nomenclature described in Concepts 3B2 and 3) have suggested that xyloglucan oligosaccharides can exist in both a twisted and a flat chain conformation. This strongly suggests that the trisaccharide side chain of XXFG, having a fucosyl-α-1,2-galactosyl-β-1,2-xylosyl structure, may regulate, or at least contribute to, xyloglucan-cellulose interactions. **Figure 2.5** shows an example of the flat and twisted conformations of xyloglucan predicted by these studies. The studies with xyloglucan oligosaccharides highlight how a considerable amount of three-dimensional structural information can be obtained using conformational studies. Further refinements of the molecular dynamics simulations will be required, however, before the three-dimensional conformation of larger polysaccharide molecules can be accurately analyzed.

NMR spectroscopy has become increasingly more important as a tool for the conformational study of oligo- and polysaccharides. Both one- and two-dimensional methods have been developed that allow us a very detailed look at the interactions between various atoms of oligosaccharides in solution, information that is essential for interpreting the conformation of the oligosaccharide. Such NMR analyses done in parallel with theoretical

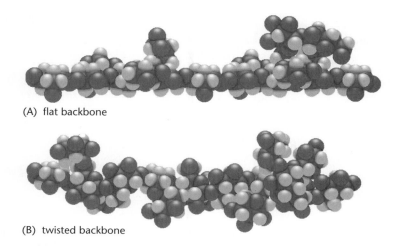

(A) flat backbone

(B) twisted backbone

Figure 2.5 Xyloglucan conformation. Conformational models of possible structures for the xyloglucan heptadecasaccharide GXXFGXXXG (see xyloglucan nomenclature described in Concepts 3B2 and 3). (A) A flat backbone conformation that is proposed to occur upon binding with the glucan chains of a cellulose microfibril. (B) A twisted backbone conformation that is proposed to occur with the molecule in solution. In all views the reducing end of the molecule is at the right of the figure. (From S. Levy et al., *Plant J.* 1, 195–215, 1991.)

molecular simulation calculations can add further information about the conformations of oligosaccharides and raise questions to challenge the data obtained from theoretical calculations. For example, NMR studies did support the twisted structure predicted earlier by the theoretical calculations of the xyloglucan oligosaccharide XXFG solution conformation.

Both theoretical and technical limits still remain on the size of oligosaccharides that can be analyzed by both NMR spectroscopy and molecular simulation calculations. Although the small size of oligosaccharins makes them ideal candidates for quick progress in conformational studies, these molecules can adopt many structures in solution, and obtaining the conformation of that structure that interacts with the putative receptor for the oligosaccharin remains a considerable challenge. The use of NMR and computational methods for the study of conformation is a rapidly growing area of research with considerable potential for cell wall structural analysis.

9. The structure of cellulose has been determined by X-ray diffraction techniques.[ref19]

Many of the physical characteristics of cellulose microfibrils, such as their high tensile strength, can be attributed to the crystalline arrangement of the β-1,4-linked glucan chains that aggregate to form microfibrils. This crystalline arrangement is stabilized by numerous regular hydrogen bonds, both within and between the β-1,4-linked glucan chains. It makes cellulose amenable to analysis by X-ray diffraction techniques. The characteristic patterns obtained by X-ray diffraction analysis of cellulose (**Figure 2.6**) have allowed many aspects of the structure of cellulose microfibrils to be determined in detail, including the locations of hydrogen bonds, the orientation of the β-1,4-glucan chains, the diameter of the microfibrils, the identification of differences between the various forms of isolated cellulose, and the identification in the microfibrils of areas of considerable crystallinity and areas with more amorphous characteristics (see Concepts 3A1 and 2). Other methods that have contributed data to cellulose structural studies include electron diffraction (performed in an electron microscope), solid-state ^{13}C-NMR, and infrared (IR) and Raman absorption spectroscopies. Electron microscopy has also been of considerable value in obtaining data on the size of cellulose microfibrils isolated from different sources.

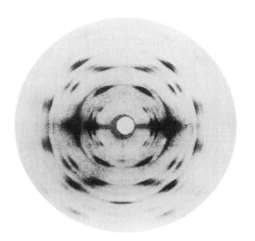

Figure 2.6 An X-ray pattern of cell wall cellulose. (From R.M. Brown Jr., ed., Cellulose and Other Natural Polymer Systems. New York: Plenum Press, 1982, p. 410. Originally from K.H. Gardner and J. Blackwell, *Biopolymers* 13, 1975–2001, 1974.)

Key Terms

arabinoxylan	mass spectrometry
cellulose	methylation analysis
conformation	nuclear magnetic resonance
degree of polymerization	pectin
glycosyl linkage composition analysis	pectic polysaccharide
glycosyl residue composition	primary structure
glycosyl sequencing	RG-I
hemicellulose	RG-II
homogalacturonan	xyloglucan

References

[1]**Primary walls isolated in a variety of ways contain the same structural polymers.**

Koller A, O'Neill MA, Darvill AG & Albersheim P (1991) A comparison of the polysaccharides extracted from dried and non-dried walls of suspension-cultured sycamore cells. *Phytochemistry* 30, 3903–3908.

Selvendran RR & Du Pont MS (1980) An alternative method for the isolation and analysis of cell wall materials from cereals. *Cereal Chem.* 57, 278–283.

Selvendran RR & O'Neill M (1987) Isolation and analysis of cell walls from plant material. *Methods Biochem. Anal.* 32, 25–153.

York WS, Darvill AG, McNeil M et al. (1986) Isolation and characterization of plant cell walls and cell wall components. *Methods Enzymol.* 118, 3–40.

[2]**Cellulose is a β-1,4-linked glucan with a disaccharide repeating unit.**

Delmer DP (1987) Cellulose biosynthesis. *Annu. Rev. Plant Physiol.* 38, 259–290.

Franz G & Blaschek W (1990) Cellulose. *Methods Plant Biochem.* 2, 291–317.

Gardner KH & Blackwell J (1974) The structure of native cellulose. *Biopolymers* 13, 1975–2001.

Mutwil M, Debolt S & Persson, S (2008) Cellulose synthesis: a complex complex. *Curr. Opin. Plant Biol.* 11, 252–257.

[3]**Pectins and hemicelluloses are the non-cellulosic polysaccharides of primary walls.**

Darvill A, McNeil M, Albersheim P & Delmer DP (1980) The primary cell walls of flowering plants. *Biochem. Plants* 1, 91–162.

Selvendran RR & O'Neill M (1987) Isolation and analysis of cell walls from plant material. *Methods Biochem. Anal.* 32, 25–153.

[4]**Wall polysaccharides are composed of 13 different sugars.**

Bacic A, Harris PJ & Stone BA (1988) Structure and function of plant cell walls. *Biochem. Plants* 14, 297–371.

Carpita NC & Gibeaut DM (1993) Structural models of primary walls in flowering plants: consistency of molecular structure with the physical properties of the walls during growth. *Plant J.* 3, 1–30.

McNeil M, Darvill AG, Fry SC & Albersheim P (1984) Structure and function of the primary cell walls of higher plants. *Annu. Rev. Biochem.* 53, 625–663.

[5]**Homogalacturonan, rhamnogalacturonan I, and rhamnogalacturonan II are three pectic polysaccharides in primary walls.**

Bacic A, Harris PJ & Stone BA (1988) Structure and function of plant cell walls. *Biochem. Plants* 14, 297–371.

McNeil M, Darvill AG, Fry SC & Albersheim P (1984) Structure and function of the primary cell walls of higher plants. *Annu. Rev. Biochem.* 53, 625–663.

Mohnen D (2008) Pectin structure and biosynthesis. *Curr. Opin. Plant Biol.* 11, 266–277.

Mort AJ, Feng Q & Niels OM (1993) Determination of the pattern of methyl esterification in pectin. *Carbohydr. Res.* 247, 21–35.

[6]**Xyloglucans and arabinoxylans are the two major hemicellulosic polysaccharides in primary walls.**

Bacic A, Harris PJ & Stone BA (1988) Structure and function of plant cell walls. *Biochem. Plants* 14, 297–371.

Carpita NC & Gibeaut DM (1993) Structural models of primary walls in flowering plants: consistency of molecular structure with the physical properties of the walls during growth. *Plant J.* 3, 1–30.

Vogel J (2008) Unique aspects of the grass cell wall. *Curr. Opin. Plant Biol.* 11, 301–307.

York WS & O'Neill MA (2008) Biochemical control of xylan biosynthesis: which end is up? *Curr. Opin. Plant Biol.* 11, 258–265.

[7]**The structural characterization of polysaccharides is difficult.**

Åman P (1993) Composition and structure of cell wall polysaccharides in forages. In Forage Cell Wall Structure and Digestibility (HG Jung, DR Buxton, RD Hatfield & J Ralph eds), pp 183–200.

McNeil M, Darvill AG, Åman P et al. (1982) Structural analysis of complex carbohydrates using high performance liquid chromatography, gas chromatography and mass spectrometry. *Methods Enzymol.* 83, 3–45.

Selvendran RR & O'Neill M (1987) Isolation and analysis of cell walls from plant material. *Methods Biochem. Anal.* 32, 25–153.

[8]Cell walls contain proteins, glycoproteins, and proteoglycans that perform enzymatic, structural, defensive, and other functions.

Bolwell GP (1993) Dynamic aspects of the plant extracellular matrix. *Int. Rev. Cytol.* 146, 261–323.

Kieliszewski MJ & Lamport DTA (1994) Extensin in repetitive motifs, functional sites, post-translational codes, and phylogeny. *Plant J.* 5, 157–172.

Showalter AM (1993) Structure and function of cell wall proteins. *Plant Cell* 5, 9–23.

[9]Primary cell walls contain an aqueous phase.

McNeil M, Darvill AG, Fry SC & Albersheim P (1984) Structure and function of the primary cell walls of higher plants. *Annu. Rev. Biochem.* 53, 625–663.

[10]Secondary walls are composed of primary walls plus additional layers of polymers.

Eriksson K-EL, Blanchette RA & Ander P (1990) Biodegradation of hemicelluloses. In Microbial and Enzymatic Degradation of Wood and Wood Components (K-El Eriksson, RA Blanchette & P. Ander eds), pp 181–222. Berlin: Springer.

Mellerowicz EJ & Sundberg B (2008) Wood cell walls: biosynthesis, developmental dynamics and their implications for wood properties. *Curr. Opin. Plant Biol.* 11, 293–300.

Shimizu K (1991) Chemistry of hemicelluloses. In Wood and Cellulosic Chemistry (DN-S Hon & N Shimaisi eds), pp 177–214. New York: Marcel Dekker.

[11]Obtaining pure polysaccharides is a prerequisite for primary structure determination.

Darvill A, McNeil M, Albersheim P & Delmer DP (1980) The primary cell walls of flowering plants. *Biochem. Plants* 1, 91–102.

Mort AJ, Moerschbacker BM, Pierce ML & Maness NO (1991) Problems encountered during the extraction, purification, and chromatography of pectic fragments, and some solutions to them. *Carbohydr. Res.* 215, 219–227.

Selvendran RR & O'Neill M (1987) Isolation and analysis of cell walls from plant material. *Methods Biochem. Anal.* 32, 25–153.

[12]Extracellular polysaccharides secreted by cultured cells are excellent models for wall polysaccharides.

McNeil M, Darvill AG, Fry SC & Albersheim P (1984) Structure and function of the primary cell walls of higher plants. *Annu. Rev. Biochem.* 53, 625–663.

Stevenson TT, McNeil M, Darvill AG & Albersheim P (1986) Structure of plant cell walls XVIII: an analysis of the extracellular polysaccharides of suspension-cultured sycamore cells. *Plant Physiol.* 80, 1012–1019.

York WS, Darvill AG, McNeil M et al. (1985) Isolation and characterization of cell walls and cell wall components. *Methods Enzymol.* 118, 3–40.

[13]Enzymatic and chemical cleavage of polysaccharides has made structural studies possible.

Selvendran RR & O'Neill M (1987) Isolation and analysis of cell walls from plant material. *Methods Biochem. Anal.* 32, 25–153.

York WS, Darvill AG, McNeil M et al. (1985) Isolation and characterization of cell walls and cell wall components. *Methods Enzymol.* 118, 3–40.

[14]Several methods are used for glycosyl composition analysis.

Selvendran RR & Ryden P (1990) Isolation and analysis of plant cell walls. *Methods Plant Biochem.* 2, 549–575.

York WS, Darvill AG, McNeil M et al. (1985) Isolation and characterization of plant cell walls and cell wall components. *Methods Enzymol.* 118, 3–40.

[15]Glycosyl linkage compositions are determined by methylation analysis.

Carpita NC & Gibeaut DM (1993) Structural models of primary walls in flowering plants: consistency of molecular structure with the physical properties of the walls during growth. *Plant J.* 3, 1–30.

Mort AJ, Parker S & Kuo M-S (1983) Recovery of methylated saccharides from methylation reaction mixtures using Sep-Pak C-18 cartridges. *Anal. Biochem.* 133, 380–384.

Selvendran RR & Ryden P (1990) Isolation and analysis of plant cell walls. *Methods Plant Biochem.* 2, 549–575.

York WS, Darvill AG, McNeil M et al. (1985) Isolation and characterization of plant cell walls and cell wall components. *Methods Enzymol.* 118, 3–40.

[16]Glycosyl sequencing of oligo- and polysaccharides is not routine.

Carpita NC & Gibeaut DM (1993) Structural models of primary walls in flowering plants: consistency of molecular structure with the physical properties of the walls during growth. *Plant J.* 3, 1–30.

O'Neill MA, Albersheim P & Darvill A (1990) The pectic polysaccharides of primary cell walls. *Methods Plant Biochem.* 2, 415–439.

Selvendran RR & O'Neill M (1987) Isolation and analysis of cell walls from plant material. *Methods Biochem. Anal.* 32, 25–153.

[17]Nuclear magnetic resonance (NMR) spectroscopy and mass spectrometry (MS) are important tools for determining the structures of polysaccharides

Hoffmann RA, Geijtenbeek J, Kamerling JP & Vliegenthart JFG (1992) ^{1}H-n.m.r. study of enzymically generated wheat-endosperm arabinoxylan oligosaccharides: structures of hepta- to tetradeca-saccharides containing two or three branched xylose residues. *Carbohydr. Res.* 223, 19–44.

Kormelink FJM, Hoffmann RA, Gruppen H et al. (1993) Characterisation by ^{1}H NMR spectroscopy of oligosaccharides derived from alkali-extractable wheat-flour arabinoxylan by digestion with endo-$(1{\rightarrow}4)$-β-D-xylanase III from *Aspergillus awamori*. *Carbohydr. Res.* 249, 369–382.

York WS, Kumar Kolli VS, Orlando R et al. (1996) The structure of arabinoxyloglucans produced by solanaceous plants. *Carbohydr. Res.* 285, 99–128.

York WS, Van Halbeek H, Darvill AG & Albersheim P (1990) Structural analysis of xyloglucan oligosaccharides by ^{1}H-n.m.r. spectroscopy and fast-atom-bombardment mass spectrometry. *Carbohydr. Res.* 200, 9–31.

[18]The conformations of cell wall oligo- and polysaccharides can be investigated using NMR and computational methods.

Levy S, Maclachlan G & Staehelin LA (1997) Xyloglucan sidechains modulate binding to cellulose during *in vitro* binding assays as predicted by conformational dynamics simulations. *Plant J.* 11, 373–386.

Levy S, York WS, Stuike-Prill R et al. (1991) Simulations of the static and dynamic molecular conformations of xyloglucan: the role of the fucosylated sidechain in surface-specific sidechain folding. *Plant J.* 1, 195–215.

Ló V-M, Hahn MG & Van Halbeek H (1994) Preparation, purification and structural characterisation of linear oligogalacturonides: a FAB-mass spectrometric and NMR spectroscopic study. *Carbohydr. Res.* 255, 271–284.

Ogawa K, Hayashi T & Okamura K (1990) Conformational analysis of xyloglucans. *Int. J. Biol. Macromol.* 12, 218–222.

[19]The structure of cellulose has been determined by X-ray diffraction techniques.

Franz G & Blaschek W (1990) Cellulose. *Methods Plant Biochem.* 2, 291–323.

Ha M-A, Apperlley D C, Evans B W et al. (1998) Fine structure in cellulose microfibrils: NMR evidence from onion and quince. *Plant J.* 16, 183–190.

McCann M, Roberts K, Wilson R H et al. (1995) Old and new ways to probe plant cell-wall architecture. *Can. J. Bot.* 73, S103–S113.

Newman RH, Davies LM & Harris PJ (1996) Solid-state ^{13}C nuclear magnetic resonance characterization of cellulose in the cell walls of *Arabidopsis thaliana* leaves. *Plant Physiol.* 111, 475–485.

Biochemistry of the Cell Wall Molecules

3

The primary cell walls of plants have structures composed of polysaccharides and often structural proteins. Secondary walls in addition contain lignin and extra amounts of cellulose. Cell walls of specialized cells (for example, epidermal cells) contain other polymers essential for the cell's function (for example, cutin). In all cases the interactions of the molecules within the wall result in the wall's architecture.

This chapter describes the structure of the most abundant cell wall structural molecules. These include cellulose, pectins, hemicelluloses, proteins, glycoproteins, lignin, cutin, suberin, and waxes. Most of these molecules have common structural features found in all walls. These structures are described together with structural modifications found in specific cell types, tissues, or particular groups of plants.

A. Cellulose

1. β-Glucan chains hydrogen-bond together to form cellulose microfibrils.[ref1]

The physical and chemical nature of cellulose is strongly influenced by the ability of the β-1,4-linked glucan chains to form both intramolecular and intermolecular hydrogen bonds. These hydrogen bonds result in the lateral aggregation and crystallization of the β-1,4-linked glucan backbones into structures called microfibrils. Approximately 30 to 50 β-1,4-linked glucan chains are held together by hydrogen bonds to form a *cellulose microfibril* of a diameter of 3 to 5 nm. **Figure 3.1** shows the potential intramolecular and intermolecular hydrogen bonds of two closely associated β-1,4-linked glucan chains in a cellulose microfibril. The many intra- and intermolecular hydrogen bonds endow the cellulose microfibrils with a well-defined structure as well as considerable strength and stiffness. It is these microfibrillar properties that are exploited in the cell wall. The crystalline arrangement of the glucan chains in the microfibrils has allowed the structures of the microfibrils to be examined in great detail using X-ray diffraction.

2. The glucan chains of cellulose microfibrils all have the same orientation.[ref2]

Cellulose exists in several crystalline forms (called I through IV in the literature) that have different X-ray diffraction patterns, ^{13}C solid-state NMR spectra, and infrared and Ramen absorption spectra. All of these methods can give information on the structure and crystallinity of cellulose microfibrils, although X-ray diffraction has contributed most to these determinations. *Cellulose I* is the native form of cellulose as it is found in

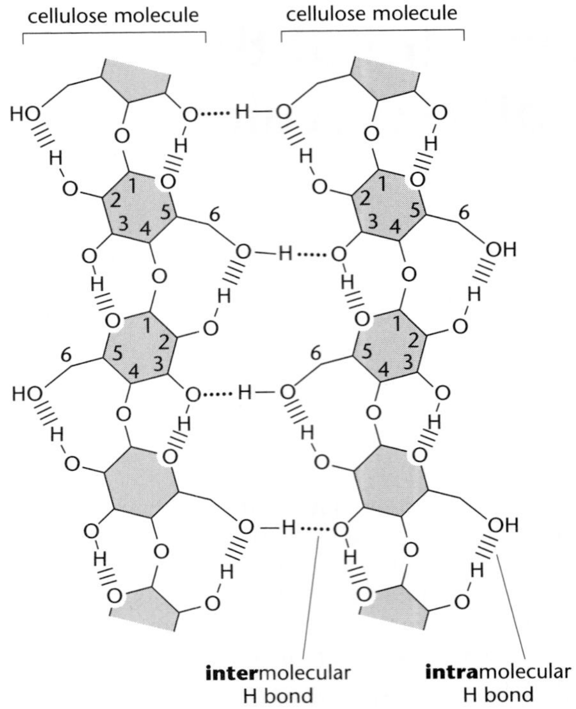

Figure 3.1 The intramolecular (parallel bars) and intermolecular (dotted lines) hydrogen bonds of two glucan chains (shaded) within a microfibril. The glucose units are shaded, and the individual carbon atoms of several glucosyl residues are labeled 1 through 6. (From G. Franz and W. Blaschek, *Methods Plant Biochem.* 2:291–317, 1990.)

plant cell walls. The characteristic feature of cellulose I is the parallel configuration of the β-1,4-linked glucan chains. A *parallel configuration* means that all of the reducing ends of the glucan chains are at the same end or pole of the microfibrils. This configuration reflects the mode of synthesis of cellulose microfibrils (see Concepts 5B1 and 5B2). Treatment of cellulose I by strong base can cause an irreversible change to *cellulose II*, in which the β-1,4-linked glucan chains are found in the antiparallel orientation.

Structural studies of cellulose microfibrils have indicated that they are composed of highly ordered crystalline regions that are interrupted by more amorphous (less crystalline) regions. The degree and distribution of the crystalline and amorphous regions in cellulose microfibrils are proposed to affect the physical properties of the cellulose microfibril within the cell wall, although the exact role of each still remains debatable.

The cellulose microfibrils within the wall are associated with hemicelluloses that are attached to the cellulose microfibrils via hydrogen bonds. It is these hemicelluloses that are the molecules that interface between the cellulose microfibrils and the rest of the matrix macromolecules. During the formation of secondary cell walls, the amount of cellulose in the wall can increase from about 20% to about 50% of the dry weight of the wall. In primary cell walls the cellulose microfibrils are seen by electron microscopy to be organized both as a random mesh and as loose parallel sheets, whereas in secondary walls they appear organized into parallel sheets and bundles. Changes in the orientation and organization of the cellulose microfibrils have been shown to affect cell shape, cell growth, and the mechanical properties of both primary and secondary cell walls.

B. Hemicelluloses

1.   Xyloglucan is a principal hemicellulose of cell walls.[ref3]

Xyloglucans are an important family of hemicellulose polysaccharides found in the primary cell walls of all higher plants. Estimates of molecular weights up to 200,000 kD have been made for primary cell wall xyloglucans. Xyloglucans account for approximately 30% of the primary cell walls of dicotyledons, up to 20% of the primary cell walls of gymnosperms, and less than 2% of the primary cell walls of the Poaceae (the grass family). Xyloglucans adhere strongly to cellulose via hydrogen bonds and attach to the surface of cellulose microfibrils in the walls of growing plant cells. This permits the formation of xyloglucan bridges between the microfibrils and may reduce the ability of the cellulose microfibrils to aggregate laterally into very large fibrils. The resulting *cellulose-hemicellulose network* appears to form a framework around which the other cell wall macromolecules become organized. A nomenclature for xyloglucan-derived oligosaccharides has been utilized by researchers in the field that greatly helps descriptions of xyloglucan structure (see Panel 3.1)

In most dicotyledons and gymnosperms the structural characteristics of xyloglucans consist of a highly branched polymer with a β-1,4-linked glucan chain backbone (like cellulose) in which approximately 75% of the β-D-glucosyl residues are substituted at C-6 with single α-D-xylosyl residue side chains. Some of the α-D-xylosyl residues have β-D-galactosyl or α-L-fucosyl-(1→2)-β-D-galactosyl units attached to C-2 (**Panel 3.1**). In sycamore cell wall xyloglucan many of the 2-linked galactosyl residues have one or two *O*-acetyl substituents. Additionally, β-D-xylosyl residues are attached to C-2 of less than 5% of the β-D-glucosyl residues in the backbone. Approximately 2% of sycamore xyloglucan also contains terminal α-L-arabinosyl residues that are attached either directly to C-2 of some of the backbone β-D-glucosyl residues or to C-3 of the β-D-xylosyl residues (see Concept 3B3).

Xyloglucans isolated from the Solanaceae (a family of dicotyledons, including tobacco and tomato) have a somewhat different structure. These xyloglucans have the same β-1,4-linked glucan backbone, but they do not contain fucosyl residues and have only minor amounts of galactosyl residues in the side chains. α-L-Arabinosyl residues are found attached to C-2 of some of the side chain xylosyl residues. Similar structures are found in olive fruit xyloglucan. Approximately 50% of the backbone glucosyl residues are branched in the Solanaceae, mostly with α-D-xylosyl residues; some glucosyl residues are acetylated at O-6; and in addition, some arabinosyl residues are also acetylated. Tomato cell walls also contain β-Ara*f* residues in their xyloglucans, the only xyloglucans so far reported to contain this glycosyl residue.

Xyloglucans isolated from Poaceae also lack fucosyl and arabinosyl residues and contain few galactosyl residues. The β-1,4-glucan backbone has much less branching than dicotyledon xyloglucan and is quite similar to the Solanaceae in this respect. In the Poaceae, nearly all of the xyloglucan side chains are single terminal α-D-xylosyl residues.

2. The structural features of xyloglucan have been defined by analysis of *endo*-glucanase-released oligosaccharides.[ref4]

Many of the structural features of xyloglucan have been determined by the detailed analysis of oligosaccharide repeating units enzymatically released from the polysaccharide. This analysis is made possible because the unbranched β-1,4-D-glucosyl residues of the backbone of xyloglucan are susceptible to hydrolysis by fungal *endo-β-1,4-glucanases*. Often, the xyloglucan oligosaccharides are released directly from primary cell walls by treatment with *endo*-glucanase; or alternatively, polymeric xyloglucans that have been released from cell walls by treatment with alkali are subsequently fragmented with *endo*-β-1,4-glucanase prior to structural analysis. The predominant oligosaccharides released from nonsolanaceous dicotyledon xyloglucans are shown in **Figure 3.2**. The nonasaccharide (XXFG), the heptasaccharide (XXXG), and the pentasaccharide (XXG) are major components of most dicotyledonous xyloglucans, accounting for approximately

XYLOGLUCAN NOMENCLATURE

A system of abbreviated names is used for xyloglucan oligosaccharides. Each differently substituted 1,4-linked β-D-glucosyl residue (or a reducing terminal D-glucose moiety*) of the xyloglucan backbone is given a one-letter code. The name of the oligosaccharide consists of these code letters listed in sequence from the nonreducing to the reducing terminus of the backbone.

code letter	structure represented
G	β-D-Glcp *-
X	α-D-Xylp-(1→6)-β-D-Glcp *-
L	β-D-Galp-(1→2)-α-D-Xylp -(1→6)-β-D-Glcp *-
F	α-L-Fucp-(1→2)-β-D-Galp-(1→2)-α-D-Xylp-(1→6)-β-D-Glcp *-
A	α-L-Araf-(1→2)⎤ α-D-Xylp-(1→6)⎦ β-D-Glcp *-
B	β-D-Xylp-(1→2)⎤ α-D-Xylp-(1→6)⎦ β-D-Glcp *-
C	α-L-Araf-(1→3)-β-D-Xylp-(1→2)⎤ α-D-Xylp-(1→6)⎦ β-D-Glcp *-
S	α-L-Araf-(1→2)-α-D-Xylp-(1→6)-β-D-Glcp *-

From Fry et al., *Physiol. Plant.* 89 : 1–3, 1993.

Using these one letter codes, below, is the abbreviated nomenclature for some commonly occurring xyloglucan oligosaccharides.

abbreviation	structure	abbreviation	structure

XLFG

```
Glc → Glc → Glc → Glc
 ↑     ↑     ↑
Xyl   Xyl   Xyl
       ↑     ↑
      Gal   Gal
             ↑
            Fuc
```

XXFG

```
Glc → Glc → Glc → Glc
 ↑     ↑     ↑
Xyl   Xyl   Xyl
             ↑
            Gal
             ↑
            Fuc
```

XLLG

```
Glc → Glc → Glc → Glc
 ↑     ↑     ↑
Xyl   Xyl   Xyl
       ↑     ↑
      Gal   Gal
```

XXLG

```
Glc → Glc → Glc → Glc
 ↑     ↑     ↑
Xyl   Xyl   Xyl
             ↑
            Gal
```

XLXG

```
Glc → Glc → Glc → Glc
 ↑     ↑     ↑
Xyl   Xyl   Xyl
       ↑
      Gal
```

XXXGol

```
Glc → Glc → Glc → Glucitol
 ↑     ↑     ↑
Xyl   Xyl   Xyl
```

XXG

```
Glc → Glc → Glc
 ↑     ↑
Xyl   Xyl
```

FG

```
Glc → Glc
 ↑
Xyl
 ↑
Gal
 ↑
Fuc
```

XXXGXXG

```
Glc → Glc → Glc → Glc → Glc → Glc → Glc → Glc
 ↑     ↑     ↑           ↑     ↑     ↑
Xyl   Xyl   Xyl         Xyl   Xyl   Xyl
```

XXFGAXXG

```
                        Ara
                         ↓
Glc → Glc → Glc → Glc → Glc → Glc → Glc → Glc
 ↑     ↑     ↑           ↑     ↑     ↑
Xyl   Xyl   Xyl         Xyl   Xyl   Xyl
             ↑
            Gal
             ↑
            Fuc
```

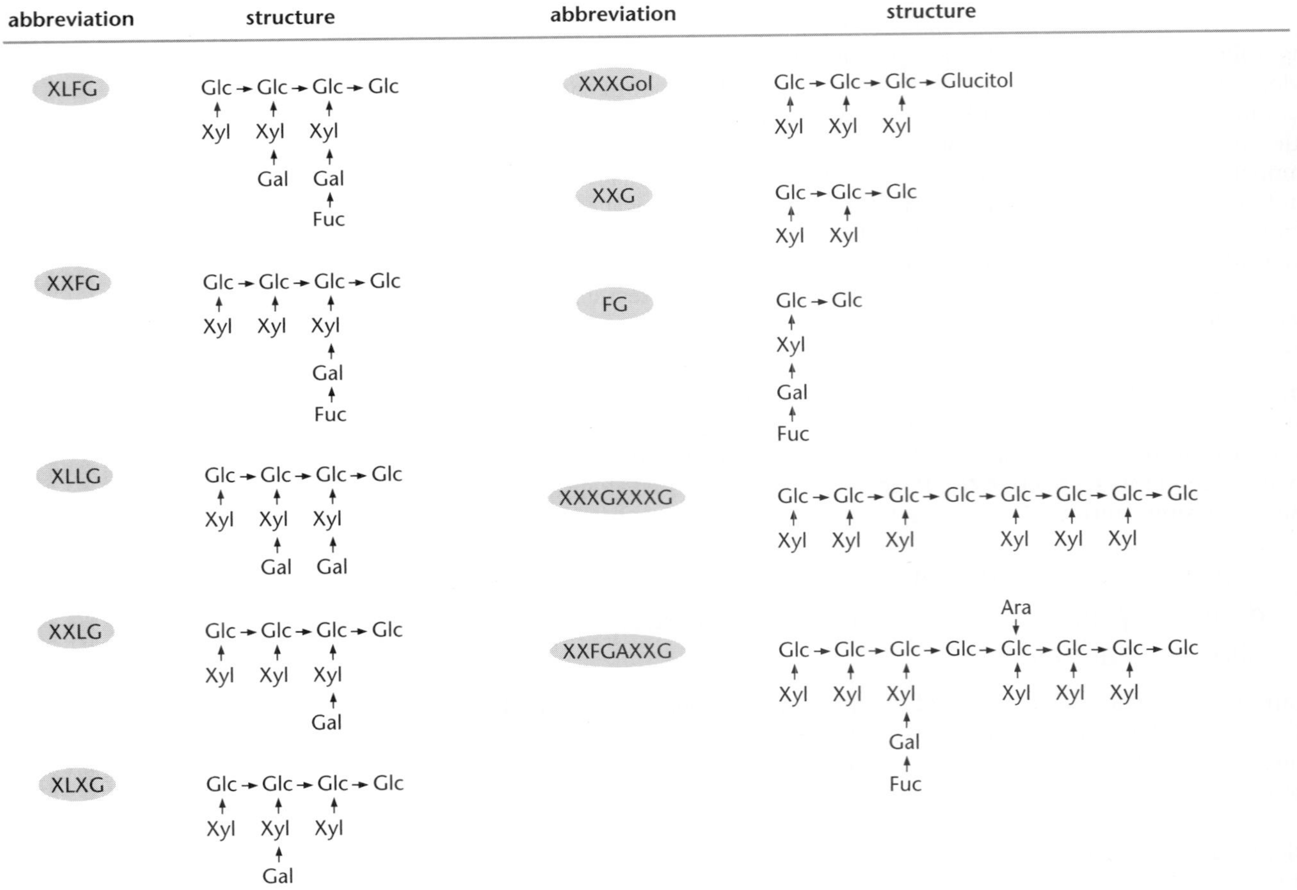

sugar residues and linkages shown are:
Ara = α-L-arabinofuranosyl (1→2), Fuc = α-L-fucopyranosyl (1→2), Gal = β-D-galactopyranosyl (1→2),
Glc = β-D-glucopyranosyl (1→4) (or reducing D-glucose), Xyl = α-D-xylopyranosyl (1→6), Glucitol = reduced glucose

75% of the xyloglucan. The more quantitatively minor decasaccharide (XLFG) and undecasaccharide (XFFG) plus oligosaccharides of sizes 17 to 21 glycosyl residues (see Concept 3B3) are also released by *endo*-β-1,4-glucanase treatment.

Oligosaccharides released from xyloglucans by treatment with *endo*-β-1,4-glucanase can be fractionated and often purified to homogeneity by a combination of gel-filtration, reversed-phase, and high-pH anion-exchange chromatographies prior to structural analysis. Often, the xyloglucan oligosaccharides are converted to their oligoglycosyl alditols by treatment with sodium borohydride prior to chromatography to allow for easier separation by removing the need to separate the reducing-end α- and β-anomers of each individual oligosaccharide (see Panel 2.3).

3. Side chain substitutions on the backbone of xyloglucan affect *endo*-glucanase cleavage of the backbone.[ref5]

Some structural features of xyloglucans highlighted by studies of sycamore cultured cell xyloglucan can dramatically affect the ability of fungal *endo*-β-1,4-glucanases to fragment the xyloglucan backbone. Normally, the glycosidic linkages of the unbranched β-1,4-linked glucosyl residues in the backbone are susceptible to cleavage by *endo*-β-1,4-glucanase. It has been observed that certain side chain structures reduce or inhibit the susceptibility of adjacent unbranched β-1,4-linked glucosyl residues to *endo*-glucanase cleavage. This resistance to *endo*-glucanase cleavage has resulted in the release and subsequent purification and structural analysis of several structurally different xyloglucan oligosaccharides consisting of 17 to 21 glycosyl residues. The structures of some of these oligosaccharides are listed in **Table 3.1**, and the structure of one of these oligosaccharides (XXXGBXFG) showing the *endo*-glucanase-resistant unbranched glucosyl residue is shown in **Figure 3.3**.

The nonsusceptibility of these oligosaccharides to *endo*-glucanase cleavage is due to the presence in these oligosaccharides of terminal α-L-arabinosyl– and/or β-D-xylosyl–containing side chains attached to C-2 of the backbone glucosyl residue adjacent to the unbranched β-D-glucosyl residue. Additionally, the presence of terminal β-D-galactosyl residues and/or α-L-fucosyl-1,2-β-D-galactosyl moieties at C-2 of two adjacent α-D-xylosyl residues in nearby side chains decreases significantly, but not totally, the susceptibility of the unbranched β-D-glucosyl backbone residue to cleavage by *endo*-glucanase.

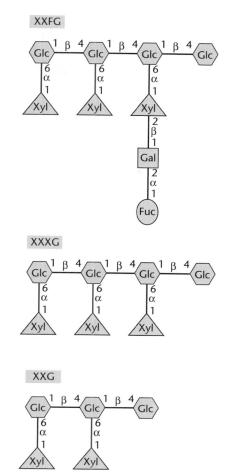

Figure 3.2 The structures of xyloglucan nonasaccharide (XXFG), heptasaccharide (XXXG), and pentasaccharide (XXG), the quantitatively major oligosaccharides released from nonsolanaceous dicotyledon cell walls by enzyme treatment.

TABLE 3.1 ABBREVIATED STRUCTURE NAMES OF OLIGOSACCHARIDES CONTAINING 17 TO 21 GLYCOSYL RESIDUES RELEASED FROM XYLOGLUCAN BY TREATMENT WITH *ENDO*-β-1,4-GLUCANASE	
XXFGAXXG	XLFGAXXG
XXFGBXXG	XLFGBXXG
XXFGCXXG	XLFGAXFG
XFFGXXXG	XLFGXLFG
XXFGAXFG	XLFGXXFG
XXFGBXFG	XXFGXLFG
XXFGCXFG	XFFGXXFG
XXBGXXFG	XXFGXFFG
XXXGBXFG	XLFGBXFG

Figure 3.3 The primary structure of the xyloglucan oligosaccharide XXXGBXFG. The arrow indicates the glycosidic bond of the *endo*-glucanase-resistant unbranched glucosyl residue.

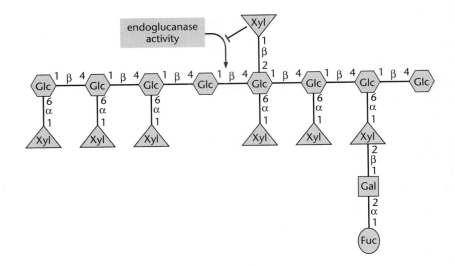

These structural features of xyloglucans have been characterized in greatest detail for the xyloglucans found in the cell walls of sycamore cultured cells and also in the extracellular polysaccharides of these cells. However, the dramatic effects of these side chain substitution patterns on the location and on the rate of *endo*-β-1,4-glucanase cleavage of xyloglucans suggest that these structural features may play a similar and important role in controlling xyloglucan metabolism *in vivo* in different plants.

4. Arabinoxylan is the predominant hemicellulose of grasses.[ref6]

Arabinoxylans constitute the majority of the hemicellulose found in the primary cell walls of the Poaceae monocotyledons. Between 30 and 40% of these walls can be accounted for by arabinoxylan, whereas less than 5% of the primary cell walls of dicotyledons and gymnosperms are accounted for by arabinoxylan. The majority of the arabinoxylans that have been structurally characterized have been released from primary cell walls either by treatment with strong alkali or by the release of oligosaccharide fragments after treatment with *endo*-xylanase.

Xylans have a backbone of β-1,4-linked xylosyl residues with side chains attached to the backbone through C-2 and/or C-3 of the xylosyl residues. Depending on the type of tissue extracted, the proportion of the backbone xylosyl residues that contains side chains can vary from 10% to approximately 90%. Side chains containing arabinosyl, galactosyl, glucosyluronic acid, and 4-*O*-methyl glucosyluronic acid residues have been identified. Those arabinoxylan polysaccharides that contain relatively few side chains have been shown to hydrogen-bond to themselves or to cellulose microfibrils. Thus, the position and amount of side chain substitution can greatly determine the structural interactions of arabinoxylans within the cell wall. Unlike xyloglucans, which are neutral, many arabinoxylans are found containing glucosyluronic acid residues or 4-*O*-methyl glucosyluronic acid residues, and thus these residues impart acidic functionality to the arabinoxylan which can affect the ability of the arabinoxylan to interact with other cell wall macromolecules.

In addition to the various glycosyl residue side chains attached to the β-1,4-linked xylosyl backbone, many arabinoxylans have been isolated that contain phenolic acids, for example, ferulic acid and coumaric acid. Such phenolic acids have been postulated to be involved in the cross linking of cell wall polysaccharides, and indeed a diferuloylarabinoxylan hexasaccharide has been isolated from bamboo shoot cell walls (see Concept 6B7).

5. The elucidation of arabinoxylan structures has been facilitated by characterizing oligosaccharides generated by *endo*-xylanases.[ref7]

Various *endo*-β-1,4-xylanases have been extremely useful for producing defined oligosaccharide fragments of arabinoxylans suitable for structural analysis, in an analogous situation to the use of *endo*-β-1,4-glucanases to release oligosaccharide fragments of xyloglucans. Many of the released xylan backbone fragments containing various side chains have been purified to homogeneity by ion-exchange, gel-filtration, and high-pH anion-exchange chromatography and structurally characterized. Furthermore, *endo*-xylanases from different organisms have been found to have slightly different specificities for where they cleave the arabinoxylan backbone thereby yielding different fragments. Some *endo*-xylanases require continuous β-1,4-linked xylosyl residues that do not contain side chains, whereas others require the presence of a terminal glucosyluronic acid residue in close proximity to the xylosidic linkage to be cleaved.

Examples of structures of arabinoxylan oligosaccharides are shown in **Figure 3.4**. Oligosaccharides isolated from arabinoxylans of Poaceae monocotyledons have been identified where backbone xylosyl residues have terminal arabinofuranosyl residues attached to C-3 and/or C-2, and contiguous backbone xylosyl residues have been identified that both have terminal arabinofuranosyl residues. When present, glucosyluronic acid

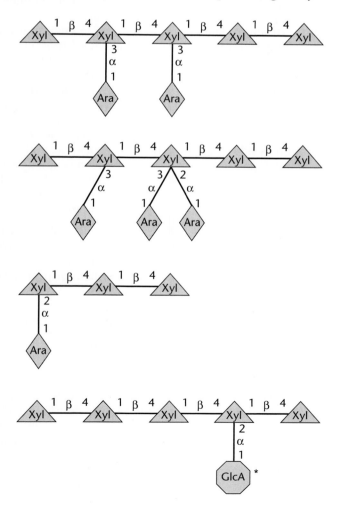

Figure 3.4 Oligosaccharides released from monocotyledon xylans using *endo*-xylanases. Polysaccharides will have the same backbone (β-1,4-linked xylosyl residues) with side chains identified here attached. The exact sequence of side chains on the backbone will vary depending on the arabinoxylan studied.

* In some glucuronoarabinoxylans the glucuronic acid is a 4-*O*-methyl glucuronic acid.

residues are located on C-2 of the backbone xylosyl residues. The arabinoxylans found in minor amounts in dicotyledonous primary cell walls have been shown to contain terminal glucosyluronic acid, terminal 4-O-methyl glucosyluronic acid, and terminal arabinosyl residues attached to C-2 of backbone xylosyl residues.

6. Secondary walls contain xylans and glucomannans.[ref8]

Xyloglucan is the quantitatively dominant hemicellulose in the primary cell walls of dicotyledons, monocotyledons, and gymnosperms, with the exception of the Poaceae. This situation changes dramatically in many walls during differentiation to secondary walls, when xylans and glucomannans often become the dominant hemicelluloses. The glucuronoxylan content in dicotyledon secondary walls can vary between 15 and 30%. The structure of glucuronoxylan is similar to that of primary cell wall xylan in that it consists of a backbone of β-1,4-linked xylopyranosyl residues. To this backbone a single terminal 4-O-methyl α-D-glucosyluronic acid residue is attached to C-2 of about 1 out of every 10 backbone xylosyl residues; arabinosyl residues have infrequently been identified. About 70% of the backbone xylosyl residues contain one O-acetyl group at C-2 or C-3. Glucomannan hemicelluloses are found in much lower amounts (2–5%) than xylans and consist of a linear 1,4-linked polysaccharide comprised of both glucosyl and mannosyl residues. The ratio of mannose to glucose can vary between 1:1 and 2:1.

Gymnosperm secondary cell walls can contain two major hemicelluloses, galactoglucomannans and glucuronoarabinoxylans. The glucuronoarabinoxylans have structures similar to those found in angiosperm primary cell walls consisting of a linear β-1,4-linked D-xylosyl residue–containing backbone with terminal α-L-arabinosyl residues attached to C-3 of some backbone xylosyl residues and terminal 4-O-methyl α-D-glucosyluronic acid residues attached to C-2 of other backbone residues. The galactoglucomannans consist of a linear β-1,4-linked glycosyl residue backbone, containing both glucosyl and mannosyl residues. The ratio of glucosyl to mannosyl residues can vary from 1:4 to 1:3. Single, terminal α-D-galactosyl residues are found as side chains attached to C-6 of some of the backbone mannosyl residues. Galactoglucomannans have been isolated that contain galactosyl residues in a ratio to backbone mannosyl residues of between 0.1:4 and 1.3:4. One O-acetyl group is found attached to either C-2 or C-3 of approximately 25% of the backbone mannosyl residues.

The structures described above for secondary cell wall hemicelluloses have resulted mostly from glycosyl residue and glycosyl linkage analyses (see Concepts 2B4 and 2B5). Elucidation of further details will have to await more detailed structural analyses as described for primary cell wall hemicelluloses.

7. A galacturonic acid–containing oligosaccharide is found at the reducing end of glucuronoxylans.[ref9]

Glucuronoxylans isolated from *Arabidopsis*, birch, and spruce have all been shown to terminate at their reducing ends with an oligosaccharide that contains galacturonic acid, rhamnose, and xylose. Glucuronoxylan solubilized from cell walls by treatment with KOH containing 1% $NaBH_4$ to convert the reducing-end sugar to an alditol was treated with *endo*-xylanase. The resulting oligosaccharide contained one structure with an alditol at its reducing end. The structure of this oligosaccharide was determined by MALDI-TOF mass spectrometry and ^1H-NMR spectroscopy to have the following structure: β-D-Xyl*p*(1→4)-β-D-Xyl*p*(1→3)-α-L-Rha*p*(1→2)-α-D-GalA*p*(1→4)-D-xylitol. The presence of this oligosaccharide at the reducing end of glucuronoxylans suggests a possible connection to other cell wall polysaccharides or proteins, although, to date, none have been identified.

C. Pectic Polysaccharides

1.   Homogalacturonans are gel-forming polysaccharides.[ref10]

A quick look through any well-stocked kitchen will show that many food products contain pectins. These pectin additives to food products are primarily used as thickening agents, taking advantage of the gel-forming properties of pectins. The pectins used predominantly in these applications are almost exclusively homogalacturonan and partially methyl esterified homogalacturonan, a major pectic polysaccharide unit of the primary plant cell wall.

The degree of methyl esterification of the homogalacturonan within the cell wall can differ quite markedly depending on the plant species and the tissue type from which the wall is removed. Values of greater than 80% esterification have been reported. The homogalacturonan regions (non-esterified) have been proposed to be the major players in gel formation within the wall. Sequences of nonesterified galactosyluronic acid residues of a size necessary for gelation occur in cell wall pectins. These homogalacturonan regions are thought to form gels by calcium cross-linking bridges between adjacent nonesterified homogalacturonan regions of two chains. It has been proposed that a minimum of seven adjacent calcium cross links are necessary for the formation of a stable calcium junction zone. Many of these junction zones would need to form between overlapping and interacting homogalacturonan molecules to form a pectin gel in the wall (**Figure 3.5**). However, it has also been shown that pectins with very high degrees of methyl esterification have the ability to form gels that do not require calcium. Such gels appear to involve hydrogen bonds and hydrophobic interactions to hold different chains together in environments where water is limiting. It remains to be seen whether gels between highly methyl esterified homogalacturonans are important in the gelation of pectins in the cell wall. It is also possible that ester linkages can hold chains of homogalacturonans together and thus contribute to pectin gelation (see Concept 6A8).

The degree of methyl esterification of homogalacturonan, the availability of the nonesterified galacturonosyl residues in blocks of a size suitable for gelation, and the presence of calcium could all affect the ability of pectins to form a gel within the cell wall. Such gels could affect not only the interactions of the homogalacturonans themselves but possibly all of the other matrix polysaccharides. In this way, the presence of a pectin gel throughout the wall or in localized regions of the wall could affect the physical properties and thus the functioning of the wall. It is interesting to observe that the nonesterified homogalacturonans are found mostly in the middle lamella and cell corner regions. The esterified pectins are present mostly throughout the cell wall surrounding the cellulose-hemicellulose network.

2.   Rhamnogalacturonan I (RG-I) is a family of large polysaccharides with a backbone of repeating disaccharide units.[ref11]

RG-I is a quantitatively important pectic polysaccharide solubilized from primary cell walls by treatment with α-1,4-*endo*-polygalacturonase. This solubilized RG-I can account for 5 to 10% of the cell walls of dicotyledons, 4 to 7% of those of gymnosperms, and about 1% of those of Poaceae. Additional RG-I (~4% of the dicotyledon cell wall) can be solubilized from the wall by sodium carbonate after treatment with *endo*-polygalacturonase. The remaining RG-I, which accounts for approximately 3% of the cell wall, remains with the wall residue. α-1,4-*Endo*-polygalacturonase–solubilized RG-I has a molecular weight of between 10^5 and 10^6 daltons and elutes as a partially included peak when chromatographed on a Bio-Gel agarose 5 M

Figure 3.5 Homogalacturonans are proposed to form gels via calcium cross-linking bridges. Many of these bridges would need to form between overlapping and interacting homogalacturonans to form a gel. (Adapted from B. Thakur et al., *Crit. Rev. Food Sci. Nutr.* 37(1):47–73, 1997.)

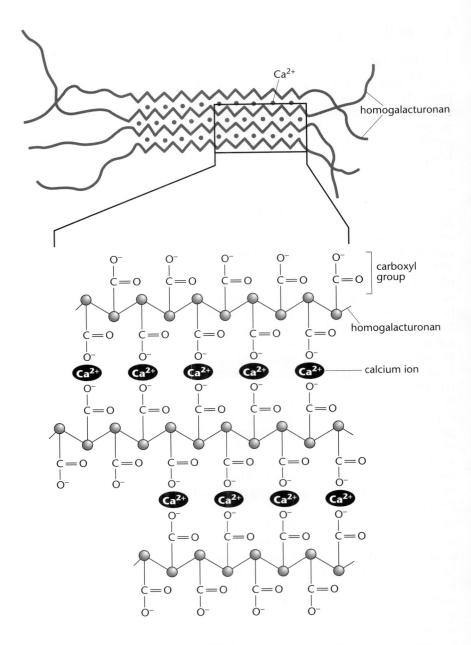

gel permeation column. The leading edge of this peak is enriched in arabinosyl residues, whereas the trailing edge is enriched in galactosyluronic acid and rhamnosyl residues. This indicates that the RG-I as isolated is not homogeneous. Indeed, RG-I is a family of structurally similar molecules.

RG-I consists of a backbone composed of alternating rhamnosyl and galactosyluronic acid residues with side chains containing arabinosyl, galactosyl, and, to a much lesser extent, fucosyl and glucosyluronic acid residues. It is variability in these side chains that explains the heterogeneity of RG-I. The backbone of RG-I is composed of the repeating unit disaccharide →4)-α-D-Gal*p*A-(1→2)-α-L-Rha*p*-(1→. In analyses performed using RG-I isolated from sycamore cultured cells, no occurrence of the disaccharide GalA-GalA or the disaccharide Rha-Rha could be detected. Between 25 and 80% of the rhamnosyl residues linked through C-2 to galactosyluronic acid residues are substituted at C-4. This is the point of attachment of the arabinosyl- and galactosyl-rich side chains.

Treatment of RG-I with a mixture of pure enzymes (*endo*-arabinase, arabinosidase, *endo*-galactanase, and *exo*galactanase) removes all of the arabinosyl residues and 90% of the galactosyl residues from the side chains of RG-I, leaving the intact repeating disaccharide backbone of RG-I. This backbone has a size equivalent to 30% of that of the original polysaccharide. These observations support the idea that RG-I has a linear backbone with multiple, largely neutral side chains attached. Analysis of this backbone from sycamore cell wall RG-I using chemical techniques and ^{1}H-NMR spectroscopy indicates that all of the galactosyluronic acid residues in the backbone contain one acetyl group at either C-2 or C-3. Structural studies of galactosyluronic acid–containing oligosaccharides released from the cell walls of cotton using anhydrous hydrogen fluoride indicates that the galactosyluronic acid residues are O-acetylated predominantly at C-3. On the other hand, structural studies on rhamnogalacturonan oligosaccharides isolated from bamboo shoot cell walls indicate that acetylation occurs primarily at C-2 of the galactosyluronic acid residues. The galactosyluronic acid residues of RG-I are not methyl esterified; this is in sharp contrast to the galactosyluronic acid residues of homogalacturonan (see Concept 3C1).

3. Rhamnogalacturonan I (RG-I) has many arabinosyl and galactosyl residue–containing side chains.[ref12]

RG-I has many (~40) structurally different side chains attached via C-4 of the backbone rhamnosyl residues. Arabinosyl and galactosyl residues are the predominant components of the side chains, although small amounts of fucosyl, glucosyluronic acid and 4-O-methyl glucosyluronic acid residues are also present. The size of individual side chains can vary from one glycosyl residue to over 30 glycosyl residues. The combination of these variously sized side chains on the backbone results in the heterogeneity of RG-I. Much is known about the structures of the side chains, but nothing about the specific location of individual side chains along the backbone.

Understanding many of the structural features of the RG-I side chains has been greatly facilitated by structural studies done on arabinans and galactans isolated from plant cell wall pectic fractions. The extraction procedures used to obtain these arabinans and galactans most likely randomly release them from covalent attachment to RG-I in the original cell wall. Studies of the arabinans and galactans, plus those of intact RG-I and oligosaccharides specifically released from RG-I, have provided detailed knowledge about the structures of the RG-I side chains.

All side chains are attached via C-4 to the rhamnosyl residues of the RG-I backbone. In the case of arabinan side chains, an arabinosyl residue forms the link to a backbone rhamnosyl residue. In contrast, in side chains containing arabinosyl and galactosyl residues, the linkage to the backbone is predominantly via a galactosyl residue. Interestingly, in any side chain containing arabinosyl residues, the arabinosyl residues are located either entirely, or almost always, at the nonreducing end of the side chain. The arabinans have conserved structures; they all contain a 5-linked backbone of arabinofuranosyl residues. Arabinosyl residues of this backbone are often branched at C-2 and C-3 and serve as points of attachment of either terminal arabinosyl residues or short oligosaccharides composed of arabinosyl residues.

The complete sequences of only a very few galactosyl residue–containing side chains have been fully determined. Linear 4-linked β-D-galactans and linear 3- and 6-linked β-D-galactans have been identified as components of RG-I side chains. Arabinosyl-containing saccharides are attached to some of these galactan side chains. The galactan-containing side chains can also be terminated by single α-L-fucosyl residues and also by single β-D-glucosyluronic acid and 4-O-methyl glucosyluronic acid residues. These three terminal residues account for approximately 3% of an RG-I molecule of dicotyledons and gymnosperms. RG-I's from the Chenopodiaceae, includ-

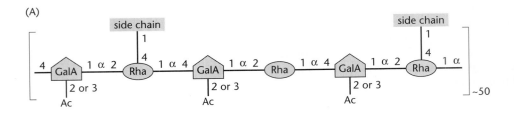

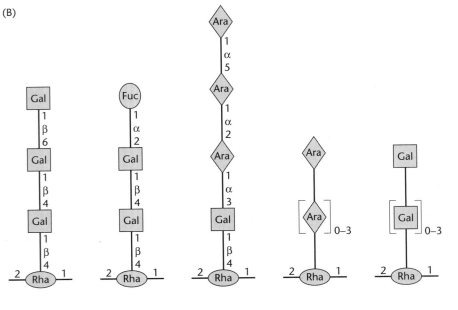

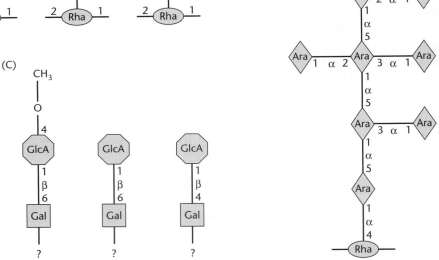

Figure 3.6 The major structural features of the pectic polysaccharide RG-I. (A) The backbone of RG-I. Depending on the source of RG-I studied, 25–80% of the rhamnosyl residues in the backbone have side chains attached at C-4. These side chains can contain 1 to 30 glycosyl residues. The sequence of attachment of the side chains to the backbone of RG-I is not known. Each backbone galactosyluronic acid residue has an *O*-acetyl group attached to it at either C-2 or C-3. (B) Side chains identified to be attached directly to C-4 of a backbone rhamnosyl residue. (C) Fragments of RG-I side chains that have been shown to contain glucuronic acid or 4-*O*-methyl glucuronic acid. In all cases these acidic sugars terminate the side chain to which they are attached. The complete structure of these side chains is not known. (D) A possible structure of an arabinan side chain of RG-I. This structure is predicted from glycosyl linkage analysis of an isolated side chain fraction of RG-I. The exact sequence of the terminal arabinosyl residues on the 5-linked arabinosyl residue backbone has not been determined.

ing spinach and sugar beet, are also esterified with phenolics such as ferulic or coumaric acid or arabinosyl or galactosyl residues (see Concept 6A7).

Considerable experimental data suggest that the structures of the RG-I's of monocotyledons, gymnosperms, and dicotyledons are extremely well conserved. The major structural features of RG-I are shown in **Figure 3.6**.

4. Rhamnogalacturonan II (RG-II) is the most complex polysaccharide known.[ref13]

The pectic polysaccharide RG-II has a structural complexity unmatched by any other plant polysaccharide so far identified. This characteristic is all the more unusual considering that the polysaccharide consists of only approximately 30 glycosyl residues. Eleven different monosaccharides are identified as components of RG-II after glycosyl residue composition analysis. No other polysaccharide contains as many different monosaccharides. All of the monosaccharides shown in Panel 2.1, except for mannose and glucose, are components of RG-II. Methylation analysis indicates that there are over 20 different glycosyl linkages in this polysaccharide. RG-II is a highly branched polysaccharide. Twenty-two percent of the glycosyl residues are singly and 6% are doubly branched. Galactosyluronic acid residues (30 mol %) and rhamnosyl residues (14 mol %) are major components of RG-II.

Many different cell wall extraction procedures have released small amounts of RG-II and/or oligosaccharide fragments of RG-II into solution. However, most if not all of the RG-II present in primary cell walls can be released by treatment with α-1,4-*endo*-polygalacturonase. This enzyme cleaves α-1,4-linked galactosyluronic acid residues and most likely cleaves homogalacturonan regions holding RG-II in covalent attachment in the cell wall with both homogalacturonan and RG-I.

RG-II purified by anion-exchange and gel-permeation chromatography contains several unusual glycosyl residue components. Aceric acid is the only branched, acidic deoxy monosaccharide to be identified in nature. The keto sugars 3-deoxy-D-manno-2-octulosonic acid (KDO) and 3-deoxy-D-lyxo-2-heptulosaric acid (DHA) have been found in plants only as components of RG-II. Apiosyl, 2-*O*-methyl fucosyl, 2-*O*-methyl xylosyl, and arabinopyranosyl residues together are diagnostic components of RG-II and are not found in any of the other major cell wall matrix polysaccharides. All of these glycosyl residues have been found in primary cell walls isolated from dicotyledons, monocotyledons, gymnosperms, pteridophytes, and lycophytes, indicating that RG-II is universally found in all these walls. Thus, with all its complexity, RG-II is a highly conserved polysaccharide. RG-II can be found in cell walls in either a monomer (m-RG-II) or dimer (d-RG-II) form. The dimer is made of two RG-II molecules held together by a borate cross link (**Figure 3.7**). d-RG-II also contains heavy metals such as Pb^{2+}, Ba^{2+}, and St^{2+}. *In vitro* m-RG-II plus borate plus heavy metal will form d-RG-II. d-RG-II is an important pectin cross link essential for normal plant growth (see Concept 6A6).

The complexity of structure of RG-II makes it resistant to degradation by microbes. For example, many commercial plant cell wall–degrading enzyme preparations are made by growing fungi (for example, *Aspergillus niger*) on plant cell walls and then isolating all of the enzymes produced by the microorganism. These enzymes can be used to degrade cell walls, for example, for use in protoplast production. Such enzyme preparations contain approximately 2% by weight of RG-II. RG-II has also been shown to be a major component of wines and fruit juices. Red wine, for example, can contain up to 100 mg/liter of RG-II, most in the d-RG-II form.

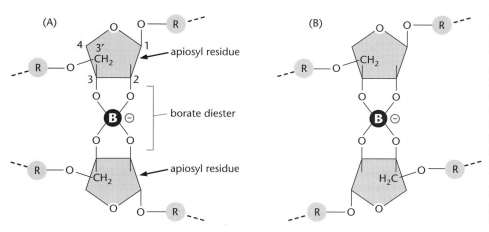

Figure 3.7 Two RG-II molecules can be cross-linked together within the cell wall to form a dimer (d-RG-II). This dimer is cross-linked via a borate ester. The apiosyl residues involved in the formation of the borate ester are those in the 2-*O*-methyl xylose–containing side chain for both partners of the dimer (see Figure 3.8). Two structures are possible to describe the relationship (structural) between the borate and the two apiosyl residues. The naturally occurring structure has yet to be determined; thus both possible structures are shown here (A and B).

5. Rhamnogalacturonan II (RG-II) has an oligogalacturonide backbone.[ref14]

Much of the structure of RG-II has been determined using mild acid hydrolysis to selectively cleave some glycosidic linkages, thus fragmenting the polysaccharide into smaller oligosaccharide fragments that have been structurally sequenced. Mild acid treatments cleave the glycosidic linkages of the apiosyl, the KDO, and the DHA glycosyl residues. Side chains containing these "acid labile" residues have been purified using anion-exchange chromatography, gel-permeation chromatography, and HPLC. The studies also resulted in the purification of the "acid resistant" backbone of RG-II. Glycosyl residue composition studies of the backbone of RG-II have shown that it contains only D-galactosyluronic acid residues. These residues are α-1,4-linked, and backbones containing 9, 10, and 11 galactosyluronic acid residues have been isolated. The variability in number of galactosyluronic acid residues in the backbone is due most likely to the solubilization of RG-II using *endo*-polygalacturonase. This results in some randomization in which α-1,4-linked galactosyluronic acid residues are cleaved, thus releasing RG-II from attachment to homogalacturonan. The backbone of RG-II is likely, therefore, a continuation of a homogalacturonan backbone; that is, it consists of a disaccharide repeating unit of α-D-galacturonosyluronic acid residues. However, in RG-II, this backbone is extensively modified by the attachment of structurally complex side chains (see Concept 3C6). It has also been determined that, as with homogalacturonans, some of the galactosyluronic acid residues in the backbone of RG-II are esterified with methyl groups.

6. Rhamnogalacturonan II (RG-II) has side chains with unique structures.[ref15]

The structural complexities of RG-II are found in the side chains that are attached to the α-1,4-linked galactosyluronic acid backbone. Four structurally different and complex side chains together with fragments of these side chains are released from RG-II by treatment with mild acid. The structures of these four side chains have been determined using a combination of carbohydrate analysis structural techniques. Their structures indicate the location in RG-II of all of the unusual and characteristic sugars of this polysaccharide. The structures of the side chains, together with the structure of the backbone of RG-II and the glycosidic linkages that attach the side chains to the backbone, are shown in **Figure 3.8**. The major structural question that has to be resolved regarding the structure of RG-II is the exact sequence of the four side chains along the backbone. As indicated in the figure, at least one of the side chains contains two *O*-acetyl substituents.

It appears that the complete glycosyl sequence of RG-II has been conserved throughout the primary walls of all dicotyledons, monocotyledons, gymnosperms, pteridophytes, and lycophytes studied. Considering the quantitatively low level but conserved nature of this polysaccharide in the cell wall and the very large amount of energy that plants expend to synthesize this complex polysaccharide, it is likely that RG-II plays an important functional role in the primary cell wall. The observation that d-RG-II (see Concepts 3C4 and 6A6) is an important pectin cross link in the cell wall may explain the important structural role this polysaccharide has in the cell wall.

The structural characterization of RG-II has also led to an important question about the synthesis of the polysaccharide. The glycosyl sequence is conserved in species as diverse as sycamore, rice, and Douglas fir and these plants likely use at least 50 enzymes to make RG-II. How these plants control to such a fine level of detail (as shown by the structural studies) the biosynthesis of RG-II is an intriguing and unanswered question of cell wall chemistry.

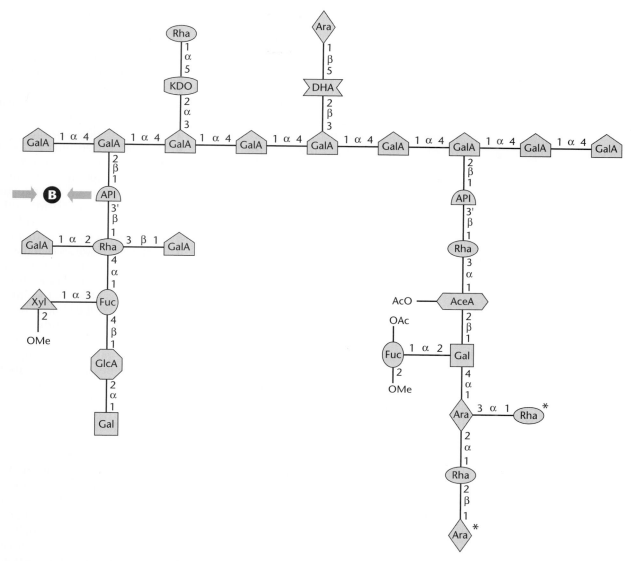

Figure 3.8 The structure of the RG-II backbone and attached side chains. The exact sequence of the side chains along the backbone of the attachment points is not known. The apiosyl residue that participates in the formation of the borate diester cross link between two RG-II molecules is labeled **B**. (*indicates glycosyl residues absent in some RG-II molecules.)

D. Other Wall Polysaccharides

1. β-1,3/1,4-(mixed-linked) glucans are found in grasses and horsetail.[ref16]

The β-1,3/1,4 glucans, like cellulose, are composed exclusively of β-linked glucosyl residues of which approximately 30% are 1,3-linked and 70% are 1,4-linked. The β-1,3/1,4 glucans, together with arabinoxylans, are the predominant noncellulosic matrix polysaccharides found in the cell walls of the Poaceae monocotyledons (grasses) and are major components of *Equisetum* (horsetail) cell walls that are also rich in pectin (**Figure 3.9**). The β-1,3/1,4 glucans show the very interesting characteristic in growing coleoptile primary walls of having a transient appearance. During periods of rapid cell growth, the amount of glucan in the wall can increase to approximately 15% but then rapidly decrease to between 0 and 4% as growth of the tissue terminates. β-1,3/1,4 glucans are also found in the cell walls of endosperm and aleurone cells of grains; for example, in barley endosperm cells the glucans can account for up to 75% of the cell wall.

These linear polysaccharides have repeated oligosaccharide sequences in the backbone β-glucosyl residues (see Concept 3D2). The molecular weights of commercially available β-1,3/1,4 glucans isolated from the cell walls of grains range from 150,000 kD to 300,000 kD.

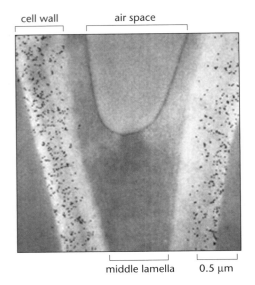

Figure 3.9 Thin-section electron micrograph of a stem cell wall of *Equisetum arvense* immunolabeled with an anti-mixed-linked glucan. (From I. Sorensen et al., *Plant J.* 54:510–521, 2008.)

2.   Enzyme-generated oligosaccharides have been used to structurally characterize mixed-linked glucans.[ref17]

Detailed structural information on β-1,3/1,4 glucans has been obtained using specific microbial enzymes to fragment the polysaccharide into oligosaccharides that can be structurally characterized. One such enzyme is β-D-glucanohydrolase from *Bacillus subtilis*. This enzyme cleaves in the mixed-linked glucan backbone a β-1,4-D-glucosyl linkage only if it is preceded by a β-1,3-D-glucosyl linkage. Major oligosaccharide products from such an enzymatic digest of Poaceae mixed-linked glucan are the trisaccharide cellobiosyl-1,3-β-D-glucose and the tetrasaccharide cellotriosyl-1,3-β-D-glucose that are found on average in a ratio of about 2:1 in most β-1,3/1,4-linked glucans studied. These two units account for approximately 90% of the total structure of the polysaccharide, although the exact sequence of the backbone of the tetra- and trisaccharides is not known. The enzyme treatment also releases oligosaccharides that contain approximately 10 contiguous β-1,4-linked glucosyl residues that have a single 3-linked glucosyl residue at the reducing terminus. It has been proposed that these contiguous stretches of β-1,4-linked glucosyl residues form sites of attachment via hydrogen bonds between chains of β-1,3/1,4-linked glucans or possibly between β-1,3/1,4 glucan chains and cellulose within the cell wall. Studies with barley β-1,3/1,4 glucans have also indicated minor amounts of contiguous β-1,3-glucosyl residues to be components of the mixed-linked glucan. It has been proposed that the single (and, in infrequent cases, contiguous) 3-linked glucosyl residues give flexibility to the structure of the backbone of the β-1,3/1,4-linked glucan. In *Equisetum*, cellotetraose-1,3-β-D-glucose is the abundant repeating unit structure of the mixed-linked glucan.

3.   Callose is a β-1,3-D-glucan.[ref18]

β-1,3-D-Glucans found in specific and often transient positions in plant cell walls are called callose. *Callose* has been found in plant walls in response to wounding or pathogen-induced stress or can be found naturally in some very specific locations, for example, in developing sieve plates, in growing pollen tubes, and in plant styles. Often callose is identified histochemically by staining with reagents, for example, exhibiting fluorescence with aniline blue, a triphenyl methane dye. Also, antibodies specific for β-1,3-D-glucans have been used to identify the location of this polysaccharide.

In those studies where callose has been isolated and structurally characterized, glycosyl residue composition analysis has shown that callose is a homopolymer containing only glucosyl residues. Glycosyl linkage analysis of a β-1,3-glucan isolated from the cell walls of *Zea mays* shoots showed that the callose consists entirely of β-1,3-linked glucosyl residues. Linkage studies of callose isolated from the cell walls of *Nicotiana alata* pollen tubes suggested that the β-1,3-linked glucan has a small amount of branching through *O*-6 and *O*-2 to some of the β-1,3-linked glucan backbone residues. Minor amounts of 3,6-linked glucosyl residues have also been identified as components of other callose fractions.

The molecular weight of the β-1,3-linked glucan isolated from *Zea* shoot cell walls was estimated to be 70,000 kD, although the problems associated with degradation of β-1,3-linked glucans by treatment with alkali may make this size determination an underestimation, as is the case with other determinations of β-1,3-linked glucan molecular weights. The β-1,3-linked glucosyl residues cause the polysaccharide to have a helical conformation that results in the formation of gels that is consistent with the translucence of callose-containing walls as shown by immuno-electron microscopy. How callose interacts with other cell wall polysaccharides when it is deposited in the wall remains undetermined.

4. Substituted galacturonans are found in some cell walls.[ref19]

Some specialized cell walls contain polysaccharides that share structural similarities with homogalacturonan and RG-II in that they have a backbone of α-1,4-linked D-galactosyluronic acid residues. In addition, like RG-II, they contain side chains attached to some of the backbone residues. These polysaccharides are called substituted galacturonans and are considered part of the pectin family.

A polysaccharide has been isolated from the cell walls of the aquatic duck-weed plant (*Lemna minor*) whose glycosyl residue composition consists of galactosyluronic acid and apiosyl residues as the only components. It has not been established whether apiogalacturonan comes from the primary and/or secondary cell walls of the plant's tissues. However, by nature of the high content of galactosyluronic acid residues, apiogalacturonan can be classified as a pectic polysaccharide.

Apiogalacturonan has side chains attached to the backbone galacto-syluronic acid residues whose structures have been identified by various chemical techniques. The side chains appear to be almost exclusively a disaccharide called apiobiose. Apiobiose contains two apiosyl residues and has the structure D-Api-1,3′-D-Api (**Figure 3.10**). It is possible that a few side chains consist of single terminal apiosyl residues. It has not been determined whether the side chains are attached to the galacturonosyl residues of the backbone through O-2 or O-3. In addition, less than 10% of the galacturonosyl residues of the backbone contain methyl esters. The structural/physiological role of apiogalacturonan and its interaction with other cell wall matrix polysaccharides in *Lemna* tissues remain to be elucidated.

The cell walls of some plant tissues (for example, apple fruit, pine pollen, bean cotyledons) contain another structurally different substituted galacturonan called a xylogalacturonan (Figure 3.10). This molecule is rich in galactosyluronic acid and xylosyl residues. Up to approximately a third of the α-1,4-linked galactosyluronic acid residues of the backbone are branched through O-3. Here terminal xylosyl residues are β-1,3 attached to the backbone. The role of this substituted galacturonan in walls remains to be determined, although it has been isolated covalently attached to other cell wall pectic components, suggesting that in those tissues where it is present this polysaccharide contributes to the overall structure/function of the pectin network.

5. Galactoglucomannans are found in Solanaceae cell walls and in the growth medium of Solanaceae cultured cells.[ref20]

Galactoglucomannans have been isolated from the cell walls of the members of the Solanaceae (for example, tobacco). These polysaccharides are usually solubilized using base extractions that also solubilize the cell wall xyloglucans. The Solanaceae xyloglucans have a structure different from most angiosperm xyloglucans (see Concept 3B1) and together with the presence of the galactoglucomannans indicate that Solanaceae walls differ quite substantially in polysaccharide structure/composition relative to other angiosperm walls. The extraction procedures used to solubilize galactoglucomannans would suggest that these molecules are hemicelluloses. Galactoglucomannans have also been identified as polysaccharides in the secondary cell walls of a variety of plant species (see Concept 3B6).

The structural features of the galactoglucomannans in the cell walls of tobacco (*Nicotiana plumbaginifolia*) have been deduced from structural studies of the galactoglucomannan secreted by suspension-cultured cells of this species. The backbone of this polysaccharide is a repeating disac-

(A) **apiogalacturonan**

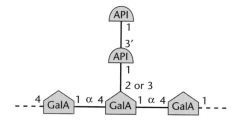

(B) **xylogalacturonan**

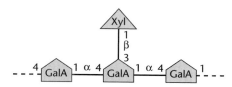

Figure 3.10 Two examples of substituted galacturonans.

charide structure of →4)-β-D-Glc-(1→4)-β-D-Man-(1→. Sixty-five percent of the backbone mannosyl residues have short side chains attached that contain D-galactosyl residues. D-Gal and β-D-Gal-(1→2)-D-Gal have been identified as monosaccharide and disaccharide side chains α-linked to C-6 of the backbone mannosyl residues. Very small amounts of terminal arabinosyl and terminal xylosyl residues plus *O*-acetyl groups are also components of this polysaccharide.

6.   Some seed cell walls contain storage polysaccharides.[ref21]

Seeds contain an embryo, together with considerable quantities of storage reserves. These reserves are stored either in the embryo itself, in the cotyledons (for example, in legumes), or in the endosperm tissue (for example, in cereals). In all cases the reserves are mobilized to provide nourishment for the growing embryo. Seed reserves often contain carbohydrates, which are usually found intracellularly (for example, starch) but are sometimes found in the cell wall itself (**Figure 3.11**).

One family of cell wall storage polysaccharides is the mannans found in the walls of endosperm cells from a variety of seeds. Three major types have been identified within this mannan family of polysaccharides (all containing β-1,4-linked mannosyl residues in the backbone). The "pure" mannans are relatively homogeneous polymers consisting of a backbone with little if any side chain substitution. The galactomannans have considerable side chain substitution (25–95%) with single terminal α-galactosyl residues attached to *O*-6 of the backbone mannosyl residues. Glucomannans have little if any side chain substitution and are comprised, depending on the plant species studied, of varying amounts of β-1,4-linked glucosyl and mannosyl residues in the backbone. The linear mannans and glucomannans, as well as functioning as an energy resource, give extremely hard physical characteristics to the seed and resist hydration. The galactomannans, on the other hand, play a role in water retention, preventing the seed embryo from desiccation. The galactomannans are found in some leguminous seeds, while examples of mannans are found in dates and glucomannans in the Liliaceae.

Xyloglucans are found as storage polysaccharides in the cell walls of the cotyledons of certain legumes and some other species. These storage xyloglucans are often referred to as amyloids, as they stain blue with the characteristic staining technique for starch that uses an iodine–potassium iodide reagent. Both tamarind and nasturtium seeds contain large quantities of amyloids in their cell walls. Amyloids have structures very similar to the cell wall xyloglucans of plants such as pea and sycamore, except that the terminal fucosyl residues are absent (Concepts 3B1–3). As well as being a food and energy reserve, amyloids function in a manner similar to galactomannans. Since the rapid mobilization of these storage polysaccharides during germination demands the expression of genes encoding a wide range of hydrolases, research in this area has resulted in the purification of these enzymes and the cloning of the genes encoding them.

7.   Gums are secreted in response to stress and are not normally considered to be wall components.[ref22]

Injury at the surface of many plants leads to the production of plant gums that are often called exudate gums. These exudate gums are produced as highly viscous fluids that after exposure to the air dry to become hard, solid masses. Most of these exudate gums are soluble in water. The major compo-

cell wall storage polysaccharides

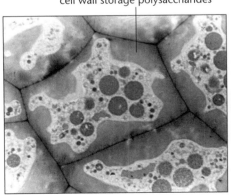

5 μm

Figure 3.11 Endosperm cell of *Myrsine laetevirens* with thickened primary cell walls due to the presence of storage polysaccharides. (Courtesy of M.S. Otegui.)

nents of these exudate gums are polysaccharides. Many angiosperms and several gymnosperms produce exudate gums, and a few of these have major industrial uses in the food and pharmaceutical industries and therefore are a major natural resource for industry. These exudate gums are utilized in industrial applications because they have physical properties that are useful for emulsifying, thickening, and gel forming (for example, gum arabic is used as an emulsifier, binder, stabilizer, and adhesive).

The exudate gum polysaccharides cover a wide variety of structures, many of which are similar to or partially resemble structures of plant cell wall polysaccharides. The structures of many exudate gums have been only partially characterized, and much of their structures remains to be determined. Although many exudate gum structures have been identified, three major groups containing molecules with related structures have been described. The first group contains branched arabinogalactans that, depending on the plant source, can contain up to about 10% protein that is covalently attached to the polysaccharide component. The polysaccharide moiety of the arabinogalactan is a branched arabinogalactan with a backbone of 3-linked β-galactosyl residues with side chains containing arabinosyl, galactosyl, rhamnosyl, xylosyl, glucosyluronic acid, and 4-O-methyl glucosyluronic acid attached to O-6 of the backbone galactosyl residues. The acacia gums (for example, gum arabic) are a major representative of this type of gum (see Concepts 3E6, 9A3).

The second exudate gum group is closely related to the cell wall pectic polysaccharide rhamnogalacturonan I. The backbone of this polysaccharide contains α-1,4-galactosyluronic acid residues interspersed with 2-linked L-rhamnosyl residues. In some gums a ratio of 1:1 occurs similar to that found in the plant cell wall RG-I. Similar to RG-I, these exudate gums have side chains attached to the backbone. These side chains contain predominantly arabinosyl, galactosyl, and glucosyluronic acid residues, but unlike RG-I, where all the side chains are attached to O-4 of the backbone rhamnosyl residues, in the exudate gums some of the side chains are reported attached to the galactosyluronic acid backbone residues. Gum karaya is an example of this group of exudate gums.

A third group of exudate gums is identified as mannoglucuronans. This group contains alternating glycosyl residues of 4-linked β-D-glucosyluronic acid residues and 2-linked α-D-mannosyl residues in the backbone. Often side chains similar in structure to those of arabinogalactan are found attached to the backbone mannosyl residues. Occasionally similar side chains are also found attached to the glucosyluronic acid residues. Gum ghatti is an example of this group of exudate gums.

The considerable industrial uses of the exudate gums (see Concept 9A3) and the potential for limited supplies due to overdemand or losses of natural resources are driving efforts for more detailed structural information about the exudate gums and investigations into the structural features of the gums that are necessary for their physical properties. Such data would help determine whether alternative sources of structurally similar polysaccharides with similar properties are available.

E. Proteins and Glycoproteins

1. Plant cell walls contain structural, enzymatic, lipid transfer, signaling, and defense proteins.[ref23]

Plant cell walls contain hundreds of different proteins, glycoproteins, and proteoglycans. These reflect both on the structural diversity of walls and on the multitude of functions they serve. In terms of mass, the structural proteins account for the bulk of these proteins, but from a functional point of view, the enzymatic, lipid transfer, signaling, and defense proteins are

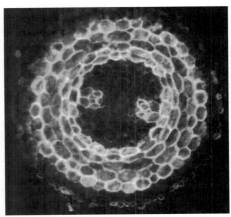

(A)

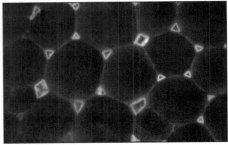

(B)

Figure 3.12 Cell wall protein distribution. (A) shows the distribution of extensin in a cryo-section of a carrot root, revealed by a monoclonal antibody to this protein. It is predominantly found in the root cortex and in some regions around the developing phloem. (B) Using a monoclonal antibody to monoamine oxidase reveals that this enzyme is found largely in the intercellular spaces of the cortex in this section of a pea epicotyl. (A, from M. Smallwood et al., *Plant J.* 5:237–246, 1994. B, from M. Laurenzi et al., *Planta* 214: 37–45, 2001.)

equally important. Most cell wall proteins identified to date have been discovered in biochemical studies of isolated cell walls. However, because of the ease with which cellular proteins can become bound to cell walls during cell disruption, it is of critical importance to verify the cell wall location of a given protein by immunolocalization techniques. Typical of secretory proteins in general, cell wall proteins possess an amino terminal signal peptide.

The synthesis and assembly of cell wall proteins is tightly controlled. Many of the proteins are developmentally regulated or expressed in a tissue-specific manner, while the synthesis of others occurs in response to wounding, infections, or environmental stresses. Immunolocalization studies have also shown that every type of protein has a unique spatial distribution (**Figure 3.12**). Although their specific targeting sequences or structures are unknown, it is probable that their size and physicochemical properties define whether they associate with the cellulose-hemicellulose network of the cell wall proper, with the homogalacturonan-rich middle lamella/cell corner regions, or with molecules at the plasma membrane–cell wall interface.

The structural proteins of plant cell walls appear to contribute to cell wall strength; to control cell wall assembly, expansion, hydration, and permeability; and to serve as possible nucleation sites for lignification and as sources of signaling molecules. The bulk of the cell wall enzymes are probably involved in cell wall metabolism, including the regulation of cell growth, the modification of other cell wall components, and the cross linking of polysaccharides, glycoproteins, and lignin precursors into networks. Other enzymes have been shown to participate in defense responses and/or in the release of cryptic regulatory molecules. The list of nonstructural, nonenzymatic cell wall proteins includes molecules involved in the transport of lipids (lipid transport proteins), the control of cell wall expansion (expansins), the suppression of ice crystal formation in cold-acclimated plants, and in the protection against pathogens (for example, thionins, glucosidases, chitinases, polygalacturonase-inhibiting proteins).

2. The repeated sequence motifs of structural proteins define their classes and their physical properties.[ref24]

Structural cell wall proteins, by definition, are proteins that contribute to the physical properties of plant cell walls and have no demonstrable catalytic properties. Their biological functions have yet to be defined precisely, but they are assumed to confer properties to the walls that cannot be duplicated by carbohydrate and polyphenolic polymers. In particular, they may contribute to cell wall strength, toughness, flexibility, elasticity, pore size, and hydration. Many, but not all, become insolubilized via ionic or covalent bonds, or by lectin-mediated cross links. Due to the presence of repeated sequence motifs, they are enriched in specific amino acids from which three classes of these proteins derive their names—the glycine-rich proteins (GRPs), the proline-rich proteins (PRPs), and the hydroxyproline-rich glycoproteins (HRGPs). Two other classes of related proteins are the solanaceous lectins and the arabinogalactan proteins (AGPs), whose sequence repeats and/or Pro/Hyp content overlap with PRPs and HRGPs and, therefore, are often classified with the latter (**Table 3.2**). Interestingly, Gly, Pro, and Hyp are also common amino acids of structural proteins of animals (for example, keratins, collagens) as well as of spiders and insects (for example, spider web and silk cocoon fibroins).

The repeated sequence motifs of structural cell wall proteins tend to be quite small (2–16 amino acids). Each repeat is a putative functional domain that alone or in conjunction with another domain imparts a specific property to the protein. Such properties include the ability to create specific geometrical conformations (helical domains, β-strands, and sheets); or certain

TABLE 3.2 FIVE CLASSES OF STRUCTURAL PLANT CELL WALL PROTEINS AND SOME DISTINGUISHING PROPERTIES

PROTEIN CLASS		% PROTEIN	%SUGAR	ABUNDANTS AAs	AA REPEATING MOTIFS
(A)	HRGPs (dicot)	~45	~55	O, S, K, Y, V, H	SOOOOSOSOOOOYYYK SOOOK SOOOOTOVYK SOOOOVYKYK
	HRGPs (monocot)	~70	~30	O, T, S, P, K	TPKPTOOTYTOSOKPO
(B)	Glycine-rich proteins (dicot)	~100	~0	G	GX
	Glycine-rich proteins (monocot)	~100	~0	G	GG, GGY
(C)	Proline-rich proteins	80–100	0–20	O, P, V, Y, K	POVYK, POVEK
(D)	Solanaceous lectins	~55	~45	O, S, C	LPSOOOOOO( )( )SOOO( )( )O
(E)	Arabinogalactan proteins	1–10	90–99	O, S, A, P	SOOAPAP, AO

protein segments may have specific physical characteristics (they may be hydrophobic, hydrophilic, charged, or stiff domains). In addition, repeated sequence motifs can serve as sites for intra- and intermolecular cross linking, as binding sites, and as defined glycosylation domains. For example, Gly-Hyp-Pro repeats account for the majority of the amino acid residues of collagen, and these repeats allow three collagen molecules to associate to form a triple helical ropelike structure. In the case of HRGPs, the blocks of Hyp, as found in typical X-Hyp-Hyp-$(Hyp)_n$ sites, favor an extended left-handed polyproline-II helix and act as major sites for O-glycosylation. This combination of inflexible, extended polyproline-II helical domains and O-glycosidically linked sugars and oligosaccharides that can conformationally wrap themselves around and H-bond to these extended protein backbones gives rise to molecules whose overall structure is dominated by their repeated, stiff, rodlike domains.

3. Sequence motifs define proline hydroxylation, glycosylation, and intra- and intermolecular cross-linking sites.[ref24]

Posttranslational modification of wall proteins is tightly controlled (**Figure 3.13**). At one level, this control resides in enzymes that react with their target amino acids only in the context of a defined sequence motif. As an example, let us consider the enzyme that converts peptide-bound proline to hydroxyproline. The rules governing the specificity of this enzymatic reaction can be deduced by comparing the amino acid sequences of peptides derived from the proteins of interest. In the case of the structural cell wall proteins it has been found that proline in the sequences -Lys-Pro-, -Tyr-Pro-, and -Phe-Pro- are never hydroxylated, while -Pro-Val- is always hydroxylated. Different isoforms of the enzyme prolylhydroxylase exist, each of which may show specificity for a different proline-containing sequence.

Analysis of the glycosylation patterns of Hyp residues in HRGPs and AGPs has revealed that not all hydroxyproline residues are automatically glycosylated. This suggests that additional amino acid sequence information plays a role in determining which Hyp residues become glycosylated and to what extent. Well-conserved sequence motifs that define glycosylation patterns are called *glycomodules* to emphasize their functional significance. The best known glycomodule is the classical Asn-Xaa-Ser/Thr motif for N-glycosylation. In contrast, the glycomodules for Hyp-O-glycosylation of cell wall proteins are still being deciphered, but some glycosylation rules

linkage	fungi	plants	animals
N-Glycosidic			
GlcNAc-Asn	●	●	●
O-Glycosidic			
Gal-Hyl	—	—	●
Ara-Hyp	—	●	—
Gal-Hyp	—	●	—
GlcNAc-Hyp	●	—	—
Glc-Hyp	—	●	—
Fuc-Ser	—	—	●
Gal-Ser	—	●	●
GalNAc-Ser	—	—	●
Glc-Ser	—	—	●
GlcNAc-Ser	—	—	●
Man-Ser	●	—	—
Xyl-Ser	—	—	●
Fuc-Thr	—	—	●
Gal-Thr	—	—	●
GalNAc-Thr	—	●	—
GlcNAc-Thr	—	—	●
Man-Thr	●	—	—
Xyl-Thr	—	●	—
C-Glycosidic			
Man-Trp	—	—	●

Figure 3.13 Occurrence of glycopeptide linkages in fungi, plants, and animals. (Adapted from M.J. Kieliszewski, *Phytochem.* 57:319–323, 2001.)

TABLE 3.3 HRGP GLYCOMODULES	
REPRESENTATIVE PEPTIDE MOTIF	**HYDROXYPROLINE OLIGOSACCHARIDE**
Xaa-(Hyp)$_2$	1-3 Ara, mostly to first Hyp
Xaa-(Hyp)$_4$	Mostly 3-4 Ara to all Hyp
(Xaa^a)$_{1-3}$-Hyp-(Xaa)$_{1-3}$-Hyp	Arabinogalactan polysaccharide

^{a}Xaa is commonly Ser, Ala, Thr, but not Hyp. (Adapted from M.J. Kieliszewski, *Phytochem.* 57:319–323, 2001.)

have already emerged. The three major rules identified to date are (**Table 3.3**): (1) peptide sequence rather than conformation determines which Hyp become glycosylated; (2) contiguous Hyp residues become arabinosylated; and (3) clustered but noncontiguous Hyp residues become modified by arabinogalactan oligosaccharides. These rules are being elucidated with the help of synthetic genes with Hyp-containing sequence motifs.

A number of protein sequences are thought to be involved in the formation of intra- and intermolecular cross links. Best characterized among these is the sequence -Tyr-X-Tyr-Lys-, which appears to be associated with the formation of stiffening intramolecular isodityrosine cross links. The sequence -Pro-Hyp-Val-Tyr-Lys- (and variants) is thought to be involved in the formation of intermolecular cross links, whose exact nature is still unclear.

4. HRGP and PRP expression is controlled by developmental programs and triggered by external stimuli.[ref25]

HRGPs and PRPs constitute large, multigene families, each member of which is synthesized in only one or a few cell types. Expression is controlled not only developmentally but also by wounding, pathogen attack, elicitors, and environmental stimuli such as water stress. For example, a specific soybean HRGP has been localized to sclerenchyma and to special hourglass cells in the seed coat. In tobacco, another HRGP is confined to the tip of newly formed lateral roots mechanically breaking through the root cortex, and a PRP to xylem vessels in stems.

In many instances, HRGP and PRP synthesis, secretion, and cross linking have been shown to be an integral part of plant defense responses. An interesting example has been reported for beans infected with the fungal pathogen *Colletotrichum lindemuthianum*. In an incompatible reaction, the HRGP gene was induced immediately after infection in the epidermal and cortical cells surrounding the infection site. In contrast, in plants displaying a compatible reaction, the induction of HRGP expression occurred only slowly and was not limited to the site of infection. H_2O_2–dependent cross linking of HRGP and PRP molecules in cell walls in response to elicitor treatments has been shown to occur within minutes of the start of the treatment. This is consistent with a role as part of the first line of a plant's defense responses.

Immunolocalization studies at the ultrastructural level in carrot storage roots have demonstrated that one HRGP, extensin, is uniformly distributed across phloem parenchyma cell walls but is excluded from the middle lamella. Indeed, the middle lamella appears to constitute a physical barrier to the passage of extensin-type HRGP molecules between adjacent cell walls. This is somewhat unexpected considering the acidic nature of the deesterified pectic polysaccharides of the middle lamella and the very basic nature of extensin. Clearly, the tightness of the pectic polysaccharide network of the middle lamella limits the ability of larger HRGP molecules to interact with the deesterified pectins *in situ*.

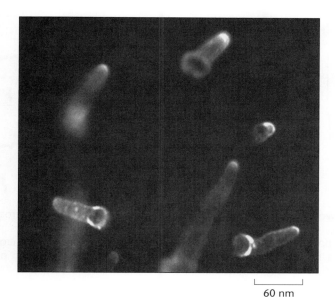

60 nm

Figure 3.14 Distribution of a proline-rich protein in root hairs. This fluorescent image demonstrates that the protein AtPRP3 is concentrated in the top and in a ringlike region around the base of *Arabidopsis* root hairs. (Courtesy of Jianhong Hu and Mary Tierney.)

The principal function of the HRGP/PRP networks appears to be to provide mechanical strength to primary cell walls (**Figure 3.14**). Such strengthening is used to increase cell wall rigidity under reduced turgor conditions, to terminate cell wall expansion during cell growth, and to make cell walls more resistant to penetration by pathogens. It is also possible that the high density of positive charges on HRGP molecules could help immobilize bacterial pathogens and disrupt tip growth of fungal hyphae.

5. Extensin-type HRGPs are rod-shaped molecules that become insolubilized by a peroxide-mediated cross-linking process.[ref26]

The *Arabidopsis* genome contains 20 genes encoding HRGPs known as *extensins*. Extensins are highly basic (H pI ~10) HRGPs that contain multiple copies of the pentapeptide sequence Ser(Hyp)$_4$ as well as significant numbers of Val, Tyr, Lys, and His residues. Most of the Hyp residues are glycosylated with one to four arabinose residues, and the serine with a single galactose residue, yielding a "sugar-coated" molecule that is highly resistant to proteolytic degradation. In the wounding-type extensin of carrots, the 25 Ser(Hyp)$_4$ repeats (45% of total amino acids), together with their O-linked arabinose and galactose-containing side chains, result in a molecule with about 90% polyproline-II helical conformation. Based on this information, it is possible to calculate the approximate length of the molecules assuming that they are 274 amino acids long, are in a 100% polyproline-II helix conformation, and have three amino acids per turn with a pitch of 0.94 nm, giving a molecular length of $274/3 \times 0.94 = $ ~85 nm. This has been confirmed by electron micrographs of purified carrot extensin monomers in which the molecules appear as about 80-nm-long, slightly kinked, rod-shaped structures (**Figure 3.15**). HRGP monomers in the wall appear to be covalently cross-linked by specific cell wall peroxidases that recognize the Val-Tyr-Lys sequence and produce di-isodityrosine- or pulcherosine-type cross-links (**Figure 3.16**). Electron micrographs of cross-linked wounding-induced extensin molecules from carrot roots show that about 60% of the cross-links involve the ends of the molecules, while the rest are scattered among internal sites. This pattern, together with the stiffness of the molecules, favors the formation of open, three-dimensional network structures capable of assembling around already formed cellulose-hemicellulose networks (see Figure 3.15), thereby stabilizing stressed or damaged mature cell walls.

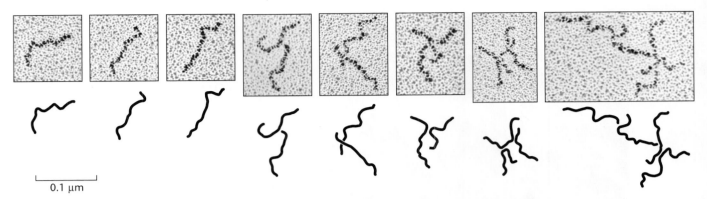

Figure 3.15 Electron micrographs of wounding-type extensin molecules and polymers. The micrographs depicts rotary shadowed, purified wounding-induced carrot extensin monomers, dimers, and oligomers in which the individual ~84-mm-long molecules appear as slightly kinked, rod-shaped structures. Note that the cross-linked molecules are frequently connected to each other by bonds that form close to or at the ends of at least one of the molecules. (From J.P. Staftstrom and L.A. Staehelin, *Plant Physiol.* 81:234–241, 1986.)

The *Arabidopsis* extensin AtEXT3 has attracted particular attention both because a knockout mutation gives rise to a lethal phenotype, which has been traced to defects in the assembly of primary walls and cell plates, and because the molecules become cross-linked into a different kind of network than the wounding-induced extensins. The unique feature of the AtEXT3 networks is that the molecules are bound to each other laterally in a staggered configuration, yielding networks that are not only more open than those of the wound extensins, but also more rigid and more planar (**Figure 3.17**). This latter network design would appear to be ideally suited to mechanically stabilize nascent primary cell walls and cell plates. In addition, it has the potential of affecting the organization of freshly secreted, negatively charged pectic polysaccharides and thereby the assembly of cell plates and growing cell walls.

6. Arabinogalactan proteins form a family of complex proteoglycans located at the cell-surface, and in the cell wall and intercellular space.[ref27]

Arabinogalactan proteins (AGPs) constitute a family of high–molecular weight proteoglycans that are associated with the plasma membrane, the cell wall, and the intercellular spaces of plants. Based on their cell-, tissue-, and organ-specific type of expression and several other properties, they have been postulated to serve regulatory and signaling roles during

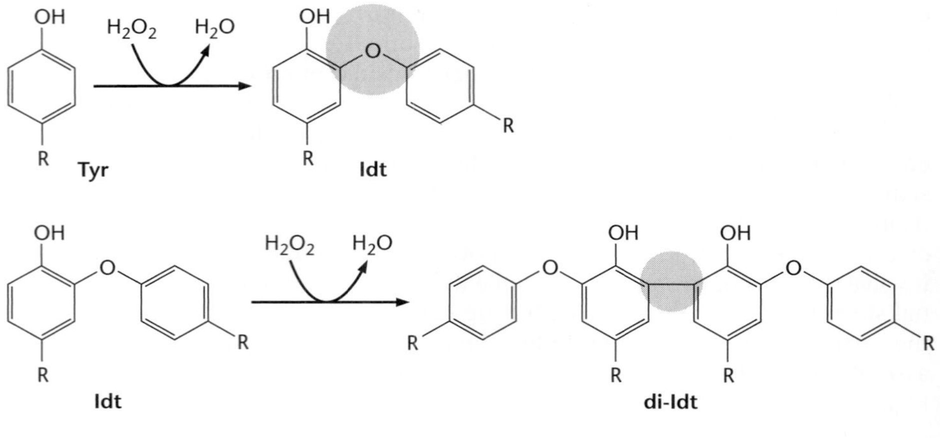

Figure 3.16 Formation of isodityrosine (Idt) and di-isodityrosine (di-Idt) cross links via a peroxidase-mediated oxidation mechanism. Tyr, tyrosine.

Figure 3.17 Atomic-force micrograph of an AtEXT3-type extensin network. The extended straight segments have been postulated to correspond to rigid network domains produced by the lateral association of aligned and staggered molecules. Such networks are formed in nascent cross walls. (From M.C. Cannon et al., *Proc. Natl. Acad. Sci. USA* 105:2226–2231, 2008.)

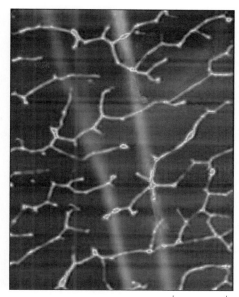

500 nm

cell differentiation, tissue development, and somatic embryogenesis. In stylar transmitting tissues of flowers, their carbohydrate domains have been implicated as an adhesive substrate for pollen tube growth as well as providers of nutrients to the pollen. AGPs are also produced in response to wounding, often in copious quantities. Gum arabic, which is used as a thickener in the food industry, is a wounding-induced mixture of AGPs of the tree *Acacia senegal* (see Concept 9A3). Finally, and in analogy with animal proteoglycans, they most likely serve physical/structural roles such as forming a protective, hydrated interface between the plasma membrane and the cell wall, maintaining the cellulose-hemicellulose network in an open conformation, and regulating the degree of hydration and porosity of cell walls and the middle lamella. It should be noted, however, that none of these functions has been firmly established.

AGPs were classically defined as cell wall proteoglycans, 1 to 10% protein and 90 to 99% carbohydrates (**Figure 3.18**), that can bind and be precipitated by a class of synthetic phenylazo dyes, the β- (but not α-) glucosyl Yariv reagents (**Figure 3.19**). The high carbohydrate content makes AGPs thermotolerant and resistant to proteolysis, two properties that add to their usefulness in the food industry. The protein backbones are typically rich in hydroxyproline/proline, alanine, serine, and threonine. Most of the hydroxyproline residues are substituted by arabino-3,6-galactans (5–25 kD), but short arabinosides have also been found (**Figure 3.20**). The AGPs appear to be encoded by a medium-sized gene family, and the diversity of the proteins is enhanced by the presence of multiple glycoforms.

Using a combined genomics and proteomics approach, about 40 genes encoding protein backbone AGPs have been identified in *Arabidopsis*. These AGPs can be subdivided into two broad classes, "classical" and "nonclassical." However, because the degree of sequence similarity within and between the classes is low and not all AGPs bind the Yariv reagent, the general definition of AGPs, and of the nonclassical subclass(es) in particular, remains somewhat controversial. Best defined is the class of classical AGPs, which contain the domain structure shown in **Figure 3.21**. The most notable features of this structure are a signal sequence, an extended hydroxyproline- (alanine-, serine-, and threonine-) rich domain, and a C-terminal GPI- (glycosylphosphatidyl inositol) anchor signal sequence consisting of an ω cleavage site, a spacer, and a hydrophobic tail. This definition implies that classical AGPs start out their life as GPI-anchored proteins (see Figure 3.23), proteins that are anchored through a GPI lipid tail to the outer

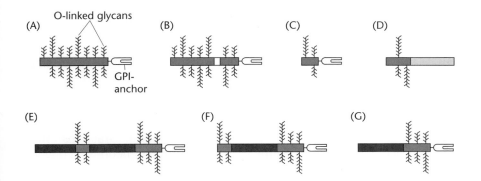

Figure 3.18 Schematic representation of the predicted structures of native AGPs after processing and posttranslational modifications. (A) Classical Hyp-rich AGP; (B) classical AGP with Lys-rich domain; (C) AG peptide (12 amino acids); (D) nonclassical AGP with Asn-rich domain; (E) protein with two AGP and two fasciclin domains; (F) protein with two AGP and one fasciclin domain; (G) protein with one AGP and one fasciclin domain. (Adapted from Y. Gaspar et al., *Plant Mol. Biol.* 47: 161–176, 2001.)

Figure 3.19 Structure of the AGP-precipitating β-D-glucosyl Yariv reagent. The reddish-brown Yariv reagent can be used to bind and precipitate certain AGPs for purification or quantification, and for staining these molecules in sections. As indicated, only certain types of sugars allow for AGP binding.

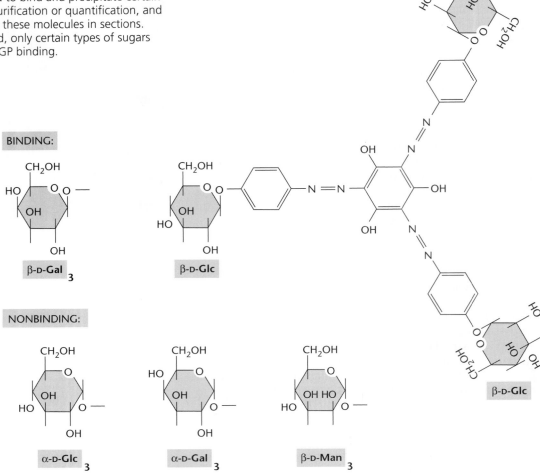

surface of the plasma membrane. As discussed in the following concept, this sequence-based prediction has been confirmed experimentally.

The nonclassical AGPs are designated AGPs based on the fact that they contain classical AGP-like hydroxyproline-, alanine-, serine-, and threonine-rich sequence motifs. What makes them nonclassical is that they also contain extensive non-proline-rich domains (~50% of protein backbone) (Figure 3.18). Some but not all of these AGPs bind the Yariv reagent, and some possess a putative GPI anchor site.

7.   Classical arabinogalactan proteins are GPI-anchored proteoglycans.[ref28]

As was predicted from the sequences of cDNA clones encoding classical AGPs, classical AGPs are GPI-anchored proteoglycans. This prediction has been verified by a number of complementary analyses. The first evidence came from the finding that AGPs purified from the styles of *Nicotiana alata* and from cell suspension cultures of *Pyrus communis* were present in buffer-soluble extracts and not in plasma membrane fractions. After deglycosylation, mass spectrometry and C-terminal peptide sequencing were used to show that the C-terminal domains of these proteins had

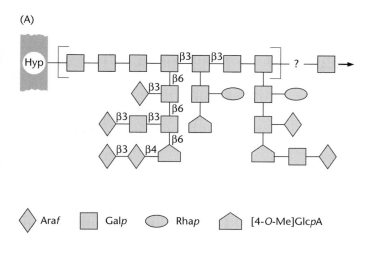

(A)

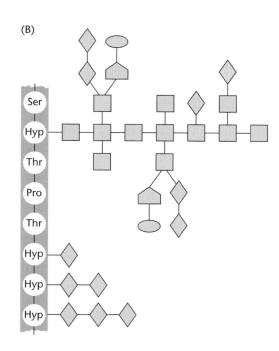

(B)

been removed by posttranslational processing as is typical of GPI-anchored proteins. Finally, the sequence of the lipid anchor from plasma membrane–bound AGPs was elucidated and its coupling to the C-terminus confirmed. This analysis yielded the structure depicted in **Figure 3.22**.

A number of GPI-anchored proteins have been characterized in animals, protozoa, and yeast, and the plant GPI anchor shows the same types of conserved residues as in these other organisms. Since several GPI-anchored proteins from animal systems have been implicated in signal transduction pathways, it is tempting to speculate that GPI-anchored AGPs could also participate in signaling pathways in plants. This signaling could involve phospholipase cleavage of the lipid anchor (**Figure 3.23**), yielding the soluble AGP and a membrane-anchored protein-free GPI from which intracellular messengers such as phosphatidyl inositol and inositol phosphoglycan could be generated. Alternatively, GPI-anchored AGPs could interact with transmembrane receptorlike proteins in adjacent cells in a manner analogous to the neuronal rat GPI-anchored protein contactin that participates in cell-cell communication. The discovery that purified AGPs from embryogenic cells can stimulate nonembryonic cells to undergo embryogenesis suggests that AGPs are indeed involved in signal transduction pathways. The challenge now is to verify these hypotheses and to elucidate the different elements of the signaling pathway(s).

Figure 3.20 Schematic models of representative carbohydrate side chains of AGPs. (A) Large, branched arabinogalactan side chain as proposed in the "wattle blossom" model of the structure of an AGP. (B) Intermediate-size, branched arabinogalactan side chain with Rha and GluA residues as proposed for the "twisted hairy rope" model of an AGP from gum arabic. Short, unbranched Ara side chains are produced where two or more contiguous Pro/Hyp residues are found. (Adapted from Y. Gaspar et al., *Plant Mol. Biol.* 47:161–176, 2001.)

Figure 3.21 (A) Domain structure of cDNA clones encoding the protein backbones of AGP: An N-terminal signal sequence, a domain rich in proline/hydroxyproline, alanine, serine, and threonine residues that contains the sites that will be glycosylated, and a C-terminal GPI-anchor signal sequence. The cleavage site defined by the ω, ω + 1, and ω + 2 residues indicates where the transamidase complex enzyme cleaves the pro-protein and attaches the GPI anchor. The hydrophobic tail corresponds to the transmembrane domain of the pro-protein. (B) The putative GPI-anchor signal sequence of a classical pear AGP (AGPPc 1). The ω residue helps define the cleavage site. The bold residue is a conserved basic amino acid located at the transition between ω + 2 and the transmembrane domain. aa, amino acid. (Adapted from C.J. Schultz et al., *Trends Plant Sci.* 3:426–431, 1998.)

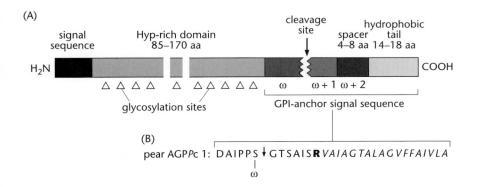

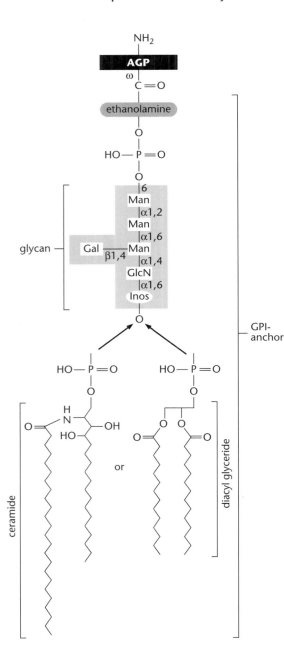

Figure 3.22 Structure of the GPI anchor of a plasma membrane AGP from a pear suspension culture. The ceramide is a lignoceroylphytosphingosine. (Adapted from D. Oxley and A. Bacic, *Proc. Natl. Acad. Sci. USA* 96:14246–14251, 1999.)

8. GRPs are structural cell wall proteins that participate in the assembly of vascular bundles and are present in root cap mucilage.[ref29]

Glycine-rich proteins (GRPs) fall into two classes that serve two different functions. One class lacks a signal peptide and accumulates within the cytoplasm most commonly under stress conditions; the other possesses signal peptides and localizes to cell walls. Here we focus only on the latter class.

Cell wall GRPs typically contain 50 to 70% glycine residues arranged predominantly in $(Gly-X)_n$ repeats, where X is frequently Gly. This sequence structure suggests that these proteins have a β-pleated sheet structure. GRPs are most often observed in vascular bundles, particularly in phloem fibers, in the modified primary walls of mature protoxylem elements, and in cell corners between xylem elements and the xylem parenchyma cells.

One of the unique features of GRPs associated with xylem wall development is that they appear to be made not only by the xylem cells themselves after the lignification of their secondary walls, but also by adjacent xylem parenchyma cells. These latter cells export the GRPs to the modified primary walls of dead xylem cells, where the proteins may help maintain the structural integrity of this tissue-stabilizing cell wall layer and possibly contribute to cell wall repair processes. Other purported functions of GRPs include providing elasticity and tensile strength during vascular development, and/or serving as nucleation sites for the oxidative polymerization chain reaction of lignin synthesis.

GRPs are also present in the mucilage secreted by root cap cells. A root cap–specific gene, *zmGRP4*, encoding a GRP has been isolated and shown to be highly expressed in root cap peripheral cells and slightly in the root epidermis. The 14-kD protein, however, is found in the mucilage that covers the root tip. After extraction from the mucilage, the proteins have an apparent size of 36 kD, suggestive of a posttranslational modification, but the nature of this modification has yet to be determined.

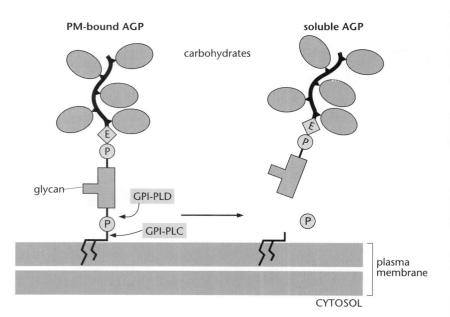

Figure 3.23 Formation of a soluble AGP from a plasma membrane–bound (GPI-anchored) AGP. The release is mediated by C- and D-type phospholipases (GPI-PLC and GPI-PLD). The soluble AGP is released into the cell wall/extracellular space.

9. AGPs, GRPs, HRGPs, and PRPs may have a common evolutionary origin.[ref30]

As discussed in the preceding sections, HRGPs, PRPs, and AGPs possess not only similar repeats, which have led to the proposal that they are members of an HRGP superfamily, but also a high degree of sequence diversity. One possible explanation for this sequence diversity is that the codons for proline (CCX), alanine (GCX), serine (TCX), and threonine (ACX) differ by only one codon in the 5′ position. Thus, during evolution, single base changes may have quickly led to interchanges between these amino acids. Even the glycine-rich proteins (GRPs) may be evolutionarily related to the PRPs by inversions and duplications between the glycine (GGX) and proline (CCX) codons. This idea has received support from the accidental discovery that HRGP clones can be used to identify GRP clones due to the nucleotide sequence homology between these two protein classes. This homology can be explained by the sequence complementarity between stretches of CCX (Pro) and of GGX (Gly), respectively, which enables the noncoding strand of an HRGP gene to share sequence homology to the coding strand of a GRP gene and vice versa.

F. Lignin

1. Lignin is a hydrophobic polymer of secondary cell walls.[ref31]

Lignin is a complex phenolic polymer that provides mechanical support and a water-impermeable surface to secondary cell walls. During lignification, the aqueous phase of the cell walls is replaced by a hydrophobic polyphenolic polymer, which surrounds all of the cellulose microfibrils and matrix polysaccharides and provides rigidity and compressive strength to the walls. Lignin also provides a barrier to pests and pathogens. *Ectopic lignification* may be induced locally as part of the plant defense response, supplementing the lignification from normal development. Thus, lignification is regulated by both developmental and environmental signals. Cell division, cell expansion, and cell elongation necessarily cease before lignin is deposited. Lignification, therefore, constitutes the last stage of differentiation (terminal differentiation), and is typically followed by programmed cell death. Lignified cell types vary greatly in size, lignin content, and location in the plant.

In contrast to most biopolymers with ordered structure, lignin primary structure is not stipulated. The polymerization process (see Concepts 5E) creates a polymer with a highly diverse series of linkages. Lignin is unusually stable in nature because the polymer is degraded slowly and only by a small number of specialized fungi. The extraordinary stability of lignin is a major factor in the terrestrial carbon cycle, resulting in large amounts of carbon in humic components of soil. Similarly, the resistance of lignin to microbial decay is likely to contribute to the long life of a great many tree species, further serving to sequester carbon in wood of living trees. The diversity of the lignin polymer may be the basis for the absence of specificity in lignin-degrading enzymes, with far-reaching consequences for life on the surface of the land.

The lignin polymer is a complex network with many different intermolecular linkages formed during its polymerization. Lignin is composed of hydrophobic subunits, polymerized from cinnamyl alcohol monomers (*monolignols*), which are synthesized in the cytoplasm from phenylalanine and then transported to the cell wall and polymerized in lignifying zones. Polymerization and deposition of lignin occurs both within the cell wall and between the walls of adjacent cells in the middle lamella (**Figure 3.24**). The

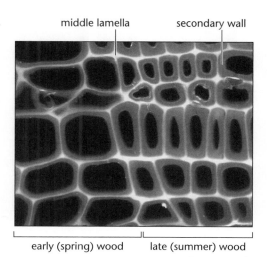

middle lamella   secondary wall

early (spring) wood   late (summer) wood

40 μm

Figure 3.24 Confocal fluorescence microscopy of pine wood showing lignin autofluorescence at 530 nm. The brightest regions contain the highest concentrations of lignin. (From L. Donaldson and J. Bond, Fluorescence Microscopy of Wood. Scion, New Zealand: New Zealand Forest Research Institute, 2005.)

Figure 3.25 Grove of giant sequoias (*Sequoiadendron giganteum*) known as the House Group in Sequoia National Park in California. The arrow indicates an adult human as a size reference. (Courtesy of C. Smith.)

hydrophobic lignin polymer serves as a barrier to lateral water transport across the lignified wall and facilitates vertical transport of water through the "waterproof" tubes of lignified tracheids and vessels.

Mechanical support and water transport were essential for the evolution of terrestrial vascular plants. Due to their size, trees have extreme needs for structural support and water transport (**Figure 3.25**). Xylem elements formed in wood may transport water more than a hundred meters in the tallest trees, and a very large tree may weigh hundreds of metric tons. Typically, 25 to 30% of the dry weight of wood is lignin, but in some trees it can be as much as 40%. In contrast, the lignin content of most herbaceous plants is much lower. Grasses, for example, contain about 8% lignin, but the amount of lignin deposited varies greatly depending on the plant species and the age of the tissue (see **Table 3.4**).

2.   Our knowledge of lignin composition is based on the chemical hydrolysis of wood.[ref32]

Lignin has a relatively amorphous, three-dimensional structure in contrast to other major abundant polymers such as cellulose or chitin that form repeating, highly ordered linear structures. Lignin is difficult to purify and characterize because of the highly heterogeneous linkages that occur between subunits within lignin and linkages to polysaccharides. Characterization of the lignin polymer is made even more difficult because it is not possible to isolate lignin in its intact state. Lignin is highly hydrophobic and is not extracted from plant tissues by either aqueous or organic solvents. Consequently, much of our knowledge of lignin structure comes from the analysis of its chemical degradation products.

Lignin is bonded to carbohydrate components and cannot be easily separated from them. Methods that extract the carbohydrate components from lignin-carbohydrate complexes may also modify lignin components. Although there are many methods for degrading lignin to quantify its content and composition, each has limitations, either in extent of solubilization, specificity of subunit detection, or modification during the process. Nonlignin phenolics, also found in the cell wall, can be difficult to distinguish from true lignin.

Purification of lignin from wood usually involves extensive mechanical grinding that fragments cells and cell walls, followed by treatment with cellulases and extraction with dioxane-water. Such preparations may contain relatively little carbohydrate and have minimally modified lignin. Severe treatment of lignin with chemicals or temperature is required to solubilize the components of lignin through oxidation or hydrolysis. The traditional estimation of lignin content (% lignin) is determined from the residual weight following sulfuric acid (72%) hydrolysis of polysaccharides, the method first applied by Klason in 1906. This estimation is known as Klason lignin.

If pulverized wood from spruce is treated with 2 M NaOH in the presence of nitrobenzene at 160°C for several hours (*nitrobenzene oxidation*), the primary phenolic product is vanillin (**Figure 3.26**). This result is typical for wood isolated from gymnosperms, which include the conifers. This group of woody plants is often described as "softwoods" (see Figure 9D3.1). In addition, a few percent of 4-hydroxybenzaldehyde is also recovered. Lignin from most woody dicots (hardwoods) and herbaceous dicots produces

TABLE 3.4 VARIATION IN LIGNIN CONTENT AND COMPOSITION

Compositions are typically determined from nitrobenzene oxidation, and content from estimation of Klason lignin. For hardwoods, nitrobenzene oxidation produces higher values for S units than other methods; therefore the values here are likely to be inflated. —, not determined.

CATEGORY	SPECIES NAME/COMMON NAME	TISSUE/CELL TYPE	% LIGNIN	COMPOSITION	REF.
Softwood	*Pinus sylvestris*/Scots pine	Wood	26.9	H/G > 0.07	a
Softwood	*Picea abies*/Norway spruce	Wood	27.5	H/G > 0.07	a
Softwood	*Pseudotsuga menziesii*/Douglas fir	Wood	27	G lignin	a
Softwood	*Pseudotsuga menziesii*/Douglas fir	Compression wood	36.5	H/G = 0.47	a
Hardwood	*Populus tremuloides*/aspen	Wood	18.0	S/G = 3.0	a
Hardwood	*Juglans nigra*/black walnut	Wood	20.9	S/G = 2.1	a
Hardwood	*Eucalyptus botryoides*/Southern mahogany	Young wood	22.9	S/G = 1.9	a
Hardwood	*Eucalyptus botryoides*/Southern mahogany	Heartwood	33.1	S/G = 1.5	a
Hardwood	*Eucalyptus botryoides*/Southern mahogany	Young bark	17.2	S/G = 1.3	a
Hardwood	Eucalyptus botryoides/Southern mahogany	Old bark	34.4	S/G = 1.5	a
Hardwood	*Belliolum haplopus*	Wood	36	S/G = 0.9	a
Hardwood	*Betula papyrifera*/white birch	Wood	22.0	—	b
Hardwood	*Betula papyrifera*/white birch	Tension wood	16.1	—	b
Hardwood	*Betula papyrifera*/white birch	Fiber cells	18	S rich	c
Hardwood	*Betula papyrifera*/white birch	Ray cells	24	S lignin	c
Hardwood	*Betula papyrifera*/white birch	Vessels	24	G lignin	c
Nonwood fiber	*Hibiscus cannabinus*/kenaf	Stem	10.9	S/G = 6.5	b, d
Monocot	*Zea mays*/maize	Corn stover	17	S/G =1.2	e
Monocot	*Oryza sativa*/rice	Internodes	15	H/G = 0.23 S/G = 0.16	h
Monocot	*Oryza sativa*/rice	Hulls	34	H/G = 0.06 S/G = 0.02	h
Monocot	*Phylostachys heterocycla*/bamboo	Shoot	2.3	H/G = 1.25 S/G = 0.37	a
Monocot	*Phylostachys heterocycla*/bamboo	Stem	26.1	H/G = 0.67 S/G = 2.0	a
Monocot	*Miscanthus sinensis*/maiden grass	Stem	23.5	H/G = 0.49 S/G = 0.46	a
Monocot	*Asparagus officinalis*/asparagus	Stem	18.8	S/G = 1.8	a
Dicot	*Arabidopsis thaliana*/mouseear cress	Inflorescence stem	—	S/G = 0.5	f
Lycophyte	*Selaginella moellendorffii*/spikemoss	Plant	—	S/G = ~2	g

[a]K.V. Sarkanen and C.H. Ludwig, Lignins: Occurrence, Formation, Structure and Reactions. New York: Wiley Interscience, 1971.
[b]C.W. Dence, The determination of lignin, in S.Y. Lin and C.W. Dence (eds.), Methods in Lignin Chemistry. Berlin: Springer, 1992, pp. 33–61.
[c]C.-L. Chen, Lignins: occurrence in woody tissues, isolation, reactions and structure, in M. Lewin and I.S. Goldstein (eds.), Wood Structure and Composition. New York: Marcel Dekker, 1991, pp. 183–261.
[d]F. Lu and J. Ralph, The DFRC method for lignin analysis 2, Monomers from isolated lignins, *J. Agric. Food Chem.* 46:547–552, 1998.
[e]G.C. Galletti, P. Bocchini, A.M. Smacchia et al. Monitoring phenolic composition of maturing maize stovers by high performance liquid chromatography and pyrolysis/gas chromatography/mass spectrometry. *J. Sci. Food Agric.* 71:1–9. 1996.
[f]C.C.S. Chapple et al., An Arabidopsis mutant defective in the general phenylpropanoid pathway, *Plant Cell* 4:1413–1424, 1992.
[g]J.-K. Weng et al., Independent origins of syringyl lignin in vascular plants, *Proc. Natl. Acad. Sci. USA* 105:7887–7892, 2008.
[h]K. Zhang et al., Gold hull and internode 2 encodes a primarily multifunctional cinnamyl alcohol dehydrogenase in rice, *Plant Physiol.* 140:972–983, 2006.

major amounts of both vanillin and syringaldehyde. Lignin from grasses (monocots) produces all three oxidation products in significant amounts.

One of many alternative methods for lignin composition analysis, called *thioacidolysis*, also provides a partial fingerprint of lignin composition (Figure 3.26). The key reaction is *solvolysis* (nonaqueous degradation) in dioxane:ethanethiol, 9:1, in a 0.2 M boron trifluoride etherate. Lignin subunits, in labile β-O-4 linkages, release thioethylated monomers with a high yield. This analytical method may be used as a test for the presence of lignins where other methods fail due to lack of specificity or sensitivity. Thioacidolysis has high specificity for β-O-4 linkages, whereas nitrobenzene oxidation provides more complete information about composition. It is always best to use a combination of methods that differ in specificity to obtain a more complete profile of lignin components and structure.

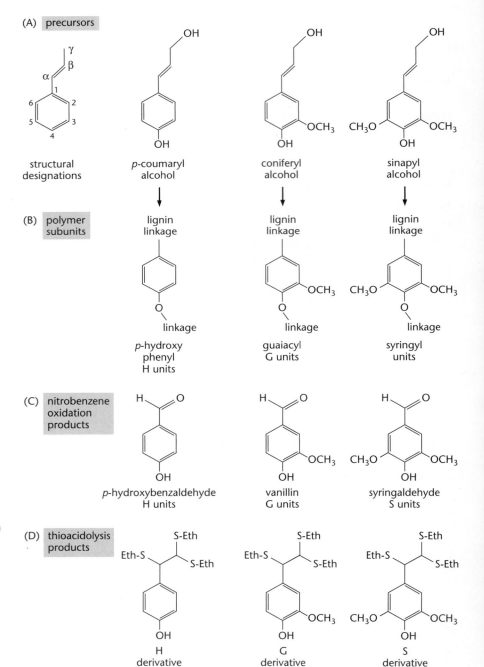

Figure 3.26 Degradation products of the major lignin subunits. Linkages are described by the numbered carbons of the phenolic ring and the α, β, and γ carbons of the propane side chain. Diverse linkages are produced when the three main lignin precursors (A) are polymerized into lignin (see Section 5E). The incorporated precursors (monolignols) produce the three types of subunits that predominate in the lignin polymer (B). The relative proportions of specific linkages and subunits may be estimated from the complete and partial degradation products by methods such as nitrobenzene oxidation (C) or thioacidolysis (D), which differ in their selectivity.

The conclusion from experiments using oxidation or hydrolysis of lignin is that the basic component of lignin is the phenylpropanoid unit (C_6C_3) with a six-carbon ring and a three-carbon side chain. The three major nitrobenzene oxidation products of native lignin are derived from three predominant types of phenylpropane units (see Figure 3.26): *hydroxyphenyl units (H units), guaiacyl units (G units),* and *syringyl units (S units).* Nitrobenzene oxidation of lignin creates soluble vanillin from guaiacyl units, syringaldehyde from syringyl units, and 4-hydroxybenzaldehyde from *p*-hydroxyphenyl units. Similarly, thioacidolysis gives rise to thioethylated derivatives of H, G, or S monomers.

Lignins are derived from more than just three predominant monomers. Many monomers of lignin are present in low amounts, and vary greatly in different species. Unusual subunits are also found. For example, lignin monomers are heavily acetylated in kenaf bast fibers, and ferulates and *p*-coumarates are found in species of grasses. Incomplete reactions during biosynthesis give rise to many precursor subunits that can also be incorporated into lignin. Where expression of genes involved in lignin biosynthesis is affected in mutants or transgenic plants, the resulting lignins often show dramatic changes in composition (see Concept 9E6).

Knowledge of the structure of the major components of lignin has provided a basis for establishing the nature of the precursors and the pathway of lignin biosynthesis (see Section 5E). The existence of the different types of lignin subunits raises fundamental questions regarding the properties of lignins, which may differ greatly in the relative abundance and linkages of the different subunits. The potential for different types of lignin to have different functions, may be important in the evolution of the major groups of higher plants. However, there is no information at present establishing a direct link between lignin composition and function.

Degradation and removal of lignin from wood or other plant material by combined mechanical, chemical, and thermal treatment is the central process of the pulp and paper industry. Plant-derived fiber, essentially cell walls depleted of lignin, is pressed and dried to produce many familiar products, including packaging materials, absorbent tissues, and different grades of paper for writing and printing (see Concept 9D).

3. Lignins in plant tissues, cells, and cell walls may be detected by histochemical staining or immunochemical localization.[ref33]

The location of lignin in plant tissues can be observed by histochemical staining. The *Weisner reaction* and the *Mäule reaction*, are color-producing chemical reactions for identifying lignin (**Figure 3.27**). Some indication of composition can be obtained *in situ* by using these reactions. The Weisner reaction is most widely used and is based on the reaction of phloroglucinol (in concentrated HCl) with aldehyde groups in lignin. The reaction is a very general one and is useful for identifying lignins in plant tissues. The purple color is caused by condensation of 4-*O*-alkylated 4-hydroxycinnamaldehyde moieties of lignin with phloroglucinol to produce a cationic chromophore that has a purple color. The aldehyde components of lignin are minor ones, and therefore the amount of staining may not be proportional to lignin quantity.

The Mäule reaction is more specific. This reaction produces a rose-red color following successive treatment of lignin with aqueous potassium permanganate, HCl, and NH_3 solutions. The reaction is specific to syringyl propane (S units) in lignin and does not react with guaiacyl propane (G units) or 4-hydroxyphenyl propane (also called *p*-hydroxyphenyl propane, H units). A chlorinated *O*-quinone has been proposed as the basis for the color. The Mäule reaction is useful for detecting lignin in angiosperms but not in gym-

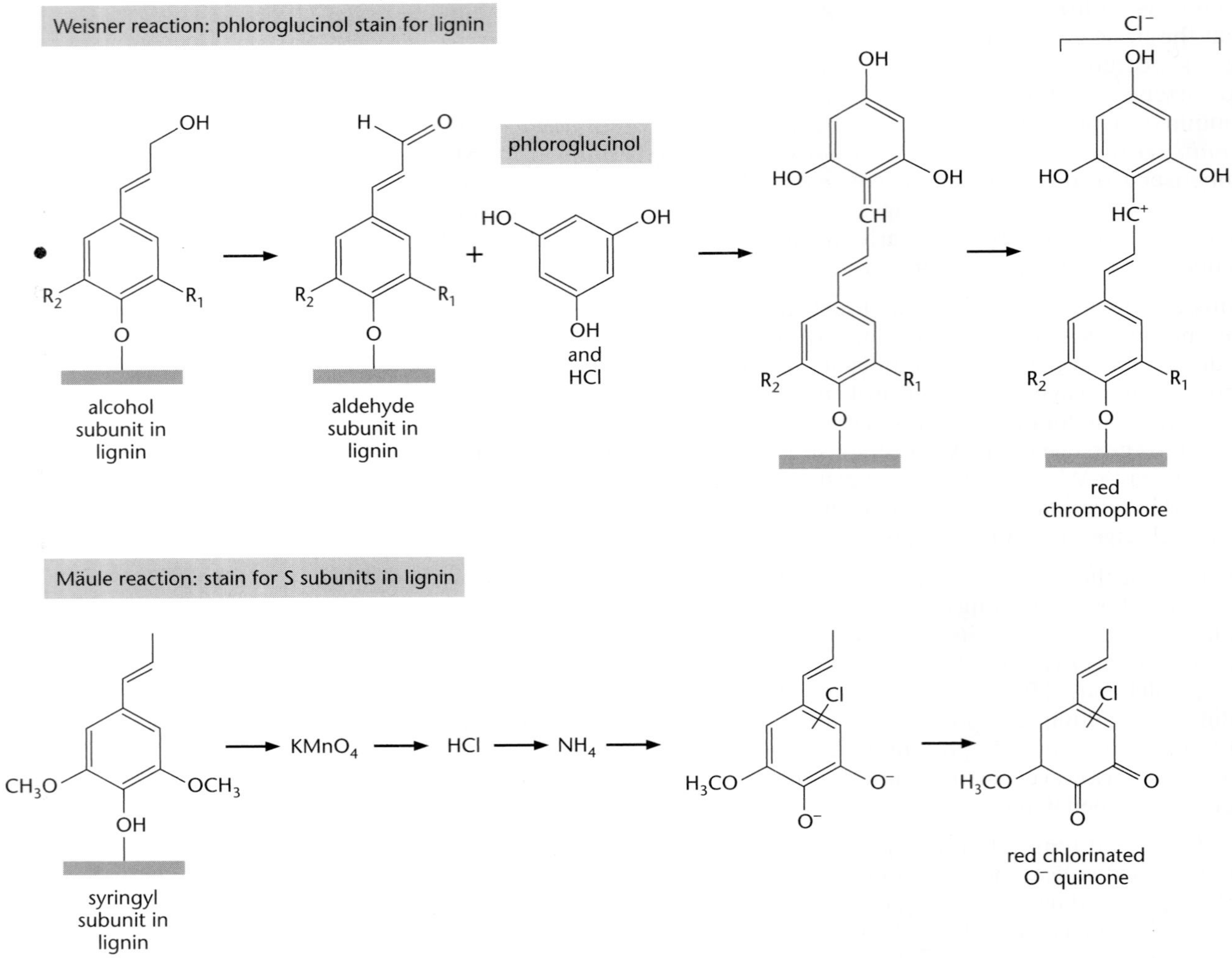

Figure 3.27 The Weisner and Mäule histochemical staining reactions.

nosperms, because gymnosperms typically contain little or no syringyl units in lignin. The Mäule reaction will understain lignin enriched in G or H units. For example, in white birch, fiber elements in the secondary wall have a syringyl type of lignin, while vessels in the same tissue contain a guaiacyl lignin. Different methods would result in differential staining. The Mäule test has been used to determine hardwood/softwood proportions in wood chip blends because hardwoods show a deep rose-red color and softwoods stain pale yellow or brown.

Recently, antibodies have been produced against synthetic lignins containing only one type of precursor, and therefore may be diagnostic for H units, G units, or S units in lignin. They have been used for labeling at the electron microscopic level. These antibodies may elucidate the distribution of different lignin subunits at high resolution.

4. The complexity of lignin is increased by the diversity of intermolecular linkages.[ref34]

Lignin substructures are known from oxidation by nitrobenzene and several other methods of oxidation or hydrolysis, although a complete structure of lignin polymers is not known. Incomplete degradation products from chemical oxidation provide insight into some of the more common intermolecular linkages.

In lignin G, S, and H units are predominantly interconnected through ether linkages and carbon-carbon linkages. By convention, linkages are described by the numbered carbons of the phenolic ring and the α, β, and γ carbons of the propane side chain (**Figure 3.28**). The major substructure in lignin is the β-*O*-4 linkage. In softwoods, the β-*O*-4 linkage represents about half of the dimeric linkages in lignin, whereas in hardwoods, the β-*O*-4 linkage

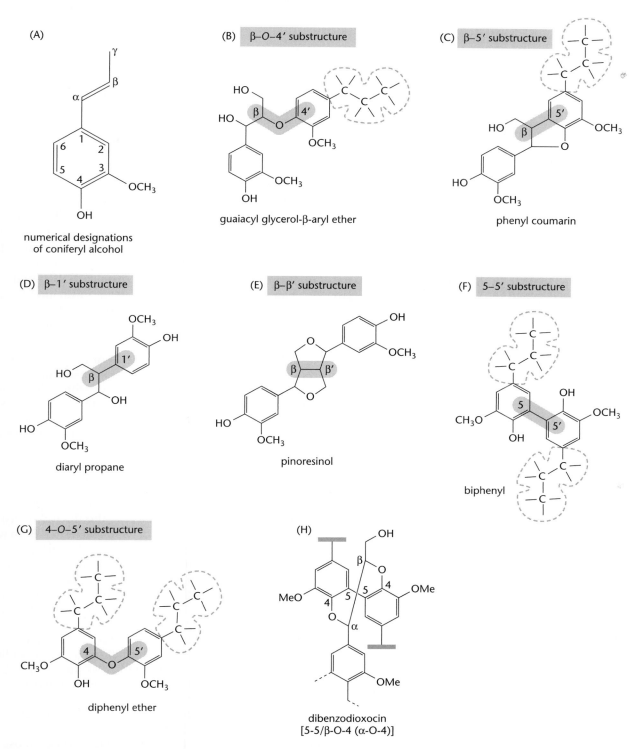

Figure 3.28 Major linkages in lignin. Numerical designation (A) and linkages of dimers (B, C, D, E, F, G). (H) One more complex linkage of three phenylpropane units. Structures are shown for coniferyl alcohol–derived guaiacyl units. Corresponding structures may occur for sinapyl alcohol–derived syringyl units, or mixed guaiacyl-syringyl units. Structures shown within dashed lines represent parts of the lignin polymer. Bonds attached to wavy lines are also connected to the lignin polymer. D and E are shown as dimers, but can also be connected to the polymer through bonds at the 4 or 5 positions. For *p*-hydroxyphenyl alcohol derivatives, R1 and R2 = H. For coniferyl alcohol derivatives, R1 = OCH₃ and R2 = H. For sinapyl alcohol derivatives R1 and R2 = OCH₃.

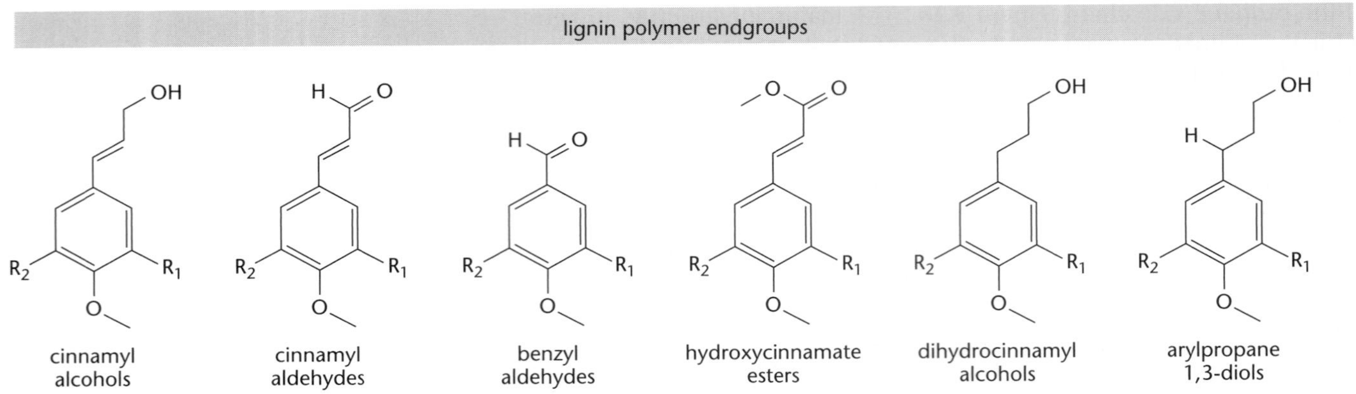

lignin polymer endgroups

cinnamyl
alcohols

cinnamyl
aldehydes

benzyl
aldehydes

hydroxycinnamate
esters

dihydrocinnamyl
alcohols

arylpropane
1,3-diols

Figure 3.29 Lignin monomer endgroups found in lignin. Each group may have alternative forms, because of differences in the 5 position of the subunits. Typically these subunits will be derived from either guaiacyl (G) or syringyl (S) subunits [with O-methylation at the 5 position (R$_2$) for the S-derived units]. The end units would be linked to the lignin polymer at the O-4 position.

represents about 80% of the linkages involving the S units (syringyl propane units). Less abundant intermolecular linkages include the α-O-4 linkage, the β-5 linkage, the β-β linkage, the 5-5 linkage, and the 4-O-5 linkage.

Recent evidence showed a more complex cyclical structure, dibenzodioxocin, which contains three phenyl rings with a 5-5/β-O-4(α-O-4) linkage (Figure 3.28H). This structure, in addition to others such as biphenyl ether (5-O-4), or 5-5 biphenyl linkages, can produce cross-links between independent growing chains. As discussed in more detail in Chapter 5, lignin polymerization is primarily due to the addition of monomers to growing polymer chains. Diverse subunits are found as endgroups on lignin (**Figure 3.29**), which can be alcohols, aldehydes, esters, or diols.

Models of higher-order structure and organization of lignin may be constructed from the nature and frequency of the intermolecular linkages (**Figures 3.30** and **3.31**). Within a cell wall, lignin exists in a three-dimensional matrix and interacts with other polymers. Physical methods of analysis (for example, NMR and infrared spectroscopy) have begun to elucidate lignin structure *in situ*. Lignin is very low in optical activity, indicating a low level of oriented structure.

5.   Models for higher-order structure of lignin.[ref35]

Physical and chemical properties of lignin do provide some insights into its higher-order structure, and preliminary models of the three-dimensional conformation of lignin have been constructed. Lignin recovered from wood cell walls following mechanical shear and cellulose degradation has a molecular weight average of 20,000 to 77,000 in different preparations, suggesting a degree of polymerization of several hundred phenylpropane subunits for particles in intact lignin. The viscosity of lignin in solution is about 3% of that of cellulose, consistent with the idea that the lignin particles have a compact spherical shape. This concept is consistent with the specific gravity of lignin, which is about 1.4 g/ml, close to that expected for space-filling spherical particles. The frequency and structure of major bonds in lignin provide information for estimating the degree of cross-linking. The β-O-4 linkage predominates in spruce lignin, and the degree of cross-linking is about 14%. In the middle lamella between cells in wood, lignin is deposited in round particles which coalesce into a compact, more uniform structure, whereas in the secondary wall lignin is associated with hemicellulose and forms a polymer in layers about 5 nm thick between layers of cellulose.

Lignin, unlike proteins, carbohydrates and DNA, is a racemic polymer. For the β-O-4 ether dimer, there are 4 optical isomers; there are 8 for a trimer and 32 for a tetramer. The complexity indicates that no two lignin polymers of substantial size are likely to be identical, within a plant, or within a species, or in all plants.

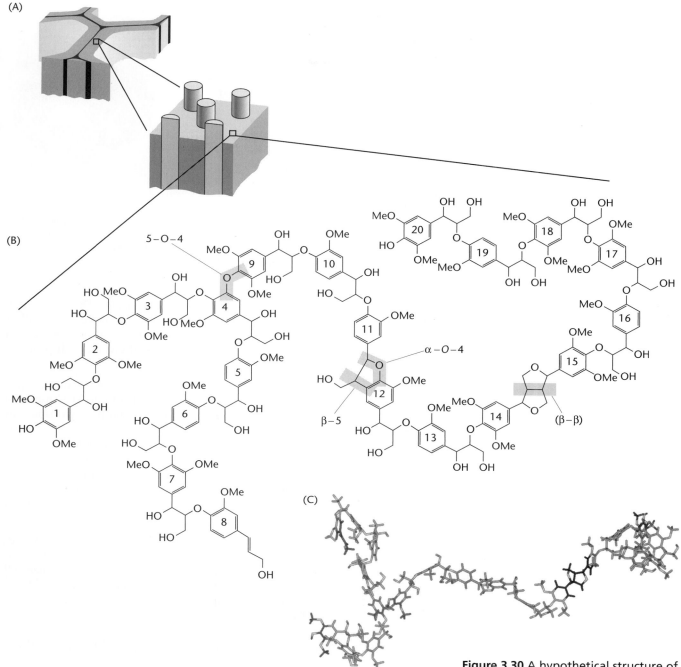

Figure 3.30 A hypothetical structure of lignin from a wood cell wall. This model is one of millions of alternative structures that could be formed from 20 monomeric lignin subunits with a composition similar to poplar. Most linkages are β-ether units. Me represents a methyl group (CH₃). Subunits 1, 2, 3, 7, 9, 14, 15, 17, 18, and 20 are S units; 4, 5, 6, 8, 10, 11, 12, 13, 16, and 19 are G units. β-*O*-4 units predominate, except for unit 4 (5-*O*-4), unit 11 (α–*O*-4 and β–5) (α-*O*-4 and β-5), and 14-15 linked as ββ. One branch point is shown as a 4-*O*-5 linkage between syringyl unit 4 (S) and guaiacyl unit 9. (After J. Ralph et al., *Phytochem. Rev.* 3:29–60, 2004.)

6.   Lignin is cross-linked to polysaccharide components in lignin-carbohydrate complexes.[ref36]

A variety of covalent cross-links couple lignin to other macromolecules in the cell wall, generally called *lignin carbohydrate complexes (LCCs)*. The strongest evidence is for hemicelluloses covalently linked to lignin in wood. Three types of cross-links have been identified. Lignin may be linked to polysaccharides through ester linkages (benzyl esters) between uronic acids of the polysaccharides and hydroxyl groups of lignin monomers. Ether bonds can link polysaccharides and lignin (benzyl ethers) through glucosyl or mannosyl residues, and similarly glycosidic linkages can bond with phenolic residues or with hydroxyls of lignin side chains (phenyl glycosides). Ferulic acid, an important component of cell walls of grasses, is frequently esterified

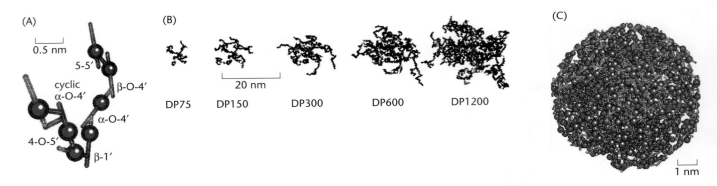

Figure 3.31 A model of a lignin polymer in spherical space. (A) A model of seven subunits connected by six different linkages (5-5, β-O-4, α-O-4, β-1, 4-O-5, cyclic β-O-4). Balls represent aromatic rings, and sticks represent side chains or groups. Short sticks are 4-*OH*, groups and propane side chains are longer. (B) Simulated development of macromolecular shape as a function of increasing polymer size. DP is the degree of polymerization ranging from 75 to 1200 subunits. (C) Lignin model of a spherical macromolecule of 1147 subunits containing 1247 bonds, 101 of which are intrachain cross-links. (Adapted from L. Jurasek, *J. Pulp Paper Sci.* 21:J274–J279, 1995.)

or linked through ether bonds to cell wall polysaccharides. Dehydroferulic acid esterified to polysaccharides can also be linked to lignin. Another potential linkage affecting the architecture of lignin is that between lignin and cell wall proteins. Both hydroxyproline-rich and glycine-rich proteins are often co-localized with lignin in secondary cell walls.

LCCs can be purified from milled wood extracted with dioxane to solublize carbohydrates not linked to lignin. The resulting insoluble LCC material from softwood has approximately 3 sugar units for 100 phenylpropanoid units. The predominant hemicelluloses in softwoods are galactoglucomannans. However, the composition of LCCs from softwood includes arabinose, xylose, and rhamnose, suggesting linkages to arabinogalactan, or arabinoglucoronoxylan. Hardwoods have xylans and glucomannans as their major hemicelluloses. Some LCCs are not labile in alkali (benzyl ethers, phenyl glycosides) and are thought to contribute to slow delignification of wood in alkaline pulping processes.

Cell walls of grasses have unusual levels of ferulate and coumarate linked to polysaccharides. Ferulate esters cross-link arabinoxylan chains through 5-5 dimers. Ferulic acid is now known to be a component of lignin. In maize cell walls, ferulate and 5-5 diferulate may be incorporated into lignin and form cross-links to xylans. Some structures of phenylpropanoid carbohydrate linkages are shown in **Figure 3.32**.

7. Lignin composition and content vary greatly among the major groups of higher plants.[ref37]

All lignins may be characterized by the relative proportion of the three major phenylpropanoid units. The most dramatic differences are found between the lignin in wood of conifers compared with the lignin of angiosperms (monocotyledons or dicotyledons). As noted earlier, a typical conifer lignin, such as lignin from pine or spruce, is composed predominantly of guaiacyl (G) units with a trace of *p*-hydroxyphenyl H units (G/H 98/2). Lignin from poplar, an angiosperm, is composed of approximately equivalent amounts of guaiacyl (G) and syringyl (S) units. However, the S/G ratio varies considerably among woody dicotyledons (Table 3.4). The lignins of grasses often have a higher proportion of H units. For example, lignin in rice internodes is 18% H units, but maize has far less. *Selaginella* (spikemoss), a lycophyte that represents a group of simple vascular plants established as a lineage (fern allies) long before the evolution of angiosperms, also contains syringyl lignin.

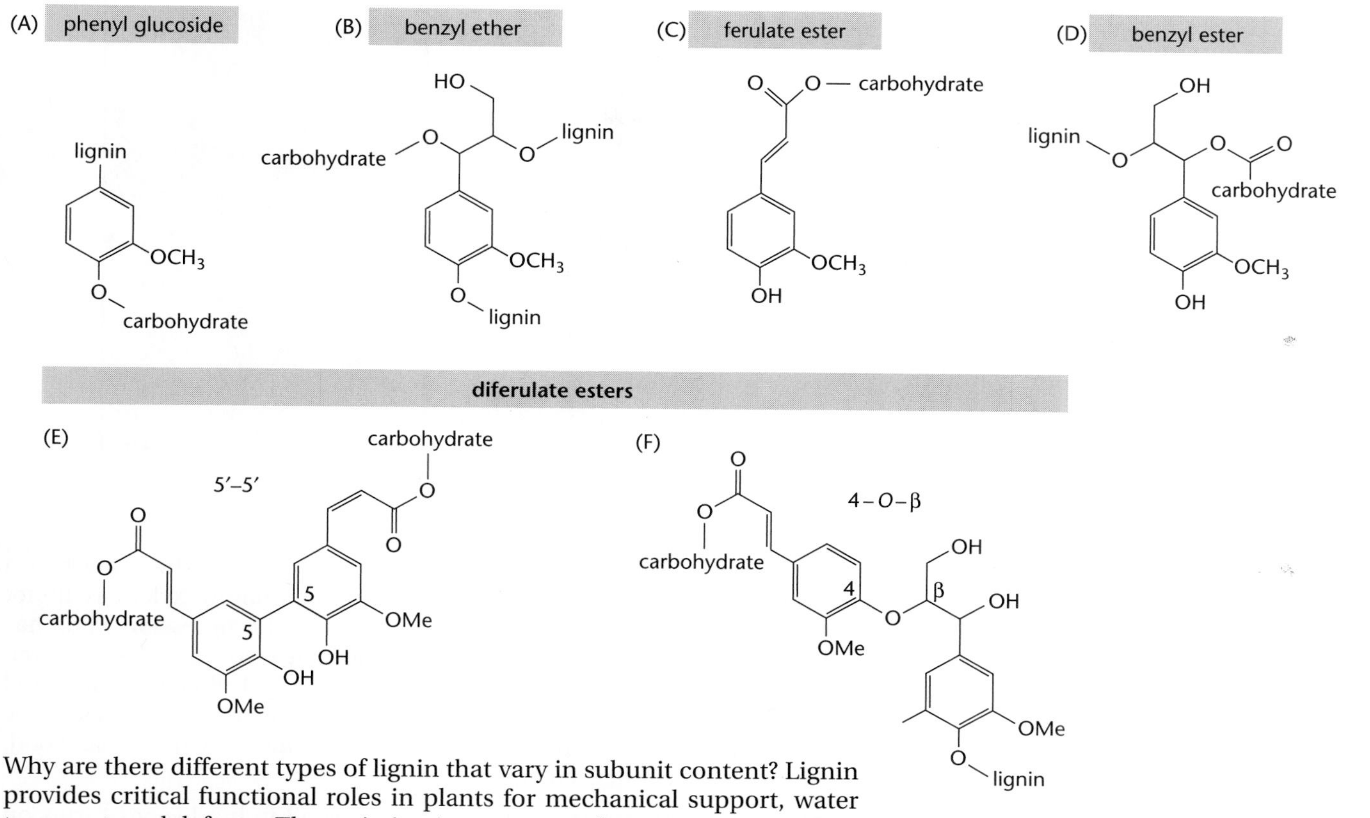

Figure 3.32 Six proposed structures for lignin carbohydrate complexes (LCCs). (A) The phenylglucoside is linked by the propane side chain to lignin and by an O-4 linkage to carbohydrate. (B) The benzyl ether has an α-O linkage to carbohydrate and a β-O linkage to lignin. (C) The ferulate ester has a γ-O linkage to carbohydrate with potential linkages to lignin at the O-4 and 5 positions on the ring. (D) The benzyl ester has a β-O-4 linkage to lignin and an α-O linkage to carbohydrate. (E) Diferulate ester with potential sites for coupling to lignin at the 4 (OH) position of both subunits. (F) Diferulate ester linked at 4-O to lignin.

Why are there different types of lignin that vary in subunit content? Lignin provides critical functional roles in plants for mechanical support, water transport, and defense. The variation in content and composition of lignin may involve adaptation to improve strength, hydrophobicity, or resistance to pests or pathogens. For example, aquatic plants have less need for support and have low lignin content. Many plants have the capacity to synthesize the three major types of lignin, but the different compositions and content result from regulation through developmental, environmental, or metabolic control of gene expression through biosynthesis and/or channeling of precursors into alternative pathways. For example, different cell types in poplar wood differ substantially in syringyl lignin content with vessels having predominantly G lignin, whereas fibers are rich in syringyl lignin.

Softwoods, with predominantly G lignin, are more resistant to decay by white rot fungi than hardwoods with S/G lignin. Similarly, in hardwoods, the G-rich vessels are more resistant to decay than the other cells in that wood. If the different lignin compositions have different susceptibilities to decay, it could explain the basis for different types of lignin in the different lineages of higher plants; however, there are many other differences in the wood of softwoods and hardwoods and between fibers and vessel cells that could also be involved in their differential decay.

8. Lignin content and composition vary in wood formed during increased mechanical stress.[ref38]

A well-studied example of an environmental effect on cell wall composition is the effect on lignin formed in response to mechanical stress. Wood produced under increased mechanical stress, called *reaction wood*, differs from wood formed during normal growth and development. Reaction wood differs in lignin content and composition and in addition may vary in morphology, color, wood density, and carbohydrate composition. Reaction wood occurs in both softwoods and hardwoods, but the mechanism of its formation is different in these major groups of woody plants. In the absence of mechanical stress, such as found in aquatic plants or in submerged internodes of rice, lignin content is greatly reduced.

Figure 3.33 Light micrograph of a section of wood from a sweetgum tree (hardwood) stained with toluidine blue showing normal xylem cells (A) and cells characteristic of tension wood (B) showing the thick G (gelatinous) layer typically formed on the upper side of a leaning stem. (From J. Andrew, J. Bowling and K.C. Vaughn, *Amer. J. Bot.* 95:655–663, 2008.)

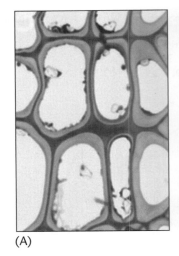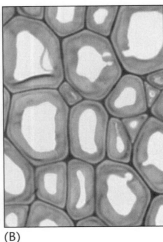

(A)   (B)

In softwoods, reaction wood forms on the underside of the branch or a tree bent toward the ground, where the differentiating xylem is under compression, and is known as *compression wood*. Compression wood has considerably higher lignin content than normal wood. In Alaskan larch (*Larix laricina*), compression wood can contain up to 40% lignin. Compared with normal wood, compression wood has shorter fibers and higher specific gravity, and is enriched in H units several times more than normal wood, yielding a more condensed lignin.

In contrast to compression wood, reaction wood in hardwoods forms on the upper side of a tree stem bent toward the ground and forms where the wood is under greater tension (*tension wood*). Tension wood of hardwoods is typically lighter in color (except in Eucalyptus, where it is darker) and contains less lignin than corresponding normal wood. In *Eucalyptus goniocalyx* (mountain gray gum) the lignin content of tension wood is only 10%, compared with normal wood, which has 25% lignin. The composition of the lignin is still a G-S type, but the relative proportion of subunits is modified, as the ratio of G units to S units is increased. The reduced lignin content of tension wood is due to the presence of the G layer, a thick, cellulose-rich *gelatinous layer* formed on the inside of the secondary cell wall (**Figure 3.33**). This *G layer* is very low and almost devoid of lignin.

Yet to be determined is how a plant perceives a mechanical stimulus, how the composition of lignin is modified in response to this stimulus, and whether more condensed lignin imparts greater strength to the wood.

G. Suberin, Cutin, Waxes, and Silica

1. Suberin is a complex hydrophobic material that forms physical and biological barriers essential for plants.[ref39]

Suberin is the insoluble material in cork that makes it impermeable to water. Hooke, in 1665, first observed the cell walls in cork and named the compartments delineated by these cell walls "cells" (**Figure 3.34**). Bottle cork is derived from the bark of the cork oak, *Quercus suber* L. (Figure 9D5.3). Suberin is an essential polymer in vascular tissues that produces insoluble barriers during normal development and in response to wounding (*wound periderm*). Suberin barriers are also laid down in response to fungal infections. In healthy plants, suberin layers form protective barriers within the epidermal walls of underground organs (roots, tubers), and in the stems of woody plants. Suberin also makes the casparian bands of the root endodermal cell layer impermeable to water (see Figures 1.35B and 1.37), and is

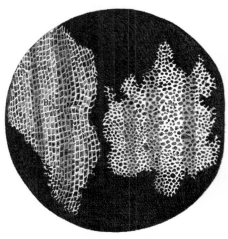

Figure 3.34 Micrograph of cross-sectioned cork cells published by Robert Hooke in 1665.

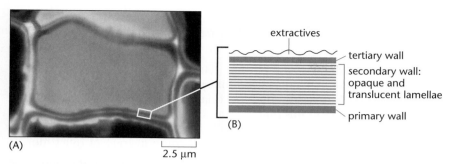

(B)

extractives

tertiary wall

secondary wall: opaque and translucent lamellae

primary wall

(A)

2.5 μm

Figure 3.35 A proposed architecture of a suberized cell wall in cork. (A) Cross-section of a suberized cork cell as seen in a UV fluorescence microscope. (B) The boxed, suberized secondary cell wall region possesses a lamellar architecture consisting of alternating opaque and translucent layers. (Adapted from J. Graça and S. Santos, *Macromol. Biosci.* 7:128–135, 2007.)

deposited in the walls around the bundle sheath cells of grasses, and in the oxalate crystal containing cells and the abscisic cells of leaves. Due to its chemical inertness, it is a residual component of terrestrial soils.

The suberin content of bark from most trees ranges from 2 to 8%, but in the bark of birch and cork oak it may reach 50%. Cork cell walls contain, in addition to suberin, 30% polyaromatics and 20% polysaccharides. One characteristic of suberin is a lamellar structure (**Figure 3.35**), as seen in the suberized surface of potatoes, the cork cell wall, and in green lint cotton fibers. In the cork cell wall, lamellar suberin is located in the secondary wall between the primary and tertiary cell wall layers. Suberinized walls are usually less than 1 μm thick, with the lamellar suberin layer accounting for most of its thickness. The lamellae are composed of regular translucent layers (~3 nm thick) and less regular opaque layers (~8 nm thick). Cork cell walls contain suberin layers consisting of 30 to 60 lamellae. Current models of the suberin lamellae propose that they consist of planar sheets of ordered aliphatic polyester molecules composed of fatty acids, which are flanked by functional groups and cross-links (**Figure 3.36**). These structural models are

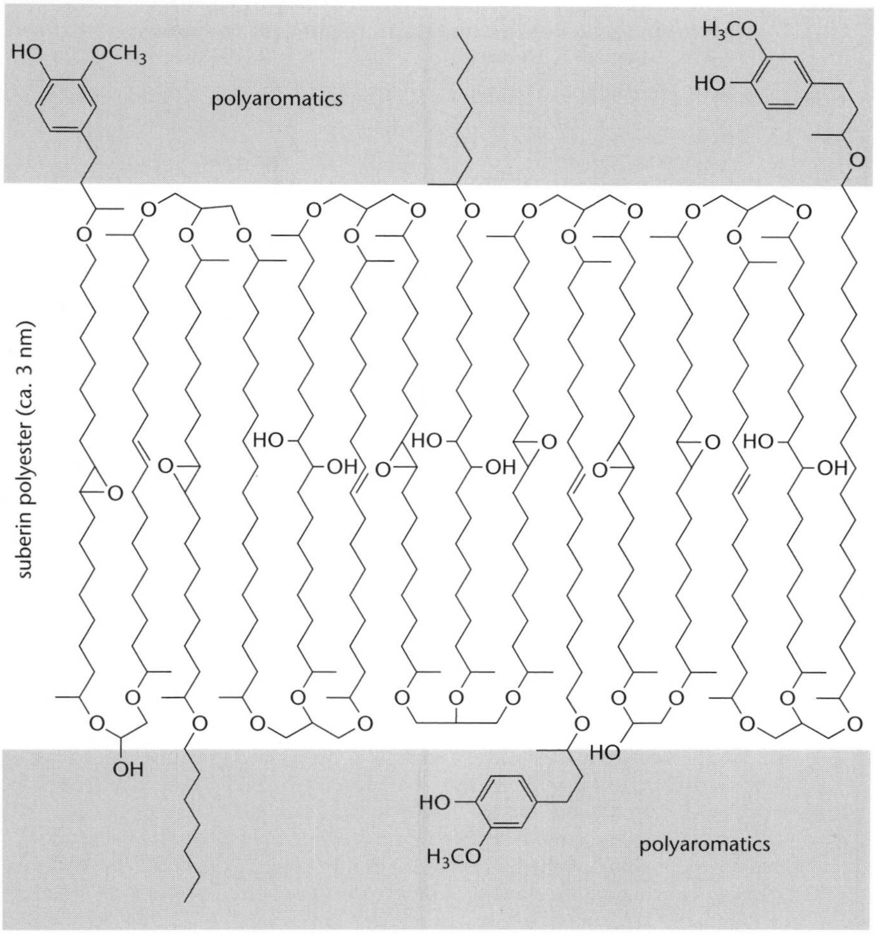

Figure 3.36 Molecular model of a suberin polyester lamella. (Adapted from J. Graça and S. Santos, *Macromol. Biosci.* 7:128–135, 2007.)

based on the monomeric composition of suberin, on the structure of oligomers recovered from partial depolymerization, and on solid-state NMR data.

Suberin is typically isolated from plant tissues by solvent extraction (dichloromethane, ethanol, water) followed by enzymatic hydrolysis with cellulases, hemicellulases, and pectinases. Purified polymeric suberin extracted from cork in this manner has an average molecular mass of 2025 g/mol. Polymeric suberin derived by alkali treatments contains a mixture of acidic and alcoholic chains with a molecular mass average of 750 g/mol.

The most diagnostic feature of suberins is the high content of monomeric aliphatic molecules (**Figure 3.37**), with C_{18} omega (ω) hydroxy acids and

Figure 3.37 Principal monomeric building blocks of suberin and cutin. (Adapted from M. Pollard et al., *Trends Plant Sci.* 13:236–246, 2008.)

	suberin	cutin
unsubstituted fatty acids 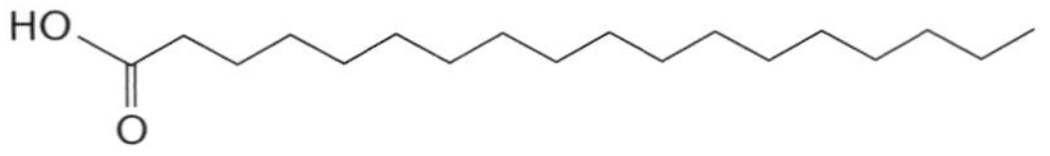	1–10% C18:0 to C24:0	1–25% C16:0, C18:0, C18:1, C18:2
ω-hydroxy fatty acids 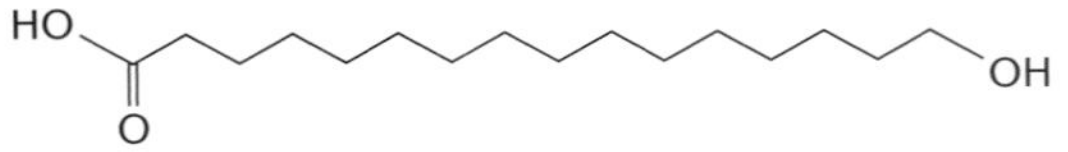	11–43% C18:1, C16:0 to C26:0	1–32% C16:0, C18:1, C18:2
α, ω-dicarboxylic acids 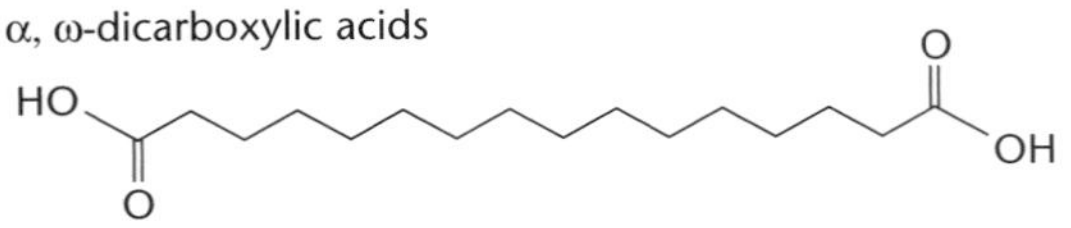	24–45% C18:1, C18:2, C16:0 to C26:0	5–50% C16:0, C18:0, C18:1, C18:2
epoxy fatty acids 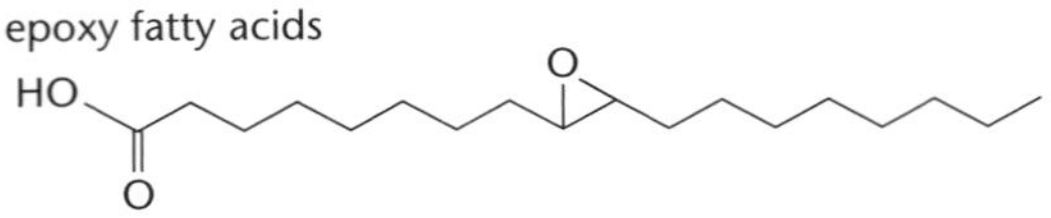	0–30% C18:1 (9, 10-epoxy-18-hydroxy) C18:0 (9, 10-epoxy-1, 18-diacid)	0–34% C18:0 (9, 10-epoxy) C18:1 (9, 10-epoxy)
polyhydroxy fatty acids 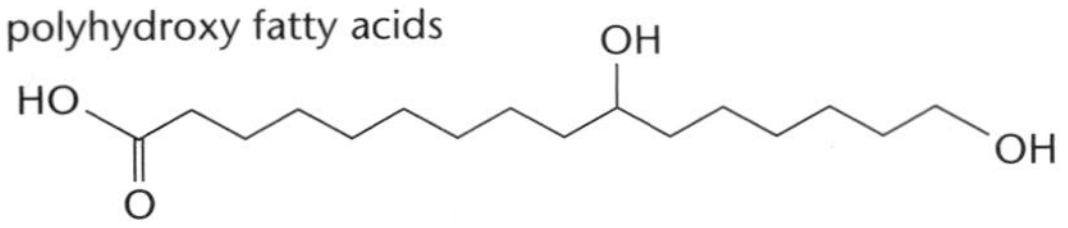	0–2% C18:0 (9, 10, 18-trihydroxy)	16–92% C16:0 (10, 16-dihydroxy) C18:0 (9, 10, 18-trihydroxy)
polyhydroxy α ω-dicarboxylic acids 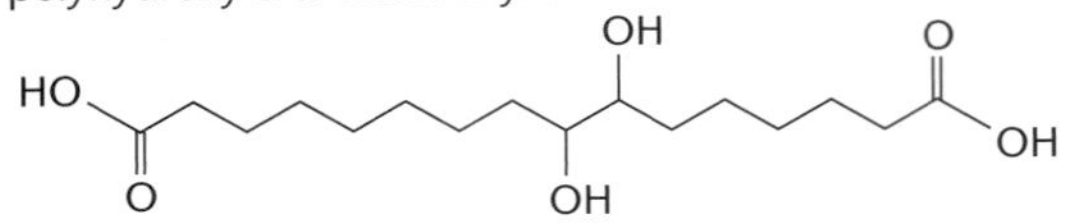	0–8% C18:0 (9, 10-dihydroxy)	traces
alkan-1-ols and alken-1-ols 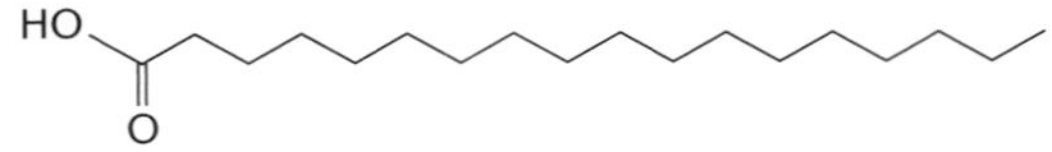	1–6% C18:0 to C22:0	0–8% C16:0, C18:1
α,ω-alkanediols and α,ω-alkenediols 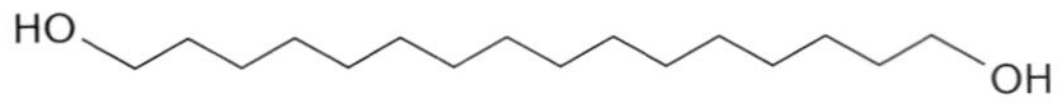	0–3% C22:0	0–5% C18:1
glycerol 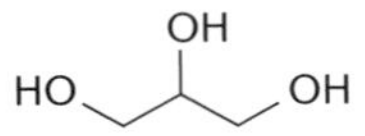	14–26%	1–14%
ferulic acid 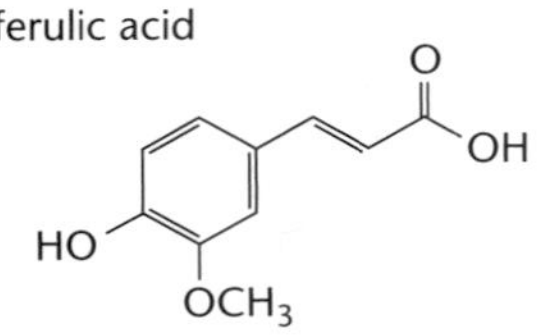	0–10% ferulate, smaller amounts of coumarate, sinapate, caffeate	0–1% ferulate

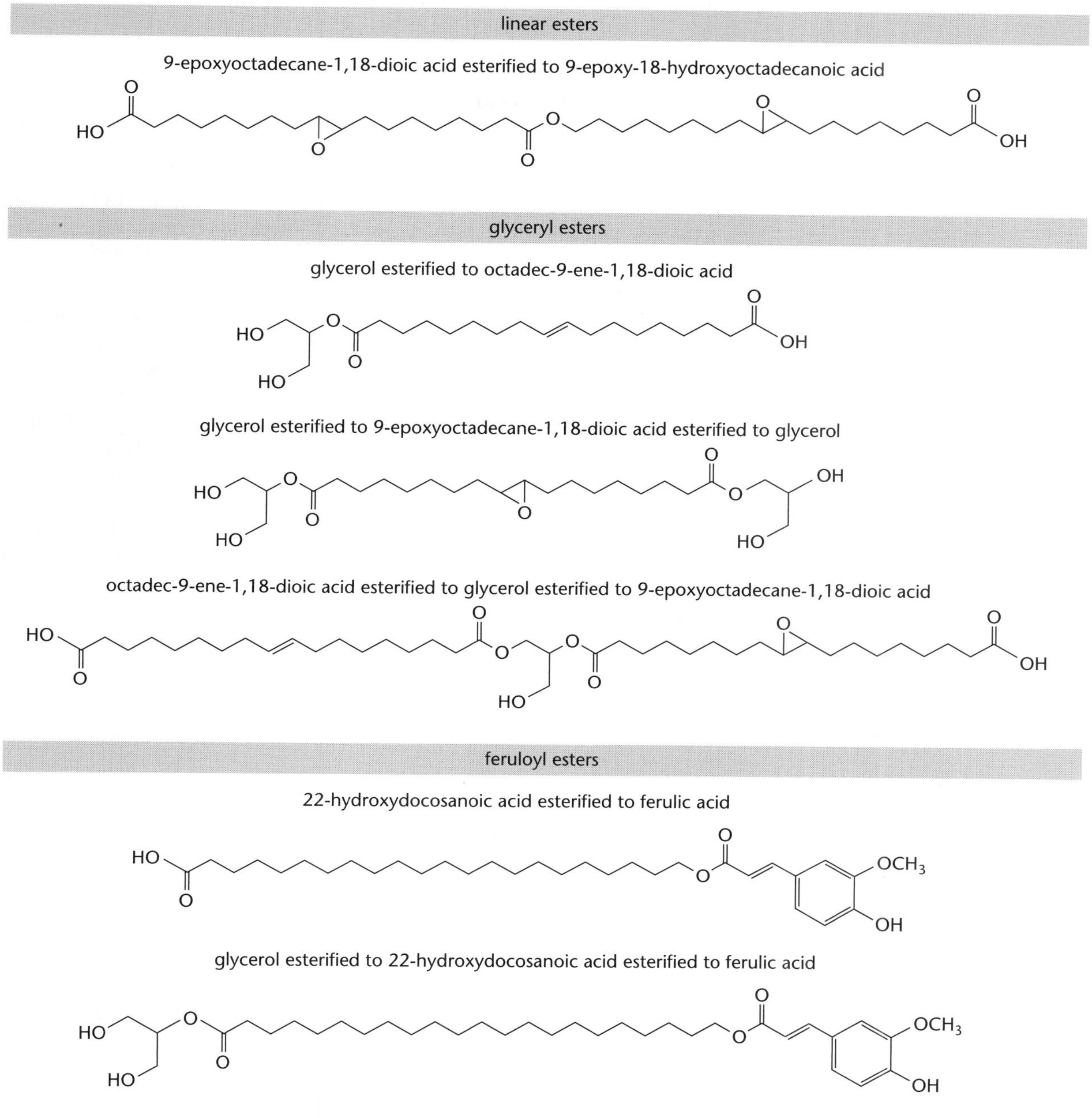

Figure 3.38 Oligomeric degradation products of cork suberin. (Adapted from J. Graça and S. Santos, *Macromol. Biosci.* 7:128–135, 2007.)

corresponding alcohols being the predominant monomer classes. Glycerol and glycerol esterified to cinnamic acids are also components of suberin. Partial depolymerization of cork suberin has resulted in the purification and characterization of oligomeric blocks, some of which are shown in **Figure 3.38**.

The phenolic components of potato periderm suberin have been characterized by thioacidolysis, which is a degradative method used to study the composition of lignin (see Concept 3F2). The phenolic material of suberin resembles the lignins of woody angiosperms in that it contains both guaiacyl (G) and syringyl (S) subunits. The yield of phenolic subunits was about one-tenth of that obtained from a corresponding amount of wood or straw cells, possibly reflecting the low proportion of phenylpropanoids in suberin. Evidence has also been obtained for ether-linked ferulic acid amides such as feruloyltyramine in potato suberin.

2. Cutin is an aliphatic polyester that is part of the plant cuticle.[ref40]

Cutin is found embedded with waxes in the cuticle of plants, which is usually an amorphous protective layer that covers leaves and stems of plants. Cutin, the major component of the cuticle, is a polymer composed of hydroxylated and esterified aliphatic acids (see Figure 3.37 and **Figure 3.39**). It is a characteristic polymer of the plant epidermis and is found on all higher land plants, mosses, and liverworts, and, curiously, even on some aquatic plants. It is also present on fruits and seed coats. It serves as a barrier to moisture loss and infection by microorganisms.

Cutin is attached to the epidermal cell wall by a pectinaceous layer and can be purified chemically or enzymatically. Relatively pure preparations of cutin are prepared by successive enzymatic treatments to remove cellulose, pectins, and hemicelluloses. The remaining, branched and esterified polymers are composed of a complex mixture of aliphatic acids but have a small number of predominant subunits, often hydroxylated derivatives of hexadecanoic acid (C_{16}) and octadecanoic acid (C_{18}), varying in the number and location of hydroxy or epoxy groups. Small amounts of phenolic acids, such as ferulic acid, are associated with cutin that may be involved in interactions with other components of the cell wall. The aliphatic component of cutin is similar to suberin and is difficult to distinguish.

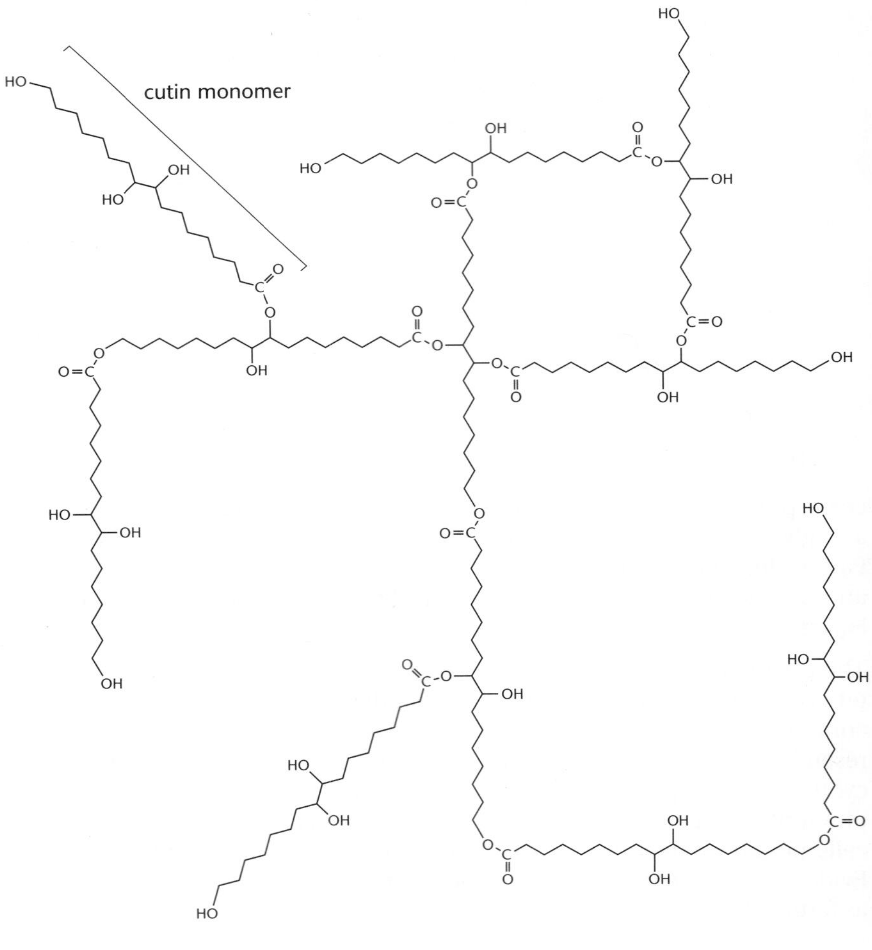

Figure 3.39 Schematic diagram of a cutin network.

TABLE 3.5 SOME OF THE TYPES OF ALIPHATIC COMPONENTS FOUND IN EPICUTICULAR WAXES Subscript numbers represent total backbone carbons; for example, C_{19}–C_{37}.		
Hydrocarbons	CH_3-$(CH_2)_n$-CH_3	C_{19}–C_{37}
Wax esters	CH_3-$(CH_2)_n$-CO-O-$(CH_2)_m$-CH_3	C_{30}–C_{60}
Primary alcohols	CH_3-$(CH_2)_n$-CH_2-OH	C_{12}–C_{36}
Fatty acids	CH_3-$(CH_2)_n$-COOH	C_{12}–C_{36}
Aldehydes	CH_3-$(CH_2)_n$-CHO	C_{14}–C_{34}
Ketones	CH_3-$(CH_2)_n$-CO-$(CH_2)_m$-CH_3	C_{25}–C_{33}
β-Diketones	CH_3-$(CH_2)_n$-CO-CH_2-CO-$(CH_2)_m$-CH_3	C_{27}–C_{35}
Secondary alcohols	CH_3-$(CH_2)_n$-CHOH-$(CH_2)_m$-CH_3	C_{21}–C_{33}

3. Epicuticular waxes form the primary interface between the outside of the plant and the aerial environment.[ref41]

Epicuticular waxes, derived from long-chain fatty acids, serve as a hydrophobic barrier to moisture loss from the plant as a barrier to gases, herbivorous insects, and pathogen attack; and as a defense against chemicals in the environment. The morphology of the wax layers, which can be smooth or organized, also affects the physical and biological interactions of the environment with the surface of the plant (see Figure 1C1.1). Natural wax is crystalline and has a similar morphology in the waxy surface of the cuticle to that of the purified and recrystallized material (see Figure 1C1.2).

The waxes vary in composition with species and tissues (**Table 3.5**) and are synthesized from elongated fatty acids in multienzyme complexes and transported by lipid transfer proteins to the surface of the cuticle, where the epicuticular wax deposits are formed. Synthesis of wax follows several steps, including elongation of fatty acids by addition of two carbon units to produce units of at least C_{26} (see Concept 5E14). The acids are reduced to the corresponding aldehydes and alcohols, and decarbonylation of the aldehydes produces alkanes. Alkanes are sequentially oxidized to secondary alcohols and ketones (Table 3.5).

Gas chromatography (GC) and mass spectrometry (MS) are the established methods for characterization of chemical composition of waxes. Normal epicuticular wax in juvenile leaves of maize has long-chain primary alcohols as the major constituent, with lesser amounts of aldehydes and esterified alcohols (C_{32} and C_{33}). Arabidopsis has shorter alkanes, ketones, and primary alcohols (C_{28} and C_{29}) as its primary constituents. Many mutations in the pathway of epicuticular wax biosynthesis have been identified in grasses and Arabidopsis. Mutations can be readily detected by the surface appearance of stems or leaves. 85 genes have been identified that affect the wax of barley. Fewer mutant genes have been identified in Arabidopsis and maize, but these mutations have been studied more extensively.

4. Silica deposits are formed on the surface of epidermal cells as well as within the cell walls of internal tissues of many herbaceous and woody plants.[ref42]

In herbaceous plants, *silica* is frequently deposited in the leaf epidermis or the cuticle associated with epicuticular wax. Silica is abundant in many grasses (**Figure 3.40**), in many woody plants, and at very high levels in the primitive horsetail, *Equisetum* (**Figure 3.41**). Many trees, including both temperate and tropical species, have silica, both in bark and in wood. The

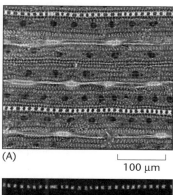
(A) |— 100 μm —|

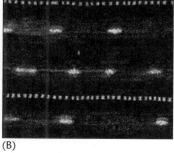
(B)

(C)

Figure 3.40 Silica on the surface of a rice leaf. (A) Scanning electron micrograph of the surface of a rice leaf. (B) The corresponding area is shown imaged with an energy-dispersive X-ray spectroscopy attachment to the microscope. This can be used to specifically identify a signal from a single element, in this case, silicon, and determine its position. The white areas are rich in silicon and appear over two special rows of hourglass, or dumbbell-shaped, cells (C), which lie over the veins, and also in more dispersed cells called bulliform cells. (Adapted from N. Yamaji, N. Mitani and J.F. Feng Ma, *Plant Cell* 20:1381–1389, 2008.)

50 µm

Figure 3.41 Silica bodies on an *Equisetum* leaf surface. A pair of stomatal guard cells is seen in this scanning electron micrograph, which, along with the neighboring epidermal cells, is encrusted with regular small bright silica bodies. (From H.A. Currie and C.C. Perry, *Ann. Bot.* 100:1383–1389, 2007.)

amount of silica in plants can vary between 0.1% and 10% of their above-ground weight. *Silica bodies* are discrete deposits of dehydrated silica on epidermal cells or parenchyma. Silica can take many forms: amorphous or crystalline, fibrillar, globular, or extended sheets. The gross texture and micro-roughness of plant leaves and surfaces may depend upon the abundance, distribution, and structure of the silica. Silica is extremely stable. Particles of plant-derived silica are known as phytoliths, which have a distinctive morphology that can be used to infer the taxa of plant material, a feature valuable for archeology, forensics, and soil science. Plant-derived silica is also known as biogenic opal.

Silica protects many different plant species from a variety of biotic and abiotic stresses. In grasses, it is a defense against herbivory and is present in diverse monocot families, including the cereals, bamboos, and orchids. Silica helps protect rice from a number of important fungal pathogens. Silica also accumulates in the leaves of soybean and in cucumbers, where it protects against fungal infection. In other species, insect damage due to the Asian corn borer is reduced as silica is increased; similarly, higher silica in sorghum reduces damage from shootflies. Silica can be differentially or uniformly distributed in the epidermis. In ryegrass, silica is located specifically on the epidermal walls of the leaf edges and trichomes, while tufted hair grass has silica bodies in all epidermal cells. Modifications of content or distribution of silica may affect other components of the plant epidermis. In sorghum, mutations affecting extracuticular wax also affect silica associated with the wax and the cuticle.

One of the traits associated with the domestication of maize from its wild ancestor teosinte is a softening of the glume, a tissue associated with the kernels. The glume epidermis of teosinte has more silica than maize. The domestication of maize has selected for a reduction in silica, which has contributed in part to an improvement in food quality. Silica-rich plant material is often used as a natural abrasive. Leaves of *Equisetum* and of the tropical plant mukunoki are traditionally used to polish wood products such as ebony. In woody tissues, silica consists of aggregate grains lying free in the lumen of tracheids, vessel elements, or fibers, or frequently in ray parenchymal cells, where they may be associated with polyphenols. Many tropical woods have high silica contents that increase their density and also render them unsuitable for sawn wood production.

Somehow silica has to be taken up from the soil and transported to the leaf surface, where it is deposited into, and on, the cell wall as a polymer of hydrated amorphous silica, forming silica-cuticle sandwiches or being deposited as discrete silica bodies. Some of the transporters, which are required to move silica through the symplast and into the xylem for transport to the leaves, have been characterized. In rice, they are members of a subgroup of the aquaporin family. LS1 is an influx transporter, carrying the silica from the wall into the endodermal cells, while LS2 is an efflux transporter, moving the silica out of the endodermis and into the xylem, where it can move in the transpiration stream to the leaves. These transporters bypass the casparian band, which prevents apoplastic solutions moving beyond the endodermis. Once in the leaves another transporter, LS6, allows the silica to leave the xylem again and enter the xylem parenchyma cells, from where it is delivered symplastically to the leaf epidermal cells that will use it.

Key Terms

callose
cellulose I
cellulose II
cellulose-hemicellulose network
cellulose microfibril
compression wood
cutin
ectopic lignification
endo-β-1,4-glucanase
epicuticular wax
extensin
gelatinous layer (G layer)
glycomodule
guaiacyl unit (G unit)
hydroxyphenyl unit (H unit)
lignin-carbohydrate complex

lipid transfer protein
Mäule reaction
monolignol
nitrobenzene oxidation
parallel configuration
reaction wood
silica
silica body
solvolysis
suberin
syringyl unit (S unit)
tension wood
thioacidolysis
Weisner reaction
wound periderm

References

[1] **β-Glucan chains hydrogen bond together to form cellulose microfibrils.**

Brown RM Jr (2003) Cellulose structure and biosynthesis: what is in store for the 21st century? *J. Polymer Sci.* 42, 487–495.

Franz G & Blaschek W (1990) Cellulose. *Methods Plant Biochem.* 2, 291–317.

Gardner KH & Blackwell J (1974) The hydrogen bonding in native cellulose. *Biochem. Biophys. Acta* 343, 232–240.

[2] **The glucan chains of cellulose microfibrils all have the same orientation.**

Franz G & Blaschek W (1990) Cellulose. *Methods Plant Biochem.* 2, 291–317.

Gardner KH & Blackwell J (1974) The structure of native cellulose. *Biopolymers* 13, 1975–2001.

Sarko A & Muggli R (1974) Packing analysis of carbohydrates and polysaccharides III: valonia cellulose and cellulase II. *Macromolecules* 7, 486–494.

[3] **Xyloglucan is a principal hemicellulose of cell walls.**

Kiefer LL, York WS, Darvill AG & Albersheim P (1989) Xyloglucan isolated from suspension-cultured sycamore cell walls is *O*-acetylated. *Phytochemistry* 28, 2105–2107.

Scheller HV & Ulvskov (2010) Hemicelluloses. *Annu. Rev. Plant Biol.* 61:10.1–10.27. (doi: 10.1146/annurev-arplant-042809-112315)

Sims IM, Munro SLA, Currie G et al. (1996) Structural characterization of xyloglucan secreted by suspension-cultured cells of *Nicotiana plumbaginifolia*. *Carbohydr. Res.* 293, 147–172.

Vierhuis E, York WS, Kumar Kolli VS et al. (2001) Structural analyses of two arabinose containing oligosaccharides derived from olive fruit xyloglucan: XXSG and XLSG. *Carbohydr. Res.* 332, 285–297.

Watanabe T, Shida M, Murayama T et al. (1984) Xyloglucan in cell walls of rice hull. *Carbohydr. Res.* 129, 229–242.

York WS, Impallomeni G, Hisamatsu M et al. (1995) Eleven newly characterized xyloglucan oligoglycosyl alditols: the specific effects of sidechain structure and location on [1]H-n.m.r. chemical shifts. *Carbohydr. Res.* 267, 79–104.

York WS, Kumar Kolli VS, Orlando R et al. (1996) The structures of arabinoxyloglucans produced by solanaceous plants. *Carbohydr. Res.* 285, 99–128.

[4] **The structural features of xyloglucan have been defined by analysis of *endo*-glucanase-released oligosaccharides.**

Hayashi T (1989) Xyloglucans in the primary cell wall. *Annu. Rev. Plant Physiol. Plant Mol. Biol.* 40, 139–168.

Hisamatsu M, York WS, Darvill AG & Albersheim P (1992) Characterization of seven xyloglucan oligosaccharides containing from seventeen to twenty glycosyl residues. *Carbohydr. Res.* 227, 45–71.

York WS, Impallomeni G, Hisamatsu M et al. (1995) Eleven newly characterized xyloglucan oligoglycosyl alditols: the specific effects of sidechain structure and location on [1]H-n.m.r. chemical shifts. *Carbohydr. Res.* 267, 79–104.

[5] **Side chain substitutions on the backbone of xyloglucan affect *endo*-glucanase cleavage of the backbone.**

Hisamatsu M, York WS, Darvill AG & Albersheim P (1992) Characterization of seven xyloglucan oligosaccharides containing from seventeen to twenty glycosyl residues. *Carbohydr. Res.* 227, 45–71.

York WS, Impallomeni G, Hisamatsu M et al. (1995) Eleven newly characterized xyloglucan oligoglycosyl alditols: the specific effects of sidechain structure and location on [1]H-n.m.r. chemical shifts. *Carbohydr. Res.* 267, 79–104.

[6]Arabinoxylan is the predominant hemicellulose of grasses.

Carpita NC & Gibeaut DM (1993) Structural models of primary cell walls in flowering plants: consistency of molecular structure with the physical properties of walls during growth. *Plant J.* 3, 1–30.

Darvill JE, McNeil M, Darvill AG & Albersheim P (1980) Structure of plant cell walls XI: glucuronoarabinoxylan, a second hemicellulose in primary cell walls of suspension-cultured sycamore cells. *Plant Physiol.* 66, 1135–1139.

Dervilly-Pinel G, Rimsten L, Saulnier L et al. (2001) Water-extractable arabinoxylan from pearled flours of wheat, barley, rye and triticale: evidence for the presence of ferulic acid dimers and their involvement in gel formation. *J. Cereal Sci.* 34, 207–214.

Ishii T & Hiroi T (1990) Isolation and characterization of feruloylated arabinoxylan oligosaccharides from bamboo shoot cell-walls. *Carbohydr. Res.* 196, 175–183.

Scheller HV & Ulvskov (2010) Hemicelluloses. *Annu. Rev. Plant Biol.* 61:10.1–10.27. (doi: 10.1146/annurev-arplant-042809-112315)

Selvendran RR & O'Neill M (1987) Isolation and analysis of cell walls from plant material. *Methods Biochem. Anal.* 32, 25–153.

[7]The elucidation of arabinoxylan structures has been facilitated by characterizing oligosaccharides generated by *endo*-xylanases.

Brogerg A, Thomsen KK & Duus JØ (2000) Application of nano-probe NMR for structure determination of low nanomole amounts of arabinoxylan oligosaccharides fractionated by analytical HPAEC-PAD. *Carbohydr. Res.* 328, 375–382.

Hoffmann RA, Geijtenbeek T, Kamerling JP & Vliegenthart JFG (1992) ^{1}H-NMR study of enzymically generated wheat-endosperm arabinoxylan oligosaccharides: structures of hepta- to tetradeca-saccharides containing two or three branched xylose residues. *Carbohydr. Res.* 223, 19–44.

Kato Y & Nevins DJ (1984) Enzymic dissociation of *Zea* shoot cell wall polysaccharides IV: dissociation of glucuronoarabinoxylan by purified *endo*-(1→4)-β-xylanase from *Bacillus subtilis*. *Plant Physiol.* 75, 759–765.

Viëtor RJ, Hoffmann RA, Angelico SAGF et al. (1994) Structures of small oligomers liberated from barley arabinoxylans by xylanase from *Aspergillus awamori*. *Carbohydr. Res.* 254, 245–255.

[8]Secondary walls contain xylans and glucomannans.

Eriksson K-EL, Blanchette RA & Ander P (1990) Biodegradation of hemicelluloses. In Microbial and Enzymatic Degradation of Wood and Wood Components (K-EL Eriksson, RA Blanchette, P Ander eds), pp 181–222. Springer –Verlag, Berlin.

Lundqvist J, Teleman A, Junel L et al. (2002) Isolation and characterization of galactoglucomannan from spruce (*Picea abies*). *Carbohydr. Polymers* 48, 29–39.

Scheller HV & Ulvskov (2010) Hemicelluloses. *Annu. Rev. Plant Biol.* 61:10.1–10.27. (doi: 10.1146/annurev-arplant-042809-112315)

Shimizu K (1991) Chemistry of hemicelluloses. In Wood and Cellulosic Chemistry (DN-S On and N Shiraishi eds), pp 177–214. Marcel Dekker, NY.

[9]A galacturonic acid–containing oligosaccharide is found at the reducing end of glucuronoxylans.

Andersson S-I, Samuelson O, Ishihara M & Shimizu K (1983) Structure of the reducing end-groups in spruce xylan. *Carbohydr. Res.* 111, 283–288.

Johansson MH & Samuelson O (1977) Reducing end groups in birch xylan and their alkaline degradation. *Wood Sci. Technol.* 11, 251–263.

Peña MJ, Zhong R, Zhou G-K et al. (2007) *Arabidopsis irregular xylem8* and *irregular xylem9*: implications for the complexity of glucuronoxylan biosynthesis. *Plant Cell* 19, 549–563.

Persson S, Caffal KH, Freshour G et al. (2007) The *Arabidopsis irregular xylem8* mutant is deficient in glucuronoxylan and homogalacturonan, which are essential for secondary cell wall integrity. *Plant Cell* 19, 237–255.

[10]Homogalacturonans are gel-forming polysaccharides.

Mort AJ, Qiu F & Maness NO (1993) Determination of the pattern of methyl esterification in pectin: distribution of contiguous non-esterified residues. *Carbohydr. Res.* 247, 21–35.

Powell DA, Morris ER, Gidley MJ & Rees DA (1982) Conformations and interactions of pectins II: influence of residue sequence on chain association in calcium pectate gels. *J. Mol. Biol.* 155, 517–531.

Ridley BL, O'Neill MA & Mohnen D (2001) Pectins: structure, biosynthesis, and oligogalacturonide-related signaling. *Phytochemistry* 57, 929–967.

Thakur BR, Singh RK & Handa AK (1997) Chemistry and uses of pectin: a review. *Crit. Rev. Food Sci. Nutrition* 37, 47–73.

Willats WGT, Orfila C, Limberg G et al. (2001) Modulation of the degree and pattern of methyl-esterification of pectic homogalacturonan in plant cell walls: implications for pectin methyl esterase action, matrix properties, and cell adhesion. *J. Biol. Chem.* 276, 19404–19413.

[11]Rhamnogalacturonan I (RG-I) is a family of large polysaccharides with a backbone of repeating disaccharide units.

Ishii T (1995) Isolation and characterization of acetylated rhamnogalacturonan oligomers liberated from bamboo shoot cell walls by Driselase. *Mokuzai Gakkaishi* 41, 561–572.

Komalavilas P & Mort AJ (1989) The acetylation at *O*-3 of galacturonic acid in the rhamnose-rich portion of pectins. *Carbohydr. Res.* 189, 261–272.

Lau JM, McNeil M, Darvill AG & Albersheim P (1985) Structure of the backbone of rhamnogalacturonan I, a pectic polysaccharide in the primary cell walls of plants. *Carbohydr. Res.* 137, 111–125.

O'Neill M, Albersheim P & Darvill A (1990) The pectic polysaccharides of primary cell walls. *Methods Plant Biochem.* 2, 415–441.

[12]Rhamnogalacturonan I (RG-I) has many arabinosyl and galactosyl residue–containing side chains.

An J, O'Neill MA, Albersheim P & Darvill AG (1994) Isolation and structural characterization of β-D-glucosyluronic acid and 4-O-methyl β-D-glucosyluronic acid-containing oligosaccharides from the cell-wall pectic polysaccharide, rhamnogalacturonan I. *Carbohydr. Res.* 252, 235–243.

Carpita NC & Gibeaut DM (1993) Structural models of primary cell walls in flowering plants: consistency of molecular structure with the physical properties of walls during growth. *Plant J.* 3, 1–30

O'Neill M, Albersheim P & Darvill A (1990) The pectic polysaccharides of primary cell walls. *Methods Plant Biochem.* 2, 415–441.

Selvendran RR & O'Neill M (1987) Isolation and analysis of cell walls from plant material. *Methods Biochem. Anal.* 32, 25–153.

[13]Rhamnogalacturonan II (RG-II) is the most complex polysaccharide known.

Darvill AG, McNeil M & Albersheim P (1978) Structure of plant cell walls VIII: a new pectic polysaccharide. *Plant Physiol.* 62, 418–422.

Kobayashi M, Ohno K & Matoh T (1997) Boron nutrition of cultured tobacco BY-2 cells II: characterisation of the boron-polysaccharide complex. *Plant Cell Physiol.* 38, 676–683.

O'Neill MA, Warrenfeltz D, Kates K et al. (1996) Rhamnogalacturonan II, a pectic polysaccharide in the walls of growing plant cells, forms a dimer that is covalently cross-linked by a borate ester. *In vitro* conditions for the formation and hydrolysis of the dimer. *J. Biol. Chem.* 271, 22923–22930.

Pellerin P, O'Neill MA, Pierre C et al. (1997) Lead present in wines is bound to the dimers of rhamnogalacturonan II, a pectic polysaccharide from the grape berry cell walls. *J. Int. Sci. Vigne Vin* 31, 33–41.

[14]Rhamnogalacturonan II (RG-II) has an oligogalacturonide backbone.

O'Neill MA, Ishii T, Albersheim P & Darvill A (2004) Structure and function of a borate cross-linked cell wall pectic polysaccharide. *Annu. Rev. Plant Biol.* 55, 109–139.

Whitcombe AJ, O'Neill MA, Steffan W et al. (1995) Structural characterization of the pectic polysaccharide, rhamnogalacturonan-II. *Carbohydr. Res.* 271, 15–29.

[15]Rhamnogalacturonan II (RG-II) has side chains with unique structures.

Matsunaga T, Ishii T, Matsumoto S et al. (2004) Occurrence of the primary cell wall polysaccharide rhamnogalacturonan II in pteridophytes, lycophytes, and bryophytes: implications for the evolution of vascular plants. *Plant Physiol.* 134, 339–351.

O'Neill MA, Eberhard S, Albersheim P & Darvill AG (2001) Requirement of borate cross-linking of cell wall rhamnogalacturonan II for *Arabidopsis* growth. *Science* 294, 846–849.

O'Neill MA, Warrenfeltz D, Kates K et al. (1996) Rhamnogalacturonan II, a pectic polysaccharide in the walls of growing plant cells, forms a dimer that is covalently cross-linked by a borate ester. *In vitro* conditions for the formation and hydrolysis of the dimer. *J. Biol. Chem.* 271, 22923–22930.

Stevenson TT, Darvill AG & Albersheim P (1988) Structural features of the plant cell wall polysaccharide rhamnogalacturonan-II. *Carbohydr. Res.* 182, 207–226.

[16]β-1,3/1,4-(mixed-linked) glucans are found in grasses and horsetail.

Gibeaut DM & Carpita NC (1991) Tracing cell wall biogenesis in intact cells and plants: selective turnover and alteration of soluble and cell wall polysaccharides in grasses. *Plant Physiol.* 97, 551–561.

Kato Y & Nevins DJ (1986) Fine structure of (1→3),(1→4)-β-D-glucan from *Zea* shoot cell walls. *Carbohydr. Res.* 147, 69–85.

Luttenegger DG & Nevins DJ (1985) Transient nature of a (1→3),(1→4)-β-D-glucan in *Zea mays* coleoptile cell walls. *Plant Physiol.* 77, 175–178.

Woodward JR, Phillips DR & Fincher GB (1983) Water soluble (1→3),(1→4)-β-D-glucans from barley (*Hordeum vulgare*) endosperm I: physiochemical properties. *Carbohydr. Polymers* 3, 143–156.

[17]Enzyme-generated oligosaccharides have been used to structurally characterize mixed-linked glucans.

Buliga GS, Brant DA & Fincher GB (1986) The sequence statistics and solution conformation of a barley (1→3),(1→4)-β-D-glucan. *Carbohydr. Res.* 157, 139–156.

Fry SC, Nesselrode BHWA, Miller JG & Mewburn BR (2008) Mixed-linkage (1→3),(1→4)-β-D-glucan is a major hemicellulose of *Equisetum* (horsetail) cell walls. *New Phytologist* 179, 104–115.

Nevins DJ & Kato Y (1984) Enzymic dissociation of *Zea* shoot cell wall polysaccharides II: dissociation of (1→3),(1→4)-β-D-glucans by purified (1→3),(1→4)-β-D-glucan 4-glucanohydrolase from *Bacillus subtilis. Plant Physiol.* 75, 745–752.

Sorensen I, Pettolino FA, Wilson SM et al. (2008) Mixed-linkage (1→3),(1→4)-β-D-glucan is not unique to the Poales and is an abundant component of *Equisetum arvense* cell walls. *Plant J.* 54, 510–521.

Staudte RG, Woodward JR, Fincher GB & Stone BA (1983) Water soluble (1→3),(1→4)-β-D-glucans from barley (*Hordeum vulgare*) endosperm III: distribution of cellotriosyl and cellotetraosyl residues. *Carbohydr. Polymers* 3, 299–312.

[18]Callose is a β-1,3-D-glucan.

Kato Y & Nevins DJ (1985) A (1→3)-β-D-glucan isolated from *Zea* shoot cell wall preparations. *Plant Physiol.* 78, 20–24.

Meikle PJ, Bonig I, Hoogenraad NJ et al. (1991) The location of (1→3)-β-glucans in the walls of pollen tubes of *Nicotiana alata* using a (1→3)-β-glucan specific antibody. *Planta* 185, 1–8.

Rae AL, Harris PJ, Bacic A & Clarke AE (1985) Composition of the cell walls of *Nicotiana alata* Link et Otto pollen tubes. *Planta* 166, 128–133.

Stone BA (1989) Cell walls in plant-microorganism associations. *Aust. J. Plant Physiol.* 16, 5–17.

[19]Substituted galacturonans are found in some cell walls.

Hart DA & Kindel PK (1970) Isolation and partial characterization of apiogalacturonans from the cell wall of *Lemna minor. J. Biochem.* 116, 569–579.

Hart DA & Kindel PK (1970) A novel reaction involved in the degradation of apiogalacturonans from *Lemna minor* and the isolation of apiobiose as a product. *Biochemistry* 9, 2190–2192.

Schols HA, Bakx EJ, Schipper D & Voragen AGJ (1995) A xylogalacturonan subunit present in the modified hairy regions of apple pectin. *Carbohydr. Res.* 279, 265–279.

[20]Galactoglucomannans are found in Solanaceae cell walls and in the growth medium of Solanaceae cultured cells.

Eda S, Akiyama Y, Kato K et al. (1985) A galactoglucomannan from cell walls of suspension-cultured tobacco (*Nicotiana tabacum*) cells. *Carbohydr. Res.* 137, 173–181.

Sims IM & Bacic A (1995) Extracellular polysaccharides from suspension cultures of *Nicotiana plumbaginifolia. Phytochemistry* 38, 1397–1405.

Sims IM, Craik DJ & Bacic A (1997) Structural characterization of galactoglucomannan secreted by suspension-cultured cells of *Nicotiana plumbaginifolia. Carbohydr. Res.* 303, 79–92.

[21]Some seed cell walls contain storage polysaccharides.

Reid JSG & Edwards ME (1995) Galactomannans and other cell wall storage polysaccharides in seeds. In Food Polysaccharides and their Application (AM Stephen ed), pp 155–187. Marcel Dekker, NY.

Reis D, Vian B, Darzens D & Roland J-C (1987) Sequential patterns of intramural digestion of galactoxyloglucan in tamarind seedlings. *Planta* 170, 60–73.

[22]Gums are secreted in response to stress and are not normally considered to be wall components.

Steven AM, Charms SC & Vogt DC (1990) Exudate gums. *Methods Plant Biochem.* 2, 483–514.

Verbeken D, Dierckx S & Dewettinck K (2003) Exudate gums: occurrence, production, and applications. *Appl. Microbiol. Biotechnol.* 63, 10–21.

[23]Plant cell walls contain structural, enzymatic, lipid transfer, signaling, and defense proteins.

Cassab GI (1998) Plant cell wall proteins. *Annu. Rev. Plant Physiol. Plant Mol. Biol.* 49, 281–309.

Knox JP (1997) The use of antibodies to study the architecture and developmental regulation of plant cell walls. *Int. Rev. Cytol.* 171, 79–120.

[24]The repeated sequence motifs of structural proteins define their classes and their physical properties.

Sequence motifs define proline hydroxylation, glycosylation, and intra- and intermolecular cross-linking sites.

Kieliszewski MJ (2001) The latest hype on Hyp-O-glycosylation codes. *Phytochem.* 57:319-323.

Kieliszewski MJ & Shpak E (2001) Synthetic genes for the elucidation of glycosylation codes for arabinogalactan-proteins and other hydroxyproline-rich glycoproteins. *Cell Mol. Life Sci.* 58, 1386–1398.

Kieliszewski MJ, O'Neill M, Leykam JF & Orlando R (1995) Tandem mass spectroscopy and structural elucidation of glycopeptides from a hydroxyproline-rich plant cell wall glycoprotein indicate that contiguous hydroxyproline residues are the major sites of hydroxyproline O-arabinosylation. *J. Biol. Chem.* 270, 2541–2549.

Showalter AM (1993) Structure and function of plant cell wall proteins. *Plant Cell* 5, 9–23.

[25]HRGP and PRP expression is controlled by developmental programs and triggered by external stimuli.

Cassab GI & Varner JE (1987) Immunocytolocalization of extensin in developing soybean seedcoat by immunogold-silver staining and by tissue printing on nitrocellulose paper. *J. Cell Biol.* 105, 2581–2588.

Schopfer P (1996) Hydrogen peroxide-mediated cell-wall stiffening *in vitro* in maize coleoptiles. *Planta* 199, 43–49.

Zimmermann P, Hirsh-Hofmann M, Hennig L & Gruissem W (2004) GENEVESTIGATOR: *Arabidopsis* microarray database and analysis toolbox. *Plant Physiol.* 136, 2621–2632.

[26]Extensin-type HRGPs are rod-shaped molecules that become insolubilized by a peroxide-mediated cross-linking process.

Cannon MC, Terneus K, Hall Q et al. (2008) Self-assembly of plant cell walls requires an extensin scaffold. *Proc. Natl. Acad. Sci. USA* 105, 2226–2231.

Held MA, Kamyab A, Hare M et al. (2004) Di-isodityrosine is the intermolecular cross-link of isodityrosine-rich extensin analogs cross-linked *in vitro. J. Biol. Chem.* 279, 55474–55482.

Stafstrom JP & Staehelin LA (1986) Cross-linking patterns in salt-extractable extensin from carrot cell walls. *Plant Physiol.* 81, 234–241.

[27]Arabinogalactan proteins form a family of complex proteoglycans located at the cell surface, and in the cell wall and intercellular space.

Du H, Clarke AE & Bacic A (1996) Arabinogalactan-proteins: a class of extracellular matrix proteoglycans involved in plant growth and development. *Trends Plant Sci.* 6, 411–414.

Gaspar Y, Johnson KL, McKenna JA et al. (2001) The complex structures of arabinogalactan-proteins and the journey towards understanding function. *Plant Mol. Biol.* 47, 161–176.

Kieliszewski MJ & Lamport DTA (1994) Extensin: repetitive motifs, functional sites, post-translational codes, and phylogeny. *Plant J.* 5, 157–172.

Nothnagel EA (1997) Proteoglycans and related components of plant cells. *Int. Rev. Cytol.* 174, 195–291.

Showalter AM (2001) Arabinogalactan-proteins: structure, expression and function. *Cell Mol. Life Sci.* 58, 1399–1417.

[28]Classical arabinogalactan proteins are GPI-anchored proteoglycans.

Schultz C, Gilson P, Oxley D et al. (1998) GPI-anchors on arabinogalactan proteins: implications for signaling in plants. *Trends Plant Sci.* 3, 426–431.

[29]GRPs are structural cell wall proteins that participate in the assembly of vascular bundles and are present in root cap mucilage.

Keller B (1993) Structural cell wall proteins. *Plant Physiol.* 101, 1127–1130.

Matsuyama T, Satoh H, Yamada Y & Hashimoto T (1999) A maize glycine-rich protein is synthesized in the lateral root cap and accumulates in the mucilage. *Plant Physiol.* 120, 665–674.

Ryser U, Schorderet M, Zhao GF et al. (1997) Structural cell-wall proteins in protoxylem development: evidence for a repair process mediated by a glycine-rich protein. *Plant J.* 12, 97–111.

[30]AGPs, GRPs, HRGPs, and PRPs may have a common evolutionary origin.

Keller B (1993) Structural cell wall proteins. *Plant Physiol.* 101, 1127–1130.

Showalter AM (1993) Structure and function of plant cell wall proteins. *Plant Cell* 5, 9–23.

[31]Lignin is a hydrophobic polymer of secondary cell walls.

Boudet A-M (2003) Toward an understanding of the supramolecular organization of the lignified cell wall. In The Plant Cell Wall. *Annu. Plant Rev.* 8, 155–182.

Eriksson K-EL, Blanchette RA & Ander P (1990) Microbial and Enzymatic Degradation of Wood and Wood Components. Berlin: Springer.

Iqbal M (1990) The Vascular Cambium. New York: John Wiley & Sons.

Lewis NG & Yamamoto E (1990) Lignin: occurrence, biogenesis, and biodegradation. *Annu. Rev. Plant Physiol. Plant Mol. Biol.* 41, 455–496.

Ralph J, Brunow G, Harris PJ et al. (2008) Lignification: Are Lignins Biosynthesized via Simple Combinatorial Chemistry or via Proteinaceous Control and Template Replication. (F Daayf, A. El Hadrami, L. Adam, GM Ballance eds), pp 36-66. Oxford UK: Wiley-Blackwell.

[32]Our knowledge of lignin composition is based on the chemical hydrolysis of wood.

Lin SY & Dence CW (eds) (1992) Methods in Lignin Chemistry. Berlin: Springer.

Lu F & Ralph J (1997) DFRC method for lignin analysis 1: new method for β–aryl ether cleavage: lignin model studies. *J. Agric. Food Chem.* 45, 4655–4660.

[33]Lignins in plant tissues, cells, and cell walls may be detected by histochemical staining or immunochemical localization.

Browning BL (1967) Methods of Wood Chemistry, vol 1, pp 278-230. New York: Wiley Interscience.

Joseleau JP & Ruel K (1997) Study of lignification by non-invasive techniques in growing maize internodes: an investigation by Fourier transform infrared, cross-polarization-magic angle spinning ^{13}C-nuclear magnetic resonance spectroscopy and immunochemical transmission electron microscopy. *Plant Physiol.* 114, 1123–1133.

[34]The complexity of lignin is increased by the diversity of intermolecular linkages.

Boerjan W, Ralph J & Baucher M (2003) Lignin biosynthesis. *Annu. Rev. Plant Biol.* 54, 519–546.

Ralph J, Lundquist K, Brunow G et al. (2004) Lignins: natural polymers from oxidative coupling of 4-hydroxyphenylpropanoids. *Phytochem. Rev.* 3, 29–60

[35]Models for the higher-order structure of lignin.

Brunow G (2001) Methods to reveal the structure of Lignin: In Lignin, Humic Substances and Coal (M Hofrichter, A Steinbuchel eds), *Biopolymer* vol. 1, pp 89–116. Wiley-VCH, Weinheim.

Jurasek L (1995) Toward a three-dimensional model of lignin structure. *J. Pulp Paper Sci.* 21, J274–J279.

Ralph J, Lundquist K, Brunow G et al. (2004) Lignins: natural polymers from oxidative coupling of 4-hydroxyphenylpropanoids. *Phytochem. Rev.* 3, 29–60.

[36]Lignin is cross-linked to polysaccharide components in lignin-carbohydrate complexes.

Balakshin MY, Capanema E & Chang HM (2007) MWL fraction with a high concentration of lignin carbohydrate linkages: isolation and 2D-NMR spectroscopic analysis. *Holzforschung* 61, 1–7.

Grabber JH, Ralph J & Hatfield R (2000) Crosslinking of maize walls by ferulate dimerization and incorporation into lignin. *J. Agric. Food Chem.* 48, 6106–6113.

Lawoko M, Henriksson G, & Gellerstedt G (2005) Structural differences between the lignin-carbohydrate complexes present in wood and in chemical pulps. *Biomacromolecules* 6, 3467–3473.

[37]Lignin composition and content vary greatly among the major groups of higher plants.

Campbell M & Sederoff R (1996) Variation in lignin content and composition: mechanisms of control and implications for the genetic improvement of plants. *Plant Physiol.* 110, 3–13.

Donaldson LA (2001) Lignification and lignin topochemistry: an ultrastructural view. *Phytochemistry* 57, 859–873.

Lewis NG & Yamamoto E (1990) Lignin: Occurrence, biogenesis, and biodegradation. *Annu. Rev. Plant Physiol. Plant Mol. Biol.* 41, 455–96.

Sauter M & Kende H (1992) Levels of β-glucan and lignin in elongating internodes of deepwater rice. *Plant Cell Physiol.* 33, 1089–1097.

[38]Lignin content and composition vary in wood formed during increased mechanical stress.

Joseleau J-P, Imai T, Kuroda K & Ruel K (2004) Detection in situ and characterization of lignin in the G layer of tension wood fibers of *Populus deltoides. Planta* 219, 338–345.

Timmel TE (1986) Compression Wood in Gymnosperms. Heidelberg: Springer.

[39]Suberin is a complex hydrophobic material that forms physical and biological barriers essential for plants.

Franke R & Schreiber L (2007) Suberin: a biopolyester forming apoplastic plant interfaces. *Curr. Opin. Plant Biol.* 10, 252–259.

Kolattukudy PE (2001) Polyesters in higher plants. *Adv. Biochem. Eng. Biotechnol.* 71, 1–49.

Pollard M, Beisson F, Li Y & Ohlrogge JB (2008) Building lipid barriers: biosynthesis of cutin and suberin. *Trends Plant Sci.* 13, 236–246.

[40]Cutin is an aliphatic polyester that is part of the plant cuticle.

Kunst L, Samuels AL & Jetter R (2004) The plant cuticle: formation and structure of epidermal surfaces. In Plant Lipids: Biology, Utilization and Manipulation (DJ Murphy ed), pp 270–302. Oxford: Blackwell.

Nawath C (2006) Unraveling the complex network of cuticular structure and function. *Curr. Opin. Plant Biol.* 9, 281–287.

[41]Epicuticular waxes form the primary interface between the outside of the plant and the aerial environment.

Samuels AL, Kunst L & Jetter R (2008) Sealing plant surfaces: cuticular wax formation by epidermal cells. *Annu. Rev. Plant Biol.* 59, 771–812.

[42]Silica deposits are formed on the surface of epidermal cells as well as within the cell walls of internal tissues of many herbaceous and woody plants.

Currie HA & Perry CC (2007) Silica in plants: biological, biochemical and chemical studies *Ann. Bot.* 100, 1383–1389.

Yamaji N, Mitatni N & Feng Ma JF (2008) A transporter regulating silicon distribution in rice shoots. *Plant Cell* 20, 1381–1389.

Membrane Systems Involved in Cell Wall Assembly

4

The focus of Chapter 4 is on the membrane systems that produce and transport the many different types of polysaccharide, glycoprotein, and proteoglycan molecules found in plant cell walls. It thus sets the stage for Chapter 5, in which a more detailed biochemical discussion of cell wall molecule assembly is presented. Considering the water-soluble nature of these cell wall molecules, it is quite remarkable that they are synthesized by membrane-associated enzymes. The membranes that contain these enzymes are commonly known as the *endomembrane system*, which includes all membranous organelles capable of exchanging membrane molecules either by lateral diffusion through continuous membrane domains or by transport vesicles that bud from one type of membrane and fuse with another. In plant cells, the membrane systems related to each other in this manner include the membranes of the *secretory pathway* (ER, nuclear envelope, Golgi apparatus, *trans* Golgi network, plasma membrane, cell plate membrane, multivesicular bodies, vacuoles, and different types of transport vesicles) and the membranes of the *endocytic pathway* (plasma membrane, cell plate membrane, clathrin-coated vesicles, *trans* Golgi network, multivesicular bodies, vacuoles, and transport vesicles). Trafficking between these compartments not only delivers cargo molecules to the cell surface and cell plates, and proteins to vacuoles, but also membrane proteins and membrane lipids to all of the endomembrane compartments (**Figure 4.1**).

Many different types of sorting, packaging, and targeting molecules are involved in regulating this trafficking and ensuring that the different products are delivered to the proper places. Because of the nature of the sorting, packaging, and trafficking mechanisms, the forward, or anterograde, traffic is always accompanied by backward, or retrograde, traffic to recycle displaced membrane proteins to where they normally operate as well as to remove and degrade old or damaged plasma membrane molecules. In addition to providing an introduction to the architecture, dynamic properties, and functional organization of the membrane systems involved in the synthesis of cell wall molecules, Chapter 4 also details some of the intricate membrane events and transformations that are involved in cell plate formation and cytokinesis in plant cells.

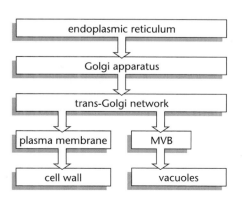

Figure 4.1 The secretory pathway of plants. This pathway produces and transports proteins, glycoproteins, proteoglycans, lipids, and complex polysaccharides for cell walls and cell plates; storage proteins and hydrolytic enzymes for vacuoles; and proteins, glycoproteins, and lipids for membranes of the endomembrane compartments. MVB, multivesicular body.

A. Sites of Cell Wall Polymer Assembly in Interphase Cells

1. Different polymers are synthesized at different locations, in the ER and the Golgi apparatus, at the plasma membrane, or in the cell wall.[ref1]

All cell wall polysaccharides and glycoproteins are synthesized by membrane-bound enzyme systems, but the assembly of each type of polymer takes place in a particular set of membrane compartments: ER and/or Golgi cisternae, or plasma membrane (**Figure 4.2**). Thus, although cell wall glycoproteins and complex polysaccharides are usually transported together in secretory vesicles to the cell surface, the actual sites of their assembly differ. The assembly of the protein backbone of N-*linked glycoproteins* is produced by membrane-bound polyribosomes on the ER, and the addition of the oligosaccharide side chain to specific asparagine residues takes place co-translationally as the nascent polypeptide enters the ER lumen. Processing of the attached *N*-glycans is initiated in the ER and is completed in the *Golgi apparatus*. In contrast, the glycosylation of O-*linked glycoproteins* as well as the assembly of *complex polysaccharides* of the cell wall matrix occurs in different cisternae of the Golgi apparatus.

Unlike the cell wall matrix polymers that are synthesized in the ER and Golgi membrane systems, cellulose, callose, and β-1,3/1,4-(mixed-linked) glucans are produced at the plasma membrane by enzymes that are made and assembled in the endomembrane system but become activated only upon reaching the plasma membrane. Activation of callose synthase in most but not all cases requires calcium; what triggers the activation of cellulose and of mixed-linked glucan synthases in the plasma membrane is not known. All of these *linear glucose polymers* are discharged directly into the cell wall.

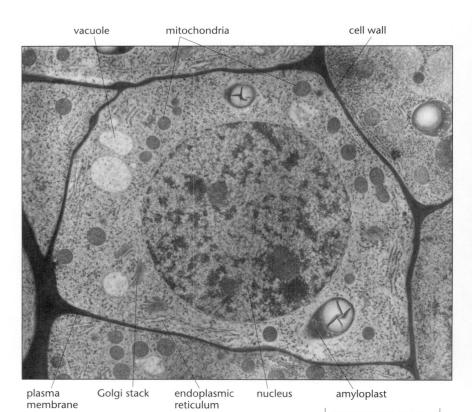

Figure 4.2 Electron micrograph of a meristematic cell in a *Nicotiana* root tip depicting the membrane systems involved in cell wall polymer synthesis: endoplasmic reticulum, Golgi stack, vacuole, and plasma membrane, as well as cell wall, nucleus, mitochondria, and amyloplast. (From L.A. Staehelin et al., *Protoplasma* 157:75–91, 1990.)

It is generally assumed that the precursors of lignin, cutin, suberin, and waxes are transported as monomers from their intracellular sites of synthesis to the extracellular matrix, where they are assembled into polymers. However, we still have only a limited knowledge about how these hydrophobic molecules are synthesized and how they are transported to the cell walls (see Concepts 5E1–14). Lignin precursors are conjugated to sugars to make them more hydrophilic prior to secretion, and there is some evidence to suggest that these precursors are secreted via transporters in the plasma membrane, but definitive cell biological data on lignin trafficking have yet to be obtained. A specific plasma membrane transporter for wax molecules has recently been identified and characterized, but transporters for cutin and suberin subunits have yet to be discovered. When fatty acid–derived cutin subunits arrive at the cell walls of epidermal cells, they become bound to lipid exchange proteins, which then mediate their transport through the extracellular matrix to the cell wall surface, where cuticle assembly takes place.

2. The endoplasmic reticulum consists of a three-dimensional membrane network that extends throughout the cytoplasm and is differentiated into specialized domains.[ref2]

The morphology of the *endoplasmic reticulum (ER)* varies tremendously depending on cell type and stage of development, as well as on the biosynthetic activity of ER-associated enzyme systems. Recent studies have shown that the ER is also subdivided into a large number of functional domains (**Figure 4.3**) including the ribosome-studded, rough ER membranes, which

Figure 4.3 Functional domains of the plant ER. The plant ER serves many different functions, and many of these activities are carried out by morphologically distinct membrane domains. This diagram distinguishes 16 functional domains, but more have been identified. MT, microtubule; PM, plasma membrane; TGN, *trans* Golgi network. (Adapted from L.A. Staehelin, *Plant J.* 11, 1151–1165, 1997.)

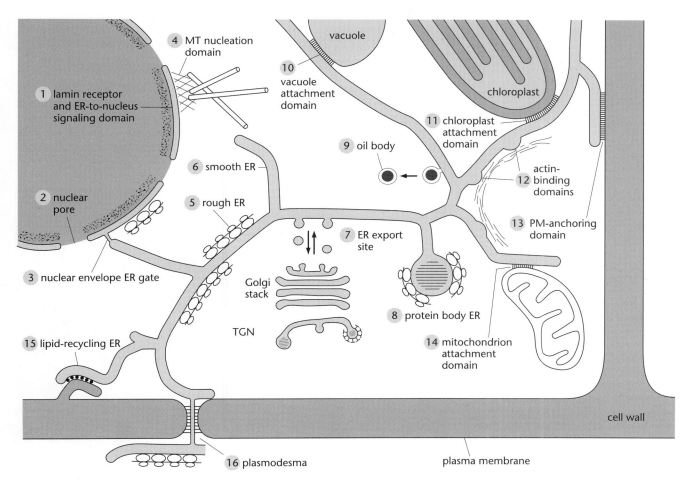

have a sheetlike structure; and the smooth ER membranes, which lack ribosomes and have a tubular conformation. Within each cell, all ER cisternae form a continuous network that is connected to the ER networks of all adjacent cells by means of the plasmodesmata. The ER-bound ribosomes are assembled into polysomes and are involved in the synthesis of proteins, which are first inserted into either the ER membrane or the ER lumen and then delivered to their destinations by transport through the secretory pathway. Targeting to specific compartments of the secretory pathway other than the cell surface or the cell wall is generally achieved through "addressing" or "signaling" domains on the proteins. To this end, ER and Golgi proteins possess *retention* and/or *retrieval sequences* that either prevent them from moving beyond a given compartment or allow them to be retrieved from a downstream compartment, respectively. Soluble vacuolar proteins contain *vacuolar targeting sequences*, which are recognized by *trans* Golgi network (TGN) receptors that "extract" them into specific clathrin-coated vesicles targeted for the vacuoles. Molecules that contain no targeting signals are delivered to the cell surface and cell wall in what is referred to as the *default pathway*. Of the different functional subdomains of the ER highlighted in Figure 4.3, three, the ER export sites (number 7), the lipid recycling ER (number 15), and the ER chloroplast attachment domains (number 11), are discussed in greater detail in later sections (Concepts 4A5, 4A11, and 5E14).

Two major functions of the ER that are important for the synthesis of cell wall glycoproteins are (1) the assembly of the 14-sugar *N*-glycans on the *dolichol* carriers prior to their transfer to specific asparagine residues on the *N*-linked glycoproteins, and (2) the enzymatic conversion of proline residues to hydroxyproline residues so that such proteins (for example, hydroxyproline-rich glycoproteins, arabinogalactan proteins) can be *O*-glycosylated. The assembly of the *O*-linked glycans appears to commence in the ER and to be completed in the Golgi.

3. Golgi stacks consist of sets of flattened cisternae that are structurally distinct and exhibit a polar architecture.[ref3]

Electron micrographs and electron tomographic images of plant *Golgi stacks* preserved by high-pressure freezing/freeze-substitution methods demonstrate that they are comprised of five to seven flattened cisternae and exhibit a distinct structural *cis-trans* polarity (**Figure 4.4** and **Figure 4.5**). This polarity is evident in a variety of morphological parameters, includ-

Figure 4.4 Thin-section electron micrograph of a plant Golgi stack and associated *trans* Golgi network (TGN). COPIa-type vesicles bud from *cis* Golgi cisternae, COPIb-type vesicles from medial and *trans* Golgi and early TGN cisternae, and clathrin-coated (CCV) and secretory (SV) vesicles from TGN cisternae. The small dots and thin lines (arrow) seen between the darkly staining *trans* cisternae correspond to intercisternal elements (see also Figure 4.20). (From L.A. Staehelin et al., *Protoplasma* 157:75–91, 1990.)

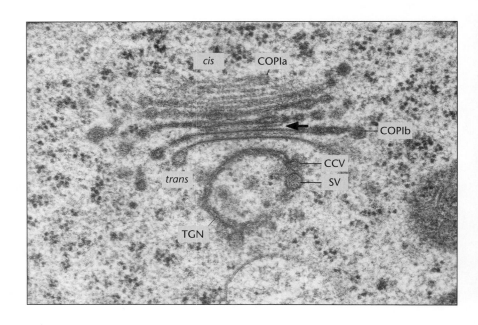

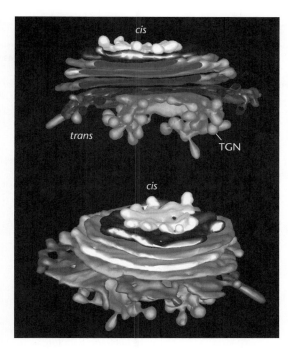

Figure 4.5 Two views of a three-dimensional model of a plant Golgi stack and TGN cisternae reconstructed from 363 electron tomography slice images 2 nm thick. Note the smaller size of the forming *cis*-side cisternae and the numerous budding vesicles on the TGN. (Courtesy of Byung-Ho Kang.)

ing diameter and width of the cisternae, intercisternal spacing, membrane and cargo staining, the types of budding vesicles formed by different cisternae, and the distribution of intercisternal elements. As will be discussed in greater detail later, shedding of the *trans*-most Golgi cisternae gives rise to trans *Golgi network (TGN)* cisternae, which undergo further structural changes as they mature.

Golgi cisternae can be divided into three classes, *cis*, medial, and *trans*, based on their position in the stack as well as their structural attributes and enzymatic functions. Typically, each Golgi stack contains roughly equal numbers of *cis*, medial, and *trans* cisternae. In thin-section electron micrographs, the cis *Golgi cisternae* are often difficult to discern because of their small size, their irregular shape, and the light staining of their luminal contents. *Cis* Golgi cisternae correspond to cisternae that are being assembled from fusing coat protein complex II vesicles, *COPII vesicles*, that arise at *ER export sites* and transport cargo and membrane molecules from the ER to the Golgi (**Figure 4.6**). The function of the *COPIa vesicles* is to recycle escaped ER proteins from the forming *cis* cisternae back to the ER. The lack of luminal staining of the *cis* cisternae is consistent with the idea that *cis* cisternae serve as assembly structures and are not involved in biosynthetic activities.

The transition from *cis* to *medial Golgi cisternae* is marked by the sudden appearance of darkly stained cisternal contents. This seems to reflect the onset of the Golgi-associated biosynthetic functions and the accumulation of sugar nucleotides, glycoproteins, and polysaccharides in the cisternal lumen. Consistent with this interpretation, immunolocalization experiments have shown that in plant Golgi stacks, the enzyme *mannosidase I*, the first enzyme involved in the processing of *N*-glycans, is absent from *cis* cisternae and confined mostly to medial cisternae (**Figure 4.7**).

Distinguishing morphological features of trans *Golgi cisternae* are their osmotically collapsed central luminal domain and the dark staining of the compressed luminal products with osmium. The low pH–dependent osmotic collapse of the *trans* cisternae seems to provide the force for sequestering the polysaccharide products into the budding vesicles at the cisternal margins. The *COPIb vesicles* that bud from the margins of the medial and *trans* Golgi cisternae recycle membrane proteins between medial and *trans* cisternae. In cells producing mucilage-type molecules, the Golgi stacks

Figure 4.6 Diagrammatic model summarizing the sites of origin and trafficking patterns of the five types of vesicles produced by ER, Golgi, and TGN cisternae. Fusion of COPII vesicles gives rise to new *cis* Golgi cisternae. COPIa vesicles recycle lost ER proteins back to the ER; COPIb vesicles recycle medial, *trans* cisterna, and early TGN proteins in a *trans*-to-medial direction. This recycling process recycles 30–35% of the cisternal surface area. Clathrin-coated vesicles collect cargo destined for multivesicular bodies and vacuoles, and secretory vesicles carry cell wall molecules to the cell surface.

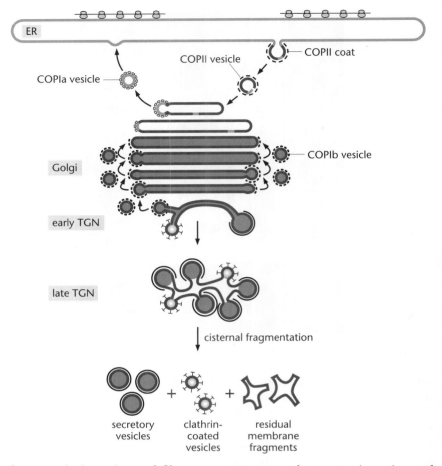

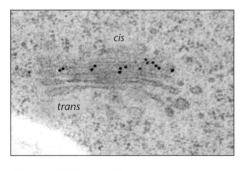

Figure 4.7 Thin-section electron micrograph of an *Arabidopsis* root meristem Golgi stack labeled with anti-mannosidase I–gold antibodies. The immunogold particles are seen exclusively over the medial cisternae, consistent with the hypothesis that the onset of Golgi-associated biosynthetic functions occurs in medial cisternae. (Courtesy of Byung-Ho Kang.)

also contain intercisternal filamentous structures known as *intercisternal elements* (see Figure 4.20). In thin-section electron micrographs these elements appear as rows of small dots and/or thin lines between the darkly staining *trans* cisternae (see Figure 4.4).

TGN cisternae are derived from the *trans*-most Golgi cisternae by a cisternal shedding mechanism, which involves peeling of the *trans*-most Golgi cisterna from the stack (see Figure 4.6). Upon separation, the detached TGN cisterna is converted into clusters of interconnected budding secretory and clathrin-coated vesicles, all of which are released simultaneously when the cisterna fragments. The main functions of the TGN compartment in plants is to complete the biosynthesis of polysaccharides, and to sort and package the products of the Golgi stacks into different vesicles that are targeted to the cell surface, to multivesicular bodies, and/or to vacuoles.

Double immunolabeling studies of plant cells with different combinations of anti-cell wall molecule antibodies have demonstrated that each Golgi stack–TGN unit contains the enzymes needed for the simultaneous assembly of *N*- and *O*-linked glycoproteins, pectic polysaccharides, and hemicelluloses. This observation lends support to the notion that each Golgi stack together with its associated TGN can function as an independent processing and assembly station for glycoproteins and cell wall polysaccharides.

4. Plant Golgi stacks are dispersed throughout the cytoplasm, travel along actin filament tracks with myosin motors, and stop at ER export sites.[ref4]

Unlike animal cells, in which the interconnected Golgi stacks are organized around the cell center and in close association with ER cisternae, the Golgi stacks of most types of plant cells are dispersed as single units or in

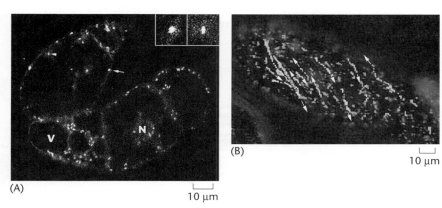

Figure 4.8 Fluorescent micrographs of living tobacco BY-2 cells expressing a Golgi-localized mannosidase I–green fluorescent protein fusion protein. (A) Optical section through the middle of a cluster of cells. The bright, disk-shaped spots correspond to individual Golgi stacks that are dispersed throughout the cytoplasm. Arrow and arrowheads point to the stacks shown in the inset. (B) Tracing of selected Golgi stacks in the cortical cytoplasm over a 30-second period. The lines of the dots correspond to sites of translocating Golgi stacks, and the irregular dots to Golgi stacks that only carried out wiggling motions. V, vacuole; N, nucleus. (From A. Nebenführ et al., *Plant Physiol.* 121:1127–1141, 1999.)

small clusters throughout the cytoplasm (**Figure 4.8**). The number of Golgi stacks per cell varies from about 35 in small *Arabidopsis* meristem cells to up to about 800 in large tobacco BY-2 cells. The evolutionary advantage of having dispersed Golgi stacks in plant cells can be traced to the need for the uniform distribution of matrix molecules to growing cell walls, even in very large and highly vacuolated cells. Furthermore, because about 80% of the secretory products of plant cells are polysaccharides that are assembled entirely in the Golgi stacks, a close, stable relationship between ER and Golgi membranes is less needed than in animal cells, where most of the secretory products are glycoproteins.

The dynamic behavior of Golgi stacks in living plant cells can be investigated with the help of *green fluorescent protein (GFP)*–tagged Golgi proteins (Figure 4.8A and B). Such fusion proteins can be produced by fusing GFP to the C-terminus of type II Golgi processing enzymes such as mannosidase I, which results in the GFP being localized to the lumina of Golgi cisternae in the transgenic cells. Golgi stacks labeled in this manner appear as small bright fluorescent disks whose movement can be followed over time. Analysis of such labeled Golgi stacks in the cortical cytoplasm of tobacco BY-2 suspension-cultured cells, in *Nicotiana* leaf cells, and in different cell types of *Arabidopsis* plants has demonstrated that individual stacks undergo "stop-and-go" movements. When stopped, the stacks carry out wiggling motions. The go phase corresponds to a linear translocation type of movement with top speeds of about 1 μm/s in *Arabidopsis* meristem cells and 4 μm/s in tobacco BY-2 cells. In cells expressing both fluorescently labeled ER export site marker proteins and resident Golgi proteins, the stopping of the Golgi stacks is seen to occur at ER export sites. ER export sites of tobacco BY-2 cells that are not associated with a Golgi stack have a half-life of about 10 s, suggesting that if the budding COPII vesicles produced at such a site are not plucked off by a passing Golgi stack within about 10 s (see Concept 4A5), they are reabsorbed by the ER and new ones are produced at the same or at a different site.

Plant Golgi stack movement is susceptible to drugs such as cytochalasin D or butanedione monoxime, which interfere with actin filament assembly and myosin motor activity, respectively. In contrast, the microtubule-disrupting drug nocodazole, which perturbs Golgi organization and trafficking in animal cells, has no effect on plant Golgi movement. These findings suggest that Golgi stacks travel along *actin tracks* by means of myosin motors.

5. COPII vesicles bud from the ER with an external scaffold that is transferred with the vesicles to the cis side of the Golgi stacks, giving rise to the Golgi scaffold (matrix).[ref5]

As discussed in the preceding concept, plant Golgi stacks are dynamic organelles that move in a stop-and-go fashion along actin tracks by means of myosin motor proteins. This raises the question as to how the cargo-

Figure 4.9 Schematic diagram illustrating the organization of the three types of scaffolds (matrices) that form on and completely encompass COPII vesicles, Golgi stacks and associated vesicles, and TGN cisternae. Note that secretory vesicles bud with a TGN-derived scaffold layer that is also transported to the vesicle destination site (plasma membrane or cell plate).

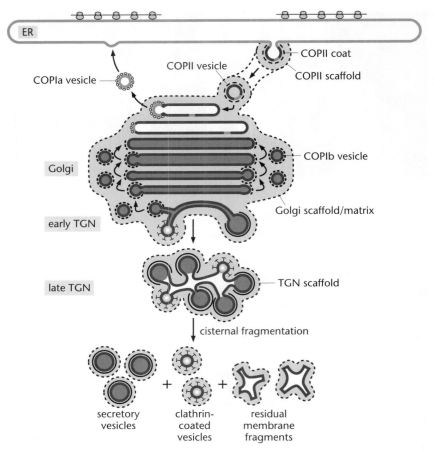

Figure 4.10 Tomographic slice image of a Golgi stack, TGN cisterna, and ER membrane in an alfalfa root meristem cell. Note the absence of ribosomes (*small black dots*) in the immediate surroundings of the Golgi stack and TGN cisterna. This ribosome-free zone corresponds to the Golgi/TGN scaffold (matrix). (Courtesy of Byung-Ho Kang.)

carrying COPII vesicles are transferred from the ER to Golgi stacks. Recent studies have demonstrated that the stopping of Golgi stacks at ER export sites is mediated by a scaffold system that originates on COPII vesicles. In particular, each COPII vesicle is born with about a 40-nm-wide scaffold layer on its surface, the *COPII scaffold* (**Figure 4.9**). This scaffold, which contains a scaffold-forming, long coiled-coil protein called p115, is capable of creating links to the *cis* side of the *Golgi scaffold/matrix* (**Figure 4.10**). This enables the ER export site to "capture" passing Golgi stacks and force them to dock with their *cis* cisternae facing the budding COPII vesicles (**Figure 4.11**). Once the COPII scaffold is firmly connected to the Golgi scaffold, the wiggling motions of the docked Golgi stacks mechanically pluck, and thereby harvest, the COPII vesicles (**Figure 4.12**). The COPII scaffold is then transferred together with the COPII vesicle to the Golgi scaffold,

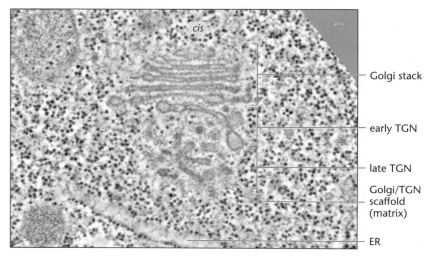

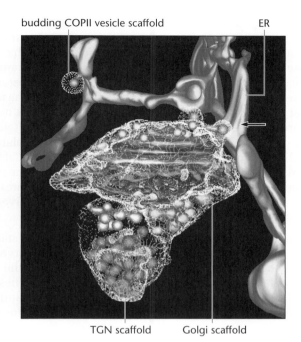

budding COPII vesicle scaffold

ER

TGN scaffold   Golgi scaffold

Figure 4.11 Three-dimensional tomographic model of a Golgi stack that is docked to an ER export site via COPII and Golgi scaffolding molecules. Note the scaffold that surrounds the free, budding COPII vesicle on the left, and the continuity between the COPII scaffold and the cis side of the Golgi scaffold (matrix) on the right (*arrow*). (Courtesy of Byung-Ho Kang.)

where the scaffold proteins become part of the *cis*-most region of the Golgi scaffold, while the COPII vesicle is induced to fuse with one of the *cis*-type cisternae. Transfer of the COPII vesicles and their scaffolds to the Golgi scaffold frees the Golgi stack from the ER export site and allows it to resume trafficking along actin filaments.

6. Intra-Golgi transport of membrane and cargo molecules is best explained by the cisternal progression/maturation model.[ref6]

How the ER-derived products move from the *cis* through the medial to the *trans* Golgi cisternae has been debated for close to 40 years. Now, however, virtually all plant Golgi researchers subscribe to the *cisternal progression/*

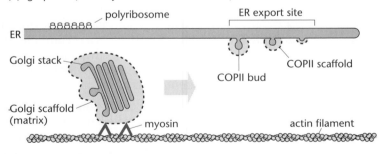

(A) go-phase (saltatory/directional movements)

polyribosome

ER export site

ER

Golgi stack

COPII scaffold

COPII bud

Golgi scaffold (matrix)

myosin

actin filament

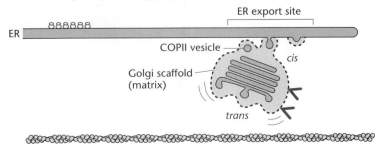

(B) dock-and-pluck phase (wiggling movements)

ER export site

ER

COPII vesicle

cis

Golgi scaffold (matrix)

trans

Figure 4.12 The dock, pluck, and go model of ER-to-Golgi trafficking in plant cells. (A) Go phase: Golgi stacks travel along actin filaments propelled by myosin motors. (B) Dock-and-pluck phase: the COPII scaffold of a budding vesicle attaches to the *cis* side of the Golgi scaffold/matrix and pulls the passing Golgi off the actin track. Once the COPII scaffold binds to the Golgi scaffold, the wiggling motion of the Golgi stack facilitates release of the COPII vesicles by plucking. After the harvesting of the COPII vesicles is complete, the Golgi is free to resume its movement along the actin track. (Adapted from L.A. Staehelin and B.-H. Kang, *Plant Physiol.* 147:1454–1468, 2008.)

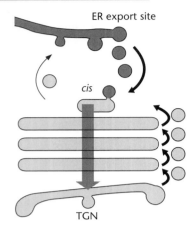

progression/maturation model

ER export site

cis

TGN

Figure 4.13 Schematic diagram of the cisternal progression/maturation model of intra-Golgi transport of cargo molecules. In this model, the cargo molecules remain within a given cisterna as this cisterna progresses in a *cis*-to-*trans* direction across the stack. During this cisternal transport, the cargo molecules are exposed to different sets of biosynthetic enzymes, which are recycled in a retrograde direction by means of COPIb-type vesicles (see Figure 4.6).

maturation model (**Figure 4.13**) of intra-Golgi transport. This model proposes that Golgi cisternae are transient structures and that they, and not the vesicles, are the main cargo carriers across the stack. To this end, upon arriving at the *cis* face of a Golgi stack, sets of ER-to-Golgi transport vesicles are induced to fuse with each other to form a new *cis* cisterna. When this cisterna reaches a critical size, assembly of the next cisterna is initiated, while the first cisterna continues to expand until it reaches a mature size. During this process, the existing cisternae appear to serve as templates for the assembly of the younger cisternae and thereby help regulate cisternal size. One of the predictions of the cisternal progression model is that the *cis* cisternae should be more variable in size and shape than the medial and *trans* cisternae, and that distinct types of assembly intermediates should be present. Such assembly intermediates have indeed been observed in tomographic reconstructions of *cis* Golgi cisternae (**Figure 4.14**).

With the formation of each new cisterna on the *cis* side of the stack and the conversion of the *trans*-most cisternae into a TGN cisterna that peels off from the stack, the first cisterna progressively moves across the stack with its cargo molecules until it, too, reaches the *trans* side, detaches, and ultimately vesiculates. Support for the cisternal progression/maturation model has also come from studies in which large, structurally distinct secretory products such as aggregates of storage proteins in developing plant embryo cells, and cell wall–forming scales in algae, have been observed to move across Golgi stacks without entering the vesicles associated with the cisternal margins. Maintenance of the sequential organization of the enzymes across the stack is postulated to involve COPIb-type vesicles that retrieve enzymes from older cisternae and then recycle them to younger cisternae (see Concept 4A3).

7. The *trans* Golgi network cisternae of plants are transient organelles that sort and package Golgi products into secretory and clathrin-coated vesicles.[ref7]

In plants, the transformation of a *trans* Golgi cisterna into a TGN cisterna starts with the onset of peeling of the *trans*-most cisterna from the stack. Such peeling-type TGN cisternae, termed early TGN cisternae, possess a flat, central domain and sometimes attached intercisternal elements, but the forming secretory vesicles along the curved cisternal margins tend to be more numerous, more spherical, and differently stained than those budding from *trans* Golgi cisternae. The transformation of a *trans*-most Golgi cisterna into an early TGN cisterna also involves a membrane surface area reduction of 30 to 35%. Upon separation from the stack, the cisterna becomes an independent organelle that undergoes a series of maturational changes before it fragments into different kinds of vesicles.

In functional terms, the TGN is defined as the compartment in which products produced in the Golgi stack are sorted and packaged into vesicles for transport to vacuoles and to the cell surface, respectively. Separation of

Figure 4.14 Face-on, *cis*-side views of electron tomography–based three-dimensional reconstructions of three different Golgi stacks. The variations in size and shape of the *cis*-most cisterna are consistent with the prediction of the cisternal progression/maturation model of Golgi transport that new cisternae are assembled on the *cis* side of the stack. The shedding of *trans* Golgi/TGN cisternae is illustrated in Figure 4.15. (Courtesy of Byung-Ho Kang.)

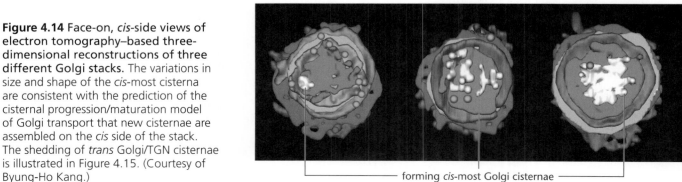

forming *cis*-most Golgi cisternae

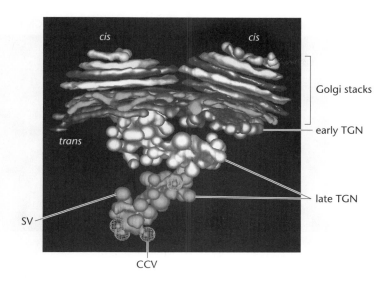

SV

CCV

Figure 4.15 Tomographic model of a dividing root apical meristem Golgi stack with several TGN cisternae positioned at different distances from the *trans*-most Golgi cisterna. "Early" TGN cisternae are TGN-type cisternae that are still attached to the *trans* side of a Golgi stack, whereas "late" TGN cisternae are free entities. In this model, clathrin-coated vesicles (CCVs) are seen on only the most mature late TGN cisterna. SV, secretory vesicle. Compare with Figure 4.16. (Courtesy of Byung-Ho Kang.)

different types of vacuolar proteins from the glycoproteins and polysaccharides destined for the cell wall involves TGN-located receptor proteins that bind to the targeting domains of vacuolar proteins and then become concentrated in *clathrin-coated vesicles* together with their cargo molecules. The ratio of secretory to clathrin-coated vesicles formed on TGN cisternae is highly variable, but in vegetative cells the secretory vesicles tend to greatly outnumber the clathrin-coated vesicles, consistent with the need for more cell wall than vacuole molecules.

The contents of *secretory vesicles* of plant cells do not appear to be concentrated beyond what occurs during packaging in the TGN. Furthermore, all secretory vesicles seem to become immediately secretory competent and do not accumulate in the cytoplasm once they separate from the TGN. In animal systems, this type of secretion is known as *constitutive secretion*, in contrast to regulated secretion, which involves secretory granules containing highly concentrated products (for example, zymogen granules of pancreatic cells) that are transiently stored in the cytoplasm and released only in response to hormonal or electrical signals. At present, there is no evidence to suggest that plants employ this type of regulated secretion.

In root meristem cells, where the movement of Golgi stacks is restricted due to the small size of the cells and the limited amount of cytoplasmic streaming, electron tomography reconstructions often show several detached TGN cisternae in a row extending from the *trans* side of the stacks (**Figure 4.15**). In contrast, in large tobacco BY-2 cells, which exhibit vigorous cytoplasmic streaming and Golgi movements, no such detached TGN cisternae are seen in the vicinity of a large percentage of stacks, even though many *trans* cisternae exhibit a peeling off type of morphology. These findings suggest that once a TGN cisterna is detached from a Golgi stack it becomes a structurally and functionally independent organelle that no longer traffics together with its originating Golgi stack.

The maturational steps of TGN cisterna development can be deduced from the sequence of structural and compositional changes seen in the meristem Golgi-TGN complexes in which the older TGN cisternae are displaced further from the *trans* side of the stacks than the younger TGN cisternae. TGN cisternae in the process of peeling off the *trans* side of a Golgi stack are known as "early" and those that have completely separated from a stack as "late" TGN cisternae. Typically, the *early TGN cisternae* produce budding vesicles only along their margins, whereas the *late TGN cisternae* become more grapelike as the number and size of the budding secretory and clathrin-coated vesicles increases (**Figure 4.16**). Also, the formation of the secretory vesicles usually starts before the onset of clathrin-coated vesi-

Figure 4.16 Tomographic models (frontal and side views) of the early and the two late TGN cisternae illustrated in Figure 4.15. Note that the early TGN cisterna has a flat central domain (*asterisk*) where it was still partly attached to the Golgi stack and has budding vesicle profiles only around its margins. In contrast, the late types of TGN cisternae have budding vesicles emanating from their entire surface and exhibit a more grapelike architecture. Budding clathrin-coated vesicles (CCVs) are seen on only the most mature late TGN cisterna. Compare with the diagrammatic model of Figure 4.6. (Courtesy of Byung-Ho Kang.)

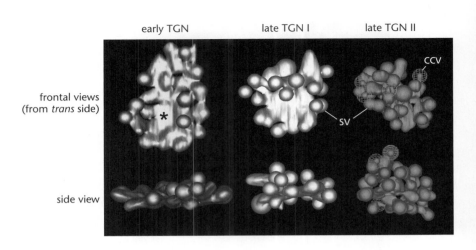

cle formation. The process of TGN cisterna fragmentation (see Figure 4.6), which releases most of the vesicles simultaneously, appears to be a rapid event and is rarely observed in cryofixed cells. Yet to be determined is the extent to which detached TGN cisternae are carried around by cytoplasmic streaming, and whether such cisternae might be recruited to specific sites of secretion before they fragment.

When a TGN cisterna detaches from a Golgi stack it also takes along a portion of the Golgi scaffold (matrix). The resulting TGN scaffold contains different types of scaffolding proteins than the Golgi scaffold and retains its integrity until the TGN cisterna fragments. During TGN maturation, the TGN scaffold differentiates in parallel with the underlying membranes, and when the cisterna fragments, the individual secretory vesicles acquire parts of the scaffold and travel to their next destination with the acquired scaffold proteins, which are then used to build a new scaffold around that membrane system. For example, the cell plate assembly matrix/scaffold, which regulates the assembly of the cell plate membranes, is brought to the cell plate region by the secretory-type cell plate–forming vesicles (**Figure 4.17**; see also Concept 4B3). Interestingly, new multivesicular body-forming vesicles also appear to arise in close proximity to late TGN cisternae and within the TGN scaffold. It is not known where the multivesicular body membranes come from, but a possible source is the residual membrane fragments of the TGN cisternae that are "left over" after all of the vesicles have been released.

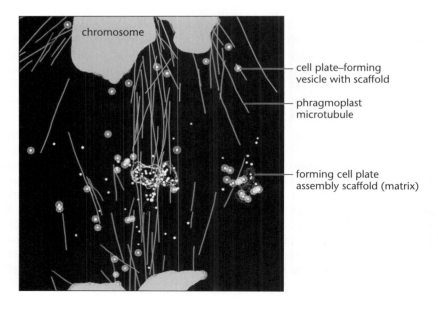

Figure 4.17 Tomographic model of cell plate–forming secretory vesicles carrying scaffold proteins to the cell plate assembly region. These scaffold molecules give rise to the cell plate assembly matrix (CPAM; see Concept 4B3). (Courtesy of Jotham R. Austin.)

8. Maintenance of an appropriate pH in different Golgi compartments is essential for their functions.[ref8]

Many cellular activities are based on transmembrane *pH gradients* or are regulated by pH. Indeed, plant cells derive most of their chemical energy from electron transport chains that produce proton gradients, and much of that energy, in turn, is used to set up and maintain a pH gradient (ΔpH) across nearly all membranes of a cell. The importance of pH and ion gradients for cellular functions is also evidenced by the fact that an average eukaryotic cell spends about one-third of its energy pumping protons and other ions across membranes.

The proton pumps that are responsible for controlling the pH of the different compartments of the secretory pathway are of two types, the vacuolar *H+-ATPase (V-ATPase)*, and the vacuolar H+-pyrophosphatase (V-PPase). Evidence to support the hypothesis that the cisternae of the Golgi stacks and TGN are acidified comes from three lines of research: biochemical fractions enriched in Golgi membranes contain H+-ATPases, immunolabeling experiments with anti-V-ATPase antibodies have demonstrated the presence of V-ATPases in Golgi and TGN cisternae, and cells treated with the ionophores monensin and nigericin exhibit swelling of Golgi and TGN compartments.

Nigericin (**Figure 4.18**) and *monensin* are *ionophores* that exchange monovalent cations across membranes and thereby affect the acidification of membrane compartments. The cisternal swelling response produced by these ionophores is thought to result from a nigericin/monensin-mediated exchange of protons with K+ (or Na+) ions across the membranes of acidified compartments, thereby causing these compartments to swell osmotically. Both swelling and maintenance of the swollen state require the presence and maintenance of a proton gradient. When sycamore suspension-cultured cells are exposed to 10 µM monensin, the first compartments to swell are the free (late) TGN cisternae. This is followed by the *trans*-most Golgi and early TGN cisternae, the remaining *trans* cisternae, and finally the *cis* and medial cisternae. These differences in rates of swelling have been interpreted in terms of a hierarchy of compartmental acidification, with the early and late TGN cisternae being the most acidic compartments. All of these changes are reversible upon removal of the ionophore.

The functional significance of the differences in cisternal acidification is still poorly understood. When the VHA-A V-ATPase of *Arabidopsis* is functionally disrupted by a T-DNA insertion in one of its catalytic subunits, complete male and partial female gametophytic lethality is observed. In the mutant male gametophytes, the first visible symptom of cell degeneration is severe changes in the architecture of the Golgi stacks, consistent with the postulated earliest site of action of the V-ATPases in the secretory pathway. It is also likely that, as in animal cells, different Golgi-localized glycosyl-

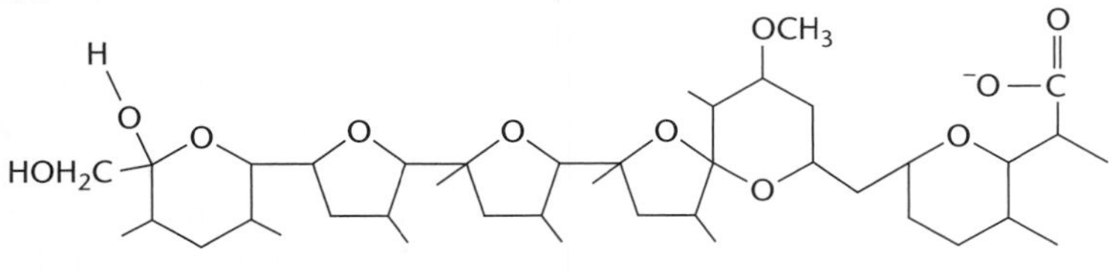

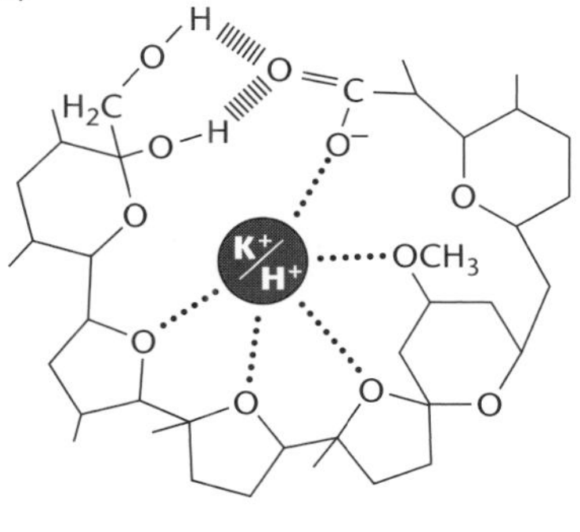

Figure 4.18 Monovalent cation-induced changes in nigericin structure.
(A) Chemical structure of the ionophore nigericin. (B) Binding of H+ and K+ (or Na+) ions to nigericin causes the molecule to assume a ring configuration by H-bond formation between the groups shown by hatched lines at the ends, while simultaneously chelating the ion at the center of the molecule. This conformational change makes the molecule more lipid-soluble and able to shuttle ions across a lipid bilayer.

transferases will have different pH optima, thereby providing the means for controlling transferase activities via pH changes. Neutralization of the TGN also affects the pH-dependent receptors that sort vacuolar proteins into clathrin-coated vesicles for subsequent transport to the vacuoles. As mentioned above, a pH-dependent osmotic collapse of *trans* Golgi cisternae may aid in the displacement of complex polysaccharides to the cisternal rims and their packaging into forming secretory vesicles. Thus, any perturbations of cisternal pH would be expected to not only change the patterns of glycosylation of cell wall polysaccharides and glycoproteins and affect the sorting of Golgi products, but also alter Golgi trafficking, the formation of secretory vesicles, and the rate of secretion.

9. Golgi stacks are structurally and functionally differentiated in a tissue- and developmental stage–specific manner.[ref9]

During the nineteenth century, early plant anatomists demonstrated that different types of plant cells could be distinguished based on their position in a tissue, their shape and dimensions, and the histochemical staining properties of their cell walls. Over 100 years later, with the help of antibody probes that recognize specific epitopes on cell wall polysaccharides and glycoproteins, a new generation of plant anatomical studies is beginning to unravel the molecular basis for the different staining patterns. In particular, immunocytochemical studies with these antibody probes have demonstrated that each cell type produces a cell wall that is uniquely tailored to its needs (see Concept 6C1). As might be expected, a major source of these differences resides with the cell type–specific forms of structural and enzymatic cell wall proteins. However, monoclonal antibodies that recognize sugar epitopes on arabinogalactan proteins and complex polysaccharides suggest that the diversity of the sugar moieties may equal or even exceed the diversity of the proteins. The ramifications of these discoveries for understanding plant development and plant cell wall function are only now beginning to be appreciated.

Since these differences in carbohydrate structures arise from changes in the activity and/or composition of the enzymatic machinery that gives rise to these molecules, one has to postulate that the Golgi apparatus of plant cells has to be enzymatically retailored in a cell type–specific manner during plant development. This hypothesis has received support from both morphological and immunocytochemical investigations. For example, during root cap development, cells derived from the apical meristem undergo a series of changes as they differentiate first into the gravity-sensing columella cells and then into the mucilage-secreting peripheral cells. Electron microscopic analyses have demonstrated that during this developmental progression, the Golgi stacks also undergo a programmed series of distinct morphological changes such that each cell type displays a characteristic type of Golgi stack (compare **Figures 4.19A and B**). These changes include alterations in the size, staining properties, and numbers of *cis*, medial, and *trans* types of Golgi cisternae. In addition, intercisternal elements associated with *trans* Golgi cisternae are seen only in the Golgi stacks of columella and peripheral cells, consistent with their postulated role in anchoring mucilage processing enzymes (**Figure 4.20**). Further support for a cell type–specific retailoring of Golgi stacks has come from immunocytochemical studies in which the distribution of specific epitopes between *cis*, medial, and *trans* cisternae has been shown to differ.

Taken together, there is now strong support for the hypothesis that each cell type produces its own specialized type of Golgi stack, which is optimized for the production of its unique cell wall. Furthermore, since cell growth and differentiation is confined to a limited period of time, additional changes in the functional organization of Golgi stacks over the life of a given cell are to

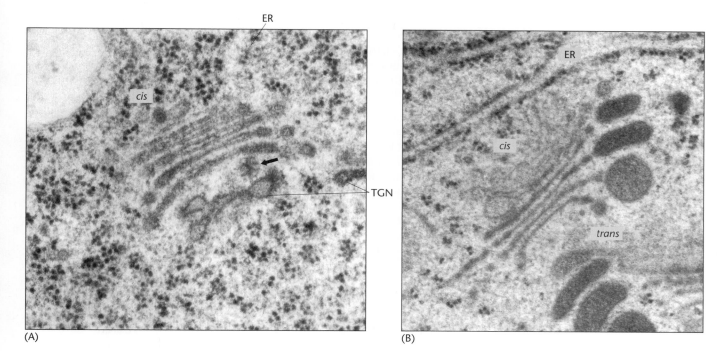

(A)

(B)

Figure 4.19 Architectural variability of plant Golgi stacks. Golgi stack in (A) a meristematic and (B) a peripheral root cap cell. Arrow, clathrin-coated vesicle associated with the TGN.

be expected. These findings highlight some of the challenges faced by Golgi researchers in the future and the problems that will have to be overcome to isolate uniform Golgi fractions from tissues for biochemical analyses.

10. The process of exocytosis delivers matrix molecules to the cell wall and membrane molecules to the plasma membrane.[ref10]

The ultimate destination of secretory vesicles is the cell surface or the cell plate, which is destined to become part of the cell surface. Delivery of the complex carbohydrates and the structural and enzymatic proteins to spe-

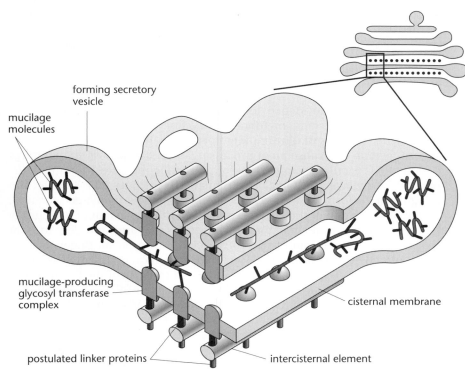

Figure 4.20 Diagrammatic model of a *trans* Golgi cisterna involved in the synthesis of polysaccharide mucilage molecules. The intercisternal elements (see Figure 4.4) associated with its surface are connected through linker proteins to glycosyl transferase complexes involved in the synthesis of polysaccharide molecules. This anchoring is postulated to serve two purposes: (1) to preserve the structural integrity of the stack, and (2) to prevent the large (mol. wt. > 10^6) mucilage molecules from dragging the enzymes into the forming secretory vesicles. (Adapted from L.A. Staehelin et al., *Protoplasma* 157, 75–91, 1990.)

cific sites in the cell requires first targeting of the Golgi-derived vesicles to defined cell surface or cell plate domains, followed by fusion of the vesicles with the plasma membrane or the cell plate membrane. In this process, the vesicle membrane becomes integrated into the target membrane, thereby allowing the latter to expand and to turn over. Excess membrane is retrieved by endocytosis and through molecular lipid transfer pathways.

Little is known about the mechanism of *exocytosis* in walled plant cells, which is difficult to study due to the fact that the actual fusion event occurs in 1 ms or less. Ca^{2+} has been implicated in three stages of exocytosis: delivery of vesicles to the cell surface (cytoskeletal effects), attachment of vesicles to the plasma membrane, and the formation of fusion pores. The exocytotic apparatus includes several types of proteins and protein complexes that perform different functions, including the GTP-binding and regulatory Rab proteins, different tethering and exocyst-type complexes that ensure proper vesicle targeting, the membrane fusion-related v- and t-SNAREs, and the ATP-dependent NSF complexes, which disassemble the membrane fusion machinery after vesicle fusion. The family of annexin-binding proteins may also participate in the fusion of secretory vesicles with the plasma membrane.

11. Turgor pressure has profound effects on vesicle-mediated secretion, membrane recycling from the plasma membrane, and the movement of secreted molecules through cell walls.[ref11]

Although the importance of *turgor pressure* for plant cell growth is well known, its profound effects on secretion and membrane recycling are not widely appreciated. Turgor pressure is defined as the osmotically generated pressure of the semipermeable plasma membrane against the cell wall. Any process that attempts to bring materials into a turgid cell has to overcome or circumvent the turgor pressure forces. This is particularly important for a process such as *endocytosis*, which is used for the bulk uptake of materials into eukaryotic cells. In plant cells, the rate of endocytosis is much smaller than in animal cells. This is due to the fact that in plants this process is used nearly exclusively for the recycling and turnover of plasma membrane molecules, whereas in animal cells it is also employed for the uptake of nutrient molecules. However, the rate of endocytosis in plant cells appears to be further reduced by molecular lipid-recycling mechanisms that bypass the endocytotic pathway, possibly to reduce the energetic cost of membrane recycling.

The most graphic evidence for unique plant mechanisms of vesicle-mediated secretion and membrane recycling has come from electron microscopic studies of ultra-rapidly frozen cells (**Figure 4.21**). Ultra-rapid freeze-fixation and subsequent processing of the samples at temperatures below –80°C is critically important for preserving the ultrastructure of turgid cells for viewing in the electron microscope. **Figure 4.22** illustrates in diagrammatic form the events associated with vesicle-mediated secretion. Stages 1 and 2 depict the events associated with the fusion of a secretory vesicle with the plasma membrane, two events that are equivalent to what has been reported for nonturgid animal cells. However, starting with stage 3, membrane configurations are seen that are unique to plant cells under turgor pressure. The turgor pressure–induced collapse of the fused vesicle into a pancake-like structure during the discharge of the vesicle contents (stage 3) appears to result from the fact that, unlike in animal cells, the turgid plant plasma membrane cannot expand laterally to incorporate the new membrane. The subsequent tipping over of the flattened membrane appendage is likely caused by movement of the cytoplasm (stage 4), and the concomitant reorganization of the interconnecting domain of the

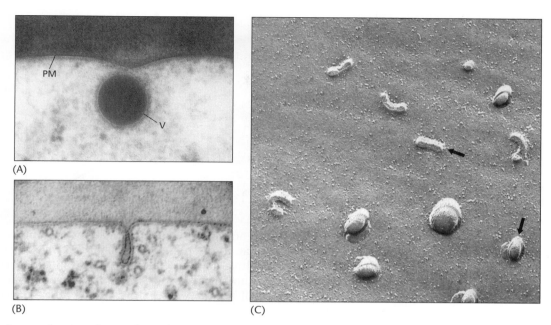

(A)

(B)   (C)

fused membranes leads to horseshoe-shaped profiles in the plasma membrane from which membrane needed for cell expansion is withdrawn and excess membrane is recycled.

Two possible mechanisms of *membrane recycling* are depicted in **Figure 4.23**. The first involves typical clathrin-coated, endocytotic vesicles, which have been well documented in plant cells. Clathrin-coated vesicles function in the retrieval of integral plasma membrane proteins and membrane lipids, and in the internalization of some molecules from the cell wall compartment. The second mechanism involves molecular uptake of membrane lipids via a lipid hopping mechanism, which allows the lipids to transfer directly from the plasma membrane to the adjacent ER membranes. Two observations support the idea of molecular recycling of lipids, the very close apposition of transiently formed "tongues" of ER against the horseshoe-shaped plasma membrane domains (**Figure 4.24**), and the very rapid transfer of externally applied, fluorescently labeled phospholipid molecules to underlying ER membranes. Clearly, much more work is needed to verify and quantify these recycling pathways, but the data clearly support the hypothesis that turgor pressure has a major effect on secretion and membrane recycling by plant cells.

Figure 4.21 Electron micrographs of different stages of vesicle-mediated secretion in a turgid plant cell (see also the interpretative diagram in Figure 4.22). (A) Cross-sectional image of a vesicle (V) initiating membrane fusion with the plasma membrane (PM). (B) Secretory vesicle that has fused with the plasma membrane and has discharged its contents. The disk-shaped membrane configuration shown here is observed only in turgid cells. (C) Freeze-fracture electron micrograph displaying different membrane configurations of secretory vesicles that have discharged their contents. The arrows point to membrane configurations that correspond to the image shown in B. (B, Courtesy of Yoshinobu Mineyuki.)

top view

stage 1   stage 2   stage 3   stage 4

cell wall

vesicle

plasma membrane

vesicle

bottom view

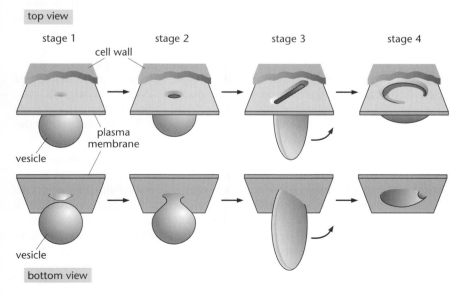

Figure 4.22 Interpretative diagrams of the membrane events associated with vesicle-mediated secretion in a turgid plant cell as depicted in Figure 4.21. Lower figures, bottom-side views of plasma membrane configurations; upper figures, top-side views. The model illustrates different stages in the fusion of a vesicle with the plasma membrane (stages 1 and 2), the discharge of the vesicle contents and the concomitant changes in vesicle and plasma membrane architecture (stage 3), and the final reorganization of the plasma membrane appendages (stage 4) that creates the horseshoe-shaped membrane configurations shown in Figure 4.21C. (Adapted from L.A. Staehelin and R.L. Chapman, *Planta* 171, 43–57, 1987.)

Figure 4.23 Three mechanisms for removing excess membrane from the horseshoe-shaped infoldings of the plasma membrane of a turgid plant cell. Mechanism 1 involves cell expansion; mechanism 2, the formation of clathrin-coated endocytotic vesicles; and mechanism 3, the direct hopping of lipid molecules from the plasma membrane via lipid-binding proteins into closely appressed ER membranes, from where they are recycled into the secretory pathway. CP, cell plate. (Adapted from L.A. Staehelin and R.L. Chapman, *Planta* 171, 43–57, 1987.)

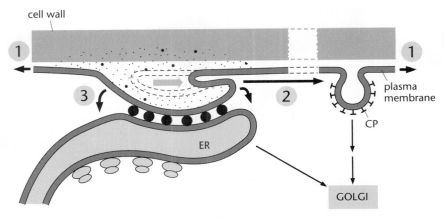

Turgor pressure also affects the structural organization and properties of cell walls. In particular, turgor pressure stretches cell walls, increases their porosity, and helps drive newly secreted molecules into the existing wall. Without turgor pressure, most freshly secreted molecules would accumulate between the wall and the plasma membrane and not become fully integrated into the cell wall.

12. Cellulose microfibrils, callose, and mixed-linked glucans are made by enzyme complexes in the plasma membrane.[ref12]

Cellulose microfibrils, callose, and mixed-linked glucans are produced by enzymes in the plasma membrane and deposited directly into the cell wall. These enzymes are made and assembled into complexes in the ER and Golgi membrane systems and then transported to the cell surface in secretory vesicles. Since neither callose, cellulose, nor β-1,3/1,4 glucans can be detected by immunolabeling experiments within intracellular compartments (**Figure 4.25**), it has been postulated that the enzymes are activated after their insertion into the plasma membrane. This contrasts with the synthesis of complex polysaccharides and glycoproteins, which are made by enzymes that reside in the ER and Golgi cisternae (see Concepts in 5B and 5C) and whose products are delivered in vesicles to the plasma membrane.

The rationale for having the synthesis of cellulose and callose confined to the plasma membrane can perhaps be found in the properties and the functions of these polysaccharides. Cellulose microfibrils are both long and stiff polymers that would be nearly impossible to package into small secretory vesicles, and also to properly deposit into the cell wall were they brought to the plasma membrane in such vesicles. Callose, on the other hand, is often used to protect or stabilize the plasma membrane during wound responses. Since time is of critical importance in containing damage after wounding, having latent callose synthases in the plasma membrane instead of delivering callose to the wound site by transport vesicles may significantly increase the effectiveness of the response. In addition, the insoluble nature of callose, once formed, would cause additional problems of delivery to the critical sites.

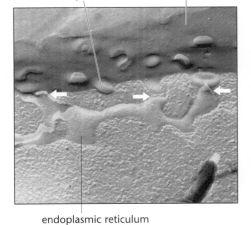

Figure 4.24 Freeze-fracture electron micrograph of the lipid-recycling membrane structures shown in Figure 4.23. Arrows, ER "tongues" that contact the collapsed, fused secretory vesicles. (From S. Craig and L.A. Staehelin, *Eur. J. Cell Biol.* 146:80–93, 1988.)

Why mixed-linked glucans are synthesized at the plasma membrane and not in the Golgi apparatus has yet to be determined. Indeed, it is somewhat ironic that we know more about the effects of mixed-linked glucans on human health (see Concepts 9B1 and 2) than on the functional role of these transiently produced and unique glucans in grasses. One common feature of cellulose, callose, and β-1,3/1,4 glucans is that they are all linear polymers of glucose, whereas the matrix polysaccharides produced in the Golgi apparatus are typically branched polysaccharides composed of at least two types of sugars. These latter features may dictate the need for synthesis within a

Figure 4.25 Electron micrograph of a Golgi stack adjacent to a cell plate in a tobacco BY-2 cell immunolabeled with anticallose antibodies. Notice that the gold particles are seen only over the cell plate and not over the Golgi cisternae. (Courtesy of A. Lacey Samuels.)

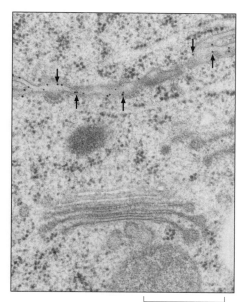

0.5 µm

closed, intracellular membrane compartment even though this organization is less energy-efficient and more difficult to maintain over evolutionary times. In the absence of such biosynthetic constraints, synthesis of linear carbohydrate polymers at the plasma membrane would be favored due to the greater simplicity and efficiency of the biosynthetic apparatus.

13. The enzymes that synthesize cellulose microfibrils in higher plants are organized into complexes that visually look like rosettes.[ref13]

Cellulose microfibrils of higher plants have been estimated to contain thirty to forty β-1,4-linked glucan chains that are laterally associated with each other in the form of a crystalline lattice (Concept 3A1). Conceptually, cellulose microfibril assembly can be divided into three steps: simultaneous initiation of the cellulose chains, chain elongation, and lateral aggregation of the chains by means of H-bonding into a crystalline lattice of the microfibril. How the proteins that perform these functions are organized in the plasma membrane is still largely a matter of conjecture.

The idea that an enzyme, located at the tip of a microfibril, might be responsible for the assembly of glucose molecules into crystalline microfibrils arose in the 1950s. Two decades later, the first experimental evidence in support of this concept was obtained in an ultrastructural study of the alga *Oocystis*. In particular, freeze-fracture electron micrographs of the plasma membrane of this alga demonstrated elongated arrays of particles, putative *cellulose synthase complexes* associated with the ends of nascent cellulose microfibrils (**Figure 4.26**). In subsequent investigations it was found that the "linear type" of cellulose synthase complex seen in *Oocystis* is present only in certain algae. Land plants and one algal group, the Chlorophyceae, which appear to be closely related to lower plants, possess "rosette type" cellulose synthase complexes (**Figures 4.27** and **4.28**). Each *rosette complex* consists of six hexagonally arranged 8-nm particles and a large cytoplasmic domain, and has an outside diameter of 24 nm. The cellulose synthase enzymes contained within each rosette produce a classic type of *elementary microfibril* with a diameter of 3 to 4 nm and consisting of an estimated 36 glucan chains. Yet to be determined is how the enzymes within a rosette complex are organized to enable the nascent cellulose molecules to aggregate laterally into a crystalline microfibril after their extrusion from the cell surface. Experimental proof for the presence of catalytic subunits of cellulose synthase in the rosette complexes has been obtained by means of a technique that involves immunolabeling of appropriately processed freeze-fracture replicas with anti-cellulose synthase antibody–gold probes (Figure 4.27 insert). A more detailed discussion of the molecular composition and architecture of rosette complexes is presented in Concept 5D7.

14. Cellulose synthase complexes are pushed forward in the plasma membrane by the growing cellulose microfibrils that they extrude into the cell wall.[ref13]

During cellulose synthesis, the synthase complexes are postulated to be pushed forward in the membrane, most likely driven by the forces associated with the polymerization of the cellulose molecules and their self-assembly into stiff crystalline microfibrils (**Figure 4.29**). As the cellulose synthase

100 nm

Figure 4.26 Freeze-fracture electron micrograph of a linear-type cellulose synthase complex and cellulose microfibril imprinting in the plasma membrane (E-fracture face) of the alga *Oocystis*. The arrow indicates the direction of microfibril growth. (Courtesy of R. Malcolm Brown Jr.)

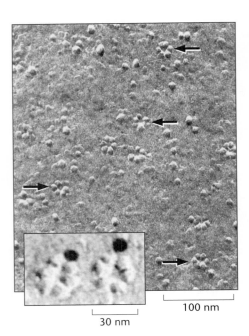

Figure 4.27 Electron micrograph of freeze-fractured rosette-type cellulose synthase complexes in the plasma membrane (P-fracture face) of a differentiating tracheary element of *Zinnia elegans*. Insert: Rosette complexes immunolabeled (*black dots*) with antibodies raised against a peptide corresponding to a domain of the catalytic subunit of the cellulose synthase complex. (Courtesy of Candace H. Haigler; insert from S. Kimura et al., *Plant Cell* 11, 2075–2085, 1999.)

complexes are translocated in the plane of the plasma membrane, the nascent microfibrils that they produce are wrapped around the cell much the way insect larvae deposit the silk thread fibers of their cocoons. This mode of cellulose synthesis explains why no cellulose microfibrils extend across the middle lamella, why the cellulose fibrils produced by neighboring cells do not interweave, and why the youngest cellulose microfibrils are found closest to the plasma membrane.

Support for the idea that the cellulose synthase complexes are pushed forward in the plasma membrane by the growing cellulose microfibrils has come from three observations. First, studies of the Gram-negative bacterium *Gluconacetobacter xylinus* (formerly *Acetobacter xylinum*) have shown that this bacterium uses the force generated by the extrusion of a stiff ribbon of laterally aggregated cellulose microfibrils into the surrounding medium to propel itself forward at a rate of up to 1 μm/min (**Figure 4.30**). This finding demonstrates that the extrusion of a microfibril from a cellulose synthase complex can generate a force on the complex capable of displacing the complex in the plane of the fluid plasma membrane. The second line of support comes from freeze-fracture studies of plant and algal cells involved in cellulose synthesis. Replica images of the plasma membranes of such cells often depict an imprint of a microfibril that appears to originate from the cellulose synthase complex (see Figure 4.26). Most recently, the expression of a yellow fluorescent protein fusion to cellulose synthase (CESA) has produced an experimental tool for directly visualizing the movements of the rosette complexes in living cells of transgenic *Arabidopsis* plants. These complexes were found to travel in straight trajectories at speeds of between 100 and 500 nm/min in the plane of the plasma membrane, which corresponds to the addition of about 300 to 1000 glucose residues per glucose chain per minute. The direction of the rosette movements typically coincided with the orientation of the cortical microtubules.

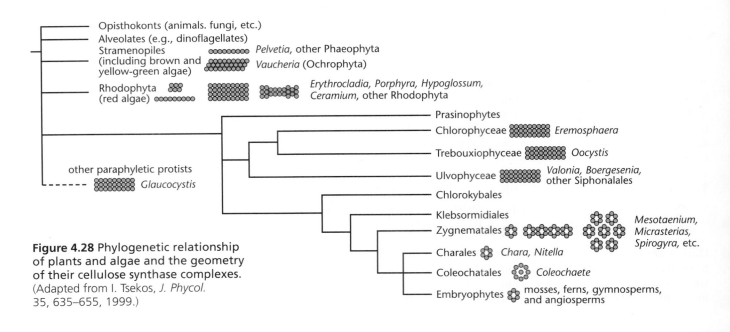

Figure 4.28 Phylogenetic relationship of plants and algae and the geometry of their cellulose synthase complexes. (Adapted from I. Tsekos, *J. Phycol.* 35, 635–655, 1999.)

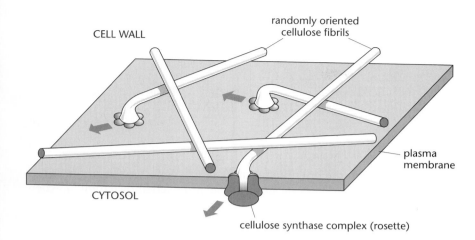

Figure 4.29 Model of cellulose "elementary" microfibril synthesis by individual rosette-type cellulose synthase complexes in a plasma membrane with no attached microtubules. Without orienting microtubules that constrain the movement of the synthase complexes in the plane of the membrane (see Figure 6.27) the microfibrils are deposited initially in a random pattern.

Two lines of evidence also suggest that the process of lateral association of cellulose molecules into crystalline microfibrils occurs after the cellulose molecules have emerged from the confines of the cellulose synthase complex. Thus, when *Calcofluor white* (**Figure 4.31**), a fluorescent molecule that can hydrogen-bond to the surface of cellulose molecules, is applied to *G. xylinus* cells producing cellulose microfibrils, the individual cellulose molecules are prevented from coalescing into microfibrils (Figure 4.30B). This suggests that the crystallization process occurs after the growing cellulose molecules have emerged from the synthase complexes. The fact that microfibrils produced by adjacent rosette complexes organized into defined rows or hexagonal arrays in the alga *Micrasterias* (**Figure 4.32A**) can also aggregate laterally to form thicker cellulose microfibrils (Figure 4.32B and **Figure 4.33**) lends further support to the idea that polymerization and crystallization are both temporally and spatially separable events.

15. Rosette complexes are concentrated in domains of the plasma membrane underlying sites of rapid cellulose synthesis.[ref14]

The diversity of plant forms arises from the control of where new cell walls are laid down and of where and how much old and new walls are allowed to expand. Cell wall growth is to a significant extent dependent on when, where, and in what amounts cellulose microfibrils are deposited in the cell

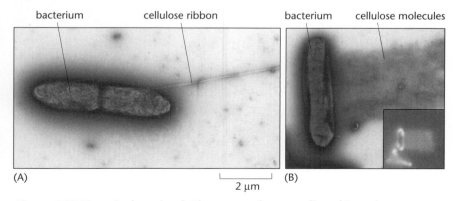

(A)   2 μm   (B)

Figure 4.30 Negatively stained *Gluconacetobacter xylinus* (*Acetobacter xylinum*) cells producing cellulose. (A) Control cell with a cellulose ribbon. (B) *G. xylinus* bacterial cell producing cellulose in the presence of 0.25 mM Calcofluor white. Note how the cellulose molecules emerge from the cell as a broad band and do not condense laterally into a crystalline ribbon as shown in Figure 4.30A. Inset: Calcofluor-labeled cellulose band associated with a *G. xylinus* cell as seen with fluorescence microscopy. (Courtesy of Candace H. Haigler.)

Figure 4.31 Structure of Calcofluor white ST. Calcofluor is a highly planar derivative of stilbene that associates with straight-chain polysaccharides with free hydroxyl groups. Calcofluor produces a blue-white fluorescence when excited with UV light and is included as an "optical brightener" in laundry detergents.

wall, as well as on the orientation of the newly deposited microfibrils (see Concept 6B9). Freeze-fracture studies have provided information on where newly formed rosette complexes are inserted into the plasma membrane, as well as the direction in which they start moving after cellulose synthesis has commenced.

In tip-growing cells such as root hairs and pollen tubes, new rosette complexes are brought to the plasma membrane in secretory vesicles that fuse with the expanding cell tip. Upon activation at the cell surface, the rosettes move in trajectories over, around, or away from the tip, thereby depositing a set of randomly oriented microfibrils (**Figure 4.34**). High concentrations of rosettes have also been observed in bandlike domains of the plasma membrane of differentiating xylem cells. However, in sharp contrast to the tip-growing cells, these rosette complexes remain confined to bandlike domains to ensure that all of the new cellulose microfibrils are deposited at the sites of the future xylem wall thickenings (**Figure 4.35**). The delivery and confinement of the rosette complexes to bandlike regions in the plasma membrane, and the concomitant orientation of the cellular microfibrils parallel to the long axis of the wall thickenings (see Concept 6B10), are dependent on cortical microtubules that are attached to the plasma membrane in the bandlike areas.

Redistribution of rosette complexes and corresponding alterations in patterns of cellulose deposition have been observed in plants subjected to changes in the direction of gravitropic forces or by subjecting cells to strong magnetic or electrical fields. Changes in the distribution, the size, the organization, and/or the number of rosettes can also be induced by chemicals such as colchicine (a microtubule-disrupting agent), 2,6-dichlorobenzonitrile (DCB, an inhibitor of the putative cellulose synthesis primer

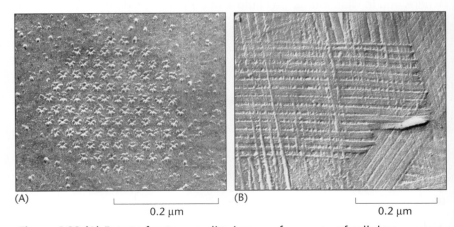

(A)   0.2 μm   (B)   0.2 μm

Figure 4.32 (A) Freeze-fracture replica image of an array of cellulose synthase complexes (*rosettes*) in the plasma membrane of the alga *Micrasterias denticulata* during secondary wall synthesis. Such arrays give rise to bands of parallel cellulose microfibrils with different widths but constant, center-to-center spacing (B). A model of how the rows of rosette complexes produce cellulose microfibrils of different widths is depicted in Figure 4.33. (From T.H. Giddings, D.L. Brower and L.A. Staehelin, *J. Cell Biol.* 84:327–339, 1980.)

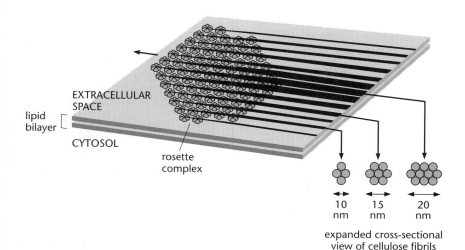

Figure 4.33 Model of the formation and deposition of parallel bands of cellulose microfibrils during secondary wall formation in *M. denticulata*. The thickness of the cellulose microfibrils depends on the number of rosettes associated with each row (see Figure 4.32), while the center-to-center distance between the microfibrils remains constant, being defined by the spacing of the rosette rows. (Adapted from T.H. Giddings, D.L. Brower and L.A. Staehelin, *J. Cell Biol.* 84:327–339, 1980.)

sistosterol-β-glucoside), the thiatriazine herbicide CGA 325′615 (an inhibitor of crystalline cellulose assembly), Calcofluor white/Tinopal LPW (an inhibitor of lateral glucan chain association), cycloheximide (a protein synthesis inhibitor), and nigericin and monensin (ionophores that perturb acidic compartments of the secretory pathway). A loss of rosette complexes has also been noted in the temperature-sensitive *Arabidopsis* mutant *rsw1*, which expresses a mutation in the cellulose synthase gene *CESA1*. Together, these results demonstrate that cellulose microfibril synthesis can be manipulated experimentally in many different ways.

16. The half-life of rosette complexes determines the length of cellulose microfibrils.[ref13, 14, 15]

Assembly of the protein subunits of the rosette complexes into the typical rosette structures occurs in the Golgi apparatus. From there, they are transported in secretory vesicles to the plasma membrane (**Figure 4.36**). Cellulose synthesis starts after insertion of the complexes into the new environment of the plasma membrane and appears to be accompanied by slight changes in the morphology of the complexes as seen in freeze-fracture replicas. These structural changes are consistent with the idea of activation of the complexes only upon their arrival at the plasma membrane. Based on observations of living cells expressing fluorescently labeled rosette complexes, the lag time between the arrival of a rosette complex in the plasma membrane and the onset of cellulose microfibril deposition is < 10 seconds.

The *half-life of rosettes* associated with the deposition of cellulose microfibrils in xylem wall thickenings has been calculated to be about 20 minutes. Rosette complexes containing fluorescently labeled CESA proteins have been shown to move at speeds of about 0.3 μm per minute, which, over a 20-minute period, would yield cellulose microfibrils with a length of about 6 μm and a degree of polymerization of ~12,000. Quite remarkably, this value essentially matches the degree of polymerization (10,000–15,000) reported for solubilized cellulose molecules from secondary cell walls. It also suggests that the length of cellulose molecules and of cellulose microfibrils is determined by the half-life of the rosette complexes. Considering the rate at which a given rosette complex is moving through the plasma membrane while spinning out a cellulose microfibril, it is highly unlikely that damaged enzymes in the complex could be repaired or replaced on the fly. Most likely, the subunits of damaged complexes become tagged for proteolytic degradation by ubiquitination, which triggers removal of the complexes from the cell surface by endocytosis and their transfer to multivesicular bodies and to lytic vacuoles. This hypothesis suggests that the microfibrils should become thinner towards their ends.

Figure 4.34 Distribution and trajectories of rosette complexes in the tip region of a tip-growing fern protonema.

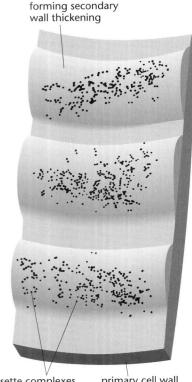

forming secondary
wall thickening

rosette complexes   primary cell wall

Figure 4.35 Pattern of rosette distribution in the plasma membrane of a developing xylem element of a young maize root. The rosettes (*black dots*) are restricted to areas underlying forming secondary wall thickenings, which are also defined by bands of parallel cortical microtubules. (Adapted from R. Schneider and W. Herth, *Protoplasma* 131:142–152, 1986.)

17. The cellulose microfibrils of primary cell walls are shorter than those of secondary cell walls.[ref16]

Assuming a length of about 6 µm and a diameter of about 4 nm, the length-to-width ratio of a secondary wall cellulose microfibril is about 1500. Translated into the dimensions of a cooked spaghetti, this would amount to 6- to 8-m-long spaghetti "microfibrils" forming the tensile elements of walls of "cells" of the size of rooms or small houses. This simple analogy highlights the fact that even the very long cellulose microfibrils of secondary cell walls are not long enough to wrap completely around a cell. This fact has important implications for understanding cell wall mechanics and in particular for understanding how cell wall expansion occurs during growth. Cellulose microfibrils produced during primary wall formation tend to be composed of cellulose molecules with a much lower degree of polymerization (50%, 200–500; 50%, 2500 to 4500 glucose moieties in tobacco) than those synthesized during secondary wall deposition (10,000–15,000). One possible reason for having shorter cellulose molecules in primary cell walls is that cell growth requires microfibril sliding and that cells can control the process of microfibril sliding and thereby the process of cell expansion only with shorter cellulose microfibrils. Production of the very short cellulose molecules may be associated with cell plate formation.

18. Callose appears transiently during growth and development and is also formed in response to biotic and abiotic stresses.[ref17]

Callose, a linear β-1,3-linked glucose polymer with occasional 1,6 branches, plays an important role during various steps of plant growth and development as evidenced by the programmed patterns of deposition and removal of callosic cell wall layers (**Figure 4.37**). In addition, it has the capacity to protect the plasma membrane, seal tissue wounds, and protect cells against pathogens. Like cellulose, callose is synthesized by enzymes located in the plasma membrane. However, in contrast to the cellulose synthase complexes, which are produced only during certain stages of cell growth, latent callose synthases appear to be present in the plasma membrane throughout a cell's life, ready for action at a moment's notice.

Fresh *callose deposits* usually appear as plug- or sheetlike structures sandwiched between the external surface of the plasma membrane and the cell wall (**Figure 4.38**). This mode of deposition can be explained by the location of the enzyme, the self-aggregating properties of the helical callose molecules, and the dense, insoluble nature of the highly hydrated but semipermeable callose gels. These physical properties make callose deposits ideal for creating transient physical and chemical barriers, structural molds, and "fillings" that produce expansive forces. Callose can be detected cytochemically, as it forms an intensely fluorescent complex with *aniline blue*. However, whereas aniline blue staining is a fairly reliable callose stain for higher plants, it is less so for other organisms, since it can also bind to other polysaccharides such as laminarin and substituted β-1,3-glucans. Removal of callose deposits *in vivo* involves callase, a β-1,3-glucanase.

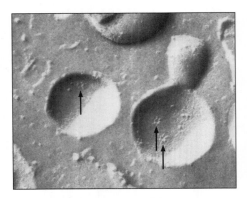

Figure 4.36 Secretory vesicles transporting rosette complexes (*arrows*) to the plasma membrane in a semicell of the alga *M. denticulata* engaged in synthesis of the primary cell wall. (From T.H. Giddings, D.L. Brower and L.A. Staehelin, *J. Cell Biol.* 84:327–339, 1980.)

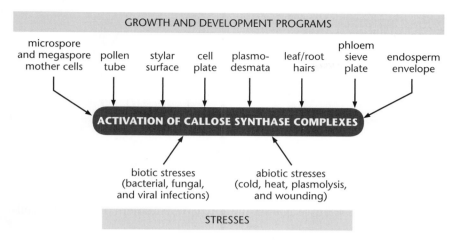

Figure 4.37 Regulation of callose synthase complexes. Many growth and development activities of plants depend on the transient expression of callose at specific locations. Callose synthase complexes also respond to signals produced by biotic and abiotic stresses. (Adapted from D.P.S. Verma and Z. Hong, *Plant Mol. Biol.* 47:693–701, 2001.)

19. Callose is a transient cell wall component that is critically important for many growth and developmental processes.[ref18]

The functions of the transiently produced callose deposits during specific stages of plant growth and development are still being explored. For example, during cell plate formation, callose is seen to form a thick, coat-like structure on the surface of the membranes during the maturation of the early tubular vesicular network into a fenestrated cell plate and then into a cell wall (see Figures 4.38A and 4.44). It has been postulated that this coat helps to mechanically stabilize the delicate membrane networks and to create a spreading force that widens the tubules and converts them into platelike structures.

Pollen formation is another process that relies on transient callose deposits for proper development. Here, the formation of a callose type of wall commences when the pollen mother cell begins its meiotic cycle. Two functions have been postulated for the microspore callose wall. It may physically protect the haploid microspores from genetic and physiological influences of the diploid sporophytic tissues and/or provide a molecular scaffold that can be dissolved by specific enzymes as the exine and intine layers are laid down with the help of surrounding tapetal cells. Mutant or transgenic plants in which callose dissolution occurs prematurely or is delayed are

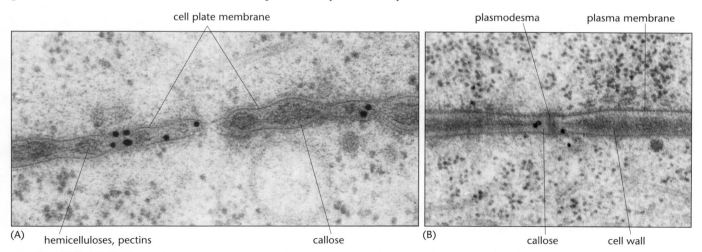

Figure 4.38 (A) Cell plate of a tobacco root tip cell immunolabeled with anticallose antibodies. The gold particles are seen over the more lightly stained callose layer adjacent to the cell plate membrane. The darkly staining material in the cell plate corresponds to complex polysaccharides (xyloglucans and esterified pectins). (B) A callose sleeve also surrounds plasmodesmata as evidenced by anticallose gold labeling. (A, from A.L. Samuels, T.H. Giddings and L.A. Staehelin, *J. Cell Biol.* 130:1345–1357. B, Courtesy of A.L. Samuels.)

Figure 4.39 Changes of β-1,3-glucan hydrolase activity during microsporogenesis in anthers of fertile and sterile Petunia lines. (Adapted from S. Izhar and R. Frankel, *Theor. Appl. Genet.* 44:104–108, 1971.)

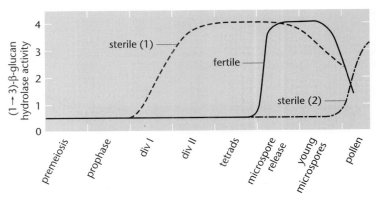

male sterile due to bursting or inhibition of development of the pollen cells (**Figure 4.39**).

Callose synthesis plays a role in two aspects of pollen tube growth. A callosic cell wall layer that accounts for about 80% w/w of total wall carbohydrates is laid down as a sheathlike structure inside the fibrillar outer wall starting about 10 μm back from the growing tip. Whether this sheath is produced to stiffen the cell wall or to serve as a protective molecular filter, as has been reported for the callose layer of the endosperm envelope of muskmelon seeds, has yet to be determined. The second type of callosic structure involved in pollen tube growth is the callosic plugs that are deposited inside the pollen tubes by the trailing end of the advancing pollen cytoplasts (**Figure 4.40**). These plugs most likely prevent the pollen cells from reversing their direction of growth.

Sporophytic self-incompatibility brought about by pollen-stigma interactions in plants such as *Brassica* also involves callose formation. Thus, when a pollen grain settles on a stigma whose genome contains an identical S-allele as the pollen parent, callose formation is observed both in the tip of the pollen and in the stigma cells underlying the pollen. Presumably the two callose deposits help prevent penetration of the stigma by the pollen tube.

The formation of *sieve plate pores* in developing phloem constitutes another system in which callose synthesis is both spatially and temporally controlled. Thus, as the forming sieve plate expands during growth, the space between the plasmodesmatal tube-lining plasma membrane and the cell wall fills with a callose collar that is subsequently removed, leaving behind a void that allows the plasmodesmatal opening to expand into a mature sieve tube pore. In deciduous trees, whose life span extends over several growing seasons, the sieve tube pores are physically closed in the fall by the deposition of callose between the plasma membrane tube of the plasmodesmata and the cell wall. In the spring, the pores reopen when the callose is enzymatically removed.

Plasmodesmata are also ensheathed by a sleeve of callose (see Figure 4.38). This callose layer appears to serve as a "valve" that can regulate the transmission of molecules through plasmodesmatal pores. Thus, activation of callose synthases around the ends of plasmodesmatal openings causes a buildup of callose between the wall and the plasma membrane, which physically constricts the pores (see Concept 8B2).

20. Stress-induced callose synthesis serves to protect cells from potentially lethal biotic and abiotic insults.[ref19]

Wound-induced callose deposits are most readily seen in sieve plates, around plasmodesmata, and adjacent to physically or chemically damaged plasma membranes. Most likely, callose synthesis in these locales is triggered by mechanical stresses associated with the sudden loss of turgor pressure and by the resulting fluxes of Ca^{2+} across the plasma membrane. In

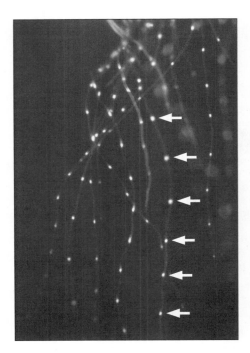

Figure 4.40 Callosic plugs produced at the trailing ends of advancing pollen cytoplasts inside growing pollen tubes. (Courtesy of Alice Y. Cheung.)

sieve plates close to cut stem regions, large amounts of callose are deposited between the plasma membrane and the cell wall of the pores, which results in their constriction and eventual closure. This closure resembles the shutdown of the sieve plate pores in the fall (see Concepts 1C7 and 4A18) and may involve the same callose synthases (see Concept 5D11). Similarly, the wound-induced closure of plasmodesmata also appears to be triggered by Ca^{2+} fluxes that activate the callose synthases around the plasmodesmatal openings.

Callose synthesis is often of critical importance in the defense against pathogens where the defense response does not involve cell death (see Chapter 8). In such instances, deposition of callose at the inside surface of the cell wall is observed at the point of penetration of the fungal haustorium through the wall. As the fungal hypha extends into the cell, the callose layer expands to form a protective interface between the plasma membrane and the hypha. Inhibition of callose synthesis has been shown to reduce the resistance of barley coleoptile epidermal cells to powdery mildew, indicating that these callose deposits do contribute to disease resistance in general. However, this is not always the case. Thus, although the *Arabidopsis pmr4* mutants have a lesion in a callose synthase gene and do not deposit callose in response to biotic or abiotic stress, they are more resistant than wild-type plants to pathogen attack. The enhanced resistance has been traced to the activation of the salicylic acid defense-signaling pathway. How the absence of callose stimulates this pathway has yet to be determined.

B. Membrane Systems Involved in *De Novo* Cell Wall Assembly during Cytokinesis

1. Plant Golgi stacks multiply by division.[ref20]

Cytokinesis in plant cells involves the formation of a cell plate from Golgi-derived vesicles and the conversion of this cell plate into a new wall. Prior to these events, all critical cell components have to be duplicated and separated into the cytoplasmic domains that become the two independent daughter cells. While these changes have been characterized in greatest detail for the genetic material, all membranous organelles can be expected to follow a similar sequence of events.

In *Arabidopsis* shoot meristem cells, doubling of the Golgi stacks (from ~34 to ~66) occurs during the G2 stage of the cell cycle, that is, just prior to mitosis. However, *Golgi stack duplication* can also occur in interphase cells in conjunction with cell growth. In both cases, duplication appears to start with the formation of two half-sized *cis* cisternae on the forming side of the stack (**Figure 4.41**). Each of these separate cisternae then serves as an independent template for the assembly of the next cisterna. As this process is repeated and the older *trans*-side cisternae are shed, the two parallel stacks of duplicated cisternae get bigger, while the number of single, *trans*-side cisternae to which they are attached decreases. Separation of the last of the nonduplicated *trans* Golgi cisternae from the stack liberates the two fully formed daughter stacks from the physical constraints of the mother stack and allows them to become both structurally and functionally independent organelles.

2. During mitosis and cytokinesis cytoplasmic streaming stops and the Golgi stacks redistribute to specific locations.[ref21]

When we think about mitosis, our thoughts invariably concentrate on the events associated with preprophase band formation, the assembly and the transformations of the diverse sets of spindle MTs, and finally chromo-

(A)

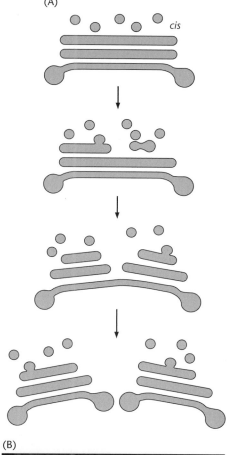

(B)

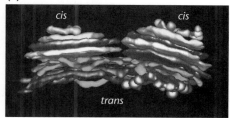

Figure 4.41 Mechanism of Golgi stack duplication in plants. (A) Schematic diagrams illustrating how plant Golgi stacks divide by initiating the formation of two separate *cis*-most cisternae that serve as separate templates for the assembly of subsequent cisternae. (B) Electron tomography–based model of a dividing Golgi stack of *Medicago sativa*. (Courtesy of Byung-Ho Kang.)

some separation. However, while all these MT-mediated events are taking place, all *cytoplasmic streaming* and all saltatory Golgi movements stop. The mechanistic basis for the cessation of cytoplasmic streaming has yet to be elucidated. Recruitment of actin to the spindle might deplete the cytoplasm of the actin filaments needed for streaming. From a functional perspective, cessation of streaming may aid in the proper sorting and positioning of organelles during mitosis, their partitioning between the two daughter cells, the assembly of the spindle microtubules, and the alignment of the chromosomes. In addition, cessation of cytoplasmic streaming during cytokinesis may be essential for ensuring the efficient delivery of Golgi-derived vesicles, of cytoskeletal proteins, and of enzymes involved in cell plate assembly to the division plane. Indeed, transport to the division plane appears to become the default trafficking pathway in dividing cells, much like the trafficking of secretory vesicles to the plasma membrane in interphase cells.

The behavior of the Golgi stacks during *mitosis* and *cytokinesis* supports the hypothesis that these organelles are recruited to sites where their products are needed for cell plate/wall assembly. This redistribution is most notable in large cells such as tobacco BY-2 cells. Thus, when the general streaming of Golgi stacks stops, about one-third of the Golgi stacks redistribute from the cortical to the perinuclear cytoplasm. By metaphase about 20% of all Golgi stacks aggregate in the immediate vicinity of the mitotic spindle, while a similar number congregate in an equatorial region under the plasma membrane (**Figure 4.42**). This latter aggregation has been termed the *Golgi belt*, since it accurately predicts the future site of cell division after the disassembly of the preprophase band of microtubules. During telophase and cytokinesis, a higher density of Golgi stacks is seen around the phragmoplast, the site of cell plate assembly. The active recruitment of the Golgi stacks to these different cellular sites is evidenced by the fact that plastids are simultaneously excluded from these locations.

To what extent might this redistribution of Golgi stacks aid in cell plate assembly? During the past century, many studies of cytokinesis in living plant cells have documented not only the striking events associated with chromosome separation but also the remarkably rapid growth of the initial *cell plate* during early telophase. For example, in tobacco BY-2 cells, a 20-μm-diameter cell plate can be observed within 10 to 15 minutes of cell plate initiation. Based on the size of the Golgi-derived vesicles and the surface area of a 20-μm-diameter disk, it can be calculated that the formation of this initial BY-2 cell plate requires between 100,000 and 200,000 vesicles. How can over 100,000 Golgi-derived vesicles be delivered to the forming cell

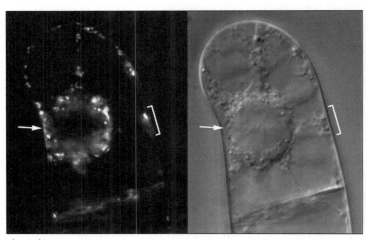

Figure 4.42 Golgi stack distribution in a dividing tobacco BY-2 cell containing GFP-labeled Golgi stacks. The fluorescence (*left*) and the differential interference contrast (DIC) microscopy image (*right*) depict the same cell during metaphase. The bright dots in the fluorescent micrograph correspond to single or small groups of Golgi stacks. Note the high density of Golgi stacks around the spindle and in the equatorial belt region. The arrow denotes the plane of the metaphase plate of chromosomes, and the bracket the position of the Golgi belt in the cortical cytoplasm. (From A. Nebenführ, J.A. Frohlick and L.A. Staehelin, *Plant Physiol.* 124:135–151, 2000.)

10 μm

plate region within less than 15 minutes? Since cell plate–forming vesicles have been shown to travel along MTs to the cell plate, it seems likely that most of the initial vesicles are produced by the spindle-associated Golgi stacks prior to metaphase and begin traveling along the spindle MTs to the future site of cell plate assembly before chromosome separation occurs. The presence of such vesicles between spindle MTs has been verified by means of electron tomography (Figure 4.17). Indirect evidence in support of this hypothesis has also come from experiments with synchronized BY-2 cells in which Golgi vesicle formation was stopped with the drug brefeldin A prior to metaphase and the initial stages of cell plate assembly still occurred. In a similar manner, the Golgi stacks that accumulate in the equatorial Golgi belt region are most likely involved in creating pools of cell plate–forming vesicles that are used during the later centrifugal phase of cell plate assembly.

3. Somatic-type cell plate formation can be divided into four distinct phases.[ref22]

Somatic-type cell plate formation begins during late anaphase as the sister chromatids congregate around the spindle poles, and yields a new cell wall by late telophase (**Figure 4.43**). The structure that produces the cell plate and the new cell wall is known as the *phragmoplast*. The three principal structural elements of the phragmoplast are two opposing arrays of MTs and actin filaments, and a ribosome-excluding, cell plate–encompassing filamentous matrix/scaffold, commonly known as the *cell plate assembly matrix* (CPAM). As its name suggests, the CPAM corresponds to a protein-aceous scaffold that contains all of the proteins (enzymatic, regulatory, structural, and motor) needed for cell plate assembly and growth.

Based on the spatial organization of the CPAM and associated MTs, the process of cell plate assembly can be divided into four phases (**Figure 4.44**). Phase I coincides with the events that lead to the assembly of the *phragmo-plast initials*. During this phase, clusters of residual polar spindle MTs serve as the initial scaffoldings for bringing the first Golgi-derived cell plate–forming vesicles and CPAM components to the site of cell plate assembly. Shortly thereafter, the first hourglass- and dumbbell-shaped fused vesicles are observed within these very early CPAM domains. Lateral expansion and coalescence of the initial CPAM domains together with the recruitment of new MTs lead to the creation of the next distinctive cell plate assembly intermediate, the solid-type phragmoplast.

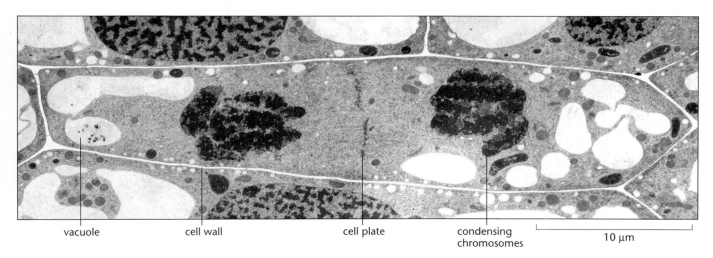

vacuole   cell wall   cell plate   condensing chromosomes   10 μm

Figure 4.43 Electron micrograph of a late anaphase, dividing onion cotyledon epidermal cell. At this early stage of cell plate formation, the cell plate still consists of several separate tubulovesicular membrane networks. (Courtesy of Ichirou Karahara.)

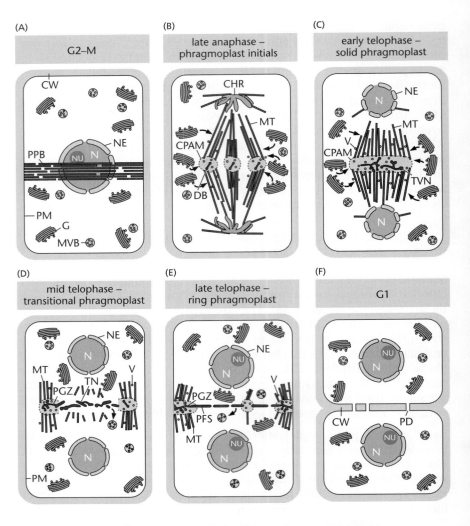

Figure 4.44 The different stages of plant somatic-type cytokinesis. (A) Formation of the preprophase band (PPB) of microtubules (MT) during the G2-M stage of the plant cell cycle. The preprophase band of microtubules defines where the cell plate will subsequently fuse with the plasma membrane (PM) and the cell wall (CW) of the mother cell. (B) Phragmoplast assembly phase during late anaphase. The phragmoplast initials arise from opposite sets of polar spindle MTs. Golgi (G)-derived vesicles (V) travel along MTs towards the assembly sites, defined by the presence of cell plate assembly matrices (CPAM). Inside the CPAMs, the first cell plate initials are produced through the formation of dumbbell-shaped (DB) vesicle intermediates. (C) Solid phragmoplast phase. After fusion of vesicles to the dumbbell ends and joining of the phragmoplast initials, a tubulovesicular network (TVN) cell plate develops within the cocoonlike CPAM, which extends across the entire interzone between the two sets of opposing MTs of the solid phragmoplast. (D) Transitional phragmoplast phase. As the central CPAM and associated MTs disassemble, a new CPAM with MTs arises at the edge of the cell plate, giving rise to the ring-shaped phragmoplast and the peripheral growth zone (PGZ). In parallel the tubulovesicular network cell plate is converted into a tubular network (TN) through callose synthesis in the cell plate lumen, and by the removal of excess membrane via clathrin-coated vesicles that are delivered to multivesicular (MVB) bodies for degradation. (E) Ring-shaped phragmoplast phase. As the central cell plate domain is converted to a planar fenestrated sheet (PFS), secondary CPAMs and associated MTs re-form over the remaining large fenestrae, focusing cell plate growth to these regions. The ring-shaped phragmoplast and CPAM define the PGZ, which expands centrifugally until the cell plate reaches and fuses with the plasma membrane. (F) Completion of the newly formed cell wall and plasmodesmata (PD) between the daughter cells. Onset of the G1 stage of the plant cell cycle. NE, nuclear envelope; CHR, chromosomes; N, nucleus. (Adapted from J.M Segui-Simarro et al., *Plant Cell* 16:836–856, 2004.)

Phase II starts with the completion of the *solid phragmoplast* and ends with the completion of the coherent tubulovesicular network stage cell plate. The solid phragmoplast consists of a flat, cocoonlike CPAM that spans the entire width of the phragmoplast, and two opposing sets of MTs, whose (+) ends terminate within the CPAM. This association of the MT (+) ends with the CPAM stabilizes the MTs and thereby ensures that the cell plate–forming vesicles are delivered to the cell plate. Assembly of the cell plate starts with the formation of the first dumbbell-shaped fused vesicles (see Concept 4B4). Later-arriving vesicles either form new dumbbell-shaped vesicles or fuse with the bulbous ends of existing ones. The end result of this vesicle-driven growth, which occurs simultaneously across the entire width of the CPAM, is the tubulovesicular membrane network.

Phase III corresponds to the *transitional phragmoplast* stage, during which the CPAM and MTs of the solid phragmoplast break down, while a new CPAM and new MTs assemble in the cell plate periphery. The resulting *ring phragmoplast* creates the centrifugally expanding peripheral cell plate growth zone, which leads to the fusion of the cell plate with the cell wall. In the central region, the CPAM-free cell plate is converted to a flattened tubular network by callose deposition.

The final phase of cell plate assembly, phase IV, involves two parallel sets of events. In the central cell plate region, the transformation of the tubular network into a CPAM-free planar fenestrated sheet can be traced to a combination of callose-mediated membrane spreading (see Figure 4.38) and removal of excess membrane via clathrin-coated vesicles. Small secondary CPAMs with attached MTs arise *de novo* over the remaining large fenestrae to focus vesicle-mediated local growth to these regions. In the cell plate periphery, the centrifugal expansion and maturation of the peripheral growth zone results in the fusion of the cell plate with the mother cell walls and to the completion of the new wall.

The sites on the plasma membrane with which the cell plate fuses correspond to the plasma membrane domains demarcated earlier by the *preprophase band* of microtubules (see Concept 1A4). Although the preprophase band breaks down before the mitotic spindle forms, it leaves behind a spatial "memory" on the plasma membrane. Creation of this altered membrane domain involves the removal of selected plasma membrane proteins via clathrin-coated vesicle-mediated endocytosis in the preprophase band region. Actin filament labeling studies have shown that as the preprophase band microtubules degrade, so too do the membrane-associated actin filaments. However, because the cortical actin filaments outside the preprophase band site do not degrade, this process leaves behind an actin-depleted zone on the cytoplasmic surface of the plasma membrane that persists throughout cytokinesis and is accessible to the fusion tubes that facilitate the fusion of the cell plate with the cell wall. The remaining microfilament-rich domains on either side of the actin-depleted zone may interact, in turn, with cell plate–associated actin filaments to help orient the cell plate, a task that is particularly important in cells undergoing asymmetric cell divisions (see Concept 1A5).

Immuno- and cytochemical labeling studies have demonstrated that the Golgi-derived vesicles deliver not only matrix polysaccharides and proteins—esterified pectins, xyloglucans, and arabinogalactan proteins—to the forming cell plate, but also callose and cellulose synthases. Thus, when cellulose synthesis begins during the later stages of cell plate maturation, the microfibrils are exuded from the cell surface into a soup of cell wall matrix molecules, where self-assembly apparently leads to the formation of the new wall. Interestingly, this cellulose microfibril-triggered assembly has also been correlated with the stiffening of the wavy cell plate into a straight cell wall. Since both daughter cells wrap themselves in their own cellulose microfibril "cocoons," the space left between the cocoons eventually becomes the new middle lamella. Enzymatic degradation of the mother cell wall microfibrils at the juncture between the new and the old walls allows the middle lamella of the new wall to connect to the middle lamella of the mother cell wall, thereby completing an expanded three-way junction region (see Figure 1.18).

4. Dynamin-GTPases create dumbbell-shaped, cell plate–forming vesicles containing dehydrated and possibly gelled cell wall–forming molecules.[ref23]

One of the central challenges faced by meristematic cells during cytokinesis is how to induce the Golgi-derived vesicles to assemble into a planar cell plate and not into a large, swollen, vacuole-like structure. The critical event that prevents the latter scenario from developing is the formation of dumbbell-shaped vesicles. *Dumbbell-shaped vesicles* are produced from freshly fused, hourglass-shaped vesicles within the CPAM with the help of mechanoenzymes known as *dynamin-GTPases* (**Figure 4.45**).

Several dynamin homologs have been localized to somatic-type cell plates, where they participate in two different activities, the formation of dumbbell-shaped vesicles and the production of clathrin-coated vesicles. Dynamin-GTP polymers constrict and elongate membrane tubules by assembling into tight, membrane-pinching spirals. Hydrolysis of dynamin-bound GTP causes the spirals to expand like springs, and the mechanical energy released during this expansion can be harnessed to do work. In the case of clathrin-coated buds, the energy is sufficient to shear the neck of these buds, whose rigid coat prevents the vesicle membrane from stretching to accommodate the dynamin-induced squeezing force.

Unlike clathrin-coated buds, the bulbous ends of the dumbbell-shaped vesicles are not covered by rigid coats. Thus, when their dynamin springs expand, the membranes encompassing the bulbous ends can stretch transiently to accommodate the fluids that are propelled into those ends. This short-lived membrane stretching would be expected to transiently increase membrane permeability, allowing water, but not the cell plate–forming polysaccharides, to exit to alleviate the pressure (**Figure 4.46**). Measurements

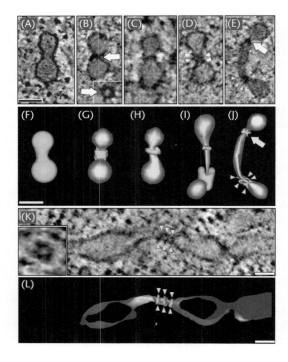

Figure 4.45 Formation of dumbbell-shaped vesicles from freshly fused hourglass vesicles. (A–E, K) Electron tomographic slice images of hourglass-shaped (A) and dumbbell-shaped (B–E, K) vesicles/tubules in *Arabidopsis* meristem cells preserved by cryofixation and freeze-substitution methods. (F–J, L) Models of structures similar to those seen in the tomographic slices. Arrows point to compact, dynamin-like springs; small arrowheads point to individual loops of expanded dynamin springs. (From J.M. Segui-Simarro et al., *Plant Cell* 16:836–856, 2004.)

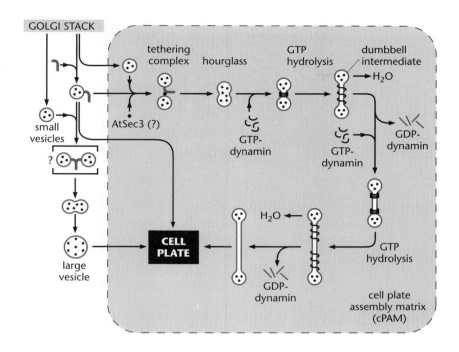

Figure 4.46 Hypothetical model explaining how Golgi-derived, cell plate–forming vesicles can give rise to large vesicles and to dumbbells. Small Golgi-derived vesicles outside the cell plate assembly matrix (CPAM) interact through an unknown tethering complex prior to fusion and give rise to large vesicles. Vesicles that enter the CPAM before fusion first become tethered through an exocyst-type complex and then give rise to an hourglass intermediate. This hourglass-type vesicle is converted into a dumbbell vesicle with the help of dynamin springs. First, GTP-dynamin monomers are assembled into tight springs around the hourglass vesicle necks. GTP hydrolysis causes expansion of the dynamin spring, which elongates the neck to form a dumbbell. New GTP-dynamin springs are assembled next to the bulbous ends of the dumbbell, and the expansion/stretching process is repeated. During each neck elongation step the volume of the vesicle is reduced, most likely by expulsion of water. The resulting elevated concentration of the polymeric vesicle contents (esterified pectins, xyloglucans, arabinogalactan proteins) has been postulated to lead to polysaccharide gelling and thereby to the stabilization of the elongated vesicle architecture. (From J.M. Segui-Simarro et al., *Plant Cell* 16:836–856, 2004.)

of the surface area and volume of dumbbell-shaped vesicles have shown that the vesicle constriction/elongation process can reduce vesicle volume by up to 50% with no change in surface area. Later-formed dynamin rings associated with the membranes of the tubulovesicular network–stage cell plates can further reduce the cell plate volume by up to 70% compared to the originating vesicles.

The functional significance of these vesicle volume-reducing events becomes evident when one considers the effects of water removal on the physical properties of the vesicle contents: esterified pectic polysaccharides, hemicelluloses, and arabinogalactan proteins. Dehydration of the esterified pectic polysaccharides is likely to produce the formation of hydrophobic pectin gels (see Concept 3C1). Gels can maintain their shape over extended periods of time, since they represent an elastic structure that does not readily rearrange. This suggests that the principal function of dumbbell formation is to induce the gelling of the cell plate polysaccharides and thereby create mechanically stable, elongated cell plate–forming vesicle intermediates. The mechanical stability of the dumbbell-shaped vesicles is evidenced by the fact that even when new vesicles fuse with the bulbous ends, the dumbbells show no signs of swelling (**Figure 4.47**). In this manner, the dumbbell vesicles can be used to create cell plate assembly foci that can grow without swelling, and which can give rise to a planar cell plate membrane network that can subsequently be converted into a cell wall.

Another type of cell wall molecule that is expressed in meristematic tissues and may play a role in cell plate assembly is an isoform of the structural hydroxyproline-rich glycoprotein extensin (AtEXT3; see Concept 3E4). Like the stress- and wounding-induced extensins, this lysine-rich HRGP is composed of repetitive amphiphilic motifs, has a rodlike structure, and can be cross-linked by extensin peroxidase. However, unlike those extensins, the AtEXT3 molecules can also aggregate laterally in a staggered configuration, which both stiffens and extends the polymeric building blocks and creates more open, rigid, and extended networks. A knockout mutation of AtEXT3 produces an embryo-lethal phenotype and displays ruptured cell walls. This suggests that the AtEXT3 network provides mechanical strength to young cell walls and/or cell plates.

Figure 4.47 Assembly of the initial cell plate membrane network from Golgi-derived vesicles and dumbbell-shaped vesicles. (A–D) Fusion of vesicles to the bulbous dumbbell ends gives rise to increasingly complex tubulovesicular-type cell plate assembly intermediates. (E) Example of an early tubulovesicular-type cell plate membrane network within which some of the assembly intermediates shown in A–D can be seen. cPAM, cell plate assembly matrix (outer edge). (From J.M. Segui-Simarro et al., *Plant Cell* 16:836–856, 2004.)

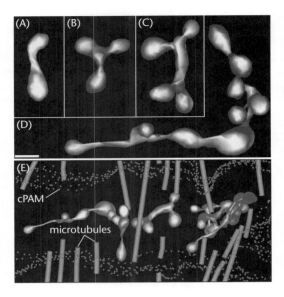

5. The association of cisternae of the endoplasmic reticulum with forming cell plates increases over time, but the functional importance of this spatial relationship has yet to be fully explored.[ref24]

How the different cell plate assembly events are regulated remains controversial. Several lines of evidence suggest that Ca^{2+} could play an important role in controlling some of these activities, including membrane fusion, the dynamics of cytoskeletal elements, and callose synthase activity. Because the ER is known to regulate local Ca^{2+} levels in other systems, the organization of cell plate–associated ER has been the subject of a number of studies. The most detailed electron tomography investigation of this relationship has shown that the number of ER cisternae in close proximity to the forming cell plate is low during the initial rapid growth phase (phragmoplast initial and solid phragmoplast stages) and increases dramatically as the cell plate matures into a fenestrated sheet (transitional and ring phragmoplast stages). These findings suggest that Ca^{2+} may be less important for regulating the initial vesicle fusion processes, and more important for controlling cell plate maturation events such as callose synthesis and membrane lipid recycling.

More clear cut is the relationship between ER and the formation of the tubular templates for the assembly of *primary plasmodesmata*. During the early stages of cell plate formation, the ER–cell plate association is ill defined. However, as the cell plate matures, some of the ER cisternae become more closely appressed to both sides of the plate and form distinct tubules that cross the fenestrae. Eventually, the cell plate fenestrae close around these bridging ER strands, giving rise to the primary plasmodesmata (see Concept 1C9).

Formation of the plasmodesmata involves compression of the ER tube into a thin membrane cylinder with a diameter of about 14 nm, as well as the assembly of cross-linking structures between this cylinder and the plasma membrane of the newly formed cross wall. The end result is a structure in which the space between the ER-derived cylinder and the plasma membrane serves as the conduit for exchanging small molecules (< 1000 mol. wt.) between adjacent cells. However, where required (for example, junctions between companion cells and phloem sieve tubes), the size-restricting channel structures can be opened up by specific sequence domains on proteins to allow for the passage of molecules that are much larger.

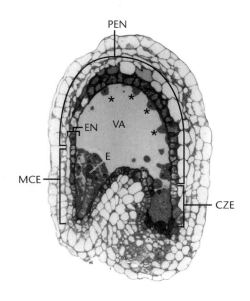

Figure 4.48 Longitudinal section view of a developing *Arabidopsis* seed. Three distinct regions can be recognized in the developing endosperm: the micropylar endosperm domain (MCE), in which the embryo (E) is located; the peripheral endosperm (PEN) domain; and the chalazal endosperm (CZE) domain. The central zone consists of a thin peripheral layer of cytoplasm with regularly spaced nuclei (*asterisks*) and a large central vacuole (VA). The endosperm is surrounded by the endothelium (EN). (From M.S. Otegui and L.A. Staehelin, *Plant Cell* 12:933–947, 2000.)

6. Formation of cell walls in the syncytial endosperm and in meiocytes involves a special kind of cell plate, the syncytial-type cell plate.[ref25]

As discussed in Concept 4B3, mitosis in somatic cells is followed immediately by cell plate formation and cytokinesis at sites defined earlier by the preprophase band of MTs. This coupling of preprophase band formation, mitosis, and cytokinesis does not occur in syncytial systems like the nuclear *endosperm, meiocytes,* and gametophytic cells. Instead, the primary nucleus initially undergoes several rounds of divisions without cytokinesis, forming a syncytium (**Figure 4.48**). In the developing endosperm, *cellularization* then begins with the simultaneous formation of cell walls between both sister and nonsister nuclei. The formation of the very first anticlinal walls between nonsister nuclei has remained controversial for the past 90 years, but a recent study of this process in the endosperm of *Arabidopsis* has led to the discovery of a new kind of cell plate, the *syncytial-type cell plate*.

The initial round of endosperm nuclear divisions is completed during the late globular embryo stage, and the newly formed nuclei then migrate to their positions in the cortical cytoplasm. The final positioning is aided by arrays of radial MTs that originate in MT-organizing centers associated with the surface of the nuclei. Each of these arrays defines a *nucleocytoplasmic domain* and thereby the boundary of the future cell (**Figure 4.49**). The interface zones between the opposing sets of MTs are where the new cell walls will be laid down.

Cell plate formation starts with the synchronous organization of small clusters of oppositely oriented MTs (< 10 MTs in each set) into phragmoplast-like structures, termed *mini-phragmoplasts*, between both sister and nonsister nuclei (Figure 4.49B). These mini-phragmoplasts give rise to multiple separate but aligned cell plates that assemble in unison and eventually become the cell walls. The details of this process are depicted schematically in **Figure 4.50**.

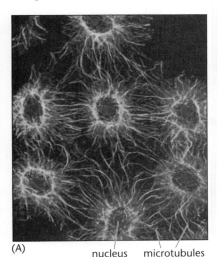

(A)   nucleus   microtubules

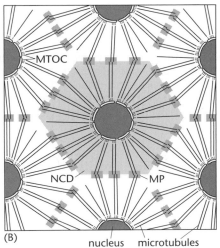

(B)   nucleus   microtubules

Figure 4.49 Spatial regulation of the sites of new cell wall formation by nuclear cytoplasmic domains (NCD) in the nuclear endosperm. (A) Light micrograph of the nuclear endosperm of *Coronopus didymus* (Brassicaceae). The microtubules (MTs) are visualized with fluorescently labeled antitubulin antibodies. (B) Explanatory diagram showing how the MTs depicted in (A) arise from MT-organizing centers (MTOCs) on the surface of the nuclear envelope. The mini-phragmoplasts (MPs), which produce the syncytial-type cell plates, form in the regions where opposing sets of MTs overlap. (A, courtesy of Roy C. Brown.)

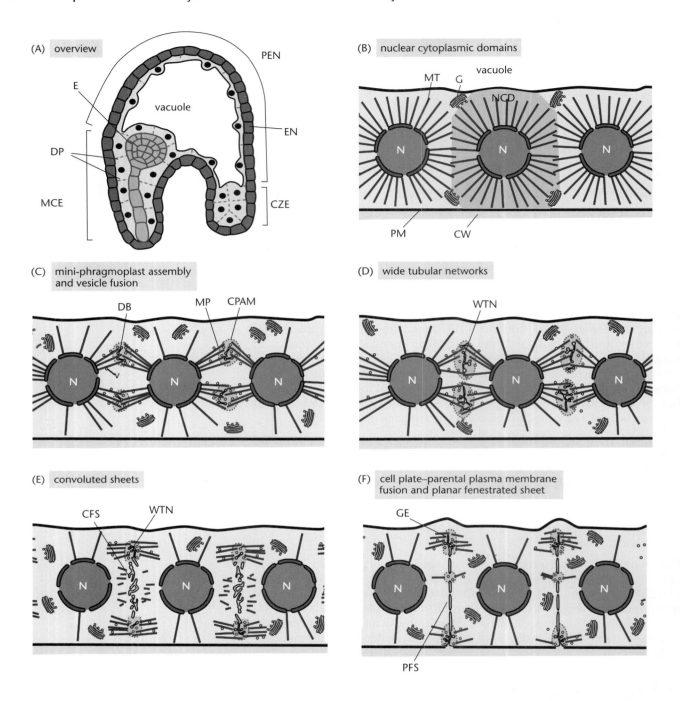

Figure 4.50 Model illustrating the main developmental stages of syncytial-type cell plate formation in *Arabidopsis* **endosperm.** (A) Overview of a developing *Arabidopsis* seed just prior to endosperm cellularization. Cellularization starts in the micropylar endosperm domain (MCE) when the embryo (E) reaches the globular stage. (B) At this stage, the nuclear endosperm is organized into nuclear cytoplasmic domains (NCD) defined by radial systems of microtubules (MT). (C) Cellularization starts with the assembly of mini-phragmoplasts (MP) at the boundaries of adjacent nuclear cytoplasmic domains, the accumulation of vesicles within the forming cell plate assembly matrices (CPAM), and vesicle fusion as evidenced by the presence of dumbbell-shaped fusion intermediates (DB) at the division plane. (D) Assembly of multiple wide tubular networks (WTN) across the division plane.
(E) Formation of a coherent cell plate by fusion of adjacent wide tubular networks and maturation of the central cell plate domains into convoluted fenestrated sheets (CFS) driven by spreading forces generated by callose deposition. (F) Fusion of the cell plate with the parental plasma membrane and conversion of the convoluted sheets into planar fenestrated sheets (PFS). Unlike in somatic-type cytokinesis, the callose deposits produced in the syncytial walls persist after completion of the walls. In *Arabidopsis*, this callose is used as a carbohydrate reserve for the developing embryo. CZE, chalazal endosperm; CW, cell wall; DP, division plane; EN, endothelium; GE, cell plate growing edge; G, Golgi stack; N, nucleus; PEN, peripheral endosperm; PM, plasma membrane. (Adapted from M.S. Otegui et al., *Plant Cell* 13:2033–2051, 2001.)

Syncytial-type cell plates, like somatic-type cell plates, are formed from Golgi-derived vesicles, which fuse with each other via hourglass-shaped intermediates to form wide tubules. These wide tubules (~45 nm in diameter) quickly become coated and surrounded by a ribosome-excluding matrix, and as they elongate they branch and fuse with each other to form wide tubular networks. Each of these aligned networks expands laterally until it meets and merges with other small networks, eventually giving rise to a coherent wide tubular network whose mini-phragmoplasts are positioned around its margins. As depicted in Figure 4.49, each nucleus in the cell cortex is surrounded by up to six neighbors, and a wide tubular network is formed between each pair of nuclei. These networks continue to expand until they meet and fuse with each other at the sites of the future cell corners.

Due to the small number of MTs associated with each mini-phragmoplast and the limited number of vesicles that can be delivered to any given cell plate per unit time, syncytial-type cell plate formation occurs at a much slower rate than somatic-type cell plate formation. For this reason, wide tubular networks have to be maintained in a stable configuration for prolonged periods of time. Aiding the stabilization of the tubular membrane conformation are small, transient, spiral-shaped protein assemblies of *dynamin*-type molecules that appear to form around and transiently pinch the wide tubules at random intervals. These dynamin spirals disappear shortly after the wide tubular networks join up in the corners and as the wide tubules are converted into convoluted, fenestrated sheets at multiple sites throughout the networks. These events are accompanied by the appearance of callose in the cell plate lumen and of clathrin-coated budding vesicles that remove excess membrane, a process that reduces the membrane surface area by about 75%. Upon fusion of the forming cell plate with the syncytial wall, the final convoluted fenestrated sheets are converted into planar fenestrated sheets and ultimately into the new cell walls. The total number of Golgi-derived vesicles required to produce the walls around each nucleus has been calculated to be about 1.5 million.

At the end of this first phase of cell wall formation, the narrow endosperm region around the suspensor and the sides of the developing embryo is completely cellularized, whereas in the central and the chalazal zones, the honeycomb-like new cell wall assemblies, termed alveoli, remain open on the side of the central vacuole. Completion of the cellularization process in these regions involves a combination of expansion of the walls along their free margins, branching and fusion of the septum-like walls with each other, and the formation of somatic-type cell plates between newly divided nuclei. In cereals, the endosperm-derived cells differentiate into five different cell types, including the starchy endosperm and the aleurone layer. In contrast, the shorter-lived cellularized endosperm of *Arabidopsis* is almost completely digested during the late stages of seed formation.

As in the nuclear endosperm, the pollen-producing, cellularizing meiocytes also undergo cytokinesis via the formation of syncytial-type cell plates (**Figure 4.51**). As groups of microtubules become organized into phragmoplasts, cell plate assembly sites consisting of a cell plate assembly matrix and vesicles form across the entire division plane. After the fused vesicles have been transformed into wide tubular networks, those networks located in the cell periphery begin to fuse with the plasma membrane before the more central assembly sites become organized into a coherent cell plate. Fusion of the peripheral cell plate domains with the parental plasma membrane triggers a rapid accumulation of callose at the fusion sites, and the concomitant transformation of the wide tubular networks into stublike projections that grow toward the center of the syncytium. As this occurs, the stublike projections expand by fusing with the remaining wide tubular networks in the division plane until the four new callosic cell walls are completed.

(A) nuclear cytoplasmic domains

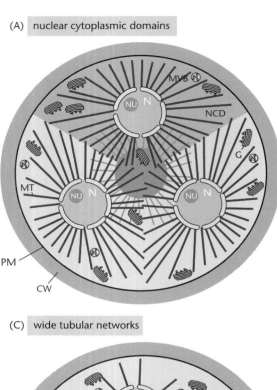

(B) mini-phragmoplast assembly–vesicle fusion

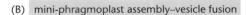
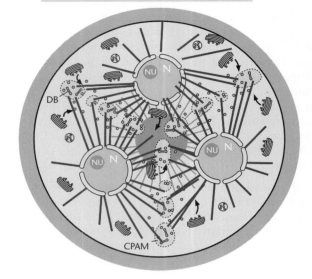

(C) wide tubular networks

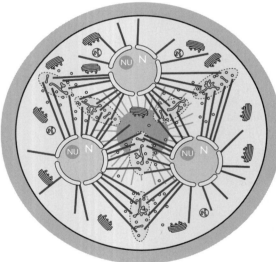

(D) convoluted sheets

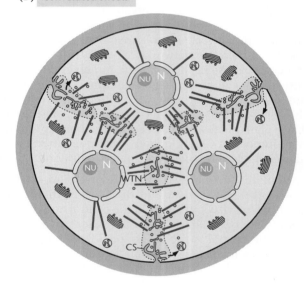

(E) fusion of cell plate with parental plasma membrane

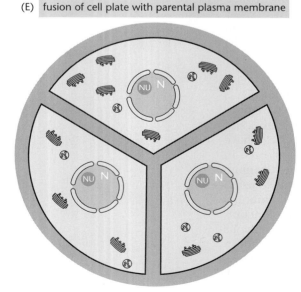

Figure 4.51 Model illustrating the main developmental stages of syncytial-type cell plate formation in *Arabidopsis* microsporocytes during pollen formation. (A) Organization of four nuclear cytoplasmic domains (NCD) defined by radial microtubule (MT) arrays after meiosis. Six division planes between the four nuclei are established. (B) Multiple phragmoplasts, each with a cell plate assembly site defined by a cell plate assembly matrix (CPAM) containing vesicles and dumbbell-shaped fusion intermediates (DB), arise across the entire division plane. (C) Assembly of multiple wide tubular networks (WTN). (D) Early fusion of the most peripheral cell plate assembly sites with the parental plasma membrane. These peripheral domains mature into convoluted sheets (CS) and stublike projections by the accumulation of callose. (E) Upon completion of cytokinesis, six callose-rich cell walls with no plasmodesmata are formed. G, Golgi stacks; MVB, multivesicular bodies; N, nucleus; NU, nucleolus; PM, plasma membrane; CW, cell wall. (Adapted from M.S. Otegui and L.A. Staehelin, *Planta* 218:501–515, 2004.)

During the formation of the postmeiotic tetrad, the callose walls break the symplastic connections and isolate the microspores from each other, a process that is essential for the expression of the gametophytic genome without interference either from the parental sporophyte or from neighboring microspores. The callose walls undergo local degradation to make room for the primexine layer, the blueprint of the exine, which is deposited on the outer surface of the microspores. Massive callose degradation by *callases* secreted from the surrounding tapetal cells releases the microspores from the tetrad during maturation.

Key Terms

actin track	elementary microfibril	nigericin
aniline blue	endocytic pathway	nucleocytoplasmic domain
Calcofluor white	endocytosis	*O*-linked glycoprotein
callase	endomembrane system	pH gradient
callose deposit	endoplasmic reticulum (ER)	phragmoplast
cell plate	endosperm	phragmoplast initial
cell plate assembly matrix	ER export sites	pollen
cellularization	exocytosis	preprophase band
cellulose microfibril	Golgi apparatus	primary plasmodesma
cellulose synthase complex	Golgi belt	retention sequence
cis Golgi network	Golgi matrix	retrieval sequence
cisternal progression/maturation	Golgi scaffold	ring phragmoplast
model	Golgi stack	rosette complex
clathrin-coated vesicle	Golgi stack duplication	rosette half-life
complex polysaccharide	green fluorescent protein (GFP)	secretory pathway
constitutive secretion	H$^+$-ATPase	secretory vesicle
COPIa vesicle	intercisternal element	sieve plate pore
COPIb vesicle	ionophore	solid phragmoplast
COPII scaffold	late TGN cisterna	sporophytic self-incompatibility
COPII vesicle	linear glucose polymer	syncytial-type cell plate
cytokinesis	mannosidase-I	TGN
cytoplasmic streaming	medial Golgi cisterna	*trans* Golgi cisterna
default pathway	meiocyte	*trans* Golgi network
dolichol	membrane recycling	transitional phragmoplast
dumbbell-shaped vesicle	mini-phragmoplast	turgor pressure
dynamin	mitosis	vacuolar targeting sequence
dynamin-GTPase	monensin	V-ATPase
early TGN cisterna	*N*-linked glycoprotein	wound-induced callose

References

[1]Different polymers are synthesized at different locations, in the ER and the Golgi apparatus, at the plasma membrane, or in the cell wall.

Johansen JN, Vernhettes S & Höfte H (2006) The ins and outs of plant cell walls. *Curr. Opin. Plant Biol.* 9, 616–620.

Staehelin LA & Newcomb EH (2000) Membrane structure and membranous organelles. In Biochemistry and Molecular Biology of Plants (BP Buchanan, W Gruissem, RL Jones eds), pp 2–50. Rockville, MD: American Society of Plant Physiologists.

Suh MC, Samuels AL, Jetter R et al. (2005) Cuticular lipid composition, surface structure, and gene expression in *Arabidopsis* stem epidermis. *Plant Physiol.* 139, 1649–1665.

[2]The endoplasmic reticulum consists of a three-dimensional membrane network that extends throughout the cytoplasm and is differentiated into specialized domains.

Vitale A, Schnell D, Raikhel NV & Chrispeels MJ (2010) Targeting and intracellular traffic of proteins. In Biochemistry and Molecular Biology of Plants (BP Buchanan, W Gruissem, RL Jones eds). Rockville, MD: American Society of Plant Physiologists, in press.

Staehelin LA (1997) The plant ER: a dynamic organelle composed of a large number of discrete functional domains. *Plant J.* 11, 1151–1165.

[3]Golgi stacks consist of sets of flattened cisternae that are structurally distinct and exhibit a polar architecture.

Jürgens G (2004) Membrane trafficking in plants. *Annu. Rev. Cell Dev. Biol.* 20, 481–504.

Nebenführ A & Staehelin LA (2001) Mobile factories: Golgi dynamics in plant cells. *Trends Plant Sci.* 6, 160–167.

Staehelin LA, Giddings T Jr, Kiss JZ & Sack FD (1990) Macromolecular differentiation of Golgi stacks in root tips of *Arabidopsis* and *Nicotiana* seedlings as visualized in high pressure frozen and freeze-substituted samples. *Protoplasma* 157, 75–91.

Staehelin LA & Kang B-H (2008) Nanoscale architecture of endoplasmic reticulum export sites and of Golgi membranes as determined by electron tomography. *Plant Physiol.* 147, 1454–1468.

[4]Plant Golgi stacks are dispersed throughout the cytoplasm, travel along actin filament tracks with myosin motors, and stop at ER export sites.

Boevinck P, Jarka K, Santa-Cruz S et al. (1998) Stacks on tracks: the plant Golgi apparatus traffics on an actin/ER network. *Plant J.* 15, 441–447.

Nebenführ A, Gallagher LA, Dunahay TG et al. (1999) Stop-and-go movements of plant Golgi stacks are mediated by the acto-myosin system. *Plant Physiol.* 121, 1127–1141.

Yang YD, Elamawi R, Bubeck J et al. (2005) Dynamics of COPII vesicles and the Golgi apparatus in cultured *Nicotiana tabacum* BY-2 cells provides evidence for transient association of Golgi stacks with endoplasmic reticulum exit sites. *Plant Cell* 17, 1513–1531.

[5]COPII vesicles bud from the ER with an external scaffold that is transferred with the vesicles to the *cis* side of the Golgi stacks, giving rise to the Golgi scaffold (matrix).

Kang B-H & Staehelin LA (2008) ER-to-Golgi transport by COPII vesicles in *Arabidopsis thaliana* involves a ribosome-excluding scaffold that is transferred with the vesicles to the Golgi matrix. *Protoplasma* 234, 51-64.

Robinson DG, Herranz MC, Bubeck J et al. (2007) Membrane dynamics in the early secretory pathway. *Crit. Rev. Plant Sci.* 26, 199–225.

[6]Intra-Golgi transport of membrane and cargo molecules is best explained by the cisternal progression/maturation model.

Glick BS (2000) Organization of the Golgi apparatus. *Curr. Opin. Cell Biol.* 12, 450–456.

Nebenführ A (2003) Intra-Golgi transport: escalator or bucket brigade? *Annu. Plant Rev.* 9, 76–89.

Storrie B & Nilsson T (2002) The Golgi apparatus: balancing new with old. *Traffic* 3, 521–529.

[7]The *trans* Golgi network cisternae of plants are transient organelles that sort and package Golgi products into secretory and clathrin-coated vesicles.

Staehelin LA, Kang B-H (2008) Nanoscale architecture of endoplasmic reticulum export sites and of Golgi membranes as determined by electron tomography. *Plant Physiol.* 147, 1454–1468.

Otegui MS, Herder R, Schulze J et al. (2006) The proteolytic processing of seed storage proteins in Arabidopsis embryo cells starts in the multivesicular bodies. *Plant Cell* 18, 2567–2581.

[8]Maintenance of an appropriate pH in different Golgi compartments is essential for their functions.

Schumacher K (2006) Endomembrane proton pumps: connecting membrane and vesicle transport. *Curr. Opin. Plant Biol.* 9, 595–600.

Zhang GF, Driouich A & Staehelin LA (1993) Effect of monensin on plant Golgi: re-examination of the monensin-induced changes in cisternal architecture and functional activities of the Golgi apparatus of sycamore suspension-cultured cells. *J. Cell Sci.* 104, 819–831.

[9]Golgi stacks are structurally and functionally differentiated in a tissue- and developmental stage–specific manner.

Lynch MA & Staehelin LA (1992) Domain-specific and cell type-specific localization of two types of cell wall polysaccharides in the clover root tip. *J. Cell Biol.* 118, 467–479.

Staehelin LA, Giddings T Jr, Kiss JZ & Sack FD (1990) Macromolecular differentiation of Golgi stacks in root tips of *Arabidopsis* and *Nicotiana* seedlings as visualized in high pressure frozen and freeze-substituted samples. *Protoplasma* 157, 75–91.

[10]The process of exocytosis delivers matrix molecules to the cell wall and membrane molecules to the plasma membrane.

Battey NH, James NC, Greenland AJ & Brownlee C (1999) Exocytosis and endocytosis. *Plant Cell* 11, 543–659.

Campanoni P & Blatt MR (2007) Membrane trafficking and polar growth in root hairs and pollen tubes. *J. Exp. Bot.* 58, 65-74.

[11]Turgor pressure has profound effects on vesicle-mediated secretion, membrane recycling from the plasma membrane, and the movement of secreted molecules through cell walls.

Geldner N & Jürgens G (2006) Endocytosis in signaling and development. *Curr. Opin. Plant Biol.* 9, 589–594.

Grabski, S.; de Freijter, A.W.; Schindler, M. Endoplasmic reticulum forms a dynamic continuum for lipid diffusion between contiguous soybean root cells. *Plant Cell* 5:25-38, 1993.

Staehelin LA & Chapman RL (1987) Secretion and membrane recycling in plant cells: novel intermediary structures visualized in ultra-rapidly frozen sycamore and carrot suspension-culture cells. *Planta* 171, 43–57.

[12]Cellulose microfibrils, callose, and mixed-linked glucans are made by enzyme complexes in the plasma membrane.

Philippe S, Saulnier L & Guillon F (2006) Arabinoxylan and 1,3/1,4-β-glucan deposition in cell walls during wheat endosperm development. *Planta* 224, 449–461.

Samuels AL, Giddings TH & Staehelin LA (1995) Cytokinesis in tobacco BY-2 and root tip cells: a new model of cell plate formation in higher plants. *J. Cell Biol.* 130, 1345–1357.

Wilson SM, Burton RA, Doblin MS et al. (2006) Temporal and spatial appearance of wall polysaccharides during cellularization of barley (*Hordeum vulgare*) endosperm. *Planta* 224, 655–667.

[13]The enzymes that synthesize cellulose microfibrils in higher plants are organized into complexes that visually look like rosettes.

Cellulose synthase complexes are pushed forward in the plasma membrane by the growing cellulose microfibrils that they extrude into the cell wall.

Giddings TH, Brower DL & Staehelin LA (1980) Visualization of particle complexes in the plasma membrane of *Micrasterias denticulata* associated with the formation of cellulose fibrils in the primary and secondary cell walls. *J. Cell Biol.* 84, 327–339.

Haigler CH, Brown RM & Benziman M (1980) Calcofluor White ST alters *in vivo* assembly of cellulose microfibrils. *Science* 210, 903–906.

Kimura S, Laosinchai W, Itoh T et al. (1999) Immunogold labeling of rosette terminal cellulose-synthesizing complexes in the vascular plant *Vigna angularis*. *Plant Cell* 11, 2075–2085.

Paredez AR, Sommerville CR & Ehrhardt DW (2006) Visualization of cellulose synthase demonstrates functional association with microtubules. *Science* 312, 1491–1495.

Saxena IM & Brown RM Jr (2005) Cellulose biosynthesis: current views and evolving concepts. *Ann. Bot.* 96, 9–21.

Tsekos I (1999) The sites of cellulose synthesis in algae: diversity and evolution of cellulose-synthesizing enzyme complexes. *J. Phycol.* 35, 635–655.

[14]Rosette complexes are concentrated in domains of the plasma membrane underlying sites of rapid cellulose synthesis.

Schneider R & Herth W (1986) Distribution of plasma membrane rosettes and kinetics of cellulose formation in xylem development in higher plants. *Protoplasma* 131. 142–152.

Wada M & Staehelin LA (1981) Freeze-fracture observations on the plasma membrane, the cell wall and the cuticle of growing protonemata of *Adiantum capillus-veneris* L. *Planta* 151, 462–468.

[15]The half-life of rosette complexes determines the length of cellulose microfibrils.

Rudolf U & Schnepf E (1988) Investigations of the turnover of the putative cellulose-synthesizing particle "rosettes" within the plasma membrane of *Funaria hygrometrica* protonema cells I: effects of monensin and cytochalasin B. *Protoplasma* 143, 63–73.

[16]The cellulose microfibrils of primary cell walls are shorter than those of secondary cell walls.

Blashek W, Koehler H, Semler U & Franz G (1982) Molecular weight distribution of cellulose in primary cell walls. Investigations with regenerating protoplasts, suspension cultured cells and mesophyll of tobacco. *Planta* 154, 550–555.

Brett CT (2000) Cellulose microfibrils in plants: biosynthesis, deposition, and integration into the wall. *Int. Rev. Cytol.* 199, 161–199.

[17]Callose appears transiently during growth and development and is also formed in response to biotic and abiotic stresses.

Bacic A, Fincher GB & Stone BA (eds) (2009) Chemistry, Biochemistry and Biology of 1–3 β Glucans and related polysaccharides. New York: Academic Press.

Verma DPS & Hong Z (2001) Plant callose synthase complexes. *Plant Mol. Biol.* 47, 693–701.

[18]Callose is a transient cell wall component that is critically important for many growth and developmental processes.

Ferguson C, Teri TT, Siikaaho M et al. (1998) Location of cellulose and callose in pollen tubes and grains of *Nicotiana tabacum*. *Planta* 206, 452–460.

Samuels AL, Giddings TH & Staehelin LA (1995) Cytokinesis in tobacco BY-2 and root tip cells: a new model of cell plate formation in higher plants. *J. Cell Biol.* 130, 1345–1357.

[19]Stress-induced callose synthesis serves to protect cells from potentially lethal biotic and abiotic insults.

Lush WM & Clarke AE (1997) Observations of pollen tube growth in *Nicotiana alata* and their implications for the mechanism of self-incompatibility. *Sex. Plant Reprod.* 10, 27–35.

Nishimura MT, Stein M, Hou BH et al. (2003) Loss of callose synthase results in salicylic acid-dependent disease resistance. *Science* 301, 969–972.

Stanghelini ME, Rasmussen SL & Vandemark GJ (1993) Relationship of callose deposition to resistance of lettuce to *Plasmopara lactucae-radicis*. *Phytopathology* 83, 1498–1501.

Worral D, Hird DL, Hodge R et al. (1992) Premature dissolution of the microsporocyte callose wall causes male sterility in transgenic tobacco. *Plant Cell* 4, 759–771.

Yim K-O & Bradford KJ (1998) Callose deposition is responsible for apoplastic semipermeability of the endosperm envelope of muskmelon seeds. *Plant Physiol.* 118, 83–90.

[20]Plant Golgi stacks multiply by division.

Segui-Simarro JM & Staehelin LA (2006) Cell cycle-dependent changes in Golgi stacks, vacuoles, clathrin-coated vesicles and multivesicular bodies in meristematic cells of *Arabidopsis thaliana*: a quantitative and spatial analysis. *Planta* 223, 223–236.

Staehelin LA & Kang B-H (2008) Nanoscale architectue of endoplasmic reticulum export sites and of Golgi membranes as determined by electron tomography. *Plant Physiol.* 147, 1454–1468.

[21]During mitosis and cytokinesis cytoplasmic streaming stops and the Golgi stacks redistribute to specific locations.

Nebenführ A, Frohlick JA & Staehelin LA (2000) Redistribution of Golgi stacks and other organelles during mitosis and cytokinesis. *Plant Physiol.* 124, 135–151.

[22]Somatic-type cell plate formation can be divided into four distinct phases.

Austin JR, Segui-Simarro JM & Staehelin LA (2005) Quantitative analysis of changes in the (+)-end geometry of microtubules involved in plant cell cytokinesis. *J. Cell Sci.* 118, 3895–3903.

Jürgens G (2005) Cytokinesis in higher plants. *Annu. Rev. Plant Biol.* 56, 281–299.

Karahara I, Suda J, Tahara H et al. (2009) The preprophase band is a localized center of clathrin-mediated endocytosis in late prophase cells of the onion cotyledon epidermis. *Plant J.* 57, 819–831.

Mineyuki Y (1999) The preprophase band of microtubules: its function as a cytokinetic apparatus in higher plants. *Int. Rev. Cytol.* 187, 1–49.

Samuels AL, Giddings TH & Staehelin LA (1995) Cytokinesis in tobacco BY-2 and root tip cells: a new model of cell plate formation in higher plants. *J. Cell Biol.* 130, 1–13, 1995.

Segui-Simarro JM, Austin JR II, White EA & Staehelin LA (2004) Electron tomographic analysis of somatic cell plate formation in meristematic cells of *Arabidopsis* preserved by high pressure freezing. *Plant Cell* 16, 836–856.

Segui-Simarro JM, Otegui MS, Austin JR & Staehelin LA (2007) Plant cytokinesis: insights gained from electron tomography studies. In Cell Division Control in Plants (DPS Verma, Z Hong eds), pp 251–287. Heidelberg: Springer.

[23]Dynamin-GTPases create dumbbell-shaped, cell plate–forming vesicles containing dehydrated and possibly gelled cell wall–forming molecules.

Cannon MC, Ternus K, Hall Q et al. (2008) Self-assembly of the plant cell wall requires an extensin scaffold. *Proc. Natl. Acad. Sci. USA* 105, 2226–2231.

Kang BH, Busse JS & Bednarek SY (2003) Members of the *Arabidopsis* dynamin-like gene family, ADL1, are essential for plant cytokinesis and polarized cell growth. *Plant Cell* 15, 899–913.

Segui-Simarro JM, Austin JR II, White EA & Staehelin LA (2004) Electron tomographic analysis of somatic cell plate formation in meristematic cells of *Arabidopsis* preserved by high pressure freezing. *Plant Cell* 16, 836–856.

Stowell MHB, Marks B, Wigge P & McMahon HT (1999) Nucleotide-dependent conformational changes in dynamin: evidence for a mechanochemical molecular spring. *Nat. Cell Biol.* 1, 27–32.

Thakur BR, Singh RK & Handa AK (1997) Chemistry and uses of pectin: a review. *Crit. Rev. Food Sci. Nutr.* 37, 47–73.

[24]The association of cisternae of the endoplasmic reticulum with forming cell plates increases over time, but the functional importance of this spatial relationship has yet to be fully explored.

Segui-Simarro JM, Austin JR II, White EA & Staehelin LA (2004) Electron tomographic analysis of somatic cell plate formation in meristematic cells of *Arabidopsis* preserved by high pressure freezing. *Plant Cell* 16, 836–856.

[25]Formation of cell walls in the syncytial endosperm and in meiocytes involves a special kind of cell plate, the syncytial-type cell plate.

Brown RC & Lemon BE (2001) The cytoskeleton and spatial control of cytokinesis in the plant life cycle. *Protoplasma* 215, 35–49.

Otegui MS (2007) Endosperm cell walls: formation, composition, and functions. In Endosperm Development and Molecular Biology Series: Plant Cell Monographs, vol 8 (O-A Olsen ed), pp 159–178. Heidelberg: Springer.

Otegui M & Staehelin LA (2000) Cytokinesis in flowering plants: more than one way to divide a cell. *Curr. Opin. Plant Biol.* 3, 493–502.

Otegui M, Mastronarde DN, Kang B-H et al. (2001) Three-dimensional analysis of syncytial-type cell plates during endosperm cellularization visualized by high resolution electron tomography. *Plant Cell* 13, 2033–2051.

Shivanna KR, Cresti M & Ciampolini F (1997) Pollen development and pollen-pistil interaction. In Pollen Biotechnology for Crop Production and Improvement (VK Sawhney ed), pp 15–39. Cambridge, UK: Cambridge Univesity Press.

Biosynthesis of Cell Wall Polymers

5

The magnitude of the task of making and maintaining plant cell walls is illustrated by the estimate that these cellular activities involve the products of well over 2000 different genes. One factor contributing to this large number of required genes is the large number of genes that code for glycosyltransferases and glycosidases. Indeed, plants contain a higher number of such genes than any other organism sequenced to date (**Figure 5.1**) due to the tremendous diversity and complexity of the carbohydrates found in the plasma membrane and cell walls. In this chapter, the assembly pathways of different cell wall polymers are organized around their sites of assembly, namely, the ER and Golgi apparatus, the plasma membrane, and the cell wall. We begin with a discussion of the assembly and processing of N- and O-linked glycans in the ER and Golgi cisternae and then progress to the synthesis of cell wall matrix polysaccharides, which involves many different types of glycosyltransferases that have also been localized to Golgi cisternae. Our discussion of cellulose synthesis focuses on the CESA proteins and their functional domains, and the role of sucrose synthase in producing UDP-glucose for cellulose synthase. The discussion of the functional organization of the catalytic subunit of callose synthases (GSL proteins) is based on the analysis of their gene sequences. We also discuss the differences between Ca^{2+}-dependent and Ca^{2+}-independent callose synthase systems. Finally, we address the environment of the cell wall in terms of polymer assembly and how this applies to the synthesis of lignin.

A. General Mechanisms of Polymer Assembly

1. Polymers are assembled from building blocks.

Cell wall polymers of plants are assembled from four types of building blocks: proteins from activated amino acids; polysaccharides and the oligosaccharide side chains of glycoproteins from activated sugars; lignin

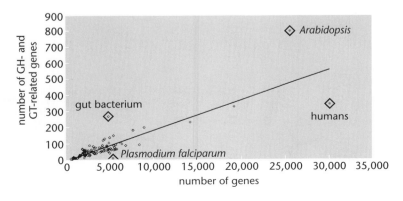

Figure 5.1 Correlation between the number of glycosidase- (GH) and glycosyltransferase- (GT) related genes and the total number of genes in 86 bacterial and 9 eukaryotic genomes. (Adapted from P.M. Coutinho et al., *Trends Plant Sci.* 8:563–565, 2003.)

from monolignols (or phenylpropanoids); and cutin, suberin, and waxes from various CoA thioesters of fatty acids and aromatic molecules. All, or most, of these building blocks are synthesized in the cytoplasm and then transported to the different intracellular and extracellular sites of polymer assembly. Each cell appears to produce the subunits needed for the synthesis of its own wall polymers; few if any of the secreted subunits or polymers pass beyond the middle lamella that separates the walls of adjacent cells. Interestingly, the bulk of the polymers are composed of subunits that contain only carbon, hydrogen, and oxygen. Only glycoproteins, which are relatively nonabundant, are made of nitrogen-(N)-containing building blocks, consistent with the N-limited, sessile lifestyles of plants. Animals, by contrast, which are not N-limited, have abundant protein in their extracellular matrix.

2. Nucleotide sugars are the source of the glycosyl residue building blocks used in the synthesis of cell wall polysaccharides and glycoproteins.[ref1]

Photosynthesis is the ultimate source of the sugars incorporated into cell wall polysaccharides and glycoproteins. Newly synthesized sugars are often stored transiently in starch, fructans, or oligosaccharides in photosynthetic or storage tissues. Subsequently the storage polymers are converted into sucrose, raffinose, or stachyose sugars for transport to their destinations, where they are converted via a series of reactions to hexose monophosphate and then usually to *nucleoside diphosphate sugars* (NDP-sugars or sugar nucleotides). Only after this activation can the sugars be utilized by glycosyltransferases for the synthesis of cell wall polysaccharides and glycoproteins. In cells undergoing rapid wall growth, such as elongating cottonseed hairs, the total cytoplasmic NDP-sugars have been estimated to be present in up to millimolar concentrations, the most abundant being UDP-Glc (75% of total NDP-sugars) and UDP-Gal.

NDP-sugars contain a monosaccharide attached to the terminal phosphate of a ribonucleoside diphosphate (**Figure 5.2**). In the case of D-sugars, the

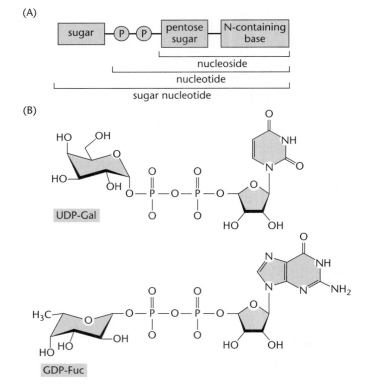

Figure 5.2 Nomenclature and structure of sugar nucleotides. (A) General structure of a sugar nucleotide. (B) Structures of UDP-Gal and GDP-Fuc.

TABLE 5.1 SUGAR NUCLEOTIDES FOUND IN PLANTS

NAME	USUAL ABBREVIATION
adenosine-diphosphate-L-arabinose	ADP-Ara
adenosine-diphosphate-D-fructose	ADP-Fru
adenosine-diphosphate-D-galactose	ADP-Gal
adenosine-diphosphate-D-glucose	ADP-Glc or ADPG
adenosine-diphosphate-D-mannose	ADP-Man
guanosine-diphosphate-L-fucose	GDP-Fuc
guanosine-diphosphate-D-galactose	GDP-Gal
guanosine-diphosphate-D-glucose	GDP-Glc or GDPG
guanosine-diphosphate-D-mannose	GDP-Man or GDPM
thymidine-diphosphate-D-galactose	TDP-Gal
thymidine-diphosphate-D-galacturonic acid	TDP-GalA
thymidine-diphosphate-D-glucose	TDP-Glc or TDPG
uridine-diphosphate-D-apiose	UDP-Api
uridine-diphosphate-L-arabinose	UDP-Ara
uridine-diphosphate-D-fructose	UDP-Fru
uridine-diphosphate-D-galactose	UDP-Gal
uridine-diphosphate-D-galacturonic acid	UDP-GalA
uridine-diphosphate-D-glucose	UDP-Glc or UDPG
uridine-diphosphate-D-glucuronic acid	UDP-GlcA
uridine-diphosphate-N-acetyl-D-galactosamine	UDP-GalNAc
uridine-diphosphate-N-acetyl-D-glucosamine	UDP-GlcNAc
uridine-diphosphate-L-rhamnose	UDP-Rha
uridine-diphosphate-D-xylose	UDP-Xyl

Note: The activated sugars for KDO (3-deoxy-D-manno-octulosonic acid) and DHA (3-deoxy-D-lyxo-2-heptulosaric acid) are not known, but may be CMP-KDO and CMP-DHA.

linkage between the sugar and the NDP is an α-glycosidic bond, whereas L-sugars form a β-glycosidic bond. Most NDP-sugar residues are in the pyranose ring form, with UDP-Api being an exception. Uridine and, to a lesser extent, guanine are the principal nucleosides of the NDP-sugars used in the synthesis of plant cell walls; adenosine is used in starch synthesis, and thymidine derivatives have also been reported. **Table 5.1** shows a list of NDP-sugars that have been found in plant cells.

The starting material for the synthesis of most NDP-sugars utilized in the synthesis of cell wall molecules is a pool of five hexose monophosphates—Glc-1-P, Glc-6-P, Fru-6-P, Man-6-P, and Man-1-P—that are produced and maintained in rapid equilibrium by a combination of phosphoglycomutases and 6-phosphate isomerases (**Figure 5.3** and **Figure 5.4**). This pool can be tapped at two points, Glc-1-P and Man-1-P, by pyrophosphorylases to form some NDP-sugars directly, and others through subsequent enzyme-medi-

Figure 5.3 Enzymes involved in sugar activation and interconversion reactions.

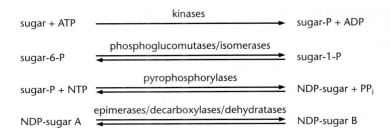

ated NDP-sugar interconversions. Most of the monosaccharides used in cell wall synthesis are derived from UDP-D-glucose via a series of dehydrogenation, decarboxylation, and 4-epimerization reactions (Figure 5.4). Also depicted in this figure are enzymatic steps that have been shown to be interrupted in cell wall mutants. For example, the *mur1* mutation of *Arabidopsis* causes a deficiency in L-fucose in the shoot due to a lesion in an isoform of GDP-D-mannose 4,6-dehydratase. Rapid progress is being made in the cloning and characterization of *Arabidopsis* genes encoding other enzymes involved in sugar interconversions. A particularly interesting finding is that in plants the synthesis of UDP-Rha involves a bifunctional 3,5 epimerase/4-keto reductase enzyme (UDP-L-Rha synthase), whereas in eubacteria these two activities are associated with two different enzymes.

Most of the enzymes involved in the *nucleotide sugar interconversions* appear to be soluble cytosolic proteins. However, the 4-epimerases specifically acting on UDP-GlcA/GalA and on UDP-Xyl/Ara as well as a synthase that produces UDP-Xyl from UDP-GlcA are located within Golgi cisternae (**Figure 5.5**). These membrane-bound enzymes are type II membrane proteins (see Figure 5.7) with their conserved C-terminal catalytic domain facing the lumen and the variable N-terminal domain in contact with the cytosol. Different isoforms of the UDP-D-xylose synthase appear to provide UDP-Xyl to different Golgi-localized biosynthetic pathways—specifically, for the assembly of *N*-glycans, xyloglucans, or xylans, respectively. Those enzymes possessing a luminal catalytic domain could conceivably funnel their activated sugars directly into luminal xylosyltransferases (for example, for *N*-glycan biosynthesis), whereas the soluble ones in the cytosol could donate such sugars to xylosyltransferases with a transmembrane disposition (possibly involved in xylan biosynthesis).

Plants also possess *salvage pathways* to reutilize monosaccharides produced by hydrolysis of storage or structural polysaccharides. These recycling pathways provide a handle for experiments, because they permit *in vivo* labeling of polysaccharides by simply using radiolabeled sugars. However, once an NDP-sugar is formed, it is subject to modification by interconverting enzymes, as illustrated in Figure 5.3. The most compelling evidence in support of the recycling pathways has come from *in vivo* labeling of intermediates. For example, in spinach, [^{3}H]Ara has been shown to be incorporated into Ara-1-P, and then into UDP-Ara, UDP-Xyl, and Xyl-1-P, respectively.

A given sugar may form NDP-sugars with more than one type of nucleotide. This may reflect the use of different types of NDP-sugars as building blocks for different synthetic pathways. For example, whereas ADP-Glc is made in plastids and used there for starch synthesis, UDP- and GDP-Glc are produced in the cytoplasm and are delivered to enzymes involved in the synthesis of cellulose, callose, and hemicellulose. However, as discussed in Concept 5D9, UDP-glucose used in cellulose and callose synthesis may also be formed by a membrane-bound form of sucrose synthase.

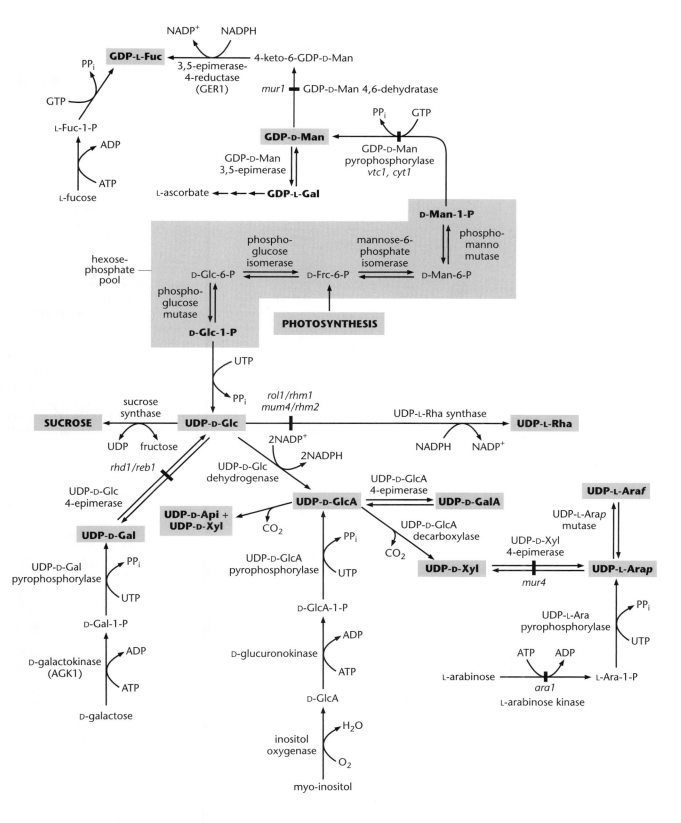

Figure 5.4 Overview of nucleotide sugar interconversion pathways and how they relate to the hexose-phosphate pool, which provides the sugar precursor molecules. For simplicity, only pathways important in the synthesis of cell wall polymers in higher plants are shown. The cloned genes and mutants refer to *Arabidopsis*. (Adapted from W.-D. Reiter, *Curr. Opin. Plant Biol.* 11:236–243, 2008.)

Figure 5.5 Biosynthesis pathways of nucleotide sugars made by Golgi-localized enzymes.

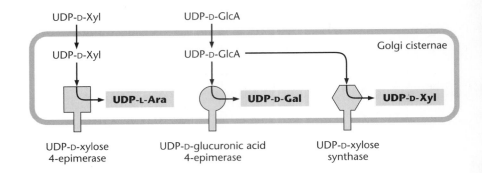

3. Transport of nucleotide sugars into ER and Golgi cisternae is mediated by specific NDP-sugar/NMP antiporters.[ref2]

Nucleotide sugar transporters (NSTs) are present in all eukaryotes and provide substrates for glycosyltransferases that have their active site in the lumina of the ER, Golgi, and TGN. As illustrated schematically in **Figure 5.6**, these cytoplasmically synthesized NDP-sugar molecules are transported into the cisternae by means of membrane-spanning transporter proteins that function as antiporters, exchanging NMP for specific NDP-sugars. Upon binding of the NDP-sugar to the catalytic domain of a given glycosyltransferase, the sugar is transferred to the nascent oligosaccharide and NDP is released into the cisternal lumen. NDP is then immediately cleaved to NMP and P_i by a nucleotide diphosphatase both to drive the uptake of new NDP-sugars and to reduce the competitive binding of NDP to the glycosyltransferase. NMP and P_i are then returned through separate carriers to the cytoplasm, where they can be recycled.

Figure 5.6 Relationship between NDP-sugar biosynthesis and glycosylation of a glycoprotein in a Golgi cisterna. The regeneration of NDP-sugars from nucleotide phosphate, ATP, and salvage sugars (and the *de novo* synthesis of NDP-sugars by the sugar interconversion pathway; see Figure 5.4) occurs in the cytosol. Transport of the NDP-sugar into the lumen of the Golgi cisterna is by means of an antiport system. The glycosyltransferase transfers the sugar from the NDP-sugar molecule to the oligosaccharide chain on the glycoprotein. The remaining nucleoside diphosphate molecule is immediately hydrolyzed by a phosphatase in the membrane. Return of the nucleoside monophosphate and the P_i to the cytosol is mediated by the NDP-sugar/NMP antiporter and a P_i channel, respectively.

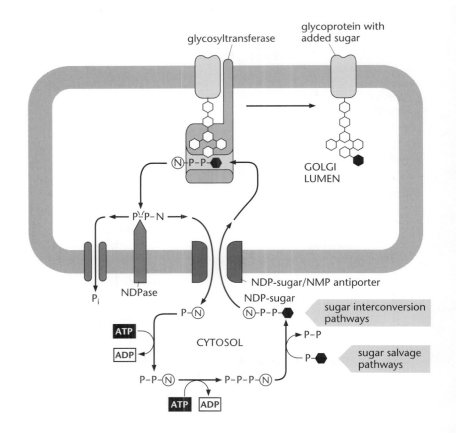

The finding that most NDP-sugar transport mutants have macromolecules with a 95 to 100% deficiency in the corresponding sugar has led to the suggestion that each type of nucleotide sugar is transported by a specific nucleotide-sugar/nucleotide antiporter, but a few transporters able to transport more than one NDP-sugar have also been identified. Hydrophobicity analysis of the amino acid sequences of NDP-sugar transporters, which are typically 300 to 350 amino acids long, predicts that they contain between 6 and 10 transmembrane domains. Several lines of evidence, including radiation inactivation data, suggest that NSTs form homodimers in the membranes. In *Arabidopsis*, NDP-sugar transporters for UDP-Gal, UDP-Gal/UDP-Glu, and GDP-Man have been identified and partially characterized.

The Golgi-localized protein known as *reversibly glycosylated protein* has attracted a lot of interest and created a lot of confusion in the field of plant cell wall research. This protein has been reported to become labeled upon incubation with UDP-[^{14}C]glucose and to be able to subsequently transfer the label to certain nucleotide acceptors in the presence of high concentrations of unlabeled UDP-glucose. However, because the nature of the postulated high-energy sugar–amino acid bond has never been established, the role of this protein in polysaccharide synthesis has remained obscure. A recent study suggests that, based on sequence homology, the reversibly glycosylated protein is a UDP-arabinose mutase that converts UDP-arabinopyranose to UDP-arabinofuranose. The rice UDP-arabinose mutase also possesses reversible glycosylating properties.

4. Wall polysaccharides are made by membrane-bound polysaccharide synthases and glycosyltransferases.

Glycosyltransferases are enzymes that catalyze the transfer of a single sugar from nucleotide sugar donors onto acceptors such as glycan chains, peptides, or lipids: NDP-sugar + acceptor-OH → acceptor-O-sugar + NDP. All measurable enzyme activities involved in the synthesis of wall polysaccharides and glycoproteins are found associated with defined membrane fractions and not with the cytosolic fraction when mechanically disrupted plant cells are fractionated on a sucrose gradient. Since these activities are not removed from the membranes by water or salt extractions, it has been concluded that glycosyltransferases are integral membrane proteins. This conclusion is supported directly by freeze-fracture images of putative cellulose synthase complexes that appear as organized clusters of particles in plasma membranes (see Figure 4.27), and by gene sequence data.

5. Glycosyltransferases are generally nonabundant proteins.[ref3]

The enormous diversity of carbohydrate structures associated with glycoproteins and polysaccharides is a reflection of the underlying diversity of the glycosyltransferase enzymes that assemble these structures. Considering that nearly all glycosyltransferases can catalyze the synthesis of only a single glycosidic linkage, together with the multitude of sugar linkages already recorded in the glycoprotein, glycolipid, and polysaccharide databases, one can easily imagine that the number of different glycosyltransferases in plant cells is on the order of hundreds (see Table 5.2 and Table 5.3). Since the majority of these enzymes are confined to the membranes of Golgi stacks, which have a finite amount of surface area, the quantities of specific enzymes per Golgi stack and even per cell are small. Indeed, simple calculations suggest that each Golgi stack may contain fewer than 10 copies of an average glycosyltransferase. This explains the difficulty encountered by biochemists in purifying such enzymes from isolated Golgi preparations

6. All glycosidases and glycosyltransferases involved in the modification of glycoproteins and polysaccharides have a common topology and common Golgi retention mechanisms.[ref4]

To date, over several hundred glycosidases and glycosyltransferases involved in the processing of glycoproteins and polysaccharides have been identified and partially characterized. The glycosyltransferases have been classified into 92 distinct, sequence-based families using the *Carbohydrate Active Enzyme (CAZy) database* (**Sidebar 5.1**). Most of the sequences predict enzymes with a common *type II membrane protein* topology (**Figure 5.7**). Typically, such enzymes have a short N-terminal cytoplasmic tail (4 to 24 amino acids for glycosyltransferases, 5 to 43 amino acids for glycosidases); a transmembrane helix of 15 to 22 amino acids; a proline-rich stem region of variable length (up to 400 amino acids long), which lacks a discernible secondary architecture but may participate in coiled-coil structures; and a relatively large C-terminal catalytic domain (300 to 1125 amino acids), which protrudes into the lumen of ER or Golgi cisternae. This ball-on-a-chain design provides the enzymes with a large amount of structural flexibility, which is probably needed to reach oligosaccharide chains in different orientations. An example of a glycosyltransferase that does not fit this mold is GlcNAc transferase, the enzyme that catalyzes the first step of dolichol-oligosaccharide synthesis (see Figure 5.13). This enzyme has four transmembrane helices. Most likely, the four hydrophobic helices are needed for the binding of the long dolichol lipid tail (see Figure 5.12) to the enzyme.

The precise mechanism of targeting of these enzymes to specific Golgi cisternae has yet to be elucidated. The two principal hypotheses of Golgi retention are (1) the bilayer thickness hypothesis, which postulates that retention is based on the length of the hydrophobic membrane-spanning domain (<18 amino acids for *cis*/medial enzymes, 18 to 22 amino acids for *trans*-side cisternae), and (2) the oligomerization-kin hypothesis, which suggests that retention involves the formation of homo-/hetero-oligomerization mediated by the hydrophilic cytoplasmic and luminal domains of the proteins. There is evidence to support both hypotheses, as well as combinations of these features. However, it is also possible that the Golgi scaffold/matrix (see Concept 4A5) could be involved in this enzyme-targeting process.

Despite the topological similarities, the degree of sequence similarity between members of distinct glycosyltransferase families that add sugars to glycoproteins is remarkably small. Only glycosyltransferases within small catalytically related families such as the α-1,3-fucosyltransferases appear to share any significant primary sequence similarity. As shown by a comparative sequence analysis of α-1,2- and α-1,3-fucosyltransferases, even the use of the same sugar nucleotide, GDP-Fuc, and the capability to act on the same oligosaccharide substrate do not guarantee any primary sequence similarity.

7. The synthesis of polysaccharide backbones involves initiation, elongation, and termination reactions.[ref5]

The synthesis of all biopolymers involves three types of reactions: chain *initiation*, chain *elongation*, and chain *termination*. These reactions have been characterized in greatest detail for the biosynthesis of polypeptides, but apply also to the biosynthesis of polysaccharides.

The need for a specific polysaccharide synthesis-initiating enzyme is evident from the fact that most polysaccharide synthases identified to date can transfer sugars only to a preexisting polysaccharide. Theoretically, the first glycosidic bond could be created by the transfer of a sugar residue from a sugar nucleotide to another, free sugar. Accumulating evidence suggests,

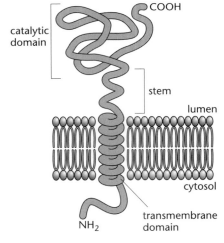

Figure 5.7 Topology of a type II integral membrane protein characteristic of Golgi glycosidases and glycosyltransferases. The protein has one transmembrane domain, and the catalytic domain is attached to a flexible stem. Note that the N-terminus is in the cytosol and the C-terminus in the lumen, which together with the single transmembrane domain, make it a type II integral membrane protein.

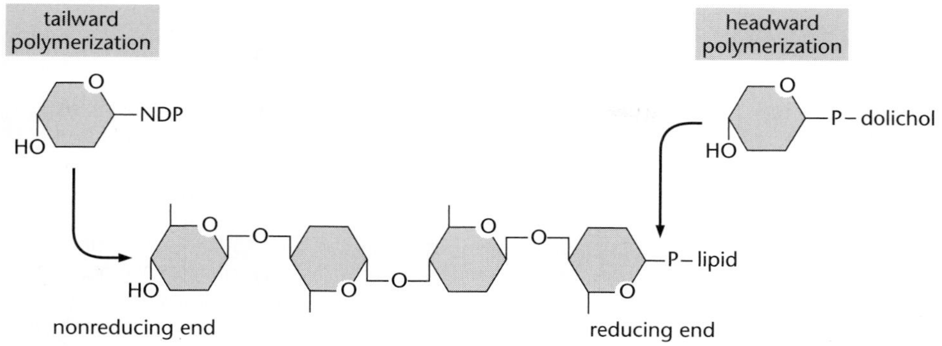

Figure 5.8 Comparison of headward versus tailward growth (polymerization) of a polysaccharide polymer. Tail polymerization occurs at the nonreducing end and involves NDP-sugar precursors. In the case of head polymerization, the addition of sugars to the reducing end of a growing lipid-linked polysaccharide involves transfer to a nonreducing position of a sugar attached to a lipid acceptor.

however, that the first sugar is instead attached through its reducing end to a *noncarbohydrate primer*, which can act as an acceptor for the transfer of further sugar residues. In the case of glycogen and starch, the primer is a protein; some bacterial polysaccharides are initiated on a lipid carrier; the primer for callose synthesis appears to be β-furfuryl-β-glucoside; and a possible primer for cellulose synthesis is sitosterol-β-glucoside.

Elongation of an initiated polysaccharide may involve either the direct transfer of a monosaccharide from an NDP-sugar, or the indirect transfer of one or several sugars from a lipid or other intermediate to the growing polysaccharide. Polymerization of polysaccharides theoretically may occur by *"tailward growth"* or by *"headward growth"* (**Figure 5.8**). Tailward growth involves the addition of saccharide units to the nonreducing end of the growing polymer. Headward growth refers to systems in which new sugars are attached to the reducing end of a growing oligosaccharide and which involve lipid intermediates. Evidence from *in vitro* studies of the assembly of β-glucans such as cellulose in bacteria points to a tailward mechanism of polysaccharide growth; that is, they grow by the addition of sugars from NDP-sugar precursors to the nonreducing end.

Termination of polymer synthesis can be achieved either by an active or by a passive process. Where the synthesis of a polymer is controlled by a template such as is the case with nucleic acids and proteins, termination is an active process that leads to polymers of a predetermined size. In contrast, enzyme-guided polymerization systems tend to operate in a rather uncoordinated termination mode, in which the enzyme complex simply disengages from the polymer after a certain period of time or the polymer reaches a size where the stability of its association with the enzyme complex is reduced. This leads to products of variable length. Plant cell wall polysaccharides are likely examples of this latter type of polymerization system. For example, xyloglucans extracted from mung bean vary in molecular weight from 50 to 200 kD. However, there is also evidence for some control in the length of cellulose molecules and in the size of the matrix polysaccharide RG-II. Thus, in cotton fibers, the average degree of polymerization (DP) of cellulose fibrils increases markedly from 500–1000 to 10,000–15,000 at the onset of secondary wall formation. In other systems, the DP for primary wall cellulose varies from 1000 to 6000. The size change in cotton occurs independently of environmental factors known to alter the rate of cellulose fibril synthesis. Most likely it is caused by a switch to a different cellulose synthase isoform, but the involvement of an active termination process cannot be ruled out.

A second example of polysaccharide polymers with controlled lengths comes from an investigation of the contour length of isolated xyloglucan molecules from onion cell walls. In that study, the xyloglucan molecules exhibited an approximately 30-nm periodicity (about 60 backbone glucosyl residues) in length. This finding has led to the suggestion that xyloglucan is synthesized using a *block polymer construction*. Where and how the 30-nm blocks are ligated into larger molecules is still unknown.

Sidebar 5.1 The CAZy Database (Peter Ulvskov, University of Copenhagen)

The CAZy database classifies Carbohydrate Active enZymes, including hydrolases and glycosyltransferases (GTs), into families, and provides links to NCBI and UNIPROT databases where the sequences are stored. The CAZy classification is more detailed and recognizes many more gene products as glycosyltransferases than the general gene annotations or the Pfam protein family database. The CAZy classification has informed reverse genetic approaches to gene discovery enabled by the genome sequence of *Arabidopsis* and other plants. Gene names usually convey information about function or mutant phenotype, but CAZy-based naming of GT genes was adopted for the *Selaginella* genome. This is expected to set a trend for new genome sequencing projects.

The methods used by the CAZy database to identify GTs and assign them to a family are primarily sequence-based and secondarily based on three-dimensional structure. Many families are quite divergent—as little as 11% sequence identity is often found between the least related amino acid sequences in a family,

Most known GTs adopt one of three protein folds, GT-A, GT-B, or GT-C. Fold information is useful when trying to decide whether a candidate gene could encode a GT or not. GTs that adopt the GT-C fold use dolichol rather than nucleotide sugar donors. Functional correlations for the other folds have not yet been identified.

CAZy provides predictions about whether candidate GTs use retaining or inverting reaction mechanisms. Retaining means that the configuration of the nucleotide sugar donor substrate is also found in the product (**Figure 1**).

Retaining glycosyltransferases produce β-glycosidic bonds from all nucleotide sugars except UDP-Ara, UDP-Rha, CMP-KDO and GDP-Fuc, where β-linkages are formed (the configurations of the nuceotides of DHA and aceric acid are

not known. The L-galactose in RG-II is probably transferred from GDP-β-L-Gal). Inverting GTs do the opposite. All plant integral membrane proteins from family GT2 (the cellulose synthase superfamily) characterized so far form β-1,4 linkages between hexoses.

CAZy GT families are rarely confined to one donor sugar, so inferring which donor sugar is used requires additional information, for example by mutant analysis. Single gene mutants are, however, unlikely to feature a strong phenotype where there is redundancy—here double mutants or RNAi transformants are called for. Whether there is redundancy can often be resolved by analysis of sequence similarity and phylogeny of all membrers of the GT family in question followed by co-expression analysis. If a mutant or RNAi-transformant is reduced in a particular cell wall epitope, then the question is: Does this epitope comprise a glycosidic linkage that is compatible with the GT family membership of the gene which is knocked out? The biochemical function is probably inferred correctly if the answer to this question is affirmative, but proof has to be provided, typically by assaying the GT produced by heterologous expression in a suitable host.

Where the mammalian members of a family are outnumbered by the plant sequences, such families are likely to include GTs for plant-specific glycans. The diversification of the angiosperm families that began some 200 million years ago was accompanied by gene duplication on a massive scale, resulting in unique lineages within some families. Poplar has, for example, about 1600 CAZy genes, compared with only 1000 in *Arabidopsis*. [B.L. Cantarel et al. The Carbohydrate-Active enZymes database (CAZy): an expert resource for glycogenomics. *Nucleic Acids Res.* 37 (Database issue): D233–238, 2009.]

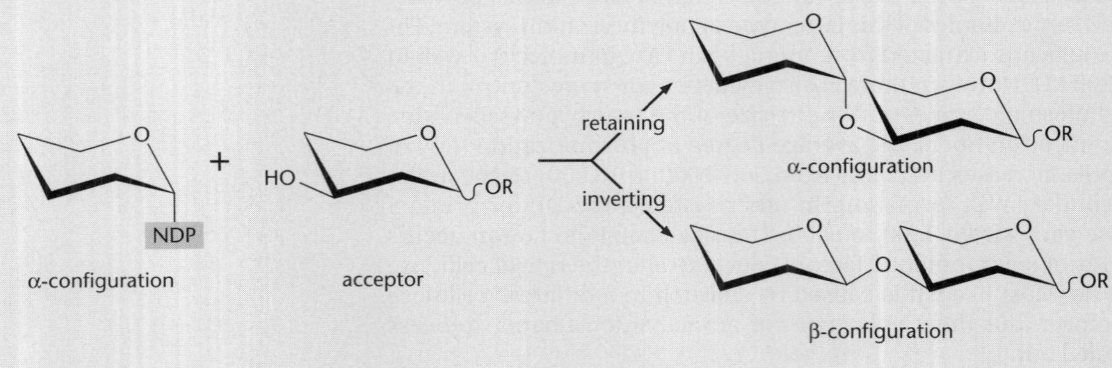

Figure 1 Diagrammatic representation of retaining versus inverting reaction mechanisms of sugars.

TABLE 1 LIST OF THE GENE FAMILIES IN THE CAZy DATABASE THAT HAVE PLANT GT ENTRIES. THE CURRENT DATABASE DEFINES 92 FAMILIES, BUT GT90 IS THE LAST THAT CONTAINS HIGHER-PLANT ACCESSIONS. THE FAMILIES OF PARTICULAR CELL WALL RELEVANCE ARE HIGHLIGHTED WITH BOLD NUMBERS.

FAMILY	MECHANISM	FOLD	*Arabidopsis*	RICE
1	Inverting	GT-B	121	202
2	Inverting	GT-A	42	47
4	Retaining	GT-B	24	25
5	Retaining	GT-B	6	11
8	Retaining	GT-A	42	39
10	Inverting	GT-B	3	3
13	Inverting	GT-A	1	1
14	Inverting	Fold not known	11	12
16	Inverting	GT-A	1	1
17	Inverting	GT-B	6	4
19	Inverting	GT-B	1	1
20	Retaining	GT-B	11	11
21	Inverting	GT-A	1	1
22	Inverting	GT-C	3	4
24	Retaining	GT-A	1	1
28	Inverting	GT-B	4	4
29	Inverting	Fold not known	3	5
30	Inverting	GT-B	1	1
31	Inverting	GT-A	33	40
32	Retaining	GT-B	6	3
33	Inverting	GT-B	1	2
34	Retaining	Fold not known	8	6
35	Retaining	GT-B	2	2
37	Inverting	Fold not known	10	18
41	Inverting	GT-B	2	3
43	Inverting	GT-A	4	10
47	Inverting	GT-B	39	35
48	Inverting	Fold not known	12	11
50	Inverting	GT-C	1	1
57	Inverting	GT-C	2	2
58	Inverting	GT-C	1	1
59	Inverting	GT-C	1	1
61	Inverting	Fold not known	7	25
64	Retaining	GT-A	3	3
65	Inverting	Fold not known	1	1
66	Inverting	GT-C	2	2
68	Inverting	Fold not known	3	1
75	Inverting	Fold not known	5	3
76	Inverting	Fold not known	1	1
77	Retaining	Fold not known	19	16
90	Inverting	Fold not known	9	10

8. Polysaccharide synthases can be identified biochemically by substrate and activator binding activities, by product entrapment techniques, and by catalytic activities detected in nondenaturing gels.[ref6]

Glycosyltransferases have binding sites for activators such as cyclic-D-GMP (bacterial cellulose synthesis), for oligosaccharide or polysaccharide acceptor substrates, and for nucleotide sugars where these are the donor molecules. These properties can be exploited for the identification of the transferases in membrane fractions or in partially purified enzyme fractions by using labeling techniques with fluorescent or radioactive compounds.

A technique known as *product entrapment,* first exploited for the purification of yeast chitin synthase, and subsequntly for callose and cellulose synthases, has proved effective for identifying proteins associated with functional polysaccharide synthase complexes. This method takes advantage of the fact that when detergent-solubilized enzyme complexes are incubated with substrates and effectors, the glycan polymers produced remain associated with the complexes and can be used to recover the complexes by low-speed centrifugation. Whether the association is by entanglement or through covalent bonds has not been determined, but effective purification is achieved.

Native gel electrophoresis in a matrix containing an acceptor polysaccharide has been used to identify a glycosyltransferase complex from pea epicotyl membranes responsible for adding fucose to the side chains of xyloglucan. Tamarind seed xyloglucan, which lacks fucosyl residues, served as the acceptor for the fucosyltransferase while entrapped in the polyacrylamide gel matrix. Following electrophoresis of the solubilized fucosyltransferase-containing extract, the gel was incubated with a radiolabeled fucosyl donor and autoradiographed. This revealed the presence of an active fucosyltransferase in a 150-kD complex.

9. The three-dimensional structure of the backbone of cell wall polysaccharides suggests synthesis by glycosyltransferases with two active sites.[ref7]

There is general agreement that the polysaccharides of plant cell walls are formed by polymerization of monosaccharide building blocks (Concept 5A2). It has been hypothesized that the assembly of the backbones of the six polysaccharides that are present in the walls of all higher plants may involve binding of NDP-sugars to two separate sites followed by the formation of disaccharide intermediates. This hypothesis is based on the fact that in terms of their three-dimensional structure all of these polysaccharides are composed of diglycosyl repeating units. Similar considerations also apply to other polysaccharides with disaccharide repeats such as chitin, hyaluronan, and heparin.

In the case of rhamnogalacturonan I, it is readily apparent that the backbone is composed of many repeats of a diglycosyl unit, namely, rhamnosyl-galactosyluronic acid. In contrast, the homogalacturonan backbone of rhamnogalacturonan II and the backbone of homogalacturonan itself appear to be formed from the single glycosyl residue, galactosyluronic acid, repeated many times. In terms of their three-dimensional structures, however, their backbones are in fact also constructed of diglycosyl repeating units. This is equally true of the cellulosic backbone of xyloglucan, of the β-1,4-glucan chains of cellulose itself, and of the β-1,4-xylan backbone of glucuronoarabinoxylan.

The backbones of xyloglucan and cellulose are linear chains of β-D-glucosyl residues, and the backbone of glucuronoarabinoxylan is a linear chain of

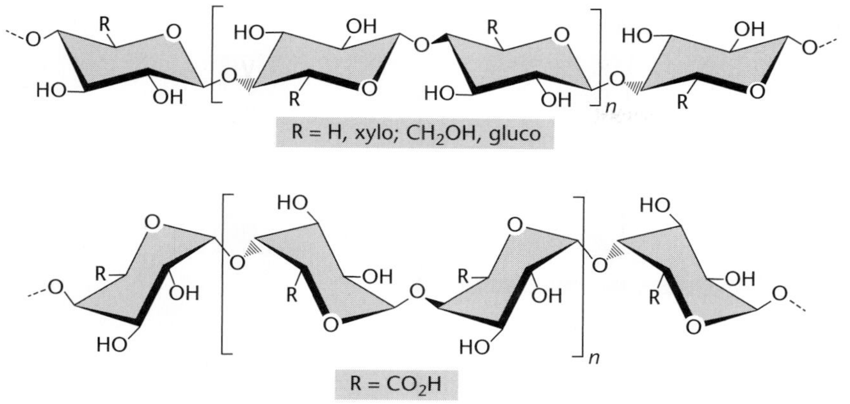

R = H, xylo; CH₂OH, gluco

R = CO₂H

Figure 5.9 Disaccharide repeating unit of the backbones of (A) cellulose, xyloglucan, and xylan, and (B) homogalacturonan and rhamnogalacturonan II. In (A) R = H for xylan, CH_2OH for cellulose and xyloglucan; in (B) R = CO_2H.

β-D-xylosyl residues. β-D-glucosyl and β-D-xylosyl residues both take the $^4C_1$ chair form when connected glycosidically in 4-linked chains (**Figure 5.9**). In the $^4C_1$ chair form, O-1 and O-4 of β-D-glucosyl and β-D-xylosyl residues are both in the equatorial configuration. Furthermore, the backbones of the corresponding polysaccharides tend to assume the energetically favored ribbonlike form that allows the glucan chains of cellulose to hydrogen-bond to one another to form crystalline fibers. Xyloglucan and arabinoxylan take this form when they hydrogen-bond to cellulose. Even xyloglucan in solution is thought to have a closely related structure. In the ribbonlike form, every glucosyl or xylosyl residue is flip-flopped in relation to its two immediate neighbors.

The backbones of rhamnogalacturonan II and homogalacturonan are linear chains of α-D-galactosyluronic acid residues. The O-1 and O-4 of the α-D-galactosyluronic acid residues of homogalacturonan are both in the *axial* configuration (Figure 5.9). Thus, when a 4-linked chain of these residues is interconnected by glycosidic bonds in the α configuration, as in homogalacturonan, a puckered ribbonlike chain is believed to form in which every galactosyluronic acid residue is inverted in relation to its neighbors, just as are the glucosyl residues of cellulose and of the xyloglucan backbone and the xylosyl residues of the arabinoxylan backbone.

The inversion of α-D-galactosyluronic acid, β-D-glucosyl, or β-D-xylosyl residues results in the corresponding face being "up" and then "down" in neighboring glycosyl residues of the backbones of these 4-linked polysaccharides. This has interesting consequences. For example, an *endo*-glycanase approaching such a chain interacts with different structures on alternating glycosyl residues. In other words, each of these polysaccharides presents a disaccharide repeating structure to *endo*-glycanases. Cellobiose is in fact the product of the *endo*-β-1,4-glucanase (cellulase) breakdown of the β-1,4-glucan chains of cellulose (see Figure 8.61).

The disaccharide nature of the repeating units of galacturonan, glucan, and xylan chains also has consequences for the mechanism by which they are synthesized. Cellulose fibrils, for example, are synthesized by a large, plasma membrane–located, multienzyme complex (rosette) that catalyzes the simultaneous synthesis of the numerous β-1,4-glucan chains that associate to form a crystalline microfibril of cellulose (see Concept 4A13). If each glucan chain synthesized by a rosette were elongated by a single UDP-glucose:glucosyltransferase that had only a single active site, the transferase would have to move from one side of the elongating glucan chain to the other to add the successive residues with flipped orientations in space. Alternatively, the transferase could have two active sites, with the sugar in the second active site rotated 180° from the sugar in the first site (**Figure 5.10**; see also Concept 5D). The two-site glycosyltransferase model has been tested for chitin synthase. Thus, to determine if this enzyme has two UDP-GlcNAc-binding sites, potential bivalent inhibitors were synthe-

Figure 5.10 A model for the synthesis of wall polysaccharides with backbones constructed of a disaccharide repeat. The assembly process is subdivided into five steps defining the growth of the polysaccharide, from its nonreducing end, by the addition of pairs of glycosyl residues transferred from two molecules of nucleoside diphosphate sugar (NDP-sugar). The enzyme complex contains two sites capable of binding NDP-sugars. The sites are loaded by transfer of the glycosyl residues from the NDP-sugars to an Asp-containing binding site in the glycosyltransferase. The model postulates that the two activated sugars are connected to form a dimer and then attached to the growing polysaccharide. The polysaccharide must translocate before two new activated glycosyl residues can initiate another elongation cycle. (Modified from N.C. Carpita, M.C. McCann and L.R. Griffing, *Plant Cell* 8:1451–1463, 1996.)

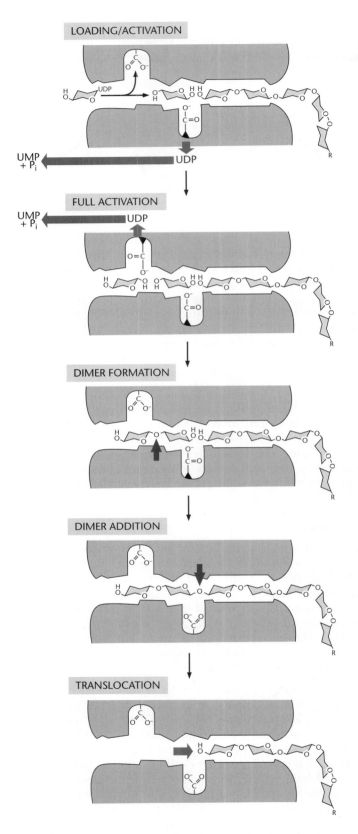

sized by linking together 5′-deoxy-5′-aminouridine residues using ethylene glycol linkers of different lengths. The finding that some of these dimers were an order of magnitude more potent inhibitors than monomeric derivatives supports the hypothesis that polysaccharides composed of disaccharide repeats are produced by glycosyltransferases with two binding sites.

10. Polysaccharide synthesis is carefully controlled, but the control points are still poorly understood.[ref8]

Plants consist of about 40 types of cells, which can be distinguished by the composition of their cell walls. Classical plant anatomists exploited this feature when they used different histochemical stains to define and distinguish the different cell types and tissues that make up a plant, and these differences have been characterized more recently by means of immunolabeling experiments with monoclonal antibodies (see Concepts 6B8 and 6C1). The observed differences in cell wall size, composition, and structure reflect on the ability of each cell to regulate the types and amounts of cell wall molecules that it makes, when these molecules are synthesized, and where they are secreted. Clearly, the proper assembly of the many cell wall types requires a variety of control systems.

The major factor controlling the qualitative composition of primary cell walls resides in the complement of synthases in the endomembrane system that produce the matrix components of each cell's walls. To what extent the synthesis of these enzymes and their activity is regulated by hormones, by transcriptional or translational controls, by allosteric conformation or by turnover of the enzymes, by environmental or mechanical stresses, and/or by the availability of nucleotide sugars has yet to be determined. Based on the recent finding that small interfering RNAs (siRNAs) derived from the coding region of the cellulose synthase gene *CESA6* also downregulate several *CSL* genes it has been postulated that different glycosyltransferases share common regulatory control points. Support for the idea that feedback regulation of levels of specific nucleoside diphosphate sugars may be important has come from the finding that UDP-dehydrogenase activity is inhibited by high levels of UDP-xylose. It is also likely that the coordinated regulation of multiple enzymes at the transition from primary to secondary wall synthesis involves the activation/deactivation of different sets of enzyme isoforms. In aspen, the bending of stems has been shown to increase cellulose synthesis in the phloem fibers on the tensionally stressed side of the stem, and to reduce cellulose synthesis on the opposite side of the stem that was subject to compression forces. Future studies will also have to address questions pertaining to the feedback mechanisms that sense and control cell size, as well as the thickness and the rate of expansion of plant cell walls during growth, development, and adaptation to environmental stresses.

B. Assembly and Processing of Glycoproteins and Proteoglycans

1. *N*-linked glycans of glycoproteins are assembled on dolichol lipids.[ref9]

A family of polyisoprenol lipids, known as *dolichols*, serve as carriers of sugars and of preassembled, "activated" oligosaccharides that, when completed, are transferred in their entirety to defined asparagine residues of nascent polypeptides in the ER (**Figure 5.11** and **Figure 5.12**). Eukaryotic dolichol molecules consist of 16 to 22 isoprene units. With the exception of polymeric isoprene (rubber and gutta-percha), dolichols are probably the longest natural hydrocarbons made up of a single repeating unit. A fully extended dolichol molecule measures close to 10 nm in length, compared to 2.5 nm for a typical fatty acid of a membrane phospholipid molecule (Figure 5.12). The large size of the hydrophobic dolichol tail appears to be essential for firmly anchoring the large hydrophilic, 14-sugar oligosaccharides during their assembly at the ER membrane and their transfer to the nascent polypeptide. Although much of the information about the use of dolichols in the assembly of N-*linked glycans* has come from studies with

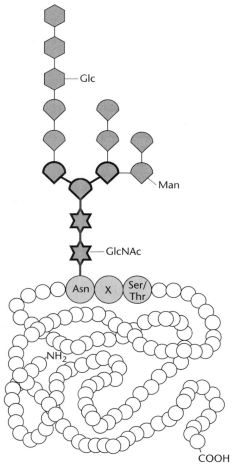

Figure 5.11 The asparagine-linked 14-sugar oligosaccharide (*N*-linked glycan) that is added to many secretory and membrane proteins in the ER. The five highlighted sugars constitute the "core region" of this glycan. In many cases only these core residues survive the extensive trimming process that occurs in conjunction with the assembly of complex *N*-glycans. *N*-glycans are attached only to asparagines in the sequences Asn-X-Ser/Thr (where X is any amino acid except proline).

(A)

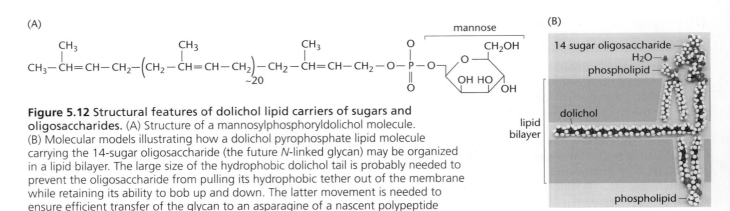

Figure 5.12 Structural features of dolichol lipid carriers of sugars and oligosaccharides. (A) Structure of a mannosylphosphoryldolichol molecule. (B) Molecular models illustrating how a dolichol pyrophosphate lipid molecule carrying the 14-sugar oligosaccharide (the future *N*-linked glycan) may be organized in a lipid bilayer. The large size of the hydrophobic dolichol tail is probably needed to prevent the oligosaccharide from pulling its hydrophobic tether out of the membrane while retaining its ability to bob up and down. The latter movement is needed to ensure efficient transfer of the glycan to an asparagine of a nascent polypeptide chain (see Figure 5.14). The long tail can also wrap around integral proteins.

yeast and animal cells, the fact that the same 14-sugar oligosaccharides are assembled on dolichols in plants demonstrates that this assembly pathway is highly conserved.

The two types of functions of dolichols in the assembly of the 14-sugar oligosaccharides of the *N*-linked glycans are diagrammed in **Figure 5.13.** The first seven sugars are added stepwise to a dolichol pyrophosphate while its polar head is protruding from the cytoplasmic side of the ER membrane. A transport enzyme then flips the seven-sugar oligosaccharide, together with the head region of the dolichol, to the opposite face of the bilayer. The oligosaccharide is further enlarged on the luminal side of the membrane by the stepwise addition of seven more sugar residues. These sugars, four mannose and three glucose residues, are individually transported from the cytoplasm to the ER lumen while attached to a dolichol phosphate and then transferred directly to the growing oligosaccharide chain. The final product is the 14-sugar oligosaccharide $Glc_3Man_9(GlcNAc)_2$ that is connected through a pyrophosphate bridge to the dolichol carrier (Figure 5.13).

The entire synthesis of the 14-sugar dolichol can be inhibited by the antibiotic *tunicamycin*, which prevents attachment of the first GlcNAc sugar

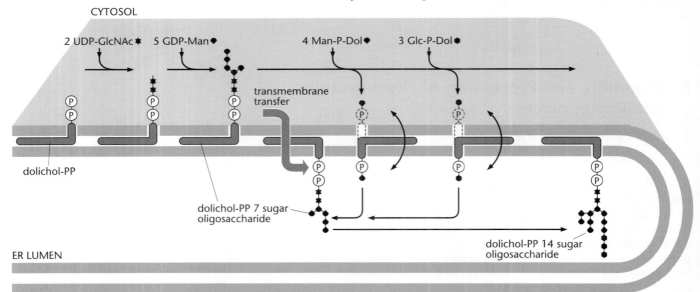

Figure 5.13 Synthesis of the dolichol-linked precursor oligosaccharide of *N*-linked glycans in the ER. The first two GlcNAc and five Man residues are added, one by one, to the dolichol pyrophosphate with its polar head facing the cytoplasm. An enzyme then flips the 7-sugar assembly across the bilayer and into the ER lumen with the dolichol tail serving as anchor. The following four Man and three Glc residues are then brought individually across the bilayer membrane while attached to a dolichol phosphate carrier. Both the Man and Glc residues are transferred to the growing oligosaccharide by glycosyltransferase enzymes in the ER lumen.

residue of the oligosaccharide to the dolichol phosphate. Glycoproteins synthesized in the presence of tunicamycin lack *N*-linked oligosaccharide side chains and can therefore be used to study the importance of such side chains for protein structure and function.

Dolichols function in bacteria in the synthesis of extracellular polysaccharides as well as in the assembly of glycoproteins and lipopolysaccharides. This has elicited extensive, but to date only partially successful, efforts to demonstrate the involvement of dolichols in the assembly of plant cell wall polysaccharides. Nevertheless, further studies of this topic seem warranted considering that the cellulose synthase CESA protein has a potential dolichol-binding domain and how little we know about the assembly of the different side chains of complex polysaccharides, in particular those associated with pectic polysaccharides and arabinogalactan proteins.

2. Newly synthesized *N*-linked glycans are transferred *en bloc* to nascent polypeptides and immediately subjected to processing by two glucosidases.[ref10]

After the assembly of the $Glc_3Man_9(GlcNAc)_2$ oligosaccharide precursor on the dolichol lipid carrier (Figure 5.13), the 14-sugar oligosaccharide is transferred *en bloc* and co-translationally from the dolicholpyrophosphate carrier to the amino side chain of an Asn in an Asn-Xaa-Ser/Thr sequence (Xaa being any amino acid other than proline) as the nascent polypeptide enters the ER lumen (**Figure 5.14**). The *oligosaccharyltransferase* that catalyzes this reaction, together with the signal sequence–removing peptidase, the dolichol-binding ribophorins, and several other proteins, has been shown to be part of a large peptide-gated channel complex that mediates the transfer of nascent polypeptides across the ER membrane. Binding of the oligosaccharide-carrying dolichol to the complex is mediated by a dolichol recognition consensus sequence on the transmembrane domain of ribophorin I. This complex is present in a ratio of 1:1 with ER-bound ribosomes, which suggests that it is part of the permanent protein translocation apparatus and can therefore monitor emerging polypeptides for potential glycosylation sites.

Not all of the potential Asn glycosylation sites become glycosylated. For example, in the enzyme β-fructosidase of carrot, only three of the six poten-

Figure 5.14 Co-translational transfer of a preassembled *N*-glycan from its dolichol carrier to a nascent polypeptide in the ER. The *en bloc* transfer is catalyzed by the enzyme oligosaccharyltransferase, which connects the 14-sugar oligosaccharide to an asparagine in an Asn-Xaa-Ser/Thr sequence of the polypeptide. The headgroup of the dolichol pyrophosphate is then flipped from the luminal back to the cytoplasmic side of the ER membrane for the assembly of a new oligosaccharide (see Figure 5.13). Two glucosidases then remove the three terminal glucose residues from the oligosaccharide.

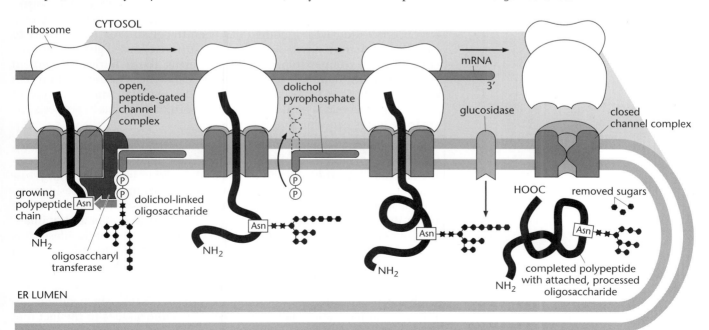

tial sites have oligosaccharide side chains attached, and in the bean E subunit of phytohemagglutinin the potential *N*-glycosylation site closest to the N-terminal end of the protein is always used, the middle site is sometimes used, and the site closest to the C-terminus is never used. The fact that the site closest to the N-terminus is the most highly glycosylated suggests that co-translational folding and previously bound glycans can cover up potential, later emerging sites before they can be glycosylated.

Shortly after the transfer of the 14-sugar oligosaccharide to the protein, the three terminal Glc residues are removed by two hydrolytic enzymes, glucosidase I and II, to yield a high-mannose-type glycan with nine Man residues (Figure 5.14). These glucosidases can be inhibited by the alkaloid castanospermine and the Glc analog 1-deoxynojirimycin. The short-lived Glc residues of *N*-glycans have been postulated to constitute a signal that indicates completion of the assembly of the 14-sugar precursor oligosaccharide and promotes its binding to the oligosaccharide transferase.

Folding and assembly of the completed proteins into functional complexes are believed to be the rate-limiting steps in transporting the proteins out of the ER. This has been demonstrated by site-directed mutagenesis of sites that promote the folding of the proteins or are critical for their assembly into complexes. Misfolded glycoproteins with *N*-glycan side chains bind to a UDP-Glc:glycoprotein glucosyltransferase that has a high affinity for improperly folded proteins. This soluble enzyme then transfers a glucose residue from UDP-Glc to the high-mannose glycans, which, in turn, mediates binding of the misfolded protein to the lectin chaperones calnexin and calreticulum. Based on studies in yeast, proteins that remain misfolded are exported from the ER into the cytoplasm via peptide channels and are then degraded in proteosomes.

3. Processing of *N*-linked glycans in the plant Golgi apparatus is similar but not identical to the processing in other organisms.[ref11]

Transfer of the glycoprotein from the ER to the Golgi initiates the next level of *N*-glycan processing, yielding *high-mannose-type, paucimannosidic-type, hybrid-type,* or *complex-type glycans* (**Figure 5.15**). The reactions that give rise to these different types of *N*-glycans are catalyzed by glycosidases and glycosyltransferases as shown schematically in **Figure 5.16**. The reason that some *N*-linked glycans remain in the high-mannose form while others become extensively modified has been traced to their accessibility to the oligosaccharide-processing enzymes; surface exposed glycans become converted to complex-type glycans, whereas those buried in protein folds remain in their high-mannose form. Differences in the local protein environment of the exposed oligosaccharides may determine the type of complex glycan that is produced on a given protein, but this hypothesis has yet to be tested experimentally.

The first enzyme to act on *N*-linked glycans in the Golgi is *mannosidase I,* which trims the four α-1,2-linked Man residues (Figure 5.16). This processing is needed for the addition of a GlcNAc residue by *GlcNAc transferase I,* which, in turn, is required for the removal of two more Man residues by mannosidase II to yield GlcNAcMan$_3$(GlcNAc)$_2$. All of the steps leading up to this six-sugar oligosaccharide are common to plants and animals. Immunoelectron microscopy has shown that α-1,2-mannosidase I is located in the medial cisternae of the Golgi stacks in *Arabidopsis* meristem cells (see Figure 4.7).

The term *complex glycan* refers to the products of the second half of the *N*-glycan processing pathway (Figures 5.15 and 5.16), which differs significantly in plants, animals, and yeast. The name alludes to the fact that these molecules are more highly branched and contain sugars (in plants,

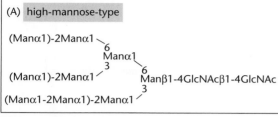

(A) high-mannose-type

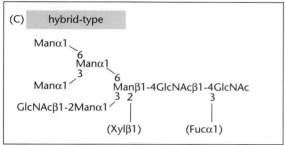

(C) hybrid-type

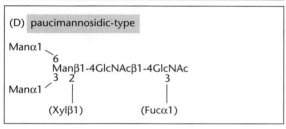

(D) paucimannosidic-type

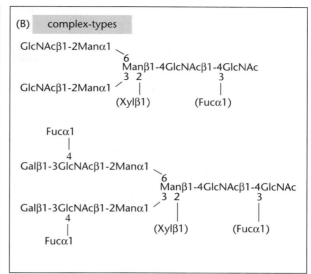

(B) complex-types

Figure 5.15 Structure of different types of N-linked glycans isolated from plant glycoproteins. (A) A high-mannose-type N-linked glycan, (B) complex-type N-linked glycans, (C) a hybrid-type N-linked glycan, and (D) a paucimannosidic-type N-linked glycan. (Adapted from P. Lerouge et al., *Plant Mol. Biol.* 38:31–48, 1998.)

fucose, xylose, and galactose) not found in the high-mannose intermediates. Hybrid-type N-linked glycans are produced by the processing of only the α-1,3-mannose branch of the intermediate $Man_5(GlcNAc)_2$, leading to oligosaccharides having α-1,3-fucose and/or β-1,2-xylose residues linked to $GlcNAcMan_5NAc_2$ (Figure 5.15). The paucimannosidic-type N-linked glycans result from the elimination of terminal residues from complex-type N-linked glycans. This type of processing occurs only in vacuoles.

Immunolabeling experiments of sycamore cells have shown the β-1,2-xylose to be added in medial and the α-1,3-fucose in *trans* Golgi cisternae. No plant glycoproteins containing an α-1,3-fucose without a β-1,2-xylose have been described, but the reverse situation is not uncommon. In contrast to the N-linked glycans of animal cells, those of plants do not contain any N-acetylneuraminic acid (sialic acid) residues.

The *clg* mutation in *A. thaliana* in which the GlcNAc transferase I enzyme is knocked out provides further support for the idea that the sequential action of the glycosyltransferases in the N-linked glycan processing pathway is based on each glycosyltransferase recognizing the sugar group added by the preceding transferase. Thus, the *clg* mutant was initially identified as a mutant that lacked β-1,2-xylose residues, and only later was it found that the mutant has a defective GlcNAc transferase I gene and is blocked at the earlier $Man_5(GlcNAc)_2$ step. When the *clg* mutant is complemented with a human GlcNAc transferase I gene, normal processing of N-linked glycans is observed.

4. The sites of O-linked glycosylation are defined in part by rules of proline hydroxylation.[ref12]

Three major classes of structural cell wall glycoproteins are heavily glycosylated. These are the hydroxyproline-rich glycoproteins (30–55% sugars), the solanaceous lectins (~45% sugars), and the arabinogalactan proteins

Figure 5.16 Processing of *N*-linked glycans in the ER, Golgi, and vacuolar compartments of plants. α-Man I/II: α-mannosidases I/II; GlcNAcT I/II: *N*-acetylglucosaminyltransferases I/II; β-1,2-XylT: β-1,2-xylosyltransferase; α-1,3-FucT: α-1,3-fucosyltransferase; α-1,3- and α-1,4-FucT: α-1,3- and α-1,4-fucosyltransferases; β-1,3-GalT: β-1,3-galactosyltransferase. (Adapted from P. Lerouge et al., *Plant Mol. Biol.* 38:31–48, 1998.)

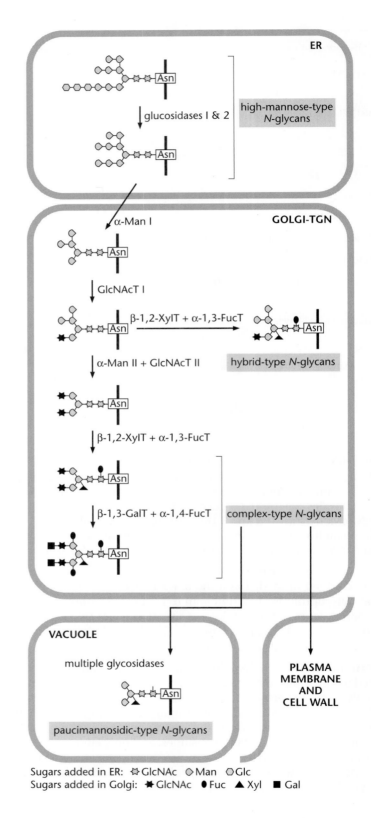

(90–98% sugars). As depicted in Figure 3.13, in plants, Ara, Gal, and Glc can be linked to hydroxyproline, Gal to serine, and GalNAc and Xyl to threonine.

Proline hydroxylation by the enzyme *prolylhydroxylase* (**Figure 5.17**) is a posttranslational event. In tobacco, prolyl-4-hydroxylase is a type II integral membrane protein with its catalytic domain exposed to the lumen of ER

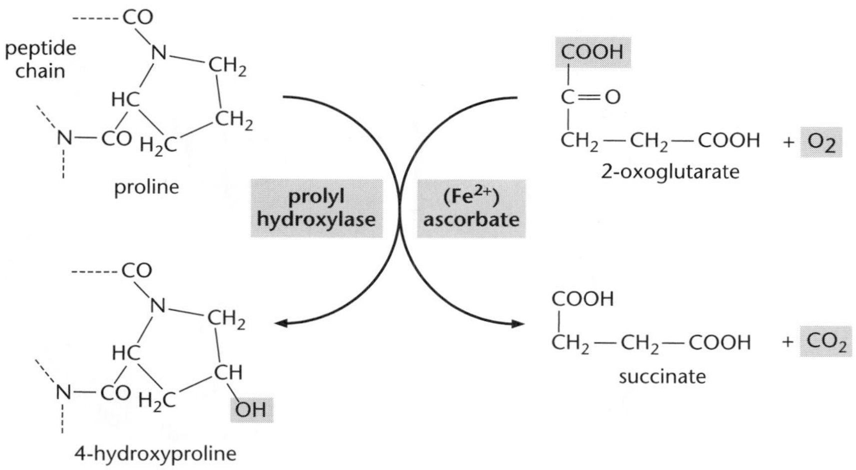

Figure 5.17 Hydroxylation of proline by prolylhydroxy-lase, a mixed-function oxygenase (prolyl peptide, 2-oxoglutarate: oxygen oxidoreductase).

cisternae. The membrane anchor appears to be unique to plants; in animals all prolylhydroxylases are soluble proteins. Proline hydroxylation may occur on some but not all prolines of a given protein. Some of the sequence "rules" that define which prolines become converted to hydroxyprolines in plants are described in Concept 3E3. For example, prolylhydroxylase has the highest affinity for series of four or more proline residues.

By using antibodies that recognize the arabinose-containing side chains of extensin in immunocytochemical labeling experiments, the initiation of O-linked arabinosylation of extensins has been located to late *cis* or medial Golgi cisternae, but it is possible that some arabinoses are already added in the ER. No information is currently available on where the galactosyltransferases that attach galactose to serine residues are localized in plant Golgi stacks. It is also not known where the more peripheral sugars of the longer and branched oligosaccharide side chains of arabinogalactan proteins are added. However, analogous to the biosynthesis of the side chains of the pectic polysaccharide RG-I, a *trans*-cisternal localization of these latter glycosyltransferases would seem likely.

5. The GPI anchors of arabinogalactan proteins are both synthesized and attached to the protein backbone in the ER.[ref13]

As described in Concept 3E7, classical-type AGPs are linked to the plasma membrane via a glycosylphosphatidylinositol (GPI) anchor. The assembly of a *GPI anchor* and its attachment to an AGP occurs in several steps (**Figure 5.18**). All of these reactions are mediated by ER-associated enzymes. GPI-anchor synthesis resembles to some extent the synthesis of a dolichol-linked *N*-glycan (see Figure 5.13) in that the synthesis of the headgroup is initiated by enzymes localized on the cytosolic membrane surface and completed by enzymes in the ER lumen. A flippase enzyme translocates the partially completed headgroup across the bilayer. The protein backbone of AGP is co-translationally inserted into the ER and is initially anchored to the membrane by means of a transmembrane domain close to the C-terminus. Coupling of the protein to the GPI anchor involves cleavage of the protein between two specific sites followed by the attachment of the GPI anchor to the protein. Hydroxylation of designated proline residues (Figure 5.18) appears to occur before the protein leaves the ER.

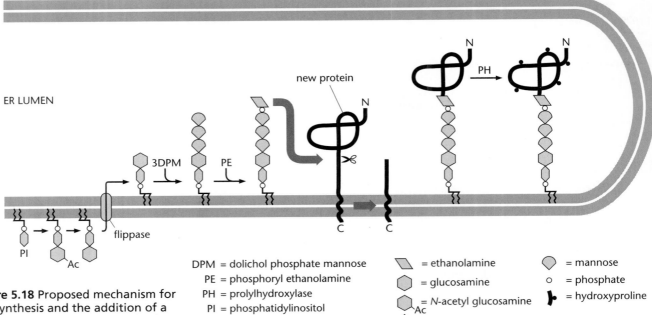

Figure 5.18 Proposed mechanism for the synthesis and the addition of a glycosylphosphatidylinositol (GPI) anchor to an AGP. (Adapted from C. Schulz et al., *Trends Plant Sci.* 3:426–431, 1998.)

DPM = dolichol phosphate mannose
PE = phosphoryl ethanolamine
PH = prolylhydroxylase
PI = phosphatidylinositol

= ethanolamine
= glucosamine
= N-acetyl glucosamine
= inositol

= mannose
= phosphate
= hydroxyproline

C. Assembly of Polysaccharides in the Endomembrane System

1. The backbones of some complex polysaccharides are synthesized by Golgi-located enzymes encoded by genes of the cellulose synthase-like (CSL) gene family.[ref14]

Analysis of the *Arabidopsis* and other plant genomes has led to the discovery of a large family of genes that show sequence similarity to the genes encoding the CESA-type cellulose synthase enzymes. Based on predicted protein sequences, these genes have been grouped together into a *cellulose synthase superfamily* (**Figure 5.19**), which consists of the *CESA* family and eight families of *cellulose synthase-like genes* (*CSLA, CSLB, CSLC, CSLD,*

Figure 5.19 Unrooted, bootstrapped tree of the *CESA/CSL* superfamily. All of these genes share the D, D, D, Q/RXXRW motif and belong to the GT2 family of glycosyltransferases (for more details see the CAZy Database Sidebar 5.1). Subfamilies highlighted by shading contain one or more members whose enzymatic function has been demonstrated. The corresponding structures of the polysaccharide products of these subfamilies are indicated. The functions of the subfamilies within the dashed ovals are unknown. The scalebar indicates the number of changes per basepair for the gene sequences compared. (Adapted from http://cellwall.stanford.edu.)

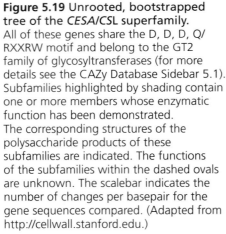

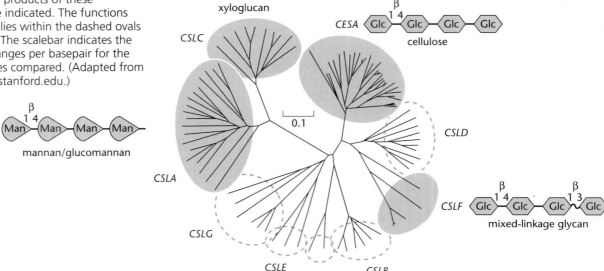

In addition to the four glucosyl residue repeats, xyloglucans also appear to be built of blocks of 30-nm repeats. Such blocks have been observed in electron micrographs of isolated XG molecules. This finding suggests that XG polymers could be assembled from repeats of 60 glucosyl residues (or 15 four-glucosyl repeat oligosaccharides). The biosynthetic origin of these larger repeats and how and where they get joined into longer molecules remains to be determined.

3. The XG-fucosyl and XG-galactosyltransferases can add fucosyl and galactosyl residues during or after synthesis of the XG backbone.[ref16]

When pea Golgi membranes are incubated with GDP-[14C]Fuc or UDP-[14C]Gal, these sugars can be incorporated into large endogenous or externally added XG molecules that do not undergo detectable elongation with time (**Figure 5.22**). This suggests that the addition of fucose and galactose to XG can occur independently of chain elongation. The functional independence of the *XG-Fuc-transferase* and the *XG-Gal-transferase* enzymes is consistent with the hypothesis that they are structurally independent enzymes and not tightly associated with the core XG-Glc/XG-Xyl transferase multienzyme complex. Both enzymes are predicted to be type II membrane proteins (see Figure 5.7). Yet to be determined is whether these transferases act as completely independent units or whether they become organized into homo- or hetero-oligomeric complexes.

Cloning of the XG-Fuc-transferase gene was preceded by the isolation and purification of the enzyme from pea epicotyls. Partial sequences of this protein led to the identification of the first Fuc-transferase of *Arabidopsis* (AtFUT1). The 60-kD protein contains three amino acid motifs that are conserved among all α-1,2-Fuc-transferases, but the other protein domains bear little resemblance to Fuc-transferases of other organisms.

The *mur2* mutant of *A. thaliana* has been shown to lack fucosylated XG because of a single point mutation in AtFUT1. This single mutation eliminates XG fucosylation in all major plant organs, indicating that AtFUT1 accounts for all of the XG-Fuc-transferase activity in *Arabidopsis*. The finding also suggests that the other putative fucosyl transferases, AtFUT2–AtFUT10, are involved in adding fucose residues to other polysaccharides such as RG-I and RG-II, since the *N*-linked glycan Fuc-transferase does not belong to the *AtFUT* gene family.

The acceptor structure that is required for the fucosylation of tamarind XG has been determined by analyzing the ability of partial hydrolysis products of tamarind XG to serve as acceptors for [14C]Fuc in *in vitro* experiments with detergent-solubilized XG-Fuc-transferase. Whereas the individual

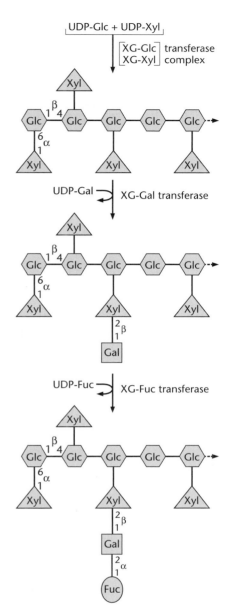

Figure 5.22 Mechanism of synthesis of XG. Note the coordinated assembly of the glucose backbone and the xylosyl side groups, and the stepwise addition of the galactose and fucose residues.

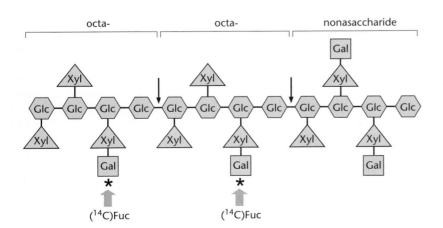

Figure 5.23 Substrate site specificity of the α-1,2-fucosyltransferase activity from pea epicotyl membranes. Fucosylation of the depicted sites (*asterisks*) requires oligosaccharides containing three repeats. The hepta-, octa-, and nonasaccharide fragments alone do not become fucosylated. Arrows, cellulase cleavage sites.

octa- and nonasaccharide fragments shown in **Figure 5.23** are unsuitable as fucosyl acceptors, oligosaccharide fragments containing all three subunits are excellent acceptor substrates for the Fuc-transferase. The two sites that become fucosylated and those that do not are also shown in Figure 5.23.

An intriguing aspect of the enzyme KAM1/MUR3, which adds galactosyl residues to XG, is that it serves as a dual-function protein. In particular, the XG galactosyltransferase activity of KAM1/MUR3 is contained in the C-terminal region of the protein, whereas the N-terminal portion is capable of interacting with the actin network. However, because the latter domain lacks any predictable actin-binding motifs, intermediary protein partners such as Golgi scaffold proteins (see Concept 4A5) might provide the actual links to the actin filaments.

4. The assembly of xyloglucan occurs in *trans* Golgi and early *trans* Golgi network cisternae.[ref17]

In cryofixed/freeze-substituted plant cells, the cisternae of the Golgi stacks can be resolved into morphologically distinct *cis*, medial, and *trans* types of Golgi cisternae (see Concept 4A3). By combining this type of specimen preparation technique with quantitative immunolabeling experiments using antibodies that recognize specific sugar groups on polysaccharides and glycoproteins, insights into the compartmental organization of the assembly pathways of such molecules can be obtained. This approach has been employed to map the spatial organization of the XG synthesis pathway within Golgi stacks using antibodies specific to the backbone and to the terminal fucose of the trisaccharide side chains of XG (Figure 5.19). Based on the observed labeling patterns, assembly of the backbone of XG occurs in *trans* Golgi cisternae. Fucosylation of the side chains also takes place in *trans* Golgi cisternae and may continue in early *trans* Golgi network cisternae. Immunolocalization of the XG-Fuc-transferase has shown that it too is associated with *trans* Golgi cisternae, consistent with the polysaccharide labeling data. Taken together, these results demonstrate unambiguously that the biosynthesis of cell wall matrix polysaccharides can occur exclusively in Golgi cisternae and does not depend on precursor or backbone molecules assembled in the ER.

5. The large number of enzymes needed to make pectic polysaccharides makes understanding their synthesis a challenge.[ref18]

The magnitude of the problem of understanding how pectic polysaccharides are assembled is highlighted by the following considerations. Pectic polysaccharides constitute the most complex polysaccharides discovered to date (see Section 3C), and complete structural characterization of the pectic polysaccharides has yet to be achieved. Nevertheless, based on the sugar residues of known pectic polysaccharides and the linkages by which they are connected, one can calculate that at least 67 transferases, including glycosyl-, methyl- and acetyltransferases, are required for their assembly in Golgi cisternae (**Table 5.2** and **Table 5.3**). Since each enzyme probably has a number of isoforms to provide for cell type–specific control of expression and to respond to environmental challenges, the total number of such enzymes in a given plant could be several-fold higher. Not to be forgotten is the potential need for different acetyl-, methyl- and feruloyltransferases to further increase the diversity of sugar residues of these truly complex polysaccharides.

Of the pectin biosynthesis enzymes identified to date, only GAUT1, the pectin backbone-synthesizing enzyme <u>GA</u>lact<u>U</u>ronic acid <u>T</u>ransferase 1, has been identified through biochemical studies. The discovery of all of

TABLE 5.2 GLYCOSYLTRANSFERASES "REQUIRED" FOR RG-I BIOSYNTHESIS

TYPE OF GLYCOSYLTRANSFERASE	PARENT POLYMER	ENZYME[a]	
		ACCEPTOR SUBSTRATE	ENZYME ACTIVITY
D-GalAT	RG-I	L-Rhaα1,4-GalA	α1,2-GalAT
D-GalAT	RG-I/HGA[b]	GalAα1,2-L-Rha	α1,4-GalAT
L-RhaT	RG-I	GalAα1,2-L-Rha	α1,4-L-RhaT
L-RhaT	HGA/RG-I[b]	GalAα1,4-GalA	α1,4-L-RhaT
D-GalT	RG-I	L-Rhaα1,4-GalA	β1,4-GalT
D-GalT	RG-I	Galβ1,4-Rha	β1,4-GalT
D-GalT	RG-I	Galβ1,4-Gal	β1,4-GalT
D-GalT	RG-I	Galβ1,4-Gal	β1,6-GalT
D-GalT	RG-I/AGP	Galβ1,3-Gal	β1,3-GalT
D-GalT	RG-I/AGP	Galβ1,3-Gal	β1,6-GalT
D-GalT	RG-I/AGP	Galβ1,6-Galβ1,3-Gal	β1,6-GalT
D-GalT	RG-I	L-Araf-1,4-Gal	1,5-GalT
L-AraT	RG-I	Galβ1,4-Rha	α1,3-L-ArafT
L-AraT	RG-I	L-Arafα1,3-Gal	α1,2-L-ArafT
L-AraT	RG-I	L-Arafα1,2-Ara	1,5-L-ArafT
L-AraT	RG-I	L-Rhaα1,4-GalA	1,4-ArafT
L-AraT	RG-I	L-Arafα1,5-Ara	α1,5-L-ArafT
L-AraT	RG-I	L-Arafα1,5-Ara	α1,2-L-ArafT
L-AraT	RG-I	L-Arafα1,5-Ara	α1,3-L-ArafT
L-AraT	RG-I	L-Arafα1,3-Ara	α1,3-L-ArafT
L-AraT	RG-I	Galβ1,4-Gal	α1,3-L-ArafT
L-AraT	RG-I	L-Araf-1,3-Gal	1,5-L-ArafT
L-AraT	RG-I/AGP	Galβ1,6-Gal	α1,3-L-ArafT
L-AraT	RG-I/AGP	Galβ1,6-Gal	α1,6-L-ArafT
L-AraT	RG-I	Galβ1,4-Gal	1,4-L-ArapT
L-FucT	RG-I	Galβ1,4-Gal	α1,2-L-FucfT
D-GlcAT	RG-I	Gal...	β1,6-GlcAT
D-GlcAT	RG-I	Gal...	β1,4-GlcAT
Methyl-RG-I:GlcA 4-O-methyltransferase	RG-I	GlcAβ1,6-Gal	
Acetyl-RG-I:GalA 3-O/2-O-acetyl-transferase	RG-I	GalAα1,2-L-Rhaα1,4$_{(n)}$	

[a]All sugars are D sugars and have pyranose rings unless otherwise indicated. Glycosyltransferases add to the glycosyl residue on the left of the indicated acceptor.
[b]Enzyme that may be required to make an HGA/RG-I junction.
Adapted from D. Mohnen, in G. Seymour and J.P. Knox (eds.), Pectins and Their Manipulation, pp. 52–96. Oxford: Blackwell/CRC Press, 2002.

TABLE 5.3 GLYCOSYLTRANSFERASE ACTIVITIES LIKELY TO BE REQUIRED FOR RG-II BIOSYNTHESIS[a]

TYPE OF GLYCOSYLTRANSFERASE	RG-II SIDE CHAIN	PARENT POLYMER	ENZYME[b]	
			ACCEPTOR SUBSTRATE	ENZYME ACTIVITY
D-GalAT	—	HGA/RG-II	GalAα1,4-GalA	α1,4-GalAT
D-GalAT	A	RG-II	L-Rhaβ1,3'-Apif	α1,2-GalAT
D-GalAT	A	RG-II	L-Rhaβ1,3'-Apif	β1,3-GalAT
L-RhaT	A,B	RG-II	Apifβ1,2-GalA	β1,3'-L-RhaT
L-RhaT	C	RG-II	KDO2,3-GalA	α1,5-L-RhaT
L-RhaT	B	RG-II	L-Araα1,4-Gal	α1,2-L-RhaT
L-RhaT	B	RG-II	L-Araα1,4-Gal	β1,3-L-RhaT
L-GalT	B	RG-II	GlcAβ1,4-Fuc	α1,2-L-GalT
D-GalT	B	RG-II	L-Acef Aα1,3-Rha	β1,2-GalT
L-AraT	D	RG-II	βDHA2,3-GalA	β1,5-L-ArafT
L-AraT	B	RG-II	Galβ1,2-L-AcefA	α1,4-L-ArapT
L-AraT	B	RG-II	L-Rhaα1,2-L-Ara	β1,2-L-ArafT
L-FucT	A	RG-II	L-Rhaβ1,3'-Apif	α1,4-L-FucT
L-FucT	B	RG-II	Galβ1,2-L-AceAf	α1,2-L-FucT
D-ApifT	A,B	RG-II	GalAα1,4-GalA	β1,2-ApifT
D-XylT	A	RG-II	L-Fucα1,4-L-Rha	α1,3-XylT
D-GlcAT	A	RG-II	L-Fucα1,4-L-Rha	β1,4-GlcAT
D-KDOT	C	RG-II	GalAα1,4-GalA	2,3-KDOT
D-DHAT	D	RG-II	GalAα1,4-GalA	β2,3-DHAT
L-AcefA	B	RG-II	L-Rhaβ1,3'-Apif	α1,3-AceAfT
Methyl-RG-II:xylose 2-O-methyltransferase		RG-II	D-Xylα1,3-L-Fuc	
RG-II:fucose 2-O-methyltransferase		RGII	L-Fucα1,2-D-Gal	
Acetyl-RG-II:fucose acetyltransferase		RG-II	L-Fucα1,2-D-Gal	
RG-II:aceric acid 3-O-acetyltransferase		RG-II	L-Acef Aα1,3-L-Rha	

[a]This list of glycosyltransferase activities is based on the most extended structure of RG-II. Note that the terminal βRhap$\rightarrow$3αAraf- in side chain B and the terminal βAraf$\rightarrow$2αRhap- in side chain B are not present in all RG-II preparations.
[b]All sugars are D sugars and have pyranose rings unless otherwise indicated. Glycosyltransferases add to the glycosyl residue on the left of the indicated acceptor.
Adapted from D. Mohnen, in G. Seymour and J.P. Knox (eds.), Pectins and Their Manipulation, pp. 52–96. Oxford: Blackwell/CRC Press, 2002.

the other glycosyltransferases involved in pectin biosynthesis identified to date has involved the use of alternative genomic methods, most notably the CAZy database (see Sidebar 5.1). In this database, GAUT1, GAUT7, and the related GATL enzymes are assigned to the GT8 family, whereas ARAD1, an enzyme that appears to synthesize the arabinan side chains of RG-I, is a member of the very large GT47 family. RGXT1 and RGXT2, two xylosyltransferases involved in the synthesis of RG-II, are members of the plant-specific family GT77. Predicting the specificity of the different glycosyltransferases is, however, very difficult, because the nucleotide sugar and acceptor substrates are not well conserved within each family. Thus, to determine the function of a given gene product requires complementation of the genomic analyses with detailed biochemical, structural, and immunocytochemical analyses of knockout and/or insertional mutants.

The GAUT1 enzyme is predicted to be a type II membrane protein with a single N-terminal transmembrane domain. Considering that the CSL-type polysaccharide backbone-synthesizing enzymes contain multiple transmembrane domains (see Figure 5.20), discovery of the type II architecture of the GAUT1 enzyme was quite unexpected. The GATL family of GAUT1-related enzymes has also been postulated to be involved in the biosynthesis of the backbones of RG-I and RG-II. Interestingly, there is evidence to suggest that specific glycosyltransferases are associated with specific nucleotide sugar-synthesizing enzymes and nucleotide sugar transporters.

6. Synthesis of pectic polysaccharides involves enzymes localized to late *cis,* medial, and *trans* Golgi cisternae.[ref19]

The general layout of the biosynthetic pathways of homogalacturonan (HG) and rhamnogalacturonan I (RG-I) has been elucidated by means of immunolabeling techniques using antibodies directed against specific groups of these polysaccharides: a deesterified HG/RG-I transition domain, methylesterified HG, and two side-chain domains of RG-I. Based on a quantitative analysis of the different labeling patterns seen on thin sections of sycamore suspension-cultured cell Golgi stacks, the following tentative biosynthesis pathway was deduced (**Figure 5.24**). Assembly of the unesterified backbone of HG/RG-I appears to be initiated in the transitional *cis*/medial Golgi cisternae and apparently completed in medial cisternae. Methylesterification of the galacturonic acid residues in the HG domains seems to follow also in the medial cisternae. Finally, epitopes of the arabinose-containing side chains of RG-I are first detected in *trans* Golgi cisternae, suggesting that these side chains are added in that compartment.

Interestingly, when clover root tip cells are immunolabeled with the anti-unesterified HG/RG-I backbone antibodies, the label is seen nearly exclusively over late *cis* and medial Golgi cisternae in cortical parenchyma cells, but mostly over *trans* Golgi cisternae and the TGN in mucilage-secreting root cap cells. This observation suggests that the organization of the assembly pathway(s) of pectic polysaccharides may differ among cell types. Immunolocalization of specific enzymes is now required to confirm the proposed spatial organization of the glycosyltransferases and modifying enzymes within Golgi stacks of a variety of cells at different stages of development.

7. Only a few steps of the pectic polysaccharide synthesis pathway have been studied biochemically *in vitro.*[ref20]

Biochemical analysis of pectic polysaccharide synthesis has been hampered by the lack of a commercial source of nucleotide sugar substrates for most of the pectin biosynthetic enzymes and of oligo/polysaccharides required to assay the enzymes. Indeed, finding the right combination of nucleotide sugar donor and acceptor substrates can be daunting, particularly in the absence of other biochemical leads. A method for producing UDP-GalA from UDP-GlcA via an epimerase from radish roots, followed by an anion-exchange chromatography purification step, provides a relatively easy

Figure 5.24 Model of the biosynthetic pathway of HG/RG-I in the Golgi apparatus of sycamore suspension-cultured cells as deduced from immunolabeling experiments. (Adapted from G.F. Zhang and L.A. Staehelin, *Plant Physiol.* 99:1070–1083, 1992.)

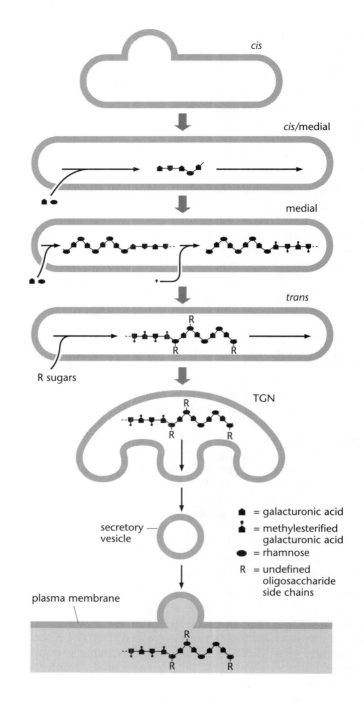

route for researchers to produce their own UDP-GalA for *in vitro* synthesis studies. This is an example of the kind of challenge that awaits researchers entering the field. The situation is being made somewhat easier through efforts such as the NSF-funded CarboSource services (http://giti.ccrc.uga.edu/web/services/carbosource/css_home.html).

In vitro synthesis of HG has been achieved with microsomal membranes isolated from a variety of plants. GalA transferase studies using tobacco suspension-cultured cells demonstrated that the product made in the presence of exogenous acceptors was partially esterified HG. The methyl ester groups are added to the C-6 carboxyl group of GalA by transmethylation from *S*-adenosylmethionine by enzymes located on the luminal side of Golgi cisternae. Preliminary data suggest that *GalA-transferase* and *HG-methyltransferase* may act together in a complex. Studies with preparations of solubilized GalA transferases have demonstrated that this enzyme adds GalA to the nonreducing end of the polymer. Based on the immunolabeling data discussed in the preceding section, it is likely that most of the *in vitro* synthesized and methylesterified HG molecules produced *in vitro* are assembled by enzymes located in medial Golgi cisternae, because only in those cisternae are the polymerization and esterification enzymes found together.

8. The biosynthesis of galactomannans has parallels to the biosynthesis of xyloglucans.[ref21]

Galactomannans assume the role of storage carbohydrates in the seeds of certain legumes and coconut palms (see Figure 9.19), where they accumulate in the secondary wall thickenings of the endosperm. They possess a backbone of 1,4-linked β-Man residues and single α-D-Gal residues attached to C-6 of the backbone. The degree of galactose substitution is species-specific, with a higher degree of substitution occurring in the more phylogenically advanced legumes (see Concept 7C6).

Biosynthesis experiments with endosperm-derived microsome fractions from fenugreek (*Trigonella foenum-graecum*) and guar (*Cyamopsis tetragaroloba*) have shown that [^{14}C]Man from GDP-[^{14}C]Man can be transferred to an endogenous substrate by a *Man-synthase* to yield a soluble galactomannan product. When UDP-Gal is included in the reaction mixture together with Mn^{2+}, significant amounts of C-6-linked galactose become attached to the mannan backbone. However, when the 1,4-linked mannan backbone is produced first and then incubated with UDP-Gal but not GDP-Man, very little galactose is incorporated into the final product. These results suggest that the *Gal-transferase* and the Man-synthase work in concert to prevent the mannan chain from assuming a folded conformation before the Gal residues can be added. This interdependence of the two enzymes resembles the interdependence of the Glc-synthase and the Xyl-transferases during the *in vitro* synthesis of xyloglucan.

Further experiments have demonstrated that *in vitro* the degree of substitution of mannan with galactose can be manipulated by adjusting the relative concentrations of GDP-Man and UDP-Gal. However, the maximum degree of substitution that can be achieved *in vitro* is dependent on the source of the Gal-transferase. Thus, microsomes derived from fenugreek, which naturally produces galactomannans with a high degree of galactose substitution (~96%), also yield *in vitro* products with a higher maximum degree of substitution than those from guar, whose rate of galactan substitution is 65%. These findings together with data from experiments with defined mannooligosaccharide acceptors have led to the proposal that the Gal transferase has a principal recognition sequence of six Man residues, with transfer of Gal to the third Man residue from the nonreducing end.

The mannan synthase gene from guar, *CtMANS*, has been cloned and shown to be a member of the *CSLA* family of genes (see Figure 5.19). The gene codes for a protein with a diagnostic D, D, D, Q/RXXRW motif of processive glycosyltransferases, and a catalytic domain on the cytosolic side of the membrane as depicted in Figure 5.20. The functional identification of the cloned gene includes the demonstration that expression of *CtMANS* in soybean somatic embryos leads to high levels of mannan synthase activities in the Golgi.

A galactomannan Gal-transferase gene from the developing endosperm of fenugreek has also been cloned. The deduced protein has a typical type-II membrane protein topology with one transmembrane domain near the N-terminus.

9. Several glycosyltransferases involved in xylan biosynthesis have been identified, but the mechanism of xylan biosynthesis remains obscure.[ref22]

The three major components of secondary cell walls of woody plant tissues are cellulose, lignin, and glucuronoxylans. Since woody tissues constitute the bulk of terrestial biomass, the principal source of bioenergy, understanding the biosynthesis of these molecules is a prerequisite for the bioengineering of cellulosic energy crops (see Concept 9E). Whereas significant progress has been made in understanding how xyloglucan is assembled (see Concepts 5C2, 3), remarkably little is known about the biosynthesis of the other major hemicellulose, glucuronoxylan.

The structural feature of glucuronoxylan that makes its biosynthesis difficult to understand is its two-domain architecture. Thus, each glucuronoxylan molecule possesses a typical polymer domain and a distinct "glycosyl sequence" domain. The basic polymer domain consists of a backbone of β-1,4-linked xylosyl residues with about 10% of these residues bearing a single glucuronic acid or methylated glucuronic acid side chain. The glycosyl sequence domain Xyl-Xyl-Rha-GalA-Xyl is attached to the reducing end of the xylan polymer domain. This glycosyl sequence domain was first identified in birch, but is now known to occur in *Arabidopsis* and spruce as well, suggesting that it constitutes an evolutionary conserved structure of glucuronoxylans. Based on this structural information, it is likely that several glycosyltransferases are involved in the initiation and elongation of the polymer backbone, and that other enzymes are required for the addition and/or modification of the side chains.

Microarray profiling and co-regulation analyses of *Arabidopsis* genes expressed during secondary cell wall formation have failed to identify any CSL candidate genes for glucuronoxylan biosynthesis. However, several other glycosyltransferases belonging to the GT8, GT43, and GT47 families of CAZy sequences (see Sidebar 5.1) are also highly expressed in secondary cell wall–forming tissues in *Arabidopsis*, poplar, and aspen. Furthermore, in the glucuronoxylan-deficient *Arabidopsis* mutants *irx8*, *irx9*, and *fra8*, all of which exhibit defective secondary walls, the mutations have been traced to enzymes of the same glycosyltransferase enzyme families. Further analysis of these mutants has demonstrated that the *irx8* mutation produces glucuronoxylan molecules with a reduced length, that the *irx9* mutation gives rise to glucuronoxylan molecules lacking the reducing terminal glycosyl sequence domain, and that the *fra8* mutation yields xylans that lack unmethylated but not methylated side chains. A common theme of current glucuronoxylan biosynthesis models is that they postulate multi-subunit enzyme complexes. The idea of interdependent biosynthetic activities is also supported by earlier *in vitro* synthesis experiments in which the incorporation of glucuronosyl residues from UDP-GlcA was increased by the addition of UDP-Xyl, suggesting that backbone and side-chain residues are added in a concerted manner, similar to the xyloglucan synthesis system.

Another interesting question is how the arabinofuranosyl precursors of glucuronoarabinoxylans are made. In most plant polymers containing this sugar, including glucuronoarabinoxylans, rhamnogalacturonan-I, arabinogalactan-proteins, and hydroxyproline-rich glycoproteins, L-arabinose is in the furanose ring conformation, whereas UDP-Ara is in the pyranose form. Yet to be determined is whether the arabinosyltransferases distinguish themselves from other glycosyltransferases by their ability to catalyze ring rearrangement before addition of the sugar to the polymer.

D. Assembly of Polysaccharides at the Plasma Membrane

1. Studies of cellulose synthesis by *Gluconacetobacter xylinus (Acetobacter xylinum)* provided many of the paradigms for similar studies in higher plants.[ref23]

Identification of the cellulose synthase genes in higher plants was made possible by using sequence motifs of the *Gluconacetobacter xylinus* (formerly *Acetobacter xylinum*) cellulose synthase genes as probes. This gene was identified using knowledge derived from biochemical studies of cellulose synthesis in this bacterium (see also Concept 4A14), which employs a more stable cellulose synthesis machinery than higher plants. The cellulose synthase proteins were identified by means of product entrapment. An 83-kD polypeptide was shown to be the substrate-binding subunit based on labeling with azido-UDP-glucose. In other experiments, the *in vitro* rate of cellulose synthesis from UDP-glucose was found to be stimulated to near *in vivo* rates (~200-fold) by the activator *cyclic-di-GMP* (**Figure 5.25**).

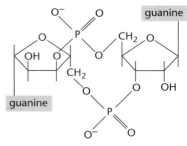

Figure 5.25 Structure of cyclic-di-GMP.

The advances in understanding of the *in vitro* synthesis of β-1,4-glucans in *G. xylinus* were then matched by the discovery of all of the genes that code for the various subunits of the entire synthase complex. (Note: Because the nomenclature for the cellulose synthesizing genes in bacteria has not been standardized, the original *A. xylinum* names are used in the following discussion.) Identification of these genes came from two approaches: cloning based on the complementation of a mutant unable to make cellulose, and sequence-based cloning from the N-terminal sequence of the 83-kD protein. Typical of genes coding for proteins involved in biosynthetic pathways, all of the cellulose synthase subunit genes of *G. xylinus* are organized in a single operon, whose genes were given different names by different groups: *BCSA, B, C, D* (*A. xylinum*), and *ACSA, B, C, D* (*A. xylinum* strain ATC 53582).

The *BCSA/ACSA* gene codes for a protein that binds the UDP-Glc substrate molecules and is therefore considered to be the catalytic subunit (CESA protein equivalent), whereas the BCSB/ACSB protein binds cyclic-di-GMP and appears to be the catalytic/regulatory subunit. The *BCSC/ACSC* gene product most likely produces the pores in the outer envelope through which the cellular microfibrils are exuded, and the BCSD/ACSD protein contributes to the crystallization of the microfibrils.

2. Hydrophobic cluster analysis played a critical role in the identification of the plant *CESA* genes.[ref24]

The discovery of the first plant cellulose synthase gene, *GhCESA1*, occurred during the course of a project that involved random sequencing of cotton cDNA clones corresponding to mRNAs present during the early stages of secondary wall formation in developing cotton fibers. Cotton fibers are single elongated epidermal cells of the ovule that grow synchronously within a boll (see Figure 9.14). At the end of elongation they initiate the synthesis of a cell wall in which cellulose constitutes more than 90% of the dry weight compared to about 25% in tissue culture cells. During the transition period, synthesis of other wall components ceases, and the rate of cellulose synthesis increases nearly 100-fold. Identification of the gene was made possible by short sequence motifs recognized during earlier sequence alignments of a number of UDP-Glc and UDP-GlcNAc-binding proteins, including the cellulose synthase gene of *G. xylinus*. A major factor contributing to

Figure 5.26 Comparison of domain structures in plant and bacterial *CESA* genes and plant *CSL* genes.

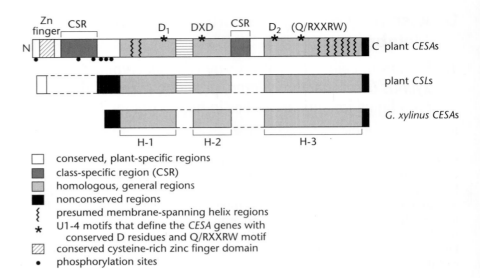

☐ conserved, plant-specific regions
▨ class-specific region (CSR)
▨ homologous, general regions
■ nonconserved regions
{ presumed membrane-spanning helix regions
* U1-4 motifs that define the *CESA* genes with conserved D residues and Q/RXXRW motif
▨ conserved cysteine-rich zinc finger domain
• phosphorylation sites

the discovery of these short but critical sequence motifs was the development of a technique known as *hydrophobic cluster analysis*. This method provides a means for relating sequence motifs to reaction mechanisms of glycosyltransferases.

The central feature of hydrophobic cluster analysis is an integration of sequence alignment with secondary structure prediction statistically centered on hydrophobic clusters. This approach yields information about the similarities in the three-dimensional organization of polypeptides that share functional but few sequence homologies. Thus, it can reveal domains of similar architecture separated by regions of high variability. In particular, it can predict which residues might perform catalytic functions and therefore would be expected to be conserved. When applied to glycosyltransferases that utilize NBD-sugars as substrates, hydrophobic cluster analysis distinguishes two groups based on the presence of A and B types of domains. Proteins that contain only an A domain correspond to transferases that add a single sugar, whereas those that contain both an A and a B domain can add two or more sugars. The A domain contains conserved aspartate (D) residues, and domain B a conserved QXXRW (where X is any amino acid) sequence motif. The spacing and location of these and other conserved residues were of critical importance for relating the two cotton *CESA* genes to the corresponding bacterial cellulose synthase genes (**Figure 5.26**).

3. Plant *CESA* genes appear to be derived from cyanobacterial precursors.[ref25]

It is generally accepted that endosymbiotic *cyanobacteria* gave rise to the chloroplasts of autotrophic eukaryotes. Less appreciated is the fact that their genomes also contributed many other genes to the genomes of plants and algae. The list of plant genes that can be traced to cyanobacteria now includes the cellulose synthase (*CESA*) and cellulose synthase-like (*CSL*) genes (see Figure 5.19). Cellulose synthase genes have been identified in several cyanobacterial species, and all of these species have also been shown to produce cellulose fibrils. The most notable feature of these cyanobacterial genes is that they share an insertion between the conserved H-1 and H-2 regions with the cellulose synthase genes of higher plants, which is missing from the corresponding genes of *G. xylinus* (*A. xylinium*) (Figure 5.26). This finding implies both an ancient origin of cellulose synthases in eukaryotes and the endocytic transfer of a *CESA* precursor gene from cyanobacteria to plants.

4. The cellulose synthase (CESA) proteins appear to correspond to the catalytic subunits of cellulose synthase complexes.[ref26]

The lack of an *in vitro* demonstration of cellulose synthesis by a purified CESA protein means that its function has not been proven yet. However, several lines of indirect evidence are consistent with CESA proteins being intimately associated with cellulose synthesis, most likely serving as the catalytic subunits of the cellulose synthase complexes. For example, antibodies raised against a domain of the CESA protein label the cellulose-synthesizing rosette complexes in the plasma membranes of cells producing cellulose fibrils (see Figure 4.27). A more direct method for providing proof of functionality of an unknown protein is to demonstrate that a lost function in a defined mutant can be restored by the transformation of the mutant with a wild-type gene of this protein. This has been done with the *Arabidopsis* root swelling mutant *rsw1*, which carries a mutation in the *AtCESA1* gene (**Table 5.4**).

rsw1 is a temperature-sensitive mutant. Shoots of *rsw1* seedlings grown at the restrictive 31°C temperature contain less than 50% of crystalline cellulose of wild-type seedlings but more noncrystalline β-1,4-linked glucans. Freeze-fracture electron microscopy demonstrates rosette disintegration within three hours of transfer of seedlings from 18°C to 31°C and a concomitant increase in clusters of aligned particles. Transformation of the mutant with a wild-type gene corrects the radial swelling and restores both rosettes and normal rates of cellulose synthesis at the nonpermissive temperature. Based on this evidence, it has been suggested that the mutant allele interferes with the assembly of cellulose molecules into crystalline microfibrils by structurally disrupting the assembly of the subunits of the rosettes into rosette complexes. A radial swollen root phenotype is also observed when the *CESA1* gene of tobacco is knocked out using virus-induced gene silencing methods.

Procuste1, another cellulose-deficient *Arabidopsis* mutant with severe swelling in regions of cell expansion, has a mutation in *AtCESA6*. Similarly, the *irx1* and *irx3* mutants, which have mutations in *AtCESA8* and *AtCESA7*, respectively, contain reduced amounts of cellulose in their secondary walls and exhibit collapsed xylem vessels due to the reduced strength of those walls. Two point mutations that render *Arabidopsis* resistant to the cel-

TABLE 5.4 MUTANTS AND GENES IMPLICATED IN CELLULOSE SYNTHESIS IN *Arabidopsis*

GENE NAME	MUTANT	PHENOTYPE(S)
CESA1	rsw1	Root swelling, stunted growth (*rsw1-1*), embryo lethality (*rsw1-2*); primary cell wall
CESA2, CESA3	ixr1	Isoxaben resistance, stunted growth; primary cell wall
CESA4	irx5	Irregular xylem; secondary cell wall
CESA5, CESA6	ixr2	Isoxaben resistance; primary cell wall
	prc1	Reduced root and hypocotyl length; primary cell wall
CESA7	irx3	Irregular xylem; secondary cell wall
CESA8, CESA9, CESA10	irx1	Irregular xylem; secondary cell wall

lulose synthesis inhibitor *isoxaben* and affect primary cell wall formation have been mapped to other *AtCESA* genes, the *ixr1-1* mutation to the *CESA3* gene, and the *ixr1-2* mutation to *CESA6*. Together, these findings provide strong support for the hypothesis that the *CESA* genes code for cellulose synthase genes.

5. The CESA protein is an integral protein with several domains characteristic of processive glycosyltransferases.[ref27]

The *CESA* genes code for proteins with a molecular weight of about 110 kD. CESA proteins appear to contain eight transmembrane helices, two of which are located close to the N-terminus and the remaining six close to the C-terminus (Figure 5.26 and **Figure 5.27**). The large, approximately 60-kD intervening domain is located on the cytoplasmic side of the membrane and contains regions that are highly conserved among all *processive glycosyltransferases*, enzymes that produce linear polysaccharides via multiple rounds of coupling without releasing the elongating product. Foremost among these regions are motifs that contain conserved aspartate residues thought to be involved in the binding of UDP-Glc, the predicted activated sugar substrate. Such motifs have also been identified in the glycosyltransferases that make chitin, hyaluronan, and heparin, all of which contain consecutive β-glycosidic linkages in which one sugar is oriented about 180° with respect to each neighboring sugar. Support for the postulated UDP-Glc-binding function has come from experiments with a glutathione *S*-transferase-CESA fusion protein containing the soluble ~60-kD domain but no transmembrane helices. This protein was capable of binding UDP-Glc in a Mg^{2+}-dependent manner. Binding was lost when one of the critical aspartate residues in the UDP-Glc binding domain was deleted. The conserved Q/RXXRW motif, which has been postulated to serve as the catalytic site of the synthase, is also found in this region. Protein modeling suggests that both the D, D, D and the Q/RXXRW motifs are located adjacent to molecular cavities suitable for the binding of UDP-Glc and for performing the catalytic reactions.

Another domain, named the *class-specific region (CSR)*, has been shown to be a diagnostic region for subclasses of CESA proteins. Thus, whereas significant sequence differences are seen between members of different

Figure 5.27 Topological model of the structure of a typical CESA protein, in the plasma membrane of a plant. The eight transmembrane domains are postulated to form a pore through which the cellulose molecule is secreted into the cell wall. Note, however, that other models have suggested that secretion pores could be formed by the transmembrane domains of dimerized protein complexes. The large, cytosolic domain contains a number of functional subdomains, including the three critical aspartate (D) residues and the Q/RXXRW motif that appears to be important for substrate binding and catalysis, respectively. The class-specific region (CSR) is also associated with the large cytosolic domain. The N-terminal domain, which is exposed to the cytosol, contains Zn finger binding motifs (see Figure 5.28). (Adapted from D. Delmer, *Annu. Rev. Plant Physiol. Plant Mol. Biol.* 50:245–276, 1999.)

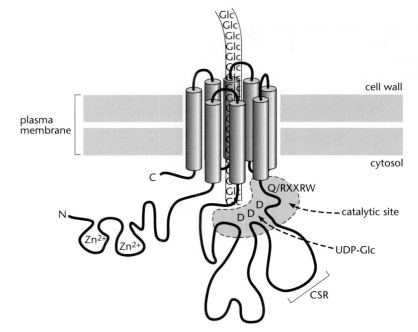

subclasses, the CSRs within a given subclass are very similar. Although no function has been ascribed to the CSRs, several motifs containing cysteine, basic, acidic, and aromatic residues suggest that the CSR may function in substrate binding specificity and catalysis.

The polar N-terminal domain of CESA proteins is much larger than in *G. xylinus*. It contains conserved cysteine residues organized in the form of Zn-binding motifs. In a Zn-binding assay, this domain has been shown to bind two Zn^{2+} atoms. The functional importance of this domain is discussed in the next section.

6. The Zn-binding domains of CESA and some CSLD proteins appear to mediate dimerization of the catalytic subunits of cellulose synthases.[ref28]

CESA and some CSLD proteins contain *Zn finger domains* located within the cytoplasmic N-terminal region of the protein (see Figure 5.27). The two Zn fingers of GhCESA1 and GhCESA2 show high homology to the RING-finger motif, which is capable of binding two Zn atoms in a unique "cross-brace" arrangement. Zn- or RING-finger domains have been found to mediate protein-protein interactions in other systems. Analysis of the N-terminal domains of the GhCESA1 and GhCESA2 proteins in the yeast two-hybrid system has shown that these domains can interact to form homo- or heterodimers via intermolecular disulfide bonds (**Figure 5.28**). Such bonds, however, can be formed only under oxidizing conditions. Under reducing conditions, as found in the cytosol, Zn^{2+} would be expected to be bound to the cysteines. This finding suggests that CESA dimerization is subject to regulation by the redox state of its local environment and that the proteins in the cellulose synthase complexes are also held together by other bonds. Oxidation of the CESA proteins may involve H_2O_2, since H_2O_2 production increases dramatically during secondary wall synthesis in cotton, and when this production is experimentally reduced, secondary wall formation stops.

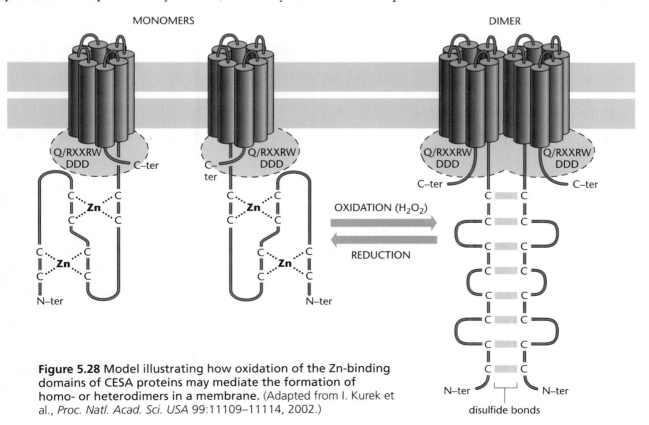

Figure 5.28 Model illustrating how oxidation of the Zn-binding domains of CESA proteins may mediate the formation of homo- or heterodimers in a membrane. (Adapted from I. Kurek et al., *Proc. Natl. Acad. Sci. USA* 99:11109–11114, 2002.)

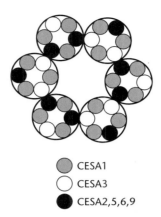

CESA1
CESA3
CESA2,5,6,9

Figure 5.29 A heteromeric model for the structure of a primary wall CESA rosette complex. The 36 CESA subunits of the rosette complex are postulated to consist of three CESA isoforms. CESA1 and CESA3 occupy positions that are exclusive for these CESAs. CESA2, 5, 6, and 9 appear to be interchangeable for the third position. CESA9 is highly expressed in pollen. Secondary wall rosettes are composed of CESA4, CESA7, and CESA8 isoforms. (Adapted from M. Mutwil, S. Debolt and S. Persson, *Curr. Opin. Plant Biol.* 11:252–257, 2008.)

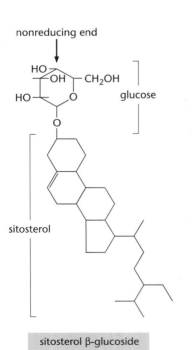

nonreducing end

HO—
OH — CH$_2$OH
HO
O
glucose

O

sitosterol

sitosterol β-glucoside

Figure 5.30 Chemical structure of sitosterol-β-glucoside, the postulated lipoglucan primer of cellulose synthesis.

7. The cellulose-synthesizing rosette complexes are composed of three different types of CESA proteins.[ref29]

The discovery of ten *CESA* genes in the *Arabidopsis* genome, and of similar numbers of *CESA* genes in other plant species, has led to a variety of studies of the functional roles of the different CESA isoforms. Techniques used to address this question have included the analysis of gene sequences and of cell wall mutants, gene expression and gene complementation studies, and co-immunoprecipitation and bimolecular fluorescence complementation experiments. As a whole, these investigations have demonstrated that the hexameric, rosette-type cellulose synthase complexes contain three types of CESA proteins, and that the complexes involved in primary and secondary cell wall synthesis are composed of different sets of proteins.

Sequence comparison of *Arabidopsis CESA* genes with those of other plant species demonstrates that the ten *CESA* isoforms belong to six different orthologous groups. These six groups have nonredundant functions as evidenced by mutational analysis. Thus, plants with mutations in the *CESA4*, *CESA7*, and *CESA8* genes show defects in secondary wall formation, whereas those with defects in the *CESA1*, *CESA3*, and *CESA6* genes exhibit defects in primary cell walls. Microarray data demonstrate that each isoform subgroup is co-regulated at the RNA level, and diverse structural studies have localized the expression of these subgroups of genes to the same cells. Indeed, when two different CESA fusion proteins with fluorescent tags and from the same subgroup are expressed together, the fluorecent complexes are seen to move together along linear trajectories at the cell surface. Finally, both co-immunoprecipitation and bimolecular fluorescence complementation experiments have provided evidence for physical interactions between members of each subset of CESA proteins.

These findings have led to the cellulose synthase complex model depicted in **Figure 5.29**, in which three different CESA subunits are needed to create the rosette-type membrane complexes seen in freeze-fracture electron micrographs. In this model, each subunit type interacts in different ways with its neighbors; this explains why three different subunits are required. It is interesting to note that in the complexes involved in primary cell wall synthesis, CESA1 and CESA3 cannot be complemented with any other CESA proteins, whereas CESA6 is partially redundant with the related isoforms CESA2, 5, and 9. CESA10 has yet to be characterized. Finally, it should be noted that not all cellulose fibrils in higher-plant cell walls are produced by CESA proteins organized in rosette complexes. For example, no rosette complexes have been observed in cell plates at the time of deposition of the first detectable cellulose polymers.

8. The involvement of KORRIGAN, a β-1,4-glucanase, in cellulose synthesis has yet to be proven conclusively.[ref30]

KORRIGAN (KOR), a membrane-anchored *endo*-β-1,4-glucanase capable of cleaving cellulose but not xyloglucan, is expressed in all cells, and some studies suggest that it associates with CESA proteins. One interesting hypothesis related to KOR function proposes that this enzyme separates a postulated cellulose primer from the nascent polymer. A potential primer was identified by digesting noncrystalline (soluble) cellulose produced by cells exposed to the herbicide CGA325′615 with cellulase, which released glucans and a lipid molecule, *sitosterol*-β-*glucoside* (**Figure 5.30**). Based on this and other evidence, sitosterol-β-glucoside was postulated to serve as a primer for cellulose synthesis (**Figure 5.31**). Since KOR mutants also accumulate noncrystalline cellulose, it has been postulated that this enzyme serves to cleave the sitosterol-β-glucoside primer from the nascent cellulose oligomer (a cellodextrin).

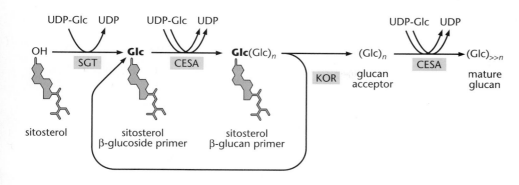

Figure 5.31 Possible pathway of cellulose synthesis in which β-sitosterol-glucoside is used as a primer. In this scheme, the membrane associated *endo*-β-1,4-glucanase KOR is needed to release the glucan acceptor from the primer. SGT, β-sitosterol glucoside transferase; CESA, cellulose synthase; KOR, Korrigan cellulase.

Two other postulated functions of the KOR protein include separating the completed cellulose molecules from the cellulose synthase complex, or cleaving noncrystalline glucan molecules from crystalline cellulose micro-fibrils to remove tensional stresses that may arise during the assembly of the large number of glucan chains into a microfibril. Lending support to this latter hypothesis is the finding that bacterial cellulose synthesis requires a related glucanase for *in vivo* but not *in vitro* activity.

9. UDP-glucose, the substrate for the cellulose and callose synthase systems, may be produced by a membrane-bound form of sucrose synthase.[ref31]

The discovery that in cells producing cellulose-rich secondary walls (for example, cotton fibers, xylem vessels) about half of the *sucrose synthase (SuSy)* enzymes are bound to the plasma membrane (**Figure 5.32**) has led to the suggestion that sucrose may be the principal source of UDP-Glc used by the cellulose synthase systems. In this scheme, UDP-Glc would be supplied to the complex via the reaction

$$\text{sucrose} + \text{UDP} \xrightarrow{\text{sucrose synthase}} \text{UDP-Glc} + \text{fructose}$$

Using sucrose as a substrate for detached and permeabilized cotton fibers, sucrose has been shown to support cellulose synthesis at rates that exceed those of callose in the presence of low calcium. One advantage of this path-way is that the high energy of the glucose-fructose bond is used efficiently for the synthesis of the glycosidic bond in cellulose. In addition, the revers-ible binding of SuSy to the plasma membrane, and most likely to cellulose synthase, suggests that UDP-Glc produced by sucrose synthase could be channeled directly to the cellulose synthase (**Figure 5.33**). This binding would also allow for the UDP released during the condensation reaction to

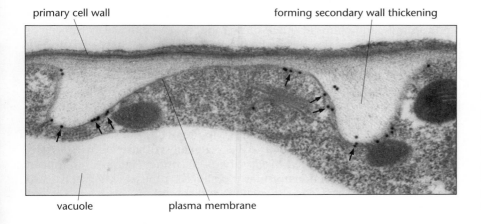

Figure 5.32 Electron micrograph of a developing *Zinnia* tracheary element immunolabeled with anti-sucrose synthase (SuSy) antibodies. Note the concentration of label (*arrows*) over the plasma membrane areas adjacent to the cell wall thickenings where cellulose synthesis is occurring. (From V.V. Salnikov et al., *Phytochemistry* 57:823–833, 2001.)

Figure 5.33 Hypothetical model of how sucrose synthase (SuSy) might produce UDP-glucose and channel it directly into a cellulose synthase enzyme in the plasma membrane. The UDP released during the condensation reaction might be channeled directly back to the SuSy to prevent inhibition of the cellulose sucrose synthase. Membrane binding of SuSy is regulated by phosphorylation.

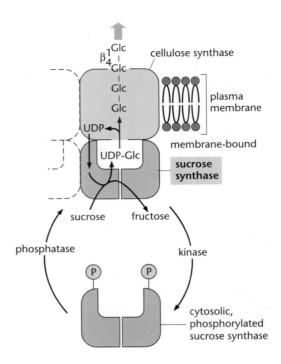

be immediately recycled by the sucrose synthase complex and thereby prevent it from inhibiting the cellulose synthase. The association of SuSy with the cellulose synthase complex appears to be controlled by phosphorylation of the SuSy proteins, with phosphorylation inducing their detachment from the membrane. A similar phosphorylation-dependent on/off relationship of SuSy appears to operate with the plasma membrane–localized enzyme callose synthase. Phosphorylated SuSy appears to have an affinity for actin, but the functional significance of this association is unclear.

10. There are two types of callose synthase systems, a Ca^{2+}-dependent and a Ca^{2+}-independent type.[ref32]

Two types of *callose synthase* systems have been identified that differ in their Ca^{2+} requirements. The most frequently studied type has an absolute requirement for micromolar amounts of Ca^{2+}. The best-known example of such a callose synthase is the one that produces wound callose in somatic cells. In contrast, the callose synthase that deposits a callose layer in pollen tube walls is not stimulated by Ca^{2+} or other cations. Since most of the research to date has focused on the wound-induced type of callose synthase of somatic cells, the emphasis here is on this form of the synthase.

The most potent inducers of callose formation by somatic cells are physical stretching and compression as well as chemicals such as digitonin, polyamines, and chitosan. All of these inducers seem to act by causing an influx of Ca^{2+} across the plasma membrane and a transient rise in the level of intracellular Ca^{2+}, since no effect is seen when Ca^{2+} is removed from the medium surrounding the cells. Furthermore, because Ca^{2+} ionophores like AZ3187 are much less potent inducers than, for example, digitonin, it has been postulated that callose synthesis requires Ca^{2+} uptake through gated channels. Calmodulin, however, does not appear to be part of the Ca^{2+} activation pathway, but phosphorylation events might be important. In addition to Ca^{2+}, the wound-induced type of β-1,3-glucan synthase requires a β-glucoside primer to get started. *β-Furfuryl-β-glucoside* (**Figure 5.34**) appears to serve as the endogenous primer of callose synthesis in *Vigna radiata*. This compound is stored in the vacuole, and conditions that lead to callose synthesis also cause an elevation of cytoplasmic β-furfuryl-β-glucoside.

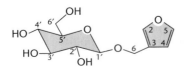

Figure 5.34 Structure of β-furfuryl-β-glucoside, an endogenous activator of callose synthase.

Ca^{2+} raises the V_{max} of β-1,3-glucan synthesis and enhances the affinity of UDP-Glc at low substrate concentrations. Recent studies suggest that the activity of a Ca^{2+}-dependent, partially purified cotton fiber callose synthase may be regulated by annexins, a class of Ca^{2+}-binding proteins that interact with membranes in a Ca^{2+}- and phospholipid-dependent manner. The callose synthase of forming cell plates is also Ca^{2+}-dependent, but little else is known about its biochemical properties. Whether callose deposition by tapetal cells in anthers requires Ca^{2+} has yet to be determined.

The callose in pollen tube walls is deposited as an inner wall layer about 30 μm behind the tube tip, where the level of free Ca^{2+} is low. *In vitro* analysis of the biochemical properties of the β-1,3-glucan synthase from pollen tubes has shown that although it also requires activation by β-glucosides and low levels of compatible detergents, which render the membrane vesicles permeable to the UDP-Glc substrate, it differs significantly in its kinetic and regulatory properties from the wound-activated, Ca^{2+}-dependent callose synthases. Most notably the pollen tube callose synthase synthesizes callose at high rates independent of the concentration of Ca^{2+}. However, the affinity for the *in vivo* substrate UDP-Glc is low (K_m 1.8–2.5 mM) compared to that of wound-activated callose synthases (K_m 0.25–0.65 mM), but consistent with the calculated concentration of UDP-Glc (3.5 mM) in the pollen tube cytoplasm.

11. Callose synthase is a large protein that differs in many ways from cellulose synthase.[ref33]

Prior to the identification of the cellulose synthase and the callose synthase genes, many cell wall researchers believed that callose was made by perturbed cellulose synthases. However, it is now clear that most if not all callose is produced by enzymes coded for by the *GSL* (glucan synthase-like) genes, which have also been named *CALS* (callose synthase) or *CFL* (cotton *FKS*-like) genes. The *GSL proteins* have a molecular weight of more than 200 kD, and, based on hydropathy plot analysis, they contain 16 transmembrane domains versus 8 for CESA (**Figure 5.35**). The large, hydrophilic cytoplasmic domain between transmembrane domains 6 and 7 contains binding sites for Rop1, annexin, UDP-Glc-transferase, and sucrose synthase. However, it lacks the D, D, D, and Q/RXXRW conserved residues of the CES and CSL

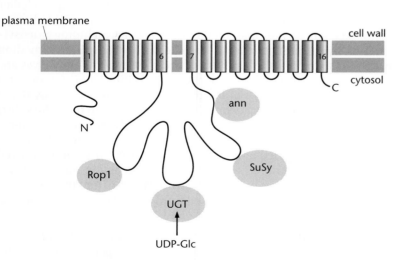

Figure 5.35 Topological model of the callose synthase showing 16 transmembrane domains and a large cytosolic domain with binding sites for Rop1, UDP-Glc transferase (UGT), sucrose synthase (SuSy), and annexin (ann). (Adapted from D.P. Verma and Z. Hong, *Plant Mol. Biol.* 47:693–701, 2001.)

proteins. The absence of the conserved D, D, D UDP-Glc binding domains might explain the need for an associated UDP-Glc transferase to feed activated sugars to the catalytic site of the enzyme.

The *GSL* gene family of *Arabidopsis* is comprised of 12 members, and the GSL enzymes are expressed in a tissue-specific manner, and/or differentially regulated in response to biotic and abiotic stresses (see Figure 4.37). For example, the *GSL5* gene has been shown to be responsible for the production of papillary and wound-induced callose in response to a variety of stresses, and none of the other 11 *Arabidopsis* callose synthases are able to compensate for the loss of this gene function (see also Concept 8B3). From a gene evolution point of view, it is interesting to note that, whereas the callose synthases of plants (GSL-type) and fungi (IKS-type) have similar sequences, neither of these proteins seems to be evolutionarily related to the CRDS protein of *Agrobacterium*, which also produces β-1,3-glucans.

12. The ratio of tri- and tetrasaccharide units in mixed-linked glucans made *in vitro* can be altered experimentally.[ref34]

β-1,3-/1,4-Linked (mixed-linked) glucans are cell wall polysaccharides that are produced in grasses (Poales) and in *Equisetum* (horsetail). They are absent in meristematic cell walls, but increase markedly in the cell walls of elongating cells, reaching a maximum that is coincident with the maximum rate of elongation. When growth ceases, the polymer is degraded.

As of this writing, there is uncertainty about where mixed-linkage glucans are assembled, in the Golgi or at the plasma membrane. Based on cell fractionation studies, it was postulated that these polysaccharides were produced by Golgi-localized enzymes and then transported in secretory vesicles to the cell surface. However, based on recent immunolabeling studies with a β-1,3-/1,4 glucan–specific monoclonal antibody, which showed strong labeling over cell walls but little cytoplasmic labeling, many cell wall researchers have come to the conclusion that this polysaccharide is synthesized at the plasma membrane and not in the Golgi.

Because of the unusual structure of the backbone of β-1,3-/1,4-linked glucans—mostly tri- and tetrasaccharide types of glucosyl repeats (cellobiosyl and cellotriosyl units connected by single β-1,3-linked glucosyl subunits)—understanding their mode of assembly constitutes a unique challenge. Initial biosynthesis studies with isolated Golgi fractions yielded very few mixed-linked glucans compared to other molecules including glucosylated flavonoids. By adding activated charcoal to the homogenization medium to remove the flavonoids, the synthesis of these latter compounds was greatly reduced, and a marked enhancement of mixed-linked glucan products was observed. However, analysis of the products was complicated by the presence of cellulose and β-1,3 glucans. UDP-Glc was shown to be the substrate for mixed-linkage glucan synthesis, and Mg^{2+} or Mn^{2+} was required as a cofactor. One of the most intriguing observations made to date is that the ratio of tri- and tetrasaccharide units produced *in vitro* can be altered by changing the concentration of UDP-Glc in the reaction mixture. Thus, by increasing the concentration of UDP-Glc from 5 to 30 μm, the ratio of tri- to tetrasaccharide units can be increased from 1.5 to 11. In native grasses, the ratio is strictly controlled at between 2 and 3.

Identification of the mixed-linkage β-1,3/1,4 glucan synthase genes has been achieved by a combination of comparative genetic and functional approaches. The breakthrough research involved correlating a quantitative trait locus (QTL) of barley that regulates the mixed-linked glucan content of the plants with a region in the rice genome that contains six CSLF genes (see Figure 5.19). *Arabidopsis* plants transformed with two of these genes gained the ability to produce mixed-linked glucans in their epidermal cells.

E. Polymer Assembly in the Wall: Biosynthesis of Lignins, Waxes, Cutins, and Suberins

1. The secondary cell wall provides a unique environment for the polymerization of lignins from cinnamyl alcohols.

Lignin is an amorphous polymer formed by an oxidative process in the cell wall and middle lamella. The mechanism of lignin polymerization differs from the polymerization of cellulose and callose and the hemicelluloses of the plant cell wall. It also differs fundamentally from the template-based polymerization of nucleic acids and the template-adaptor mechanism for the assembly of proteins. Whereas cellulose and callose are formed by enzymes embedded in the plasma membrane from phosphorylated (activated) sugar precursors (see Section 5D), lignin is assembled from monolignols by *oxidative polymerization* outside the plasma membrane, that is, in the extracellular space, and in the cell wall. The environment in the cell wall is quite different from that of the cytoplasm and greatly restricts the types of reactions that can take place. Secondary walls, where lignin deposition occurs, are multilayered, and contain short hemicelluloses and highly polymerized cellulose in roughly a 1:2 ratio. The morphological features of secondary walls are clearly visible before there is histochemical evidence of most lignin deposition. Most lignin biosynthesis occurs during the terminal differentiation of the cells following the formation of the cell layers of the secondary wall. Lignin deposition also occurs in the middle lamella, where there is little or no cellulose or hemicelluloses.

Compared to the cytoplasm, the cell wall environment in which lignin synthesis occurs has a low pH, contains low amounts of solutes, proteins, and lipids, and generally has no direct source of free energy such as activated monomers. The pores in nonlignified cell walls are about 5 to 10 nm in diameter, which allows for the passage of proteins of up to 100 kD. However, because primary walls are dynamic and flexible, transient larger pores can develop, allowing for the passage of larger molecules over time. The secondary wall is less porous, with the crystalline cellulose microfibrils being closely packed into thick, ordered layers. The accumulation of free phenolic precursors of lignin in the cell walls is also less likely to have toxic effects on the cells.

There are many unanswered questions about the biosynthesis of lignin. We know that the precursors for lignin are cinnamyl alcohols, which are derived from phenylalanine; however, the precise pathway for the series of modifications is not fully known. We still know relatively little about the mechanism of polymerization. We also do not know how the precursors are delivered to the lignifying zone. Lignins are diverse polymers and it is likely that biosynthesis varies within and between the major groups of higher plants. Lignin biosynthesis is initiated at specific sites, typically starting at the cell corners, and progresses slowly through and between the secondary walls, suggesting different mechanisms for initiation and for polymer growth.

2. Cinnamyl alcohols are the predominant precursors for lignin biosynthesis.[ref35]

Lignin biosynthesis is a part of the general phenylpropanoid (C_6C_3) pathway that shares enzymatic steps with the synthesis of many other phenolic products, including suberin, *lignans*, flavonoids, and anthocyanins (**Figure 5.36**). Lignin is typically synthesized from cinnamyl alcohols that correspond to the main subunits of the lignin polymer (see Concept 3F). These lignin precursors, called *monolignols*, are phenylpropanoids derived from phenylalanine through the shikimic acid pathway, and the different monolignols are characterized by side group additions to the phenolic ring. Monolignols are likely to be synthesized by membrane-bound enzymes of

Figure 5.36 Metabolic pathways related to the biosynthesis of monolignols and lignin. The dark arrows represent the major flow of carbon for lignin biosynthesis.

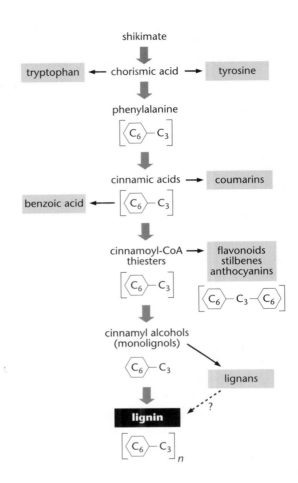

the secretory pathway and, following glycosylation, transported to the lignifying cell wall or stored in vacuoles (see Concepts 4A1–11).

Lignin is typically composed of three types of subunits, guaiacyl, sinapyl, and *p*-hydroxyphenyl subunits. The most common subunit of lignin, the guaiacyl (G) subunit, is derived from *coniferyl alcohol*, whereas syringyl (S) units are derived from *sinapyl alcohol*, and the less abundant H units from p-*coumaryl alcohol* (**Figure 5.37**). The relative abundance of different lignin subunits varies greatly in different groups of plants. Lignin composition is largely determined by the amount and composition of the monolignols delivered to the lignifiying zone. In conifers, and many other gymnosperms, coniferyl alcohol is the predominant monolignol, whereas in grasses, *p*-coumaryl alcohol, coniferyl alcohol, and sinapyl alcohol are all major precursors. In woody angiosperms coniferyl alcohol and sinapyl alcohol are both synthesized in substantial proportions.

The monolignols are derived from phenylalanine, a product of the shikimic acid pathway. Chorismic acid (chorismate) is also derived from shikimate, and is a precursor common to the three aromatic amino acids, phenylalanine, tyrosine, and tryptophan. Monolignols are usually derived from phenylalanine through cinnamic acid followed by modification of the phenyl ring and reduction of the γ carbon of the propanoid side chain. Many diverse phenylpropanoids and their derivatives are produced from this pathway.

A general monolignol biosynthetic pathway has been inferred from studies of many different plant species (**Figure 5.38** and **Table 5.5**). Recently, the role of some major enzymes has been reevaluated and the pathway has been substantially modified. Some of the uncertainties may be a result of metabolic redundancy where the same precursor could be produced in different

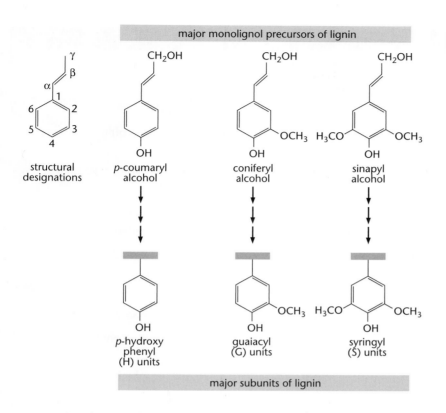

Figure 5.37 Major subunits in lignin and their corresponding monolignol precursors. The H, G, and S subunits are shown linked to the lignin polymer (*gray bars*). Two sets of designations are used for the propane side chain: α, β, and γ carbons are also often designated with numbered positions 7, 8, and 9.

ways through a metabolic grid, rather than through a specific pathway. Although alternative pathways are likely, they are not mutually exclusive and some routes are used preferentially. The majority of the metabolic flux might follow one route in a particular cell or tissue or species, and follow a different route in other circumstances.

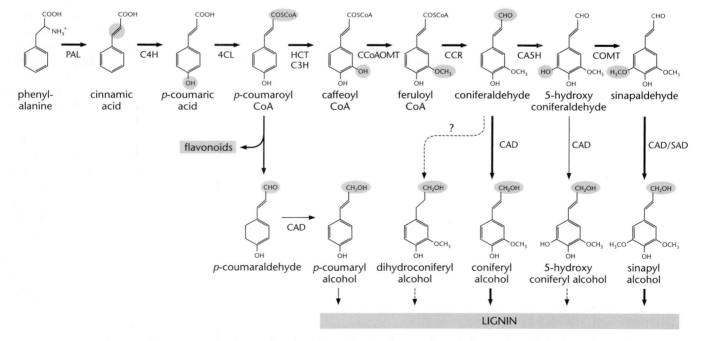

Figure 5.38 A general scheme for the biosynthesis of monolignols from phenylalanine. The pathway shown is based on the inferences for the predominant flux through the pathway in lignifying tissues. The enzymes for each step are described in more detail in Table 5.5. Shaded areas indicate modifications resulting from the previous step. An alternative pathway for 3-hydroxylation of the aromatic ring in monolignol biosynthesis is presented in Figure 5.39.

TABLE 5.5 ENZYMES OF MONOLIGNOL BIOSYNTHESIS

PAL, phenylalanine ammonia lyase	Deamination of phenylalanine
C4H, cinnamate-4-hydroxylase	4-hydroxylation of cinnamate
4CL, 4-coumarate CoA ligase	CoA ester formation of cinnamic acids
C3H, p-coumarate/p-coumaroyl-shikimate/quinate-3-hydroxylase	3-hydroxylation of p-coumaroyl shikimate/quinate esters
HCT, p-hydroxycinnamoyl-CoA: shikimate/quinate p-hydroxycinnamoyltransferase	Shikimate/quinate reversible esterification of p-coumaroyl-CoA
CCoAOMT, caffeoyl-CoA-O-methyltransferase	3-O-methylation of caffeoyl-CoA
CCR, hydroxy cinnamoyl-CoA reductase	Reduction of feruloyl-CoA to coniferaldehyde
CAld5H, ferulate/coniferaldehyde-5-hydroxylase	5-hydroxylation of coniferaldehyde
COMT, caffeic acid/5-hydroxyconiferaldehyde-O-methyltransferase	O-methylation of 5-OH coniferaldehyde
CAD, cinnamyl alcohol dehydrogenase	Reduction of cinnamylaldehydes to cinnamyl alcohols
SAD, sinapyl alcohol dehydrogenase	Reduction of sinapylaldehyde to sinapyl alcohol

3. Monolignols are formed by successive enzymatic modifications of phenylalanine.[ref36]

The first step in the biosynthesis of monolignols is the deamination of phenylalanine by *phenylalanine ammonia-lyase (PAL)* to produce cinnamic acid (see Figure 5.38). The ammonium ion released by PAL is recycled into the shikimate-chorismate pathway for the synthesis of phenylalanine and tyrosine by three enzymes, glutamine synthase, glutamine:α-ketoglutarate amino transferase (GOGAT), and prephenate aminotransferase, to make arogenate, which is the precursor for both tyrosine and phenylalanine. Following deamination of phenylalanine, the resulting propanoid side chain of cinnamate is unsaturated between the α and β carbons. This unsaturated bond is characteristic of monolignols, and these α and β carbons are reactive sites for many of the linkages made during formation of the lignin polymer.

PAL is one of the most intensively studied enzymes in plants because it is a major entry point into secondary metabolism. At least 20% of the carbon fixed by woody plants is shunted by PAL into secondary metabolites. PAL expression is regulated by many environmental and developmental stimuli, and the regulation of the level of PAL may affect the biosynthesis of all phenylpropanoid-derived metabolites. PAL is typically found in multiple isoforms encoded by a small family of four genes. In lignin-forming cells and tissues, one or more predominant isoforms may be expressed. Similarly, multiple isoforms may exist for other monolignol biosynthetic enzymes, and in some cases only one of those isoforms may be predominantly expressed. PAL enzymes from monocots can utilize tyrosine in addition to phenylalanine, resulting in the synthesis of *p*-coumaric acid from tyrosine.

A series of subsequent enzymatic reactions convert cinnamic acid (cinnamate) into diverse monolignols. The aromatic ring undergoes hydroxylation and *O*-methylation. The terminal carbon (γ) of the propanoid side chain is activated by a coenzyme A (CoA) ligation and subsequently reduced by two NADPH-dependent reductases to form the terminal alcohol group of the monolignols. Hydroxylation may occur at the 3, 4, and 5 positions, but only the 3 and 5 positions are *O*-methylated. Most, if not all, of the aromatic ring modified cinnamate derivatives have a corresponding reduced propanoid derivative (monolignol).

Cinnamate is converted to *p*-coumarate by cinnamate-4-hydroxylase (C4H), a cytochrome P-450 that hydroxylates the 4 position of cinnamate. Many coumarins, a large family of plant defense metabolites, are derived from *p*-coumarate. *p*-Coumarate is activated by 4-coumarate CoA ligase (4CL) to form 4-coumaroyl CoA. A subsequent reaction, catalyzed by chalcone synthase, using 4-coumaroyl CoA and three molecules of malonyl-CoA, is the branch point for a complex pathway leading to flavonoids and anthocyanins. *p*-Coumaroyl CoA is also the last metabolite in the pathway common to the biosynthesis of all monolignols.

For decades, it was presumed that the formation of monolignols first involved hydroxylation and *O*-methylation of the cinnamic acids. It was thought that cinnamic acids were subsequently converted to the corresponding CoA esters, and then enzymatically reduced stepwise to aldehydes, and then to alcohols to form the corresponding monolignols. Studies of substrate specificity, discovery of new enzymes, and results from transgenic plants modified at different steps in the pathway have required a revision of this concept. Major uncertainties remain about which aromatic ring modifications occur at the level of the cinnamic acids, or at the CoA esters, the aldehydes, or the alcohols. All reactions appear possible, but enzyme substrate specificity and lignin composition in transgenic plants indicate a preferred route to the major monolignols through the CoA esters and aldehydes. It is also not yet certain that all of the enzymes involved in monolignol biosynthesis have been found or even correctly identified.

The proposed pathway to *p*-coumaryl alcohol from *p*-coumaric acid is the least controversial. *p*-Coumaroyl CoA is thought to be reduced to *p*-coumaraldehyde by cinnamoyl CoA reductase (CCR) and reduced in turn by a cinnamyl alcohol dehydrogenase (CAD) to the monolignol *p*-coumaryl alcohol. The phenyl ring of *p*-coumaryl alcohol is the least modified of the monolignols, and therefore, it is able to form linkages at the 3′-position that do not occur in the more modified monolignols.

4. Hydroxylation and *O*-methylation at the 3 and 5 positions on the aromatic ring are unlikely to take place at the level of cinnamic acids.[ref37]

It has now become clear how *p*-coumaryl CoA is converted to coniferaldehyde and how the aromatic ring is hydroxylated at the 3 position (see Figure 5.38). In *Arabidopsis* a mutation in the gene *ref8* was found to block 3-hydroxylation of monolignols. The *ref8* gene encodes a cytochrome P-450–dependent monooxygenase, which was presumed to be the *p*-coumarate-3-hydroxylase (C3H) for monolignol biosynthesis; it was also presumed that C3H used either *p*-coumarate or its CoA ester as substrate. Surprisingly, neither *p*-coumarate, its CoA ester, *p*-coumaraldehyde, nor *p*-coumaryl alcohol is an adequate substrate for the enzyme *in vitro*. Similarly, ferulic acid has proved to be a poor substrate for 5-hydroxylation, *in vitro*, by another cytochrome P-450, known by genetic studies to be essential for 5-hydroxylation *in vivo*. Therefore, the hydroxylation and *O*-methylation reactions of the monolignol aromatic rings are unlikely to take place at the level of the cinnamic acids.

In studies of substrate specificity for C3H, shikimate and quinate esters of *p*-coumarate were found to be preferred substrates for C3H. *p*-Coumaroyl CoA is converted to *p*-coumaroyl CoA-shikimate or *p*-coumaroyl CoA-quinate by a *p*-hydroxycinnamoyl-CoA:shikimate/quinate *p*-hydroxy cinnamoyl transferase (HCT) (**Figure 5.39**). HCT from tobacco is able to hydroxylate both shikimate and quinate esters.

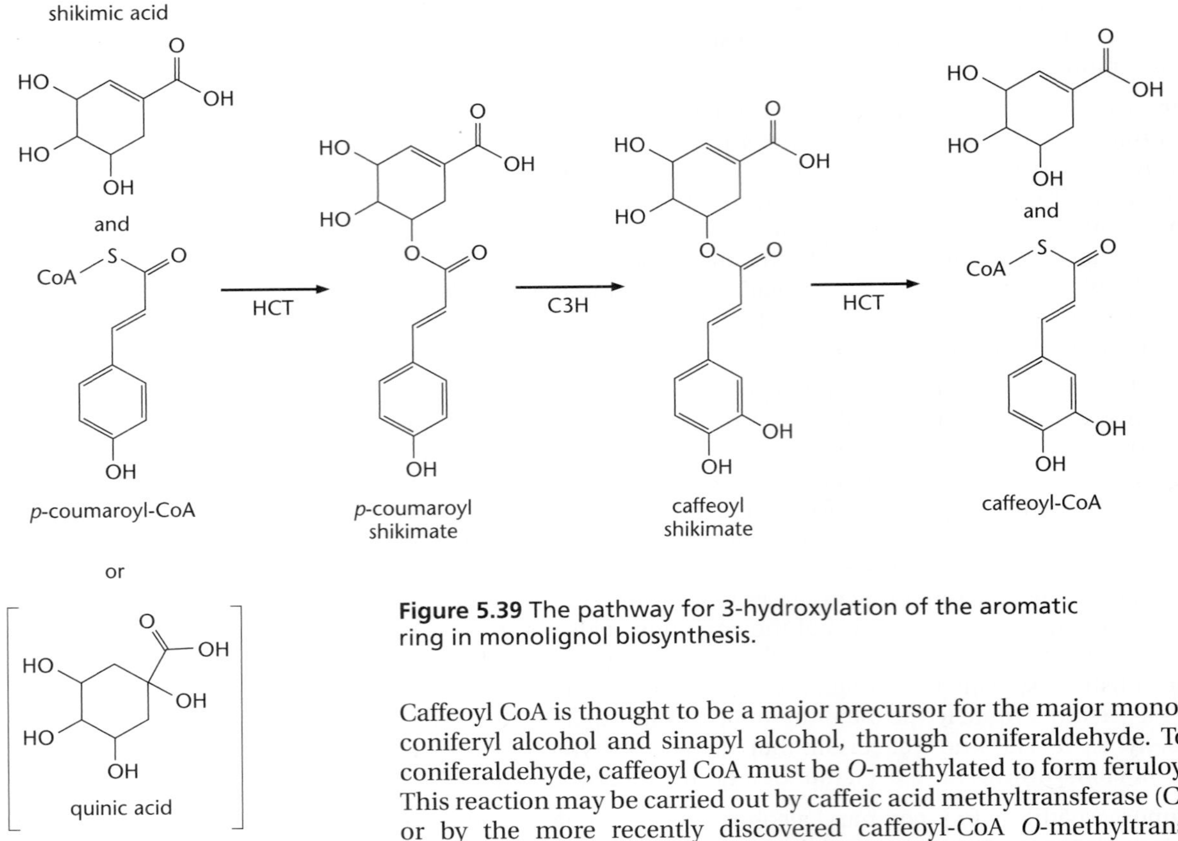

Figure 5.39 The pathway for 3-hydroxylation of the aromatic ring in monolignol biosynthesis.

Caffeoyl CoA is thought to be a major precursor for the major monolignols coniferyl alcohol and sinapyl alcohol, through coniferaldehyde. To form coniferaldehyde, caffeoyl CoA must be *O*-methylated to form feruloyl-CoA. This reaction may be carried out by caffeic acid methyltransferase (COMT), or by the more recently discovered caffeoyl-CoA *O*-methyltransferase (CCoAOMT) to produce feruloyl-CoA. Caffeoyl-CoA is the preferred substrate for CCoAOMT. Feruloyl-CoA is efficiently reduced by cinnamoyl CoA reductase to coniferaldehyde.

5.   Coniferaldehyde is the branch point for the biosynthesis of coniferyl alcohol and sinapyl alcohol.[ref38]

In gymnosperms, coniferaldehyde is reduced by CAD to coniferyl alcohol, the major precursor for gymnosperm lignin. In angiosperms coniferaldehyde is reduced to coniferyl alcohol and is also converted to 5-hydroxy coniferaldehyde by a specific hydroxylase, coniferaldehyde-5-hydroxylase, and subsequently into sinapaldehyde. 5-Hydroxy coniferyl alcohol is a significant component of lignin only in mutants or transgenic variants with reduced *O*-methyltransferase activity (see Figure 5.38). In normal plants most of the 5-hydroxy coniferyl alcohol is converted to sinapaldehyde by an *O*-methyltransferase. Sinapaldehyde may be subsequently converted to sinapyl alcohol by a novel enzyme, sinapyl alcohol dehydrogenase (SAD). Immunolocalization in wood forming tissues of aspen showed enzymatic specificity of cell types and formation of syringyl or guaiacyl lignin. CAD was found exclusively in xylem elements where only coniferyl alcohol–derived guaiacyl lignin was deposited, particularly vessels. In cells producing guaiacyl-syringyl lignin, where both coniferyl alcohol and sinapyl alcohol are required, both CAD and SAD were present.

6.   Monolignols are glucosylated, stored, transported, and deglycosylated before polymerization.[ref39]

Monolignols may be produced by the cells undergoing lignification (cell autonomous derivation) or by other cells, even cells of nonlignifying tissues. During xylem development, the biosynthesis of monolignols may also be initiated prior to the onset of lignification. At present, very little is known about the mechanisms that regulate allocation, storage, and transport of monolignols, but glycosylation of the monolignol precursors plays an important role in these processes.

Free monolignols are typically present at low concentrations in the cytoplasm, most likely because they are toxic, insoluble, and unstable molecules. For lignification, monolignols are needed at high concentrations; therefore, plant cells store and transport monolignols as soluble conjugates with glucose, called *monolignol glucosides*, which are both nontoxic and water-soluble. These monolignol glucosides are formed by glucosylation of the monolignols at the 4 OH of the phenyl ring, by UDP-glucose coniferyl alcohol glucosyltransferase (**Figure 5.40**). *p*-Coumaryl alcohol, coniferyl alcohol, and syringyl alcohol glucosides are designated as *p*-hydroxycinnamyl alcohol glucoside, coniferin, and syringin, respectively.

Glucosides accumulate to high concentrations (mM range) in immature xylem in the spring before cambial reactivation, that is, before lignification begins. Coniferin is the major glucoside found in conifers. Protoplasts derived from immature pine xylem have no cell walls but retain most if not all of the coniferin present in the intact cells. Coniferin must therefore be stored within the cell, with vacuoles and vesicles the most likely locations.

Where these glucosides are formed and how they are transported to vacuoles for storage, and to the cell wall for lignin synthesis, is poorly understood (**Figure 5.41**). Developing xylem cells possess prominent Golgi and associated *trans* Golgi network cisternae. The function of the *trans* Golgi network cisternae is to sort and package Golgi-derived products into vesicles that transport these products to either vacuoles or the cell surface (see Concept 4A7). If monolignols were synthesized by enzymes associated with the *trans* Golgi network, they could be immediately glucosylated and released (or transported) into the lumen of the *trans* Golgi network and packaged into vesicles for delivery to vacuoles or to the cell surface. Upon release into the cell wall, β-glucosidases located at the cell membrane would deglycosylate the monolignols for polymerization in the lignifying zone. A coniferin β-glucosidase has been identified in immature xylem of pine, and its expression coincides in time and place with lignification.

If glucosides were produced by cytosolic enzymes, then they could pass through membranes into *trans* Golgi network cisternae or secretory vesicles by means of membrane-bound transporters such as ABC transporters. Such transporters are abundantly expressed in developing xylem and represent a large gene family that encodes transporters for diverse molecules including lipids and glucose conjugates. Where the different types of ABC transporters operate in plant cells remains to be determined. For example, if they were present in *trans* Golgi network cisternae, where vesicles destined for the cell surface and for vacuoles are formed, then vesicular transport would

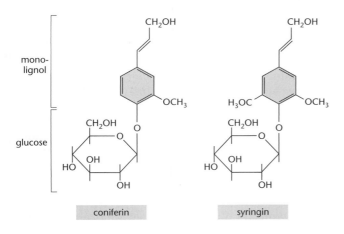

Figure 5.40 The structures of the glucosides of the major monolignols. Coniferin is the glucoside of coniferyl alcohol, and syringin is the corresponding glucoside of sinapyl alcohol.

Figure 5.41 A model for the glucosylation, storage, and transport of monolignol glucosides. Monolignol glucosides may exist as free molecules or in vesicles. Vesicles may be targeted to the vacuole for storage or released at the membrane into the secondary wall. Alternatively, monolignols may be released from the Golgi and be transported directly across the plasma membrane via a transporter into the lignifying zone. Glucosidase activity releases monolignols for polymerization in the wall or the middle lamella. Monolignols could also come from other cells. In all cases, monolignols become part of a polymer in the lignifying zone. TGN, *trans* Golgi network.

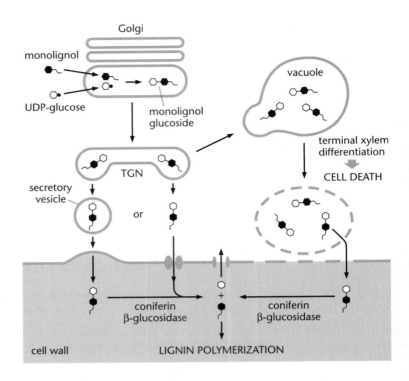

be an integral part of monolignol glucoside transport to the site of lignin synthesis. However, if such transporters were present in both vacuolar and plasma membranes, then there might be no need for a vesicular transport mechanism for delivering monolignol glucosides to the lignifying cell walls. Radioactive tracer evidence and advanced microscopy argue for transport of monolignols through membranes by yet unidentified transporters rather than by export through Golgi vesicles.

7. Lignin polymers are formed through enzymatic oxidation of monolignols.[ref40]

It has long been proposed that lignin is polymerized from monolignols through an oxidative mechanism involving free radical intermediates. Polymerization is thought to occur through the enzymatic oxidation of monolignols resulting in the formation of free radicals, formed by enzymes in or near the lignifying zone. The free radicals represent a higher oxidation state, with an unpaired electron and a propensity for coupling with other monolignols or with a lignin polymer. A monolignol radical can transfer an electron to another monolignol, or to a lignin oligomer or polymer. The radicals are relatively stable due to the electron delocalization characteristic of conjugated bonds. In a monolignol, the radical may be delocalized across the phenylpropane structure. In the polymer, the radical may be restricted to the phenolic ring. Cross-coupling results from radicals on both coupling partners and at specific positions favored by the chemical structure.

Plant cells contain many oxidative enzymes including peroxidases and laccases that have been postulated to oxidize monolignols to form free radicals (**Table 5.6**). For example, a specific coniferyl alcohol oxidase has been identified and characterized from lignin forming tissues. Alternatively, a nonenzymatic oxidation mechanism has been proposed for lignin biosynthesis involving a protein called a dirigent or guide protein with no enzymatic activity of its own, to create dimers from monolignols.

To be implicated in the polymerization of lignin, proteins should satisfy several criteria. They should be cell wall–associated, correlated in time and place with lignification, and able to polymerize monolignols *in vitro*.

TABLE 5.6 ENZYMES AND PROTEINS ASSOCIATED WITH LIGNIN POLYMERIZATION

ENZYME OR PROTEIN	METAL OXIDATION CENTER	OXIDIZING SOURCE	SUBSTRATE SPECIFICITY	MASS	FEATURES
Peroxidase	Iron	H_2O_2	Diverse phenolics	~45 kD	Diverse peroxidases implicated
Laccase	Copper	O_2	o-diphenols; p-diphenols	~60 kD	Blue copper glycoprotein
Coniferyl alcohol oxidase	Copper	Unknown	Coniferyl alcohol	Apparent M_r 84–120 kD	Glycoprotein
Dirigent protein	—	Laccase/O_2; peroxidase/H_2O_2	E-coniferyl alcohol	~24 kD	Nonenzymatic glycoprotein; stereospecific coupling

Genetic support for function *in vivo*, through mutation or transgenic suppression or activation, is also essential. Biochemical characterization of the purified candidate protein should be consistent with the proposed mechanism of action. More than one isoform and many types of enzymes may be involved in a process as complex as the formation of lignin polymers.

Peroxidases use hydrogen peroxide to generate free radicals (**Figure 5.42**). The source of H_2O_2 is not known. Enzymes in the cell wall such as oxalate oxidase may be involved in the generation of H_2O_2. Peroxidases can form H_2O_2 in reactions with reducing substrates such as ascorbate. Peroxidases have been implicated in cell cultures forming lignin where inhibition of peroxidase activity reduced lignin formation. Also, overexpression of peroxidase can result in ectopic lignin formation. Mutations or transgenics in some cases have affected lignin content. A deficiency in a maize peroxidase, and down-regulation of peroxidase in aspen or tobacco, have resulted in reduced lignification. Laccases use O_2 to oxidize monolignols; however, down-regulation of individual laccases did not affect lignin content or composition in transgenic poplar.

Model systems for lignin polymerization *in vitro* have been developed where the mixture of an oxidase and monolignols results in the formation of lignin-like polymers known as *dehydrogenation polymers* (DHP). DHPs resemble lignin in many ways, but the degree of polymerization is low and the relevance to polymerization *in vivo* is yet to be established. Nevertheless, DHP formation represents a valuable *in vitro* system where the components can be readily manipulated. Classical studies showed that the linkages formed at high concentrations of monolignols differed from those linkages formed by slow addition of monolignols to a polymerizing reaction mixture. The slow addition oligomer products are more like those of native lignin. Lignin-like linkages are formed in the presence of a poplar cell wall–associated peroxidase at high monolignol concentrations, and therefore the specificity of the cell wall–associated oxidases may be an important factor in the

Figure 5.42 Formation of coniferyl alcohol free radicals from H_2O_2 and O_2 for lignin polymer formation.

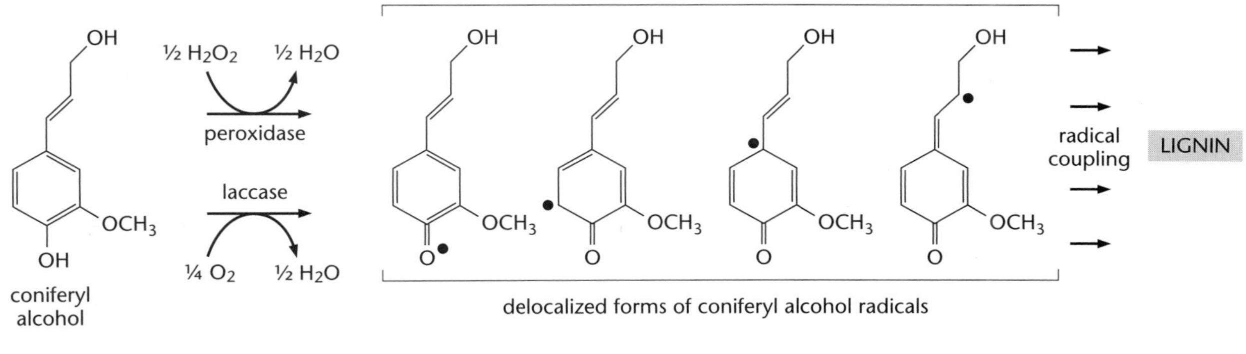

delocalized forms of coniferyl alcohol radicals

specificity of linkage formation. DHP experiments also show that coniferyl alcohol transfers electrons to a lignin oligomer at low efficiency, suggesting the need for an additional enzyme or mechanism for more efficient free radical formation in the polymer.

8. Lignin polymerization is primarily due to the addition of a monolignol to a lignin polymer.[ref41]

The most significant proposed reactions in the formation of lignin polymers are the coupling of a monolignol to a polymer and the coupling of two oligomers, resulting in a growing lignin polymer. For a coniferyl alcohol monomer, there are two favored reactions at the β position of a monolignol with a polymeric lignin subunit: formation of a β-O-4 linkage (**Figure 5.43**), and formation of a β-5 linkage. For sinapyl alcohol or a sinapyl alcohol–derived lignin subunit, with CH_3O at the 5 position, further linkages at the 5 position are precluded. Other linkages are made, based on similar principles, but the β-O-4 and β-5 are the main products, and these linkages produce linear growing polymers. Linkages of 5-5 (biphenyl), and 5-O-4 (biphenyl ether) linkages, are less frequent, but are important because they are the major linkages that form between two oligomers (or polymers) resulting in cross-links between the lignin "chains." Linkage of two subunits at the 5-5 position may be followed by addition of a third monolignol linking through both a β-O-4 linkage and an α-O-4 linkage to produce the cyclic ether structure called dibenzodioxocin (see Concept 3F4.1).

In contrast to the general evidence that lignin biosynthesis is predominantly the result of coupling of a monolignol to a polymer, addition of dilignols to lignifying *Zinnia* tracheary elements resulted in incorporation into lignin. Incorporation was observed when the supply of monolignols was blocked by inhibition of phenylalanine ammonia lyase. At least under some circumstances, dilignols could be intermediates in lignin biosynthesis.

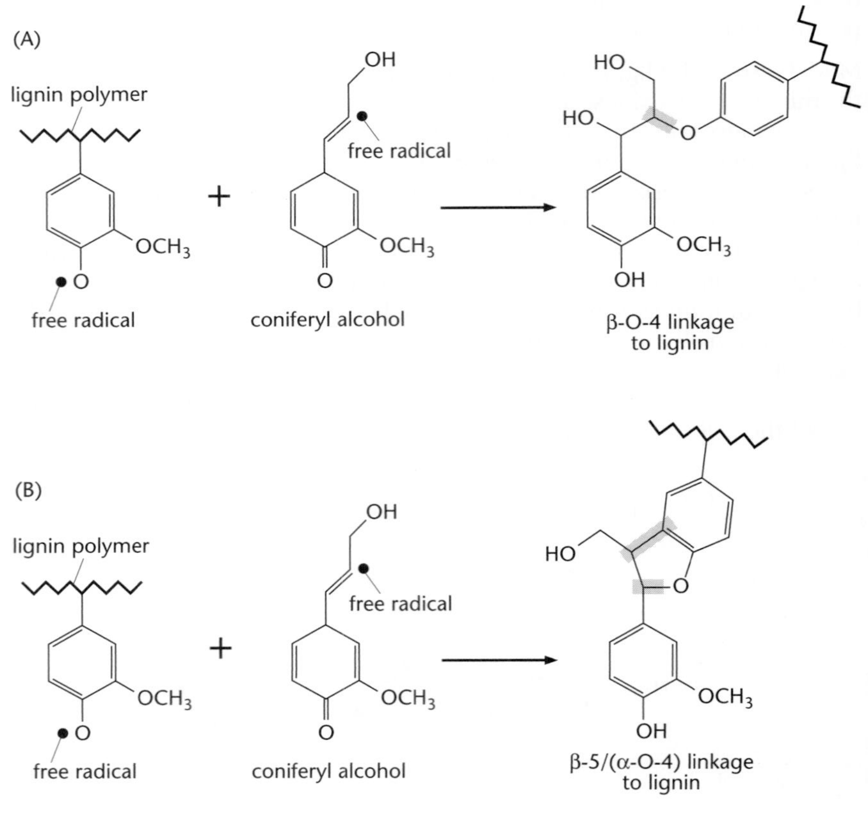

Figure 5.43 Coupling of coniferyl alcohol to the lignin polymer. Formation of the two different linkages β-O-4 and β-5 is shown. (A) Coniferyl alcohol with a free radical at the β position forms a linkage to the polymer. (B) The free coniferyl alcohol monomer forms a link to the 5 position, and an α-O-4 linkage follows. Shaded regions show new bonds formed.

In the current version of the classical model of lignin biosynthesis, now called the *combinatorial model*, the polymerization step is controlled by monolignol availability, coupling propensities, reactant concentrations, and the conditions in the lignifying zone and the cellulosic matrix. There is no fixed sequence of units in the lignin polymer. While there are different probabilities for specific reactions, the polymer changes with each new linkage, resulting in a significant component of "randomness." As the polymer grows, the probability that any two polymers will have the same structure diminishes rapidly. Second, subject to coupling propensities and concentration at any time in the lignifying zone, any monolignol may be involved in coupling. As a consequence of the chemical reactions, the units formed are racemic, lacking optical activity (in contrast to polysaccharides and proteins). The combinatorial model easily accounts for the diversity of linkages in lignin, which provides lignin with a structure that is difficult to degrade by any known enzymes, thereby resulting in the exceptional resistance of lignin to microbial decay (see Concept 8C4). The model also readily accounts for variation in lignin composition and content, and relegates the variation in lignin to the composition and concentration of monolignols delivered to the lignifying zone.

9. Oxidative carriers or radical mediators may be involved in the polymerization process.[ref42]

If lignin polymerization requires an oxidizing protein, it has been argued that radical-forming oxidases could not be readily delivered to the lignifying zone through the thickened cell wall. Similarly, a monolignol radical might easily react before it reaches the lignifying zone. A model of oxidative polymerization (**Figure 5.44**) illustrates how these problems might be solved through $2Mn^{3+}$ as a diffusible, oxidative carrier. Mn ions are less reactive than monolignol radicals and could therefore diffuse more readily to the sites of lignin polymerization. At those sites the $2Mn^{3+}$ ions would help generate monolignol radicals and polymer radicals, both of which are needed to form covalent bonds. The reduced $2Mn^{2+}$ then shuttles back to a location where it can be reoxidized. Following transfer and oxidation of the polymer and the monolignol, a subunit is added to the preexisting lignin polymer to form a β-*O*-4 linkage.

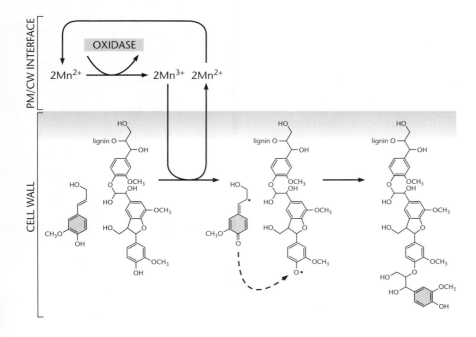

Figure 5.44 Redox shuttle model for lignin biosynthesis. In this model, an enzyme oxidizes Mn^{2+} to $2Mn^{3+}$, which is able to penetrate the pores in the cell wall and reach the lignifying zone where free monolignols and growing chains of lignin are present. Mn^{3+} oxidizes monolignols and phenolic residues in the lignin polymer to produce a monolignol radical and a radical on the polymer. Covalent bonds form when two radicals meet. Mn^{2+} regenerated and returns to the site of oxidation. (Adapted from H. Önnerud et al., *Plant Cell* 14:1953–1962, 2002.)

10. A nonenzymatic model for lignin polymerization has been proposed involving "dirigent" or guide proteins.[ref43]

The traditional model for lignin polymerization has been challenged on the basis of a mechanism discovered for the formation of lignans. Lignans are dimers of coniferyl alcohol, and it has been argued that lignin polymerization is not random, but instead involves bimolecular stereospecific coupling reactions similar to that proposed for lignan biosynthesis. For example, (+)-pinoresinol, a β-β linked lignan, can be formed *in vitro* through stereospecific reactions of a nonspecific oxidase and a *dirigent*, or guide, protein (**Figure 5.45**). Only the production of a specific optical form is permitted. This proposed reaction of lignan formation is a model for specific linkage formation in lignin through highly regulated stereospecific reactions where each type of linkage depends on the action of a specific dirigent protein. The model is attractive because it eliminates "randomness" from the polymerization mechanism and substitutes a novel protein-mediated linkage-specific reaction.

The dirigent model makes several testable predictions. First, there should be large numbers of different dirigent proteins with high specificity. This specificity should lead to fixed, repeating sequences within the polymer. Second, nontraditional monomers would be excluded, and directed modification of monomer concentration or composition would not affect composition of lignin. Third, lignin would be optically active. The key issue is the extent to which relatively nonspecific reactions are involved in linkage specificity, using few enzymes and determined by the inherent chemistry of diverse phenolics delivered to the lignifying zone, or whether lignification is a highly ordered stereospecific process controlled by many dirigent proteins.

At present, most predictions of the dirigent model have not been satisfied. Only one dirigent protein has been adequately identified and characterized. That protein is implicated in the formation of pinoresinol, which represents only a minor component of lignin. Predicted structures formed from coupling of two coniferyl alcohol radicals to form pinoresinol that is then incorporated into lignin have not been detected or are rare. No genetic test showing a "loss of function" effect on lignin has been shown for a dirigent protein. Dirigent proteins for the major linkages such as β-*O*-4 have not been identified. To account for all dimeric linkages, about 50 dirigents would be needed. If the formation of higher-order linkages also required dirigent proteins, the number of dirigents required would represent a large fraction of all the genes in the plant genome. No such genome fraction has been identified, and no fixed sequence of linkages has been found for oligolignols or polymers. Lignins are not optically active. Nontraditional monomers and conjugates of monolignols abound in native lignins. Lignin content and composition varies dramatically in different species, and in mutants and transgenics of the same species. Metabolic plasticity is a significant characteristic of natural lignins.

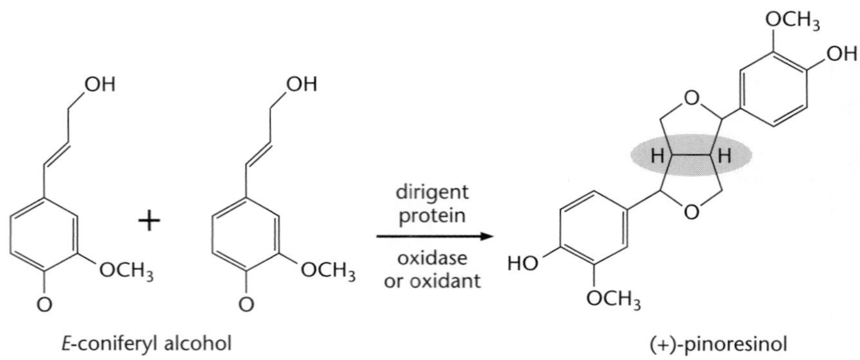

Figure 5.45 Dirigent protein based formation of pinoresinol: Two molecules of *E*-coniferyl alcohol form (+)-pinoresinol by β-β coupling. (Adapted from L.B. Davin and N.G. Lewis, *Plant Physiol.* 126:453–461, 2000.)

E-coniferyl alcohol

(+)-pinoresinol

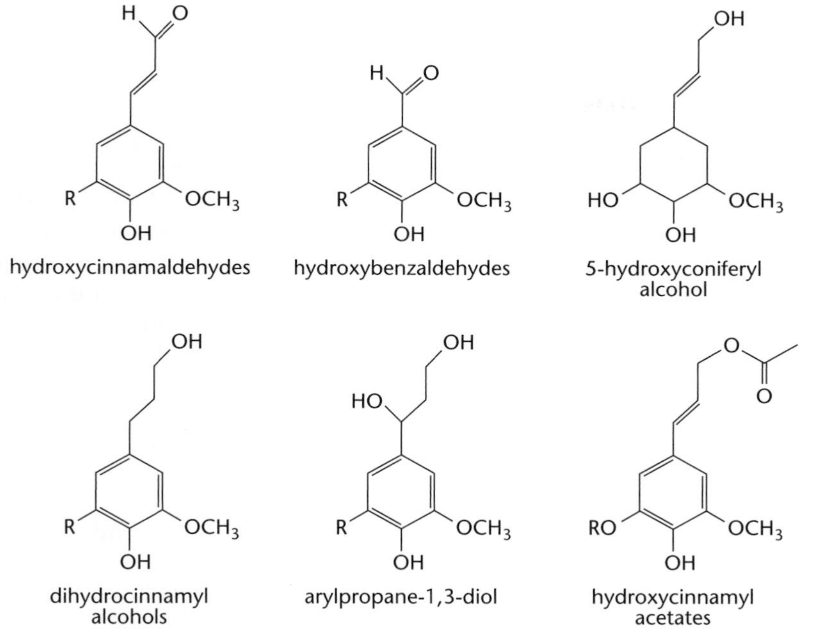

hydroxycinnamaldehydes

hydroxybenzaldehydes

5-hydroxyconiferyl alcohol

dihydrocinnamyl alcohols

arylpropane-1,3-diol

hydroxycinnamyl acetates

Figure 5.46 Some nontraditional monomers that are incorporated into natural lignins. The 5 position may have alternative linkages. Where R = H, the compounds are coniferyl alcohol, coniferaldehyde, and derived monomers. Where R = OCH_3 the derived compounds are sinapaldehyde and sinapyl alcohol derived monomers. (Adapted from J. Ralph et al., *Annu. Rev. Plant Biol.* 54:519–564, 2004.)

How might a dirigent system evolve? It is not obvious how a fixed sequence of a lignin polymer would evolve, because lignin is inert, and the order of subunits would not have apparent selective value. On the contrary, the "randomness" of the lignin polymer has adaptive value in increased resistance to microbial enzymes.

11. Nontraditional monomers are readily incorporated into lignin, indicating a high level of metabolic plasticity.[ref44]

Lignin has been traditionally considered to be a product of three types of monolignols, *p*-coumaryl alcohol, coniferyl alcohol, and sinapyl alcohol, which differ in proportions during differentiation, development, and evolution. However, lignin can incorporate subunits derived from a considerable variety of other monomers (**Figure 5.46**). Many are intermediates of monolignol biosynthesis, and others are derivatives of the traditional monomers. These results imply that many types of phenolics may be formed in the cell, transported to the lignifying zone in the cell wall, and incorporated into lignin.

The hydroxycinnamaldehydes, coniferaldehyde and sinapaldehyde are integrally incorporated into lignins, and are likely to be the basis for the classical *phloroglucinol* reaction used for staining lignin (see Concept 3F3). Cinnamaldehydes increase when there is a reduction or inhibition of cinnamyl alcohol dehydrogenase (CAD) activity. CAD is one of the major enzymes involved in formation of cinnamyl alcohols from cinnamaldehydes (Concept 5E3). Hydroxybenzaldehydes, derived from cinnamaldehydes by cleavage, may also be incorporated as monomers. Similarly, 5-hydroxyconiferyl alcohol is found in lignin of plants reduced in the expression of caffeic acid/5-OH coniferaldehyde *O*-methyl transferase (COMT), resulting in the accumulation of 5-hydroxyconiferaldehyde, which is converted by CAD to 5-hydroxyconiferyl alcohol.

Dihydroconiferyl alcohol is lacking the conjugated bond between the α and β carbons of the propane side chain, and is present in normal softwood lignins at low levels. A related subunit with an additional hydroxyl group added to the α carbon (arylpropane-1,3-diol) is also found in lignin of most softwoods. In a pine deficient for CAD, dihydroconiferyl alcohol–derived subunits can be a predominant component of the lignin (**Figure 5.47**). It

Figure 5.47 Photograph of wood fragments from normal loblolly pine (*left*) and from a CAD null mutation (*right*) that accumulated dihydroconiferyl alcohol subunits in its lignin. The wood is brown, much like the brown midrib phenotype produced by CAD mutations in grasses, presumably due to a by-product of the accumulation of coniferaldehydes.

is likely that dihydroconiferyl alcohol is derived from coniferaldehyde because its abundance is directly associated with a reduction in CAD, and because coniferaldehyde added to pine microsomes results in the formation of both dihydroconiferyl alcohol and coniferyl alcohol.

Subunits in lignin may be also derived from monolignol conjugates. Among these are the pre-acetylated monomers found in kenaf. The γ-hydroxyl of the monomers is acetylated, and this modification may be found in 50% of the subunits of lignin in kenaf bast fibers. Grasses may have as much as 17% of their lignin in γ-*p*-coumarate conjugates. Similarly, γ-*p*-benzoate conjugates are found in lignin of poplar, willow, and palms. Esters and amides of *p*-hydroxycinnamates are incorporated by radical coupling into lignin. Ferulic acid (ferulate) and derived ferulate dimers provide cross-links of lignin and polysaccharides (see Concept 3F4).

12. The deposition of lignin is both temporally and spatially controlled.[ref45]

Deposition of lignin is highly regulated in time and place, and has been intensively studied in the formation of the secondary cell walls in tracheids and fibers during wood formation. Lignification proceeds in three stages, each preceded by deposition of specific carbohydrates in specific locations (**Figure 5.48**, **Figure 5.49**, and **Figure 5.50**). The first stage is the initiation of lignification, which follows the radial expansion of cell walls, the deposition of pectin in the middle lamella, and the formation of the first cellulosic layer of the secondary wall (S1 layer). Lignification is then initiated at the cell corners and in the middle lamella. The second stage is a slow lignification stage, during the formation of the major layer of the secondary wall, the S2 layer. This pattern suggests that polymerization is initiated by local nucleation events.

Most lignification occurs after formation of the S3 layer. Lignification is not just a matter of filling in space in or between the cell walls. Lignin polymerization is likely to be specifically initiated, because intracellular spaces in normal wood are not filled with lignin. Conifer wood formed under gravitational stress (compression wood; see Concepts 3F8, 7C8) is characterized by large intercellular air spaces, and these spaces, also, are not filled with lignin. Displacement of water and solutes results from the deposition of lignin as lignification gradually converts the cellulosic hydrophilic gel into a hydrophobic phenolic polymer. Shrinkage is anisotropic and perpendicular to the direction of the cellulose microfibrils.

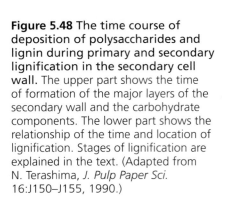

Figure 5.48 The time course of deposition of polysaccharides and lignin during primary and secondary lignification in the secondary cell wall. The upper part shows the time of formation of the major layers of the secondary wall and the carbohydrate components. The lower part shows the relationship of the time and location of lignification. Stages of lignification are explained in the text. (Adapted from N. Terashima, *J. Pulp Paper Sci.* 16:J150–J155, 1990.)

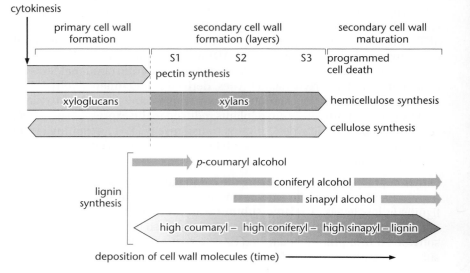

Figure 5.49 Model of the lignifying zone showing monolignol subunit linkages (lignin-lignin bonds) lignin carbohydrate linkages, carbohydrate (cellulose-hemicellulose) hydrogen bond linkages, and diferulate crosslinks. Lignin monomers are shown as filled hexagons with propane side chains. Hemicelluloses are shown as thick lines. Ferulate is shown as a circle. Cellulose microfibrils are vertical lines. Hemicelluloses and cellulose microfibrils are connected by hydrogen bonds shown as dots. (Adapted from N. Terashima, *J. Pulp Paper Sci.* 16:J150–J155, 1990.)

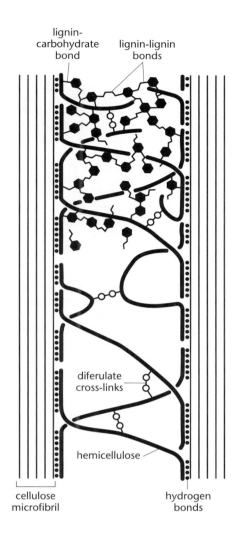

13. The composition of lignin differs between cell wall domains and cell types.[ref46]

The composition of lignin changes as different regions of the wall lignify, with the early lignifying middle lamella having a higher *p*-hydroxyphenyl content and the later lignifying cell wall layers having a guaiacyl lignin (in conifers) and a guaiacyl-syringyl lignin in woody angiosperms. Different cell types in the same wood may have different types of lignin; for example, in angiosperms with diffuse porous woods, vessels have only guaiacyl lignin, while fibers have guaiacyl-syringyl lignin (**Figure 5.51**; see also Table 3.4). Similarly, during compression wood formation, the outer region of the secondary wall has a higher concentration of lignin, and in tension wood formation, the innermost layer of the wall is gelatinous, with almost no lignin (see also Figure 3.33). In such cases, the composition is uniform around the circumference of the cell, suggesting an intracellular origin of monolignols and regulation of composition through monolignol biosynthesis. The individual cell, therefore, largely controls the temporal and spatial deposition of lignin, in addition to the subunit composition of the lignin in the specific layers of the cell walls.

These conclusions lead to an unresolved dilemma. In earlier sections on transport, it was argued that lignification could occur in dying cells and that monolignols could come from extracellular sources as cells and vacuoles lyse. However, the distribution of the variation in lignin composition in layers of the cell wall argues that the major supply of monolignols should come from inside the cell. Perhaps the regulation of biosynthesis and the source of monolignols could change during differentiation.

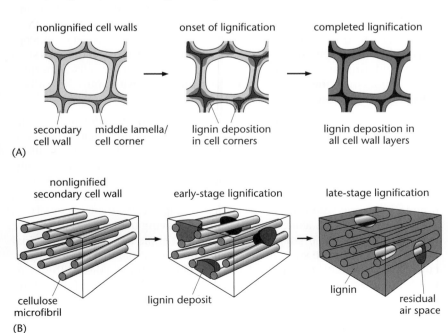

Figure 5.50 Lignin deposition in the secondary wall. (A) A cross section of lignifying secondary cell walls is shown. Lignin polymerizes around cellulosic fibers in the wall and in the middle lamella. Polymerization begins at the cell corners and in the middle lamella and spreads through the secondary wall. Lignin concentration is highest in the middle lamella because of the absence of the cellulose and hemicellulose fibers. (B) Lignin polymerization begins at nucleation sites and expands as polymerization continues to fill space around the cellulosic scaffold. (Adapted from J. Nakashima et al., *Plant Cell Physiology* 38:818–827, 1997; and L.A. Donaldson, *Phytochemistry* 57:859–873, 2001.)

Figure 5.51 Confocal micrographs of lignin fluorescence in normal and reaction wood. (A) Wood from *Populus nigra* stained with acriflavin shows differential distribution of lignin. The G-lignin label marks a highly lignified vessel cell wall, known to be rich in guaiacyl lignin adjacent to less lignified fiber cells containing guaiacyl-syringyl (G-S) lignin. Length of side ~75 μm (see also Figure 3.3). (B) Tension wood from *P. nigra*, stained with acriflavin and congo red to highlight the unlignified G layer adjacent to the fiber lumen. ML marks lignified middle lamella. Length of side ~92 μm. (C) Fluorescence of lignin in normal *Pinus radiata* wood stained with acriflavin showing high lignin concentration in the middle lamella (ML). Length of side ~80 μm. (D) Fluorescence of lignin in compression wood of *Pinus radiata* stained with acriflavin showing high lignin concentration in the S2 layer (layer) compared to normal wood. Length of side ~76 μm. (Courtesy of L.A. Donaldson and J. Bond, © Scion, NZ.)

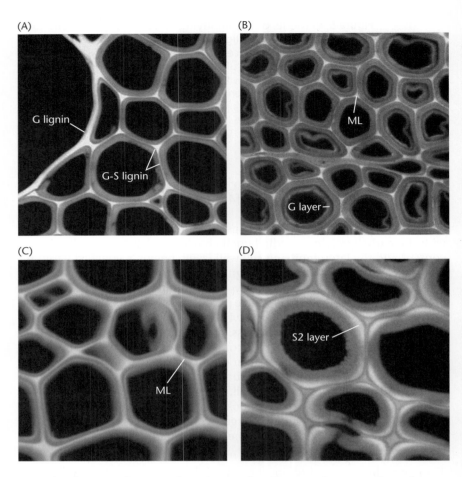

14. The biosynthesis of cutins, suberins, and waxes involves enzymes in chloroplasts, the endoplasmic reticulum, and cell walls.[ref47]

Plants produce cutins, suberins, and waxes to create hydrophobic cell wall layers capable of controlling the movement of gases, water, and solutes, as well as to impart pathogen resistance (see Concepts 3G1–3). Cutins and waxes are usually found together in the *waxy cuticle*, a water-repelling layer that covers the outer walls of epidermal cells of aerial organs (see Concept 1C1). Suberin-impregnated cell walls are formed in specific root tissues such as the exodermis and in the casparian bands of endothelial cells. Suberin is also a major component of bark cell walls and is deposited into the walls of cells bordering wounded tissues.

The waxy cuticle consists of two parts: cutin networks, two-dimensional, polyester-type networks of oxygenated C_{16} and C_{18} fatty acids that confer mechanical strength to the cuticle (see Figure 3.39); and crystallized wax molecules, long chain (C_{20}–C_{36}) hydrocarbons, fatty acids, alcohols, ketones, aldehydes, and esters that both impregnate the cutin networks (to form *intracuticular waxes*) and form surface coats (*epicuticular waxes*), thereby reducing water losses by orders of magnitude (see Concept 1C1). The waxes can be extracted with chloroform, leaving behind the cutin layer. Very-long-chain (>C_{18}) fatty acids and fatty alcohols are also the principal building blocks of the suberin layers, but unlike cutins, the suberin molecules do not contain secondary alcohols or epoxy groups. These differences make suberin more hydrophobic than cutin. Nevertheless, all of these fatty acid–derived molecules have a common origin and are synthesized by the same types of enzymes shown in the schematic wax biosynthesis pathway (**Figure 5.52**).

In epidermal cells, wax biosynthesis occurs in three stages. First, C_{16} and C_{18} fatty acids are synthesized in nonphotosynthetic plastids. Second, following transfer to the ER, the C_{16} and C_{18} fatty acids are elongated to produce *very-long-chain fatty acids* (VLCFAs) with lengths of C_{20} to C_{36}. During the final stage, the VLCFAs are converted to the major wax products, alcohols, esters, aldehydes, alkanes, and ketones.

A complex of water-soluble proteins, *the fatty acid synthase complex*, located in the plastid stroma, produces C_{16} and C_{18} fatty acids for both membrane and cuticular, wax and suberin lipid biosynthesis. During fatty acid biosynthesis, the growing acyl chain is attached to an *acyl carrier protein* (ACP), a cofactor of the fatty acid synthase complex. Fatty acid synthesis starts with the condensation of acetyl CoA with malonyl-ACP (Figure 5.52). After the

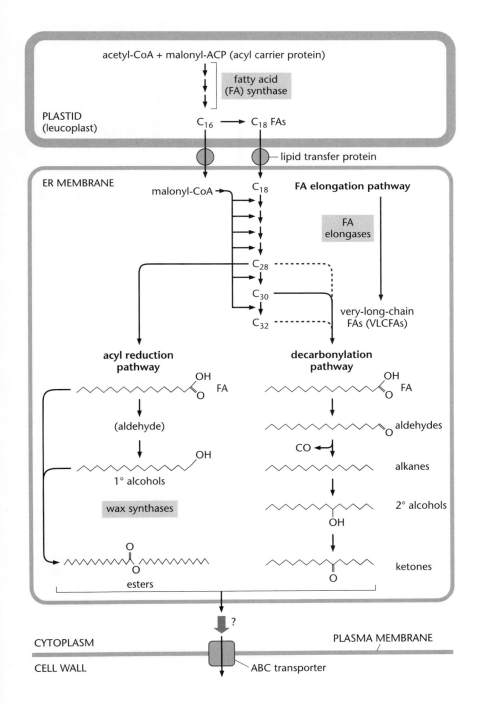

Figure 5.52 The wax biosynthesis pathway of *Arabidopsis* that provides the principal building blocks of waxes, cutins, and suberins. C_{16}- and C_{18}-type fatty acids are synthesized in colorless plastids, the leucoplasts. Export of these fatty acids to the endoplasmic reticulum (ER) membrane occurs by means of a molecular transfer mechanism mediated by lipid transfer proteins at sites of close contact between the outer chloroplast envelope membrane and the ER. Elongation of the C_{18} fatty acids to very-long-chain fatty acids (VLCFAs) and conversion of the VLCFAs to the different kinds of wax molecules is catalyzed by ER membrane enzymes. The mechanism of transport of the wax molecules to the plasma membrane is unknown, but transport of the molecules across the plasma membrane involves an ABC-type transporter. (Adapted from L. Kunst and A.L. Samuels, *Prog. Lipid Res.* 42:51–80, 2003.)

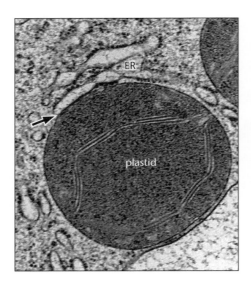

Figure 5.53 Thin-section electron micrograph of an ER-plastid contact site (*arrow*) in an epidermal cell of *Arabidposis thaliana.* Such membrane contact regions have been postulated to serve as direct lipid transfer sites between the two organelles. (Courtesy of A. Lacey Samuels.)

condensation step, a series of reactions yields an acyl-ACP product lengthened by two carbons. The assembly of C_{16} and C_{18} fatty acids requires multiple rounds of condensation reactions catalyzed by three different condensing enzymes, each of which has a strict acyl chain length specificity. In rapidly expanding epidermal cells up to 50% of these fatty acids are channeled to cuticular lipid biosynthesis.

How the C_{16} and C_{18} fatty acids are exported from the plastids to the ER, and how they are partitioned between the membrane glycerolipid and the wax biosynthesis pathways, is poorly understood. Release of the fatty acids from ACP involves a thioesterase, and prior to export, the fatty acids are esterified to acyl CoAs in the outer plastid envelope membrane. Transfer of the acyl CoAs from the outer envelope membrane to the ER occurs most likely at sites of close contact between the two membrane systems via a molecular lipid hopping type of transfer mechanism (Figure 5.52; **Figure 5.53**).

Extension of the C_{16} and C_{18} fatty acids to VLCFAs is catalyzed by *fatty acid elongases*, multienzyme complexes in the ER. The C_2 donor for this process is also malonyl-CoA, and, as for the synthesis of C_{16} and C_{18} fatty acids, multiple elongation cycles involving several distinct elongases are required to produce the VLCFAs. The specificity of each elongation step lies in the condensing enzymes, β-ketoacyl-CoA synthases (KCS), of the elongation complex. The *Arabidopsis* genome contains 21 KCS-like genes. One of these, a wax-specific KCS protein called CER6, has been shown to be involved in the lengthening of C_{22} fatty acyl-CoAs in *Arabidopsis*. In *cer6* mutants, wax accumulation on the stems is nearly abolished.

Primary wax alcohols are typically 26 or 28 carbons long. Formation of these alcohols in the acyl reduction pathway (Figure 5.52) involves a reduction of corresponding-length VLCFAs via aldehyde intermediates. In *Arabidopsis,* this reaction is mediated by the fatty acid reductase CER4. An alternative pathway, the decarboxylation pathway (Figure 5.52), converts the VLCFAs into alkanes, secondary alcohols, and ketones. The relative abundance of each of the final products varies between different tissues of a given plant and between plant species.

Enzymes called *wax synthases* catalyze the formation of wax esters from long-chain alcohols and VLCFAs. In jojoba, a desert plant that accumulates liquid waxes in its seeds and has been developed as a commercial crop, the wax synthases are able to combine saturated and unsaturated, C_{14} to C_{24} long, acyl-CoAs into C_{36} to C_{46} straight-chain wax esters. Twelve homologs of the jojoba-type wax synthases have been identified in *Arabidopsis*.

Many mysteries still surround the transport of the waxes, cutins, and suberins from the ER to the cell wall. Due to their hydrophobicity, these molecules probably accumulate in the interior of the lipid bilayer of the ER membranes. Whether they are transported in Golgi-derived secretory vesicles to the cell surface or via direct molecular transfer at ER-plasma membrane contact sites is unknown. However, a recent study has demonstrated that wax transport across the plasma membrane involves *ABC (ATP binding cassette)-type transporters*, CERT5 and WBC1, which may function as heterodimer complexes in the plasma membrane. Upon entering the hydrophilic cell wall, the different types of hydrophobic molecules are assumed to bind to lipid transfer proteins, enabling them to reach the outer cell wall layers by diffusion. Lipid transfer proteins accumulate to high levels in epidermal cell walls, contain a hydrophobic pocket capable of binding VLCPAs and related hydrophobic molecules, and are small enough to traverse the pores of the cell wall. However, definitive proof for this postulated transport function is still lacking. Polymerization of cutin occurs within the cuticular region of the cell wall and yields extremely stable cutin polyesters. Due to their insolubility and chemical stability, both cutin and suberin polyesters break down very slowly and comprise a significant proportion of forest litter.

The *wax mutants* in *Arabidopsis* and barley are called eceriferum (*cer* mutants), and in maize, glossy (*gl*) mutants. Mutants have as little as 15% of the normal level of wax. Several *cer* mutations have been characterized sufficiently to establish the steps in the pathway that are blocked. For example, *cer1* in *Arabidopsis* blocks the conversion (decarbonylation) of a C_{30} aldehyde to a C_{29} alkane, *cer8* blocks the reduction of various acids to aldehydes, and *cer4* blocks the reduction of various long-chain aldehydes to their corresponding alcohols. In maize, glossy mutations corresponding to *cer1* (aldehyde decarbonylase) and *cer2* (fatty acid elongation) have been identified.

Epicuticular wax production and deposition is tightly regulated both during development and in response to environmental changes. This regulation has been traced, in part, to transcription factor *WAX INDUCER 1 (WIN1)*. Overexpression of *WIN1* in *Arabidopsis* triggers increased rates of wax and cutin synthesis in both vegetative and reproductive organs. Such plants exhibit enhanced drought tolerance, suggesting that manipulation of *WIN1* expression might provide a means for producing more drought-resistant crops.

Key Terms

ABC (ATP binding cassette)-type transporter
Acetobacter xylinum
acyl carrier protein
atty acid synthase complex
block polymer construction
callose synthase
Carbohydrate Active Enzyme (CAZy) database
cellulase synthase superfamily
cellulose synthase-like gene
CESA protein
combinatorial model
complex-type glycan
coniferyl alcohol
CSL protein
cyanobacterium
cyclic-di-GMP
D, D, D, Q/RXXRW motif
dehydrogenation polymer
dirigent
dolichol
elongation
epicuticular wax
fatty acid elongase
Gal-transferase
GalA-transferase
GlcNAc transferase I
Gluconacetobacter xylinus
glycosyltransferase
GPI anchor
GSL protein
"headward growth"
herbicide 2,6-dichlorobenzonitrile (DCB)
HG-methyltransferase
high-mannose-type
hybrid-type glycan
hydrophobic cluster analysis
initiation
intracuticular wax
isoxaben

Korrigan protein
lignan
Man-synthase
mannosidase I
monolignol glucoside
monolignol
N-linked glycan
non-carbohydrate primer
nucleotide diphosphate sugar
nucleotide sugar interconversion
nucleotide sugar transporter
oligosaccharyltransferase
oxidative polymerization
p-coumaryl alcohol
paucimannosidic-type glycan
phenylalanine ammonia-lyase (PAL)
phloroglucinol
processive glycosyltransferase
product entrapment
prolylhydroxylase
reversibly glycosylated protein
salvage pathway
sinapyl alcohol
sitosterol-β-glucoside
sucrose synthase (SuSy)
"tailward growth"
termination
tunicamycin
type II membrane protein
very-long-chain fatty acid (VLCFA)
wax mutant
wax synthase
waxy cuticle
XG-Fuc-transferase
XG-Gal-transferase
XG-Glc-synthase
XG-Xyl-transferase
Zn finger domain
β-furfuryl-β-glucoside

References

[1]**Nucleotide sugars are the source of the glycosyl residue building blocks used in the synthesis of cell wall polysaccharides and glycoproteins.**

Konishi T, Takeda T, Miyazaki Y et al. (2007) A plant mutase that interconverts UDP-arabinofuranose and UDP-arabinopyranose. *Glycobiology* 17, 345–354.

Reiter WD (2008) Biochemical genetics of nucleotide sugar interconversion reactions. *Curr. Opin. Plant Biol.* 11, 236–243.

Seifert GJ (2004) Nucleotide sugar interconversions and cell wall biosynthesis: how to bring the inside to the outside. *Curr. Opin. Plant Sci.* 7, 277–284.

Sharples SC & Fry SC (2007) Radioisotope ratios discriminate between competing pathways of cell wall polysaccharide and RNA biosynthesis in living plant cells. *Plant J.* 52, 252–262.

[2]**Transport of nucleotide sugars into ER and Golgi cisternae is mediated by specific NDP-sugar/NMP antiporters.**

Reyes F & Orellana A (2008) Golgi transporters: opening the gate to cell wall polysaccharide biosynthesis. *Curr. Opin. Plant Biol.* 11, 244–251.

Rollwitz I, Santaella M, Hille D et al. (2006) Characterization of AtNST-KT1, a novel UDP-galactose transporter from *Arabidopsis thaliana. FEBS Lett.* 580, 4246–4251.

[3]**Glycosyltransferases are generally nonabundant proteins.**

Lerouxel O, Cavalier DM, Liepman AH & Keegstra K (2006) Biosynthesis of plant cell wall polysaccharides: a complex process. *Curr. Opin. Plant Biol.* 9, 621–630.

Shoreibah MG, Hindsgaul O & Pierce M (1992) Purification and characterization of rat kidney UDP-N-acetylglucosamine: alpha-6-D-mannoside beta-1,6-N-acetylglucosaminyltransferase. *J. Biol. Chem.* 267, 2920–2927.

[4]**All glycosidases and glycosyltransferases involved in the modification of glycoproteins and polysaccharides have a common topology and common Golgi retention mechanisms.**

Breton C, Mucha J & Jeanneau C (2001) Structural and functional features of glycosyltransferases. *Biochimie* 83, 713–718.

Perrin R, Wilkerson C & Keegstra K (2001) Golgi enzymes that synthesize plant cell wall polysaccharides: finding and evaluating candidates in the genomic era. *Plant Mol. Biol.* 47, 115–130.

[5]**The synthesis of polysaccharide backbones involves initiation, elongation, and termination reactions.**

Iiyama K, Lam TBT, Meikle PJ et al. (1993) Cell wall biosynthesis and its regulation. Forage Cell Wall Structure and Digestibility (H Jung, D Burton, R Hatfield, J Ralph eds), pp 621–683. Madison, WI: American Society of Agronomy.

McCann MC, Wells B & Roberts K (1991) Complexity in the spatial localization and length distribution of plant cell wall matrix polysaccharides. *J. Microscopy* 166, 123–136.

Ohana P, Delmer DP, Steffens JC et al. (1991) β-Furfuryl-β-glucoside: an endogenous activator of higher plant UDP-glucose (1-3)-β-glucan synthase. *J. Biol. Chem.* 266, 13742–13745.

[6]**Polysaccharide synthases can be identified biochemically by substrate and activator binding activities, by product entrapment techniques, and by catalytic activities detected in nondenaturing gels.**

Hanna R, Brummell DA, Camirand A et al. (1991) Solubilization and properties of GDP-fucose-xyloglucan 1,2-α-L-fucosyltransferase from pea epicotyl membranes. *Arch. Biochem. Biophys.* 290, 7–13.

Kudlicka K & Brown RM Jr (1997) Cellulose and callose biosynthesis in higher plants I: solubilization and separation of (1-3)- and (1-4)-β-glucan synthase activities from mung bean. *Plant Physiol.* 115, 643–656.

[7]**The three-dimensional structure of the backbone of cell wall polysaccharides suggests synthesis by glycosyltransferases with two active sites.**

Albersheim P, Darvill A, Roberts K et al. (1997) Do the structures of cell wall polysaccharides define their mode of synthesis? *Plant Physiol.* 113, 1–3.

Saxena IM, Brown RM Jr, Fevre M et al. (1995) Multidomain architecture of β-glycosyl transferases: implications for mechanism of action. *J. Bacteriol.* 177, 1419–1424.

Yeager AR & Finney NS (2004) The first direct evaluation of the two-active site mechanism for chitin synthase. *J. Org. Chem.* 69, 13–18.

[8]**Polysaccharide synthesis is carefully controlled, but the control points are still poorly understood.**

Pilling E & Höfte H (2003) Feedback from the wall. *Curr. Opin. Plant Biol.* 6, 611–616.

Wu L, Joshi CP, & Chiang VL (2001) A xylem-specific cellulose synthase gene from aspen (*Populus tremuloides*) is responsive to mechanical stress. *Plant J.* 22, 495–502.

[9]*N*-**linked glycans of glycoproteins are assembled on dolichol lipids.**

Abeijon C & Hirschberg C (1992) Topography of glycosylation reactions in the endoplasmic reticulum. *Trends Biochem. Sci.* 17, 32–36.

Elbein AD & Kaushal GP (1990) Lipid-linked saccharides in plants: intermediates in the synthesis of *N*-linked glycoproteins. Methods in Plant Biochemistry, vol 2: Carbohydrates (PM Dey, JB Harborne eds), pp 79–110. London: Academic Press.

[10]**Newly synthesized *N*-linked glycans are transferred *en bloc* to nascent polypeptides and immediately subjected to processing by two glucosidases.**

Helenius A & Aebi M (2004) Roles of N-linked glycans in the endoplasmic reticulum. *Annu. Rev. Biochem.* 73, 1019–1049.

Yan Q & Lennatz WJ (1999) Oligosaccharyltransferase: a complex multisubunit enzyme of the endoplasmic reticulum. *Biochem. Biophys. Res. Commun.* 266, 684–689.

[11]Processing of *N*-linked glycans in the plant Golgi apparatus is similar but not identical to the processing in other organisms.

Lerouge P, Cabanes-Machebeau M, Rayon C et al. (1998) *N*-glycoprotein biosynthesis in plants: recent developments and future trends. *Plant Mol. Biol.* 38, 31–48.

von Schaewen A, Sturm A, O'Neill J & Chrispeels MJ (1993) Isolation of a mutant *Arabidopsis* plant that lacks *N*-acetyl glucosaminyl transferase I and is unable to synthesize Golgi-modified complex N-linked glycans. *Plant Physiol.* 102, 1109–1118.

[12]The sites of *O*-linked glycosylation are defined in part by rules of proline hydroxylation.

Moore PJ, Swords KM, Lynch MA & Staehelin LA (1991) Spatial organization of the assembly pathways of glycoproteins and complex polysaccharides in the Golgi apparatus of plants. *J. Cell Biol.* 112, 589–602.

Yuasa K, Toyooka K, Fukuda H & Matsuoka K (2005) Membrane-anchored prolyl hydroxylase with an export signal from the endoplasmic reticulum. *Plant J.* 41, 81–94.

[13]The GPI anchors of arabinogalactan proteins are both synthesized and attached to the protein backbone in the ER.

Schultz C, Gilson P, Oxley D et al. (1998) GPI-anchors on arabinogalactan-proteins: implications for signalling in plants. *Trends Plant Sci.* 3, 426–431.

Udenfriend S & Kodukula K (1995) How glycosylphosphatidyl-inositol-anchored membrane proteins are made. *Annu. Rev. Biochem.* 64, 563–591.

[14]The backbones of some complex polysaccharides are synthesized by Golgi-located enzymes encoded by genes of the cellulose synthase-like (CSL) gene family.

Doblin MS, Vergara CE, Read S et al. (2003) Plant cell wall biosynthesis: making the bricks. *Annu. Plant Rev.* vol 8: The Plant Cell Wall (JKC Rose ed), pp 183–222. Oxford: Blackwell/CRC Press.

Lerouxel O, Cavalier D, Liepman AH & Keegstra K (2006) Biosynthesis of plant cell wall polysaccharides: a complex process. *Curr. Opin. Plant Biol.* 9, 621–630.

Suzuki S, Li L, Sun YH & Chiang VL (2007) The cellulose synthase gene superfamily and biochemical functions of xylem-specific cellulose synthase-like genes in *Populus trichocarpa*. *Plant Physiol.* 142, 1233–1245.

Yin Y, Huang J & Xu Y (2009) The cellulose synthase superfamily in fully sequenced plants and algae. *BMC Plant Biol.* 9, 99. (doi:10.1186/1471-2229-9-99)

[15]Synthesis of the glucan backbone of xyloglucan in *trans* Golgi cisternae is enhanced by the cooperative assembly of glucosyl and xylosyl residues.

Faik A, Price N, Raikhel NV & Keegstra K (2002) An *Arabidopsis* gene encoding an α-xylosyltransferase involved in xyloglucan biosynthesis. *Proc. Natl. Acad. Sci. USA* 99, 7797–7802.

Fry SC (2004) Primary wall metabolism: tracking the careers of wall polymers in living plant cells. *New Phytol.* 161, 641–675.

Gordon R & Maclachlan G (1989) Incorporation of UDP-[^{14}C] glucose into xyloglucan by pea membranes. *Plant Physiol.* 91, 373–378.

McCann MC, Wells B & Roberts K (1992) Complexity in the spatial localization and length distribution of plant cell wall matrix polysaccharides. *J. Microscopy* 166, 123–136.

[16]The XG-fucosyl and XG-galactosyltransferases can add fucosyl and galactosyl residues during or after synthesis of the XG backbone.

Faik A, Price NJ, Raikhel NV & Keegstra K (2002) An *Arabidopsis* gene encoding an α-xylosyltransferase involved in xyloglucan biosynthesis. *Proc. Natl. Acad. Sci. USA* 99, 7797–7802.

Madson M, Dunand C, Li X et al. (2003) The *MUR3* gene of *Arabidposis* encodes a xyloglucan galactosyltransferase that is evolutionarily related to animal exostosins. *Plant Cell* 15, 1662–1670.

Perrin RM, De Rocher AE, Bar-Peled M et al. (1999) Fucosyltransferase, an enzyme involved in plant cell wall biosynthesis. *Science* 284, 1976–1979.

Tamura K, Shimada T, Kondo M et al. (2005) KATAMARI1/MURUS3 is a novel Golgi membrane protein that is required for endomembrane organization in *Arabidopsis*. *Plant Cell* 17, 1764–1776.

[17]The assembly of xyloglucan occurs in *trans* Golgi and early *trans* Golgi network cisternae.

Zhang GF & Staehelin LA (1992) Functional compartmentalization of the Golgi apparatus of plant cells. *Plant Physiol.* 99, 1070–1083.

[18]The large number of enzymes needed to make pectic polysaccharides makes understanding their synthesis a challenge.

Mohnen D (2008) Pectin structure and biosynthesis. *Curr. Opin. Plant Biol.* 11, 266–277.

Scheller HV, Jensen JK, Sørensen SO et al. (2007) Biosynthesis of pectin. *Physiol. Plantarum* 129, 283–295.

Sterling JD, Atmodjo MA, Inwood SE et al. (2006) Functional identification of *Arabidopsis* pectin biosynthetic homogalaturonan galacturonosyltransferase. *Proc. Natl. Acad. Sci. USA* 103, 49–52.

[19]Synthesis of pectic polysaccharides involves enzymes localized to late *cis*, medial, and *trans* Golgi cisternae.

Lynch MA & Staehelin LA (1992) Domain-specific and cell type-specific localization of two types of cell wall matrix polysaccharides in clover root tip. *J. Cell Biol.* 118, 467–479.

Zhang GF & Staehelin LA (1992) Functional compartmentalization of the Golgi apparatus of plant cells. *Plant Physiol.* 99, 1070–1083.

[20]Only a few steps of the pectic polysaccharide synthesis pathway have been studied biochemically *in vitro*.

Liljebjelke K, Adolphson R, Baker K et al. (1995) Enzymatic synthesis and purification of uridine diphosphate [^{14}C] galacturonic acid: a substrate for pectin biosynthesis. *Anal. Biochem.* 225, 296–304.

Mohnen D (2002) Biosynthesis of pectins. Pectins and Their Manipulation (GB Seymour, JP Knox eds), pp 52–98. Oxford: Blackwell/CRC Press.

Scheller HV, Jensen JK, Sørensen SO et al. (2007) Biosynthesis of pectin. *Physiol. Plantarum* 129, 283–295.

Sterling JD, Atmodjo MA, Inwood SE et al. (2006) Functional identification of *Arabidopsis* pectin biosynthetic homogalaturonan galacturonosyltransferase. *Proc. Natl. Acad. Sci. USA* 103, 49–52.

[21]The biosynthesis of galactomannans has parallels to the biosynthesis of xyloglucans.

Dhugga KS, Barreiro R, Whitten B et al. (2004) Guar seed β-mannan synthase is a member of the cellulose synthase super gene family. *Science* 303, 363–366.

Edwards ME, Marshall E, Gidley MJ & Reid JSG (2002) Transfer specificity of detergent-solubilized fenugreek galactomannan galactosyltransferase. *Plant Physiol.* 129, 1391–1397.

[22]Several glycosyltransferases involved in xylan biosynthesis have been identified, but the mechanism of xylan biosynthesis remains obscure.

Peña MJ, Zhong R, Zhou GK et al. (2007) *Arabidopsis irregular xylem8* and *irregular xylem9*: implications for the complexity of glucuronoxylan biosynthesis. *Plant Cell* 19, 549–563.

York WS & O'Neill MO (2008) Biochemical control of xylan biosynthesis: which end is up? *Curr. Opin. Plant Biol.* 11, 258–265.

Zhong R, Peña MJ, Zhou GK et al. (2005) *Arabidopsis Fragile Fiber8*, which encodes a putative glucuronyltransferase, is essential for normal secondary wall synthesis. *Plant Cell* 17, 3390–3408.

[23]Studies of cellulose synthesis by *Gluconacetobacter xylinus (Acetobacter xylinum)* provided many of the paradigms for similar studies in higher plants.

Römling U (2002) Molecular biology of cellulose production in bacteria. *Res. Microbiol.* 153, 205–212.

[24]Hydrophobic cluster analysis played a critical role in the identification of the plant *CESA* genes.

Callebaut I, Labesse G, Durand P et al. (1997) Deciphering protein sequence information through hydrophobic cluster analysis (HCA): current status and perspectives. *Cell. Mol. Life Sci.* 53, 621–645.

Pear JR, Kawagoe Y, Schreckengost W et al. (1996) Higher plants contain homologs of the bacterial *celA* genes encoding the catalytic subunit of cellulose synthase. *Proc. Natl. Acad. Sci. USA* 93, 12637–12642.

Saxena IM, Brown RM Jr, Fèvre M et al. (1995) Multidomain architecture of β-glycosyltransferases: implications for mechanism of action. *J. Bacteriol.* 177, 1419–1424.

[25]Plant *CESA* genes appear to be derived from cyanobacterial precursors.

Nobles DR, Romanovicz DK & Brown RM Jr (2001) Cellulose in cyanobacteria: origin of vascular plant cellulose synthase? *Plant Physiol.* 127, 529–542.

[26]The cellulose synthase (*CESA*) proteins appear to correspond to the catalytic subunits of cellulose synthase complexes.

Arioli T, Peng L, Betzner AS et al. (1998) Molecular analysis of cellulose biosynthesis in *Arabidopsis*. *Science* 279, 717–720.

Burton RA, Gibeaut DM, Bacic A et al. (2000) Virus-induced silencing of a plant cellulose synthase gene. *Plant Cell* 12, 691–705.

Desprez T, Juraniec M, Crowell EF et al. (2007) Organization of cellulose synthase complexes involved in primary cell wall synthesis in *Arabidopsis thaliana*. *Proc. Natl. Acad. Sci. USA* 104, 15572–15577.

Mutwil M, Debolt S & Persson S (2008) Cellulose synthesis: a complex complex. *Curr. Opin. Plant Biol.* 11, 252–257.

[27]The CESA protein is an integral protein with several domains characteristic of processive glycosyltransferases.

Delmer D (2000) Cellulose biosynthesis: exciting times for a difficult field of study. *Annu. Rev. Plant Physiol. Plant Mol. Biol.* 50, 245–276.

Mutwil M, Debolt S & Persson S (2008) Cellulose synthesis: a complex complex. *Curr. Opin. Plant Biol.* 11, 252–257.

Saxena IM & Brown RM (2005) Cellulose biosynthesis: current views and evolving concepts. *Ann. Bot.* 96, 9–21.

[28]The Zn-binding domains of CESA and some CSLD proteins appear to mediate dimerization of the catalytic subunits of cellulose synthases.

Doblin MS, DeMelis L, Newbigin E et al. (2001) Pollen tubes of *Nicotiana alata* express two genes from different β-glucan synthase families. *Plant Physiol.* 125, 2040–2052.

Kurek I, Kawagoe Y, Jacob-Wilk D et al. (2002) Dimerization of cotton fiber cellulose synthase catalytic subunits occurs via oxidation of the zinc-binding domains. *Proc. Natl. Acad. Sci. USA* 99, 11109–11114.

[29]The cellulose-synthesizing rosette complexes are composed of three different types of CESA proteins.

Desprez T, Juraniec M, Crowell EF et al. (2007) Organization of cellulose synthase complexes involved in primary cell wall synthesis in *Arabidopsis thaliana*. *Proc. Natl. Acad. Sci. USA* 104, 15572–15577.

Persson S, Paredez A, Carroll A et al. (2007) Genetic evidence for three unique components in primary cell-wall cellulose synthase complexes in *Arabidopsis*. *Proc. Natl.Acad. Sci. USA* 104, 15566–15571.

Taylor NN, Howells RM, Huttly AK et al. (2003) Interactions between three distinct CESA proteins essential for cellulose synthesis. *Proc. Natl. Acad. Sci USA* 100, 1450–1455.

[30]The involvement of KORRIGAN, a β-1,4-glucanase, in cellulose synthesis has yet to be proven conclusively.

Nicol F, His I, Jaunau A et al. (1998) A plasma membrane-bound putative endo-1,4-β-glucanase is required for normal wall assembly and cell elongation in *Arabidopsis*. *EMBO J.* 17, 5563–5576.

Peng L, Kawagoe Y, Hogan P & Delmer D (2002) Sitosterol-β-glucoside as primer for cellulose synthesis in plants. *Science* 295, 147–150.

[31] UDP-glucose, the substrate for the cellulose and callose synthase systems, may be produced by a membrane-bound form of sucrose synthase.

Haigler CH, Ivanova-Datcheva M, Hogan PS et al. (2001) Carbon partitioning to cellulose synthesis. *Plant Mol. Biol.* 47, 29–51.

Salnikov VV, Grimson MJ, Delmer DP & Haigler CH (2001) Sucrose synthase localizes to cellulose synthesis sites in tracheary elements. *Phytochemistry* 57, 823–833.

[32] There are two types of callose synthase systems, a Ca^{2+}-dependent and a Ca^{2+}-independent type.

Bacic A, Fincher GB & Stone BA (eds) (2009) Chemistry, Biochemistry and Biology of 1-3 Beta Glucans and Related Polysaccharides. London: Academic Press.

Li H, Bacic A & Read SM (1997) Activation of pollen tube callose synthase by detergents: evidence for different mechanisms of actions. *Plant Physiol.* 114, 1255–1265.

Ohana P, Delmer DP, Volman G & Benziman M (1992) β-Furfuryl-β-glucoside, an endogenous activator of higher plant UDP-glucose: (1-3)-β-glucan synthase: Biological activity, distribution, and *in vivo* synthesis. *Plant Physiol.* 98, 708–715.

[33] Callose synthase is a large protein that differs in many ways from cellulose synthase.

Doblin MS, De Melis L, Newbigin E et al. (2001) Pollen tubes of *Nicotiana alata* express two genes from different β-glucan synthase families. *Plant Physiol.* 125, 2040–2052.

Verma DPS & Hong Z (2001) Plant callose synthase complexes. *Plant Mol. Biol.* 47, 693–701.

[34] The ratio of tri- and tetrasaccharide repeats in mixed-linked glucans made *in vitro* can be altered experimentally.

Burton RA, Wilson SM, Hrmova M et al. (2006) Cellulose synthase-like *CslF* genes mediate the synthesis of cell wall (1,3; 1,4)-β-D-glucans. *Science* 311, 1940–1942.

Gibeaut DM & Carpita NC (1994) Improved recovery of (1-3), (1-4)-β-D-glucan synthase activity from Golgi apparatus of *Zea mays* (L.) using differential flotation centrifugation. *Protoplasma* 180, 92–97.

Wilson SM, Burton RA, Doblin MS et al. (2006) Temporal and spatial appearance of wall polysaccharides during cellularization of barley (*Hordeum vulgare*) endosperm. *Planta* 224, 655–667.

[35] Cinnamyl alcohols are the predominant precursors for lignin biosynthesis.

Buchanan BB, Gruissem W & Jones RL (eds) (2001) Biochemistry and Molecular Biology of Plants, ch 24: Natural Products (Secondary Metabolites). Rockville, MD: American Society of Plant Physiologists.

Freudenberg K & Neish AC (eds) (1968) Constitution and Biosynthesis of Lignin. Berlin: Springer.

Higuchi T (1997) Biochemistry and Molecular Biology of Wood. New York: Springer.

[36] Monolignols are formed by successive enzymatic modification of phenylalanine.

Anterola AM, Jeon J-H, Davin LB & Lewis NG (2002) Transcriptional control of monolignol biosynthesis in *Pinus taeda*: Factors affecting monolignol ratios and carbon allocation in phenylpropanoid metabolism. *J. Biol. Chem.* 277, 18272–18280.

Boerjan W, Ralph J & Baucher M (2003) Lignin biosynthesis. *Annu. Rev. Plant Biol.* 54, 519–546.

Boudet AM (2000) Lignins and lignification: selected issues. *Plant Physiol. Biochem.* 38, 81–96.

Dixon RA, Chen F, Guo D & Parvathi K (2001) The biosynthesis of monolignols: a metabolic grid or independent pathways to guaiacyl and syringyl units? *Phytochemistry* 57, 1069–1084.

Ralph J, Lundquist K, Brunow G et al. (2004) Lignins: natural products from oxidative coupling of 4-hydroxyphenylpropanoids. *Phytochem. Rev.* 3, 29–60.

[37] Hydroxylation and *O*-methylation at the 3 and 5 positions on the aromatic ring are unlikely to take place at the level of cinnamic acids.

Franke R, Humphreys JM, Hemm MR et al. (2002) The *Arabidopsis REF8* gene encodes the 3-hydroxylase of phenylpropanoid metabolism. *Plant J.* 30, 33–45.

Hoffman L, Maury S, Martz F et al. (2002) Purification, cloning and properties of an acyltransferase controlling shikimate and quinate ester intermediates in phenylpropanoid metabolism. *J. Biol. Chem.* 278, 95–103.

Nair RB, Xia Q, Kartha CJ et al. (2002) Arabidopsis CYP98A3 mediating aromatic 3-hydroxylation. Developmental regulation of the gene, and expression in yeast. *Plant Physiol.* 130, 210–220.

Ulbrich B & Zenk MH (1980) Partial purification and properties of *para*-hydroxycinnamoyl-CoA-shikimate-*para*-hydroxycinnamoyl transferase from higher plants. *Phytochemistry* 19, 1625–1629.

[38] Coniferaldehyde is the branch point for the biosynthesis of coniferyl alcohol and sinapyl alcohol.

Humphreys JM & Chapple C (2002) Rewriting the lignin roadmap. *Curr. Opin. Plant Biol.* 5, 224–229.

Li L, Cheng XF, Leshkevich J et al. (2001) The last step of syringyl monolignol biosynthesis in angiosperms is regulated by a novel gene encoding sinapyl alcohol dehydrogenase. *Plant Cell* 13, 1567–1585.

Li L, Popko JL, Umezawa T & Chiang VL (2000) 5-hydroxyconiferyl aldehyde modulates enzymatic methylation for syringyl monolignol formation: a new view of monolignol biosynthesis in angiosperms. *J. Biol. Chem.* 275, 6537–6545.

[39] Monolignols are glucosylated, stored, transported, and deglycosylated before polymerization.

Dharmawardhana P, Ellis BE & Carlson JE (1995) A β-glucosidase from lodgepole pine xylem specific for the lignin precursor coniferin. *Plant Physiol.* 107, 331–339.

Kaneda M, Rensing KH, Wong JCT et al. (2008) Tracking monolignols during wood development in Lodgepole Pine. *Plant Physiol.* 147, 1750–1760.

Samuels AL, Rensing KH, Douglas CJ et al. (2002) Cellular machinery of wood production: differentiation of secondary xylem in *Pinus contorta* var. latifolia. *Planta* 216, 72–82.

Steves V, Forster H, Pommer U & Savidge R (2001) Coniferyl alcohol metabolism in conifers I: glucosidic turnover of cinnamyl aldehydes by UDPG: coniferyl alcohol glucotransferase from pine cambium. *Phytochemistry* 57, 1085–1093.

[40] Lignin polymers are formed through enzymatic oxidation of monolignols.

Ranocha P, Chabannes M, Chamayou S et al. (2002) Laccase down-regulation causes alterations in phenolic metabolism and cell wall structure in poplar. *Plant Physiol.* 129, 145–155.

Ros Barceló A, Gómez Ros LV, Gabaldón C et al. (2004) Basic peroxidases: the gateway for lignin evolution? *Phytochem. Rev.* 3, 61–78.

Sasaki S, Nishida T, Tsutsumi Y & Kondo R (2004) Lignin dehydrogenative polymerization mechanism: a poplar cell wall peroxidase directly oxidizes polymer lignin and produces in vitro dehydrogenative polymer rich in β-*O*-4 linkage. *FEBS Lett.* 562, 197–201.

[41] Lignin polymerization is primarily due to the addition of a monolignol to a lignin polymer.

Ralph J, Lundquist K, Brunow G et al. (2004) Lignins: natural products from oxidative coupling of 4-hydroxyphenylpropanoids. *Phytochem. Rev.* 3, 29–60.

Tokunaga N, Sakakibara N, Umezawa T et al. (2005) Involvement of extracellular dilignols in lignification during tracheary element differentiation of isolated *Zinnia* mesophyll cells. *Plant Cell Physiol.* 46, 224–232.

[42] Oxidative carriers or radical mediators may be involved in the polymerization process.

Li K, Helm R & Eriksson K-EL (1998) Mechanistic studies of the oxidation of a nonphenolic lignin model compound by the laccase/1-hydroxybenzotriazole redox system. *Biotechnol. Appl. Biochem.* 27, 239–243.

Onnerud H, Zhang L, Gellerstedt G & Henriksson G (2002) Polymerization of monolignols by redox shuttle-mediated enzymatic oxidation: a new model in lignin biosynthesis I. *Plant Cell* 14, 1953–1962.

[43] A nonenzymatic model for lignin polymerization has been proposed involving "dirigent" or guide proteins.

Davin LB & Lewis NG (2000) Dirigent proteins and dirigent sites explain the mystery of specificity of radical precursor coupling in lignan and lignin biosynthesis. *Plant Physiol.* 123, 453–461.

Davin LB, Wang H-B, Crowell AL et al. (1997) Stereoselective biomolecular phenoxyradical coupling by an auxiliary (dirigent) protein without an active center. *Science* 275, 362–366.

Hatfield R & Vermerris W (2001) Lignin formation in plants. The dilemma of linkage specificity. *Plant Physiol.* 126, 1351–1357.

[44] Nontraditional monomers are readily incorporated into lignin, indicating a high level of metabolic plasticity.

Kim H, Ralph J, Yahiaoui N et al. (2000) Crosscoupling of hydroxycinnamyl aldehydes into lignins. *Org. Lett.* 2, 2197–2200.

Lu F & Ralph J (2002) Preliminary evidence for sinapyl acetate as a lignin monomer in kenaf. *J. Chem. Soc. Chem. Commun.* 1, 90–91.

Ralph J, Lundquist K, Brunow G et al. (2004) Lignins: natural products from oxidative coupling of 4-hydroxyphenylpropanoids. *Phytochem. Rev.* 3, 29–60.

Sederoff RR, MacKay JJ, Ralph J & Hatfield R (1999) Unexpected variation in lignin. *Curr. Opin. Plant Biol.* 2, 145–152.

[45] The deposition of lignin is both temporally and spatially controlled.

Donaldson LA (2001) Lignification and lignin topochemistry: an ultrastructural view. *Phytochemistry* 57, 859–873.

Nakashima J, Mizuno T, Takabe K et al. (1997) Direct visualization of lignifying secondary wall thickenings in *Zinnia elegans* cells in culture. *Plant Cell Physiol.* 38, 818–827.

Terashima N (1990) A new mechanism for formation of a structurally ordered protolignin macromolecule in the cell wall of tree xylem. *J. Pulp Paper Sci.* 16, J150–J155.

[46] The composition of lignin differs between cell wall domains and cell types.

Donaldson LA (2001) Lignification and lignin topochemistry: an ultrastructural view. *Phytochemistry* 57, 859–873.

Joseleau J-P, Imai T, Kuroda K & Ruel K (2004) Detection in situ and characterization of lignin in the G-layer of tension wood fibers of *Populus deltoides*. *Planta* 219, 338–345.

[47] The biosynthesis of cutins, suberins, and waxes involves enzymes in chloroplasts, the endoplasmic reticulum, and cell walls.

Kannangara R, Branigan C, Liu Y et al. (2007) The transcription factor WIN1/SHN1 regulates cutin biosynthesis in *Arabidopsis thaliana*. *Plant Cell* 19, 1278–1294.

Kunst L & Samuels L (2009) Plant cuticles shine: advances in wax biosynthesis and export. *Curr. Opin. Plant Biol.* 12, 721–727.

Pighin JA, Zheng H, Balakshin LJ et al. (2004) Plant cuticular lipid export requires an ABC transporter. *Science* 306, 702–704.

Pollard M, Beisson F, Li Y & Ohlrogge JB (2008) Building lipid barriers: biosynthesis of cutin and suberin. *Trends Plant Sci.* 13, 236–246.

Samuels L, Kunst L & Jetter R (2008) Sealing plant surfaces: cuticular wax formation by epidermal cells. *Annu. Rev. Plant Biol.* 59, 771–812.

Principles of Cell Wall Architecture and Assembly

6

In the previous five chapters, we described in detail the structure and chemistry of the different macromolecules found in plant cell walls, and we saw how the various polymers are made by the cell and then deposited outside the plasma membrane. In this chapter we examine how these polymers, once outside the cell, are assembled into a functional cell wall. We discuss the general architectural principles involved in wall construction, starting with the "glue": the different ways that wall polymers can be attached to each other through both covalent and noncovalent cross-linkages. We then see how two fundamental networks underlie the construction of all primary cell walls, and how this can be elaborated by the addition of other polymer networks such as protein and lignin. Just as we saw that different cells have appropriately specialized walls, it will become clear that different domains within the wall of an individual cell can also have different compositions and structures. Finally, we show, on a larger scale, how a combination of cell shape, internal turgor pressure, and wall properties contribute to the structural and mechanical properties of plant parts. The differences in texture between a ripe tomato and a crisp apple reside in cell wall properties!

A. Cross-Links between Wall Polymers

1. Wall polymers are cross-linked by covalent bonds as well as by noncovalent bonds and interactions.[ref1]

The cellulose microfibrils, matrix polysaccharides, and structural proteins (glycoproteins) constitute the structural framework of the wall, where the cellulose microfibrils are embedded in a matrix provided by the other wall macromolecules. The wall provides the strength for cells to withstand external and internal forces, and yet at the same time must be flexible and extensible enough to allow for changes in cell shape and size during growth and development. At first sight these physical properties of the cell wall may seem paradoxical if we consider those of its macromolecular components. Many of the individual matrix molecules of the wall are soluble in aqueous solution. This reflects the need for these components to be soluble during their synthesis and delivery to the cell wall. Yet the wall itself, which contains these components, is insoluble. This paradox can be resolved if we suggest that the wall macromolecules interact with each other and that cross-linking between the interacting wall macromolecules is the basis of wall insolubility. In addition, a contribution from the direct physical "interweaving" of some wall macromolecules may contribute to the insolubility of the wall. Information about cross-links between the various components of the wall has emerged from two sources: from the methods used to solubilize specific components (for example, enzymes, chemical extractions, and chelators), and from studies of matrix macromolecules (or components of these) that have been isolated from the wall.

The data from both sources support the existence of both covalent and noncovalent bonds and interactions between cell wall matrix polymers, which together can be considered to play important roles in achieving the functional cohesion of both primary and secondary cell walls.

2. Hemicelluloses bind strongly to cellulose microfibrils by noncovalent bonds.[ref2]

A major noncovalent cross-link that occurs in the cell wall is the binding of hemicelluloses, via multiple hydrogen bonds, to cellulose microfibrils. In eudicots and nongrass monocots, xyloglucan is the predominant hemicellulose hydrogen-bonded to microfibrils, whereas in the grasses, arabinoxylans fulfill this role. Small amounts of arabinoxylan in dicots and xyloglucan in grasses are also found hydrogen-bonded to cellulose. Thus, the hemicelluloses provide the contact point in the cell wall between the cellulose microfibrils and the rest of the matrix polymers.

There is considerable evidence to support the hypothesis that cellulose microfibrils and hemicelluloses bind strongly together via multiple hydrogen bonds. Often, after sequential extraction of cell walls with various chemical treatments, the residual wall is found to contain glycosyl residues that are representative of cellulose and hemicelluloses, suggesting a strong association between these macromolecules. Incubation of cell walls with strong alkali releases hemicelluloses into solution, and treatment of walls with strong chaotropes, for example, MMNO (4-methyl-morpholine-N-oxide-hydrate), solubilizes hemicellulose and cellulose microfibrils. Together these two observations suggest hydrogen bonding between the molecules. Additionally, both arabinoxylan and xyloglucan have been shown to bind strongly to cellulose in *in vitro* experiments.

Conformational studies of both arabinoxylan and xyloglucan have shown that they can adopt structures that partially resemble the structure of the individual cellulose molecule and are thus able to hydrogen-bond to the flat-ribbon shapes of cellulose microfibrils (**Figure 6.1**). Interestingly, the fucosyl-containing side chains of xyloglucan can influence whether xyloglucan can bind to cellulose (see Figure 2.5), as can the number of side chains attached to a xylan backbone. In this case the less substituted the backbone, the greater the potential for binding to cellulose. In many walls the hemicelluloses may form a complete monolayer coating the microfibrils, although some of the hemicellulose molecules are believed to penetrate, and possibly disrupt, the microcrystalline arrangement of cellulose molecules within the microfibril. Other hemicellulose molecules are believed to span the gap between two cellulose microfibrils, hydrogen-bonding to each to create a cross-link between them, as we now discuss (**Figure 6.2**).

Figure 6.1 Hydrogen bonding of a hemicellulose molecule to a cellulose microfibril. This schematic diagram shows a small portion of a xyloglucan molecule, modeled in the flat-ribbon conformation. The β-glucan backbone in this conformation is exposed and is thought to be able to hydrogen-bond to the similar β-glucan chains of cellulose on the surface of a microfibril.

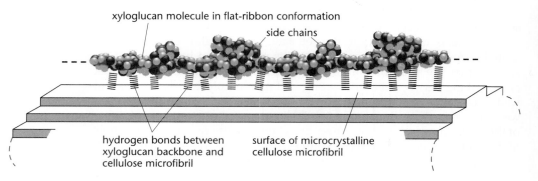

xyloglucan molecule in flat-ribbon conformation

side chains

hydrogen bonds between xyloglucan backbone and cellulose microfibril

surface of microcrystalline cellulose microfibril

Figure 6.2 A model accounting for different xyloglucan domains in a xyloglucan/cellulose network. The model shows possible arrangements of xyloglucan domains in relation to cellulose microfibrils (A, B, and C), but does not represent the mass ratio of these domains. The model shows a xyloglucan domain accessible to possible enzyme modification that could also cross-link cellulose microfibrils (domain 1), a xyloglucan domain bound to the surface of cellulose microfibrils (domain 2), and a xyloglucan domain that is trapped within or between cellulose microfibrils (domain 3).

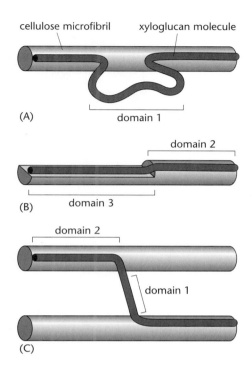

3. Hemicelluloses cross-link cellulose microfibrils.[ref3]

The observation that hemicelluloses hydrogen-bond to cellulose microfibrils suggests that they may function to connect, or cross-link, cellulose microfibrils. One hemicellulose molecule could bind to more than one microfibril, thus both connecting and separating neighboring microfibrils. In this manner, cross-linking hemicellulosic molecules could both keep the cellulose microfibrils apart from each other and potentially influence the ability of the microfibrils to slip past one another, thus playing an important role in the control of expansion growth of the cell wall.

Evidence for cross-links has been obtained by direct visualization of the wall. Using electron microscopy of rapidly frozen, deep-etched primary cell walls, xyloglucans were shown to form cross-links, 20 and 40 nm long, between cellulose microfibrils (**Figure 6.3**). Treatment of the tissues with base (1 N and 4 N KOH) resulted in the removal of the cross-links and the collapse of the microfibrils onto one another. It has been proposed that arabinoxylans play a similar role in grass cell walls.

Additional evidence that hemicellulose molecules can create cross-links between cellulose microfibrils comes from artificial composites formed from pure bacterial cellulose and plant xyloglucan or other hemicelluloses. Cross-links are observed that look remarkably similar to those in the plant cell wall itself (Figure 6.3B).

4. Pectic polysaccharides are probably covalently interconnected by glycosidic bonds.[ref4]

The pectic polysaccharides homogalacturonan, partially methylesterified homogalacturonan, rhamnogalacturonan I, and rhamnogalacturonan II constitute the pectin network of the plant cell wall. A vast literature describes the release of part of this pectic network from the wall by chemical extractions using cold water, hot water, buffers, or chelators. The various pectic polysaccharides released in this way cannot be separated easily using gel-filtration or ion-exchange chromatography unless glycosidic bonds holding them together are broken, suggesting that the pectic polysaccharides are glycosidically attached to each other.

Figure 6.3 Hemicellulose cross-links. (A) Electron micrograph showing a shadowed replica of a primary cell wall that has had most of its pectin and some of its hemicellulose removed. What remain are long cellulose microfibrils, cross-linked to each other by links, about 20 to 40 nm long, that are thought to be xyloglucan molecules. (B) Very similar images of cross-links are obtained from artificial wall-like composites made by incubating pure, newly-synthesized bacterial cellulose microfibrils in a solution of purified plant cell wall xyloglucan.(C) Schematic representation of how hemicellulose molecules may form cross-links, about 30 nm long, between cellulose microfibrils. (A, courtesy of Maureen McCann. B, from S. Whitney, E. Wilson et al., *Amer. J. Bot.* 93:1402–1414, 2006.)

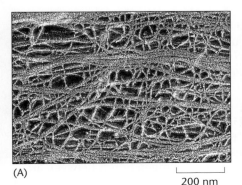

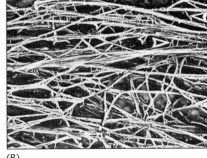

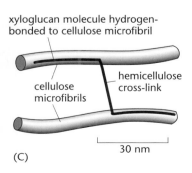

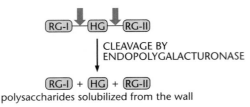

CLEAVAGE BY
ENDOPOLYGALACTURONASE

polysaccharides solubilized from the wall

Figure 6.4 Cross-links between pectic polysaccharides. The pectic polysaccharides RG-I, RG-II, and homogalacturonan (HG) can all be released from the wall after its treatment with *endo*-polygalacturonase. This enzyme cleaves short unesterified stretches of α-1,4-linked galactosyluronic acid residues that are thought to connect the three pectic polysaccharides to each other in the wall. Their exact sequence and structural organization within the wall is not known, and a conjectural scheme is shown here for simplicity.

Treatment of cell walls with enzymes offers the most compelling evidence that some pectic polysaccharides are covalently attached to each other via *glycosidic bonds*. Much of the data has come from experiments where cell walls are treated with an α-1,4-*endo*-polygalacturonase that releases on average 50% of the pectic polysaccharides of the wall. For example, all of the RG-II and approximately 50% of the RG-I and homogalacturonan are solubilized from sycamore cell walls by *endo*-polygalacturonase. The substrate specificity of *endo*-polygalacturonase requires two or three consecutive nonesterified α-1,4-galactosyluronic acid residues for hydrolysis of an α-1,4-galactosyluronic acid glycosidic linkage. Chromatography of the solubilized pectic polysaccharides by ion exchange and/or gel filtration results in the separation of RG-I and RG-II and homogalacturonan in the form of oligogalacturonides of various sizes, depending on the degree of esterification of the homogalacturonan in the originating wall. Methyl esters of the carboxyl groups will limit the ability of the enzyme to cleave the homogalacturonan backbone. All oligo- and polysaccharide products have a galactosyluronic acid residue at the reducing end indicating that all solubilized products were released by cleavage of an α-1,4-galactosyluronic acid glycosidic bond.

If the solubilized products are deesterified by base treatment and then incubated with *endo*-polygalacturonase, the "homogalacturonan tails" attached to the ends of the RG-I and RG-II polysaccharides are removed. This also results in cleavage of oligogalacturonides (homogalacturonan fragments) into galacturonic acid and di- and trigalactosyluronic acid oligosaccharides, the final products of treating nonesterified homogalacturonan with the enzyme. These data provide very strong evidence that some of the pectic polysaccharides in the cell wall are joined together via glycosidic linkages whose cleavage results in the solubilization of all three components of the network (**Figure 6.4**). The total amount of pectic material solubilized in this way is dependent on the source of the cell wall material and is likely to be influenced by the "cross-linking" of the pectic polysaccharides by various covalent and noncovalent linkages.

5.   Homogalacturonans and partially methylesterified homogalacturonans form gels.[ref5]

In vitro studies on homogalacturonans with various degrees of methylesterification, isolated from different plant species, strongly suggest that homogalacturonans within the cell wall can form *gels*. Much of this work has been driven by the desire to understand gel formation in homogalacturonans as it relates to gelling in industrial applications, where these macromolecules are used in enormous quantities, for example, in jams and jellies.

How multiple homogalacturonan polysaccharides can be aligned and cross-linked together to form a gel is still not entirely clear, but two major mechanisms are proposed (**Figure 6.5**). The first is that homogalacturonans, with a high degree of methylesterification (>60%), can form gels via noncovalent hydrophobic interactions and hydrogen bonds between areas of the molecules with a high concentration of methylesterified carboxyl

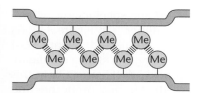

(A) noncovalent hydrophobic interactions and hydrogen bonds between methylesterified homogalacturonans

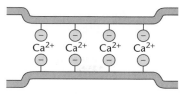

(B) calcium cross-links between adjacent carboxyl groups of homogalacturonan

Figure 6.5 Cross-links between pectic homogalacturonans. Two ways in which noncovalent bonds act to create stable interactions between homogalacturonan molecules. (A) Regions with a high degree of methylesterification can interact with a similar region on a neighboring molecule via a combination of hydrophobic and hydrogen bonds, while (B) unesterified regions interact via calcium cross-links between free carboxyl groups.

groups. The second is that nonesterified homogalacturonans, in the presence of divalent cations, for example, calcium, can form noncovalent ionic cross-links between adjacent carboxyl groups. Details of the exact orientations and of the required sizes of the component molecules involved in the formation of these gels remain to be fully characterized, but there is a high probability that both such kinds of gels exist in cell walls and exert profound effects on the physical characteristics and properties of the wall (see Figure 3.5).

Localization studies with various staining techniques, and in particular with monoclonal antibodies, indicate that in many tissues the nonesterified homogalacturonans are particularly abundant in the middle lamella and three-way junctions (**Figure 6.6**; see also Figure 3.7), while partially methylesterified units are more commonly distributed throughout the body of the wall (see Section 6C). Local regulation either of calcium concentrations or of the activity of pectin methylesterase, whose activity would remove methyl esters from wall homogalacturonans, could affect the local gel-forming properties and characteristics of the homogalacturonans within the wall. How, or if, such regulation occurs is currently unknown. When considering the gel-forming properties of homogalacturonans, it is also important to remember that these molecules are probably glycosidically linked to the other pectic polysaccharides RG-I and RG-II and that these rhamnogalacturonans could have a dramatic influence on the gelation properties of homogalacturonans in a particular cell wall.

6.   Borate esters cross-link RG-II dimers in the wall.[ref6]

RG-II is released from walls (for example, from sycamore or pea), by treatment with α-1,4-*endo*-polygalacturonase, mostly in the form of a dimer (d-RG-II), although some monomer (m-RG-II) is also found (see Concept 3C4). The two RG-II molecules in the dimer are held together by a single *borate diester* cross-link (**Figure 6.7**, see also Figure 3.7). The diester is stable at pH 5, the normal pH of the wall; and *in vitro* the amount of d-RG-II formed at pH 5 is increased by the presence of certain divalent cations (for example, Sr^{2+}, Pb^{2+}, and Ba^{2+}). Indeed, d-RG-II isolated from cell walls contains significant amounts of these cations. These observations suggest that the divalent cations may contribute to the conformation of RG-II to optimize the formation of d-RG-II.

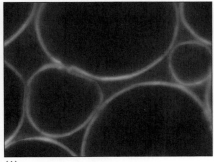

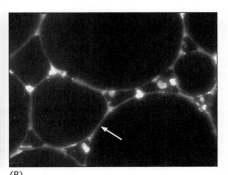

(A)   (B)

Figure 6.6 Homogalacturonan in the middle lamella and three-way junctions. (A) Cross section of parenchyma cells in the petiole of a fern, stained with calcofluor to show the cellulose in the cell walls. (B) The same section stained with a monoclonal antibody to homogalacturonan, showing its concentration in the middle lamella (arrow) and three-way junctions (Courtesy of P. Knox, from F. Leroux et al., *Ann. Bot.* 100: 1165–1173, 2007.)

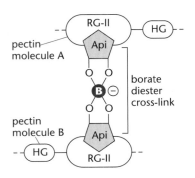

Figure 6.7 Borate diester cross-links. This schematic diagram shows how, in principle, a single borate diester cross-link, between apiosyl residues in two different RG-II molecules, could help cross-link larger pectic polysaccharide networks in the wall, since RG-II molecules are themselves glycosidically linked to other pectic polysaccharides, in this case homogalacturonan (HG).

The borate diester is attached to C-2 and C-3 of one specific apiosyl residue in each of the two RG-II molecules in the dimer (see Figure 3.8). As discussed earlier, RG-II is likely to be covalently connected to both RG-I and HG via glycosidic bonds, and d-RG-II, therefore, is a strong candidate as a covalent cross-link of the cell wall pectin network (Figure 6.7). In this capacity, d-RG-II has very important effects on the physical properties and expansion characteristics of the wall and in particular in the organization of the cell wall pectin network. This has been shown with *Arabidopsis* mutants deficient in d-RG-II that are stunted in their growth. Recovery of these mutants to normal wild-type size is achieved by watering with additional borate, which results in normal amounts of d-RG-II in the primary cell wall.

7. Diferulic acids probably cross-link wall polysaccharides.[ref7]

Oxidative coupling between two phenolic groups, each attached to a separate polysaccharide, is the source of another important cross-link in the wall. Ferulic acid and *p*-coumaric acid, two hydroxycinnamic acids, have been found ester-linked, in a number of plant species, to terminal nonreducing glycosyl residues. Several of these points of attachment have been structurally identified. In the grasses, ferulic acid can be ester-linked to terminal arabinofuranosyl residues of arabinoxylans and to a terminal xylosyl residue of xyloglucan. In addition, *p*-coumaric acid is found ester-linked to an arabinoxylan trisaccharide from bamboo. Ferulic acid has also been identified ester-linked to terminal arabinopyranosyl and galactopyranosyl residues in pectin (likely RG-I) of sugar beet and spinach.

These phenolic groups, ester-linked to cell wall polysaccharides, provide potential sites for cross-linking. The phenols can be oxidatively coupled to form a biphenyl (C-C) bond or a diphenyl (C-O-C) ether bond. Such cross-links between two phenol groups, for example, *diferulic acid* (see **Figure 6.8**), may be formed in a reaction catalyzed in the wall by peroxidase, in the presence of hydrogen peroxide; both H_2O_2 and peroxidase are found in the walls of many species.

Diferulic acid has been isolated from several plant species, and its presence has resulted in the structural identification of a diferulic acid arabinoxylan hexasaccharide from bamboo cell walls. The hexasaccharide contains two trisaccharides, each trisaccharide attached via an ester to one of the ferulic acid units of the diferuloyl cross-link. This suggests strongly that, at least in some cell walls, diferuloyl (and di-*p*-coumaryl) units could cross-link arabinoxylan molecules. However, even the structural characterization of the diferuloyl arabinoxylan hexasaccharide did not prove that the cross-link was intermolecular, and it is possible that the unit identified originated from an intramolecular cross-link, as both of the component trisaccharides could have originated from different regions of the same polysaccharide. Whether inter- or intramolecular (resulting in anchoring one polysaccharide to another wall polymer), phenol cross-links could be important contributors to the structure of the wall by controlling the interactions between various cell wall polysaccharides. In many grass secondary walls, the attached feruloyl units become incorporated into subsequently deposited lignin.

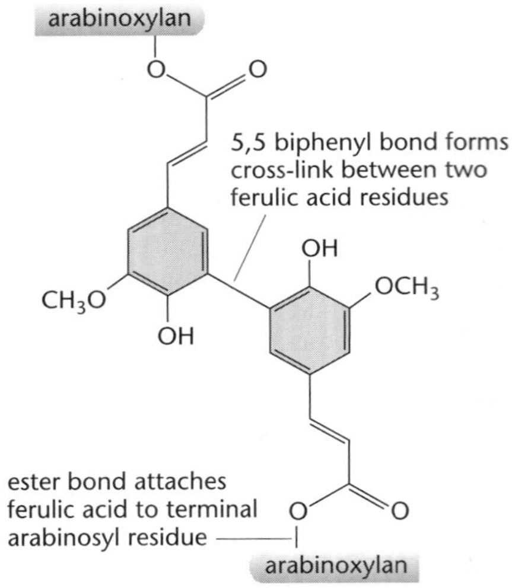

Figure 6.8 Diferulic acid cross-link. Schematic diagram to show the biphenyl bond that is formed by an oxidative reaction between two ferulic acid residues, each of which in turn is ester-linked to an arabinoxylan molecule from bamboo cell walls. As with many wall cross-links, this diferulic acid link might be formed either between separate polysaccharides (intermolecular) or within a single polysaccharide (intramolecular).

8. Transesterification may produce other possible cross-links.[ref8]

Another possible cross-link between wall polysaccharides is a direct ester linkage between the carboxyl group of a glycosyluronic acid of one polysaccharide and a hydroxyl group of a glycosyl residue on a neighboring polysaccharide. All of the pectic components and many arabinoxylans of the cell wall contain glycosyluronic acid residues that could participate in such direct ester linkages. The formation of such an ester requires that the participating carboxyl group be activated (energized) prior to cross-linking. In the cell wall it is likely that *sugar-ester cross-links* are formed by the transesterification of an already methylesterified (energized) carboxyl group, for example, the methylesterified galactosyluronic acid residues of homogalacturonan (**Figure 6.9**).

Although no such direct ester cross-links between sugar residues have been isolated and properly characterized from a cell wall, there is circumstantial evidence that supports their existence. Walls from several species, including tobacco and maize, have been analyzed by chemical procedures for the percentage of galactosyluronic acid residues that are esterified and for the percentage that are specifically methylesterified. These analyses indicate that methyl esters account for between 25 and 66% of the total esterified galactosyluronic acid residues in tissues of different species under various physiological conditions, supporting the conclusion that esters other than methyl esters occur in these walls. Researchers have failed to find other low–molecular weight alcohols that could explain these esters, and therefore the possibility of interpolysaccharide ester cross-links within the wall is considered likely.

9. Some wall proteins become insolubilized by covalent cross-links.[ref9]

Many different proteins become insolubilized in the wall at some point after they have been deposited there, but how is still poorly understood. Work to date suggests that this occurs either by the formation of covalent cross-links between tyrosine residues or by the formation of disulfide bonds. The most likely candidate for the cross-linking of the structural protein extensin was identified several years ago as isodityrosine (**Figure 6.10A**), two tyrosines joined together by a diphenyl ether link. *Isodityrosine* is released from extensin using either acid or alkaline treatments, and it has been isolated as a component of a small polypeptide fragment of extensin by treatment with trypsin. The two cross-linked tyrosines, however, were separated by only a

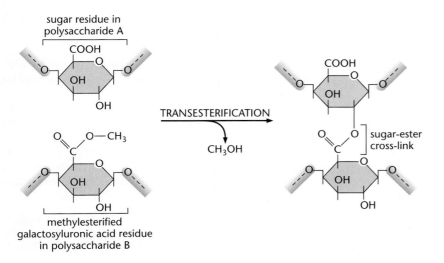

Figure 6.9 Sugar esters as putative wall cross-links. This schematic shows how, in principle, an esterified uronic acid residue in a pectin molecule (in this case a methylesterified galactosyluronic acid residue in homogalacturonan) could engage in a transesterification reaction with a hydroxyl group from a sugar residue in another polysaccharide (in this case an unesterified galactosyluronic acid residue in a neighboring homogalacturonan). The resulting sugar-ester cross-link could in theory cross-link separate polysaccharides or form intramolecular cross-links.

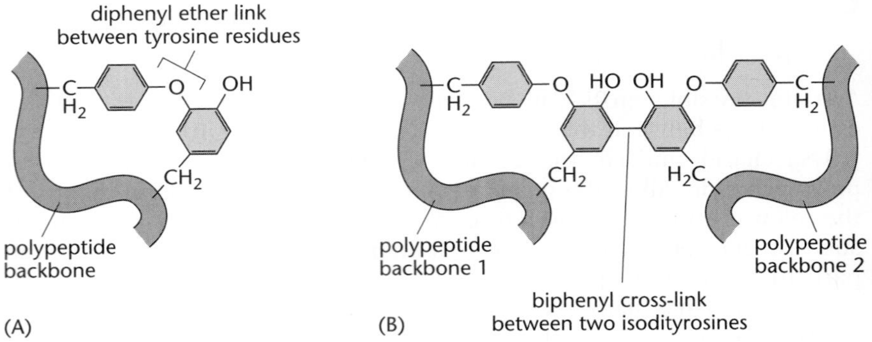

Figure 6.10 Isodityrosine cross-links. (A) The diphenyl ether bond between peptidyltyrosine residues that creates the intramolecular isodityrosine cross-link. Although this structure has been found as an intramolecular cross-link, it has not so far been found as an intermolecular cross-link. (B) Di-isodityrosine is an intermolecular cross-link between two isodityrosines on separate polypeptide chains. These have been demonstrated to cross-link the hydroxyproline-rich glycoprotein extensin.

(A)   (B)

single amino acid (lysine), and the cross-link was therefore intramolecular rather than intermolecular.

The formation of isodityrosine in the cell wall is probably catalyzed by peroxidase in the presence of hydrogen peroxide in a manner similar to the formation of diferulic acid. Although cell wall structural proteins contain many tyrosine residues, each potentially a site for cross-linking, the search for an intermolecular isodityrosine has been inconclusive. However, several accounts now suggest that rather than the isodityrosine itself being the cross-link, it is a biphenyl link between two separate intramolecular isodityrosines that is instead the main intermolecular cross-link, a reaction that forms *di-isodityrosine* (Figure 6.10B). This oxidative reaction can form a cross-linked extensin network in the wall, a reaction that has some value in pathogen protection.

Since the environment of the cell wall is oxidative, it is an obvious thought that proteins might also be immobilized through the formation of intermolecular disulfide bonds. This idea has received a major boost from studies of a class of tyrosine- and lysine-rich proteins (TLRP), cross-linked proteins that localize to the sites of lignification in protoxylem cells. These TLRP wall proteins share in their C-terminal region a highly conserved structural domain rich in cysteine residues, named the cysteine domain (CD), with the consensus sequence $CXYXCC(X)_6C(X)_{2-3}CCSY$.

CD sequences have also been identified in other cell wall proteins and may form the basis for intermolecular cystine links that stabilize appropriate protein conformations and thereby promote interactions with other wall components. The role of the CD sequence in insolubilizing cell wall proteins has been investigated by means of transgenic plants expressing fusion proteins consisting of the cysteine-rich domain and a secreted, soluble protein. These studies have shown that the CD sequence alone is sufficient to anchor soluble proteins in the cell wall. Yet to be determined are which specific amino acids are responsible for the insolubilization, since both cysteine and tyrosine are possible candidates.

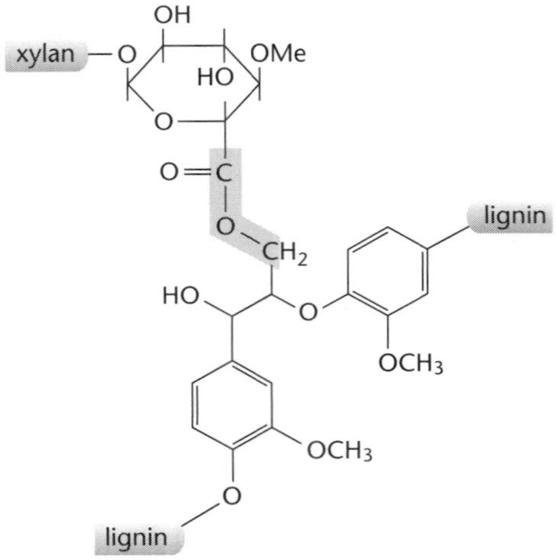

10. Lignin is covalently attached to hemicelluloses, and their interaction adds strength to the secondary cell wall.[ref10]

There is strong evidence that hemicelluloses are covalently linked to lignin. Lignin-carbohydrate complexes are readily extracted from wood of many species by a variety of solvents including water, dilute alkali, or dimethylformamide, but separating them, either by gel-filtration, electrophoresis, centrifugation, or hydrophobic-interaction chromatography has proved very difficult. Four kinds of cross-link have been suggested: ester (**Figure 6.11**), benzyl ether, glycosidic, and acetal (or hemiacetal).

Figure 6.11 A structure for a benzyl ester between xylan and lignin. Other lignin-carbohydrate cross-links include benzyl ethers and phenyl glycosidic linkages. Such components are difficult to both separate and characterize.

Lignin hemicellulose complexes are amphipathic structures with hydrophobic and hydrophilic components linking the carbohydrate and phenolic components of the lignified cell wall. Lignin has a higher affinity for hemicellulose than for cellulose, and the complex of lignin with hemicellulose may contribute significantly to the strength of the secondary cell wall. Additional noncovalent interactions may occur between lignin and hemicellulose, giving the interface adhesive properties. According to this view, much of the lignin in the secondary wall does not interact with cellulose directly but derives strength from the layered interactions.

Cellulase and hemicellulase treatment of a lignin-hemicellulose complex indicates that spruce lignin is linked to an L-arabinosyl side chain in a xylan and to a D-galactosyl side chain in a galactoglucomannan. In similar experiments with pine, lignin is found linked to D-galactosyl, D-xylosyl, and L-arabinosyl units in hemicellulose with two sugars linking a single lignin fragment. In wood from eucalyptus, for example, xylose, arabinose, galactose, and glucose are involved in lignin linkages. NMR characterization of a water-soluble lignin-carbohydrate complex indicates that attachment is likely to be to the propane side chain.

B. Architectural Principles: Putting the Polymers Together

1. Two coextensive polysaccharide networks underlie the structure of the primary cell wall.[ref11]

Having discussed the polymers isolated from the wall and the ways in which they may be associated with one another, it is now time to consider how, in the plant, they are put together outside the plasma membrane, in three dimensions, to construct a wall. We have emphasized the complexity and diversity of cells and their walls, and it will therefore come as no surprise that one single simple wall model is unlikely to account satisfactorily for all aspects of the variety of walls found in nature.

We can, however, derive some general underlying architectural principles that are common to all primary walls. All growing plant parts are subjected, either through their own weight or through the forces of the wind, to alternating compressive and stretching forces. It follows that the cell walls, and also the middle lamella that glues the cells together, must be able to withstand adequately compression, tension, and shearing forces. Engineers, to solve similar problems in structural components, use *fiber composite materials*, in which fibrous elements, capable of resisting tension, are embedded in a matrix of more amorphous material that can withstand compressive and shear forces (**Figure 6.12**). Examples include reinforced concrete and fiberglass. Plants have solved the problem, if a little earlier, in a similar way! Three lines of evidence—biophysical and mechanical measurements of cell wall properties, structural analysis in the electron microscope, and chemical information about the polymers and their cross-links—all support the idea that the cell wall can be regarded as a two-component fiber composite material.

The polysaccharides of the wall are used to construct two networks, independent of but intermeshed with each other. The first network is constructed from the ordered cellulose microfibrils, coated with, and cross-linked by, hemicellulose molecules. This network provides tensile strength. Woven into this is another network, coextensive with it, but constructed from the pectic polysaccharides. This water-retentive, gel-like phase provides compressive stiffness and resistance to shear. Although there is little evidence at present for any covalent links between elements of the two networks, their coextensiveness ensures that each is tightly related to the other and each is probably responsive to stresses and strains in the other.

Although plants will differ, not only in the relative proportions of each network but also in their chemical composition, it seems very likely that the

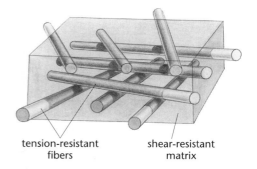

tension-resistant fibers   shear-resistant matrix

Figure 6.12 Construction principles of a fiber composite material. Fiber composites, for example, fiberglass or reinforced concrete, consist of fibrous or rodlike elements that resist stretching forces (tension), embedded in a matrix material that resists compressive or shear forces. The resultant composite material combines both these desirable properties.

Figure 6.13 Schematic diagram to show the construction of a single lamella within the cell wall. The cellulose microfibrils are roughly parallel both to each other and to the plane of the plasma membrane. Each lamella is only one cellulose microfibril thick.

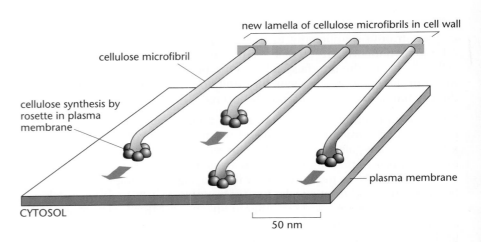

primary walls of all plants are constructed using this same general architectural principle, of two coextensive polysaccharide networks forming a fiber composite material.

2. Walls are constructed of lamellae that are one cellulose microfibril thick.[ref12]

The components of the *cellulose-hemicellulose network* are not just randomly intertwined within the wall. Instead considerable order is imparted to the network by the biosynthetic machinery. Cellulose microfibrils, spun out by adjacent rosettes in the plasma membrane, are spatially restricted to the space between the existing wall and the plasma membrane that is pushed firmly against it by the cell's turgor pressure. The result is that each new set of cellulose microfibrils invariably lies in a plane parallel to the outer surface of the membrane. In addition, adjacent rosettes appear to travel through the membrane along roughly parallel tracks, with the consequence that adjacent microfibrils in the innermost layer of the wall form a layer, or *lamella*, of roughly parallel microfibrils rather like the vapor trails left by a line of aircraft flying in formation (**Figure 6.13**; see also Figure 4.32).

Electron micrographs of intact walls from a wide variety of plant species have confirmed the general conclusion that the primary walls of all higher plants are composed of a series of such lamellae, each one cellulose microfibril thick, stacked one on top of the other, rather like the different floors in a large multistory building or the different strata in sedimentary rocks (**Figure 6.14**). In most walls, the lamellae are not as clear-cut as in Figure 6.14, often having more meandering microfibrils and rather variable gaps between them. However, there is no evidence for the extensive weaving or intertwining of microfibrils within a lamella. In isodiametric cells, the orientation of each lamella with respect to its neighbors varies, with the result that there is no net microfibril orientation in the wall. In elongating cells, by contrast, adjacent lamellae are more or less aligned with each other, resulting in a net orientation of microfibrils in the wall at right angles to the growth direction (**Figure 6.15**). In thickened secondary walls there can be more complex ordering of adjacent lamellae, and these are discussed in more detail later.

3. The primary wall is usually composed of only a small number of lamellae.[ref13]

Primary cell walls are surprisingly thin. Careful measurements of the cell walls in meristematic tissues suggest that, with very few exceptions, the distance between the plasma membranes of adjacent cells is between 150 nm and 250 nm. This means that the wall of a single cell, that is, the distance between the plasma membrane and the middle lamella, is only about half of

0.5 µm

Figure 6.14 Cell wall lamellae. This electron micrograph shows the unusually regular layers of cellulose microfibrils in the cell wall of the single-celled alga *Oocystis apiculata*. Although the cellulose microfibrils are much larger than those in higher plants, they demonstrate clearly the general principle of lamella construction in which each lamella consists of a single layer of roughly parallel microfibrils, oriented at an angle to those in adjacent lamellae. In this case about five successive lamellae can be distinguished, each more or less at right angles to the next. (Courtesy of D. Montezinos and R.M. Brown Jr.)

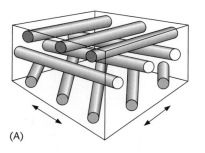
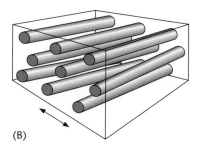

Figure 6.15 Arrangement of successive lamellae to construct a wall. This may contain (A) randomly oriented lamellae in isodiametric cells or (B) directionally oriented lamellae as in an elongating cell. The arrows show the direction of cell elongation. In both cases only three lamellae are shown.

this, that is, 75 to 125 nm. How many individual lamellae of cellulose microfibrils does this correspond to? To answer this question we need to know the distance between microfibrils in adjacent lamellae. Direct visualization of shadowed replicas of primary walls from a variety of species suggests that this distance is between 20 and 40 nm, much the same distance, in fact, as between microfibrils within a single lamella. If we include the fact that microfibrils themselves vary between 5 and 12 nm in diameter, we can see that simple geometrical considerations limit the possible number of lamellae within a primary wall to about three or four (**Figure 6.16**).

While this conclusion holds for the newly formed primary walls of most meristematic cells, primary walls in more mature tissues, such as fruits and vegetables, can often be much thicker. The expanding cell walls of the flesh of young processing tomatoes, for example, are very thick and may be built of a hundred or more individual lamellae (**Figure 6.17**).

4. The spacing of cellulose microfibrils is determined by matrix polysaccharides.[ref14]

We have discussed the relatively constant distance that separates cellulose microfibrils within and between lamellae, and it is now time to consider how these spacings are generated. The thin primary wall is under considerable tension, generated by the turgor pressure of the cell; in a typical spherical parenchyma cell the tension in the wall may be equivalent to

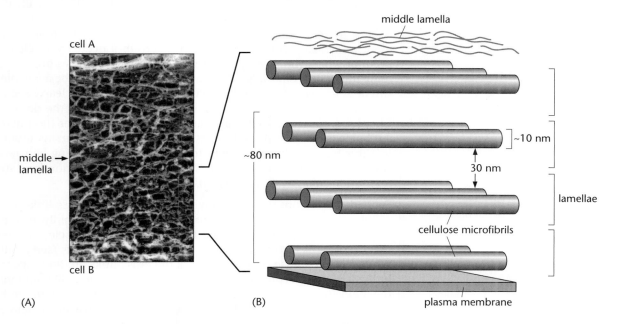

Figure 6.16 (A) Electron micrograph of two primary *Arabidopsis* cell walls and their intervening middle lamella. Each wall is about 100 nm thick. (B) Schematic diagram, drawn to scale, showing the probable arrangement of cellulose microfibrils and lamellae within the walls shown in (A). Each wall can probably accommodate only around four lamellae.

Figure 6.17 A section of the thick primary cell wall of a processing tomato. These walls can be between 1 and 4 μm thick and may contain, therefore, more than 100 lamellae. The section is labeled with a colloidal gold probe for relatively deesterified pectin (*black dots*). One wall is more labeled than the other, emphasizing that the middle lamella can act as a functional barrier in the wall.

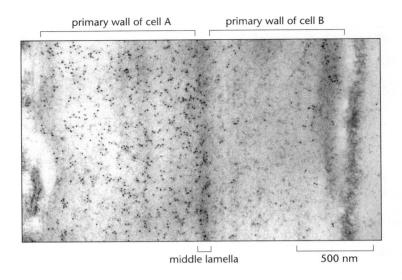

primary wall of cell A   primary wall of cell B

middle lamella   500 nm

between 100 and 1000 times atmospheric pressure. To withstand this, the cellulose microfibrils have to be very securely attached to each other to create a strong, resistant, and coherent network. It is generally accepted that the molecules responsible for this are the hemicelluloses (arabinoxylans in the case of grasses and xyloglucans in other plants).

Evidence for the noncovalent interaction of xyloglucan with cellulose comes from at least two different classes of experiment. First, *in vitro* binding assays suggest tight hydrogen bonding exists between the flat planar β-1-4-glucan or xylan backbones of hemicellulose and the cellulose molecules available at the surface of the microfibrils. Computer modeling of xyloglucan conformations suggest that steric hindrance from the side chains of hemicellulose restrict to about eight or so the number of xyloglucan molecules that can contribute to coating the microfibril at any one point along its length. The fact that strong alkali treatment is needed to remove all the xyloglucan from a cell wall suggests that at least some of the molecules are woven more tightly into the texture of the microfibril, locally disturbing the strict crystallinity of the cellulose (see Figure 6.2). This situation is hard to achieve *in vitro*, and the conclusion must be that the attachment of xyloglucan molecules to the microfibril occurs very close to the point at which the newly synthesized cellulose molecules are self-assembling into a microfibril. Newly secreted xyloglucan must be available at the site of active cellulose synthase complexes, and this is clearly seen in the early stages of cell plate formation (see **Figure 6.18** and the discussion in Chapter 4, Concept 4B3). Calculations suggest that, after the microfibrils are coated, perhaps three fourths of the xyloglucan remains unattached. It is this fraction that is thought to be involved in bridging between adjacent microfibrils within lamellae. Xyloglucans and arabinoxylans are often referred to collectively as *cross-linking glycans.*

Second, direct observation provides evidence for cross-links of hemicellulose. High-resolution electron micrographs of rapidly frozen walls from which the pectin has been largely removed show clear evidence of short cross-links, 20 to 40 nm long, between microfibrils (see Figure 6.3). When hemicellulose is also removed, the cross-links are no longer visible. Individual xyloglucan molecules can be between 60 and 500 nm long, with the majority about 200 nm long, plenty long enough, in fact, to bind at one end to a microfibril and then make a jump of 30 nm on average before binding to another (Figure 6.3).

The length of the cross-links, at around 30 nm, seems to be conserved across a wide range of plant species, including the grasses (**Figure 6.19**), and this raises the question of what controls the point at which a hemicellulose

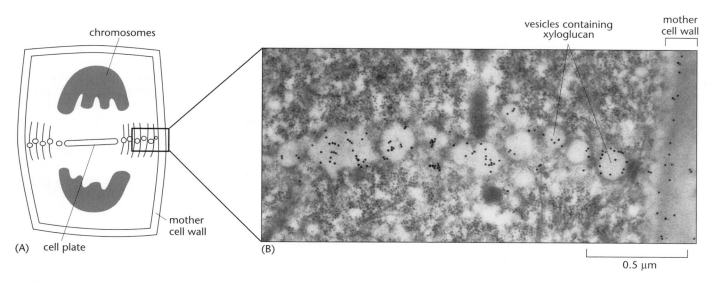

vesicles containing
xyloglucan

mother
cell wall

mother
cell wall

(A)   cell plate   (B)

0.5 µm

molecule lifts off the microfibril to which it is bound in order to form a link. One possible explanation would be the presence of some structural feature along the backbone molecule that interfered with its hydrogen bonding to cellulose. This would mean some sort of perturbation about every 60 sugar residues of the xyloglucan or xylan backbones, and although there is no direct chemical evidence at present for such long-range order, length measurements of xyloglucan molecules do suggest that they show a length preference for multiples of 30 nm.

Efforts to assemble xyloglucan and cellulose *in vitro* into a wall-like cellulose-hemicellulose network have been made using the pure cellulose ribbons spun out by *Acetobacter* cells grown in a medium containing purified plant xyloglucan. Remarkably the network that is assembled from this bacterial cellulose looks rather like a normal primary plant cell wall with the pectic network removed, including the presence of regular 30-nm cross-links (Figure 6.3). The images strongly suggest that in the plant this network could be self-assembled in the wall and that the size and frequency of the hemicellulose cross-links is an intrinsic function of the hemicellulose molecules themselves.

Figure 6.18 Xyloglucan deposition in the cell plate. Following nuclear division a new cell wall is constructed between the two sets of chromosomes. (A) The cell plate grows through vesicle accumulation and fusion until it meets the mother cell wall, and cellulose synthesis begins only after the two cells are completely separated. (B) In this electron micrograph the vesicles involved in cell plate growth are immunogold-labeled with an antibody that recognizes xyloglucan. The cell plate thus contains abundant xyloglucan before cellulose synthesis commences. (Courtesy of G. Freshour and M.G. Hahn.)

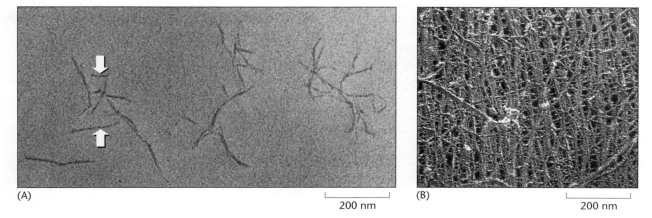

(A)   200 nm   (B)   200 nm

Figure 6.19 Hemicellulose and wall architecture. (A) Aggregates of laterally associated, isolated hemicellulose molecules as seen in the electron microscope. The individual molecules (arrows) are long enough to form the cross-bridges seen in (B). (B) This electron micrograph of a cell wall from a maize root tip, at the same magnification as (A), clearly shows the parallel cellulose microfibrils together with cross-links of hemicellulose, in this case probably arabinoxylan. Although grass cell walls have a different composition from those of other plants, their underlying molecular architecture is similar. (A, courtesy of Maureen McCann. B, courtesy of Adrian Turner.)

The links discussed so far have been lateral ones between the individual microfibrils within a single lamella. Micrographs suggest, however, that similar-length links appear to exist between one lamella and the next, and it is tempting to assume that these too are based on hemicellulose. In many cases this may well be true, but in some secondarily thickened cellulosic walls, for example, in collenchyma, there are suggestions that the pectic polysaccharide network may in some way mediate the interaction between adjacent lamellae. Mild treatments of these walls, designed to mobilize pectin, appear to allow adjacent lamellae to separate easily from one another while still retaining their structural integrity as lamellae. Weaker bonds between, rather than within, lamellae may be necessary to allow the reorientation of a lamella with respect to its neighbors, an idea we shall return to later.

5. The walls of cells in which cellulose synthesis has been inhibited are composed largely of a pectin network.[ref15]

We suggested earlier that in the two-network model of cell wall architecture it is the cellulose-hemicellulose network that provides much of the tensile strength while the pectic network largely contributes shear strength. Pectin is not, however, simply a passive jelly in the wall. Even though the complete extraction of the pectic network from the wall has no effect on the remaining architecture of the cellulose-hemicellulose network, there is now persuasive evidence for the existence of a coherent *pectin network* in the wall with considerable tensile strength of its own. This evidence includes a suspension culture of tomato cells that has been slowly adapted to grow in concentrations of the herbicide 2,6-dichlorobenzonitrile (DCB) that inhibit the biosynthesis of cellulose. Hemicellulose is still made, but lacking cellulose to bind to, the majority ends up in the culture medium. The wall that remains consists almost entirely of pectic polysaccharides, with only about 1% cellulose, and yet, remarkably, it can withstand the huge forces in the wall generated by turgor pressure, and can also successfully accommodate cell division and subsequent cell expansion (**Figure 6.20**). Equally remarkable, it would appear that the Ca^{2+} cross-links between free carboxyl groups of the pectic polysaccharides are load-bearing. Indeed, if the cells are treated with the calcium chelator CDTA, the bonds are broken and the cells then burst. Similar experiments have been done with the cellulose synthesis inhibitor isoxaben. While these cell suspensions offer us a privileged glimpse of the pectin network in isolation, it may not represent the full complexity of the network in other cell types where other cross-links, such as the various esters mentioned earlier, may also play an important role in its integrity. It is, however, a clear illustration of the need to consider the pectin network, not simply as a gel, but as a coherent fibrous structure in its own right with considerable tensile strength.

Figure 6.20 Light micrographs of sections of tomato cells that either (A) have not or (B) have been slowly adapted to grow in cell culture on 8-μM DCB, an herbicide that blocks the formation of the cellulose-hemicellulose network in the cell wall. The cell wall of the adapted cells consists almost entirely of pectic polysaccharides, with only about 1% cellulose present. (From E. Shedletzky et al., *Plant Physiol.* 94:980–987, 1990.)

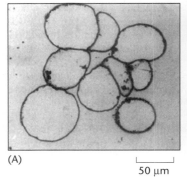

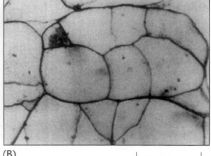

(A)   50 μm

(B)   50 μm

6. The pectin network limits the porosity of the wall.[ref16]

The two wall networks together produce a fibrous structure with characteristic pores in it. These aqueous channels through the wall are vital for the life of the cell within. Not only does the wall provide a passage for water and dissolved ions and nutrients to reach the cell, but its porosity plays a key role in determining the access of enzymes and other proteins to different parts of the wall, and for polysaccharides and proteins to reach specialized regions such as three-way junctions and air spaces. Conversely, it also determines the cell's accessibility to enzyme attack by microorganisms, and for animals it affects the digestibility of plant foods by restricting access to hydrolytic enzymes.

Both networks within the wall contribute to the effective porosity of the wall, but the main determinant is the pectic network. The main framework of the cellulose-hemicellulose network has cross-links about 30 nm long that are also spaced about 30 nm apart, thus creating a meshwork with relatively large holes. The pectic network consists of polymers that thread through this meshwork, forming junction zones whose frequency depends at least partly on the local patterns of pectin deesterification. Regions of extensive Ca^{2+} cross-linking depend on runs of at least 10 to 15 nonesterified consecutive galactosyluronic acid residues. Borate diesters formed between RG-II molecules that are in turn attached to other pectin molecules (see Concept 6A6) provide another key cross-linking mechanism that regulates the net pore size of the wall. This second cross-linked network, together with associated RG-I and RG-II polysaccharides, means that the original pores in the cellulose-hemicellulose network become considerably occluded. The size of the remaining pores has been measured in two different ways. The ability of molecular probes of different sizes to penetrate the wall has been used to derive values of pore diameters that range from 4–5 nm using a solute-exclusion method (corresponding to a globular protein of about 20 kD), through 5.7 nm using colloidal gold (see Panel 6.1 in Section 6C) particles, to 6.6–9.2 nm using fluorescent proteins and dextrans. Gentle treatment with enzymes to remove some of the pectic polysaccharides from the wall increases the apparent pore diameters. A second method to estimate pore size is to directly examine cell walls that have been carefully prepared to preserve their architecture (**Figure 6.21**). Measurements from these preparations suggest a pore diameter of between 6 and 13 nm (corresponding to a globular protein of up to 150 kD), which, when the narrowing effects

Figure 6.21 Wall porosity. These electron micrographs show a small portion of the epidermal cell wall from an elongating pea epicotyl. The material has been rapidly frozen and the ice etched away to reveal the wall architecture (A) before and (B) after the removal of pectic polysaccharides. The pectin network that occupies the spaces between the cellulose microfibrils restricts the pore size in (A) to about 6 nm. The removal of pectin creates larger pores that are about 30 nm in diameter. (C) A similar result can be seen in an onion cell wall that has been extracted with CDTA and sodium carbonate to remove much of the pectin. (A, B, courtesy of T. Itoh. C, courtesy of M. McCann.)

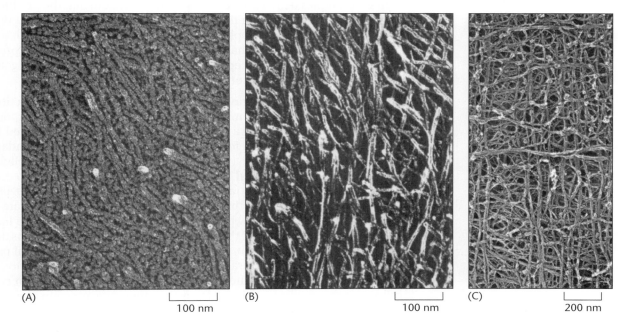

(A)   100 nm

(B)   100 nm

(C)   200 nm

of bound water are taken into account, is in very good agreement with the other methods. In fact somewhat larger molecules can eventually permeate the wall, given time (large wall polysaccharides do pass through the walls of cultured plant cells and into the medium), and this might be for two reasons. The first is that many large molecules in the wall are fibrous, and these, with their relatively small cross-sectional area, may be able to move snakelike through the pores. The second is that many of the pectin chains that obstruct the pores are very mobile. Studies of walls using NMR spectroscopy reveal that while the cellulose-hemicellulose network is relatively rigid, both the neutral galactans in RG-I side chains and the highly esterified homogalacturonans are very mobile. The size of the pores may not, therefore, be static but may instead be dynamic.

The conclusion is that the pores in the primary cell wall are in the same size range as medium-sized proteins and will, therefore, directly affect the accessibility of polymers in the wall to many enzymes, for example. Many wall proteins are in fact either fairly small (for example, lipid transfer proteins) or are long, rodlike molecules (for example, HRGPs).

7. The cell controls the thickness of its wall.[ref17]

It is a notable feature of plant anatomy that, if you compare any particular cell with its exact counterpart in another plant of the same species, then the thickness of its cell wall is surprisingly constant. The absolute value may vary a lot; meristematic cell walls may be only 75 nm thick while collenchyma walls may be several microns thick, but this thickness will depend on the cell type, its developmental history, and the species involved. Indeed, it can even vary between genotypes of the same species, since, for example, the walls of tomato fruit cells from processing varieties are consistently very much thicker than those of closely related salad tomatoes (see Figure 6.17). Since wall thickness is so species- and cell type–specific, its control must be at the genetic level and be developmentally regulated. What is the molecular basis of this control? A primary control point must be at the level of biosynthesis. For a cell at constant final size and with constant wall architecture, the thickness of the wall will vary directly with the amount of wall polymer made and deposited in the wall. This principle is likely to operate, for example, in the production of the very thick walls of collenchyma cells. The maintenance of wall thickness in expanding cells is another example that will be discussed in detail later (see Concept 7B4). All of these examples demand feedback systems to ensure that the production of matrix materials made in the endomembrane system is coupled tightly to cellulose synthesis at the plasma membrane. The transcriptional, translational, and posttranslational controls on these processes are at present largely unknown (see Concept 5A10). There are, however, some general principles that bear on the control of wall thickness.

At one level we might imagine that wall thickness is simply a function of the number of cellulose/hemicellulose-based lamellae that are produced by the cell, three or four for a thin primary wall and several hundred for a collenchyma wall. It is, therefore, instructive to note that the DCB-adapted cell walls described earlier (see Figure 6.20), which have no cellulose-based lamellae, are identical in thickness to the walls of unadapted cells (**Figure 6.22**). This suggests rather that it is the pectic network (or possibly the structural proteins) that provides information on appropriate wall thickness and that somehow the correct number of cellulose-based lamellae is then produced to "fill" the available volume. How this happens at the molecular level is not known, but it should be noted that individual pectic polysaccharides and structural proteins are more than long enough to comfortably span the width of the primary wall or even to stretch from cell to cell across the middle lamella, and they may, therefore, function as molecular rulers.

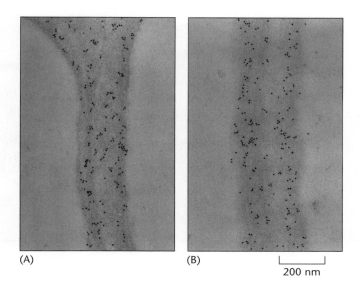

(A)   (B)

|—————| 200 nm

Figure 6.22 Pectin and wall thickness. (A) Section of a cell wall from tomato cells adapted to grow in DCB (see Figure 6.20). This wall, which lacks cellulose and hemicellulose, is the same thickness as the wall (B) from a normal unadapted cell. Both sections have been stained with an antibody to pectin, using a colloidal gold probe (*black dots*). (Courtesy of Brian Wells.)

Pectic polysaccharides may also be involved in controlling wall thickness by helping tether together consecutive lamellae within the wall. Experiments on collenchyma cell walls support this idea. Treatments, such as pectin deesterification or calcium removal, that increase the free-charge density of pectin (an increase in free carboxyl groups) cause the pectic polysaccharide network of collenchyma cells to swell, exerting an isotropic or equal stress on the wall structure in all directions. The result is that the wall responds by swelling only in a radial direction (**Figure 6.23**). This suggests that the internal integrity of an individual lamella does not depend on pectin, but that a stack of lamellae are held together by cross-links that are pectin-dependent and further that the radial spacing of each lamella is also pectin-dependent.

8. How proteins are integrated into wall architecture is unclear.[ref18]

Protein accounts for a very small fraction of the primary cell wall. Although there may be several hundred different kinds of protein represented, in total they amount to no more than 1 to 2% of the dry weight of a typical primary wall. Exceptionally this may rise to levels as high as 10%, for example, in cells grown in culture. In Chapter 3 we described the complex family of evolutionarily related wall proteins—the HRGPs, the PRPs, the GRPs (together often referred to as structural proteins), and the AGPs—and it is now appropriate to consider where these are found, how they may relate to the "two network" model of the wall described earlier, and what structural functions they may serve.

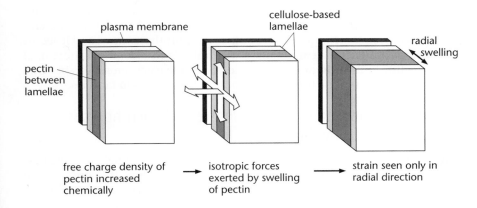

Figure 6.23 Schematic diagram to show that forces exerted by pectin between lamellae in the wall can push the lamellae apart and cause radial swelling of the wall, but that bonds within each lamella, holding it together, are not affected and do not depend on the pectin network. This experiment was done with the thick primary walls of collenchyma.

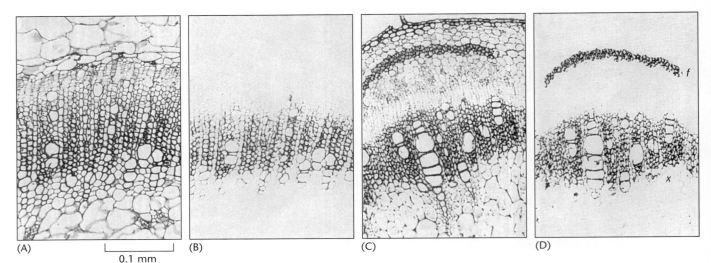

(A)   (B)   (C)   (D)

0.1 mm

Figure 6.24 Tissue localization of PRP and GRP. A thin section of a tobacco stem (A) is shown together with the consecutive section (B) that has been stained using an antibody to bean GRP1.8. The protein is localized exclusively in the walls of xylem vessels and tracheids. (C) A section of a soybean stem together with (D) the consecutive section stained using an antibody to soybean PRP. The protein is confined to the lignified walls of phloem fiber cells (*f*) and xylem vessels and tracheids (*x*). (Courtesy of Zheng-Hua Ye.)

With the exception of the ubiquitous AGPs, a key feature of the "structural" proteins is their relatively restricted expression pattern. GRPs and PRPs are commonly found in young vascular tissue, and in xylem in particular (**Figure 6.24**). Both start out as relatively soluble molecules. One, GRP1.8, is actually made in neighboring xylem parenchyma cells and is transported to the wall of the vessel elements. Later they often become insolubilized in the wall, either through oxidative cross-linking or because they become marooned during lignin deposition. Subsets of PRP- and HRGP-encoding genes are expressed specifically in root hairs, where they may function in the tip growth characteristic of these cells. The expression of both GRP and PRP gene families is also regulated in response to wounding and pathogen attack. Their distribution and regulation both suggest a role in strengthening the wall in some way in response to signals, but so far there is no direct experimental support for this idea, and the molecular or genetic basis for such a function remains unclear.

Closely related to PRPs is the family of HRGPs, highly basic rodlike molecules with characteristic Ser(Hyp)$_4$ repeats. Expression of these proteins is particularly strong in cells subjected to mechanical stress during growth, including the petiole-stem junction and phloem tissue. Again, wounding and pathogens enhance their expression. Stresses, particularly those involving the generation of H_2O_2 in the wall through an oxidative burst, rapidly insolubilize HRGPs in the wall. This probably occurs enzymatically through the action of a specific cell wall peroxidase that catalyzes the formation of cross-links between Val-Tyr-Lys sequences. In some cases this may even result in a coherent network, formed from the rodlike HRGPs, that extends through and between the cellulose and pectin-based networks in the wall. A cross-linked HRGP network may help prevent further cell enlargement and may strengthen the wall locally against pathogen ingress. This idea is supported by the observation that when cells grown in suspension culture are treated with H_2O_2, their walls rapidly become resistant to enzymes that previously could digest the wall and produce protoplasts. As with the GRPs and PRPs, however, there is little direct data to support these ideas, and it seems likely that proteins have no major structural role to play in the functional architecture of the primary cell wall. Considerable work remains to be done to elucidate the precise functions of these cell wall protein families.

9. The orientation of newly synthesized microfibrils is determined by the cell.[ref19]

We have seen how cellulose microfibrils are deposited parallel to each other within each lamella of the primary wall, but how does this order come about? Since the net orientation of cellulose microfibrils in the wall has a

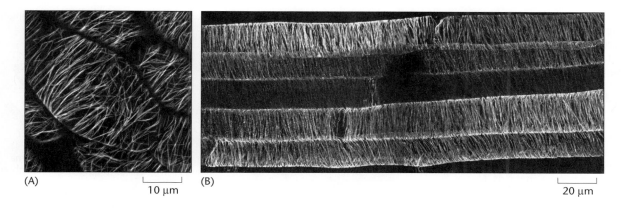

(A)   10 μm   (B)   20 μm

profound effect on the mechanical properties of the wall, and can dictate in which direction(s) the cell may grow, it is vital that each cell, relative to its position within the plant, can exert appropriate controls over this orientation. The cellular machinery responsible for spatial order, in almost all eukaryotic cells, is the cytoskeleton, and in particular the *microtubules*. We have seen earlier how microtubules are responsible for defining the plane of cell plate formation in dividing cells, and how they guide Golgi-derived vesicles containing cell wall building blocks to the growing cell plate (see Concept 1A2). Later, during interphase, a network of cortical microtubules is reestablished running roughly parallel to one another, just beneath the plasma membrane (**Figure 6.25**). Numerous structural studies of young elongating cells suggest that the internal order of the microtubules is mirrored by a similar ordering of newly deposited cellulose microfibrils outside the cell. In other words, the arrangement of cortical microtubules and the links between them and the plasma membrane have a direct bearing on the texture of the wall (**Figure 6.26**).

A simple model, to understand how this curious coupling comes about, starts with the rosettes in the plasma membrane that spin out microfibrils. Microfibril ends rapidly become integrated into the wall causing the rosettes to be pushed away in the plane of the membrane. In principle a rosette might move away in any direction in the lipid bilayer, but if constrained on either side by plasma membrane domains, defined by the tight anchoring of underlying cortical microtubules, then its net movement will be parallel to

Figure 6.25 The interphase cortical array of microtubules. (A) A confocal image of living hypocotyl epidermal cells from *Arabidopsis* reveals the cortical microtubules by their incorporation of a translational fusion between tubulin and a fluorescent protein. (B) The very elongated cells in the root epidermis, whose microtubules have been stained by immunofluorescence, show a highly organized, net-transverse cortical array. The microtubules form an array just beneath the plasma membrane in which their net orientation is at right angles to the direction of cell elongation. (A, from D. Ehrhardt, *Curr. Opin. Cell Biol.* 20:107–116, 2008. B, courtesy of Keiko Sugimoto.)

cortical microtubule
plasma membrane
cell wall   cortical microtubule   cell wall

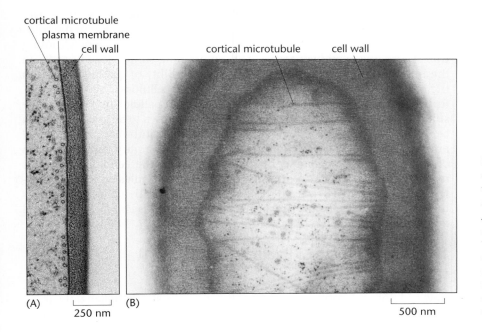

(A)   250 nm   (B)   500 nm

Figure 6.26 Cortical microtubules and cellulose deposition in the primary cell wall. These two electron micrographs are both of longitudinal sections of an *Arabidopsis* trichome, or leaf hair. (A) In a section through the middle of the hair, the cortical microtubules can be seen at a constant distance beneath the plasma membrane. (B) In a glancing section, the microtubules can be seen to parallel the orientation of the fibers within the wall. (Courtesy of Heinz Schwarz.)

Figure 6.27 Model for oriented cellulose microfibril deposition. This simple schematic shows one possible model of how the orientation of cellulose microfibrils produced by the cellulose synthase complexes (rosettes) moving through the plane of the plasma membrane might be determined by plasma membrane–anchored microtubules. In the model the microtubules that are anchored to the plasma membrane serve as physical guides that confine rosette movement to defined membrane channels. Newly deposited cellulose microfibrils, therefore, will roughly parallel the net microtubule orientation at the time. Since there seem to be reciprocal interactions between the microtubules and the cellulose microfibrils, it is equally plausible that the rosettes interact more directly with the microtubules.

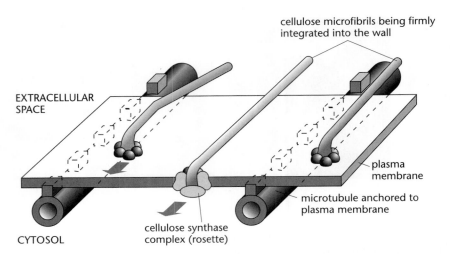

them. Thus, the orientation of groups of growing cellulose microfibrils will passively come to reflect the orientation of the microtubule-defined "canal banks" in the membrane. This, conceptually, is probably the simplest model (**Figure 6.27**), but others, in which there is more direct coupling between the rosettes and microtubules or their associated proteins, are also possible. Either way, the model holds well for almost all young cells that leave the meristem where they were born and embark on a period of expansion or elongation. Drugs that disassemble cortical microtubules, or growth factors that alter their orientation, also provide strong support for the idea of a general congruence between cellulose and cytoskeleton. There are, however, certain exceptions. In cells that have stopped growing, where cell growth controls become less important, the orientation of new cellulose lamellae, for example, in secondary wall thickening, may become uncoupled from their cytoskeletal controls. Tip-growing cells (pollen tubes and root hairs) also appear to uncouple microtubule distribution and ordered cellulose deposition. In this case the ordered pattern of microfibrils within the wall may arise entirely extracellularly through a combination of biomechanical constraints and self-assembly processes.

With plants expressing a translational fusion of tubulin and GFP, or by using micro-injected, fluorescently labeled tubulin, it is possible to follow the behavior of the cortical microtubules in living plant cells. Perhaps surprisingly, the whole microtubule array is spectacularly dynamic, constantly disassembling and reassembling sufficiently rapidly to turn over completely in less than 20 minutes. In addition, the orientation of microtubules against one wall of a cell may differ from those against a neighboring wall. These observations suggest that maintaining some sense of order in the microtubule array must depend on the stability of the links between the microtubules and the plasma membrane. If cellulose synthesis is perturbed in living cells, either by drugs that inhibit cellulose synthase or by subtle mutations in genes that encode proteins known to be associated with cellulose synthesis, the precise orientation of microtubules in the cortical array is disturbed. This strongly suggests that the stability and orientation of cortical microtubules is correlated tightly with the cellulose synthesis machinery itself. In other words, the cortical microtubules direct the trajectory of the cellulose synthase rosettes, while their activity, in turn, affects the organization of the cortical array. We return to this issue in more detail in Chapter 7.

10. The orientation of microfibrils within a wall may change during growth.[ref20]

New cellulose microfibrils are laid down in successive lamellae only at the innermost face of the primary wall, but what happens to the older outer lamellae as the cell begins to grow? In principle they could either retain their original net orientations or they could become reoriented. Examples of both cases exist.

Stem and root growth are largely the result of cell elongation with little increase in cell girth. In such cells it is common to find the cellulose microfibrils laid down in a flat helical array, rather like a compressed toy Slinky, with the net orientation of each microfibril more or less at right angles to the axis of cell elongation. In many cases elongation is accompanied by the longitudinal stretching of the helices in older lamellae, with the result that the helical angle changes (**Figure 6.28**). This is termed *passive reorientation*. The steepness of the angle will depend on the number of turns in the original helix and the degree of cell elongation. In other cases of cell elongation the net orientation of cellulose appears to remain more or less constant, typically at right angles to the axis of cell elongation. This is easily seen in micrographs of elongating cells in culture (**Figure 6.29**). Both of these examples raise unanswered questions about the spacing of microfibrils within old lamellae as they expand, and whether or not microfibrils from one lamella can be incorporated into another. In addition there are topological problems with uniformly reorienting the lamellae on all the faces of a cell together, and much more work needs to be done in this area.

Discussion so far has rested on the assumption that individual wall lamellae are discrete, independent elements, with relatively labile cross-links between them, and that they have the ability to shift orientation with respect to their neighbors. In many cases this assumption may be wrong, and lamellae with differing orientations are found in some walls, not because they have reoriented passively, but because they were laid down that way to start with and have not shifted since, a process of *directed reorientation*. A clear example can be found in the xylem vessels and tracheids of wood, where a complex, three-layered cellulosic secondary wall is laid down after cell expansion has finished but before lignification starts. The cellulose microfibrils in the first and last of these layers (S1, S3) are nearly transverse, while those in the mid-

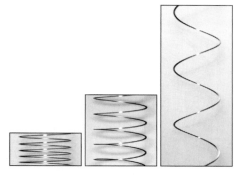

Figure 6.28 Passive reorientation of cellulose microfibrils. This image of a child's Slinky toy, before and after stretching, is used to illustrate the shift in orientation of the outer layers of cellulose microfibrils that might occur passively during cell elongation. The shift will depend both on the number of turns in the Slinky and the extent to which it is stretched.

direction of cell elongation

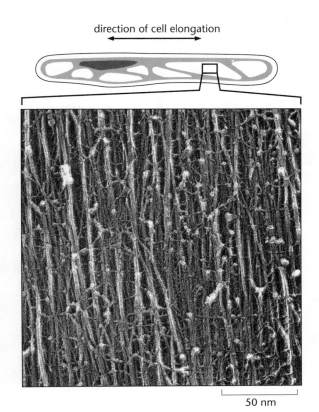

50 nm

Figure 6.29 Cellulose microfibril orientation in an elongating cell. In this electron micrograph of an intact isolated wall from a rapidly elongating carrot cell, all of the microfibrils visible are oriented at right angles to the direction of cell elongation. (Courtesy of Brian Wells.)

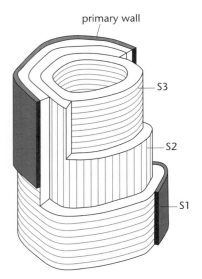

primary wall

S3

S2

S1

Figure 6.30 Cellulose orientation in xylem vessels and tracheids. This schematic cutaway diagram shows the S1, S2, and S3 secondary cell wall layers laid down successively inside the primary cell wall layer during the differentiation of xylem vessels and tracheids in wood. The net orientation of cellulose microfibrils laid down in each layer before lignification is shown by gray lines. This is an example of directed reorientation.

dle layer (S2) are nearly longitudinal (**Figure 6.30**). This has clear functional significance, since tensile strength is greatest parallel to the microfibrils and, by varying their orientation, strength is maintained in multiple directions, a principle embodied in plywood construction.

11. The self-ordering properties of polylamellate walls can lead to high degrees of structural order.[ref21]

The considerable order that we have seen in the construction of lamellae and their assembly into a wall has been attributed to the cell that made them. However, mixtures of different long-chain polymers can, on their own, often display a variety of self-ordering phenomena such as *phase separation*. Concentrated solutions containing a mixture of esterified and nonesterified pectic polysaccharides, or of pectin and hemicellulose, can resolve in a test tube into separate phases, each enriched in one or another polymer. Whether this property is used by the cell to organize the wall, however, remains unclear.

Better documented is the ability of cellulose microfibrils and hemicellulose to self-assemble into transient liquid crystal-like assemblies. This has been shown for soluble cellulose and glucuronoxylan that have been extracted from quince slime and then reassembled *in vitro*, but more important, similar assemblies have been found in normal cell walls from a variety of cell types. These complex ordered structures show *helicoidal architecture*, in which each successive lamella in the wall is rotated by a small fixed angle with respect to the next one. When sectioned at an oblique angle, helicoidal walls display a characteristic series of arced bands, an optical effect resulting from the particular geometry (**Figure 6.31**). Such helicoids are readily detectable only if large numbers of lamellae are involved, and consequently they have been most often described in relatively thick cellulosic walls. They are found, for example, in the thickened regions of the epidermal and cortical parenchyma cells of rapidly elongating mung bean hypocotyls. In this case the structures are transient and, as elongation ceases, the strict helicoidal architecture dissipates. In collenchyma cells, however, which also deposit local cellulose-rich thickenings in their walls as they elongate, helicoidal structures are found that persist and indeed enlarge as the cells stop elongating (**Figure 6.32**). Their plywoodlike nature strongly suggests that helicoidal walls possess considerably enhanced strength, able to withstand stresses in many directions.

Figure 6.31 Helicoidal cell wall architecture. (A) When successive wall lamellae are rotated with respect to each other by a small fixed angle, the resulting architecture is helicoidal. (B) When sectioned at an oblique angle, walls with helicoidal lamellae show alternating dark and light bands at low magnification, as seen in this section of a thick collenchyma cell wall. (B, from B. Vian & J-C Roland, *New Phytol.* 105:345–357, 1987.)

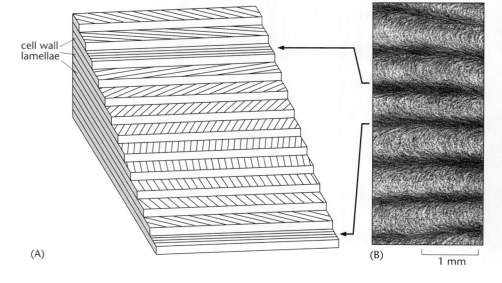

cell wall lamellae

(A)

(B)

1 mm

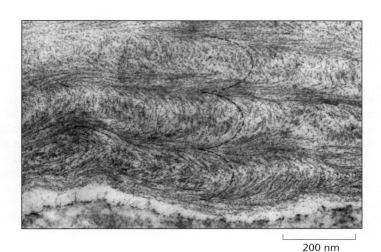

200 nm

Figure 6.32 The outer epidermal wall of a mung bean hypocotyl. At one stage of cell expansion this thick wall shows the clear helicoidal architecture seen here. Later this organization dissipates. (From J-C Roland et al., *J. Cell Sci.* 56:303–318, 1982.)

It is not yet clear whether the helicoidal architecture in these examples is originally established by laying down consecutive lamellae at a defined angle that reflects changing microtubule orientations within the cell cortex (that is, is cell-directed) or whether instead the lamellae self-assemble into a dynamic and transient array with no direction from the cell. If the latter self-assembly model is the case, then a molecular mechanism would be required to establish a preferred angle between the microfibrils in one lamella and those in the next. At present we have no way of predicting when or why a cell forms a helicoidal wall, and much remains to be learned about these remarkable structures.

It is also of interest to note that components of the pectin network have also been shown to self-assemble into larger molecular complexes. *In vitro* studies with RG-II have shown that this polysaccharide will form dimers in the presence of borate (see Concept 6A6). This observation strongly suggests that dimerization of RG-II can occur in the cell wall in the presence of borate, thus "self-assembling" a pectin complex that contributes to the structure of the cell wall pectin network.

12. Cell wall assembly is a hierarchical process.[ref22]

Plants, in common with other biological structures, are hierarchically organized, and their architecture is controlled on length scales from the molecular to the macroscopic. Many of the macroscopic features of plants, for example, their mechanical properties, depend on the architecture and properties of their cell walls, which are themselves hierarchically structured materials. Organization over such a large range of length scales (**Figure 6.33**) is the basis for the unique performance of natural materials such as wood and fibers. The properties of plant cell walls are typical of other hierarchical materials systems in biology. For example, a wide range of biological properties is attained from a relatively small set of basic building blocks. There is very precise control over the orientation of the key structural elements, cellulose at one level, elongated cell types at another. There are durable interfaces between the tensile and compressive elements in the wall—microfibrils and matrix materials, respectively. Wall architecture is resilient and resistant to fatigue, and can vary its properties in response to the local requirements of the plant. Its properties are dependent on, and sensitive to, the presence of water, and last, it shows a capacity for self-repair.

These complex and remarkable properties derive in large part from the way the wall is built, concurrent synthesis and assembly in a series of hierarchical construction steps. Most, but not all, of these steps are carefully controlled by the cell, by making them contingent on preexisting structures, a process called *template-based assembly* (in contrast to the less controllable

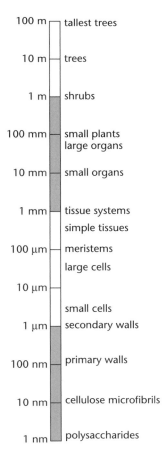

100 m	tallest trees
10 m	trees
1 m	shrubs
100 mm	small plants large organs
10 mm	small organs
1 mm	tissue systems simple tissues
100 µm	meristems large cells
10 µm	
	small cells
1 µm	secondary walls
100 nm	primary walls
10 nm	cellulose microfibrils
1 nm	polysaccharides

Figure 6.33 Plant architecture and mechanics operate at every length scale over 11 orders of magnitude, from the nanoscale to the tallest trees.

mechanism of self-assembly discussed earlier). We can illustrate the idea by using cellulose as an example. One of the first steps in the hierarchy is cellulose biosynthesis, the assembly of the glucan chain, and its adoption of a rigid twofold helical conformation. The next step is the parallel packing of an appropriate number of chains, an entropically unfavorable process, and their assembly into a cellulose microfibril. This probably depends on the structural order within the cellulose synthase complexes themselves and the rosettes that they form in the membrane (see Concept 5D8). Microfibril structure and cross-linking becomes rapidly stabilized by interaction at the cell surface with appropriate amounts of secreted hemicellulose, a process controlled by the cell. The orientation of the microfibrils is again under cellular control, and involves templating by intracellular microtubules. The lateral ordering of microfibrils into lamellae is probably dependent on both of the previous processes. Higher elements in the hierarchy of wall construction involve lamella reorientation during growth; localized deposition of new wall material, for example, secondary cell wall cellulose; and the more global control of all these processes at a cell-specific level.

This example of *hierarchical construction* applies not only to cellulose, but to all of the wall components we have described earlier, and it is the integration of all these hierarchies, and their responsiveness to internal and external controls, that gives plants their rich variety of properties and behaviors.

While discussing wall construction it is important to appreciate what the cell can and cannot accomplish with macromolecules once they have been deposited outside the cell. While some ATP is found in the wall, it is most likely there as a signaling molecule, and there is no ready supply of usable free energy outside the confines of the plasma membrane. The classes of covalent bonds that can be formed within the wall, therefore, are largely limited to those that involve no net gain in free energy. Thus, reactions that break and then re-form the same kind of bond, such as a transglycanase or a transesterase, are in principle permitted, as are oxidative cross-linking reactions, since they are energetically favorable. Forming new glycosidic bonds in the wall, however, is not, and this is why all the basic wall polymers are constructed within the cell, where there is a ready supply of activated nucleotide-sugars, and why wall construction outside the cell is largely by self-assembly.

C. Architectural Variations: The Mosaic Wall

1. Polymers are not evenly distributed within a wall.[ref23]

So far in this chapter we have considered the general architectural principles that are common to all primary cell walls. Some walls are thicker, some are thinner, some walls enclose very small cells, some enclose gigantic cells, but the assumption has been that at a molecular level they are all rather similar. While in general construction terms this may be true, this section will question that assumption by emphasizing instead the differences that exist within and between walls. We have already seen hints of this heterogeneity in the discussion of different cell types in Chapter 1, and stained histology slides of plant tissues are an obvious reminder that in plant anatomy it is often by their differing walls that cells are identified. More recently, the use of specific probes for particular cell wall components, in light microscopy, electron microscopy, and tissue printing, has dramatically increased our appreciation of the heterogeneity within walls. Widely used probes include monoclonal antibodies, fluorescent stains, and wall-degrading enzymes coupled to colloidal gold. Using these and other techniques, three main kinds of heterogeneity have been found in cell walls, and each kind may involve chemical differences, structural differences, or both (**Figure 6.34**).

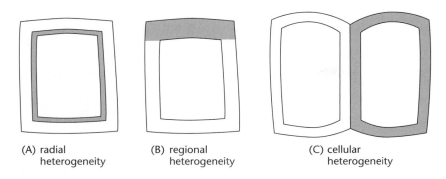

(A) radial heterogeneity

(B) regional heterogeneity

(C) cellular heterogeneity

Figure 6.34 Three kinds of heterogeneity in cell walls. (A) Radial heterogeneity: compositional and/or architectural differences within a wall, or a single wall facet, in a radial direction. (B) Regional heterogeneity: structural and/or compositional differences between the wall on one cell face and the rest of the wall. (C) Cellular heterogeneity: differences between the two neighboring walls that separate adjacent cells.

First, it is common to find *radial heterogeneity*, differences associated with lamellae near the plasma membrane compared with those further away near the middle lamella. Radial heterogeneity may be confined to a single cell face, for example, the layering of waxes, cutin, and cellulose in the outer wall of a leaf epidermal cell; or it may extend to the wall surrounding an entire cell. For example, different polysaccharides occupy different radial domains within the wall of root hairs in *Arabidopsis* (**Figure 6.35**).

A second kind of wall heterogeneity exists in which one region of a cell wall becomes distinguished from the remainder. This *regional heterogeneity*, common in both primary and secondary walls and often involving local wall thickening, can be seen clearly (**Figure 6.36**) in many of the cell types described earlier (see Section 1C). In epidermal cells, the wall facing the outside is usually thickened and then elaborated with a cuticle and surface waxes. In xylem and tracheary elements and in collenchyma cells, localized regions of the wall are thickened and, in the former case, lignified. In the endodermal cells of the root, a narrow strip of wall surrounding the cell is impregnated with waterproofing materials. In this case the regional special-ization in one wall lines up precisely with a similar region in its neighbors, forming an interconnected chain mail of casparian bands. Other examples include the elaborate wall ingrowths of transfer cells and the sieve plates at the end of phloem sieve tube elements. In some cases the regional speciali-zation of walls may depend less on obvious physical features than on the deposition of particular molecules, including proteins, at particular sites, a feature not easily detected microscopically without the appropriate specific probes (see **Figure 6.37**) .

The third kind of wall variation, *cellular heterogeneity*, is found where the neighboring walls between two adjacent cells have different structures or compositions. This often occurs where different cell types abut, but it can also be found between cells that otherwise appear similar. In some cases the differences are obvious, for example the thick walls of xylem vessels or phloem sieve tube elements and the thin walls of their associated paren-chyma cells. In other cases the differences are less obvious, for example,

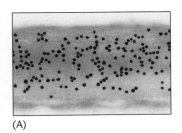

(A)

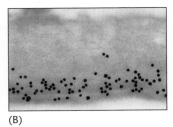

(B)

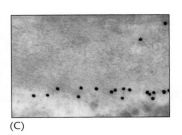

(C)

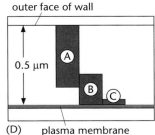

(D)

Figure 6.35 Immunogold labeling of the root hair cell wall. (A) Monoclonal antibodies that recognize different epitopes on xyloglucan, (B) RG-I, and (C) AGPs have been used to demonstrate radial heterogeneity in the distribution of these epitopes within a specific primary cell wall. (Courtesy of Glenn Freshour and Michael Hahn.)

Figure 6.36 Examples of regional heterogeneity in the walls of different cell types: (A) leaf epidermal cells, (B) xylem vessel elements and tracheids, (C) collenchyma, (D) endodermis with casparian band, and (E) transfer cell.

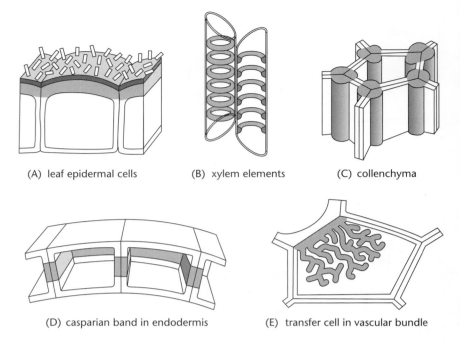

(A) leaf epidermal cells   (B) xylem elements   (C) collenchyma

(D) casparian band in endodermis   (E) transfer cell in vascular bundle

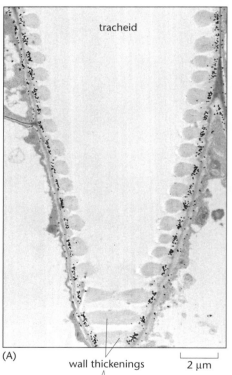

(A)

tracheid

wall thickenings   2 μm

(B)

tracheid

wall thickenings

xylem parenchyma cell

2 μm

the subtle distribution of pectin between cells in a root tip (**Figure 6.38**). Antisera, monoclonal antibodies, and other cell wall probes for cell wall polysaccharides and cell wall proteins have revealed numerous examples of the cell-specific expression of wall polymers, and they are increasingly valuable tools for studying plant cell walls (**Panel 6.1**).

In many respects the most obvious example of heterogeneity in the cell wall is the region, called the middle lamella, which exists between neighboring walls of all the cells in the plant, holding them together. This region is of great importance to the life of the plant, as we now discuss.

2.   The middle lamella forms an adhesive boundary between adjacent cells.[ref24]

We have seen earlier how, during the later stages of cytokinesis, cell wall material is laid down in the cell plate from either side by each daughter cell (see Figure 4.44). A region is created between the two young walls that functions to define the boundary of each cell and, in addition, to cement the two daughters together. This region is called the *middle lamella*, and it persists in one form or another, as part of the cell wall, for the life of each cell. The middle lamella can usually be seen in electron micrographs as a thin layer of increased electron density, about 20 nm wide (**Figure 6.39**). There are certain general features that are likely to be true of any middle lamella. For example, cellulose microfibrils do not appear to cross the middle lamella from the outermost lamella of one wall to the outermost lamella of its neighbor. This means that the adhesive forces that hold plant cells together are more likely to reside in the pectin network of the wall than in

Figure 6.37 Regional heterogeneity in xylem. (A) Tracheid from a small vein in a potato that has been probed with a monoclonal antibody that recognizes the 1-4-β-D-galactan side chains on RG-I. (B) Specific local deposition of this polymer is revealed solely in the region of the primary wall that underlies the secondary wall thickenings of the tracheid, a remarkable example of regional heterogeneity. (Courtesy of Max Bush.)

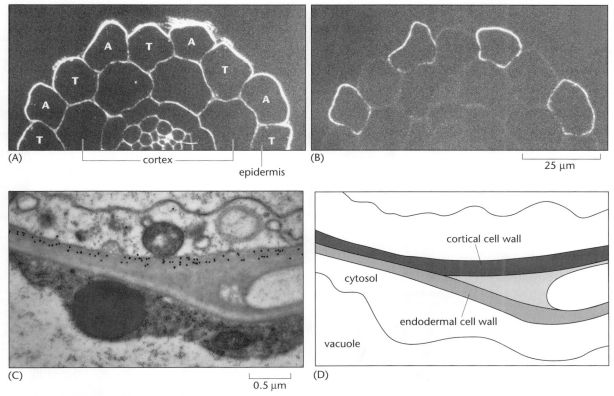

Figure 6.38 Cellular heterogeneity in the root. (A) A cross section of the seedling root, just above the meristem, reveals two kinds of cells in the *Arabidopsis* root epidermis: the future hair cells, or trichoblasts (T), that tuck into a cleft between two underlying cortical cells; and the atrichoblasts (A), which will not develop hairs, sitting on an outer wall of the cortical cell beneath. (B) In a consecutive section, treated with a fluorescent antibody to stain RG-I, it can be seen that only the atrichoblast walls contain appreciable amounts of RG-I, a clear example of cellular heterogeneity of cell walls. Further back from the root tips shown above, sections stained with the same antibody show that both the epidermis and the cortex now contain RG-I, while the endodermis and stele do not. (C) Electron micrograph following staining by immunogold labeling and (D) schematic diagram showing that the RG-I is confined to the cortical wall at the point where it and the neighboring endodermal wall are apposed. (Courtesy of Michael Hahn and Glenn Freshour.)

the cellulose-hemicellulose network. Indeed, high levels of pectic polysaccharides, rich in unesterified homogalacturonan, are commonly found in the middle lamella.

Adhesive forces that are both tissue-specific and developmentally regulated with time hold cells together. These adhesive forces are the product of a variety of different covalent and noncovalent cross-links between pec-

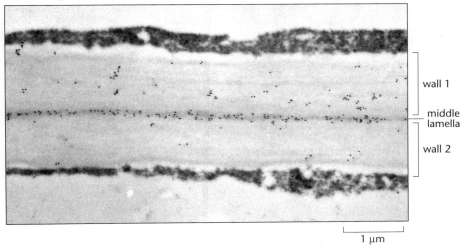

Figure 6.39 Electron micrograph of thin section of two tomato fruit cell walls, labeled with an antibody for relatively unesterified pectin. The electron-dense middle lamella cementing the thick walls of the two cells together is visible, and is enriched in unesterified pectin. (From Blumer et al., *Can. J. Bot.* 78:607–618, 2000.)

INTRODUCTION

A key problem in cell wall research is location; or where each of the molecules of the cell wall are located and how they are developmentally regulated, both at a tissue level and within individual walls. To address this problem we need probes that are specific to each molecule, and methods for revealing them at both the light and electron microscope level.

Over the last thirty or so years there has been an explosion of new methods and probes, which have transformed our ideas about the complexity and heterogeneity of the different walls within plants. This panel briefly describes some of these, albeit without the originals dramatic colors!

STAINS

A simple way to show the presence of a particular component in a section of a plant is to use a stain that specifically binds to it. Common stains in regular use include Phloroglucinol-HCl (lignin) and the Periodic Acid-Schiff stain (carbohydrate). Many stains fluoresce when bound to their substrate e.g. Aniline Blue (callose) and Calcofluor White (cellulose). Some wall components fluoresce naturally. Lignin autofluorescence in a section of the petiole of an aquatic fern. (*Courtesy of Paul Knox*)

Since plant cell walls contain cellulose, Calcofluor White is often used to reveal the positions of all cells, often as a background stain for other probes. When bound to cellulose, it is excited by near UV light and emits blue light. Here, a *Zinnia* cell in culture has developed into a tracheid (see Figure 1.25). The cellulosic wall thickenings are stained brightly with Calcofluor White. Lignin will be deposited in the same region later on.

CARBOHYDRATE-BINDING MODULES

Carbohydrate-binding modules, or CBMs, (described in Concept 8B10) are small protein domains, from microbial wall degrading enzymes, that bind to specific polysaccharides. The most common is cellulose, but others bind to mannan, xylan etc. When attached to a fluorescent marker, their small size makes them ideal probes for cell wall structures. (A) A section of a tobacco stem, stained with Calcofluor White, shows the walls of all the cells. (B) The same section probed with a his-tagged CBM that binds to crystalline cellulose. The CBM bound to the xylem vessels is detected with fluorescent anti-his antibodies. (*Courtesy of Paul Knox*)

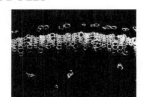

A section of tobacco stem (*above*) has been fluorescently stained with a CBM that binds specifically to xylans, present in phloem fibers and xylem cells. (*Courtesy of Paul Knox*)

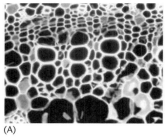

(A)

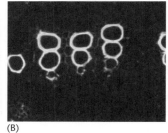

(B)

GREEN FLUORESCENT PROTEIN (GFP)

As important as the wall polymers themselves are the proteins involved in their manufacture and assembly. A key way of following where these are, in living cells, is to create a fluorescently tagged protein by introducing its gene, coupled to that encoding a fluorescent protein. Visualization of cellulose synthase, microtubule dynamics and Golgi trafficking, have all been followed in this way. Most probes are variants on green fluorescent protein (GFP).

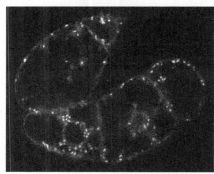

Optical section through living tobacco cells in culture. They are expressing a fusion protein between GFP and Mannosidase I, located in the Golgi stacks, which appear as white dots (see Figure 4.8).

RADIOLABELING

Feeding cells with a radioactive wall precursor, followed by autoradiography, allows the distribution of the product to be monitored. The transit of material through Golgi stacks and to the wall was first shown in this way. The electron micrograph shows silver grains over a tracheary element-like wall thickening in a cell that has been fed H^3-proline. Most of the radioactive label is incorporated into hydroxyproline-rich glycoprotein in the thickenings.

ENZYMES

Many microbial enzymes have specificity for particular wall polysaccharides, and therefore, if coupled to colloidal gold particles, they will act as good EM probes for their substrate. *Endo*xylanase, coupled to colloidal gold, is used here to reveal xylan in secondarily thickened walls from the giant cane, *Arundo donax*. (*Courtesy of Katia Ruel*)

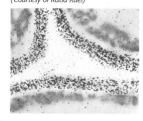

RNA LOCALIZATION

To determine which cells in a tissue express a gene of interest, RNA *in-situ* hybridization is used, usually on tissue sections. This has been used widely to determine the patterns of wall-related gene expression. Another way to do this is hybridization on a *tissue print*; a nitrocellulose sheet, on which a cut plant surface is pressed, to replace the tissue section. (A) Glycine-rich protein (GRP) mRNA distribution on a tissue print of a soybean stem. (B) A similar print showing a HRGP mRNA distribution. Both genes are expressed in different parts of the vascular system. (*Courtesy of Joe Varner*)

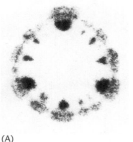

(A)

(B)

ANTIBODIES

Nature's most versatile and sensitive probes for a particular molecule of interest are antibodies. When made against plant wall related molecules, such antibodies are important probes in both the light and electron microscope. They can be applied to intact tissue, on thin sections or even to isolated molecules, and their presence can then be detected, in immunocytochemistry, by using secondary antibodies that are coupled in turn to either a fluorescent marker for use in the LM, or to colloidal gold for use in the EM. Greater specificity and reproducibility can be obtained by using monoclonal antibodies, and many of these are now available, specific to a variety of wall macromolecules or antigens.

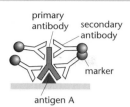

Indirect immunocytochemistry. A primary antibody recognises immobilized antigen A. Secondary antibodies, directed to the primary antibody, amplify the signal as each is coupled to a marker, either fluorescent, for LM, or colloidal gold, for EM.

MONOCLONAL ANTIBODIES

Many wall related monoclonal antibodies, derived from either rat or mouse cell lines, are now available as LM or EM probes for the wall. There are two main depositories of such reagents, one in the USA at http://cell.ccrc.uga.edu/~mao/wallmab/Home/Home.php and the other in the UK at www.plantprobes.net.

Some examples are listed here:

Monoclonal antibody	Epitope recognized
LM15	XXXG motif of xyloglucan
LM10	xylogalacturonan
LM5	1,4-β-D-galactan
LM1	HRGP
JIM4	AGP glycan
JIM7	Methyl-esterified homogalacturonan
CCRC-M1	α-L-fucosylated xyloglucan

IMMUNOFLUORESCENCE

Used widely in cell wall research, the use of fluorescent immunocytochemistry has widened our appreciation of the complexity, variety and dynamic properties of wall components.

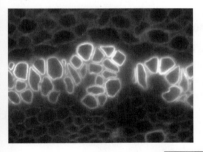

In this young hemp stem, the plasma membranes of the developing primary fibers in the cortex contain specific arabinogalactan proteins, an epitope of which is revealed here by the monoclonal antibody JIM14. All the cell walls in the section are also stained more faintly with Calcofluor White.
(Courtesy of Paul Knox)

This section of the young primary root of an *Arabidopsis* seedling has been probed with the monoclonal antibody CCRC-M1. This recognises an epitope in xyloglucan, which is particularly abundant in the outer epidermal wall.
(Courtesy of Georg Seifert)

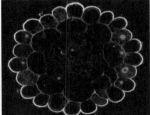

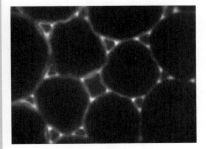

These pith cells from a tobacco stem have been probed with the xyloglucan specific monoclonal antibody LM15. Some of the epitopes are masked unless the section is first treated with endopolygalacturonase. The XG is abundant at three-way cell junctions.
(Courtesy of Paul Knox)

WHOLE MOUNTS

Whole cells or organs, fixed or alive, can be immersed in an antibody solution and their surface chemistry probed by direct or indirect immunocytochemistry.

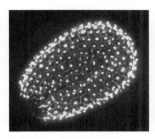

An intact seed of *Arabidopsis* labelled with the anti-xyloglucan monoclonal antibody LM15.
(Courtesy of Paul Knox)

Whole mount preparation of the moss *Physcomitrella patens* immunolabelled with **LM6**, showing the presence of arabinogalactan-proteins at the growing tip of protonemal filaments.
(Courtesy of Paul Knox)

ELECTRON MICROSCOPY

To look more closely at the precise distribution of wall macromolecules within tissues and walls, electron microscopy must be used. Colloidal gold beads, which are very electron dense, are used instead of fluorescent markers, but with the same probes.

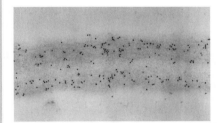

This section of a wall from a tomato cell in culture has been immunogold labeled with JIM5, a monoclonal antibody that recognises a partially methylesterified homogalacturonan, evenly distributed throughout the wall.

A section of a wall thickening of a developing *Zinnia* tracheary element has been immunogold labeled with a monoclonal antibody to an AGP, which is concentrated in the wall thickenings.
(Courtesy of Maureen McCann)

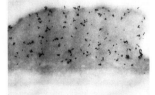

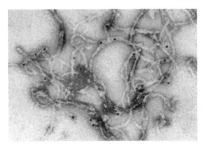

Immunogold staining with JIM5 of isolated pectin molecules. The long molecules are made visible by negative staining and the gold particles are 5 nm in diameter.
(Courtesy of Maureen McCann)

tic polysaccharides in the middle lamella. Thus, depending on the tissue, cells can be released by gentle mechanical disruption, by CDTA or other Ca^{2+}-chelating agents, by breaking ester cross-links using dilute alkali, or by combinations of these treatments. Cells in other tissues will fail to separate even after treatment with hot, strong alkali. Cells are released from the flesh of a red tomato by CDTA but not from a green tomato. We can conclude that the glue that holds cells together is combinatorial and that the combination can change with the developmental age of the cell. Cells in the green tomato, for example, may be held together by both Ca^{2+} cross-linked pectin and by ester cross-links. As the fruit ripens, the ester cross-links are broken, and the remaining Ca^{2+} cross-links make cell separation easier. The result is softer fruit. All the pectic polysaccharide cross-links, and possibly the protein cross-links, discussed at the beginning of this chapter are possible contributors to cell-cell adhesion at the middle lamella. Depending on the cell types, borate diesters between RG-II molecules, diferulate cross-links between pectic polysaccharides, Ca^{2+} cross-linked pectic polysaccharides, and ester linkages between galactosyluronic acid residues and glycosyl residues of neighboring polysaccharides may all contribute to the combinatorial "glue" that holds cells, and hence plants, together.

3. Three-way junctions are rich in protein, pectin, and phenolics.[ref25]

When cells begin their life in a meristem, they are firmly attached to their neighbors by the middle lamella at every point on their surface, like soap bubbles in foam; tightly packed with no space between them. Geometrical and energetic considerations mean that the boundaries between cells are, as in soap bubbles, a series of facets that each separate any two adjacent cells. Each facet is connected at its edges by a series of interconnected struts that form *three-way junctions* between three adjacent cells. Three-way junctions can be seen in any section of plant tissue; four plant cells almost never meet at the same junction (**Figure 6.40**).

The sum total of the three-way junctions in a tissue forms an interconnected physical structure that has features in common with a geodesic dome. Most of the tissue's strength resides, as in these domes, in the struts, formed in the plant by the three-way junctions. It is unsurprising, therefore, to find that there are concentrations of molecules in these regions that help

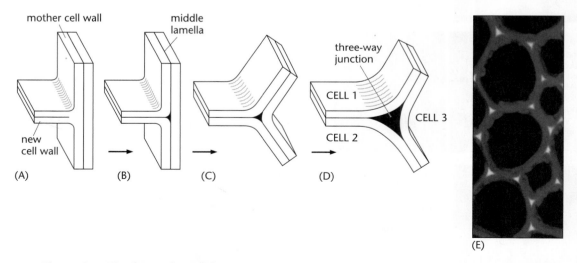

Figure 6.40 The formation of three-way junctions. (A) Following cell division, the new cell plate fuses with the mother cell wall and (B) continuity is established between the middle lamellae. As the daughter cells grow, along with their neighbors (C), a clear three-way junction is established (D). Such junctions can be seen in all plant tissues, and are illustrated here in a cross section of young fiber cells in an *Arabidopsis* stem (E). The three-way junctions are preferentially stained with an antibody to low methylesterified pectin.

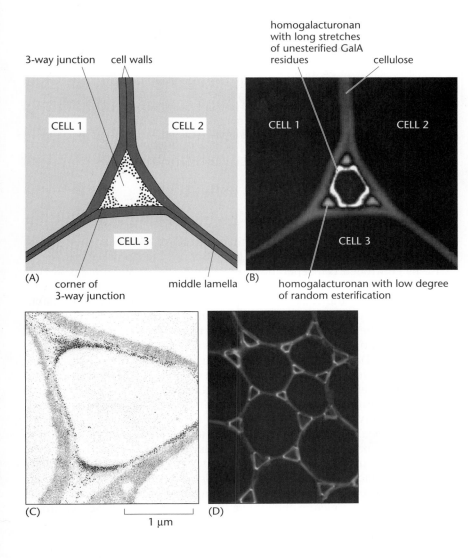

Figure 6.41 Pectin in three-way junctions. (A) As cells expand, the corners of three-way junctions take much of the strain in keeping cells stuck firmly to their neighbors. (B) Homogalacturonans, with subtly different patterns of methylesterification, are laid down in a remarkably precise pattern of microdomains in this three-way junction between cortical cells in a tobacco stem. The two microdomains are revealed using two different monoclonal antibodies that react with differently methylesterified homogalacturonan epitopes. (C) In the electron microscope, immunolabeling with a monoclonal antibody reveals concentrations of low methylesterified pectin in the corners of a three-way junction. (D) A similar picture seen in the light micrograph of cells in the pith of a tobacco stem. (B, courtesy of Bill Willats; C, courtesy of Paul Linstead; D, courtesy of Paul Knox.)

to provide particularly strong adhesion between cells at their junctions. Immunocytochemical results, using antibodies to a range of molecules, suggest that pectic polysaccharides, including RG-I and homogalacturonan, are often preferentially deposited in the corners of three-way junctions, as are some classes of hydroxyproline-rich glycoproteins (**Figure 6.41** and **Figure 6.42**). Phenolic ester cross-links are also thought to reinforce cell adhesion in several species, and this is seen particularly clearly in the Chinese water chestnut (which, despite its name, is a monocot). The crisp, crunchy texture of this edible corm results from heat-stable ferulic acid dimers that reinforce adhesion along the three corners of three-way junctions between the close-packed parenchyma cells (**Figure 6.43**).

4.   The middle lamella is involved in cell separation.[ref26]

As cells leave the meristems where most are born, they grow in volume. As they differentiate into mature cell types, their connecting three-way junctions can become modified in a variety of ways. In some cases they may enlarge a little but remain occluded or filled with electron-dense material, rich in proteins and pectin (see Figure 6.42). More commonly, however, cells may separate at their three-way junctions to an extent determined precisely by a cell's position in a tissue and its developmental age (**Figure 6.44**).

The driving force for cell separation is usually turgor, but the degree to which cell adhesion will allow cells to separate is vital in determining the

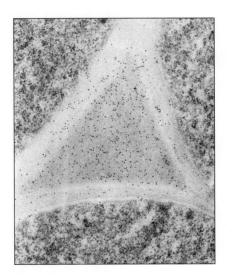

Figure 6.42 Closed three-way junction between cells in a maize root tip showing localization of a threonine-rich, hydroxyproline-rich glycoprotein (THRGP) by immunogold labeling. (From K. Roberts, *Curr. Opin. Cell Biol.* 2:920–928, 1990.)

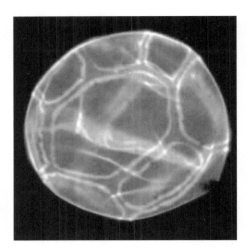

Figure 6.43 Three-way junctions define cell faces. The corners of three-way junctions seen in cross section also extend in three dimensions to define the different faces of a cell (see Figure 1.4). These corners represent particularly strong adhesion lines between cells, and they can often be seen when cells are separated from each other. The lines left by the three-way junctions of cells can be seen on the surface of an isolated cell from a Chinese water chestnut. Here the lines are seen in the light microscope by the autofluorecence of wall phenolics, concentrated at three-way junctions and involved in the strong cell-cell adhesion that is reflected in the crisp texture of water chestnuts. (Courtesy of Mary Parker and Keith Waldron.)

final functional morphology of cells in tissues. The most obvious example of controlled cell separation is in the formation of *air spaces*, not only in leaves but also throughout the plant body. Spongy mesophyll cells in leaves arise by controlled cell separation, while maintaining small areas of cell-cell contact that act as localized regions of cell growth, thereby increasing the separation of the cells and creating the large interconnected system of air spaces that facilitates gas exchange in the leaf (**Figure 6.45**). While it is clear that the precise amount by which two adjacent cell walls may be pried apart is tightly controlled, the molecular basis for these controls is still obscure. Data from young developing pea seed cotyledons suggest that small regions of the middle lamella are reinforced at an early stage and that these act effectively as sets of rivets that prevent the further separation of cells beyond them. These regions can be seen as zones of increased electron density, but to date no candidate molecules have been localized to them. A local zone of cell separation is also seen during the formation of the stoma, or pore, between a pair of stomatal guard cells. In this case separation occurs not at the three-way junctions but at a small, specified region of the face between two cells (see Figure 1.31). Three-way junctions that are reinforced with cross-links between pectic polysaccharides may be disrupted by heat. Calcium chelation and β-elimination of galactosyluronic acid residues during the cooking of potatoes lead to partial cell separation and consequent softening.

Insight into the molecular basis of cell separation and its control may come through the use of genetic approaches. Mutants in the dehiscence or abscission pathways of *Arabidopsis* may prove revealing. One mutant, called *things fall apart* (*tfa*), shows a phenotype in which cells of the hypocotyl lose adhesion with their neighbors and begin to round up and slough off. The *TFA* gene encodes a protein with no obvious homology to known proteins in the databases, but such mutants offer promise for better understanding cell-cell adhesion.

5.   The intercellular spaces form a continuum.[ref27]

The two-dimensional images we have shown of three-way junctions between cells are derived from thin sections of plant material. In three dimensions, however, within the plant, these junctions form an interconnected set of three-sided tubes, surrounding every cell in a complex integrated labyrinth of load-bearing struts (**Figure 6.46**). Each strut forms one common edge for three different wall facets, and the struts are joined at their ends at points where four cells touch in a tetrahedral junction. In

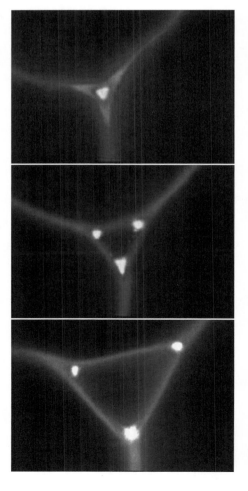

Figure 6.44 The gradual separation of walls at a three-way junction creates an intercellular space. In these cells from a pea stem the cellulose in the walls is seen using a fluorescent stain, while the bright spots represent homogalacturonan that is enriched along the corners of the junction. (Courtesy of Bill Willats and Paul Knox.)

Figure 6.45 Air spaces in the leaf. This scanning electron micrograph of spongy mesophyll cells in a leaf clearly shows the remaining points at which the cells are still attached to each other, and the elaborate labyrinth of air spaces that have been generated by a combination of localized cell separation and growth. (Courtesy of Kim Findlay.)

many tissues, for example in apical meristems, the three-way junctions are zones of tight adhesion between the three cells involved. As we have seen earlier, however, as the tissues grow the junctions can expand and become open channels surrounding the cells. These low-resistance passages, which facilitate gaseous diffusion, have obvious advantages in leaf architecture (Figure 6.45). They also facilitate access for fungi, providing ideal routes for hyphae to penetrate cortical tissues, both in pathogen attack and in the establishment of a beneficial symbiosis by endomycorrhiza.

Many wetland plants have developed strategies for ensuring that their roots, living in conditions of low oxygen tension (*hypoxia*), can obtain enough oxygen to sustain respiration and grow. During development, three-way junctions between certain cortical cells in the roots enlarge, often to the point where cells become completely detached from each other, to create extensive air spaces that act to deliver oxygen to the root tips from the aerial organs. Cortical tissues that develop in this way are called *aerenchyma*.

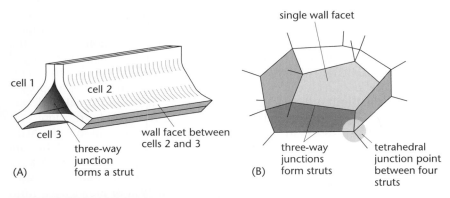

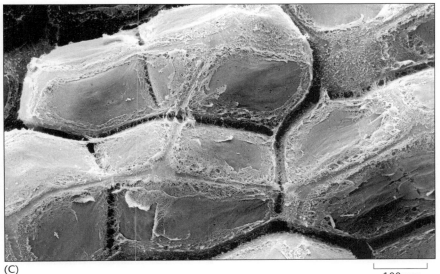

Figure 6.46 A three-dimensional network of three-way junctions surrounds each cell. (A) Each three-way junction extends in three dimensions as a strut that is triangular in cross section and (B) interconnects with other struts at tetrahedral junction points to enclose the cell. An interconnected labyrinth of three-way junctions thus surrounds and links all the cells in the plant. (C) Scanning electron micrograph of cells in a cooked potato. The lines left by the adhesive three-way junctions of cells that have been removed during cooking are clearly visible, as are the positions of the tetrahedral junction points. (C, courtesy of Mary Parker and Keith Waldron.)

Figure 6.47 Aerenchyma. (A) The large air spaces between the rows of cortical cells in this section of the underwater shoot of a water lily arise by schizogeny, the separation of living cells during shoot development. (B) Transverse section of the rhizome of the sweet flag that grows submerged in mud. Aerenchyma has developed by schizogeny, leaving thin strands of cells that have grown to surround the air spaces that conduct oxygen from the leaves to the rhizome. (A, courtesy of www.biologie.uni-hamburg.de/b-online. B, courtesy of J. Mauseth.)

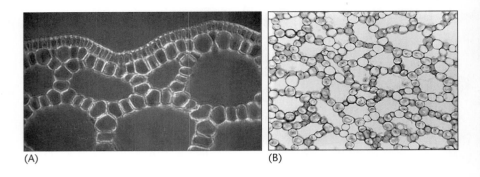

(A)   (B)

In principle, the air-filled channels of aerenchyma can be formed in two different ways, by the separation of living cells during organ growth and development (called *schizogeny*), or by the subsequent death and dissolution of groups of cortical cells (called *lysigeny*). The former is typical of plants, such as rice, water lilies, and rushes, that are well adapted to hypoxic conditions (**Figure 6.47**, and see also Figure 1.34). The latter, lysigeny, is common in plants that tolerate flooding conditions (for example, maize) by developing aerenchyma as an adaptive response, mediated by the plant growth factor ethylene.

6.   The expression of arabinogalactan proteins is developmentally regulated.[ref28]

Up to now most of our discussion of wall architectural variation has revolved around relatively stable variations in the amount or composition of the structural polymers of the wall. There are other molecules at the cell surface, however, which show more dynamic variations and generate more complex patterns of cellular heterogeneity. Notable among such molecules are the *arabinogalactan proteins* (AGPs) that were discussed in Chapter 3. Judged by studies of the fate of radioactively labeled arabinose fed to cells in culture, AGPs show a relatively high rate of turnover, and this view is supported by numerous immunocytochemical studies using monoclonal antibody probes to AGP glycan side chains, which show remarkable transient patterns of AGP expression in a range of plants and organs (**Figure 6.48**). AGPs are found at the surface of all plant cells, from mosses to dicots; and the core proteins are encoded by a large and divergent set of genes. Some AGPs have GPI anchors that attach them to the outer face of the plasma membrane, while others are either secreted directly into the wall or are cleaved from their GPI anchors. Heterogeneity exists, not only at the level of the core pro-

Figure 6.48 Arabinogalactan protein patterns in a carrot root. (A) Two sections of a carrot primary root, just above the meristem, after staining to show all of the cells. (B) The same sections stained with an antibody that recognizes an oligosaccharide side chain that is expressed only on AGPs in the epidermis (e) and in future xylem related cells in the central stele. (From Knox et al., *Plant J.* 1:317–326, 1991.)

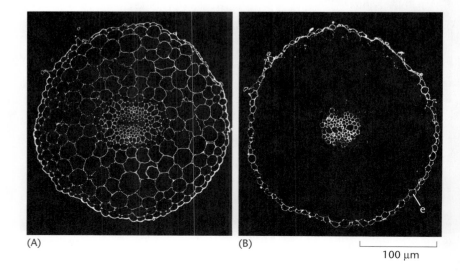

(A)   (B)

100 µm

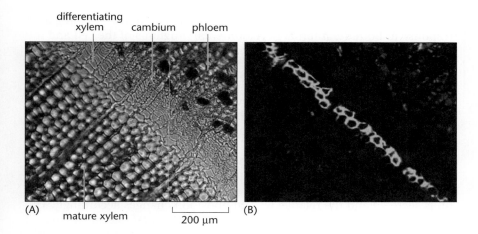

differentiating
xylem   cambium   phloem

(A)   mature xylem   200 µm   (B)

Figure 6.49 Arabinogalactan protein expression in developing xylem. (A) Cells that arise by cell divisions in the cambium of this loblolly pine differentiate into phloem tissue on one side and into xylem tissue on the other. (B) Future xylem vessel elements express a cell wall–located arabinogalactan protein, stained here with a fluorescent antibody, for a short time period during the differentiation process. (Courtesy of Yi Zhang.)

tein, but also among the complex type II arabinogalactan side chains that comprise the bulk of these molecules. It is thought that dynamic cellular heterogeneity exists in the expression patterns of both the core proteins and the glycan side chains. This is further complicated as the same core protein may bear different glycan side chains, and the same glycan side chain may be present on different core proteins. The patterns of AGP expression are dynamic and sometimes quite complex and often appear to be related to cell position rather than to particular cell types. Thus, in Figure 6.48, an antibody that recognizes a specific AGP glycan side chain binds to epidermal cells of the carrot root (which did not label at an earlier stage of development) and a subset of vascular cells (protoxylem and xylem parenchyma). This pattern will change again in the older root. An antibody that recognizes one specific AGP core protein in loblolly pine reveals that its expression is transient, appearing only at one particular stage of a xylem vessel element's formation from a cambial cell (**Figure 6.49**). Experiments in which an artificial gene construct for an AGP core peptide module was expressed in all aerial tissues of *Arabidopsis* provide a remarkable insight into the way in which these molecules are processed. The way in which the prolines in the peptide were hydroxylated and subsequently glycosylated depended on the tissue in which it was expressed. Thus the particular set of glycan epitopes decorating an AGP is dictated by the particular cell it is expressed in rather than the sequence of the particular AGP gene. In addition posttranslational and developmental effects dictate when and where such products survive.

The appearance and disappearance of particular AGPs at the cell surface during development suggests strongly that they are more likely to be involved in cell-cell interactions, or in transient stages of cell wall assembly, than in a structural role. These mysterious wall molecules have yet to yield up all their secrets.

7. A cell wall is the product of both a cell's internal developmental program and its environmental history.[ref29]

The AGPs discussed above are a rather conspicuous example of dynamic changes in the extracellular matrix that occur as a function of developmental time. But cells within a plant also undergo other substantial changes in their walls with the passage of time. Many of these changes are treated in more detail elsewhere, for example, fruit ripening, abscission, pathogen responses, or lignification; but looked at in total they fall into two main categories, those changes that occur as part of normal vegetative or reproductive development, and those that occur as part of adaptive responses to changes in the plant's environment. These two general categories are helpful in analyzing the wall changes observed in individual cell types, and it is often true that the final wall a cell ends up with is a complex and integrated response to both the internal developmental programs of the cell and the external constraints imposed by the environment.

Since all cell walls start life in dividing cells, it is clear that to form, for example, a fully differentiated dead xylem vessel element, considerable changes must be made to the wall in the course of cell differentiation; but other changes, even within the primary wall, are less obvious. The pectic polysaccharides secreted during the construction of the primary wall are usually heavily methylesterified. With time, some of these methylesters may be converted to other esters (see Concept 6A8) and some are progressively lost by deesterification, particularly at three-way junctions and in the middle lamella. Generally, during most forms of secondary thickening, less pectin is secreted and the relative wall composition changes accordingly. The thickening of collenchyma and sieve tube walls, for example, are all part of the normal developmental program of these cell types, but, sadly, remarkably little is known about the molecular basis for how these changes, as the downstream targets of the networks of signals and regulatory genes that determine cell fate in the first place, are effected. Following cell expansion, cell fate decisions have to be coupled to the machinery of cell wall biosynthesis, deposition, and modification that is appropriate to that fate. We have much yet to learn in this area.

On top of these more global temporal changes, other more subtle modifications to the wall may occur. For example, plasmodesmata spanning the wall progressively shift from simple to more complex branching forms with time. Lignin, in both xylem and fibers, shows slow but subtle shifts in monolignol composition as the cells fully differentiate.

Overlaid on these developmental and genetically determined programs are changes in the wall occasioned by the impact of the environment. At one level, these may be manifested as simple growth regulation. Cold, nutrient–starved, and water-stressed plants will be smaller than their siblings in the greenhouse! This does, however, reflect the activity of complex signaling pathways that again have to trigger changes in the cell wall production machinery essential for growth. Other wall changes are more complex and can be quite local. Wounds, inflicted by wind, herbivores or pruning shears, result in local adaptive responses in cell walls. Local suberin and lignin secretion, wound callus formation, and protective gums are all responses that involve reprogrammed production of cell wall materials. Pathogens also often elicit local wall-related defense responses such as callose and lignin production (see Chapter 8). We can see, therefore, that any particular cell wall's final size, composition, and heterogeneity constitute a complex product that reflects both the cell's developmental history and its environmental history.

8. Wall composition and architecture, together with anatomy, contribute to the mechanical properties of plants and their products.[ref30]

Throughout this chapter we have focused on how walls are assembled outside the cell as functional fiber composites that show considerable heterogeneity both within and between different cells. The final end product of all this activity is the body of the growing and developing plant itself, and we, as human beings, have always had a vital interest in the properties of plants and their parts, whether as essential fruit and vegetables, or timber for construction and papermaking. The properties we use and value, and indeed select for, are mostly mechanical. Crisp apples, juicy ripe tomatoes, crunchy carrots, load-bearing wood, and the tough, flexible fibers of cotton and linen all have complex and desirable mechanical properties that in large part can be traced back to the cell walls they contain.

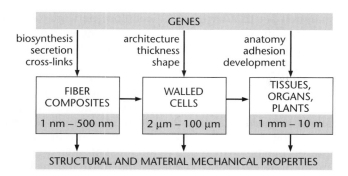

Figure 6.50 Plant mechanical properties relate to scale. Genes involved in growth and development help direct the construction of the plant at many levels of organization, from the nanoscale to the everyday. The mechanical properties determined by structure and composition at one level feed into and help determine the mechanical properties at each succeeding higher level, through cells to tissues and organs, and on to the whole plant.

To explain these *material mechanical properties* in terms of the structural mechanical properties of individual wall molecules such as cellulose has turned out to be immensely difficult, and the main reason is one of scale. Plants are constructed in a hierarchical manner, from single polysaccharide molecules in the nanometer range at one end, through walled cells in the micrometer range, to tissues and organs in the millimeter range, and finally to the whole plant, often in the meter range, at the other end (see Figure 6.33). Mechanical properties at each of these different levels can be measured, but integrating these, and trying to predict properties at scales over 10 orders of magnitude, has proved very difficult, since properties at the polymer level do not always relate directly to properties at the material level (**Figure 6.50**). This is a large and active field of research, and here we shall focus on just a few instructive examples.

Measurements on isolated cell walls or on single cells, while technically feasible, are difficult, and it is not easy to look at wall compositional variation. Luckily, however, there is an ingenious way to fabricate large artificial wall-like composites on which to perform simple mechanical tests. These composites are formed by growing the bacterium *Acetobacter*, strains of which spin out long ribbons of pure cellulose (see Figure 4.30), in flasks of growth medium that contain one or more purified, and well-characterized, plant cell wall polysaccharides. This situation mimics, at least crudely, the situation at the surface of a living plant cell, where cellulose is spun out into a region rich in secreted wall matrix polymers such as xyloglucan and pectin. The cellulosic "pellicle" that is produced in the flask can then be removed and subjected to a variety of physical tests. Some clear, and very significant, results have emerged.

The first is that the cellulose network dominates the major material properties of the composites. The physical arrangement of cellulose fibers and their level of entanglement both affect these properties, but overall, cellulose always takes the load. However, rather remarkably, their material properties can be greatly modified by the presence, and state of, other polymers (**Figure 6.51**). Experiments using hemicelluloses show that xyloglucan

Figure 6.51 The effect of pectin and xyloglucan on cellulose mechanics. Composites, made by growing *Acetobacter* cells in a medium that contains wall polymers, were subjected to uniaxial tensile tests. The stress-strain relationships of the various composites are shown here and demonstrate the large effect that matrix polysaccharides have on the mechanical properties of the composites. (From M.J. Gidley, E. Chanliaud and S. Whitney, in D. Renard, G. Della Valle and Y. Popineau, eds., Plant Biopolymer Science: Food and Non-Food Applications, pp. 39–47. Cambridge, UK: Royal Society of Chemistry, 2002.)

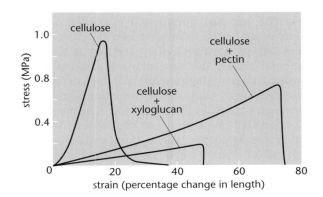

and glucomannan act to cross-link the cellulose network and that in so doing they weaken the composite but also greatly increase its extensibility and lower its stiffness. This strongly suggests that in real walls the cellulose/hemicellulose network provides a desirable balance of strength with extensibility that is not achievable with cellulose alone. Real primary walls, however, have two separate polymer networks, and growing the composites in the presence of pectin has revealed how one network may influence the mechanical properties of the other. With appropriate concentrations of pectin, with a low degree of methylesterification, and in the presence of Ca^{2+}, composites can be formed that retain most of their strength but that again are much more extensible and less stiff. Unlike hemicellulose, however, there is no direct interaction of pectin with the cellulose network, and this raises the question of how the pectin can exert such a major effect on the load-bearing cellulose network. When pectin is removed from the composite, the cellulose network that was laid down in its presence surprisingly retains its altered properties (Figure 6.51). This suggests that depositing cellulose into a preformed pectin network helps to orient the cellulose fibrils and reduce their level of entanglements, thus making it easier to pull them apart on stretching. The dependence of this effect on the degree of pectin esterification and Ca^{2+} level suggests how plants may indirectly influence the mechanical properties and extensibility of cell walls without compromising strength.

In the living plant it is important to remember that some of its mechanical properties, for example, the crispness of a lettuce, depend on turgor pressure. Turgor exerts a considerable force on the surrounding cell wall. In a tissue, turgor forces will tend to separate cells by rounding them up, and these forces are resisted by the strength of the middle lamella "glue" holding cells together and the resistance of the wall to bending. Turgor acts to compress the wall, driving water out. Primary walls commonly have about 75% water content, but pectin makes the wall a hydrated polyelectrolyte network and hydration forces act on this to create considerable network swelling forces that counteract turgor. The drive to swell is dependent on the salt concentration, being highest at low salt, and on the nature of the cation. Isolated walls, at low salt, may swell radially to more than twice their thickness *in vivo*. This process may have important mechanical and physiological consequences. For example, cells in a ripening tomato show a dramatic drop in turgor (from 0.13 MPa to 0.03 MPa), and this is accompanied by a corresponding swelling of the wall that is thought to increase wall porosity and the movement through it of ripening-related wall-degrading enzymes. Wall swelling forces in the absence of turgor also explain why the walls in conventionally fixed plant material always appear thicker in the microscope than those of rapidly frozen material that maintains the plasma membrane closely pushed against the cell wall (**Figure 6.52**).

9. Models of the cell wall have helped us to think more clearly about their construction and function.[ref31]

Throughout this chapter, our emphasis has been on the complexity and heterogeneity of walls. We have seen how heterogeneity arises both between different cell types within a plant and during the different stages of the developmental history of a particular wall. So it is certainly hard to think about constructing meaningful models of the wall that would help us understand their complexity while at the same time showing us how they might be constructed. Clearly, in this context, it is impossible to provide a single, easily digested take-home model of "the cell wall"! However, models are often didactically and pedagogically helpful, and so we present a series of diagrams in **Panel 6.2** that we hope bring together, and model, some of the ideas about wall architecture presented in this chapter.

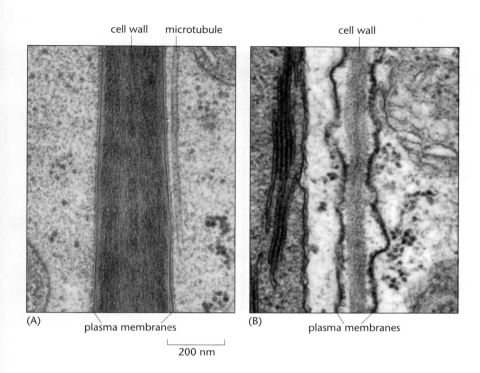

cell wall   microtubule

cell wall

(A)

(B)

plasma membranes

plasma membranes

200 nm

Figure 6.52 The well-fixed wall. These sections of plant material have been fixed either (A) by high pressure freezing/freeze substitution or (B) more conventionally in glutaraldehyde. Only in the cryofixed cells is the plasma membrane still pushed firmly against the wall, as it would be in the turgid living cell.

Historically, the first serious model of the plant cell wall was presented in the literature over 35 years ago, and was based on deductions from the enzymatic deconstruction of the wall of suspension culture cells, together with chemical information about the chemical nature of the cross-links that might be present. This model, framed around the reasonable idea that the wall was a macromolecular complex, suggested that the pectin, hemicellulose, and HRGPs were all cross-linked through covalent bonds and that the cellulose-hemicellulose network was hydrogen-bonded together. Although wrong then in details, all subsequent models have essentially been refinements on this, varying in the extent of covalent cross-links shown, in the way in which they were presented graphically, and in the degree to which issues of architecture and scale were incorporated. The first real attempts to convey something of the complexity and difference between walls were models that emphasized the differences between the walls of grasses and the walls of all other flowering plants, although even here it was clear that the underlying architectural and structural principles are the same.

The models we have shown here in Panel 6.2 reflect the core principles of previous models and build on the structural ideas presented earlier in the chapter, trying as far as possible to draw the various key wall components to scale. The focus is on a young thin primary wall and emphasizes the construction principle of two-component fiber composites. In addition, the heterogeneity emphasized elsewhere is reflected in a representation of the specialized outer wall of a shoot epidermal cell (see also Figure 1.28). It is clear that static two-dimensional drawings will never get us close to the dynamic complexity of real walls, but the thinking behind having to draw such models has been helpful to wall researchers by clarifying some of the questions about scale and assembly of the two underlying wall networks.

What we sadly lack at present is a more sophisticated, computer-generated, iterative model that would take into account the exact size and mechanical properties of the wall components, their stoichiometry, the known cross-links, and the ways in which they are all secreted and assembled. The hope is that such models may in the future be predictive and will provide us with new insights into the relationships between local wall architecture,

THE PRINCIPLE OF THE MODEL

Although they display considerable heterogeneity, all cell walls are built using the same structural or architectural principles. They are fiber-reinforced composite materials. The cellulose-hemicellulose network resists tension, while the coextensive, intermeshed pectin network resists compression and shearing forces. The middle lamella is an adhesive layer sticking adjacent cells together.

In this schematic diagram the cellulose microfibrils are shown aligned at right angles to the direction of cell elongation (*double arrow*).

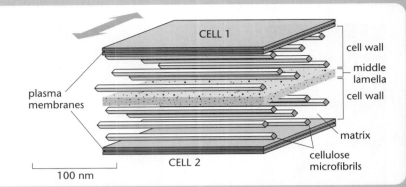

100 nm

THE PRIMARY CELL WALL

The cellulose-hemicellulose network

This is a schematic section through a wall similar to that shown above. Each lamella of cellulose microfibrils is represented by a single microfibril, and all are oriented in the same direction. They are interconnected within each wall by cross-linking hemicellulose molecules that are hydrogen-bonded to their surfaces. The resultant network is resistant to tensions in the wall. Each lamella can be oriented differently with respect to its neighbors depending on the developmental stage and fate of the cell. In each wall shown here, there are only three lamellae, whose total thickness is about the minimum needed to maintain a primary cell wall. In this diagram all the various components are drawn approximately to scale.

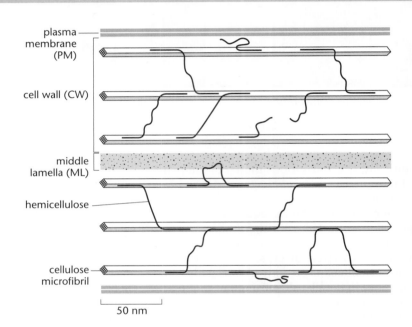

50 nm

The pectic polysaccharide network

The pectic polysaccharides assemble into a coextensive network with the cellulose-hemicellulose network, intermeshing with it to create a water-retentive matrix that resists the forces of compression and shear and controls wall permeability. The pectin is probably anchored to the plasma membrane in some way, through such molecules as Wak proteins. At the middle lamella they form an adhesive mesh involving calcium cross-bridges between homogalacturonans. Further cross-links include the borate diester bridges between RG-II molecules. It is entirely possible that the pectin molecules are far more ordered than is schematically shown here.

THE PRIMARY CELL WALL: STRUCTURAL WALL PROTEINS

Arabinogalactan protein (AGP) distribution

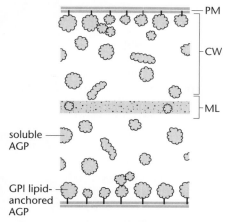

soluble AGP

GPI lipid-anchored AGP

There is an enormous heterogeneity in AGPs, both in kind and in which cells they are expressed. They form a hydrated interface between plasma membrane and cell wall.

Extensin 3 networks (cell plate/early cell wall)

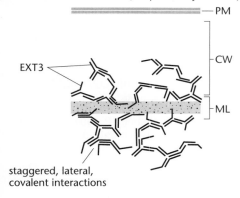

EXT3

staggered, lateral, covalent interactions

Extensin is a hydroxyproline-rich glycoprotein (HRGP).

Wound-extensin network

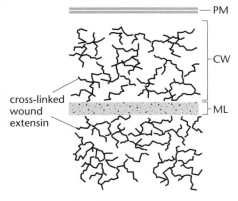

cross-linked wound extensin

Extensins are not found in all cell types, but can be induced in walls by wounding or stress to provide mechanical protection.

THE OUTER EPIDERMAL WALL

All walls are different, some subtly but others more dramatically. While respecting the same overall architectural principles, they vary in the proportions and nature of the individual components to produce considerable heterogeneity such as the lignified walls of xylem vessels. One example of wall specialization is shown here, the primary wall of the outer shoot epidermis, in which this wall, unlike the basal and side walls, is thickened and encrusted with cutin and waxes.

The cellulose-hemicellulose distribution

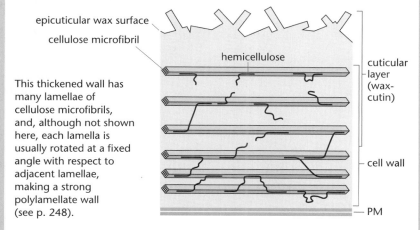

epicuticular wax surface

cellulose microfibril

hemicellulose

cuticular layer (wax-cutin)

cell wall

PM

This thickened wall has many lamellae of cellulose microfibrils, and, although not shown here, each lamella is usually rotated at a fixed angle with respect to adjacent lamellae, making a strong polylamellate wall (see p. 248).

The distribution of AGPs and pectic polysaccharides

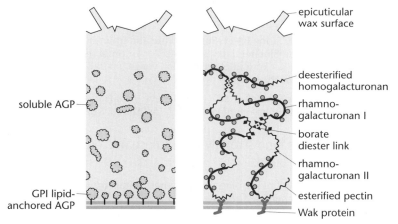

epicuticular wax surface

soluble AGP

deesterified homogalacturonan

rhamno-galacturonan I

borate diester link

rhamno-galacturonan II

esterified pectin

GPI lipid-anchored AGP

Wak protein

The deposition of cutin and waxes

In the outer layers of the epidermal wall, cutin polymers are formed. Subsequently, secreted epicuticular and intracuticular waxes crystallize in distinctive patterns, providing a hydrophobic and waterproof covering for the plant.

cutin

wax

epicuticular wax layer

cutin matrix

intracuticular wax layer

PM

cell-cell adhesion, and whole-plant mechanical properties, and into the dynamic ways in which walls develop and change during cell growth and differentiation.

Key Terms

aerenchyma	isodityrosine
air space	lamella
borate diester	material mechanical property
cellular heterogeneity	microtubule
cellulose-hemicellulose network	middle lamella
cross-linking glycan	passive reorientation
di-isodityrosine	pectin network
diferulic acid	phase separation
directed reorientation	radial heterogeneity
fiber composite material	regional heterogeneity
gel	sugar-ester cross-link
glycosidic bond	template-based assembly
helicoidal architecture	three-way junction
hierarchical construction	

References

[1]Wall polymers are thought to be cross-linked by covalent bonds as well as noncovalent bonds and interactions.

Bacic A, Harris PJ & Stone BA (1988) Structure and function of plant cell walls. The Biochemistry of Plants: A Comprehensive Treatise. (Stumpf PK, Conn EE, Preiss J eds), vol 14, pp 297–371. New York: Academic Press.

McCann MC, Wells B & Roberts K (1990) Direct visualization of cross-links in the primary plant cell wall. *J. Cell Sci.* 96, 323–334.

[2]Hemicelluloses bind strongly to cellulose microfibrils by noncovalent bonds.

Fry SC (1986) Cross-linking of matrix polymers in the growing cell walls of angiosperms. *Annu. Rev. Plant Physiol.* 37, 165–186.

Hayashi T (1989). Xyloglucans in the primary cell wall. *Annu. Rev. Plant Physiol. Plant Mol. Biol.* 40, 139–168.

Levy S, York WS, Stuike-Prill R et al. (1991) Simulation of the static and dynamic molecular conformations of xyloglucan. *Plant J.* 1, 195–215.

Pauly M, Albersheim P, Darvill A & York WS (1999) Molecular domains of the cellulose/xyloglucan network in the cell walls of higher plants. *Plant J.* 20, 629–639.

[3]Hemicelluloses cross-link cellulose microfibrils.

Chanliaud E, De Silva J, Strongitharm B et al. (2004) Mechanical effects of plant cell wall enzymes on cellulose/xyloglucan composites. *Plant J.* 38, 27–37.

Hayashi T (1989) Xyloglucans in the primary cell wall. *Annu. Rev. Plant Physiol. Plant Mol. Biol.* 40, 139–168.

McCann MC, Wells B & Roberts K (1990) Direct visualization of cross-links in the primary plant cell wall. *J. Cell Sci.* 96, 323–334.

Whitney SEC, Brigham JE, Darke AH et al. (1995) *In vitro* assembly of cellulose/xyloglucan networks: ultrastructural and molecular aspects. *Plant J.* 8, 491–504.

[4]Pectic polysaccharides are probably covalently interconnected by glycosidic bonds.

McNeil M, Darvill AG, Fry SC & Albersheim P (1984) Structure and function of the primary cell walls of plants. *Annu. Rev. Biochem.* 53, 625–663.

Mohnen D (2008) Pectin structure and biosynthesis. *Curr. Opin. Plant Biol.* 11, 266–277.

Vincken JP, Schols HA, Oomen RJ et al. (2003) If homogalacturonan were a side chain of rhamnogalacturonan I: implications for cell wall architecture. *Plant Physiol.* 132, 1781–1789.

[5]Homogalacturonans and partially methylesterified homogalacturonans form gels.

Knox JP, Linstead PJ, King J et al. (1990) Pectin esterification is spatially regulated both within cell walls and between developing tissues of root apices. *Planta* 181, 512–521.

Thakur BR, Singh RK & Handa AK (1997) Chemistry and uses of pectin: a review. *Crit. Rev. Food Sci. Nutr.* 37, 47–73.

Willats WGT, Orfila C, Limberg G et al. (2001) Modulation of the degree and pattern of methyl-esterification of pectic homogalacturonan in plant cell walls: implications for pectin methyl esterase action, matrix properties, and cell adhesion. *J. Biol. Chem.* 276, 19404–19413.

[6]Borate esters cross-link RG-II dimers in the wall.

Kobayashi M, Ohno K & Matoh T (1997) Boron nutrition of cultured tobacco BY-2 cells II: characterisation of the boron-polysaccharide complex. *Plant Cell Physiol.* 38, 676–683.

Matoh T (1997) Boron in plant cell walls. *Plant and Soil* 193, 59–70.

O'Neill MA, Eberhard S, Albersheim P & Darvill AG (2001) Requirement of borate cross-linking of cell wall rhamnogalacturonan II for *Arabidopsis* growth. *Science* 294, 846–849.

O'Neill MA, Ishii T, Albersheim P & Darvill AG. (2004) Rhamnogalacturonan II: structure and function of a borate cross-linked cell wall pectic polysaccharide. *Annu Rev. Plant Biol.* 55, 109–139.

O'Neill MA, Warrenfeltz D, Kates K et al. (1996) Rhamnogalacturonan-II, a pectic polysaccharide in the walls of growing plant cells, forms a dimer that is covalently cross-linked by a borate ester. *J. Biol. Chem.* 271, 22923–22930.

Ryden P, Sugimoto-Shirasu K, Smith AC et al. (2003) Tensile properties of Arabidopsis cell walls depend on both a xyloglucan cross-linked microfibrillar network and rhamnogalacturonan II-borate complexes. *Plant Physiol.* 132, 1033–1040.

[7]Diferulic acids probably cross-link wall polysaccharides.

Fry SC (1986) Cross-linking of matrix polymers in the growing cell walls of angiosperms. *Annu. Rev. Plant Physiol.* 37, 165–186.

Iiyama K, Lam TB-T & Stone BA (1994) Covalent cross-links in the cell wall. *Plant Physiol.* 104, 315–320.

Ishii T (1991) Isolation and characterization of a diferuloyl arabinoxylan hexasaccharide from bamboo shoot cell walls. *Carbohydr. Res.* 219, 15–22.

Lindsay SE & Fry SC (2008) Control of diferulate formation in dicotyledonous and gramineous cell-suspension cultures. *Planta* 227, 439–452.

[8]Transesterification may produce other possible cross-links.

Brown JA & Fry SC (1993) Novel *O*-D-galacturonoyl esters in the pectic polysaccharides of suspension-cultured plant cells. *Plant Physiol.* 103, 993–999.

Kim J-B & Carpita NC (1992) Changes in esterification of the uronic acid groups of cell wall polysaccharides during elongation of maize coleoptiles. *Plant Physiol.* 98, 646–653.

McCann MC, Shi J, Roberts K & Carpita N (1994) Changes in pectin structure and localization during the growth of unadapted and NaCl-adapted tobacco cells. *Plant J.* 5, 773–785.

[9]Some wall proteins become insolubilized by covalent cross-links.

Brady JD, Sadler IH & Fry SC (1996) Di-isodityrosine, a novel tetrametric derivative of tyrosine in plant cell wall proteins: a new potential cross-link. *Biochem. J.* 315, 323–327.

Cannon MC, Terneus K, Hall Q et al. (2008) Self-assembly of the plant cell wall requires an extensin scaffold. *Proc. Natl. Acad. Sci. USA* 105, 2226–2231

Domingo C, Saurí A, Mansilla E et al. (1999) Identification of a novel peptide motif that mediates cross-linking of proteins to cell walls. *Plant J.* 20, 563–570.

Iiyama K, Lam TB-T & Stone BA (1994) Covalent cross-links in the cell wall. *Plant Physiol.* 104, 315–320.

[10]Lignin is covalently attached to hemicelluloses, and their interaction adds strength to the secondary cell wall.

Balakshin MY, Capanema EA & Chang H (2007) MWL fraction with a high concentration of lignin-carbohydrate linkages: isolation and 2D NMR spectroscopic analysis. *Holzforschung* 61, 1–7.

Koshijima T & Watanabe T (2003) Association between Lignin and Carbohydrate in Wood and Other Plant Tissues. Berlin: Springer.

Lawoko M, Henriksson G & Gellerstedt G (2005) Structural differences between the lignin-carbohydrate complexes in wood and in chemical pulps. *Biomacromolecules* 6, 3467–3473.

Persson S, Caffall KH, Freshour G et al. (2007) The *Arabidopsis* irregular *xylem8* mutant is deficient in glucuronoxylan and homogalacturonan, which are essential for secondary cell wall integrity. *Plant Cell* 19, 237–255,

[11]Two coextensive polysaccharide networks underlie the structure of the primary cell wall.

Carpita NC & Gibeaut DM (1993) Structural models of primary cell walls in flowering plants: consistency of molecular structure with the physical properties of the walls during growth. *Plant J.* 3, 1–30.

Fry SC (1986) Cross-linking of matrix polymers in the growing cell walls of angiosperms. *Annu. Rev. Plant Physiol.* 37, 165–186.

Fry SC & Miller JG (1989) Towards a working model of the growing plant cell wall: phenolic cross-linking reactions in the primary cell walls of dicotyledons. In The Biosynthesis and Biodegradation of Plant Cell Wall Polymers (NG Lewis, MG Paice eds), pp 33–46. Washington, DC: American Chemical Society.

Jarvis MC (1992) Self assembly of plant cell walls. *Plant Cell Environ.* 15, 1–5.

Kerstens S, Decraemer WF & Verbelen J-P (2001) Cell walls at the plant surface behave mechanically like fiber-reinforced composite materials. *Plant Physiol.* 127, 381–385.

McCann M & Roberts K (1991) Architecture of the primary cell wall. In The Cytoskeletal Basis of Plant Growth and Form (CW Lloyd ed), pp 109–129. London: Academic Press.

Vincent J (1990) Structural Materials, 2nd ed. Princeton, NJ: Princeton University Press.

[12]Walls are constructed of lamellae that are one cellulose microfibril thick.

Carpita NC (1996) Structure and biogenesis of the cell walls of grasses. *Annu. Rev. Plant Physiol. Plant Mol. Biol.* 47, 445–476.

McCann MC & Roberts K (1991) Architecture of the primary cell wall. In The Cytoskeletal Basis of Plant Growth and Form (CW Lloyd ed), pp 109–129. London: Academic Press.

[13]The primary wall is usually composed of only a small number of lamellae.

McCann MC, Wells B & Roberts K (1990) Direct visualization of cross-links in the primary plant cell wall. *J. Cell Sci.* 96, 323–334.

[14]The spacing of cellulose microfibrils is determined by matrix polysaccharides.

Fry SC (1989) The structure and functions of xyloglucan. *J. Exp. Bot.* 40, 1–11.

Hayashi T, Marsden MPF & Delmer DP (1987) Pea xyloglucan and cellulose V: xyloglucan-cellulose interactions *in vitro* and *in vivo. Plant Physiol.* 83, 384–389.

Hayashi T, Ogawa K & Mitsuishi Y (1994) Characterization of the adsorption of xyloglucan to cellulose. *Plant Cell Physiol.* 35, 1199–1205.

McCann MC & Roberts K (1991) Architecture of the primary cell wall. In The Cytoskeletal Basis of Plant Growth and Form (CW Lloyd ed), pp 109–129. London: Academic Press.

Valent BS & Albersheim P (1974) The structure of plant cell walls V: on the binding of xyloglucan to cellulose fibers. *Plant Physiol.* 54, 105–108.

Whitney SEC, Gothard MGE, Mitchell JT & Gidley MJ (1999) Roles of cellulose and xyloglucan in determining the mechanical properties of primary plant cell walls. *Plant Physiol.* 121, 657–663.

[15]The walls of cells in which cellulose synthesis has been inhibited are composed largely of a pectin network.

Hématy K, Sado PE, Van Tuinen A et al. (2007) A receptor-like kinase mediates the response of *Arabidopsis* cells to the inhibition of cellulose synthesis. *Curr Biol.* 17, 922–931.

Manfield LW, Orfila C, McCartney L et al. (2004) Novel cell wall architecture of isoxaben-habituated *Arabidopsis* suspension-cultured cells: global transcript profiling and cellular analysis. *Plant J.* 40, 260–275.

Shedletzky E, Shmuel M, Delmer DP & Lamport DTA (1990) Adaptation and growth of tomato cells on the herbicide 2,6-dichlorobenzonitrile leads to production of unique cell walls virtually lacking a cellulose-xyloglucan network. *Plant Physiol.* 94, 980–987.

Wells B, McCann MC, Shedletzky E et al. (1994) Structural features of cell walls from tomato cells adapted to grow on the herbicide 2,6-dichlorobenzonitrile. *J. Microscopy* 173, 155–164.

[16]The pectin network limits the porosity of the wall.

Baron-Epel O, Gharyal PK & Schindler M (1988) Pectins as mediators of wall porosity in soybean cells. *Planta* 175, 389–395.

Carpita NC, Sabularse D, Montezinos D & Delmer DP (1979) Determination of the pore size of cell walls of living plant cells. *Science* 205, 1144–1147.

Fleischer A, O'Neill MA & Ehwald R (1999) The pore size of non-graminaceous plant cell walls is rapidly decreased by borate-ester cross-linking of the pectic polysaccharide rhamnogalacturonan II. *Plant Physiol.* 121, 829–838.

Fujino T & Itoh T (1998) Changes in pectin structure during epidermal cell elongation in pea (*Pisum sativum*) and its implications for cell wall architecture. *Plant Cell Physiol.* 39, 1315–1323.

McCann MC, Wells B & Roberts K (1990) Direct visualization of cross-links in the primary plant cell wall. *J. Cell Sci.* 96, 323–334.

O'Driscoll D, Read SM & Steer MW (1993) Determination of cell-wall porosity by microscopy: walls of cultured cells and pollen tubes. *Acta Bot. Neerl.* 42, 237–244.

[17]The cell controls the thickness of its wall.

Carpita NC (1996) Structure and biogenesis of the cell walls of grasses. *Annu. Rev. Plant Physiol. Plant Mol. Biol.* 47, 445–476.

Jarvis MC (1992) Control of thickness of collenchyma cell walls by pectin. *Planta* 187, 218–220.

Somerville C, Bauer S, Brininstool G et al. (2004) Toward a systems approach to understanding plant cell walls. *Science* 306, 2206–2211,

[18]How proteins are integrated into wall architecture is unclear.

Brisson LF, Tenhaken R & Lamb C (1994) Function of oxidative cross-linking of cell wall structural proteins in plant disease resistance. *Plant Cell* 6, 1703–1712.

Fowler TJ, Bernhardt C & Tierney ML (1999) Characterization and expression of four proline-rich cell wall protein genes in *Arabidopsis* encoding two distinct subsets of multiple domain proteins. *Plant Physiol.* 121, 1081–1091.

Keller B (1993) Structural cell wall proteins. *Plant Physiol.* 101, 1127–1130.

Ryser U (2003) Protoxylem: the deposition of a network containing glycine-rich cell wall proteins starts in the cell corners in close association with the pectins of the middle lamella. *Planta* 216, 854–864.

Showalter A (1993) Structure and function of cell wall proteins. *Plant Cell* 5, 9–23.

Ye Z-H, Song Y-R, Marcus A & Varner JE (1991) Comparative localization of three classes of cell wall proteins. *Plant J.* 1, 175–183.

[19]The orientation of newly synthesized microfibrils is determined by the cell.

Akashi T, Kawasaki S & Shibaoka H (1990) Stabilization of cortical microtubules by the cell wall in cultured tobacco cells. *Planta* 182, 363–369.

Chan J, Sambade A, Calder G & Lloyd C (2009) *Arabidopsis* cortical microtubules are intiated along, as well as branching from, existing microtubules. *The Plant Cell* 21, 2298–2306.

Ehrhardt D (2008) Straighten up and fly right: microtubule dynamics and organization of non-centrosomal arrays in higher plants. *Curr. Opin. Cell Biol.* 20, 107–116.

Giddings TH & Staehelin LA (1991) Microtubule-mediated control of microfibril deposition: a re-examination of the hypothesis. In The Cytoskeletal Basis of Plant Growth and Form (CW Lloyd ed), pp 85–89. London: Academic Press.

Lloyd CW, Drobak BK, Dove SK & Staier CJ (1996) Interactions between the plasma membrane and the cytoskeleton in plants. In Membranes: Specialized Functions in Plants (M Smallwood et al. eds), pp 1–20. Oxford: BIOS Scientific Publishers.

Paredez AR, Persson S, Ehrhardt DW & Somerville CR (2008) Genetic evidence that cellulose synthase activity influences microtubule cortical array organization. *Plant Physiol.* 147, 1723–1734.

[20]The orientation of microfibrils within a wall may change during growth.

Chan J, Calder G, Fox S & Lloyd C (1994) Cortical microtubule arrays undergo rotary movements in *Arabidopsis* hypocotyl epidermal cells. *Nature Cell Biol.* 9, 171–175.

Hogetsu T (1986) Orientation of wall microfibril deposition in root cells of *Pisum sativum* L. var. Alaska. *Plant Cell Physiol.* 27, 947–951.

Lloyd C & Chan J (2004) Microtubules and the shape of plants to come. *Nat. Rev. Mol. Cell Biol.* 5, 13–22.

McCann MC & Roberts K (1994) Changes in cell wall architecture during cell wall elongation. *J. Exp. Bot.* 45, 1683–1691.

Preston RD (1982) The case for multinet growth in growing walls of plant cells. *Planta* 155, 356–363.

Roelofsen PA (1965) Ultrastructure of the wall in growing cells and its relation to the direction of growth. *Adv. Bot. Res.* 2, 69–149.

Somerville C (2006) Cellulose synthesis in higher plants. *Annu. Rev. Cell Dev. Biol.* 22, 53–78.

Suslov D & Verbelen JP (2006) Cellulose orientation determines mechanical anisotropy in onion epidermis cell walls. *J Exp. Bot.* 57, 2183–2192.

[21]The self-ordering properties of polylamellate walls can lead to high degrees of structural order.

Jarvis MC (1992) Self assembly of plant cell walls. *Plant Cell Environ.* 15, 1–5.

MacDougall AJ, Rigby NM & Ring SG (1997) Phase separation of plant cell wall polysaccharides and its implications for cell wall assembly. *Plant Physiol.* 114, 353–362.

Reis D, Vian B, Chanzy H & Roland J-C (1991) Liquid crystal-type assembly of native cellulose-glucuronoxylans extracted from plant cell wall. *Biol. Cell* 73, 173–178.

Roland J-C, Reis D, Vian B & Roy S (1989) The helicoidal plant cell wall as a performing cellulose-based composite. *Biol. Cell* 67, 209–220.

Vian B, Roland J-C & Reis D (1993) Primary cell wall texture and its relation to surface expansion. *Int. J. Plant Sci.* 154, 1–9.

[22]Cell wall assembly is a hierarchical process.

Chanliaud E & Gidley MJ (1999) *In vitro* synthesis and properties of pectin/*Acetobacter xylinus* cellulose composites. *Plant J.* 20, 25–35.

Gidley MJ, Chanliaud E & Whitney S (2002) Influence of polysaccharide composition on the structure and properties of cellulose-based composites. In Plant Biopolymer Science: Food and Non-Food Applications (D Renard, GD Valle, Y Popineau eds), pp 39–47. Cambridge: Royal Society of Chemistry.

National Research Council (1997) Hierarchical Structures in Biology as a Guide for New Materials Technology, pp 1–130. Washington, DC: National Academy Press.

[23]Polymers are not evenly distributed within a wall.

Freshour G, Clay RP, Fuller MS et al. (1996) Developmental and tissue-specific structural alterations of the cell-wall polysaccharides of *Arabidopsis thaliana* roots. *Plant Physiol.* 110, 1413–1429.

Knox JP (2008) Revealing the structural and functional diversity of plant cell walls. *Curr. Opin. Plant Biol.* 11, 308–313.

Knox JP, Linstead PJ, King V et al. (1990) Pectin esterification is spatially regulated both within cell walls and between developing tissues of root apices. *Planta* 181, 512–521.

McCann MC, Wells B & Roberts K (1992) Complexity in the spatial localization and length distribution of plant cell-wall matrix polysaccharides. *J. Microscopy* 166, 123–136.

Roberts K (1989) The plant extracellular matrix. *Curr. Opin. Cell Biol.* 1, 1020–1027.

Swords KMM & Staehelin LA (1993) Complementary immunolocalization patterns of cell wall hydroxyproline-rich glycoproteins studied with the use of antibodies against different carbohydrate epitopes. *Plant Physiol.* 102, 891–901.

Ye Z-H, Song Y-R, Marcus A & Varner JE (1991) Comparative localization of three classes of cell wall proteins. *Plant J.* 1, 175–183.

[24]The middle lamella forms an adhesive boundary between adjacent cells.

Fry SC (1986) Cross-linking of matrix polymers in the growing cell walls of angiosperms. *Annu. Rev. Plant Physiol.* 37, 165–186.

Jarvis MC, Briggs SPH & Knox JP (2003) Intercellular adhesion and cell separation in plants. *Plant Cell Environ.* 26, 977–989.

Mazz Marry M, Roberts K, Jopson SJ et al. (2006) Cell-cell adhesion in fresh sugar-beet root parenchyma requires both pectin esters and calcium cross-links. *Physiologia Plantarum* 126, 243–256.

[25]Three-way junctions are rich in protein, pectin, and phenolics.

Knox JP, Linstead PJ, King J et al. (1990) Pectin esterification is spatially regulated both within cell walls and between developing tissues of root apices. *Planta* 181, 512–521.

Roberts K (1990) Structures at the plant cell surface. *Curr. Opin. Cell Biol.* 2, 920–928.

Waldron KW, Smith AC, Parr AJ et al. (1997) New approaches to understanding and controlling cell separation in relation to fruit and vegetable texture. *Trends Food Sci. Technol.* 8, 213–221.

Willats WGT, McCartney L, Mackie W & Knox JP (2001) Pectin: cell biology and prospects for functional analysis. *Plant Mol. Biol.* 47, 9–27.

[26]The middle lamella is involved in cell separation.

Jeffree CE, Dale JE & Fry SC (1986) The genesis of intercellular spaces in developing leaves of *Phaseolus vulgaris* L. *Protoplasma* 132, 90–98.

Kato Y, Yamamouchi H, Hinata K et al. (1994) Involvement of phenolic esters in cell aggregation of suspension-cultured rice cells. *Plant Physiol.* 104, 147–152.

Knox JP (1992) Cell adhesion, cell separation, and plant morphogenesis. *Plant J.* 2, 137–141.

Kolloffel C & Linssen PWT (1984) The formation of intercellular spaces in the cotyledons of developing and germinating pea seeds. *Protoplasma* 120, 12–19.

Roberts JA & Gonzalez-Carrenza Z eds. (2007) Plant Cell Separation and Adhesion Ann. Plant Reviews vol. 25. Oxford: Blackwell.

Waldron KW, Smith AC, Parr AJ. et al. (1997) New approaches to understanding and controlling cell separation in relation to fruit and vegetable texture. *Trends Food Sci. Technol.* 8, 213–221.

[27]The intercellular spaces form a continuum.

Mühlenbock P, Plaszczyca M, Plaszczyca M et al. (2007) Lysigenous aerenchyma formation in *Arabidopsis* is controlled by *LESION SIMULATING DISEASE1*. *Plant Cell* 19, 3819–3830.

Saab IN & Sachs MM (1996) A flooding-induced xyloglucan *endo*-transglycosylase homolog in maize is responsive to ethylene and associated with aerenchyma. *Plant Physiol.* 112, 385–391.

Seago JL Jr, Peterson CA & Enstone DE (2000) Cortical development in roots of the aquatic plant *Pontederia cordata* (Pontederiaceae). *Am. J. Bot.* 87, 1116–1127.

[28]The expression of arabinogalactan proteins is developmentally regulated.

Estévez JM, Kieliszewski MJ, Khitrov N & Somerville C (2006) Characterization of synthetic hydroxyproline-rich proteoglycans with arabinogalactan protein and extensin motifs in *Arabidopsis. Plant Physiol.* 142, 458–470.

Knox JP (2006) Up against the wall: arabinogalactan-protein dynamics at cell surfaces. *New Phytol.* 169, 443–445.

Knox JP, Linstead PJ, Peart JM et al. (1991) Developmentally regulated epitopes of cell surface arabinogalactan proteins and their relation to root tissue pattern formation. *Plant J.* 1, 317–326.

Nothnagel EA, Bacic A & Clarke AE (eds) (2001) Cell and Developmental Biology of Arabinogalactan-Proteins. Dordrecht, Netherlands: Kluwer Aademic Publishers/Plenum.

Pennell RI, Janniche L, Kjellbom P et al. (1991) Developmental regulation of a plasma membrane arabinogalactan protein epitope in oilseed rape flowers. *Plant Cell* 3, 1317–1326.

Seifert GJ & Roberts K (2007) The biology of arabinogalactan proteins. *Annu. Rev. Plant. Biol.* 58, 137–161.

[29]A cell wall is the product of both a cell's internal developmental program and its environmental history.

Knox JP (2008) Revealing the structural and functional diversity of plant cell walls. *Curr. Opin. Plant Biol.* 11, 308–313.

Verhertbruggen Y, Marcus SE, Haeger A et al. (2009) Developmental complexity of arabinan polysaccharides and their processing in plant cell walls. *Plant J.* 59, 413–425.

[30]Wall composition and architecture, together with anatomy, contribute to the mechanical properties of plants and their products.

Abasolo W, Eder M, Yamauchi K et al. (2009) Pectin may hinder the unfolding of xyloglucan chains during cell deformation: implications of the mechanical performance of *Arabidopsis* hypocotyls with pectin alterations. *Molec. Plant* 2, 990–999.

Chanliaud E & Gidley MJ (1999) *In vitro* synthesis and properties of pectin/*Acetobacter xylinus* cellulose composites. *Plant J.* 20, 25–35.

Jarvis MC (1998) Intercellular separation forces generated by intracellular pressure. *Plant Cell Environ.* 21, 1307–1310.

MacDougall AJ, Rigby NM, Ryden P et al. (2001) Swelling behavior of the tomato cell wall network. *Biomacromolecules* 2, 450–455.

Neville AC (1993) Biology of Fibrous Composites: Development beyond the Cell Membrane. Cambridge, UK: Cambridge University Press.

Preston RD (1974) The Physical Biology of Plant Cell Walls. London: Chapman & Hall.

Vincent JFV (2007) Plant Biomechanics. Handbook of Plant Science. Chichester, UK: John Wiley & Sons.

Wei C & Lintilhac PM (2007) Loss of stability: a new look at the physics of cell wall behavior during plant cell growth. *Plant Physiol.* 145, 763–772.

Whitney SEC, Gothard MGE, Mitchell JT & Gidley MJ (1999) Roles of cellulose and xyloglucan in determining the mechanical properties of primary plant cell walls. *Plant Physiol.* 121, 657–663.

[31]Models of the cell wall have helped us to think more clearly about their construction and function.

Baa K (2006) Models of plant cell walls. The Science and Lore of the Plant Cell Wall: Bioynthesis, Structure and Function (T Hayashi ed), pp 3–10. Boca Raton, Florida: BrownWalker Press.

Carpita NC & Gibeaut DM (1993) Structural models of primary cell walls in flowering plants: consistency of molecular structure with the physical properties of the walls during growth. *Plant J.* 3, 1–30.

Hayashi T (1989) Xyloglucans in the primary cell wall. *Annu. Rev. Plant Physiol. Plant Mol. Biol.* 40, 139–168.

Keegstra K, Talmadge KW, Bauer WD & Albersheim P (1973) The structure of plant cell walls III: a model of the walls of suspension-cultured sycamore cells based on the interconnections of the macromolecular components. *Plant Physiol.* 51, 188–197.

McCann M & Roberts K (1991) Architecture of the primary cell wall. In The Cytoskeletal Basis of Plant Growth and Form (CW Lloyd ed), pp 109–129. London: Academic Press.

The Cell Wall in Growth and Development

7

In Chapter 6 we examined how the various polymers of the wall are assembled outside the cell to form a coherent structure. And we saw how different kinds of walls are built using similar architectural principles. The rich diversity of walls is made even more complex when we take into account the heterogeneity within and between walls, even within a single cell or tissue type. The picture presented there was a relatively static one, but of course plants are living organisms whose key feature is their amazing dynamic growth and development. Bamboo can have a growth rate of four feet a day!

In this chapter, we now consider how the architecture of the wall accommodates the dynamic processes of cell growth and differentiation. We start with a consideration of the complex but vital interactions between the cytoskeletal structures within the cell and the wall without. It is these interactions, initiated at the cell's behest, together with subtle and still not well-understood changes in the wall, which underpin cellular growth and expansion. Once laid down, and influenced by both internal and external signals, the wall of the fully grown cell may subsequently change dramatically during cell differentiation, as it constructs an appropriate functional enclosure for the demands of different cell types. The wall's dynamic behavior does not rest there; we then discuss how walls are remodeled, turned over, and finally broken down. Finally, we look briefly at a topic we shall return to in the next chapter, the ways in which plant cells have adapted to, recognize, and respond to the appearance of fragments of carbohydrate polymers in their walls.

A. Interactions Between the Cytoskeleton and the Wall

1. Cortical microtubule reorientation changes the orientation of cellulose microfibril deposition during growth.[ref1]

The final shape of a growing cell is determined by the turgor-driven expansion of its primary cell walls. As already described in Chapter 6, this process is controlled primarily by the innermost layers of cellulose fibrils in the growing walls, whose direction at the time of their deposition generally appears to be correlated with the orientation of cortical microtubules. Support for this hypothesis has come from studies in which a genetically or experimentally induced loss of microtubules has led to the formation of cell walls containing more randomly oriented cellulose fibrils and to isodiametrically shaped cells. Similarly, when the orientation of microtubules is altered by growth factors or light treatments, corresponding changes in cellulose fibril

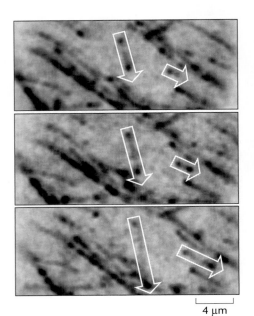

4 µm

Figure 7.1 Microtubule dynamics.
The growth of microtubules from cortical nucleation sites is shown here in a sequence of images from a living *Arabidopsis* cell. The cell is expressing a microtubule end-binding protein (AtEB1a) fused to GFP. Two growing microtubules are indicated. The relatively static end, where the microtubule is nucleated, is the minus end, while the rapidly growing end is the plus end. The time between successive frames is 70 seconds. Such probes have revealed the dynamics of the plant cytoskeleton. (From J. Chan et al., *Nat. Cell Biol.* 5:967–971, 2003.)

orientation and cell shape are seen. These findings raise the question of how cells control the orientation and reorientation of cortical microtubules during growth, and whether microtubules are required throughout cellulose fibril biosynthesis to ensure their organized deposition.

During cell wall formation, microtubule guidance of rosette complexes in the plasma membrane is needed to reorient fibril deposition in successive cell wall lamellae. Once a given set of rosette complexes has begun depositing cellulose fibrils in a new direction, subsequent deposition of oriented fibrils may no longer require the continued presence of microtubules, at least over short periods of time. However, as the rosette complexes turn over (with about a 10-min half-life), microtubules may be required for the rechanneling of new complexes as new lamellae, with differently oriented fibrils, are initiated.

While no information is yet available on the mechanism(s), either molecular or biophysical, that control the absolute orientation of cortical microtubules, some intriguing clues have emerged about the events associated with the reorientation of microtubule arrays. The most direct approach is to study microtubule orientations in living cells, by injecting fluorescently labeled tubulin subunits (for example, rhodamine-conjugated tubulin) into cells and following the behavior over time of microtubules containing these markers. The use of transgenic plants expressing translational fusions, between either tubulin or microtubule associated proteins and a fluorescent protein (YFP, GFP), has greatly facilitated the study of the cytoskeletons of large numbers of cells over time (**Figure 7.1**). Due to the dynamic instability and rapid turnover of cortical microtubules, the fluorescent tubulin molecules are rapidly incorporated into the microtubules, whose dynamics can then be monitored over several hours. In pea epidermal cells, cortical microtubule reorientation is brought about by the formation of local groups of discordant microtubules that anticipate the new alignment. These increase in number as the existing microtubules depolymerize and are replaced by microtubules oriented in the new direction. During this process several alignments may coexist before the final new alignment is achieved, and individual microtubules appear to be nucleated at random over the cell surface. It is likely that simple self-assembly mechanisms play a role in the local ordering of the microtubule array, probably reinforced through cross-bridges of a *microtubule-associated protein (MAP)* such as MAP65 (**Figure 7.2**).

It is important to emphasize that although some microtubules can be found deeper within the plant cell, it is the cortical array alone that appears to be intimately involved in wall deposition. The majority of the cell's microtubules reside in this dynamic *cortical microtubule array*, and they are defined by being closely associated with the inner face of the plasma membrane, maintaining a relatively constant distance from it of between 10 and 20 nm. This suggests the involvement of linker proteins that connect the microtubules with membrane-associated proteins, but we are currently short of candidates for this role (**Figure 7.3**).

Two remarkable and unexpected features of cortical microtubule behavior have emerged from the careful monitoring of their dynamics in live cells as a function of time. The first is that what appears superficially to be a single, coherent, and well-ordered cortical array may in fact be a complex mosaic of individually polarized domains, somewhat like the panels on the surface of a soccer ball. Indeed, considering that it is physically impossible to cover the surface of a cell with a single coherent set of parallel lines, without introducing dislocations, we should perhaps have been more alert to this notion! The second is that the whole microtubule array undergoes slow rotary movements, both clockwise and counterclockwise.

The images in **Figure 7.4** illustrate both of these new principles. Microtubules are imaged here by following growth at their plus ends using a GFP-labeled

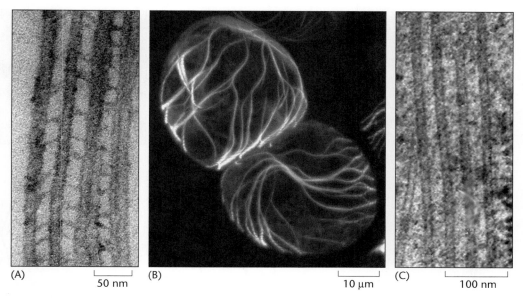

(A)   (B)   (C)
          50 nm   10 μm   100 nm

Figure 7.2 Microtubules can be cross-bridged to form parallel groups of cortical microtubules. (A) Purified brain microtubules have been incubated with the filamentous microtubule-associated protein MAP65 from carrot. Parallel groups of microtubules are seen, connected by 25- to 30-nm cross-bridges of MAP65. (B) If MAP65, as a fusion protein with GFP, is expressed at high levels in *Arabidopsis* cells in suspension culture, the microtubules are cross-linked by the MAP65 to form long, thick bundles. (C) Grazing section of cortical microtubules just below the plasma membrane in an *Arabidopsis* root cell. The clear parallel arrangement of microtubules *in planta,* and the spacing between them, closely resembles the *in vitro* result shown in (A). (A, courtesy of J. Chan and C. Lloyd. B, courtesy of J. Chan.)

microtubule end-binding protein, EB1. Spinning-disk confocal microscopy allows images to be recorded over long periods of time. Before describing the observations, an important distinction must be made between, on the one hand, the dynamic growth and disassembly of individual microtubules and, on the other hand, the higher-order feature of self-sustaining tracks along which the individual microtubules appear to move, usually all in the same direction. The microtubules underlying the outer epidermal wall of an *Arabidopsis* hypocotyl appear to form one coherent array (Figure 7.4A). The paths of individual microtubules, however, show that the array is in fact a mosaic of distinct domains, within which microtubule tracks are reasonably well aligned, and in which most microtubules are polarized or moving in the same direction. These domains are shown schematically in Figure 7.4C. These domains and their constituent tracks, along which individual microtubules are growing at their plus ends at about 6 μm/min, themselves migrate, slowly rotating around the wall at about 0.3 μm/min, in one direction or the other. It takes between about 4 and 8 hours to make a full 360° turn, but the direction of rotation can also periodically reverse.

If cellulose microfibrils are being deposited in line with the tracks of microtubules as they slowly migrate around, this might help to explain the

Figure 7.3 Cortical microtubules. (A) In this section of a freeze-substituted plant cell, a cortical microtubule can be seen, closely following the contour of the plasma membrane and maintaining a constant distance of about 20 nm from it. (B) Two microtubules are seen here in cross section. There are no centrosome-based microtubule organizing centers in plants, and microtubules appear to be initiated at many sites at the cell surface during interphase by nucleating sites thought to contain centrin and other plant-related versions of animal centrosomal proteins. (A, courtesy of Byung-Ho Kang.)

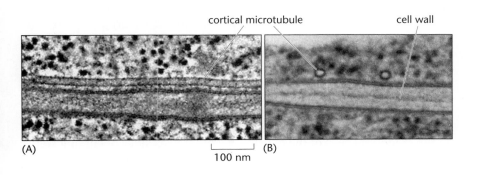

cortical microtubule   cell wall

(A)   (B)
          100 nm

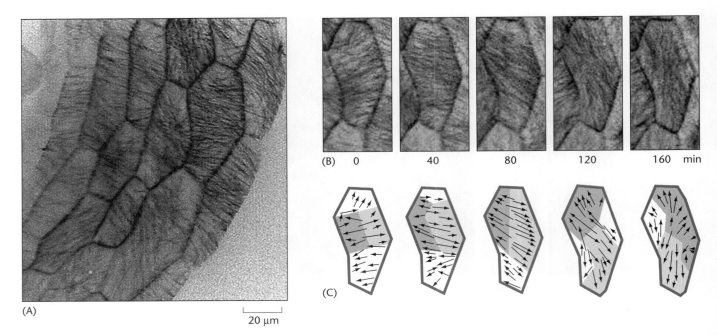

Figure 7.4 Microtubule group behavior over time. (A) Cortical microtubules underlying the outer epidermal wall of a hypocotyl are imaged, using AtEB1-GFP, as a time projection over 6 min. (B) Subsequent images of a selected cell are shown at time intervals of 40 min. (C) Interpretative diagrams record the presence of a mosaic of polarized domains. Each domain consists of tracks in which the vast majority of microtubules grow, and tracks migrate, in a common direction (*arrows*). Two domains, shown in gray, can be seen to steadily rotate clockwise around the cell over the period of 160 min recorded. (Adapted from J. Chan et al., *Nat. Cell Biol.* 9:171–175, 2007.)

complex arrangement of successive lamellae found within the helicoidal walls discussed earlier (see Figure 6.31). The outer epidermal wall is indeed a wall with helicoidal architecture.

2. Intracellular factors can alter the orientation of cortical microtubules.[ref2]

Growth factors are intimately involved in all aspects of plant growth and differentiation. They also serve as mediators of environmental stimuli that affect growth. Since they act at the cellular level, this implies that the orientation of cortical microtubules may also be under the control of growth factors, and this is indeed the case. *Gibberellic acid (GA₃)*, ethylene, and auxin all exert their effects on cell expansion, at least in part, by altering cortical microtubule-dependent orientation of cellulose microfibrils, as discussed in previous sections.

For example, downstream of the GA_3 signal transduction pathway that leads to enhanced elongation growth is the reorientation of cortical microtubules into a transverse array and the subsequent deposition of cellulose fibrils in the same direction (**Figure 7.5**). Drug-induced disruption of microtubules with colchicine, and the inhibition of cellulose synthesis, both prevent plants from displaying the GA_3-dependent growth response. The stunted growth of ethylene-treated seedlings caused by enhanced radial cell expansion has been correlated with the realignment of cortical microtubules parallel to the seedling's longitudinal axis (**Figure 7.6**). Similarly, the auxin-mediated tropic responses (gravitropism, phototropism) have been shown to correlate with reciprocal reorientations of cortical microtubules in the epidermal cells of the concave versus convex sides of curving stems and roots.

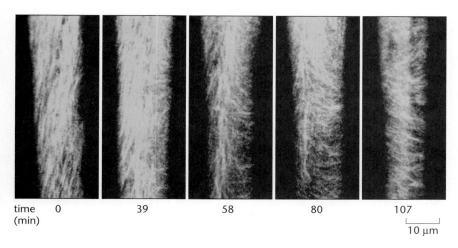

time (min)   0   39   58   80   107

10 μm

Figure 7.5 Gibberellic acid–induced microtubule realignment. In these sequential images of a single live pea epidermal cell, the rearrangement of fluorescently labeled microtubules is seen to proceed from net longitudinal to net transverse during the couple of hours following addition of GA₃. (Adapted from C. Lloyd et al., *J. Microsc.* 181:140–144, 1996.)

Little is known about the molecular mechanisms that bring about these *microtubule reorientation responses*. In principle, growth factor signaling could alter the molecules that affect microtubule dynamics; could change the properties or types of microtubule-associated proteins (MAPs) that mediate binding of microtubules to each other, to the plasma membrane, and to other membranous structures; or could even modify microtubules themselves. In one study of the effects of GA₃ on pea microtubules, it was found that the elongation response involved the detyrosination of one of the isoforms of α-tubulin. In animal cells, tyrosination or detyrosination of α-tubulin has been linked to changes in microtubule stability, with loss of tyrosine increasing stability.

3. Bundles of microtubules can define distinct domains of the wall that will thicken during cell development.[ref3]

Xylem vessel and tracheid differentiation is an elaborate example of cellular morphogenesis that has attracted much interest as a model for cell differentiation. Probably the most striking event in this process is the deposition in tracheary elements of secondary cell wall thickenings as annular rings, helical bands, or reticulate or pitted wall layers (see Figure 1.40). One of the first recognizable changes in cell morphology is the clustering of cortical microtubules and actin filaments into bands that precisely predict the location and orientation of the secondary cell wall thickenings (**Figure 7.7**). Simultaneously, cortical ER membranes become closely associated with the plasma membrane in domains between the microtubule bundles, thereby restricting secretory vesicle access solely to the plasma membrane regions where the secondary cell wall layers are to be deposited. Since these vesicles also deliver the rosette complexes to the cell surface, cellulose synthesis becomes confined to the plasma membrane domains defined by the cortical microtubule bundles (**Figure 7.8**). In turn, the parallel-oriented cellulose fibrils appear to serve as templates for the sequential assembly of all of the other secondary cell wall components in the thickenings, including xylans and lignin.

The local positioning of wall thickenings by the cytoskeleton can be seen in primary walls as well as in secondary walls. Hoops of thicker wall are laid down early in developing leaf mesophyll cells in positions that are defined

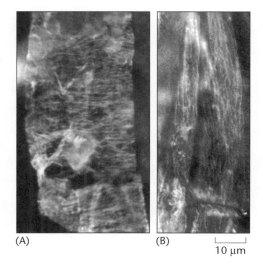

(A)   (B)

10 μm

Figure 7.6 Cortical microtubules in young pea epicotyls adopt a new orientation in response to ethylene treatment. In this case microtubules without ethylene are net transverse (A), but after 30 minutes' treatment with ethylene begin to polymerize at an angle and finally end up net longitudinal (B). (From I. Roberts et al., *Planta* 164:439–447, 1985.)

Figure 7.7 Microtubule arrangement in differentiating xylem elements.
(A) Secondary wall thickenings in a developing xylem element are overlaid by a bundled array of cortical microtubules.
(B) Immunofluorescent staining of the bundles of microtubules in a young tracheary element isolated from the stem of a plant. (C) Isolated leaf cells from some plants, in this case *Zinnia*, can be induced to transdifferentiate into tracheids in liquid culture. This fully differentiated and dead cell has deposited lignin, seen here by its autofluorescence, in exactly the same areas as the cellulosic secondary wall thickenings were deposited, just as tracheids in intact plants do.

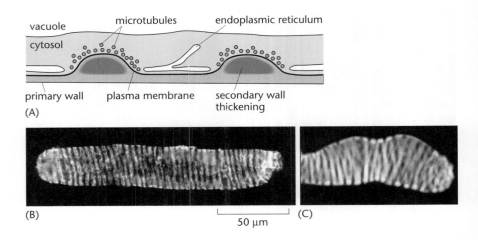

(A)

(B)   50 μm   (C)

by cortical bundles of microtubules and a fine actin meshwork (**Figure 7.9**). These bands act rather the way strips of sticky tape do if stuck around a balloon that continues to be inflated. As the cell continues to expand, bulges appear between the wall thickenings, separating the cells at these points and helping to create the air spaces that surround the characteristically bulging mature mesophyll cells. It is noteworthy, as can be seen in Figure 7.9A, that the position of the microtubule bundle in one cell is accurately reflected in the adjacent cell. This ensures, of course, that the expanding bulges remain connected with those of neighbors, and that the cells separate at the wall thickenings to create the air spaces. This remarkable cooperation between mesophyll cells is an example of a more general phenomenon, as we now discuss.

4. The arrangement of cytoskeletal elements in one cell often relates to that in a neighboring cell.[ref4]

Since plant cells are firmly attached to their neighbors, they are obliged to cooperate with each other as they expand and differentiate. It is not surprising, therefore, that the arrangement and dynamics of the cytoskeleton in one cell are often coordinated with those in its neighbors. This can be true for both the microtubule-based and the actin-based cytoskeleton.

A good example of the former can be seen in the wound-induced differentiation of xylem cells in a cut stem (**Figure 7.10**). Long before microtubules were discovered, it was observed that as the cortical cells began to transdifferentiate into tracheary elements, thick bands of cytoplasm formed

Figure 7.8 Cellulose-synthesizing rosettes are restricted to regions of wall thickening in differentiating xylem elements. Freeze fracture images of developing xylem elements in roots of cress seedlings were used to map the position of rosettes in relation to regions of wall thickening. (A) Distribution of rosettes in one region of a cell's plasma membrane. The gray shading marks the band of wall thickening, defined by the bundled cortical microtubules, in which all the rosettes are found. (B) The density of rosettes can be very high, almost 200 in 1 μm², and the grouping of them shown here confirms that they can be touching each other. (Adapted from W. Herth, *Planta* 164:12–21, 1985.)

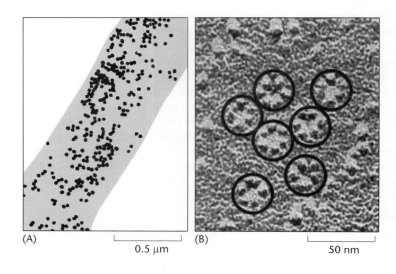

(A)   0.5 μm   (B)   50 nm

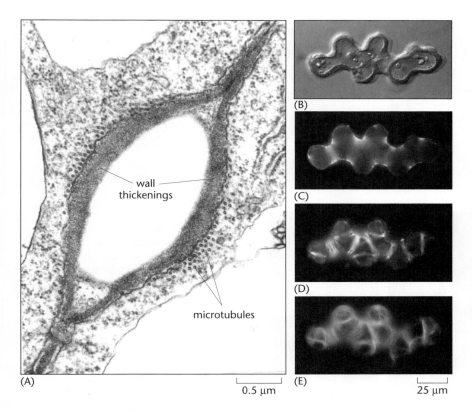

(A)

0.5 μm

(B)

(C)

(D)

(E)

25 μm

Figure 7.9 Microtubules and the shaping of leaf mesophyll cells.
(A) As the mesophyll cells of a wheat leaf develop, bundles of cortical microtubules, seen in this electron micrograph, define regions of the growing wall that will thicken and resist wall expansion. The thickened regions of the wall separate from their neighbors at this point, as the adjacent wall continues to expand resulting in the distinctive shape of mature spongy mesophyll cells (see Figure 6.45). (B) An isolated mesophyll cell, at a later stage of development than (A). The cell has been stained to reveal the microtubules that are shown in an optical section through the center of the cell (C) and at the top of the cell (D). (E) A combined set of optical sections. The bands of microtubules can be seen that define the constraining wall thickenings. (Courtesy of Wolfgang Wernicke.)

around the cells to reestablish vascular continuity and predicted where the lignified wall thickenings would later be deposited. We now appreciate that these bands contain bundles of microtubules, but it was already clear that the bands in one cell often aligned exactly with the bands in the neighboring cells. The anchoring of the resultant strong lignified wall thickenings to those in the next cells achieves a mechanical advantage, a robust vascular tissue; but how the cells communicate with one another to achieve this is still not clear.

The response of cells to wounding also provides us with an example of coordinated actin rearrangements. If an epidermal cell in a leaf is pricked with a needle and killed, the surrounding cells appear to be able to respond very rapidly by dividing, with their new division walls aligned tangentially with respect to the wounded cell (**Figure 7.11**). The earliest response to the wound is a rapid rearrangement of the actin filaments that connect the nucleus to the cell cortex. Cables of actin become aligned, from cell to cell, and surround the wound in the center. Their location subsequently predicts the location of the preprophase band of microtubules and the position of the new division walls.

Figure 7.10 Xylem elements with aligned cell wall thickenings. If a *Coleus* stem is cut, adjacent pith cells will differentiate into xylem elements and restore vascular continuity around the wound. These careful drawings (A) show how defined bands of cytoplasm form early on and (B) predict where cellulosic thickenings will form that will later be lignified. As we saw in the last section, the cytoplasmic bands contain bundles of aligned microtubules, and it is clear that the position of bands in one cell is reflected in the bands of neighboring cells, implying some form of transcellular ordering of cytoskeletons. (From E.W. Sinnott and R. Bloch, *Am. J. B.* 34:151–156, 1945.)

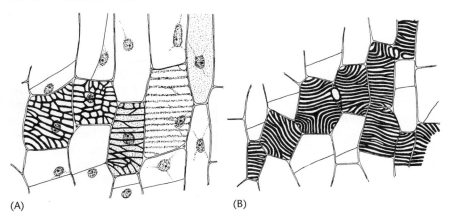

(A)   (B)

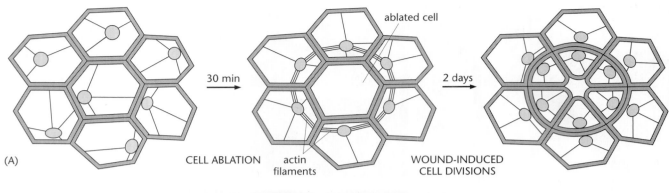

(A)   CELL ABLATION   actin filaments   WOUND-INDUCED CELL DIVISIONS

ablated cell

30 min   2 days

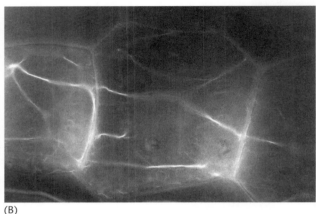

(B)

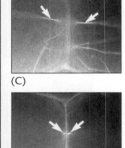

(C)

(D)

Figure 7.11 The actin cytoskeleton responds to wounding. (A) An ablated leaf epidermal cell rapidly triggers the rearrangement of actin filaments in the neighboring cells. Actin cables predict the subsequent division planes that are precisely oriented with respect to the ablated cell. (B) The actin cables in epidermal cells bordering a cut in the leaf of a *Tradescantia* plant clearly line up with those in adjacent cells. (C, D) Actin cable alignment at two planes of focus. (From K.C. Goodbody and C.W. Lloyd, *Protoplasma* 157:92–101.1990.)

The two examples above both involve a cytoskeletal element whose arrangement is directly mirrored in the neighboring cell, but there are examples where the opposite occurs. A well-documented example is the spatial arrangement of both actin and microtubules during the development of the lobes found on many leaf epidermal cells. In the mature leaves of many grasses, for example, the side walls of the epidermal cells are deeply scalloped (**Figure 7.12**). This pattern is established earlier on by groups of microtubules that are focused at regular intervals along the wall and that alternate with corresponding groups in the neighboring cell. A fine actin meshwork is located in the regions between the groups of microtubules in one cell, but directly opposite a group of microtubules in the adjacent cell, and it is this arrangement which appears to determine where new wall material is deposited as the wall gradually grows in a concertina-like manner (Figure 7.12).

These examples again raise the problem of how information about the position of the cytoskeleton in one cell might be transmitted to its neighbor. At least three models have been proposed to account for this transcellular signaling. The first suggests that there might be direct molecular links from the cytoskeleton in one cell, via transmembrane adapter proteins and molecules that span the wall, to the cytoskeleton of another. The second suggests that local signals, chemical or mechanical, might be transmitted through the plasmodesmata that connect adjacent cells. The third evokes the idea that more global mechanical stresses within tissues and organs can be transduced at the level of the constituent cells to reinforce the orientation or distribution of both cytoskeletal elements and net cellulose orientation. Which one, or more, of these models turns out to be operating is still unclear.

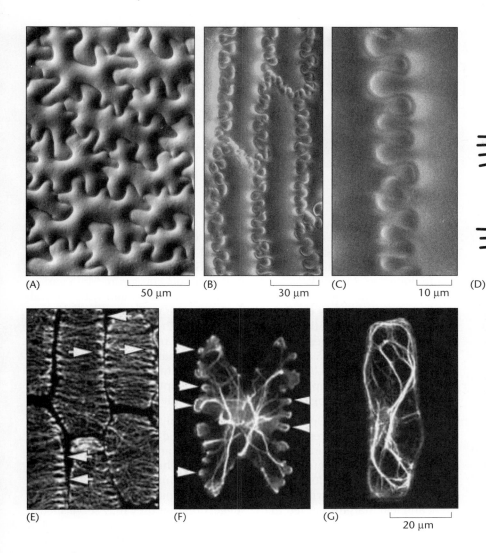

(A)   50 μm   (B)   30 μm   (C)   10 μm   (D)

(E)   (F)   (G)   20 μm

Figure 7.12 The role of the cytoskeleton in the formation of lobes on leaf epidermal cells. Leaf epidermal cells of many plants, both dicots (A) and monocots (B, C), are deeply lobed. Presumably, these formations, like jigsaw puzzles, increase the mechanical strength of the sheet of cells. (D) During the formation of these lobes, each of which corresponds to an invagination in the neighboring cells, an alternating pattern of focused groups of microtubules on one side of the wall and a fine actin meshwork on the other are set up. (E) The alternate bundles of microtubules in a developing maize leaf; and (F) the actin network in a living leaf cell. (G) Disabling mutations in genes that encode the plant Arp2/3 complex, which is required for appropriate local actin polymerization, abolish the lobes entirely, resulting in smooth-walled bricklike cells. (E–G, courtesy of Laurie Smith.)

5. The wall is attached to receptor-like proteins in the plasma membrane.[ref5]

Plasmolysis experiments, performed close to 100 years ago by Hecht, provided the first evidence for physical links between the plasma membrane and its adjacent cell wall. Tiny regions of the plasma membrane were seen to remain attached to the wall when the remaining protoplast contracted during plasmolysis (**Figure 7.13**). The resulting delicate cytoplasmic threads became known as *Hechtian threads* or *strands*, but the molecules underlying the plasma membrane–cell wall adhesion are only now being elucidated.

There are several plasma membrane–associated proteins that could, in principle, function as mechanical or signaling bridges between the wall and the interior of the cell. These include the large family of *GPI-anchored AGPs* (see Concept 3E7 and Figure 7.13) and the *LRR-extensin proteins*. In animal cells, the details of the molecular connections between the extracellular matrix and the internal cytoskeleton are very well documented, and it is curious that almost nothing is currently known about such connections in plants.

The best-characterized protein family known to provide a physical and signaling connection between the wall and the cytoplasm is that of the *WAK proteins (wall-associated kinases)*. In *Arabidopsis* there are five such kinases, all of which contain a highly conserved cytoplasmic serine/threo-

**Figure 7.13 Plasmolysis and Hechtian
threads.** (A) Living, suspension-cultured
tobacco cells that have been plasmolyzed
in a solution of 3% salt. This cell line
carries a gene encoding a plasma
membrane–associated AGP (from tomato)
fused to green fluorescent protein.
(B) The plasma membrane of the
contracted protoplast is brightly
fluorescent, as are the numerous Hechtian
threads that span the gap between the cell
wall and the protoplast. Some Hechtian
threads end at plasmodesmata, but as is
clear from those seen here on outer walls
(*arrowhead*), which do not connect to
another cell, many do not. (Adapted from
W. Sun et al., *Physiol. Plantarum*
120:319–327, 2004.)

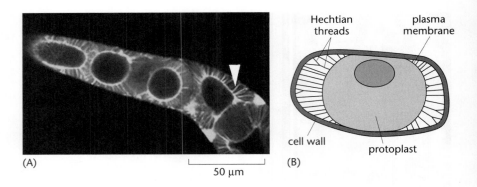

(A)   50 µm   (B)

nine kinase domain and a transmembrane domain that extends into the extracellular region to bind tightly to the wall (**Figure 7.14**). The extracellular domains, which contain two cysteine-rich EGF-like repeats, diverge significantly between the five WAKs. When plant tissues are ground in a variety of buffers and detergents, most of the WAK proteins remain with the wall fraction. However, they can be released from the wall into a microsomal fraction by digestion of the cells with pectinase. Chemical analysis of these microsomal WAKs suggests that the putative WAK receptor domain is covalently linked to pectin. Yeast two-hybrid analyses have also identified a class of glycine-rich proteins (GRPs) that bind strongly to at least one of the WAK proteins. This binding appears to activate the receptor, but the downstream elements of the signaling cascade have yet to be identified.

The five different *WAK* genes, which are tandemly arranged on the same chromosome, are expressed in an organ-specific manner. For example, *WAK*s 1, 3, and 5 are expressed primarily in leaves and stems of *Arabidopsis*. *WAK 4* expression is detected only in siliques, while *WAK 2* expression is seen in leaves, stems, flowers, and siliques. The finding that *WAK*s 1 and 2 are prominently transcribed at junctions of organs, the root apical meristem, and leaf margins has led to the suggestion that they may serve special functions in cells subjected to compression or expansion. Several WAK proteins are also induced by pathogen infection and wounding. The phenotype of plants with reduced *WAK* expression has led to the suggestion that WAK function is required for cell expansion, but the mechanism remains obscure, as does the nature of the extracellular ligand and the details of the downstream signaling.

B. Cell Expansion

1. A key driver of plant growth is postmitotic cell expansion.[ref6]

One of the striking features of plant growth is the massive, yet precisely regulated, accompanying increase in cell size. This contrasts sharply with the situation in animals, where, in general, cells usually end up not much more than double their starting size. There are exceptions, of course. Muscle cells are huge, but they are syncytia, containing lots of nuclei; and egg cells can also be very large, although much of their bulk is accounted for by stored food. By and large though, most of the cells in our own bodies, for example, have a fairly small volume range. Plants are very different. If postmitotic cell size increases in an oak tree were more like those in an animal, the giant tree would be reduced to the size of a small shrub. Plant cells arise in meristems and commonly have a volume of about 200 to 1000 μm^3. However, as cells leave the meristem and expand, they can easily reach volumes, for example, in storage tissue, of 10^6 μm^3 or more, an increase of at least 1000-fold (**Table 7.1**).

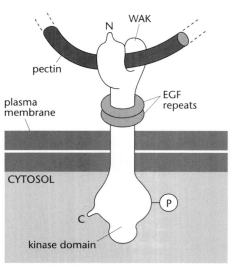

Figure 7.14 Schematic diagram of a WAK protein at the cell surface. This class of plasma membrane–associated protein is tightly bound to pectin components of the cell wall.

TABLE 7.1 RELATIVE PLANT CELL VOLUMES. Typical cell volumes of several common cell types are shown in relation to the volume of the small phloem parenchyma cells in an *Arabidopsis* root, arbitrarily assigned a volume of 1.

CELL TYPE	RELATIVE VOLUME
Phloem parenchyma (root apex)	1
Shoot apical meristem cell	100
Leaf palisade mesophyll cell	300
Storage parenchyma cell	50,000–500,000
Large metaxylem vessel element	100,000

What is it about plant cells that can accommodate such large increases in cell size? To answer this, we first need to make a distinction between cell growth and cell expansion. Animal cells get bigger largely by the production of more cytoplasm. This increase in macromolecular mass, and the corresponding increase in size, is commonly called *cell growth*. In plants, however, much of the cell size increase occurs without a corresponding increase in the amount of cytoplasm. Instead most of the cell volume increase is accounted for by an increase in the size of the vacuole, and the cytoplasm is simply pushed out to form an increasingly thin layer just inside the cell wall. In the meristem, the situation is more like that in yeast or animal cells. Cells grow, double in size, and divide. Only when cells leave the meristem, and the cell cycle, does the massive increase in vacuolar volume accompany controlled *cell expansion*. For plants, limited by nitrogen availability, such a strategy makes sense. Building large structures like leaves to capture light is more economically achieved by using large, inflated, water-filled compartments, constrained by walls made of "'cheap" carbohydrate, than by making an equivalent volume of nitrogen-rich cytoplasm. Proteins account for rather little of the mass of a watermelon! The distinction between cell growth, as the irreversible addition of macromolecular mass, and cell expansion, driven by vacuolation, can be seen clearly in root tips. Cells in the meristem region grow at a constant rate, have few vacuoles, and divide when they reach an appropriate size. As the cells stop dividing and begin vacuolation and expansion growth, the rate of expansion is consistently fourfold higher. Presumably it takes more time to manufacture cytoplasm than to deposit wall and take up water as the vacuole expands.

The final size of a plant organ is thus a complex balance between cell proliferation, cell growth, and cell expansion through vacuolation (**Figure 7.15**). It is clear, from the observed relative constancy of plant organ size and form, that these processes are both tightly controlled and genetically underpinned. One simple mechanism that plants use to regulate cell expansion is to successively double the total amount of chromosomal DNA in the nucleus, without undergoing mitosis, in a process called *endoreduplication*. Auxin levels are thought to control the postmitotic switch to endoreduplication, and there appears to be a rough correlation between increases in nuclear ploidy and the amount of cytoplasm in the cell (**Figure 7.16**). A combination of endoreduplication and vacuolation together allows very large cells to be made. We can now see that the considerable postmitotic, turgor-driven cell expansion underpinning plant growth must involve at least two key processes. The first is the increase in surface area of the membrane—the tonoplast—that surrounds the vacuole. How the production of this membrane—and the important proteins it contains including the water channels, or *aquaporins*, and the K+ uptake channels—is regulated is not well understood, although mutants that are compromised in vacuole

Figure 7.15 Two processes contribute to the postmitotic increase in cell size. (A) When a cell leaves the meristem it may grow by increasing its cytoplasmic mass, a process that can be enhanced by DNA endoreduplication cycles, and it may also expand by increasing the volume of its vacuole. A cell's final size is tightly controlled and can be thousands of times its starting size. (B) This sequence of events is seen clearly in the files of cells leaving the root meristem, migrating back, enlarging, and differentiating in this scanning electron micrograph of an *Arabidopsis* root. (B, from L. Dolan et al., *Development* 120:2465–2475, 1994.)

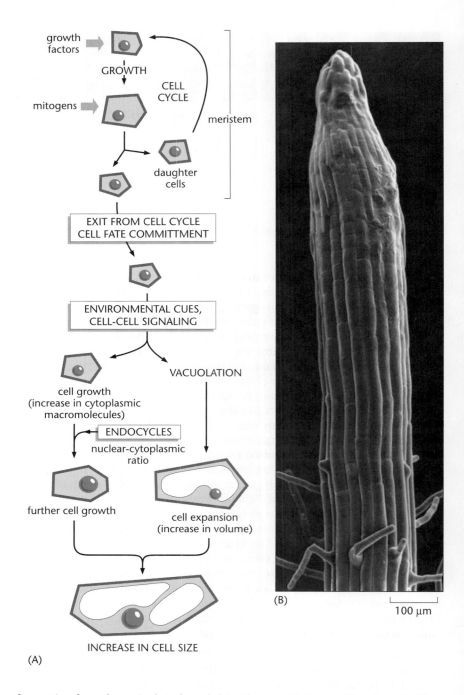

formation have been isolated, and these have cells incapable of expanding. The second process is the production and deposition of new cell wall material, which we shall discuss next. The plasma membrane, of course, must also increase its area proportionately, which it does as a consequence of vesicles delivering new material to the wall.

2. Cell expansion is usually accompanied by the deposition of new wall material.[ref7]

Cells in the shoot apical meristem of a large tree have undergone hundreds or even thousands of rounds of growth and division. Yet they, and their walls, look almost identical to when they started out decades earlier in the seedling. We must conclude that at a general level, at least, the process of cell wall deposition is very tightly correlated with the process of cell growth and expansion. If the two processes were uncoupled to even a small degree,

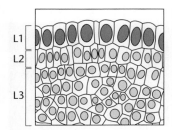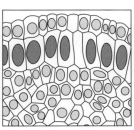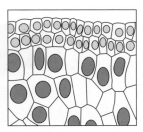

Figure 7.16 The relationship between cell ploidy and cell size. These sketches of vertical sections through floral apices of *Datura stramonium*, in which different cell layers have different ploidies, show a clear relationship between nuclear DNA content, nuclear volume, and cell size. The periclinal chimeras were induced by colchicine treatment, and the dark gray nuclei in each case are 8C, whereas the pale gray nuclei are 2C. (*Left to right*) The apices have 8C nuclei in cell layers L1, L2, and L3, respectively. Cell division and cell expansion in the different layers mutually adjust to avoid distortion of meristem anatomy, indicating that local cell-cell signaling is involved in the overall control of organ size (Adapted from S. Satina and A.F. Blakeslee, *Am. J. Bot.* 28:862–871, 1941.)

walls would get progressively thicker or thinner (**Figure 7.17**). This raises the question of whether the wall deposition process itself can drive growth, or whether cellular growth decisions instead drive the production and deposition of wall material. Although the tree example above might suggest that either model could be true, there are sufficient examples at the cellular level to suggest that the two processes are indeed separately regulated, but that they are often in practice tightly coordinated.

The large and shapely trichomes that we have seen earlier (see Figure 1.23) deposit new wall material in a carefully regulated manner to achieve their final spiky form. The nuclei in these large cells have endoreduplicated, usually to a DNA content of 32C. However, *Arabidopsis* mutants that reduce the ploidy of the trichome nucleus also reduce the size and branching of the trichome, and correspondingly the amount of wall deposited. Conversely, mutants with increased trichome ploidy make larger, and more elaborately branched, trichomes (**Figure 7.18**), and also more wall. This example emphasizes that internal cellular controls are the key regulators of cell growth and expansion. However, it is also true that if the secretion and deposition of new wall material is blocked, for example, with brefeldin A, then cell expansion immediately ceases. This suggests that the intrinsic and extrinsic signal pathways that control cell growth and expansion act in part by controlling wall deposition, which is then necessary but not by itself sufficient to drive expansion.

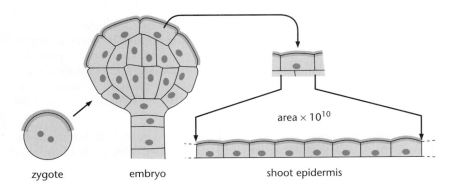

zygote   embryo   shoot epidermis

area × 10^{10}

Figure 7.17 Cell wall deposition can be tightly coupled to cell expansion. Part of the cell wall that is deposited around the initially naked plant zygote goes on to become the outer epidermal wall surrounding the globular embryo, and over time this in turn becomes the outer epidermal wall of the leaves in a mature tree, all by a gradual process of wall expansion and new wall deposition. This wall is unique in that it does not arise, as other walls do, as a cell plate during cell division. Millions of new anticlinal walls will have arisen as the cell divides repeatedly, and all of these will have been carefully pasted into the single expanding outer wall as the plant grows to maturity. Since the final surface area of a tree may be as much as 10^{10} times the area of the original zygote wall, and yet its epidermal wall thickness remains pretty much the same, this provides a remarkable example of the degree to which the processes of wall deposition and cell expansion can be coupled.

Figure 7.18 Effect of ploidy on the size of trichomes. (A) Normal trichomes usually have three points and a ploidy of 32C—that is, each trichome has 32 times the normal haploid amount of DNA in its nucleus. (B) Mutations that reduce the amount of DNA (8C) in the trichome nucleus result in much smaller trichomes with only a single point. (C) The trichomes on mutants that increase the normal amount of DNA per nucleus (64C) are correspondingly larger and have many points. (Courtesy of Keiko Sugimoto-Shirasu.)

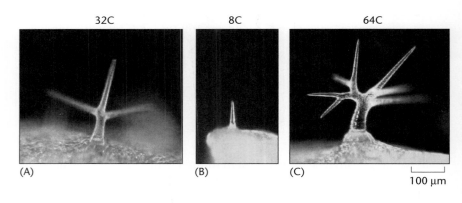

There are several examples, however, where expansion and wall deposition appear to be uncoupled. The *Arabidopsis* hypocotyl undergoes considerable extension growth when the seed germinates in the dark, and all of this 30-fold or so increase in length occurs without any cell division. Rather surprisingly, it seems that much of the wall the hypocotyl requires is actually deposited as the seed germinates, before the hypocotyl is even 1 mm long. Subsequent cell elongation is then accompanied by the progressive thinning of the thick wall laid down earlier, and is only partly offset by further wall deposition (**Figure 7.19**). Hypocotyl elongation will even take place in the presence of isoxaben, a chemical inhibitor of cellulose synthesis, provided it is added after the early stage of wall deposition has been completed. This is rather a special case, since seedling growth is limited by the carbon available in the seed, and the sole function of the hypocotyl is to raise the two cotyledons into the light, where they can fix more carbon. A counterexample can be seen in the walls of the elongating pedicel, the short stem that supports the flower and eventually the seed pod. Unlike the hypocotyl cell walls, the walls of the pedicel start out thin and then gradually thicken at the same time as the pedicel elongates, presumably to cope with the increased load of the developing silique. In both these cases, the rate of deposition of new wall material is uncoupled from the rate of cell elongation.

An extension of this idea is seen in the far more common case where cells that have ceased expanding may still continue to deposit new wall material, a process referred to as *secondary thickening* (see Concept 1B4). In conclusion, although wall deposition usually accompanies, and is necessary for, cell expansion, the two processes can be uncoupled. Cell expansion can take place with little wall deposition, and extra wall deposition can take place both during and after cell expansion.

Postmitotic cell expansion can occur through two entirely different mechanisms, *tip growth* and *diffuse growth*, which differ principally in the way

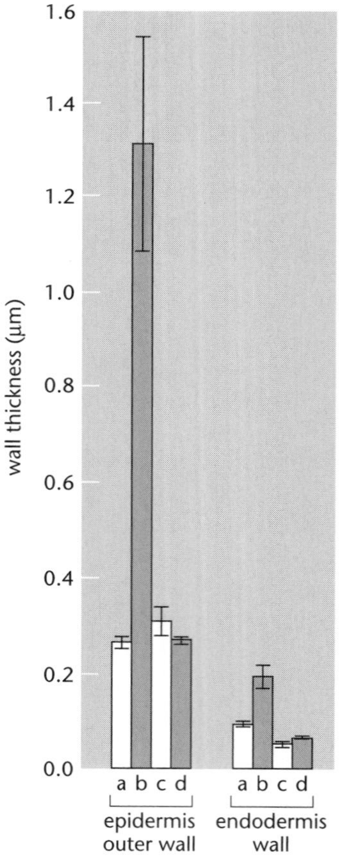

Figure 7.19 Cell wall thinning during hypocotyl elongation. This chart shows the remarkable dynamic changes in wall thickness that occur, in two different cell types, in a dark-grown *Arabidopsis* hypocotyl at four stages of seedling emergence and growth (a, in embryo prior to germination; b, emerging hypocotyl; c, hypocotyl at 50% final height; d, fully grown hypocotyl). The outer epidermal wall is the key constraint on organ growth, containing up to half of the total wall material in the hypocotyls, and at emergence it has grown to five times its earlier thickness (over 1 μm thick) before it then rapidly thins as the organ grows. Later, growth and wall deposition become coupled. A similar picture is seen in the inner walls, exemplified here by the endodermis. In dark-grown material, struggling to reach the light, these inner walls reach their minimum thickness, around 50 nm, about as thin as a wall can possibly be while still retaining integrity! (Adapted from P. Derbyshire et al., *J. Exp. Bot.* 58:2079–2089, 2007.)

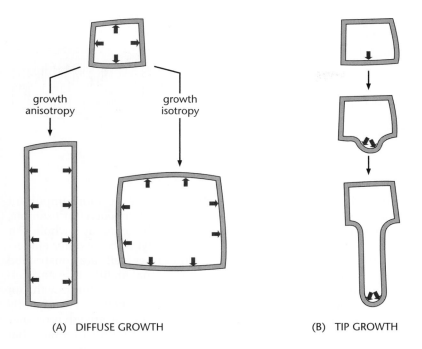

(A)   DIFFUSE GROWTH   (B)   TIP GROWTH

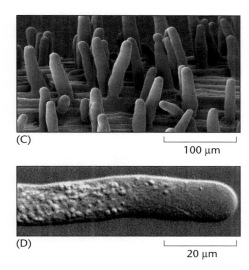

(C)

|————————| 100 µm

(D)

|————————| 20 µm

Figure 7.20 Diffuse growth and tip growth. These schematic diagrams comparing diffuse growth (A) and tip growth (B) illustrate the general concept that where and how new wall is deposited has important consequences for cell shape. Growth anisotropy depends on the precise arrangement of cellulose lamellae in the lateral walls as well as where new wall components are delivered (*arrows*). Two examples of tip-growing cells are illustrated here: (C) an SEM of young tip-growing root hairs on a growing root of *Medicago truncatula*; and (D) a light micrograph of the tip of a pollen tube. (C, courtesy of Kim Findlay. D, from L. Vidali et al., *Mol. Biol. Cell* 12:2534–2545, 2001.)

new cell wall material is deposited. Tip-growing cells extend by delivering new wall material to the growing tip, a bit like building a factory chimney. In plants such cells are rather rare, with only two common examples, pollen tubes and root hairs (Figure 7.19). Some cells—for example cotton fibers (see Concepts 9C2 and 9C3) and pointed trichomes, which look as though they might grow by tip growth—in fact grow by diffuse growth. In diffuse growth, new wall material is laid down evenly throughout the extending wall. In an isodiametric parenchyma cell this involves vesicular delivery of wall material equally to all points on the cell surface. In cells that extend preferentially along one axis, the vesicles of wall-related material are delivered preferentially along the walls that are extending (**Figure 7.20**), as are the active cellulose synthases, and the result is referred to as *anisotropic cell expansion*. To expand by a factor of 10 in all directions, a cell has to produce about 600 times as much new wall material!

3.   Wall architecture underpins anisotropic cell expansion.[ref8]

Growth and cell expansion can result in large increases in cell volume, but for most cells in a plant body these increases are not equal in all directions. Rather, the increase in size is usually greater along one axis of the cell than another—in other words, is *anisotropic*. As we saw earlier, growth anisotropy is the result of both localized wall deposition and wall mechanical properties that derive from the orientation of the cellulose microfibrils. In seedling growth, for example, the cells of the growing root and hypocotyl elongate dramatically longitudinally, but undergo almost no radial expansion. It is tempting to assume that this is part of some intrinsic developmental program, but in fact anisotropy is cell-autonomous and furthermore is not "hardwired" at the cellular level. Intrinsic factors can exert their effects during cell expansion, radically reorienting the growth axis in response to a developmental challenge, and this means that vesicle delivery in the cell has to be redirected from one set of walls to another (**Figure 7.21**). Cell polarity in plant organs is related more generally to the polarized influx and efflux across the cell of the signal molecule auxin, a process reinforced by the selective endocytosis and recycling of plasma membrane components that include auxin efflux carriers.

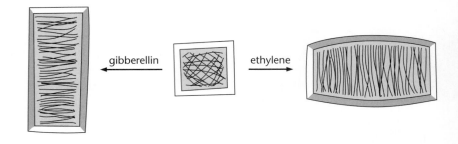

Figure 7.21 Wall architecture and anisotropic growth are not "hardwired." This schematic diagram shows how, in principle, a meristematic cell is flexible and can behave in different ways in response to different growth factors, in this case gibberellin and ethylene. The resultant microtubule rearrangements in each case mean that the net orientation of cellulose microfibrils (shown here) is different, and the degree of growth anisotropy is different. Extension is cell-autonomous, and its polarity is not wholly fixed by developmental decisions about cell identity. This property is useful for plants in adapting to changing environmental conditions. In each case new wall material is delivered to a different set of walls (shaded *gray*).

In constraining cell and organ growth, some walls are more important than others. The outer epidermal wall of root and hypocotyl, for example, exerts an overriding influence on the growth of the organ as a whole. For a start, nearly half of the total wall in the hypocotyl is in the outer epidermal wall, and internal cells with their much thinner walls are constrained in their growth anisotropy by this wall. The internal architecture and cellulose orientation of this wall alone both drives and restricts the growth, not only of the hypocotyl, but of the whole shoot. This conclusion is elegantly demonstrated by experiments in which a receptor for the growth promoter brassinosteroid, expressed only in the epidermis of a dwarf mutant that lacks the receptor, is sufficient to rescue it. In a wild-type plant, expressing an enzyme that breaks down brassinosteroids, again just in the epidermis, confers a dwarf phenotype. The nature of the signal from the "master" epidermis that directs the growth of the "slave" ground tissue remains unknown.

We have seen, in a growing cell, how a net transverse microtubule array can result in a net transverse cellulose microfibril array, which in turn leads to growth anisotropy. This simplistic account, however, does not account for all the observations. The exact nature of the interaction between microtubules and the cellulose-synthesizing rosettes is still very unclear, and several observations are less easy to reconcile with our general picture. Several studies have shown that chemically interfering with cellulose biosynthesis—for example, with DCB or oryzalin—can lead in turn to a disorganization of the cortical microtubule array. Conversely, in some cases the cortical microtubule array has been perturbed without any corresponding effect on cellulose deposition.

The general paradigm, nevertheless, appears to hold in most cases, and it is reinforced by recent experiments on helical growth patterns. Several mutants have been identified in *Arabidopsis* that cause the shoot to adopt a spiral growth pattern, some growing in a left-handed spiral and others in a right-handed spiral. Similar growth patterns can be induced by certain drug treatments, such as taxol. Cloning of the corresponding genes has directly implicated tubulin itself, and some microtubule-associated proteins, in this spiral growth process. Coupled with a careful analysis of what is happening to microtubule dynamics and cellulose deposition in the mutants, an intriguing hypothesis has emerged, which states that when the microtubule catastrophe rate is increased and shorter, more destabilized microtubules result, left-handed growth occurs. When microtubule assembly is promoted and microtubules are stabilized, then right-handed growth results. Since the spiral growth process is probably mediated by strains in wall architecture, this suggests that normal, straight organ growth is a subtle and dynamic balance between opposing forces within the wall, each pushing growth in a different direction, each of which is ultimately dictated by the dynamics of the underlying cortical microtubules. That the phenomenon of spiral growth in these mutants can be seen in a single trichome strongly suggests that these growth processes are cell-autonomous and do not depend on organized cells in tissues.

Perhaps the most significant advance in our understanding of this confusing field is the recent direct visualization of cellulose synthase complexes in living cells. By expressing in the same plant a fluorescent version of the CESA6 protein (CESA6 is one of the primary cell wall cellulose synthases) along with fluorescent tubulin, it has been possible to follow the movement of rosettes within the plasma membrane and to track them in relation to the underlying microtubules (**Figure 7.22**). These experiments finally provide proof, at least in epidermal cells, that microtubules provide a guidance mechanism for cellulose deposition. The tracks of the cellulose synthase closely followed the general tracks of the microtubules, and would readjust if the microtubule orientation were changed. When the microtubules were removed, the cellulose synthases also soon dissipated. This therefore suggests that MTs do provide guide rails for the movement of cellulose synthases and, together with the finding that the MT guide rails rotate with time, does suggest a role for MTs in regulating the variable alignment of cellulose microfibrils.

A major long-term problem is, what then determines the orientation of the cytoskeletal arrays themselves? It has been suggested that supracellular tensions and strains within whole organs or tissues can have a profound effect on growth and its direction—in other words, that strain direction could act as an alignment cue for shifts in microtubules and microfibrils. We have seen hints of this in our earlier discussion of how cytoskeletons often appear to be organized at a more global level in tissues, but we have no clear idea yet how such biophysical forces can be sensed and transduced into microtubule organization. This biophysical model might in turn suggest that the wall itself could also play an active role in tissue remodeling and growth direction, helping to influence the dynamics of the cortical microtubules. This is an attractive idea, although it still awaits experimental evidence. It has been demonstrated, however, that global alignments in cellulose microfibrils can affect growth anisotropy, in that cellulose can be equally well aligned at a cellular level but less well aligned at a global level, and that this can reduce growth anisotropy (**Figure 7.23**).

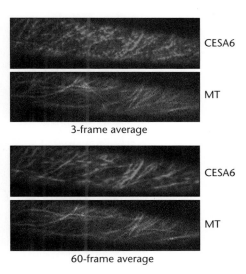

Figure 7.22 Tracking cellulose synthesis in living cells. These micrographs show the co-localization of cellulose synthase (CESA6) and microtubules (MT), both fluorescently tagged, in an epidermal cell from the *Arabidopsis* hypocotyl. A frame was collected every 10 s on a spinning-disk confocal microscope. Averages of 3 frames and 60 frames show the correspondence, over time, between the trajectories of the cellulose synthases through the plane of the membrane and the underlying cortical microtubules. (Adapted from A. Paredez et al., *Curr. Opin. Plant Biol.* 9:571–578, 2006.)

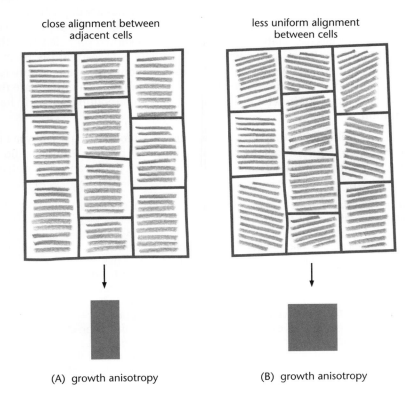

(A) growth anisotropy

(B) growth anisotropy

Figure 7.23 The global alignments of wall microfibrils affect growth anisotropy. While the net alignment of microfibrils in the wall of a single cell may be high, the resulting growth anisotropy of the tissue may depend more on the degree to which the global alignment of adjacent cells is the same or different. Tissues with cells whose wall arrays are well aligned with each other (A) will result in greater growth anisotropy than tissues with equally well-aligned microfibril arrays that are not in turn well aligned cell to cell (B). (Adapted from T.I. Baskin, *Annu. Rev. Cell Dev. Biol.* 21:203–222, 2005.)

4. Dynamic remodeling of wall architecture facilitates cell expansion.[ref9]

Increasing the surface area of the wall during expansion growth is not like just blowing up a balloon. As we have seen, cellulose microfibrils are spun out in precise order and new wall matrix materials have to be deposited in the appropriate place and the macromolecules faithfully incorporated into the texture of the enlarging wall. But the wall is under tension, and growth is a constant challenge to its integrity. The yield threshold must be exceeded to drive expansion, and yet at the same time wall integrity and strength must be constantly maintained to avoid an aneurysm. Since, by and large, wall thickness keeps pace with expansion, the new wall macromolecules have to be cut and pasted into the existing wall architecture, and that architecture has to be maintained. How is this managed?

From a consideration of its basic fiber-composite architecture we might deduce that the cellulose/hemicellulose network and the pectic polysaccharide network would both have to be remodeled during growth, as new molecules of each class are deposited. Since both contribute, albeit unequally, to the strength of the wall, it is likely that this is so, and in the case of xyloglucan, a key tether in the wall network, several candidates that may be involved in its remodeling have been described.

The first of these is the large family of *xyloglucan transglucosylase-hydrolases* (the XTH family, around 30 members), some of which, depending on the exact substrate, will cleave xyloglucan (the *xyloglucan endohydrolase*, or *XEH*, activity), and some of which will transglucosylate xyloglucan (the *xyloglucan endotransglycosylase*, or *XET*, activity). In some cases the same enzyme may be able to perform both tasks, depending on the exact substrate availability and other conditions. Both classes of activities could, in principle, assist in the remodeling of load-bearing xyloglucan as the existing cross-linked cellulose microfibrils have to move apart to incorporate new ones and their new cross-links. Even in secondary wall deposition, where no expansion growth occurs, the existing wall still has to accommodate the introduction of new polymers, and XTH family members have indeed been located in such walls. It seems that these enzymes are deposited in the wall at the appropriate time and place before the subsequent deposition of the wall materials associated with growth. This is shown very clearly in the case of developing root hairs, where XET activity is present at the precise spot on the outer root epidermal wall where the root hair will later emerge and start tip growth (**Figure 7.24**). This is presumably to accommodate the wall remodeling required for the change in growth direction implicit in hair formation.

It is also possible that other, more conventional *endo*-glucanases may act to loosen, tighten, or remodel the wall during growth, but less is known about these. Another candidate, however, for cutting xyloglucan, and indeed

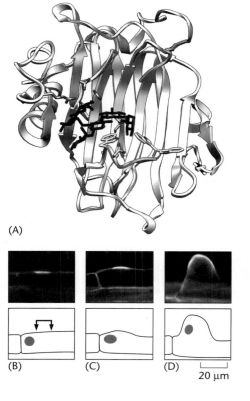

(A)

(B)   (C)   (D)
20 μm

Figure 7.24 Xyloglucan *endo*-transglycosylase (XET). (A) Shows the structure of the XET enzyme molecule from poplar. The curved β-sandwich protein structure is shown in association with a bound XXLG xyloglucan nonasaccharide. (B–D) Localized XET activity is associated with the initiation of a tip-growing root hair. In this experiment, fluorescently labeled xyloglucan (XG) oligosaccharides are supplied to walls, where they act as acceptor substrates for endogenous XETs acting on XG as a donor substrate. The local presence of XET is then revealed as insoluble fluorescent polymer deposited in the wall. This clever technique allows the visualization of localized XET activity in the outer root epidermal wall (*arrows*), at a stage in root hair initiation that precedes any other visible marker (B). The tip-localized activity persists only through the very early stages of subsequent root hair growth (C,D).

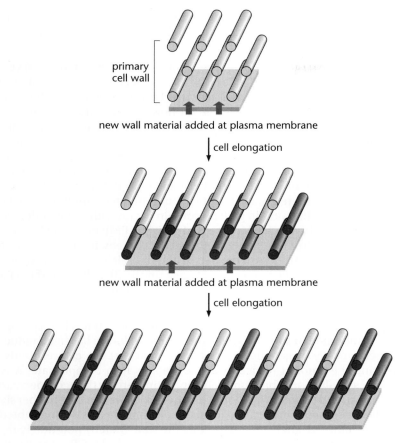

primary cell wall

new wall material added at plasma membrane

↓ cell elongation

new wall material added at plasma membrane

↓ cell elongation

NET RESULT: WALL MATERIAL IS DISPLACED TOWARDS THE MIDDLE LAMELLA
DURING CELL ELONGATION

Figure 7.25 Growth in area of the wall is accompanied by the gradual displacement of both old and new wall materials. This schematic diagram shows that wall material laid down early on at the plasma membrane (cellulose microfibrils in the case shown) will be steadily displaced to the outer lamellae of the wall, adjacent to the middle lamella, as cell expansion takes place. The diagram is based on two assumptions, both supported by experimental observations, that the primary wall stays much the same thickness during the fourfold increase in expansion shown here, and that the microfibrils maintain more or less the same separation. One conclusion from this drawing is that the newly laid-down microfibrils must move between lamellae and that they have to be "cut and pasted" into the spaces between existing microfibrils. The same conclusions also probably hold for the components of the pectin network in the wall.

possibly the pectic polymers, is the hydroxyl radical. Although this highly reactive and short-lived oxygen species can be generated in various ways in the wall, and although it is capable of cutting polysaccharide molecules, it is not easy to see how its action *in vivo* could be precisely regulated in either time or space, and at present there is no direct evidence that this molecule has a real role in wall remodeling in growth. Indeed there are very few candidates at all for remodeling the pectic network during expansion growth, although in principle there may well be transglycanases and transesterases that could act in this way. One curious consequence of adding and incorporating new wall material at the inner face of the existing wall is that if the increase in area of the wall during expansion is very much more than the number of separate lamellae within the wall, which it commonly is, then the wall material laid down earlier on will eventually be displaced to the outermost lamella of the wall, adjacent to the middle lamella, as the initial cell wall essentially disappears due to thinning (**Figure 7.25**). This has interesting consequences for regulating the wall radial heterogeneity discussed in Chapter 6 (see Concept 6C1). It also emphasizes that cellulose microfibrils have to be completely separated from their earlier neighbors as new microfibrils are constantly inserted between them during growth.

During elongation growth—for example in the hypocotyls or primary root—the orientation of cellulose microfibrils, as we have seen, is initially net transverse to the growth axis. As growth slows and eventually stops, however, the various lamellae of cellulose may rearrange and end up in other orientations. Whether these rearrangements are passive, the result of mechanical strains, or driven by self-assembly mechanisms is unknown. What is clear though is that we are a very long way from being able to translate wall composition data into wall architecture and to work from that to mechanical properties and functional behavior. Plants are exquisite but secretive engineers.

We have discussed many aspects of cell growth, but have not touched yet on what it takes to stop growing. It is clear that plant cells are able to slow and then cease growth to end up at a very precise final cell size; pollen grains from any one species are remarkable for how constant their size is, as are stomatal guard cells. Cells, therefore, must have intrinsic mechanisms to stop cell expansion at the appropriate time and place, but we know little of the molecular details. One candidate is the oxidative cross-linking of wall polysaccharides through associated phenolic groups, likely to be important in the type II walls of grasses. Another candidate is *pectin methylesterase (PME)*, for which there are over a hundred annotated family members in *Arabidopsis*. Growing walls in general contain pectin with a higher degree of esterification than walls that have ceased growing. PME might act by removing methyl esters from middle lamella pectin, thereby increasing both the charge density and stiffness of the pectin and the calcium cross-linking and rigidification of the wall. Some data suggest that artificially expressing PME does indeed prematurely arrest hypocotyl elongation.

The plant cell wall shows several signs of being an integrated system that, like the cell wall of yeast, may have a way of sensing when it has been compromised or modified in some way. In cases where cellulose has been reduced, either chemically or genetically, it is noticeable that the cell responds by increasing the amount of pectin in the wall, as if to compensate. A wall integrity signaling mechanism is likely to underpin this compensatory mechanism, and although it is not well understood, we shall see further signs of it when we discuss the wall in defense (see Concept 8B11). Undoubtedly, other mechanisms surely exist to rigidify the wall, thereby potentially regulating cell expansion. The cross-linking of extensin-HRGPs discussed in Chapter 3 (Concept 3E4) is an example of one such mechanism.

5. Proteins that enhance wall expansion have been identified.[ref10]

The incorporation of new wall material, and the remodeling of the expanding cell wall, are key elements in the control of expansion growth, but so also is the relaxation, often called loosening, of the existing *load-bearing components* of the wall. These components are called load-bearing because they provide the constraint offered by the wall to the turgor pressure of the cell. The stress relaxation of the wall in response to turgor pressure is an important feature of extension growth.

We have already discussed some of the molecules that may be involved in wall loosening or remodeling, but one family of proteins that does mediate cell wall relaxation, both *in vitro* and *in vivo*, is the expansin family. *Expansins* were first identified in reconstitution experiments in which proteins extracted and purified from plant cell walls restored acid growth responsiveness to tissues that had lost it through heat inactivation.

The exact mechanism by which expansins mediate cell wall relaxation has yet to be determined. However, it is thought that expansins work at the interface between cellulose and hemicellulose when the wall is under tension, thus relaxing a key cross-link between polymers within the wall structure, probably the hydrogen bonding between cellulose and hemicellulose.

Much of the initial protein work on expansins was done on materials isolated from cucumber hypocotyls, and although the first cDNA sequences were identified in cucumber, it is now clear that expansins form a large and divergent gene family in all plants. *Arabidopsis* alone has over 30 expansin genes. Remarkably, it now seems that expansin on its own can exert complex and dramatic effects, not only on cell shape but also on morphogenesis

5 mm   1 cm

Figure 7.26 Local induction of expansin expression is sufficient to induce a leaf developmental program. If expansin gene expression is induced locally (*arrows*) in a tobacco shoot apical meristem, this is sufficient to trigger a developmental program that recapitulates the entire process of leaf formation. In addition, as shown here, if expansin expression is induced on one side of a new leaf primordium, the leaf that develops 2 to 4 weeks later has increased local growth of the leaf blade on that flank only. (From S. Pien et al., *Proc. Natl. Acad. Sci. USA* 98:11812–11817, 2001.)

more generally. Tobacco plants harboring an expansin gene that is under the control of an inducible promoter have been used to investigate the effects of locally inducing expansin expression within the apical meristem. The results strongly suggest that expansin alone can trigger a developmental program that leads to the production of a new leaf, which appears normal in all respects. Induction on one flank of existing leaf primordia leads to the ectopic production of extra leaf blade material on that side (**Figure 7.26**).

It is unlikely that all wall loosening is mediated by expansins, and the search for other candidates that may play a role is ongoing. Although free radicals have been invoked, most of the candidates to date have been wall proteins. Among these are two very different families. The first is a small family of proteins called *yieldins*. These have structural similarity to acidic chitinases, and the recombinant protein indeed has weak enzyme activity. In an extension assay similar to that used to discover expansin, they appear to be able to reduce the yield threshold of hypocotyl walls, but after a considerable incubation time. This is a surprising result since no substrate for chitinase is known in plant walls, but it may be that the glycan binding of the protein is what is important rather than its enzyme activity. The yieldin expression pattern does not always correlate with plant parts that are expanding. The second candidate is a member of the very large family of *lipid transfer proteins (LTPs)*, with over 70 members in *Arabidopsis*. These are small basic proteins with eight conserved cysteine residues linked together by four disulfide bridges. The structures of both expansin and LTP2 are shown in **Figure 7.27**. A hydrophobic cavity binds a variety of small hydrophobic molecules such as lipids, and this cavity appears to be required in tobacco LTP2 for its novel wall-loosening activity. How the LTP acts to cause nonhydrolytic disruption of the wall to facilitate extension is not known, and neither is the putative hydrophobic wall ligand. Doubtless there will be many more candidates discovered that will contribute to our understanding of this complex area!

Figure 7.27 Structures of a lipid transfer protein and a β–expansin. (A) TobLTP2 is a small protein with four disulfide bonds and a hydrophobic cavity. (B) The β-expansin (EXPB1) shown is also a maize pollen antigen and has two domains. Domain 1 has structural similarity to a family-45 glycoside hydrolase, but has no enzyme activity, while domain 2 is an Ig-like domain. The two domains together provide an extended potential polysaccharide-binding surface. It is proposed that the expansin acts by promoting local movement and stress relaxation in the cellulose-hemicellulose network by noncovalent rearrangement of their interactions. (A, from J. Nieuwland et al., *Plant Cell* 17:2009–2019, 2005. B, from N.H. Yennawar et al., *Proc. Natl. Acad. Sci. USA* 103:14664–14671, 2006.)

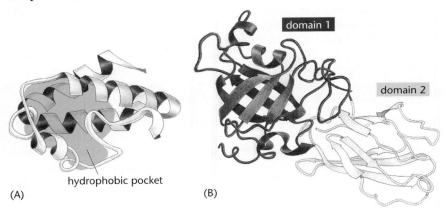

domain 1

domain 2

hydrophobic pocket

(A)   (B)

6. Acidification of the wall, enhanced by auxin, may promote cell expansion.[ref11]

Since the discovery in the 1930s that the plant hormone auxin can stimulate cell expansion in certain plant tissues, for example, pea epicotyls and maize coleoptiles, scientists have tried to determine the mechanism by which auxin can control cell expansion. It is clear that to maintain cell expansion over a period of hours auxin must stimulate the synthesis of new cell wall material for incorporation into the cell wall to maintain both its thickness and its strength. In addition, auxin-stimulated acidification of the cell wall has formed the basis of a model used to describe both the early (less than 30 min) and some of the later effects of auxin on cell expansion. In this so called *acid growth hypothesis*, auxin causes the acidification of the cell wall via the stimulation of plasma membrane ATPases that pump protons into the cell wall and lower its pH. In turn, the lower pH either directly lowers the yield threshold of the wall or optimizes the activity of cell wall–localized proteins that loosen the wall, either or both resulting in turgor-driven cell expansion.

Results obtained with fusicoccin, a fungal phytotoxin that mimics some of the effects of auxin and also causes cell elongation and cell wall acidification in isolated plant tissues, have been used to support the acid growth hypothesis. On the other hand, other researchers have pointed to data obtained, both with fusicoccin and with the addition of pH buffers on plant tissues, to question some aspects of the hypothesis. Thus, although the hypothesis is still used to explain some aspects of auxin-induced cell expansion, questions remain about the exact way in which auxin promotes the rapid expansion of plant cells. Certainly recent data, suggesting that one action of auxin is to modify vesicular traffic to the plasma membrane by its suppression of endocytosis, means that the picture is likely to be very much more complex.

C. Turnover, Remodeling, and Breakdown of the Wall

1. Although plant cell walls are relatively stable compartments, many matrix polysaccharides and proteoglycans do turn over.[ref12]

Walls are typically an end product of biosynthesis, and are usually considered to be static structures once formed. However, there are many examples of macromolecular turnover both during the formation of cell walls and after they are fully formed. There is little evidence, however, for the routine turnover of either cellulose or lignin in the wall. Walls may undergo dynamic remodeling by glycosyltransferases such as xyloglucan *endo*-transglycosylase-hydrolase (XTH; see Figure 7.24), but remodeling should be distinguished from turnover, since there is no net removal and replacement of material. Plant cells in excised tissue exhibit autolysis of wall components, and there is considerable metabolic activity during growth due to cell wall expansion. Cells in liquid culture have also provided useful insights into the assimilation and metabolism of macromolecules in the cell wall.

Transient changes in macromolecular structure or composition of cell walls may be necessary for the normal multistep progress of cellular differentiation, for mobilization of storage materials, or for responses to environmental stimulation such as wounding or pathogen attack. In rapidly growing tissues, where most cells have primary cell walls, cell division and cell elongation are highly dynamic processes. There are a number of examples of turnover of particular cell wall polysaccharides in both monocots and dicots.

The turnover of cell wall polysaccharides has been well illustrated in flax, where pulse labeling in intact plants shows extensive turnover of monosaccharides in the matrix polymers of phloem fibers. Flax plants are rich in pectic substances and xyloglucan. After pulse labeling with $^{14}CO_2$, followed by a chase with unlabeled CO_2, the relative radioactivity in glucose and other cell wall sugars changed at substantially different rates. For example, specific polysaccharides such as galactans turn over during the differentiation of phloem fibers in a cell- and tissue-specific manner.

Similarly, in proso millet (*Panicum millaceum* L.) cells in liquid culture, which exhibit little cell expansion beyond that associated with the cell cycle, arabinose and glucose are constantly recycled during the turnover of polysaccharides from the cell wall. Arabinose, hydrolyzed from glucoronoarabinoxylans, and glucose, hydrolyzed from mixed-linked β-1,3/1,4-glucans (see Concept 3D1), represent most of the turnover.

Cotyledons of cabbage seedlings, germinated in darkness, show a decrease in dry weight, and a slight increase in insoluble cell wall carbohydrate during the first week. During this time, arabinose, which begins as the most abundant sugar residue, decreases by 70%. This decrease is accompanied by an increase in glucose and uronic acid residues consistent with turnover in at least some cell wall carbohydrates.

There is also evidence for turnover of cell wall polysaccharides during postharvest storage of white asparagus. White asparagus, a seasonal speciality, is grown in sand with very low light to produce spears with tender tips. During postharvest storage, there is mobilization or turnover of cell wall polysaccharides in both the apical and basal regions of the spear, including homogalacturonans, galactans, xyloglucans, and cellulose. The increase in cellulose and xylan in the basal region has been blamed when the asparagus becomes tougher to eat (**Figure 7.28**).

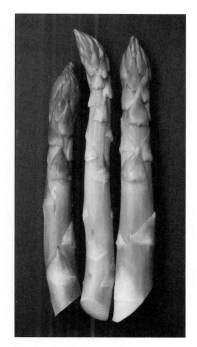

Figure 7.28 Asparagus spears.

Arabinogalactan proteins (AGPs) also appear to be rather dynamic polymers in the cell wall (see Concept 3E6). AGPs often appear transiently at the cell surface in the course of development and differentiation, often in a cell type–specific manner (**Figure 7.29**, and see also Figures 6.48 and 6.49). In pollen tube growth, AGPs in the style may provide a source of sugars to provide energy for pollen tube wall biosynthesis.

Callose is a particularly dynamic component of cell walls at many stages of growth and development (see Figure 4.25). Callose deposits typically appear between the plasma membrane and the cell wall, as transient barriers. Callase (a β-1,3-glucanase) removes callose deposits. During cell division callose appears as a major early component laid down in the cell plate, but it is later removed, and during pollen grain formation and pollen tube growth it is again produced transiently. Formation and removal of callose is important in the differentiation of sieve plate pores, where the space between the plasma membrane lining the plasmodesma and the cell wall is filled with callose, which is then later removed. In deciduous trees, callose is deposited in the fall to close sieve tube pores but is enzymatically removed again in the spring.

A striking example of synthesis and degradation of cell wall polysaccharides is found in the storage structures of the seeds of many plants, for example, nasturtium (*Tropaeolum majus*) and fenugreek (*Trigonella foenumgraecum*; **Figure 7.30**). These seeds contain large reserves of cell wall–derived storage carbohydrates, deposited as secondary thickenings in the cell walls of the endosperm. The major carbohydrate component is a galactoman-

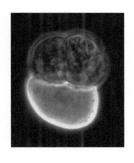

Figure 7.29 AGPs are dynamic cell surface markers. The basal cell of a somatic carrot embryo is fluorescently labeled by a monoclonal antibody that reacts with a carbohydrate epitope on an AGP molecule produced exclusively by the lower cell.

Figure 7.30 Fenugreek. The amber-colored seeds of this legume, containing large amounts of galactomannan, are widely used in Indian and Near East cuisines. (Courtesy of Wikipedia Open Commons.)

nan, which in mature fenugreek seeds can be as much as 30% of the weight of the seed. The galactomannan is rapidly broken down to mannose and galactose in the endosperm before being quickly adsorbed by the embryo.

2. The *de novo* insertion of plasmodesmata across established walls requires wall remodeling.[ref13]

Plasmodesmata, described in Chapter 1, can be subdivided into two classes based on their structure and on the time of their formation. Those inserted in the new cross-wall during cytokinesis consist of single, unbranched channels and are known as *primary plasmodesmata* (see Figure 1.46). Those inserted later, across established or expanding walls, often have a branched morphology and are referred to as *secondary plasmodesmata*. Secondary plasmodesmata appear to play a role in the regulation of metabolic trafficking between cells and in the passage of macromolecules that regulate cell differentiation and tissue development.

For example, during early stages of *Arabidopsis* leaf development there is a requirement for molecular transport into the young leaf to sustain its growth. At this stage, simple primary plasmodesmata are mostly present, which allow the free passage of molecules up to a size limit of about 50 kD. Later, as the leaf finishes growing and in turn becomes a source of raw materials for the rest of the plant, the primary plasmodesmata are largely replaced by secondary plasmodesmata, while at the same time their size exclusion limit drops to well below 20 kD. In other developmental situations the size exclusion limits of plasmodesmata may vary considerably (see Figure 1.47). Fully differentiated cells are often in communication with each other via a mixture of primary and secondary plasmodesmata, although this is controlled at some supracellular level that is not yet understood. For example, phloem sieve tube elements are connected by plasmodesmata only to their companion cell and not to the other touching phloem parenchyma cells.

How do more complex secondary plasmodesmata arise? There are currently two theories to account for their origin. The first is based on the observation that plasmodesmata can arise *de novo* in walls that were never the product of a cytokinetic event, including walls between cells of graft unions and chimeras as well as between cells of parasite and host plant. This suggests that, at least in these circumstances, they can be generated from scratch in the absence of any preexisting primary plasmodesmata. It is thought that the first step in creating a new plasmodesma is the dynamic probing of the wall by small, fingerlike projections of cortical ER on either side of the intervening wall, each carrying in a sleeve of plasma membrane around itself. The force required for such probing of the wall may come from the actin and myosin that are often found associated with plasmodesmata. If two such fingers happened to meet up in the region of the middle lamella and then fuse, the result would be a plasma membrane–lined channel, with a tube of ER now spanning from one cell to the other across the wall—in other words, the essential structure of a plasmodesma (**Figure 7.31A**). Unsuccessful fingers will be constantly resorbed.

However, in more conventional walls, there is now strong evidence that new plasmodesmata often arise in a second way, in very close association with preexisting ones, resulting in paired plasmodesmata (**Figure 7.32**). Possible ways in which this close association might arise are shown in Figure 7.31B and C, and in both cases these again depend on the production of desmotubule-like processes from cortical ER penetrating the wall space. One way involves a new desmotubule forming within the enlarged pore of an existing plasmodesma, before then separating into two daughter plasmodesmata; while the other involves the *de novo* production of a new plasmodesma, but inserted in a region of wall immediately adjacent to an existing plasmodesma. As daughters remain close to their parent this can lead to large

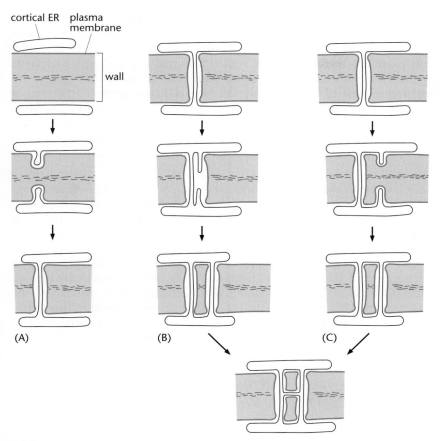

Figure 7.31 The formation of plasmodesmata. (A) The *de novo* insertion of plasmodesmata into a wall involves dynamic fingers of plasma membrane and associated ER pushing into the wall, possibly using actin and myosin to supply the force, and fusing with a similar finger from the adjacent cell. The plasma membrane fusion creates the pore itself across the wall, while the ER fusion event forms the desmotubule within the new plasmodesma.

(B, C) Secondary plasmodesmata often form in association with an existing plamodesma in a process that involves similar membrane fingers and fusion events. This may occur either through new fingers entering an expanded existing pore, followed by their separation (B), or by a new pore forming in tight association with an existing plasmodesma (C). In either case, secondary plasmodesmata often become branched and interconnected by membrane fusion events in the region of the middle lamella.

local accumulations of plasmodesmata. Commonly seen in expanding walls, such groups are often associated with a thinner region of wall that is cellulose-poor and pectin-rich. Such groupings of pores are called *primary pit fields* (Figure 7.32). Pit fields are prominent in both primary and secondary walls, although during secondary thickening the wall in the region of the pit field is often left unthickened, resulting in deep pits in the wall on either side, with the original pit field lying between the two cells at the base of the opposing pits (Figure 7.32D).

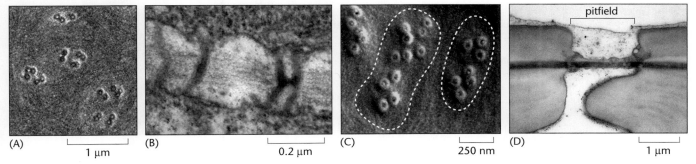

Figure 7.32 Secondary plasmodesmata are often paired. (A) As shown in the previous figure, new plasmodesmata are usually inserted adjacent to preexisting ones, creating obvious pairs of pores. The clustering of pores, shown here in a freeze fracture replica of the wall between a tobacco leaf outer epidermal cell and a trichome, is an early stage in the formation of primary pit fields. (B) The paired plasmodesmata also begin to show membrane fusion in the region of the middle lamella, creating branched plasmodesmata with a central cavity seen in a thin section. (C) Groups of related secondary plasmodesmata often exist in specialized regions of wall, called primary pit fields, which are rich in pectin and relatively poor in cellulose. An early stage is shown here in this shadowed replica of a maize cell wall. The doughnut-like rings are callose in the neck region of the plasmodesmata, and in some pores the snapped-off remains of the desmotubule can be seen. (D) Cross section of a pit field between two xylem vessel elements. The primary wall with a group of plasmodesmata remains at the base of deep pits in the secondary walls on either side. (A and B, from C. Faulkner et al., *Plant Cell* 20:1504–1518, 2008; D, courtesy of Kim Findlay.)

Secondary plasmodesmata, and particularly those in primary pit fields, are usually characterized by cross-connections between them, in the plane of the middle lamella. These connections can become enlarged and may form extensive median cavities within the wall (Figure 7.31). The creation of plasmodesmata and their subsequent elaboration both involve the local remodeling of wall components. New cellulose deposition seems to be inhibited during secondary plasmodesma and pit field formation, while pectin appears to be enriched. The erosion of middle lamella material during branched plasmodesma formation also involves pectin removal. With one exception, how any of these wall events are orchestrated is not yet known.

The exception is our emerging understanding of the role of callose in regulating the size exclusion limit of plasmodesmata. Callose is deposited rapidly in the wall in response to a variety of signals, including transient calcium rises or mechanical trauma. It forms a tight collar around the entrance to the plasmodesma, and in the process reduces its capacity to traffic molecules intercellularly. This would be to no avail if the process were irreversible, but specific β-1-3-glucanase, or callase, molecules are located around the neck of the plasmodesma. When activated, these can remove the callose collar. A balance in the wall is thereby struck between callose deposition (reducing traffic) and β-1-3-glucanase activity (enhancing traffic), but how this balance is regulated at the molecular level is not fully understood (see also Figure 8.22).

3. Local removal of wall material is used to create conducting elements from files of cells.[ref14]

A more extreme example of localized wall removal is seen during the differentiation of the vessel elements and tracheids that make up the conducting elements of xylem (see Concept 1C6). These cells, arranged in long, continuous files, differentiate sequentially into dead and empty elements. A final step in this process is the removal of the end wall between one cell and the adjacent fully differentiated cell, thus connecting the two cells by a hole or set of holes called a *perforation plate* (**Figure 7.33A**). This process is repeated cell by cell along the file, finally creating a hollow water-conducting tube.

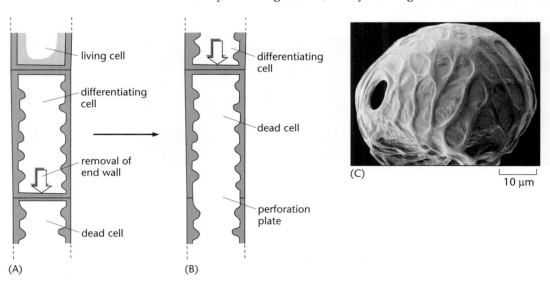

Figure 7.33 Polar end wall removal during tracheary element differentiation. Conducting vessels are made from files of cells that differentiate, one cell at a time, in a polar fashion. (A) A differentiating cell has an already dead element at one end, and it will selectively remove the common wall to form a clear channel while at the same time leaving the wall at the other end intact. (B) This remaining wall will be removed by the next cell along, when it too differentiates. (C) This process of polar end wall removal is a cell-autonomous event and can even be seen in a cell culture system in which single isolated leaf cells are induced to transdifferentiate into dead tracheary elements. (C, courtesy of Preeti Dahiya and Kim Findlay.)

The remarkable part of this whole process is its polarity: completely digesting away one end wall while leaving the other end wall, next to a still living cell, intact. Polar end wall removal is maintained even when a single cell in culture differentiates into a tracheary element (Figure 7.33B). *In vivo*, the precise pattern of end-wall dissolution can vary in a species-specific manner. Removal of a disk of wall material from the end wall creates a *simple perforation plate*, but in some vessels more complex patterns of wall removal are found—for example, scalariform, or ladderlike, perforation plates (**Figure 7.34**). In addition, areas may be removed from the side walls between vessels to create *pits* that allow lateral water flow in the xylem (**Figure 7.35**). In both cases the precise region of wall removal is determined by where the secondary, lignified wall is laid down earlier on. It is the primary wall, left unprotected, that is attacked by a combination of secreted hydrolytic enzymes and removed. A mutant tomato plant, in which secondary wall is laid down where the perforation plate should form, wilts, because it fails to remove the end walls, resulting in xylem that cannot properly transport water.

The precise pattern of secondary wall deposition during xylem cell differentiation appears to be under the control of a plant-specific family of transcription factors that are characterized by a conserved region called a *NAC domain*. Ectopic expression of different NAC family members can drive the formation of secondary wall (but not the subsequent cell death process) in the distinctive patterns seen in both tracheids and vessels. How these NAC factors act both to switch on genes for secondary wall biosynthesis and to regulate the cytoskeletal rearrangements at the cell surface—which, we saw earlier (see Figure 7.7), predict where the wall will be deposited—is not yet understood.

4. Abscission of leaves and fruit is an active process that involves controlled cell separation.[ref15]

Abscission is an active process that leads to the detachment of organs (usually senescent) from the main body of the plant. These organs commonly include leaves, flowers (and floral organs), fruit, seeds, and diseased organs. Although often associated with senescent organs, the abscission process itself does not require cell death. Instead, it is an active process that involves the controlled separation, at the middle lamella, of two groups of densely cytoplasmic cells on either side of the break, in a specialized region called the *abscission zone* (**Figure 7.36**). This group of cells is specified early at the base of all organs that will eventually abscise (**Figure 7.37**). At some point these cells acquire the competence to respond to the abscission signals that will later activate the zone of cell separation (**Figure 7.38**). Following organ

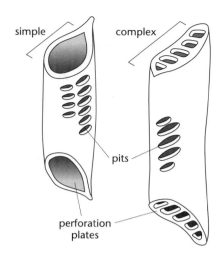

Figure 7.34 Drawing of isolated xylem vessel elements, showing simple and complex perforation plates on their end walls, and pits in their lateral walls.

simple perforation plate

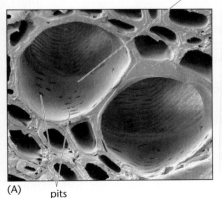
(A)   pits

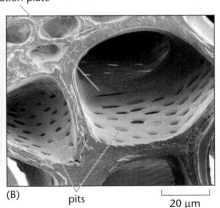
(B)   pits
20 μm

Figure 7.35 SEMs of metaxylem vessels in a *Zinnia* stem, showing the simple perforation plates connecting adjacent cells in the vessel. (Courtesy of Preeti Dahiya and Kim Findlay.)

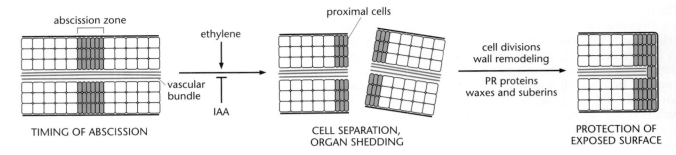

TIMING OF ABSCISSION

CELL SEPARATION, ORGAN SHEDDING

PROTECTION OF EXPOSED SURFACE

Figure 7.36 Schematic diagram of stages in the abscission process. IAA, Indoleacetic acid. PR, Pathogenesis-related proteins.

abscission, the exposed cells revealed on the plant side form a protective layer and often divide and produce suberin and waxes for waterproofing and protection.

The regulatory controls on organ abscission are not well understood, but they include the antagonistic activities of two growth factors, ethylene and auxin. Ethylene actively promotes abscission by inducing the expression of wall-degrading enzymes, while auxin in turn inhibits this, thus delaying abscission. Other key genes have been identified that are likely to be involved in these signaling pathways, including a specific MADS box transcription factor and a leucine-rich repeat receptor kinase, but their precise role remains uncertain. A range of cell wall enzymes that are up-regulated during the abscission process have been identified, including expansins, those involved in pectin breakdown (PME and PG) and those involved in xyloglucan remodeling. The latter may be involved in remodeling the walls of the proximal cells of the abscission zone as they grow and divide following organ abscission. For cell separation at the middle lamella to occur, pectic enzymes are vital, and one specific *endo*-polygalacturonase gene, among a very large gene family, appears to be implicated. Identified in oilseed rape, its expression is restricted to abscission zones and is induced by ethylene. It is not yet known if other pectin-mobilizing enzymes are also involved, and there is still much to be learned about the mechanics of how such a highly localized cell separation event can be achieved.

5.   Fruit softening depends upon the expression and activity of cell wall–modifying proteins.[ref16]

Fruit ripening is an adaptive biological process that attracts herbivores and facilitates seed dispersal. Ripening is important for our diet and is the focus of considerable attention because of its effect on fruit production, transport, and marketing. There are two main categories of ripening fruit, based on the control of ripening by ethylene. Those that require ethylene for ripening are called *climacteric* and include tomato, cucurbits, peach, plum, and apple; while *nonclimacteric fruits*, including strawberry, grape, and citrus, do not (**Figure 7.39**). Fruit ripening is usually accompanied by softening of the fleshy tissue, the pericarp or endocarp. During ripening the primary walls of the cells in these tissues are thickened and enriched in pectic substances, primarily homogalacturonan and rhamnogalacturonan (RG-I). The wall structure becomes increasingly hydrated as the cohesion of the pectin gel changes, resulting in wall swelling and a loss of cell-cell cohesion in the pectin-rich middle lamella. Softening is associated with a progressive disassembly of the network formed by cross-linking of cellulose microfibrils and matrix glucan, particularly xyloglucan. Synthesis and incorporation of new wall components also continues throughout ripening.

Softening associated with ripening fruit is determined by the loss of cell-cell adhesion, and the extent of wall degradation. Variation in the composition of the wall and the expression of wall-modifying enzymes contributes to the variation observed in the texture and quality of different fruits during ripening. Softening is evident in strawberry, which acquires a soft melting texture.

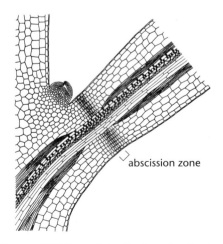

Figure 7.37 Drawing of a longitudinal section through a typical abscission zone. (From F.T. Addicott, *Abscission*. Berkeley: University of California Press, 1982.)

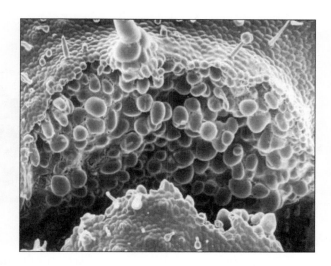

Figure 7.38 SEM of proximal face of abscission zone in tomato. It is assumed that the swollen rounded cells on the proximal surface of the zone provide part of the pressure needed for organ separation. (Courtesy of Jeremy A. Roberts.)

Softening of the cell walls is relatively delayed in apple. The cells of the apple cortex remain turgid and firmly attached to each other, giving rise to a crisp texture upon ripening. In contrast, ripe red tomato pericarp becomes soft through loss of cell-cell adhesion. Some cells dissolve completely.

The process of ripening in tomato has been studied in the greatest detail. The analysis of transgenic plants, in which the levels of wall-modifying proteins have been increased or decreased, has been particularly useful in understanding the mechanism of fruit softening. For example, polygalacturonase (PG) activity is typically undetectable before ripening, but increases during the ripening process. Although there is a large gene family for PGs, one family member is expressed at very high levels during ripening. Surprisingly, suppression of PG expression only slightly affects fruit softening in tomato, but extends the shelf life. PG is largely responsible for pectin depolymerization, subject to its prior demethylesterification by pectin methylesterase (PME). These pectin-modifying enzymes, which also include pectate lyase, affect the integrity of the middle lamella and the three-way junctions (see Figure 6.41), in turn reducing cell-to-cell adhesion, tissue integrity, and consequently fruit texture. Methylesterification of cell wall pectin declines from 90% in green tomato fruit to 35% during ripening, and removal of the methylesters makes the polyuronides more susceptible to degradation by PG and pectate lyase. Suppression of PME activity by antisense reduces the loss of methylesters in walls during ripening, but it does not affect softening during normal ripening. Another major change in tomato cell walls during ripening is the loss of galactosyl residues, most likely through the activity of a specific β-galactosidase (*exo*-β-D-galactosidase). Loss of cell wall galactose may contribute substantially to fruit softening, perhaps affecting the porosity of the wall and increasing access to other hydrolases.

The viscoelasticity of tomato paste can be increased by reducing either PG or PME activity, but the detailed roles and effects of these enzymes remain

climacteric fruit	nonclimacteric fruit
(require ethylene)	(do not require ethylene)

Figure 7.39 Climacteric and nonclimacteric fruit.

Figure 7.40 Ripe tomatoes.
(A) Cross section of a ripe, salad tomato. (B) Tomatoes on their way to a processing plant in the Philippines.

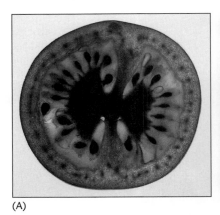

(A)

(B)

unclear. One of the greatest contributions to the viscosity of tomato paste, however, is initial wall thickness; the thicker the wall, the thicker the paste. Processing tomatoes, destined for paste (**Figure 7.40**), have walls that are several times as thick as salad tomatoes.

Expansins are a family of proteins that, as we have seen, affect cell wall extensibility (see Concept 7B5), but they may also play an important role in fruit ripening. Ectopic overexpression of one member of the expansin gene family in tomato led to enhanced softening, even in mature green tomatoes, while repression of expansin reduced softening. Its mode of action is most likely indirect, through its general role in wall remodeling.

Much remains to be learned about the processes of cell wall modification leading to fruit softening, largely because the key cell wall polymer substrates and products have yet to be identified. There are many enzymes involved in this complex process, and despite a wealth of experimental data, there is still no clear correlation between wall changes and the larger-scale mechanical properties. Other wall-localized enzymes have been investigated for their roles in fruit ripening, including pectate lyase, pectin esterase, *endo*-(1→4)-β-D-glucanase, and xyloglucan *endo*-transglucosy-lase-hydrolase, but direct evidence of their roles in softening has not yet been obtained. The sense of confusion about how wall changes relate to ripening changes is highlighted by the isolation of an intriguing tomato cultivar. The Delayed Fruit Deterioration (DFD) cultivar bears fruit that ripens normally and appears to undergo all the usual observed wall changes described above, but stubbornly refuses to soften. In this case the key difference appears to be compositional and architectural changes in the cuticle overlying the DFD fruit, which minimize water loss during ripening and result in enhanced turgor in the cells within. It is clear that ripening and softening processes are more complex than we thought, and that ripening-related physiological processes other than wall dynamics are likely to be very important. There is considerable value in understanding more about these processes, because ripening affects the quality of fruit in many ways, particularly fiber content and composition, lipid metabolism, vitamin content, and levels of antioxidants. A better understanding should lead in the future to the development of fruit with enhanced quality and nutritional value.

6. In some seeds, wall polysaccharides can form a food reserve to be used during germination.[ref17]

All seeds require an internal store of food that can be used by the growing embryo during germination. As the seed germinates, these reserves are mobilized to support the emergence and growth of the young seedling. In most seeds the polymer used to store carbon is starch, but in some instances wall polysaccharides can perform the same storage function. In the same way that starch is broken down on germination to form monosaccharides that can be mobilized and reused by the seedling, so too wall storage polymers can be hydrolyzed and the products used for seedling growth. Plants have recruited two main kinds of wall polymers, both hemicelluloses, as storage materials: mannan and xyloglucan. Pure mannans, in α-1,4 linkage, form the insoluble, crystalline, and very hard walls that are found in the endosperm of date palms and ivory nuts. Decoration of the mannan backbone with β-1,6-galactose side chains produces a variety of galactomannans that are found as storage polysaccharides in the endosperm of several legume seeds including guar, carob, and locust bean (**Figure 7.41**). These polymers are hydrophilic materials that form useful gums, stabilizers, and thickeners, which have found a wide range of uses, in both industry and food technology (see Chapter 9). In addition, their water-absorbing and retaining properties are thought, in nature, to help seed hydration and to avoid subsequent desiccation (**Figure 7.42**). The precise rheological properties, and value, of each seed galactomannan depend critically on the degree of galactose substitution. In the seeds of tamarind and nasturtium, a form of xyloglucan is deposited as a storage polysaccharide.

Like other hemicelluloses, the wall storage polymers are all assembled in the Golgi apparatus and secreted. During seed formation, these storage polymers are steadily deposited on the inner face of the primary walls of the endosperm, resulting eventually in cells with a very reduced cytoplasmic volume and enormously thick walls. On germination, an appropriate suite of hydrolytic enzymes, secreted by the aleurone layer, digest these thick walls, and the embryo then uses the resultant sugars as a primary carbon source (Figure 7.42). Considerable progress has been made in characterizing the enzymes involved in both the synthesis and the breakdown of cell wall storage polymers. Genes encoding the mannan synthase and the galactomannan galactosyltransferase (see Concept 5C8) required to make galactomannan have both been described (the galactosyltransferase being the first plant cell wall glycosyltransferase gene to be cloned). Overexpressing the mannan synthase gene in soybean cells in culture led to the production of β-1,4-mannan in their cell walls, the final proof that the mannan synthase had been successfully cloned. The gene belongs to one of the families of cellulose synthase-like genes, *CslA*, the first member of these families (see Figure 5.19) to be assigned a precise function.

7. Lignin and suberin provide physical barriers to the turnover and degradation of secondary walls.[ref18]

Macromolecular turnover is generally a feature of living cells, and therefore there is little wall turnover in cells that have formed secondary walls and completed programmed cell death. Secondary walls, formed after the primary walls have stopped expanding, are characterized by thick layers of densely packed and highly oriented crystalline cellulose. Cellulose fibrils in secondary walls are abundant and inert, and there is little indication that they are turned over.

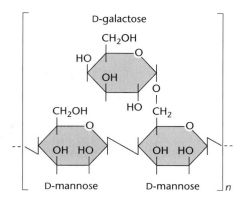

Figure 7.41 The structure of guar seed gum. The guar seed storage galactomannan has a backbone of β-1,4-mannan with side branches of α-1,6-galactose on alternate mannose residues. The polymer is used as a stabilizer in a wide variety of food and drink applications.

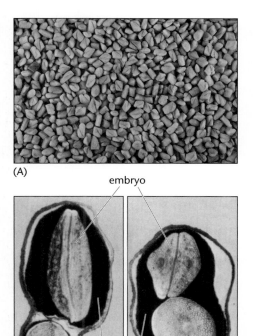

Figure 7.42 Fenugreek seeds store galactomannan.
(A) Hard amber-colored fenugreek seeds. (B) Galactomannan in the walls of the endosperm layer accounts for 30% of the mass of this legume seed. On imbibition, this reserve is rapidly mobilized and then reused to form starch in the cotyledons of the germinating embryo. In the sections of fully hydrated seeds shown here the galactomannan in the endosperm has been stained black. (A, from *Wikipedia* Creative Commons. B, from J.S.G. Reid and J.D. Bewley, *Planta* 147:145–150, 1979.)

Figure 7.43 Lignin encrustation of the cellulose network. This model of a bundle of cellulose microfibrils in the lignified S2 layer of a secondary xylem wall has been obtained by electron microscopic tomography and three-dimensional reconstruction. Ninety tomographic slices were used, each 1 nm thick, to produce the model, which shows both the digitally isolated microfibrils (A) and the encrusting lignin and residual hemicellulose (B) that completely fills up the remaining space around the cellulose microfibrils. (Adapted from P. Xu et al., *Wood Sci. Technol.* 4:101–116, 2007.)

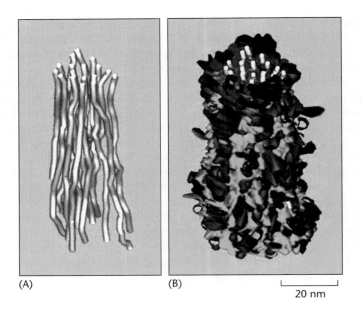

(A)   (B)

20 nm

In the formation of wood, lignification creates a phenolic polymer that extends throughout the middle lamella, the primary wall, and the secondary wall to form a dense, waterproof phenolic network and a smooth hydrophobic surface on the interior of the thickened secondary cell wall (see Figures 3.24, 3.33, and 5.50). The lignin polymer, which is formed for mechanical support, particularly in water-transporting cells, protects the secondary wall from degradation. Enzymes of even modest size do not penetrate the lignified wall readily, and the embedded polysaccharides are poorly accessible to enzymes that could act to turn over carbohydrates in the primary walls (**Figure 7.43**, see also Figure 5.50). Lignin is an effective hydrophobic barrier, and therefore lateral movement of water and solutes through the walls of tracheids and vessels is greatly restricted. Its hydrophobicity and diversity of linkages allow lignin to form an effective barrier against degradation by microorganisms. Bark is rich in suberin as well as lignin, and this tissue too is therefore highly resistant to degradation.

In secondary walls, following lignification, additional wall-associated macromolecules may be fixed in place and remain immobilized indefinitely. Cell wall–associated proteins, and some enzymes, often abundant and associated with the plasma membrane and/or the cell wall, appear to be immobilized in lignified cell walls. Most woods contain about 0.05% nitrogen. In woody plants, where mature xylem is retained, these immobilized and insoluble components are resistant to elution by the transpiration stream and remain fixed in cell walls for the life of a tree and even afterward in solid wood.

8. Gravity sensing and mechanical stress lead to compensatory changes in cell wall synthesis and architecture.[ref19]

Cell walls can be dramatically remodeled when plant stems react to the mechanical stresses imposed by bending or gravity. These forces are common in the face of environmental factors such as physical obstacles, wind, ice, and snow. Trees sense their position and balance their weight against gravity by differential growth to maintain the large mass of wood in a relatively upright position. Trees respond to mechanical stress and gravity by adjusting their height, mass, crown shape, branch density, and branch angle. Wood formed under bending is known as *reaction wood* and is distinct from

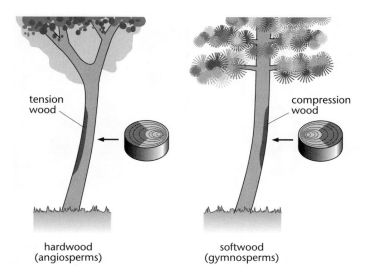

Figure 7.44 Two kinds of reaction wood. Tension wood forms on the upper side of leaning trunks and branches of angiosperms (hardwoods). Compression wood forms on the lower side of trunks and branches of gymnosperms (softwoods). Both act to help correct the deviation from vertical growth in the trunk, and to support the branches.

normal wood. Softwoods and hardwoods (see Figure 9.27) both respond to bending, but are each remodeled in different ways. These responses change the structure and chemistry of the walls and the morphology of the cells in the xylem, and these in turn affect the physical properties of the wood.

In softwoods (gymnosperms), the response to bending or mechanical stress occurs on the underside of the bent branch or stem, and the result is called *compression wood*. By contrast, the reaction wood of hardwoods (angiosperms) forms on the upper side of the bent branch, and the resultant wood is called *tension wood* (**Figure 7.44**). Both responses are adaptations that attempt to right bent stems or to maintain a proper branch angle. Compression wood is stronger in longitudinal compression but weak in tension and toughness (resistance to splitting or to sudden loads), whereas tension wood is stronger under tension, has increased toughness, and is weak under compression.

Environmental signals induce both kinds of reaction woods, but the tendency to form it may be inherited and in some cases may be associated with straightness, a desirable trait in timber and therefore for tree breeders. Trees of similar straightness may vary greatly in their response to gravity, some producing large amounts of reaction wood, which is an undesirable trait in wood used as a building material. Both types of reaction wood differ from normal wood in both cellular morphology and chemical composition.

Induction of compression wood has been attributed to inhibition of auxin transport, since auxin transport inhibitors such as morphactin can induce compression wood. Compression wood is darker in color, heavier, harder, and denser than normal wood. The tracheids are shorter, have thicker walls, and have a round rather than a rectangular shape in cross section (**Figure 7.45**). Along with the round shape, there are air spaces in the three-way junctions between tracheids, while normal tracheids do not have air spaces (Figure 7.45). The S1 layer is thicker than normal in compression wood, and the S3 layer is absent. The S2 layer in compression wood tracheids contains distinctive narrow, branched, helical cavities (see Concept 3F8).

Tension wood is formed on the upper side of the leaning branches or stems of angiosperms (hardwoods). It is characterized by thick-walled fibers, containing an S1 and an S2 layer, but the innermost layer of the wall is a gelatinous layer (G layer), consisting largely of highly crystalline cellulose oriented in the direction of the fiber axis (**Figure 7.46**). Tension wood has a reduced content of lignin.

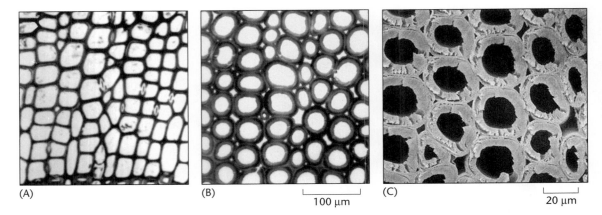

(A)   (B) └────┘ 100 μm   (C) └────┘ 20 μm

Figure 7.45 Compression wood. (A, B) A seedling stem of Douglas fir (a gymnosperm) displaying normal wood (A) and compression wood (B). The tracheids in compression wood have thicker walls, are more rounded, and have larger intercellular spaces than those in normal wood. (C) Scanning electron micrograph of compression wood. (A and B, from M. Kwon et al., *Phytochemistry* 57:847–857, 2001. C, from S.N. Kartal and S.T. Lebow, *Wood Fiber Sci.* 33:182–192, 2001.)

D. Cell Wall–Derived Signals in Growth and Development

1. Many signals combine to regulate plant growth.[ref20]

The growth and development of plants is carefully controlled by a combination of both intrinsic and extrinsic factors that are intimately linked. Intrinsic factors, for example, the constraints on growth imposed by plant growth factors, are in turn linked to extrinsic factors such as gravity, nutrients, and light. Growth factors made by cells in one part of an organism regulate processes in cells in another part of the organism and are examples of *signal molecules.*

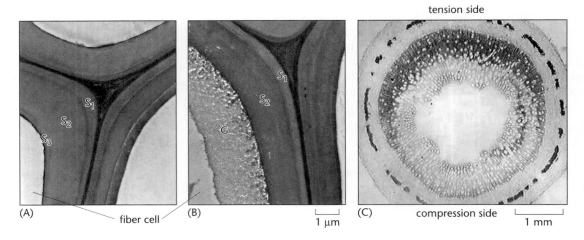

tension side

(A)   fiber cell   (B) └────┘ 1 μm   (C)   compression side   └────┘ 1 mm

Figure 7.46 Tension wood. (A, B) Electron micrographs of sections through wood from poplar, *Populus deltoides*, reveal clear differences between the fibers in normal wood (A) and those in tension wood (B). In tension wood, the S2 layer is smaller, the S3 layer appears to be missing, and, most obvious, a large G layer (gelatinous layer) has been deposited on the inner face of the wall. This G layer contains some lignin, but it is particularly rich in cellulose. (C) The increased cellulose production can be seen in this transverse section of an aspen stem that has been kept bent for 7 days. The section has been hybridized with an antisense *PtrCesA1* RNA probe, and the increase in transcript level of this secondary cell wall–specific cellulose synthase can be clearly seen on the tension stress side (*top*) compared with its relative suppression on the opposite side of the stem (*bottom*). (A and B, from J.-P. Joseleau et al., *Planta* 219:338–345, 2004. C, from S. Bhandari et al., *Planta* 224:828–837, 2006.)

Signal molecules, in both animals and plants, come in many different classes and sizes, and include the aromatic heterocycles (epinephrine in animals, and the auxins, brassinosteroids, and cytokinins in plants), lipophilic terpenoids (such as estrogen in animals and the gibberellins in plants), and a wide variety of polypeptides (insulin in animals, systemin and phytosulfokine in plants). The most studied signal molecules of plants are the classic so-called *plant growth factors*, or *phytohormones*. These include abscisic acid, auxin, brassinosteroids, ethylene, gibberellins, cytokinin, jasmonate, and salicylate. Most plant signal molecules move intercellularly to regulate gene expression and enzyme activities and thereby regulate processes associated with the physiology, development, growth, defense, and stress responses of plants (**Figure 7.47**). Many signal molecules interact with receptors situated on the outer surface of cells, although some signal molecules, such as auxin and ethylene, traverse the plasma membrane to reach their receptors. Other signal molecules, such as salicylic acid, which systemically activates defense processes in plants have equally profound effects on target tissues (see also Concept 8A7).

The binding of signal molecules to receptor proteins initiates intracellular signal pathways leading to biological responses. Although the binding of a signal molecule to its receptor resembles the binding of a substrate to an enzyme, the signal molecule is usually bound more tightly to its receptor. Binding tightness is measured as the concentration of the signal molecule or substrate necessary to occupy half of the receptor's or enzyme's binding sites (the K_m). In the case of receptors for signal molecules, the K_m is usually between 10^{-6} and 10^{-9} M (tight binding), whereas enzymes generally operate with K_m values between 10^{-3} and 10^{-6} M (more loosely bound).

The strong binding of signal molecules to receptors also means that there are stringent structural requirements for a molecule to possess signaling activity. The more exact the structural requirements for molecules to activate receptors (**Figure 7.48**), the less likely that inappropriate molecules will activate them. Most of the controls on plant cell growth and differentiation revolve around the classic plant growth factors and key external signals such as light and gravity, and this is too big a field to discuss here. It is important to realize also that the various signal inputs have to be integrated by the cell, often by core transduction pathways. For example, the gibberellin signaling pathway also acts in both abiotic and biotic stress signaling, a situation that makes sense in that plants subjected to stresses may also want to divert resources away from growth. Drought-stressed plants grow more slowly than well-watered ones. Biotic stresses or wounding often stimulate the production of ethylene, an auxin antagonist, with consequences again for growth and development. It is also clear that these controls can be subtly modulated in many cases by components of, or derivatives of, cell wall macromolecules themselves. In some cases it could be the wall itself acting as a feedback signal to the cell (see also Concept 8B11), but in other cases it is fragments released from wall carbohydrates. The actions of such wall fragments, called *oligosaccharins*, have been most clearly characterized in the plant's response to pathogen interactions (see Chapter 8), but they may also affect normal growth processes, as we now discuss.

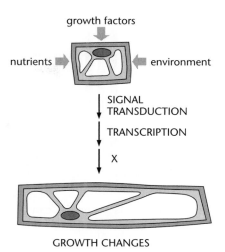

growth factors

nutrients → | | ← environment

SIGNAL TRANSDUCTION

TRANSCRIPTION

X

GROWTH CHANGES

Figure 7.47 Signals on the way to growth. Schematic diagram showing the range of signaling inputs that contribute to cell growth as the output. Environmental signals are particularly important in plants and include light, water potential, temperature, pH, gravity, and mechanical stresses. All these inputs are integrated into changes in gene expression, which finally result in the production of more wall and cytoplasm. It is the mechanics of this last step that remain inscrutable.

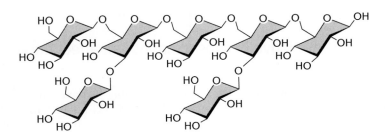

Figure 7.48 Structural specificity of the heptaglucoside elicitor. Glucan fragments released from a fungal pathogen cell wall by the action of a plant host *endo*-glucanase act as potent signals to the plant to mount a defense response. The structure of the most active fungal wall fragment, a heptaglucoside, is shown here. Small structural modifications made to this molecule have large effects on its biological activity.

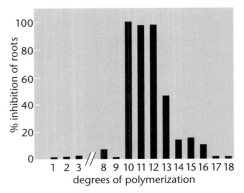

Figure 7.49 Oligogalacturonides can inhibit the formation of roots on tobacco leaf explants. Auxin can induce root formation on such explants, and oligogalacturonides with a degree of polymerization in the range from 10 to 15 can inhibit the process. Increasing the auxin concentration can overcome the inhibition. (From D. Bellincampi et al., *Plant J.* 4:207–213, 1993.)

2. Oligogalacturonides can modulate development in tobacco explants.[ref21]

Oligogalacturonides, released from the host wall's pectin by enzymes secreted by a variety of pathogens, trigger a wide range of different host defense responses (see also Concept 8B7). They can elicit phytoalexin accumulation in soybean (*Glycine max*), castor bean (*Ricinus communis*), and parsley (*Petroselium crespum*); lignification in cucurbits (Cucurbitae); β-glucanase in parsley; chitinase in tobacco (*Nicotiana tabacum*); proteinase inhibitors in tomato (*Lycopersicum esculentum*); isoperoxidases in castor bean; hydrogen peroxide in cucumber (*Cucumis sativus*); membrane responses in soybean and tobacco; and ethylene in tomato and pear. A wide variety of plants respond to these bioactive oligogalacturonides and in turn activate one or more of a variety of defense responses.

But oligogalacturonides (short linear α-1,4-homogalacturonides) also appear to exert an effect on plant growth and development. For example, oligogalacturonides with DPs of 10 to 15 can inhibit the ability of stem and leaf explants to form roots (**Figure 7.49**). Yet in leaf explants the ability of approximately 1 μM oligogalacturonide to inhibit root formation can be overcome by a tenfold increase in auxin concentration (to 6 μM). The same-sized oligogalacturonides (DPs 10–15) induce the formation of vegetative shoots as well as inhibit the formation of roots in stem explants of vegetative tobacco plants, and they induce the formation of inflorescences as well as inhibit the formation of roots in stem explants from flowering tobacco plants. Oligogalacturonides exhibit half their maximum ability to induce organogenesis at about 0.1 μM, which is 10- to 100-fold lower than the concentration required for the elicitation of defense responses.

Oligogalacturonides thus appear to have profound, if varied, effects on plants, and very little is known about how they elicit their effects, although it is noticeable that most but not all could be interpreted as being the opposite of the effect of adding auxin. The size requirement for bioactive oligogalacturonides, a minimum of ten GalA residues, suggests that a specific conformation of oligogalacturonide is important for its recognition by the plant. Although the results from some of these experiments are quite dramatic, it is important to be aware that the experimental systems involved the ectopic application of oligosaccharides, together with plant growth factors, to excised plant parts.

In principle there are two main frameworks within which the results could be interpreted. The first might include a specific oligogalacturonide receptor and a downstream signaling pathway that modulated growth, although to date, receptors for oligogalacturonides have not been identified. The second would interpret their effects as being a downstream consequence of a defense response where ethylene and auxin could be involved. Ethylene is an auxin antagonist, and some of the effects observed in response to oligogalacturonides might be attributed to auxin antagonism. Since many of the key developmental experiments were performed before the ready availability of characterized growth factor mutants, and have not been repeated in such backgrounds, it is still hard to be sure whether the observed effects of oligogalacturonides on development are primarily to elicit a defense response that has secondary downstream growth effects, or whether they act as direct signal molecules in growth and development.

3. Xyloglucan-derived oligosaccharins can affect the rate of elongation growth.[ref22]

Fragments from xyloglucan have also been found to have an effect on growth. The xyloglucan-derived nonasaccharide XXFG (at 10^{-8} to 10^{-9} M) inhibits that portion of the growth of pea stem segments that is stimulated by the auxin analog 2,4-D (2,4-dichlorophenoxyacetic acid) at 10^{-6} M. (See Table 3.1 for the structures and shorthand nomenclature of the xyloglucan oligosaccharides.) The L-fucosyl residue of XXFG is essential for activity, as its removal with an α-fucosidase abolishes its growth-inhibiting activity. Furthermore, the octasaccharide XXLG and the heptasaccharide XXXG, both lacking the L-fucosyl residue, are unable to inhibit 2,4-D-stimulated growth of pea stem segments. The rare xyloglucan fragment XFFG (which has two terminal L-fucosyl residues α-2-linked to D-galactosyl residues) is an even more potent inhibitor than XXFG of 2,4-D-stimulated growth of pea stems.

The fucosyl, galactosyl, and xylosyl residues, and probably the glucosyl residue of the fucosyl-bearing side chain of XXFG, are the only ones required to provide the oligosaccharide with growth-inhibiting activity. The ability of the reduced xyloglucan pentasaccharide FGol to inhibit 2,4-D-stimulated growth of pea stems, albeit requiring a tenfold higher concentration than is required of XXFG, supports the idea that the fucosyl-containing tetrasaccharide is the key portion of XXFG required for the oligosaccharide to exhibit growth-inhibiting activity. The inability of the trisaccharide L-fucosyl-α-1,2-D-galactosyl-α-1,6-D-glucose to inhibit 2,4-D-stimulated growth of pea stems demonstrates that the fucosyl-galactosyl disaccharide of XXFG by itself is not sufficient to possess demonstrable growth-inhibiting activity. The ability of XXFG to inhibit growth is eliminated by the action of a cell wall–localized α-fucosidase.

A variety of xyloglucan-derived oligosaccharides (with or without the fucosyl residue required for growth-inhibiting activity) have been reported to slightly stimulate the growth of pea stem segments in the absence of added 2,4-D at oligosaccharide concentrations above 10^{-6} M, which is 100-fold greater than that required to inhibit growth.

All of the xyloglucan oligosaccharides that promote growth (**Figure 7.50**) possess an unsubstituted α-D-xylosyl-1,6-β-D-glucosyl moiety at the nonreducing terminus. These same glycosyl residues are required for the xyloglucan oligosaccharides to function as acceptor substrates for xyloglucan *endo*-transglycosylase, a putative growth-regulating enzyme (see Concept 7B4); and indeed, there is evidence that XXFG *in muro* is formed by xyloglucan degradation and can be incorporated into wall xyloglucan by XET activity. Their role as acceptors for XTHs, and their participation in XET reactions, may thus account for the observations of the growth-promoting action of xyloglucan oligosaccharides at high concentrations.

This does not, however, explain the inhibitory effects on growth at low concentrations. Neither, in this case, can we explain them as being secondary effects of an elicited defense response, since these oligosaccharides are not known to act as elicitors. And, as no receptors have been identified either, the molecular basis for the growth inhibition, and its *in vivo* relevance, remain unclear. The inhibitory effects of XXFG and XFFG on 2,4-D-stimulated growth of pea stems exhibit concentration optima (**Figure 7.51**). The growth-stimulating properties of XFFG should equal those of XXFG, because both oligosaccharides possess the fucosyl-containing tetrasaccharide required (at 10^{-8} to 10^{-9} M) to inhibit growth of pea stems and both oligosaccharides possess the unsubstituted α-D-xylosyl-1,6-β-D-glucosyl disaccharide at the nonreducing terminus reportedly required (at 10^{-6} M) to stimulate stem growth. If these oligosaccharins inhibit growth at low concentrations and stimulate growth at high concentrations, it would explain the observed concentration optimum exhibited for their growth-inhibiting activity.

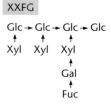

Figure 7.50 Xyloglucan fragments that can inhibit the auxin-stimulated growth of pea stems.

Figure 7.51 The effect of the xyloglucan fragment XXFG on the 2,4-D-stimulated elongation of pea stem segments has a concentration optimum. (Adapted from C. Augur et al., *Plant Physiol.* 99:180–185, 1992.)

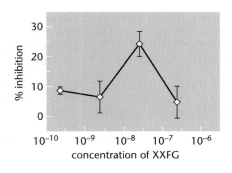

4. Lipo-oligosaccharides synthesized by rhizobia regulate nodule development in host plants.[ref23]

Some plants, and legumes in particular, are able to fix atmospheric nitrogen through a remarkable symbiosis with bacteria in the soil. To set up this mutually beneficial interaction, the two partners need to communicate. The roots secrete particular flavonoids into the soil, where specific rhizobial bacteria in the vicinity recognize them as signal molecules. In response, they then synthesize and secrete modified linear β-1,4-linked tetra- and pentasaccharides of N-acetylglucosamine that cause root hair deformation, cortical cell divisions, formation of thick short roots, depolarization of root hair plasma membranes, and, in some symbiont legumes, formation of pseudo-root nodules that are free of bacteria. These modified oligosaccharides are called *Nod factors*. Nod factors produced by a particular *Rhizobium* species have an effect on the roots of only those host legumes with which they form a nitrogen-fixing symbiosis.

Each *Rhizobium* strain secretes Nod factors that are characteristically modified N-acetylglucosamine oligosaccharides (**Figure 7.52**). Almost all the modifications that have been identified so far are made to the reducing and nonreducing glucosamine residues of the oligosaccharides. The reducing N-acetylglucosamine residue is generally modified at O-6 by acetylation, sulfation, or fucosylation, and if present, the fucosyl residue can be substituted with a methyl ether or a sulfate. The O-6 of the nonreducing N-acetylglucosamine residue can be acetylated, and the N-acetyl group of this residue is invariably removed and substituted with a fatty acyl group. The structure of the fatty acid is specific to both the *Rhizobium* strain and the legume host. Nod factors of *Rhizobium* and other nitrogen-fixing bacterial

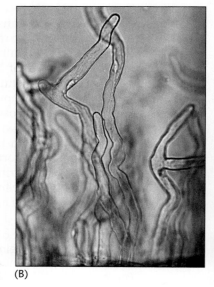

Figure 7.52 Nod factor structure and host response. (A) The structure of the specific Nod factor made by *Sinorhizobium meliloti*, during its specific symbiosis with *Medicago truncatula*, a small, cloverlike plant used widely as a genetic model organism for legume biology. The backbone oligosaccharide is attached to a fatty acyl group, and regions of the molecule where other modifications have been characterized are shaded in gray. It is the specific combination of decorations that confer specificity on the symbiotic interaction. (B) Root hairs of the legume *Medicago truncatula*, grown in the presence of the Nod factor shown here, respond by changing their morphology from their usual straight shape and become distorted and branched. These hairs may be compared with the same but untreated ones shown in Figure 7.20C. (B, courtesy of Giles Oldroyd.)

(A)

(B)

fatty acyl group

symbionts that have other host plants are being characterized at a rapid rate, and invariably the Nod factors are uniquely modified N-acetylglucosamine oligosaccharides. The Nod factor alone causes root hair deformation (Figure 7.52), downstream "spiking" of intracellular calcium, and the induction of plant target gene expression, all at a concentration of only 1 pM. The Nod factor signal is perceived by the extracellular domain of a receptor-like kinase that is located in the plant plasma membrane.

The genes responsible for the biosynthesis of the *Rhizobium* Nod factors are well characterized. Only the highly conserved *Rhizobium* nodulation genes *nodABC* are required to produce the oligosaccharide backbone of these molecules. The *nodC* gene is thought to encode the N-acetylglucosaminyltransferase, the *nodB* gene encodes the enzyme responsible for N-deacetylation of the terminal nonreducing N-acetylglucosamine residue, and the *nodA* gene encodes the enzyme responsible for N-fatty acylation of the free amino group formed by the Nod B protein on the terminal N-acetylglucosamine residue of the chitin oligosaccharides.

Each of these genes has been selectively inactivated by transposon mutagenesis, and all mutations that impair the synthesis of its Nod factor also prevent that *Rhizobium* from forming a symbiotic relationship with its host. Furthermore, transferring into a *Rhizobium* strain one or more genes responsible for synthesis of the host-specific structural features of Nod factors predictably alters the host range of the *Rhizobium*. Nod factors are by far the best-characterized oligosaccharins that regulate growth, development, and morphogenesis in plants. Many of the signaling intermediates in the nodulation pathway are shared by another symbiosis, that between the host and mycorrhizal fungi that develop a close relationship with its roots. Whether or not oligosaccharide-based signal molecules are also involved upstream in this interaction remains to be seen.

5. Chitinases may function during normal plant development.[ref24]

Chitinases are proteins encoded by a large family of genes, about 25 in *Arabidopsis* and 44 in rice. The best-characterized family members are inducible enzymes that can degrade chitin, most often in the cell walls of fungal pathogens. The plant's innate immune response recognizes the chitin oligosaccharide elicitors released and mounts a defense response. Indeed, the recognition of chitin oligosaccharides seems to play a key role in the establishment of basal resistance to potential pathogens in all plants. Plants do this through a membrane-glycoprotein receptor kinase that has some structural similarity to the Nod factor receptor mentioned earlier, in that both have extracellular ligand-binding motifs called LysM (see Figure 8.7). This is unsurprising, since the Nod factors and the chitin oligosaccharides are structurally similar. Signaling downstream is through a MAP kinase cascade.

But other chitinase family members are not pathogenesis-related and are more intriguing. One chitinase gene in *Arabidopsis*, and more than half those in rice, actually encode the protein described earlier called yieldin (see Concept 7B5), a protein with chitinase homology involved in reducing the wall yield threshold during cell elongation. Yieldin has only very limited catalytic activity, and its true substrate and mode of action are unknown. Yet other family members, encoding both acidic and basic chitinases, are constitutively expressed in the plant, appear to be developmentally regulated, and are not induced by stress signals of any kind.

For example, acidic chitinases are present in particularly high concentrations in pedicels and sepals, and chitinase gene expression also correlates with floral initiation. Although it is a possibility that chitinases have a primary function in defense, this is by no means clear. The expression of another *Arabidopsis* gene, related to the class II chitinases, is regulated during normal plant development, rather than by pathogens. If the protein, AtCTL1, which appears to have no enzyme activity, is deleted by mutation, the resultant plant has a variety of severe defects. Among these are features such as shorter, fatter hypocotyls and increased root hairs, a phenotype that can be attributed to increased ethylene production in the mutant. But there is also a considerable degree of ectopic lignin deposition in cells, such as pith parenchyma, that would never normally make lignin, and this cannot be attributed to ethylene. How this putative chitinase defect can lead to ethylene production and the ectopic lignin phenotype remains obscure.

An acidic *endo*-chitinase has been reported to function in early somatic embryo development of carrot (*Daucus carota*). A temperature-sensitive carrot cell line was selected that does not proceed beyond the globular stage at the nonpermissive temperature of 32°C. The arrested development is overcome in a small but significant proportion of the embryos by the addition to the culture medium of a mixture of the proteins secreted by wild-type carrot cells. A 32-kD protein was isolated from the secreted proteins by following the ability of the protein to rescue somatic embryos. The pure 32-kD protein was shown to be an acidic *endo*-chitinase. Restoration of the embryo's protoderm is the principal morphological effect of 10^{-8} M *endo*-chitinase on globular stage-arrested embryos. The corresponding gene in *Arabidopsis*, *AtEP3*, has also been characterized (**Figure 7.53**).

While searching for a putative product of the 32-kD carrot *endo*-chitinase that enables the temperature-sensitive carrot cell line to complete embryogenesis, it was discovered that adding *Rhizobium* Nod factors (the *N*-acetylglucosamine–containing lipo-oligosaccharides) to the cultures mimics the effect of adding carrot *endo*-chitinase. The *N*-acetylglucosamine-containing lipo-oligosaccharides (at 10^{-8} M) are as effective as the 32-kD acidic *endo*-chitinase, in rescuing the development of ts11 carrot embryos. It may be relevant that the *Rhizobium* Nod factors are *endo*-chitinase substrates. The two or three unsubstituted *N*-acetylglucosamine residues in the center of each Nod factor have the same structure as chitin and represent the point at which *endo*-chitinase cleaves these molecules.

It is important to remember that no plant substrate for *endo*-chitinases has yet been identified, either in walls or in any other part of uninfected plants, although cytochemical studies do provide some support for their existence in plant cell walls. In an attempt to localize putative chitin-related molecules in plant cells, ultrathin sections of healthy and fungus-infected plant tissues were treated with microbial chitinase-gold complexes or wheat germ agglutinin (WGA)-ovomucoid-gold complexes. WGA is a plant lectin specific for *N*-acetylglucosamine–containing oligosaccharides. Gold-labeled chitinase and WGA both bind abundantly to secondary cell walls of healthy

Figure 7.53 The localized expression of an *Arabidopsis* chitinase gene, *AtEP3*. The gene promoter has been fused to an enzyme reporter gene, *GUS*, whose presence can be detected histochemically, here seen as a dark deposit. (A) In the ovule, at this stage of seed development, the chitinase gene is expressed only in the degenerating synergid cells in the embryo sac. (B) In the root, the gene is expressed in elongating root hairs and other epidermal cells in a narrow region of the zone of cell differentiation. (Adapted from P.A. Passarinho et al., *Planta* 212: 556–567, 2001.)

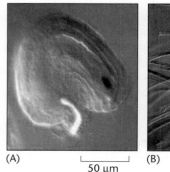

(A)   50 µm   (B)   50 µm

and fungus-infected elm (*Ulmus* spp.) wood, tomato root, potato root, and eggplant (*Solanum melongene*) stem. Enzymatic digestion of plant tissues with proteinase K did not interfere with the labeling. However, the labeling was not observed following lipase digestion of the tissues, suggesting that the putative *N*-acetylglucosamine residues to which the chitinase and WGA were binding could be linked to lipid-containing molecules, in much the same way as the lipo-oligosaccharide Nod factors. Some arabinogalactan proteins, which can also be GPI-anchored, are found in both primary and secondary walls, and have been reported to contain *N*-acetylglucosamine, so they also have been claimed as potential substrates.

We have touched in this section on how carbohydrate fragments can act as signals for plants, but in the next chapter we discuss the much more detailed and complex picture that emerges in the battleground between plants and their pathogens, where a constant interchange of molecular signals, including carbohydrate fragments, occurs between host and pathogen in their evolving battle for survival.

Key Terms

abscission
abscission zone
acid growth hypothesis
anisotropic cell expansion
aquaporin
cell expansion
cell growth
climacteric fruit
compression wood
cortical microtubule array
diffuse growth
endoreduplication
expansin
fruit ripening
gibberellic acid (GA₃)
GPI-anchored AGP
Hechtian thread (strand)
lipid transfer protein (LTP)
load-bearing component
LRR-extensin protein
microtubule reorientation response
microtubule-associated protein (MAP)

NAC domain
Nod factor
nonclimacteric fruit
oligosaccharin
pectin methylesterase (PME)
perforation plate
plant growth factor (phytohormone)
primary pit field
primary plasmodesmata
reaction wood
secondary plasmodesmata
secondary thickening
signal molecule
simple perforation plate
tension wood
tip growth
WAK protein (wall-associated kinase)
xylem vessel and tracheid differentiation
xyloglucan endohydrolase (XEH)
xyloglucan endotransglycosylase (XET)
xyloglucan transglucosylase-hydrolase (XTH)
yieldin

References

[1]Cortical microtubule reorientation changes the orientation of cellulose microfibril deposition during growth.

Azimzadeh J, Nacry P, Christodoulidou A et al. (2008) *Arabidopsis* TONNEAU1 proteins are essential for preprophase band formation and interact with centrin. *Plant Cell* 20, 2146–2159.

Baskin TI (2001) On the alignment of cellulose microfibrils by cortical microtubules: a review and a model. *Protoplasma* 215, 150–171.

Chan J, Calder G, Fox S & Lloyd C (2007) Cortical microtubule arrays undergo rotary movements in *Arabidopsis* hypocotyl epidermal cells. *Nat. Cell Biol.* 9, 171–175.

Ehrhardt DW & Shaw SL (2006) Microtubule dynamics and organization in the plant cortical array. *Annu. Rev. Plant Biol.* 57, 859–875.

Hepler PK & Hirsh JM (1996) Behavior of microtubules in living plant cells. *Plant Physiol.* 112, 455–461.

Lucas J & Shaw SL (2008) Cortical microtubule arrays in the *Arabidopsis* seedling. *Curr. Opin. Plant Biol.* 11, 94–98.

Yuan M, Shaw PJ, Warn RM & Lloyd CW (1995) Direct observation of reorientation of the cortical microtubule array, from transverse to longitudinal, in living plant cells. *Proc. Natl. Acad. Sci. USA* 91, 6050–6053.

[2] Intracellular factors can alter the orientation of cortical microtubules.

Duckett CM & Lloyd CW (1994) Gibberellic acid-induced microtubule reorientation in dwarf peas is accompanied by rapid modification of an α-tubulin isotype. *Plant J.* 5, 363–372.

Whittington AT, Vugrek O, Wei KJ et al. (2001) MOR1 is essential for organizing cortical microtubules in plants. *Nature* 411, 610–613.

[3] Bundles of microtubules can define distinct domains of the wall that will thicken during cell development.

Fukuda H (1996) Xylogenesis: initiation, progression and cell death. *Annu. Rev. Plant Physiol. Plant Mol. Biol.* 47, 299–325.

Herth W (1985) Plasma-membrane rosettes involved in localized wall thickening during xylem vessel formation of *Lepidum sativum* L. *Planta* 164, 12–21.

McCann MC (1997) Tracheary element formation: buildup to a dead end. *Trends Plant Sci.* 2, 333–338.

Oda Y, Mimura T & Hasezawa S (2005) Regulation of secondary cell wall development by cortical microtubules during tracheary element differentiation in *Arabidopsis* cell suspensions. *Plant Physiol.* 137, 1027–1036.

Wernicke W & Jung G (1992) Role of cytoskeleton in cell shaping of developing mesophyll of wheat (*Triticum aestivum* L.). *Eur. J. Cell Biol.* 57, 88–94.

[4] The arrangement of cytoskeletal elements in one cell often relates to that in a neighboring cell.

Deeks MJ & Hussey PJ (2003) Arp2/3 and "The Shape of things to come." *Curr. Opin. Plant Biol.* 6, 1–7.

Goodbody KC & Lloyd CW (1990) Actin filaments line up across *Tradescantia* epidermal cells, anticipating wound-inducing division planes. *Protoplasma* 157, 92–101.

Hamant O & Traas J (2010) The mechanics behind plant development. *New Phytol.* 185, 369–385.

Hashimoto T (2003) Dynamics and regulation of plant interphase microtubules: a comparative view. *Curr. Opin. Plant Biol.* 6, 1–9.

Overall RL, Dibbayawan TP & Blackman LM (2001) Intercellular alignments of the plant cytoskeleton. *J. Plant Growth Regul.* 20, 162–169.

Sinnott EW & Bloch R (1945) The cytoplasmic basis of intercellular patterns in vascular differentiation. *Am. J. Bot.* 32, 151–156.

Smith LG (2003) Cytoskeletal control of plant cell shape: getting the fine points. *Curr. Opin. Plant Biol.* 6, 63–73.

[5] The wall is attached to receptor-like proteins in the plasma membrane.

Baluska F, Samaj J, Wojtaszek P et al. (2003) Cytoskeleton-plasma membrane-cell wall continuum in plants: emerging links revisited. *Plant Physiol.* 133, 482–491.

He Z-H, Cheeseman I, He D & Kohorn BD (1999) A cluster of five cell wall-associated receptor kinase genes, *Wak 1-5*, are expressed in specific organs of *Arabidopsis. Plant Mol. Biol.* 39, 1189–1196.

Riese J, Ney J & Kohorn BD (2003) WAKs: cell wall associated kinases. *Annu. Plant Rev.* 8, 223–236.

[6] A key driver of plant growth is postmitotic cell expansion.

Dolan L & Davies J (2004) Cell expansion in roots. *Curr. Opin. Plant Biol.* 7, 33–39.

Rojo E, Gillmor CS, Kovaleva V et al. (2001) VACUOLELESS1 is an essential gene required for vacuole formation and morphogenesis in *Arabidopsis. Dev. Cell* 1, 303–310.

Sugimoto-Shirasu K & Roberts K (2003) "Big it up": endoreduplication and cell size control in plants. *Curr. Opin. Plant Biol.* 6, 544–553.

[7] Cell expansion is usually accompanied by the deposition of new wall material.

Derbyshire P, Findlay K, McCann MC & Roberts K (2007) Cell elongation in *Arabidopsis* hypocotyls involves dynamic changes in wall thickness. *J. Exp. Bot.* 58, 2079–2089.

Hématy K & Höfte H (2007) Cellulose and Cell Elongation. Plant Cell Monographs 6: The Expanding Cell (Verbelen J-P & Vissenberg K eds), pp 33–56. Heidelberg: Springer-Verlag Berlin.

Refregier G, Pelletier S, Jaillard D & Hofte H (2004) Interaction between wall deposition and cell elongation in dark-grown hypocotyl cells in *Arabidopsis. Plant Physiol.* 135, 959–968.

[8] Wall architecture underpins anisotropic cell expansion.

Baskin TI (2005) Anisotropic expansion of the plant cell wall. *Annu. Rev. Cell Dev. Biol.* 21, 203–222.

Dhonukshe P, Tanaka H, Goh T et al. (2008) Generation of cell polarity in plants links endocytosis, auxin distribution and cell fate decisions. *Nature* 456, 962–966.

Kerstens S & Verbelen JP (2003) Cellulose orientation at the surface of the *Arabidopsis* seedling: implications for the biomechanics in plant development. *J. Struct. Biol.* 144, 262–270.

Kutschera U (2008) The growing outer epidermal wall: design and physiological role of a decomposite structure. *Ann. Bot. (Lond.)* 101, 615–621.

Mathur J & Hülskamp M (2002) Microtubules and microfilaments in cell morphogenesis in higher plants. *Curr. Biol.* 12, R669–R676.

Paredez AR, Persson S, Ehrhardt DW & Somerville CR (2008) Genetic evidence that cellulose synthase activity influences microtubule cortical array organization. *Plant Physiol.* 147, 1723–1734.

Paredez A.R, Somerville CR & Ehrhardt DW (2006) Visualization of cellulose synthase demonstrates functional association with microtubules. *Science* 312, 1491–1495.

Savaldi-Goldstein S, Peto C & Chory J (2007) The epidermis both drives and restricts plant shoot growth. *Nature* 446, 199–202.

[9] Dynamic remodeling of wall architecture facilitates cell expansion.

Fry SC (2004) Primary cell wall metabolism: tracking the careers of wall polymers in living plant cells. *New Phytologist* 161, 641–675.

Nishitani K & Vissenberg K (2007) Roles of the XTH Protein Family in the Expanding Cell. Plant Cell Monographs 6: The Expanding Cell. (Verbelen J-P & Vissenberg K eds), pp 89–116. Heidelberg: Springer-Verlag Berlin.

Obel N, Neumetzler L & Pauly M (2007) Hemicelluloses and Cell Expansion. Plant Cell Monographs 6: The Expanding Cell. (Verbelen J-P & Vissenberg K eds), pp 57–88. Heidelberg: Springer-Verlag Berlin.

Rose JKC, Braam J, Fry SC, Nishitani K (2002) The XTH family of enzymes involved in xyloglucan endotransglucosylation and endohydrolysis: current perspectives and a new unifying nomenclature. *Plant Cell Physiol.* 43, 1421–1435.

Rose JKC, Catala C, Gonzalez-Carranza ZH & Roberts JA (2003) Cell wall disassembly. In The Plant Cell Wall (JKC Rose ed), pp 264–324. Oxford: Blackwell.

[10] Proteins that enhance wall expansion have been identified.

Cosgrove DJ (1997) Assembly and enlargement of the primary cell wall in plants. *Annu. Rev. Cell Dev. Biol.* 13, 171–201.

Cosgrove DJ (2000) Loosening of plant cell walls by expansins. *Nature* 407, 321–326.

Hematy K & Hofte H (2008) Novel receptor kinases involved in growth regulation. *Curr. Opin. Plant Biol.* 11, 321–328.

McQueen-Mason S, Le NT, Brocklehurst D (2007) Expansins. Plant Cell Monographs 6: The Expanding Cell. (Verbelen J-P & Vissenberg K eds), pp 117–138. Heidelberg: Springer-Verlag Berlin.

Nieuwland J, Feron R, Huisman BAH et al. (2005) Lipid transfer proteins enhance cell wall extension in tobacco. *Plant Cell* 17, 2009–2019.

Pien S, Wyrzykowska J, McQueen-Mason S et al. (2001) Local expression of expansin induces the entire process of leaf development and modifies leaf shape. *Proc. Natl. Acad. Sci. USA* 98, 11812–11817.

Rose JKC, Saladie M & Catala C (2004) The plot thickens: new perspectives of primary cell wall modification. *Curr. Opin. Plant Biol.* 7, 296–301.

Sampedro J & Cosgrove DJ (2005) The expansin superfamily. *Genome Biol.* 6, 242.

Yennawar NH, Li L-C, Dudzinski DM et al. (2006) Crystal structure and activities of EXPB1 (Zea m 1), a beta-expansin and group-1 pollen allergen from maize. *Proc. Natl. Acad. Sci. USA* 103, 14664–14671.

[11] Acidification of the wall, enhanced by auxin, may promote cell expansion.

Kutschera A (1994) The current status of the acid growth hypothesis. *New Phytol.* 126, 549–569.

Rayle DL & Cleland RE (1992) The acid growth theory of auxin-induced cell elongation is alive and well. *Plant Physiol.* 99, 1271–1274.

[12] Although plant cell walls are relatively stable compartments, many matrix polysaccharides and proteoglycans do turn over.

Gibeaut DM & Carpita NC (1991) Tracing cell wall biogenesis in intact cells and plants: selective turnover and alteration of soluble and cell wall polysaccharides in grasses. *Plant Physiol.* 97, 551–561.

Gorshkova TA, Chemikosova SB, Lozovaya VV & Carpita NC (1997) Turnover of galactans and other cell wall polysaccharides during development of flax plants. *Plant Physiol.* 114, 723–729.

Qouta LA, Waldron KW, Baydoun EAH & Brett CT (1991) Changes in seed reserves and cell-wall composition of component organs during germination of cabbage (*Brassica oleracea*) seeds. *J. Plant Physiol.* 138, 700–707.

Reid JSG (1985) Cell wall storage carbohydrates in seeds: biochemistry of the seed "gums" and "hemicelluloses." *Adv. Bot. Res.* 11, 125–155.

Rodriguez R, Jimenez A, Guillen R et al. (1999) Turnover of white asparagus cell wall polysaccharides during postharvest storage. *J. Agric. Food Chem.* 47, 4525–4531.

Rose JKC, Catala C, Gonzalez-Carranza ZH & Roberts JA (2003) Cell wall disassembly. In The Plant Cell Wall (JKC Rose ed), p 264–324. Oxford: Blackwell.

[13] The *de novo* formation of plasmodesmata across established walls requires wall remodeling.

Ding B & Lucas WJ (1996) Secondary plasmodesmata: biogenesis, special functions and evolution. In Membranes: Specialised Functions in Plants (M Smallwood, JP Knox, DJ Bowles eds), pp 489–506. Oxford: BIOS Scientific.

Faulkner C, Akman OE, Bell K et al. (2008) Peeking into pit fields: a multiple twinning model of secondary plasmodesmata formation in tobacco. *Plant Cell* 20, 1504–1518.

Kollman R & Glockmann C (1991) Studies on graft unions III: on the mechanism of secondary formation of plasmodesmata at the graft interface. *Protoplasma* 165, 71–85.

Levy A, Erlanger M, Rosenthal M & Epel BL (2007) A plasmodesmata-associated beta-1,3-glucanase in *Arabidopsis. Plant J.* 49, 669–682

Maule A (2008) Plasmodesmata: structure, function and biogenesis. *Curr. Opin. Plant Biol.* 11, 680–686.

Oparka KJ, Roberts AG, Boevink P et al. (1999) Simple, but not branched, plasmodesmata allow the nonspecific trafficking of proteins in developing tobacco leaves. *Cell* 97, 743–754.

[14] Local removal of wall material is used to create conducting elements from files of cells.

Butterfield BG & Meylan BA (1974) Scalariform perforation plate development in *Laurelia novae-zelandiae* A. Cunn.: a scanning electron microscope study. *Aust. J. Bot.* 20, 253–259.

Chaffey N, Barlow P & Barnett J (2000) A cytoskeletal basis for wood formation in angiosperm trees: the involvement of microfilaments. *Planta* 210, 890–896.

Kubo M, Udagawa M, Nishikubo N et al. (2005) Transcription switches for protoxylem and metaxylem vessel formation. *Genes Dev.* 19, 1855–1860.

Mitsuda N, Seki M, Shinozaki K & Ohme-Takagi M (2005) The NAC transcription factors NST1 and NST2 of *Arabidopsis* regulate secondary wall thickenings and are required for anther dehiscence. *Plant Cell* 17, 2993–3006.

Nakashima J, Takabe K, Fujita M & Fukuda H (2000) Autolysis during *in vitro* tracheary element differentiation: formation and location of the perforation. *Plant Cell Physiol.* 41, 1267–1271.

[15]Abscission of leaves and fruit is an active process that involves controlled cell separation.

Addicott FT (1982) Abscission. Berkeley, CA: University of California Press.

Cho SK, Larue CT, Chevalier D et al. (2008) Regulation of floral organ abscission in *Arabidopsis thaliana*. *Proc. Natl. Acad. Sci. USA* 105, 15629–15634.

González-Carranza ZH, Whitelaw CA, Swarup R & Roberts JA (2002) Temporal and spatial expression of a polygalacturonase during leaf and flower abscission in oilseed rape and *Arabidopsis*. *Plant Physiol.* 128, 534–543.

Jinn T-L, Stone JM & Walker JC (2000) HAESA, an *Arabidopsis* leucine-rich repeat receptor kinase, controls floral organ abscission. *Genes Dev.* 14, 108–117.

Mao L, Begum D, Chuang H et al. (2000) JOINTLESS is a MADS-box gene controlling tomato flower abscission zone development. *Nature* 406, 910–913.

Patterson SE (2001) Cutting loose: abscission and dehiscence in *Arabidopsis*. *Plant Physiol.* 126, 494–500.

Roberts JA, Elliott KA & González-Carranza ZH (2002) Abscission, dehiscence, and other cell separation processes. *Annu. Rev. Plant Biol.* 53, 131–158.

Stenvik G-E, Butenko MA, Urbanowicz BR et al. (2006) Overexpression of INFLORESCENCE DEFICIENT IN ABSCISSION activates cell separation in vestigial abscission zones in *Arabidopsis*. *Plant Cell* 18, 1467–1476.

[16]Fruit softening depends upon the expression and activity of cell wall–modifying proteins.

Blumer JM, Clay RP, Bergmann CW et al. (2000) Characterization of changes in pectin methylesterase expression and pectin esterification during tomato fruit ripening. *Can. J. Bot.* 78, 607–618.

Brummell DA & Harpster MH (2001) Cell wall metabolism in fruit softening and quality and its manipulation in transgenic plants. *Plant Mol. Biol.* 47, 311–340.

Giovannioni J (2001) Molecular biology of fruit maturation and ripening. *Annu. Rev. Plant. Mol. Biol.* 52, 725–749.

Saladie M, Matas AJ, Isaacson T et al. (2007) A reevaluation of the key factors that influence tomato fruit softening and integrity. *Plant Physiol.* 144, 1012–1028.

[17]In some seeds, wall polysaccharides can form a food reserve to be used during germination.

Dhugga KS, Barreiro R, Whitten B et al. (2004) Guar seed β-mannan synthase is a member of the cellulose synthase super gene family. *Science* 303, 363–366.

Edwards ME, Dickson CA, Chengappa S et al. (1999) Molecular characterization of a membrane-bound galactosyltransferase of plant cell wall matrix biosynthesis. *Plant J.* 19, 691–697.

Reid JSG (1985) Cell wall storage carbohydrates in seeds: biochemistry of the seed "gums" and "hemicelluloses." *Adv. Bot. Res.* 11, 125–155.

Reid JSG & Edwards M (1995) Galactomannans and other cell wall storage polysaccharides in seeds. In Food Polysaccharides and Their Applications (AM Stephen ed), pp 155–186. New York: Marcel Dekker.

[18]Lignin and suberin provide physical barriers to the turnover and degradation of secondary walls.

Bao W, O'Malley D & Sederoff RR (1992) Wood contains a cell wall structural protein. *Proc. Natl. Acad. Sci. USA* 89, 6604–6608.

Passardi F, Penel C & Dunand C (2004) Performing the paradoxical: how plant peroxidases modify the cell wall. *Trends Plant Sci.* 9, 534–540.

[19]Gravity sensing and mechanical stress lead to compensatory changes in cell wall synthesis and architecture.

Andersson-Gunneras S, Mellerowicz EJ, Love J et al. (2006) Biosynthesis of cellulose-enriched tension wood in *Populus*: global analysis of transcripts and metabolites identifies biochemical and developmental regulators in secondary wall biosynthesis. *Plant J.* 45, 144–165.

Bhandari S, Fujino T, Thammanagowda S et al. (2006) Xylem-specific and tension stress-responsive coexpression of KORRIGAN endoglucanase and three secondary wall-associated cellulose synthase genes in aspen trees. *Planta* 224, 828–837.

Hamant O & Traas J (2010) The mechanics behind plant development. *New Phytol.* 185, 369–385.

Joseleau J-P, Imai T, Kuroda K & Ruel K (2004) Detection *in situ* and characterization of lignin in the *G*-layer of tension wood fibres of *Populus deltoides*. *Planta* 219, 338–345.

Pilate G, Chabbert B, Cathala B et al. (2004) Lignification and tension wood. *C R Biol.* 327, 889–901.

Yamashita S, Yoshida M, Takayama S & Okuyama T (2007) Stem-righting mechanism in gymnosperm trees deduced from limitations in compression wood development. *Ann. Bot.* 99, 487–493.

[20]Many signals combine to regulate plant growth.

(Each year, issue 5 of *Current Opinion in Plant Biology* covers the whole area of plant cell signaling.)

Côté F & Hahn MG (1994) Oligosaccharins: structures and signal transduction. *Plant Mol. Biol.* 26, 1379–1411.

[21]Oligogalacturonides can modulate development in tobacco explants.

Bellincampi D, Salvi G, De Lorenzo G et al. (1993) Oligogalacturonides inhibit the formation of roots on tobacco explants. *Plant J.* 4, 207–213.

Eberhard S, Doubrava N, Marfa V et al. (1989) Pectic cell wall fragments regulate tobacco thin-cell-layer explant morphogenesis. *Plant Cell* 1, 747–755.

Mathieu Y, Kurkdjian A, Xia H et al. (1991) Membrane responses induced by oligogalacturonides in suspension-cultured tobacco cells. *Plant J.* 1, 333–343.

[22] Xyloglucan-derived oligosaccharins can affect the rate of elongation growth.

Augur C, Yu L, Sakai K et al. (1992) Further studies of the ability of xyloglucan oligosaccharides to inhibit auxin-stimulated growth. *Plant Physiol.* 99, 180–185.

Bellincampi D, Salvi G, De Lorenzo G et al. (1993) Oligogalacturonides inhibit the formation of roots on tobacco explants. *Plant J.* 4, 207–213.

Fry SC (1994) Oligosaccharins as plant growth regulators. *Biochem. Soc. Symp.* 60, 5–14.

Lorences EP & Fry SC (2006) Xyloglucan oligosaccharides with at least two α-D-xylose residues act as acceptor substrates for xyloglucan endotransglycosylase and promote the depolymerisation of xyloglucan. *Physiol. Plant* 88, 105–112.

McDougall GJ & Fry SC (1990) Xyloglucan oligosaccharides promote growth and activate cellulase: evidence for a role of cellulase in cell expansion. *Plant Physiol.* 93, 1042–1048.

[23] Lipo-oligosaccharides synthesized by rhizobia regulate nodule development in host plants.

Denarie J & Cullimore J (1993) Lipo-oligosaccharide nodulation factors: a new class of signaling molecules mediating recognition and morphogenesis. *Cell* 74, 951–954.

D'Haeze W & Holsters M (2002) Nod factor structures, responses, and perception during initiation of nodule development. *Glycobiology* 12, 79R–105R.

Geurts R, Federova E & Bisseling T (2005) Nod factor signaling genes and their function in the early stages of *Rhizobium* infection. *Curr. Opin. Plant Biol.* 8, 346–352.

Lerouge P, Roche P, Faucher C et al. (1990) Symbiotic host-specificity of *Rhizobium meliloti* is determined by a sulphated and acylated glucosamine oligosaccharide signal. *Nature* 344, 781–784.

Oldroyd GE & Downie JA (2006) Nuclear calcium changes at the core of symbiosis signaling. *Curr. Opin. Plant Biol.* 9, 351–357.

Schultze M, Quiclet-Sire B, Kondorosi E et al. (1992) *Rhizobium meliloti* produces a family of sulfated lipo-oligosaccharides exhibiting different degrees of plant host specificity. *Proc. Natl. Acad. Sci. USA* 89, 192–196.

Spaink HP (1992) Rhizobial lipo-oligosaccharides: answers and questions. *Plant Mol. Biol.* 20, 977–986.

[24] Chitinases may function during normal plant development.

Benhamou N & Asselin A (1989) Attempted localization of a substrate for chitinases in plant cells reveals abundant *N*-acetyl-D-glucosamine residues in secondary walls. *Biol. Cell* 67, 341–350.

De Jong AJ, Cordewener J, Lo Schiavo F et al. (1992) A carrot somatic embryo mutant is rescued by chitinase. *Plant Cell* 4, 425–433.

De Jong AJ, Heidstra R, Spaink HP et al. (1993) *Rhizobium* lipooligosaccharides rescue a carrot somatic embryo mutant. *Plant Cell* 5, 615–620.

Kasprzewska A (2003) Plant chitinases: regulation and function. *Cell Mol. Biol. Lett.* 8, 809–824.

Neale AD, Wahleithner JA, Lund M et al. (1990) Chitinase, β-1,3-glucanase, osmotin, and extensin are expressed in tobacco explants during flower formation. *Plant Cell* 2, 673–684.

Passarinho PA, van Hengel AJ, Fransz PF & de Vries SC (2001) Expression pattern of the *Arabidopsis thaliana AtEP3/AtchitIV* endochitinase gene. *Planta* 212, 556–567.

Spaink HP, Wijfjes AHM, Van Vliet TB et al. (1993) Rhizobial lipo-oligosaccharide signals and their role in plant morphogenesis: are analogous lipophilic chitin derivatives produced by the plant? *Aust. J. Plant Physiol.* 20, 381–392.

Zhong R, Kays SJ, Schroeder BP & Ye Z-H (2002) Mutation of a chitinase-like gene causes ectopic deposition of lignin, aberrant cell shapes, and overproduction of ethylene. *Plant Cell* 14, 165–179.

Cell Walls and Plant–Microbe Interactions

8

Introduction

So far in this book we have treated plants largely in isolation, as discrete, living beings, and we have discussed their cell walls in the context of the plant's own growth and development. But in the real world, on Darwin's famous "tangled bank, clothed with many plants of many kinds," plants grow in complex communities with other plants, and, more important, in close association with a multitude of microbes and other organisms (**Figure 8.1**). In most cases these associations are of little consequence; in some there is mutual benefit, while in fewer cases there is a deleterious effect on the plant. In this chapter we will look more closely at the ways in which certain microbes have evolved to become successful pathogens of plant hosts, causing disease, and how plant cell walls come to the plant's defense.

The whole area of plant pathology has immense repercussions for the future of our planet. In an era of growing population and fragile food security we still lose over 40% of our world's agricultural food produce to weeds, pests, and disease. In this chapter we shall focus on the warfare between plants and bacteria and fungi, and examine the complex roles the cell wall plays in this evolving battleground. We start by discussing the general principles of plant pathology and how plants have learned to recognize and respond to a variety of danger signals. We then look at the ways in which the preformed defense mechanisms of the plant depend largely on wall-related structures and molecules. We then discuss the complex systems of plant and microbial enzymes, and their inhibitors, that populate the apoplast, and last we

Figure 8.1 Plants and microbes coexist. In among the dense, healthy garden plants, the stump of an ancient beech tree, now felled, is slowly being recycled by the mycelium of the white rot fungus *Meripilus giganteus*, whose large fruiting bodies can be seen encircling the dead tree stump.

look at the ultimate fate of all cell walls and how, when plants die, they are at last broken down by microbes and their polymers finally recycled to the biosphere.

A. How Plants Detect and Respond to Microbes

1. Most plants are immune to most pathogens.[ref1]

Neither bacteria nor fungi can photosynthesize, and they therefore need to obtain their nutrients from elsewhere. Many adopt a *saprophytic* lifestyle, living heterotrophically on the digested products of dead and decaying plant remains, but others have coevolved with their plant hosts and have acquired ways of getting their raw materials directly from living plants. Such microbes are called *pathogens* [in contrast to *pests*, a term usually reserved for insects (Figure 8.2E) and herbivores, animals that attack or eat plants]. Some microbial pathogens, called *biotrophs*, infect living plants, spreading extensively throughout the plant's tissues, without killing cells, and at the same time evading the host's defenses. Others do kill cells to obtain food and may even kill the whole plant. These are called *necrotrophs*, and they tend to have a broader host range than the biotrophs. Many pathogens, the *hemibiotrophs*, may start off living biotrophically but end up by killing cells and living necrotrophically (**Table 8.1**). These formal definitions, however, are not hard and fast. Many plant pathogens are very adaptable and can even live as saprophytes when a suitable host is not available.

Plants generally have thickened outer epidermal walls that form a natural barrier to microbial attack. These limit the easy entry of potential pathogens either through open wounds, usually caused by insects, herbivores, or wind damage; or through the natural openings provided by stomata. Even so, most plants are resistant to most pathogens, because of a combination of preformed defenses and the plant's inducible basal defense, or immunity, that is triggered in response to the potential microbial threat. This combination of preexisting structural and chemical defenses, together with the inducible basal defenses, is usually more than sufficient to overcome microbial attack, and is described as *nonhost resistance*.

A successful pathogen can overcome these basal and/or preformed defenses only if it can either avoid triggering the basal defenses altogether or actively silence or slow the ability of the host to mount these induced basal defenses. We shall discuss how pathogens do all this later in this section, but the net result of a compatible host-pathogen interaction is a plant disease with characteristic symptoms. The particular disease is usually named after the plant affected and its associated symptoms, which historically have vivid descriptive names that include wilts, specks, rusts, scabs, mildews, blasts, rots, smuts, cankers, and blights. (**Figure 8.2A–D**). Whether or not a pathogen is successful depends not only on its invasion strategies

TABLE 8.1 PLANT PATHOGEN LIFESTYLES

PATHOGEN LIFESTYLE	BACTERIA	FUNGI	OOMYCETES
Biotrophic	*Agrobacterium tumefaciens* (crown gall)	*Cladosporium fulvum* (leaf mold) *Blumeria graminis* (cereal powdery mildew)	*Hyaloperonospora* (downy mildew of Arabidopsis)
Necrotrophic	*Pectobacterium carotovorum* (soft rot)	*Botrytis cinerea* (mold)	*Pythium* spp. (root rot, wilts, damping off)
Hemibiotrophic	*Pseudomonas syringae* (soft rots)	*Magnaporthe grisea* (rice blast)	*Phytophthora infestans* (potato late blight)

Figure 8.2 Plant diseases and pests. (A) The round, sooty lesions caused by peacock spot disease of olive trees. The growth of the fungus causing the disease, *Spilocea oleaginea*, is favored by cool humid conditions with high rainfall. (B, C) Advanced powdery mildew infection on zucchini (courgette) leaves. Different species of fungi can cause similar powdery mildew symptoms on different plants; the white eruptions are the spore-bearing fungal hyphae. (D) Leaves with rose black spot disease caused by the fungus *Diplocarpon rosae*. (E) A plant pest. This iridescent rosemary leaf beetle (*Chrysolina americana*) and its larvae both eat the young shoots of woody shrubs like rosemary, thyme, and lavender.

but also, crucially, on both a particular susceptible host plant and appropriate environmental conditions. Many fungal leaf pathogens, for example, require cool and moist conditions to establish an infection, even on a susceptible host.

2. Signals from both pathogen and host can elicit a common defense response.[ref2]

Vertebrate animals have two different systems by which they generate immunity to pathogens: an *innate immune system*, which recognizes common features of foreign organisms and then rapidly triggers a general defense response; and an *adaptive immune system*, which in time can produce custom-made, precision antibodies to recognize and destroy specific foreign molecules and organisms over a long period. Plants, however, make do with just the one immune system, which, seemingly by convergent evolution, functions in much the same way as animal innate immunity. Both systems depend on the recognition by living cells of *molecular patterns*, structural motifs present in a wide range of molecules that are often referred to as *general elicitors*. Such molecular patterns signal danger to the plant, which in turn activates a stereotypical set of responses to defend itself. For this reason, plant disease is the exception rather than the rule.

Two main classes of molecular patterns have been described, but each activates a plant's immune response in much the same way. The first are molecular patterns that are common to a whole class of microbes, some of which may be potential pathogens even while others are benign. These are called *microbe-associated molecular patterns*, or *MAMPs*, and examples include chitin and glucans from the cell walls of fungi, and flagella and peptidoglycan from bacteria. All these act as elicitors that trigger an active defense response that confers resistance to potential pathogens. This response is mounted not only to molecular patterns from microbes but also

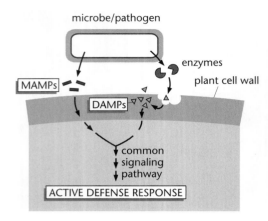

Figure 8.3 Molecular patterns elicit a common signaling pathway and an active defense response. Both MAMPs, of microbial origin, and DAMPs, of plant host origin, activate signaling pathways that converge on a common active defense response.

to certain molecular patterns from the plant itself. This second class of molecules, which includes fragments of the plant cell wall released by microbial enzymes—for example oligogalacturonides and cutin monomers—were originally described as endogenous elicitors, but since they trigger a similar set of responses to the MAMPs, they are referred to as *DAMPs*, or *damage-associated molecular patterns*. We should note now, in passing, that molecular patterns that derive from pathogens are often called pathogen-associated molecular patterns, or PAMPs, but since most—for example, chitin—are not specific to pathogens, we prefer here to use the broader term MAMPs. In addition, since plants do not have both an innate and an adaptive immune system, it is also worth noting that in this chapter we shall refer just to *plant immunity* rather than to plant innate immunity.

The plant perceives both MAMPs and DAMPs as tangible evidence that it is in potential danger; MAMPs confirm the local presence of a potential microbial threat, while DAMPs alert the plant to local damage or breakdown of its own tissues. The plant therefore treats both as danger signals and mounts accordingly a common and rapid local response (**Figure 8.3**). This process has often been referred to as pattern-triggered immunity or PTI. In an evolutionary context, the molecular components of the plant's immune response, or PTI, are both ancient and very stable. For example, almost all plants respond in the same way to almost all bacterial flagella. The strength, duration, and location of the PTI may, however, vary widely, probably depending on the local, integrated sum of all the molecular patterns and their receptors that are brought to bear in responding to the danger signals. With functional redundancy between the many DAMPs and MAMPs likely, it has been hard experimentally to dissect the contributions of any one molecular pattern, its signaling pathway, and its precise role in resistance and immunity.

3. Cell surface receptors recognize common molecular patterns.[ref3]

The first step in triggering basal, or broad-spectrum, PTI in plants is crossing of the cell wall by the danger signals—MAMPs and DAMPs—so that they can reach the plant plasma membrane, where their molecular patterns can be recognized by a family of *pattern-recognition receptors*, or *PRRs*, on the cell surface. Recognition then initiates a downstream signaling cascade, described later, that results in a range of events at the plasma membrane and eventually in defense gene activation. Very few MAMP/PRR pairs have been fully characterized, so instead of using a cell wall–derived pattern as an example, we will use the MAMP found in bacterial flagella as a paradigm to illustrate the essential features of MAMP signaling.

The filament of the bacterial flagellum is made up of thousands of copies of the protein flagellin (**Figure 8.4**). A short conserved sequence, flg22, within the flagellin molecule acts as a potent elicitor that triggers a defense response

Figure 8.4 Bacterial flagellin contains a potent MAMP. The filament of a bacterial flagellum is composed of thousands of copies of the protein flagellin. Within the flagellin molecule is a short conserved peptide, flg22, whose amino acid sequence is shown below. This is a potent MAMP that is recognized by a cell surface receptor in almost all plants. A separate peptide, D1, within the same molecule is recognized by a related receptor in animals, a case of convergent evolution.

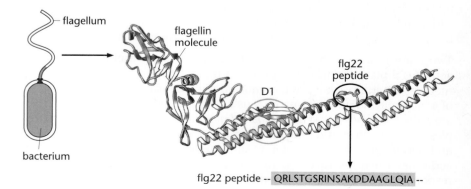

flg22 peptide -- QRLSTGSRINSAKDDAAGLQIA --

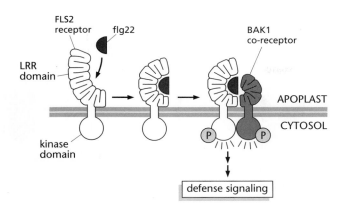

Figure 8.5 FLS2 is an example of a pattern-recognition receptor (PRR). The leucine-rich repeat receptor kinase FLS2 specifically recognizes the bacterial flagellar MAMP, the peptide flg22, and is positively regulated by its co-receptor, BAK1, to activate defense responses. Both FLS2 and BAK1 contain leucine-rich repeat domains facing the apoplast.

in *Arabidopsis* at sub-nanomolar concentrations. The flg22 epitope is recognized by the pattern-recognition receptor FLAGELLIN-SENSING 2 (FLS2) located in the plant plasma membrane of almost all higher plants. The FLS2 receptor is encoded by a member of a very large gene family in plants called *leucine-rich repeat receptor kinases* (*LRR-RKs*). The receptor kinase has a typical membrane receptor structure with three main domains: an extracellular recognition domain containing about 28 leucine-rich repeats, a transmembrane domain, and a cytosolic kinase domain. On binding the flg22 MAMP, the receptor rapidly associates with a co-receptor, another LRR-RK called BAK1; the kinase domains cross-phosphorylate each other and the signal is propagated downstream (**Figure 8.5**).

Another well-studied MAMP-receptor pair involves the bacterial MAMP EF-Tu. This is a conserved and abundant bacterial elongation factor that is recognized by plants of the family Brassicaceae by another plasma membrane LRR-RK, in this case called EFR. These two examples suggest this is a common mechanism of pattern recognition, and in fact there are more than 200 related LRR receptor-kinases in the *Arabidopsis* genome along with another 400 or so other unrelated receptor-like kinases. Needless to say most of these are "orphans" without known ligands to bind them, but many are likely to recognize other MAMPs and DAMPs.

The secreted fungal enzyme ethylene-inducing xylanase, or EIX, provides us with a different version of the same paradigm. This protein, from *Trichoderma viride*, is a potent elicitor in tomato, triggering a strong defense response. At first it was assumed that this was an indirect effect and that the true elicitor was an oligosaccharide product of the xylanase activity. However, an engineered xylanase that has no residual enzymatic activity still has strong elicitor activity, showing that it is in fact the EIX protein itself that is the true elicitor. EIX appears to be recognized at the cell surface (**Figure 8.6**) by two related LRR-containing receptors, LeEix1 and LeEix2, but only the latter is capable of transmitting the signals into the cell. LeEix1 positively associates with LeEix2 when bound to EIX, but exactly how, with no kinase domain, the signal is conveyed into the cell is not understood (**Figure 8.7**). A similar situation exists with the recognition of the ubiquitous fungal MAMP chitin. This polysaccharide, a long-chain polymer of β-1-4-linked *N*-acetylglucosamine, is found in all fungal cell walls as well as in insect and crustacean exoskeletons and is a potent MAMP, recognized by all plants. Although not all the details are fully known in any one plant, another class of plasma-membrane–located receptors, with an extracellular domain containing LysM repeats instead of LRR repeats, recognizes chitin (or fragments derived from it by chitinase activity). Receptors with similar repeats are used by the innate immune system of animals to recognize the same MAMP. Again, only one member of the receptor dimer contains the cytosolic protein kinase domain required to relay the signal to the cell interior. This associates with its co-receptor, which has no kinase domain but binds to the chitin MAMP (Figure 8.7).

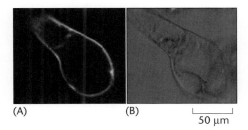

(A)   (B)   50 μm

Figure 8.6 A MAMP, the xylanase EIX, binds to a receptor at the plasma membrane. In this experiment the xylanase receptor from tomato, LeEix2, has been expressed in suspension-cultured tobacco cells. The cells have then been exposed to fluorescently labeled xylanase EIX for 30 minutes. (A) Fluorescence micrograph, showing binding of the fluorescent protein to the plasma membrane of the two cells. (B) Bright field image of the same two cells. There is no binding if the cells are not transformed with the *LeEix2* gene. (From M. Ron and A. Avni, *Plant Cell* 16:1604–1615, 2004.)

Figure 8.7 How two pattern-recognition receptors might recognize wall-related MAMPs. In addition to the LRR-RKs already described, other classes exist. (A) Schematic representation of the two receptors for the ethylene-inducing xylanase. Both contain extracellular LRRs that bind the xylanase MAMP (see Figure 8.6), but only LeEix2 is able to propagate the signal intracellularly. (B) Both partners in this putative chitin receptor contain LysM repeats. A chitin-binding protein, on binding the chitin MAMP, activates the receptor kinase which can then propagate the signal into the cell. How the two partners interact, whether directly or indirectly, is not known and the model shown here is a provisional schematic that has not been confirmed experimentally.

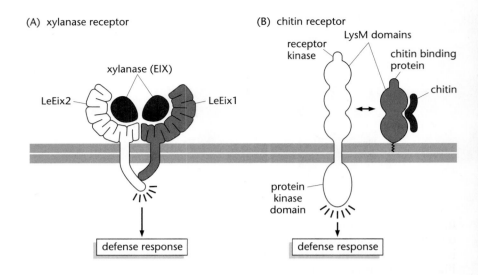

(A) xylanase receptor

xylanase (EIX)

LeEix2   LeEix1

defense response

(B) chitin receptor

LysM domains

receptor kinase

chitin binding protein

chitin

protein kinase domain

defense response

Compared to the situation with MAMP perception, described above, the situation with DAMPs is much less clear. Oligogalacturonides, or OGAs, which act as elicitors of the defense response, are important DAMPs. They are released from the host cell wall by secreted *endo*-polygalacturonases during almost all fungal attacks, and these oligosaccharides are most active as elicitors when they have a degree of polymerization between about 10 and 15. Although a lot is known about their structure and activity, no receptor that recognizes them has yet been identified [although several lines of evidence suggest that the WAK receptors described earlier (Figure 7.14) are involved]. Despite much research, this is also true for other oligosaccharide DAMPs derived from plant cell walls and also for cutin monomers, another potent elicitor released from the epidermis of plants by fungal cutinases. Many MAMPs that are recognized in animals, including bacterial lipopolysaccharide and peptidoglycan, are also recognized in plants and trigger immunity, but their corresponding receptors also remain unknown. The search to identify and characterize more receptor-elicitor pairs is an active research priority, and our understanding of their role in disease and resistance is crucially dependent on it.

The expression of many PRRs is increased by MAMP signaling itself, suggesting that feedback loops help reinforce the defense system. In the wild it seems likely that the mere presence of a potential pathogen will bring with it not one but multiple MAMPs, as well as releasing many DAMPs from the host plant. The cocktail of danger signals this represents, and the complexity of the receptors and all their corresponding inputs to the downstream signaling pathways, represents a real experimental challenge. It is already clear, however, that some combinations of molecular patterns act synergistically while others antagonize each other. Others result in the activation of different sets of target genes. Much will depend on local conditions in the cell wall that the elicitors must cross, with factors like ionic effects, wall hydration, and permeability ultimately having important indirect effects on disease resistance.

4. The plant's responses to danger involve a common sequence of events.[ref4]

So far we have discussed the various kinds of danger signals, both MAMPs and DAMPs, and how they alert plants to mount a defense response. It is now time to describe the nature of that response in more detail and how it is activated. In general terms, the cell responds to danger in two distinct ways, first by relatively rapid events at the plasma membrane, and second by rather slower events within the cell, including dynamic rearrangements

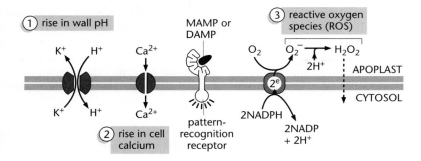

Figure 8.8 Early events in the response to danger signals. Within about 5 minutes of DAMP/MAMP recognition by a pattern-recognition receptor, three molecular events can be detected at the plasma membrane: a rise in the wall pH, a rise in intracellular calcium, and the production of reactive oxygen species in the apoplast.

within the cytosol and gene activation in the nucleus. It is worth reiterating that these general features of the basal defense system are stereotypical; they recur as a common set of responses to almost all general elicitors, and in addition they emerge as core components of the plant's response to wounding and even stress conditions such as cold and drought. Adding elicitors to cells in suspension culture, and then observing when particular changes occur, is an easy way to study the time frame of these responses. In *Arabidopsis* it has been found that addition of most MAMPs and DAMPs results in reproducible transcriptional changes in around 800 genes.

Within five minutes or so of adding an elicitor—for example, the fungal heptaglucoside described earlier (Figure 7.48)—three events can be detected at the plasma membrane, summarized in **Figure 8.8**. The first is the activation of ion channels that depolarize the membrane. An efflux of potassium is accompanied by an influx of protons and an efflux of anions. The resultant rapid rise in the pH of the medium is easily measured experimentally and forms a convenient assay for elicitation, despite the fact that the pumps and channels involved in this ion flux are not yet well understood. The second event is a net influx of calcium from the apoplast into the cytosol. If calcium is removed from the culture medium, cells cannot mount an effective defense response, suggesting that this is a crucial signaling event. The third event, detectable within a couple of minutes, is referred to as the *oxidative burst*. NADPH oxidase is an integral plasma membrane enzyme that transfers electrons from cytosolic NADPH to extracellular oxygen to create superoxide, O_2^-, in the wall space. This is rapidly converted to hydrogen peroxide, H_2O_2, which acts as a second messenger. These two very reactive molecules are collectively called *reactive oxygen species*, or *ROS*. Apart from having antimicrobial activity in their own right, ROS also act as secondary stress signaling molecules that help activate further gene expression to reinforce the defense response. Although these three rapid responses in the plasma membrane have been known for some time, exactly how the receptor kinase activation is coupled to them remains a mystery.

Onward transmission of the signal from the PRR into the cell, which results in gene transcription, is a more time-consuming process, taking between about five minutes and an hour. Further downstream events, such as wall rearrangements, may take hours or even days. The rise in intracellular calcium activates calcium-dependent kinases, and possibly other ion channels, but remarkably little is known about this branch of the signaling pathway, or indeed about how the ROS signal downstream. Much better understood is the main link between PRRs and the nucleus, through a cascade of MAP kinases. A complex chain involving several different MAP kinase family members conveys the signal finally to a set of transcription factors, notably a group in the WRKY family, which then activate the transcription of genes encoding a multitude of proteins required to defend the cell (**Figure 8.9**). Among these genes are some that are destined to ensure that the receptor is endocytosed and the response in due course curtailed.

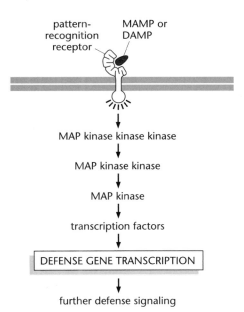

Figure 8.9 Downstream signaling in the defense response. Onward signaling from the pattern-recognition receptors is through a cascade of MAP kinases that activate a set of transcription factors. Not shown here are additional inputs to this central pathway, from calcium and reactive oxygen species, and additional outputs to further defense responses.

The final repertoire of defense responses varies with the initial stimulus and the plant that is in danger, but it would typically include reinforcement of the cell wall, the secretion of antimicrobial compounds and enzymes that will attack surface components of the microbial attacker, and the production of other important signaling molecules such as ethylene, jasmonic acid, and salicylic acid. Signaling will also often involve interactions with conventional plant growth regulators, since one response to the sustained perception of danger is to slow growth to conserve resources while the plant is under attack. Some more detailed examples of these stereotypical defense responses are described later in Section B of this chapter.

5. Some specialist pathogens deliver effectors to suppress the basal defense response.[ref5]

Thus far we have focused on the plant's basal defenses, which are usually enough to ward off the majority of potential pathogens. Indeed this is why most plants are immune to most pathogens. The existence of a single core plant defense pathway, however, has put selection pressure on ambitious microbes to find ingenious ways of disabling it, in order to gain a foothold in the plant and enhance their access to nutrients. In fact, pathogens have developed a whole range of molecules that they either secrete or directly introduce into plant cells to block, disable, interfere with, or destroy one or more components of the basal defense machinery. This large class of molecules is referred to in general terms as *effectors*. Some effectors, for example, proteases and inhibitors of microbial wall-degrading enzymes, are secreted by the pathogen.

Bacterial pathogens also make a group of effectors that they can inject directly across the cell wall and into the plant cell. For this they use a pilus, a small organelle that acts like a tiny hypodermic syringe, in a process called type III secretion. Fungi and oomycete pathogens also make such effectors, but how some of these are transferred to the interior of plant cells is not well understood. In both cases, however, these effector molecules target key components of the basal defense system, compromising general plant immunity and allowing such well-adapted pathogens a chance to succeed in what is now a susceptible host. Some of these effectors are kinases or phosphorylases that interfere with the MAP-kinase signaling cascade; yet others are targeted to the nucleus and affect transcription. Fungal and oomycete effectors are less well characterized, but one of the secreted ones provides a good example of an effector. *Cladosporium fulvum* is a fungal pathogen that causes leaf mold on tomatoes. It evades the basal defense response in its susceptible host by secreting a suite of effectors, among them a small chitin-binding lectin called Avr4. Normally the plant secretes a chitinase that digests wall material at the tips of the fungal hyphae. This releases chitin fragments, which then act as potent MAMPs that signal the plant to defend itself. The effector, Avr4, binds tightly to the chitin in the wall of the fungal tip, making it inaccessible to the plant chitinase. By preventing the release of chitin fragments, which are MAMPs, the fungus avoids triggering the plant's defenses (**Figure 8.10A and B**).

But this is a battleground that escalates with time, and the pathogen's acquisition of effector molecules means that there is now a selection pressure on the plant to fight back. The various effector molecules that pathogens have developed can all be regarded as new MAMPs, signaling to the plant that it is once again in danger. And so the plant has, in its turn, developed a new class of proteins that now recognize the presence of the effector and disable it, render it ineffective, or respond to it. These proteins are encoded by so called *resistance genes*, or *R-genes*. Most of these R-genes are relatively recent evolutionary developments. They are directed against the specific pathogen that acquired the corresponding effector molecule and are present in restricted plant species or varieties. Despite the fact that R-gene–based

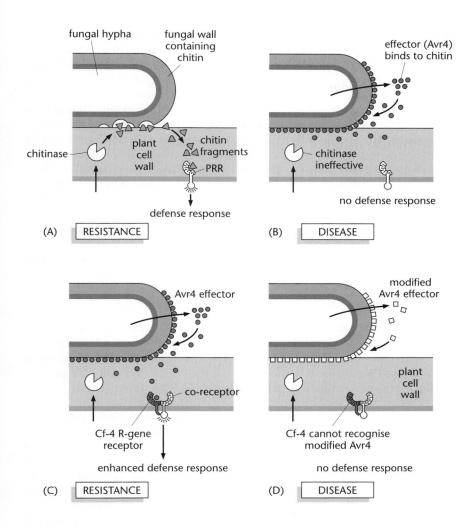

Figure 8.10 Coevolution of host and pathogen. This schematic drawing illustrates how the battle between host and pathogen swings back and forth. The example shown uses the fungal pathogen *Cladosporium fulvum*, a pathogen of tomato. (A) The plant secretes a chitinase that releases chitin fragments from the fungal cell wall. These MAMPs are recognized as danger signals by a plant pattern-recognition receptor and trigger a defense response. (B) The fungus now secretes a small effector molecule, Avr4, which protects the chitin in its wall from such an attack. The PRR is no longer activated, no defense response is mounted, and the result is that the pathogen is now successful and causes disease. (C) The plant now develops a resistance protein, Cf-4, that, probably in association with a co-receptor, somehow recognizes the presence of the effector Avr4 and triggers an enhanced defense response, including in this case cell suicide to prevent the spread of the fungus. (D) In a further twist, the fungus then develops a modified Avr4 effector, which still protects its cell wall but is no longer recognized by the plant's receptor, Cf-4, so no defense is mounted and disease results. The next step might be for the plant to provide a modified receptor that does recognize the modified Avr4, and so the cycle of attack and counterattack continues as a coevolutionary struggle. Steps C and D in this particular example are at this stage not fully understood.

resistance is soon broken down by ambitious pathogens, R-genes have long played an important part in selective plant breeding.

Many R-genes have now been characterized, and they come in many shapes and sizes. Many, though, are members of a large family of related proteins (about 150 members in *Arabidopsis*) that, like many of the pattern-recognition receptors described earlier, have LRRs. These may act either directly, by binding to the effector molecule, or indirectly, by binding to and guarding a molecular component of the basal defense machinery targeted by the effector. In either case, the basal defense pathway is reactivated, usually rather more strongly than it is in response to MAMPs and DAMPs. This response is commonly referred to as *effector-triggered immunity*, or *ETI*. A common feature of this enhanced defense response is for the targeted cell to commit suicide, which prevents the fungus from gaining nutrients and restricts its spread. This particular response is discussed later in more detail, but it is important to note that although it is more common as part of R-gene–mediated defense responses, plant cell suicide can also form part of the basal response to MAMPs and DAMPs.

In the *Cladosporium fulvum* case introduced above, some tomato varieties have acquired an R-gene that reestablishes resistance to the fungus that produces the chitin-binding lectin effector Avr4 (**Figure 8.11**). The R-gene is called Cf-4, and its structure resembles the LRR-containing pattern-recognition receptors that recognize some MAMPs and DAMPs. Since it has no kinase domain of its own, it may act in association with a co-receptor, such as BAK1, to somehow recognize the presence of the Avr4 effector produced by the fungus and to signal downstream to the defense pathway.

Figure 8.11 Tomato resistance to leaf mold. The fungal pathogen *Cladosporium fulvum* causes leaf mold on susceptible varieties of tomato. Leaves have been inoculated with spores of the fungus for 14 days. On the left is a diseased leaf from a plant that lacks the R-gene Cf-4, which confers resistance to the Avr4-producing fungal strain used. It shows extensive fungal growth and sporulation. On the right is a leaf from a plant that does have the corresponding Cf-4 resistance receptor. Recognition of the Avr4 effector triggers a strong defense response and the leaf shows no symptoms of disease. (From F.L.W. Takken et al., *Plant J.* 20:279– 288, 1999.)

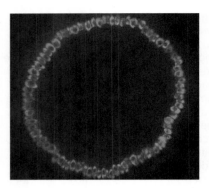

Figure 8.12 Avenacin is an antifungal toxin. The presence of this glycosylated triterpene, strategically placed in the outer epidermal cells of an oat root, is revealed by virtue of its autofluoresence. (From J.P. Morrisey and A.E. Osbourn, *Microbiol. Mol. Biol. Rev.* 63:708–724, 1999.)

Both host plant and adapted pathogen are destined to coevolve over time, with first one gaining the upper hand and then the other. The general theme is that plants, like animals, have learned to recognize danger signals and mount a stereotypical defense response. They recognize MAMPs produced by microbes and DAMPs produced by damage. Microbes are then selected that make effectors to interfere with the defense program, and that in turn leads the plant to make resistance proteins that block the activity of the effectors. And so the cycle continues, as this now places a selective pressure on the microbe to produce modified effectors that evade the resistance genes, and so on, in a continuing cycle (see Figure 8.10). We have seen then that plants have developed a single common defense response to three kinds of *danger signal*, the microbial MAMPs, the host-derived elicitors or DAMPs, and the microbial effector molecules. These are all perceived either in the wall by membrane-associated pattern-recognition receptors, or inside the cell by resistance protein receptors. The resulting defense response may vary in strength or duration, and may or may not involve cell suicide, but the core set of responses remains the same.

6.   Some defenses are preformed.[ref6]

Although we have discussed the plant in responsive mode, able to respond to danger signals by mounting a rapid defense response, the plant does not just sit there unprepared. The key barrier to microbial infection, of course, is the plant cell wall itself. If it is too thick, or too strong for pathogens to penetrate, it presents a constitutive barrier; and if it is reinforced at the point of attack it can again ward off attack; but we shall discuss all this in more detail in the next section. While a plant's major effort goes into the induced defense response, all plants have some level of additional preformed or constitutive defenses. Some of these are pre-prepared enzymes able to detoxify or metabolize dangerous microbial molecules, while others are antimicrobial compounds, sometimes called phytoanticipins. A good example is found in the root epidermis of oats (**Figure 8.12**), where a fluorescent triterpene glycoside called avenacin is found (*Avena* is the genus that includes oats and is also the Latin name for oats). Avenacin is toxic to most fungi that get as far as damaging epidermal cells and releasing it from the vacuole and is a key feature of oat's pathogen resistance. But one specialized fungus has developed a glycosidase that can remove all or part of the oligosaccharide attached to avenacin, thus rendering it nontoxic. This fungus now becomes an effective pathogen of oat roots.

It is not just secondary metabolites that are part of the plant's preformed arsenal; there are proteins active too. A large group of small basic proteins is found in the cell walls of almost all groups of plants, where they act as effective antimicrobials. These are evolutionarily ancient molecules, related to similar proteins in the animal kingdom called defensins. Although they show remarkable structural conservation (**Figure 8.13**), they are not closely related by sequence. There are about a dozen *defensins* in the *Arabidopsis*

Figure 8.13 Defensins are antimicrobial peptides. The structures of three members of the defensin family are compared to show the remarkable conservation of their structure. (A) Rs-AFP1, a cell wall defensin found in the seeds of *Raphanus sativus*. (B) NaD1, a vacuolar defensin found in flowers of *Nicotiana alata*. (C) PhD1, a vacuolar defensin from *Petunia hybrida*. The C and N termini are shown, as are the disulfide bonds as solid black lines. (From F.T. Lay and M.A. Anderson, *Curr. Protein Pept. Sci.* 6:85–101, 2005.)

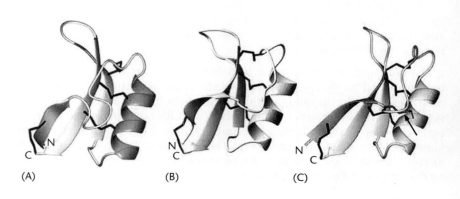

(A)   (B)   (C)

genome, each with eight disulfide-bonded cysteine residues. They are often expressed in surface cell layers, but they are also abundant in seeds. Most are secreted to the apoplast, where they have a broad range of activity against bacteria and fungi while remaining harmless to plants and animals. They are thought to act by interfering with membrane function, possibly by binding to sphingolipids, but may also enter the microbial invader and target intracellular components. In transgenic plants they have been shown to give increased protection against a wide range of pathogens, confirming their importance for preformed defenses.

Although defensins are found preformed, it has also been found that their expression is further triggered both by pathogen attack and by biotic stress signaling. In fact many components of plant defense are found at low pre-formed levels but at much higher levels after the perception of danger signals. A key class of compounds that are invariably associated with the basal defense response is the *phytoalexins*. These form a disparate group of phytochemicals: some are phenylpropanoids, some are flavonoids, and some are terpenes, but all are antimicrobials that are produced as a direct response to microbial attack, by enzymes that are downstream targets of the basal defense pathway described earlier. Mutants that cannot make a par-ticular phytoalexin are usually less resistant to pathogen attack, but plants make several such compounds and it is unlikely that any one phytoalexin exerts a dominant effect in resistance to a pathogen. The structures of three different phytoalexins are shown in **Figure 8.14**.

7. Defense can also be mounted at a distance.[ref7]

All the defense responses discussed so far take place either in the cell that is detecting the immediate danger or in its near neighbors. But it can be advantageous to a plant that is under serious attack to mount defenses further afield, for example, in distant leaves that might be threatened next. To do this the plant must take signals generated during the local response to danger and relay them to distant parts of itself, which must in turn respond in an appropriate manner. Such a distributed defense response, which provides a degree of immunization throughout the plant against fur-ther attacks, is called systemic resistance. Its effects may last from days to months, and they are most effective against biotrophic or hemibiotrophic pathogens and pests. Broadly speaking there are two distinct kinds of sys-temic resistance. The first, called *systemic acquired resistance*, or *SAR*, is triggered by a strong local defense response to a pathogen, often involving effector-triggered immunity and the suicide of cells to locally contain the pathogen. The second, the *systemic wound response*, is triggered by local mechanical wounding or cellular damage. This might be caused either by a necrotrophic pathogen or by insect pests. We shall briefly describe both kinds, starting with SAR.

One product of the general defense response, particularly when local cell suicide occurs, is the production of the small phenolic compound salicylic acid (SA) (**Figure 8.15**). This molecule acts locally as a signal to induce the expression of a set of proteins that include enzymes designed to degrade fungal cell walls. These proteins, which include sets of chitinases and β-1,3-

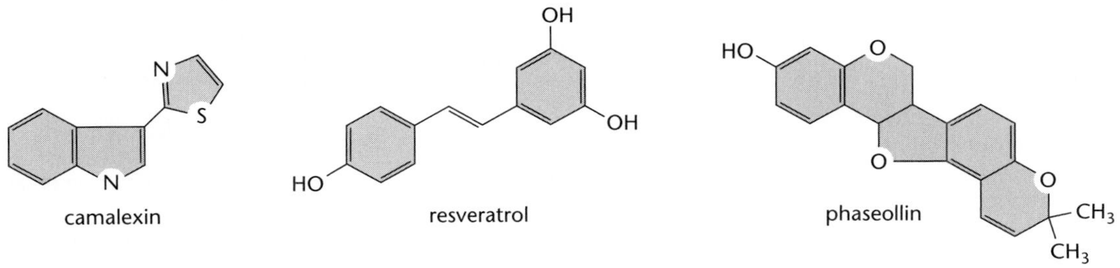

Figure 8.14 Phytoalexins. The chemical structures of three phytoalexins are shown. The sulfur- and nitrogen-containing camalexin is the main phytoalexin of Arabidopsis, resveratrol is a phenylpropanoid phytoalexin from grapevines, and phaseollin is an isoflavonoid phytoalexin from beans.

camalexin

resveratrol

phaseollin

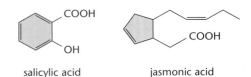

Figure 8.15 Two key components of systemic signaling. Salicylic acid is made from chorismate, a product of the shikimate pathway. The carboxyl group can be methylesterified to form the volatile signal molecule methylsalicylate. Jasmonic acid is a fatty acid derivative of linolenic acid that is finally synthesized in peroxisomes.

glucanases, are called *pathogenesis-related proteins*, or *PR proteins*, and they are a characteristic response to SA. This response can be propagated to other parts of the plant in a process that is dependent on SA. However, it is not the SA itself that moves through the plant to distant leaves, but the volatile derivative, methylsalicylate. This can travel through the phloem, and even diffuse through the air, to distant leaves, where it is then deesterified to SA, which triggers a systemic response, an SAR (**Figure 8.16**), that is similar to the local response. The systemic induction of some of the basal defense responses, together with the PR proteins, confers a relatively broad-spectrum defense against a range of pathogens including fungi, bacteria, oomycetes, and even viruses, although the range is not identical to the resistance provided by the normal basal defense response. Plants that cannot make SA show a reduced resistance to pathogens, both locally and systemically, while adding SA to a healthy plant triggers a defense response. This demonstrates that SA is the relevant signal in SAR.

The systemic wound response (see Figure 8.16), on the other hand, depends on the synthesis of another small signal molecule, jasmonic acid (JA) (see Figure 8.15). Wounds such as those caused by insects, or cellular damage caused by necrotrophs, trigger the production of ROS and calcium fluxes, which in turn promote the synthesis of JA and also ethylene. These two stimulate each other's production and cooperatively induce the expression of a specific set of defense genes that will depend on the type of plant. In potato, for example, extensin, proteinase inhibitors, and phenylalanine ammonia lyase are produced, while in *Arabidopsis* the phenylpropanoid pathway and defensin genes are activated. Additionally, oligogalacturonides (OGAs) are common products from damaged cells, and these may form part of the mobile signal that triggers a similar set of responses in distant tissues. In some plants a small peptide called systemin is made, and this may also act as a mobile signal to trigger the systemic wound response. *Arabidopsis*, at least, does not make systemin, and JA itself may act as the mobile signal in this case. A key feature of the wound response is the production of proteinase inhibitors (PINs). These act as inhibitors of digestion in herbivores such as chewing insects (**Figure 8.17**) and provide an effective defense against them. Plants that make no JA have a weakened resistance to insect pests. An important part of the local response to the initial wounding is to repair the damage. Most plants can reinforce and seal wounded surfaces by depositing on them suberin, a tough waterproof polymer of phenolic acids esterified with long-chain fatty acids and oxidatively cross-linked in a free radical–

Figure 8.16 Two systemic pathways that result in enhanced broad-spectrum defense. The systemic wound response requires jasmonic acid (JA) and ethylene (ET), both locally and in distant tissue, to enhance the defense response, including the synthesis of proteinase inhibitor proteins (PIN). The mobile signals may include oligogalacturonides (OGAs) and systemin. The latter is present in only a restricted set of plants, however. The systemic acquired resistance pathway requires the synthesis of salicylic acid (SA). The mobile signal is methylsalicylate (MSA), which is converted in distant tissue back to SA, where it enhances the defense response, including the production of pathogenesis-related (PR) proteins.

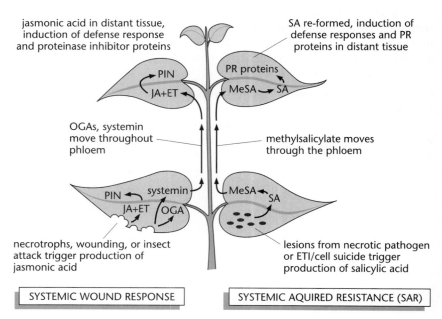

dependent reaction (see Concept 3G1). We encountered suberin earlier in the casparian band. An easily observed example is the suberized cell layer that forms at the surface of a cut potato. The suberin here prevents further loss of water and is relatively resistant to microbial attack.

Both classes of systemic resistance show a certain degree of overlap in the final defense products, although they also have the clear distinguishing features listed above. However, their effective activation depends on integrating signals from different parts of the plant, and considerable crosstalk takes place between the two pathways and between them and the basal defense pathway that responds to MAMPs, DAMPs, and effectors.

B. The Wall as a Battleground

1. Pathogens enter either by force, through wounds, or through natural openings.ref8

Pathogens, and biotrophic and hemibiotrophic pathogens in particular, need to gain access to the interior of the plant organ they are attacking. The primary obstacle to entry is the cell wall itself, which presents not only a physical barrier but also a sensory barrier, alarmed and sensitized to detect intruders and mount a defense response. Additional wall layers such as bark, suberized cells, or cork layers present further obstacles. Many invaders choose the easy route, entering through stomata, natural openings that grant access to the interior of the leaf. Many bacterial and fungal pathogens choose this method of entry. Once inside, bacteria can multiply in the moist intercellular spaces of the leaf and attach to the walls of cells, while fungi can explore the same spaces with their hyphae and enter cells to form intracellular feeding structures called haustoria (**Figure 8.18**). In some cases the germinated hypha can sense the small surface ridges on the leaf surface

Figure 8.17 Damage caused by herbivorous insects. A group of caterpillars has caused chewing damage to this nasturtium leaf, triggering a systemic wound response in neighboring leaves mediated by jasmonic acid.

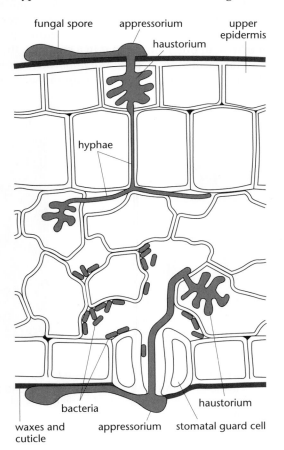

Figure 8.18 Entry of biotrophic pathogens into a plant host. Fungi enter either through natural openings such as stomata and wound damage or by forcing their way through the outer epidermal wall. Bacteria enter largely through natural openings.

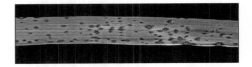

Figure 8.19 Rice blast disease. The disease lesions on this rice leaf, six days after the leaf was inoculated, are caused by the fungus *Magnaporthe oryzae*. (From P. Skamnioti and S.J. Gurr, *Plant Cell* 19:2674–2689, 2007.)

and track across them to search out the lip of a stomatal complex. There are other natural openings in plants, however, and ambitious microbes exploit them all. They include the water-secreting hydathodes on leaves, the gap where a lateral root pushes through the tissues of the primary root, the lenticels that allow gas exchange through bark, and open wounds that have not been sealed with suberin.

The outer wall of the epidermis, apart from generally being quite thick, is overlaid with layers of cutin and waxes. Yet many groups of fungi have evolved the capacity to physically penetrate this resistant covering (**Figure 8.19**). Fungi that physically push through the cuticle and wall do so with the assistance of an elaborate infection structure known as an appressorium (**Figure 8.20**). The *appressorium* is a suction-cup-like structure that physically anchors the hypha that emerges from the fungal spore tightly to the leaf and gives the hyphal tip sufficient purchase so that when it emerges from the appressorium with considerable force it can penetrate the cuticle and wall. Appressorium formation in some fungal pathogens is induced by the long-chain fatty alcohols typical of the surface waxes or cuticle of its host plant. Cutinases secreted by fungi may serve as "sensors." When a fungal spore lands on a plant surface, the cutinase cleaves esters of cutin, liberating long-chain fatty alcohols that stimulate fungal growth. Fungi have many cutinase genes, but the expression of particular ones is often highly upregulated on contact with a leaf surface. To understand appressorium function we shall use, as an example, the fungus *Magnaporthe oryzae*, which causes rice blast disease. In global economic terms, this is one of the most important of all plant pathogens. The force required to puncture the outer epidermal wall is generated osmotically by a high concentration of glycerol in the appressorium and is contained by a fungal wall reinforced by a tough layer of melanin except in a small region above the host wall (**Figure 8.21**). In a successful infection an infection peg bursts through the appressorium wall and the underlying host wall, effecting entry into the epidermal cell. Mutants that cannot make the melanin layer cannot make an effective infection peg. The pressure generated within an appressorium can reach about 8 MPa (about 30 times that in a car tire!), the highest pressures known in a biological structure. In *Magnaporthe*, a specific cutinase is essential for the proper development of the appressorium and for pen-

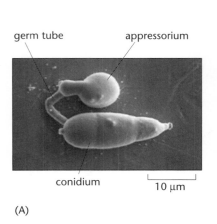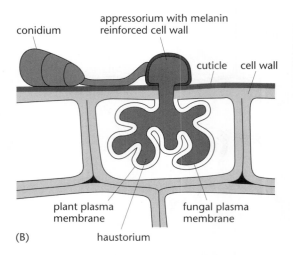

Figure 8.20 Infection by *Magnaporthe oryzae*. (A) Scanning electron micrograph of a conidium that has successfully germinated on a hydrophobic plastic surface and formed an appressorium. (B) Schematic diagram showing the appressorium and the infection peg that has enabled the hypha to cross the host cell wall and form a haustorium. (C) Successful infection occurs when the fungus invades a living rice epidermal cell and forms extensive invasive hyphae, or a haustorium, within the cell. This picture was taken 36 hours after infection. (A, from P. Skamnioti and S.J. Gurr, *Plant Cell* 19:2674–2689, 2007. C, courtesy of Prasanna Kankanala and the Barbara Valent laboratory.)

Figure 8.21 The appressorium. (A) The electron micrograph shows a cross section of an appressorium from *Magnaporthe oryzae*, formed on a hydrophobic plastic layer. The dark line around the wall is a tough layer of melanin that resists the high pressures generated in the structure. The melanin-free area can be seen; it is where the infection peg will burst through. (B) A similar section shows the infection peg after it has pushed out of the appressorium and through the plastic sheet. (Courtesy of Barbara Valent.)

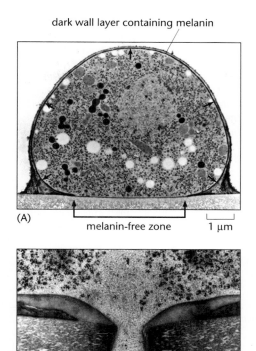

dark wall layer containing melanin

(A)   melanin-free zone   1 μm

(B)   infection peg   1 μm

etration, and mutants that lack that cutinase are markedly less pathogenic. Having entered the host cell, the fungal hypha forms a feeding structure, the haustorium, before spreading to neighboring cells.

2. Plasmodesmata provide a route for pathogen spread.[ref9]

Having finally gained a foothold in an epidermal cell, the successful pathogen must have a reliable means of spreading from cell to cell, and that means being able to cross the internal walls of the plant. Viruses, another major group of plant pathogens, generally gain access to a plant host through the intermediary of an insect vector, or through a wound, but they face the same problem: how to spread systemically through the plant. Viruses have developed two very different strategies, and both involve exploiting the natural channels between plant cells, the plasmodesmata (see Concept 1C9). Plant viruses encode movement proteins, which vary greatly between different viruses but are absolutely required for their trafficking from cell to cell. When viruses reach the phloem tissues they can then travel through sieve tubes, and the infection becomes systemic.

Plants in general need to regulate the traffic of molecules between cells, and they do this by gating the size exclusion limit of the plasmodesmata, either as a response to developmental cues or in response to stresses of various kinds. How gating is effected is still not entirely clear, but it seems likely that the deposition and removal of callose in the neck regions of the pore are a major factor. Stresses, including viral infection and the general defense response, result in callose deposition and a corresponding reduction in plasmodesmatal trafficking (see also Figures 4.37 and 4.38). For viruses, that means a reduction in the movement of viral nucleic acid through the plasmodesmata and reduction in viral spread. The location and regulation of callose deposition is connected in some way to members of the PDCB family, callose-binding, GPI-anchored, plasma membrane proteins that are located at the necks of plasmodesmata, attaching the membrane tightly to the callose deposits. If levels of these proteins are raised, more callose is deposited and less viral trafficking is seen (**Figure 8.22**).

Gating plasmodesmata implies that callose can be removed as well as deposited, in order for the channels to open again. Plants make many different β-1,3-glucanases, and we have seen that they, and chitinases, are major components of the set of PR proteins made as part of the defense

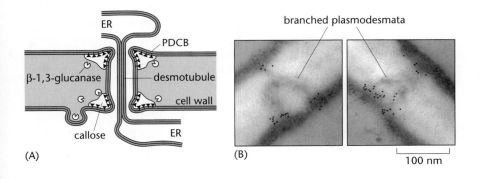

ER
PDCB
β-1,3-glucanase — desmotubule
cell wall
callose   ER
(A)

branched plasmodesmata
(B)   100 nm

Figure 8.22 Callose and the gating of plasmodesmata. (A) Schematic diagram showing the deposition of callose around the neck of a plasmodesma, constricting it and decreasing the size exclusion limit. The callose is anchored to the plasma membrane by PDCB, a callose-binding GPI-anchored protein. A balance between callose synthesis, and secreted β-1,3-glucanases that will hydrolyze it, accounts, at least in part, for the regulation of plasmodesmatal aperture and the trafficking of viral nucleic acids. (B) Two electron micrographs that show the location of PDCB at the neck of secondary, branched plasmodesmata in *Arabidopsis*, using gold-labeled antibodies to the callose-binding protein. (B, courtesy of Clare Simpson.)

Figure 8.23 Tubule-forming viruses and plasmodesmata. (A) Schematic diagram showing formation of a movement protein tubule, which can displace the central desmotubule in a plasmodesma and then transport intact spherical virus particles from one cell to another. Members of the plasmodesmatal-located protein family (PDLPs) facilitate this process. These are transmembrane proteins, lining the plasmodesma, that bind to the viral movement proteins making the tubule. Cauliflower mosaic virus (CaMV) is an example of a virus transported in this way. (B) A cultured insect cell that is expressing the viral movement protein from cowpea mosaic virus (CPMV). The cell has been stained with fluorescent antibodies to the movement protein, revealing that the movement protein on its own is sufficient to produce long tubules that extend from the surface of the cell. A similar result is found when the protein is expressed in plant protoplasts. (C) A row of small spherical virus particles being transported in a plasmodesmatal tubule in a *Chenopodium quinoa* leaf cell infected with grapevine fanleaf virus. (B, from D.T.J. Kasteel et al., *J. Gen. Virol.* 77:2857–2864, 1996. C, courtesy of Christophe Ritzenthaler.)

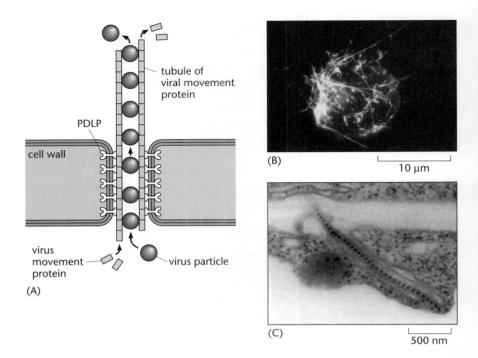

response, and in particular in systemic acquired resistance. Some of these glucanases are vacuolar, some target fungal and oomycete cell walls; but some class 1 β-1,3-glucanases are secreted and hydrolyze callose. In some viral infections such enzymes are specifically targeted to plasmodesmata in a movement protein–dependent manner; here they appear to enhance the pore size and viral traffic. Artificially increasing β-1,3-glucanase levels increases the size of lesions caused by tobacco mosaic virus, while β-1,3-glucanase deficient plants are less susceptible to virus infection and show smaller lesions.

Other virus pathogens, for example, cauliflower mosaic virus and cowpea mosaic virus, have adopted an entirely different strategy to get through plasmodesmata and into the neighboring cell. These viruses are collectively called tubule-forming viruses, because they use their movement proteins to assemble a hollow tubule that displaces the ER-related desmotubule in the center of the plasmodesma to create a free passage through which entire encapsulated virus particles can actively pass across the wall (**Figure 8.23**). The movement protein, when expressed on its own, either in plant cells or in animal cells, can directly form tubular outgrowths from the plasma membrane, demonstrating that tubule formation itself does not require a plamodesma. A class of proteins called PDLPs (plasmodesma-located proteins) line the channels of plasmodesmata, probably in association with lipid rafts or specialized membrane domains that line the pore. These proteins have receptor-like domains on their outer face (for which the corresponding ligands are not known), but on their inner face is a small domain that binds to the viral movement proteins, helping them to displace the desmotubule and to assemble into a tubule across the wall. The original function of the PDLPs, cleverly subverted by the viruses, is not known, but they are likely to be important regulators of normal plasmodesmatal gating. The membrane-spanning domain alone is sufficient to target other proteins to plasmodesmata.

Plasmodesmata are emerging as complex organelles that cross the cell wall, and their regulation is looking equally complex, involving, as we have seen, callose synthase, β-1,3-glucanase, calcium, PDLPs, callose-binding proteins, and lipid rafts. This makes them busy microdomains within the wall, and their presence has not gone unnoticed by fungal pathogens either.

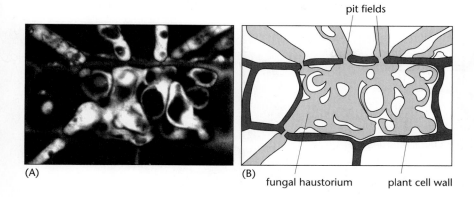

pit fields

(A)   (B)   fungal haustorium   plant cell wall

Figure 8.24 Fungal hyphae move using pit fields. (A) This fluorescence micrograph shows the rice blast fungus, *Magnaporthe*, 32 hours after a spore was inoculated on a rice leaf. The bright signal is from the fungal haustorium, which already fills the invaded epidermal cell. The fainter signal is from the cell walls of the rice epidermal cells. (B) Drawing showing the cell boundaries. New fungal hyphae have already invaded neighboring cells, passing through pit fields in the host cell wall, by loosening or partially digesting the wall around the plasmodesma, tightly constricting themselves and pushing through the narrow gap into a next-door cell, where they can expand again. (A, from P. Kankanala, K Czymmek and B. Valent, *Plant Cell* 19:706–724, 2007.)

Magnaporthe, the rice blast fungus described earlier, has also learned to exploit the presence of plasmodesmata. When the first colonized leaf epidermal cell begins to fill up with haustorium, invasive fungal hyphae begin to scan the inner surface of the cell looking for pit fields where clusters of plasmodesmata are found. The whole haustorium is encased in a host cell–derived plasma membrane and therefore separated from the plant cell by an intervening extrahaustorial matrix. The hyphal cell that detects a pit field constricts dramatically and then pushes through the thinner wall in the region of the pit field, still encased in the host cell plasma membrane (**Figure 8.24**). This invasion event does not need the high pressures generated in the original appressorium, but it presumably does need the secretion of wall-loosening and -degrading enzymes at the hyphal tip. Once the fungus is established in a neighboring cell, the first cell then dies, providing more nourishment for the growing fungus.

3. Callose deposition is a local response to both pathogens and wounding.[ref10]

Whether or not a plant succumbs to a pathogen attack is often a question of timing: of whether the attack can be rapidly mounted and overwhelming before the host plant has a chance to effectively marshal its own defenses. To overcome the resistance of the host wall a prospective pathogen must either exert enough force or deliver enough enzymes to sufficiently weaken the wall. Conversely, the host has an interest in making the wall both as mechanically strong as possible and also as resistant as possible to enzyme attack. The plant meets both of these ends by the common strategy of depositing a tough, reinforced wall apposition immediately below the point of pathogen attack. Such *papillae* are made in response to the presence of both pathogens and nonpathogens. Papillae are reinforced in a variety of ways to make them resistant to physical and enzymatic assault, and their function is to delay infection long enough for the plant's other defenses to become effective. It is a key reason why most pathogens do not succeed (**Figure 8.25**).

The papillae invariably contain deposits of callose, laid down about 16 hours after inoculation by GSL5, the relevant callose synthase (**Figure 8.26**). Callose in papillae is abundant and easily visible by its fluorescence when stained with aniline blue. The exact role of callose in strengthening the wall is unclear, but some successful pathogens, for example, *Pseudomonas*, have developed effector proteins that inhibit the deposition of callose and other papilla components. In addition to callose, there are many other molecules delivered to the papilla that increase the strength of the wall, including phenolics and hydroxyproline-rich glycoproteins that are oxidatively cross-linked. In addition lignin polymers are often deposited, increasing resistance to enzymes of wall degradation. Papillae also contain a conglom-

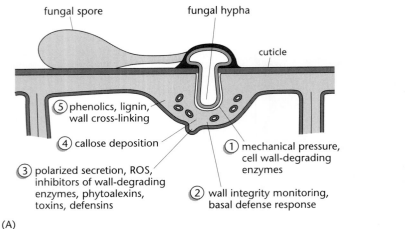

(A)

(B)

Figure 8.25 Papilla formation as a local response to pathogen attack. (A) Diagram showing the construction of a papilla immediately beneath the fungal infection peg that has penetrated the outer wall of an epidermal cell. The MAMPs and DAMPs from the interaction lead to a range of downstream events that are shown sequentially around the papilla. (B) Electron micrograph of a similar encounter shows the penetration of the outer wall by the infection peg and the complex wall material, more electron-dense than the normal wall, comprising the papilla beneath. (C) Light micrograph of a fungal germling of *Blumeria graminis* on a barley leaf cell. The spore has produced an appressorial lobe, which has been repelled by the production of an effective papilla beneath its tip. (B, From Zeyen RJ and Bushnell BR, *Can. J. Bot.* 57:898–913, 1979. C, adapted from R. Hückelhoven, *Annu. Rev. Phytopathol.* 45:101–127, 2007.)

(C)

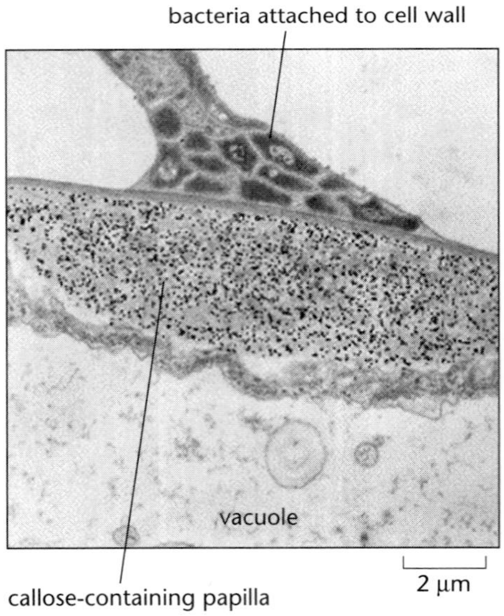

Figure 8.26 Callose deposition in the papilla. In this electron micrograph, a thickened region of cell wall material, the papilla, can be seen sitting immediately below a cluster of bacteria sticking to the cell wall in the intercellular space of an *Arabidopsis* leaf. The section has been immunogold-labeled to reveal the presence of callose, which is concentrated throughout the papilla. (Courtesy of John Mansfield.)

erate of chemical weapons: inhibitors of the cell wall–degrading enzymes made by the pathogen, antimicrobial toxins, defensins, and phytoalexins, all mentioned earlier in connection with the basal defense response.

The plasma membrane beneath the papilla is considered to be a specialized membrane microdomain containing many defense-related proteins specifically located adjacent to the site of pathogen attack. Many papillae also contain so-called paramural bodies, small membranous bodies delivered into the matrix of the papilla by the fusion of multivesicular bodies with the plasma membrane. The large number of different molecular components located in the papilla means that its effectiveness is the net integrated product of all the defenses concentrated in the one organelle.

In some cases, though, the papilla is not enough to prevent a determined pathogen from entering a susceptible host. When this happens, the papilla mounts a rear guard action trying to slow the spread and growth of the invader. For example, when a fungus breaches the papilla, the ring of remaining papilla material, now effectively a collar around the growing haustorium, is extended by secretion of further papilla wall materials in an attempt to surround and encase the intruding haustorium (**Figure 8.27**). Sometimes this works, but most times it merely slows the infection and can often promote the suicide of the cell involved after the fungus overcomes the plant's preinvasion defenses.

4. Wall strengthening helps contain the pathogen.ref11

We saw earlier (Figure 8.8) that one of the early events in the plant's perception of danger is the oxidative burst and the associated production of reactive oxygen species (ROS) through the action of membrane-located NADPH oxidases. In some cases wall-located peroxidases may also generate hydrogen peroxide. These ROS have many functions, among which are further downstream signaling in the defense program, complex interactions with the salicylic acid–dependent SAR program, and the programmed

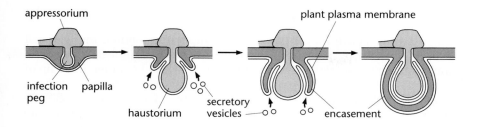

Figure 8.27 Haustorial encasement. In this diagrammatic sequence, the developing fungal haustorium has triggered formation of a papilla, but this is breached and the papilla now acquires more material in an attempt to encase the haustorium and slow its growth and spread. (Adapted from D. Meyer et al., *Plant J.* 57:986–999, 2009.)

cell death of target host cells. In addition, however, ROS have an additional and immediate function, to help reinforce the cell wall. Wounding, DAMPs, MAMPs, and effectors all trigger the production of O_2^- and H_2O_2 in the cell wall, and these can oxidatively cross-link a variety of molecules in the wall (**Figure 8.28**). Phenolics of various sorts, including monolignols, can be rapidly cross-linked to provide resistant, waterproofing materials. The suberin deposited at cut surfaces is one example, but polyphenolics and lignin are commonly deposited in walls as a result of physical insult or pathogen presence. Polyphenolics, formed by polyphenol oxidase, are the basis of the typical "browning" reaction seen at the surface of cut vegetables. Lignin and other phenolics are found as regular components of the papillae mentioned above, where one feature of such polymers, their hydrophobicity, makes it a tougher environment for wall-degrading enzymes to act. Experiments that reduce lignin production, or enzymically remove H_2O_2, both result in plants that show a reduced basal resistance to pathogens. It is worth adding that many pathogenic fungi also depend for their success on producing ROS from their own NADPH oxidases. If these enzymes are removed from the rice blast fungus described earlier, it cannot make an effective appressorium, and the fungus is rendered nonpathogenic.

Adding elicitors directly to cells in culture allows a timescale of events around the oxidative burst to be constructed. Wall reinforcement is a very rapid response. It starts around 2 min after a fungal elicitor is added and is completed by about 10 min. Key to this rapid event is the oxidative cross-linking of proteins in the wall, and in particular members of the hydroxyproline-rich glycoprotein family. Adding elicitor to soybean cells in culture causes at least two such proteins to become insolubilized in the wall, and at the same time it becomes more difficult to digest the wall with enzymes to make protoplasts. Similar protein cross-linking has been seen widely in resistance gene–mediated defense responses, including during H_2O_2-induced cell suicides. How the cross-linking of cell wall proteins affects wall digestibility and resistance to pathogens is not entirely clear, but it is likely that it functions in two related ways. The first is that the cross-linked protein meshwork may physically restrict access to wall-degrading enzymes. The second relates to the observation that the oxidative cross-linking of HRGPs in walls of cultured grapevine cells significantly decreases the wall's thickness. This cell wall shrinking is a function of reduced hydration of the wall, and it requires interactions with other wall proteins. Hydration is sensitive to wall matrix charge, and reduced wall hydration and thickness will have an additional effect on the access for degradative enzymes. This reinforces, for example, the dense and hydrophobic structure of papillae, which become increasingly resistant to enzymic attack.

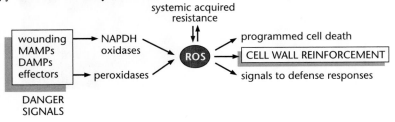

DANGER SIGNALS

Figure 8.28 The oxidative burst and wall reinforcement. This very simplified diagram shows some of the steps between elicitor, reactive oxygen species (ROS), and wall reinforcement.

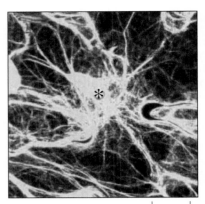

20 µm

Figure 8.29 Actin rearrangements in response to an attempted fungal attack. This *Arabidopsis* epidermal cell, expressing GFP-actin, has been inoculated with a spore of barley powdery mildew. A dense network of actin filaments, focused on the penetration site (asterisk), can be seen. (From D. Takemoto and A.R. Hardham, *Plant Physiol.* 136:3864–3876, 2004.)

5.   Cell polarization is a common response to a pathogen.[ref12]

When a prospective fungal pathogen attempts to penetrate the host cell wall, it does so by polarized secretion, delivering a molecular arsenal, including wall-degrading enzymes, in a highly focused manner at the tip of the invasion hypha. We have seen that this triggers a preinvasion resistance response in the plant that involves the assembly of a papilla immediately beneath the attempted invasion site. The plant's response, too, is highly polarized, designed to deliver all the defensive materials needed to the region of the papilla. This process is analogous to the generation of regional wall heterogeneity that we saw earlier (see Figure 6.34), and it depends on the polarized redistribution of key components within the cell. This polarized preinvasion response is typical of nonhost resistance; is triggered by danger signals emanating from bacteria, oomycetes, and fungi and even wounding; and is largely cell-autonomous.

Local signals, which might include calcium fluxes, mechanical signals, and the oxidative burst, act to trigger the basal defense machinery, a key component of which is a rearrangement of the actin cytoskeleton, creating a network of actin filaments focused tightly on the attempted penetration site (**Figure 8.29**). The actin filaments act to direct vesicles and other membrane materials toward the developing papilla. Golgi bodies, endoplasmic reticulum, and other cytoplasmic organelles are concentrated around the papilla, in a crowded area of cytoplasm, in an actin-dependent manner. The microtubule-based cytoskeleton appears to have little role in these polarized movements. Treatments that disassemble the actin network cannot assemble a papilla, and are no longer resistant to penetration.

The main secretion pathway to the plasma membrane is by Golgi-derived vesicular traffic. This will deliver, among other things, the enzymes needed to make the wall polysaccharides such as callose and cellulose, matrix materials of the wall including HRGPs and polysaccharides, toxins, phytoalexins, defensins, plasma membrane materials, and inhibitors of fungal enzymes (**Figure 8.30**). The specialized area of plasma membrane beneath the papilla contains defensive proteins and other membrane-associated molecules.

Other molecules reach the papilla in other ways. How the phenolics and monolignols needed for reinforcement of the papilla are exported to the wall is not known, but an unusual endosome-mediated secretion pathway is thought to account for the delivery of some wall-remodeling materials, some membrane and lipid materials, and defense-related plasma membrane proteins. This is thought to be how the small membrane vesicles, called paramural bodies, end up embedded in the matrix of the papilla. Endocytosis of plasma membrane materials, including membrane fusion machinery and PRR receptors, delivers these molecules to late TGN cisternae, which also serve as early/sorting-type endosomes in plants. From the TGN, the membrane molecules are recycled to to the plasma membrane via

Figure 8.30 Basal immunity involves the local synthesis of callose. (A) In response to the appressorium and infection peg of the powdery mildew fungus, the nonhost, *Arabidopsis*, forms a defensive papilla in which callose is locally deposited by the callose synthase, GSL5 (B). (A and B, from A.K. Jacobs et al., *Plant Cell* 15:2503–2513, 2003.)

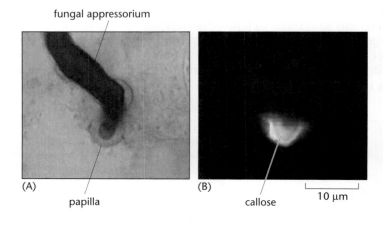

fungal appressorium

(A)

papilla

(B)

callose

10 µm

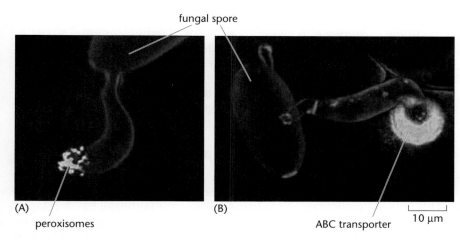

fungal spore

(A)
peroxisomes

(B)
ABC transporter   10 μm

Figure 8.31 Focal secretion at papilla during preinvasion resistance. These confocal micrographs show two GFP-fusion proteins at the site of an attempted fungal penetration. Both are *Arabidopsis* leaves inoculated with spores of powdery mildew. (A) Peroxisomes, with their associated GFP-labeled PEN2, accumulated around the area of attempted fungal penetration. (B) A GFP-labeled ABC transporter, called PEN3, is located in the plasma membrane underlying the fungal interaction site. This transporter is thought to move a glucosinolate metabolite into the wall to signal callose synthesis. The metabolite is produced by a thioglucosidase, called PEN2, associated with the surface of peroxisomes. (A and B, from U. Lipka, R. Fuchs and V. Lipka, *Curr. Opin. Plant Biol.* 11:404–411, 2008.)

secretory vesicles, or sent to the multivesicular bodies (MVBs, late endosomes) on their way to lytic vacuoles. During papilla formation the MVBs are induced to fuse with the plasma membrane, delivering their internal vesicles (often called exosomes) and lytic enzymes to the papilla wall.

A third focused delivery system to the papilla is through a membrane-located transporter. The synthesis of callose, a key component of the papilla, depends on at least one complex signaling pathway involving the breakdown of glucosinolates. These are toxic sulfur-containing compounds, mostly found in the family Brassicaceae, and usually associated with a plant's defense against insects. A thioglucosidase, PEN2, removes the glucose residue from a glucosinolate metabolite in the cytoplasm, releasing a product that is then transported through a plasma membrane ABC transporter located in the specialized membrane underlying the matrix of the papilla. It is this compound, or its breakdown products, that are required as a local signal to trigger callose biosynthesis. The PEN2 enzyme is attached to the surface of peroxisomes that are rapidly relocated to the area around the papilla in an actin-dependent manner (**Figure 8.31**). These complex and focused secretory machineries, which together deliver a wide range of host defense materials to construct the dense structure of the growing papilla, are summarized schematically in **Figure 8.32**.

We have seen that multiple strategies are used to assemble an effective papilla, making it a robust means of defense. However, as we saw earlier, the basal defense response is a target for pathogens. There is always a selection pressure to develop effector molecules that will suppress aspects of papilla construction, and such cases are known, but in general it is an effective

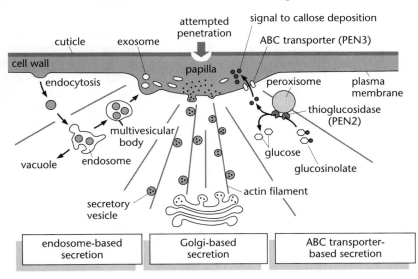

Figure 8.32 Polarized secretions at the site of a papilla. This schematic diagram illustrates the three main focused secretory events that ensure that the defensive papilla receives all the molecular weaponry that it needs. The majority of secretion is by Golgi-derived vesicular transport.

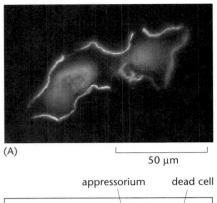

(A)

50 μm

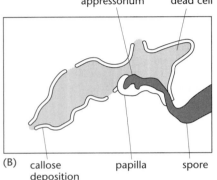

appressorium   dead cell

(B)   callose   papilla   spore
      deposition

Figure 8.33 Hypersensitive cell death.
This epidermal cell of an *Arabidopsis* leaf
has succumbed to penetration by the
barley powdery mildew fungus and has
triggered the hypersensitive response
48 h after inoculation. The programmed
cell death will constrain spread of the
fungus. The wall of the dead cell has been
reinforced by callose deposition, seen here
as fluorescence staining with aniline blue;
and the neighboring cells will mount a
strong systemic defense response to ensure
no further attack succeeds. (Adapted from
U. Lipka et al., *Eur. J. Cell Biol.,* 2009. doi:
10.1016/j.ejcb.2009.11.011)

multilayered local defense strategy. The many signaling inputs that contribute to the triggering of papillae are a source of redundancy and strength. In addition to MAMPs, DAMPs, and effectors—the usual danger signals described earlier that trigger defense responses, including papillae—the physical pressure alone that is exerted by a fungal penetration peg can also trigger them. Indeed, even a small glass needle pushing on the leaf surface can cause the rapid cytoskeletal rearrangements and the focused secretory apparatus.

6.   Localized cell death restricts pathogen spread.[ref13]

We have seen how attempted attacks by pathogens are usually resisted by the basal defenses, triggered by the plant's perception of danger signals. This immunity can break down in cases where the pathogen has developed a panel of effectors that interfere with components of the normal immune response—and sometimes just because the basal defenses were inadequate or too slow. Once a pathogen has penetrated the wall and gained entry to a cell, there is one last resort defense that can be mounted: the cell can act "altruistically" and commit suicide, for the greater good of the plant as a whole. The rapid collapse of the cell during such a programmed cell death removes access to nutrients and physically isolates and traps the pathogen, reducing the likelihood of spreading the infection (**Figure 8.33**). This rapid plant cell death response associated with restricting the growth of the pathogen has been referred to as the *hypersensitive response*, or *HR*. An extreme response like HR is understandably under very strict controls and is normally actively suppressed. While the HR can be triggered solely by a strong basal defense response, particularly after a pathogen has gained entry, it most commonly occurs where a resistance gene product recognizes an effector protein delivered by the pathogen. This recognition triggers a stronger immune response, including HR and the systemic response mediated by salicylic acid.

HR-associated cell death is a cell suicide event and is really effective only against biotrophic pathogens. Necrotrophic pathogens, on the other hand, also cause death in host cells, but in that case it is cell "murder" rather than cell suicide, a messy death that benefits the pathogen by leaking nutrients. Motive is important! Biotrophic viruses, fungi, oomycetes, and bacteria can all lead to HR and cell death, which is a rapid local event that isolates the pathogen from further access to nutrients and prevents its spread. The dying cell is rapidly dehydrated to remove accessible liquid. At the same time the systemic signals generated ensure that toxins and antimicrobials accumulate locally, that a strong defense response is mounted in neighboring cells, and that the wall of the dead cell is reinforced with materials that will make it resistant to leakage of nutrients. Callose is commonly deposited in the wall, as are polyphenols and lignin. The polyphenols can be seen as a brown staining under the light microscope (**Figure 8.34**). If wheat is inoculated with *Puccinia graminis* spores, the germlings can trigger an HR response. The resultant dead cells autofluoresce brightly as a result of lignin deposition in their walls. In this case the lignin that is deposited is entirely S-lignin, a response that can also be induced by the presence of chitosan. The programmed cell death process can be distinguished from other pro-

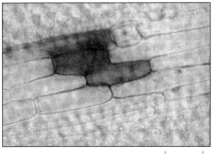

100 μm

Figure 8.34 Polyphenolics and hypersensitive cell death. The barley powdery mildew fungus, *Blumeria graminis,* has triggered hypersensitive cell death in a barley epidermal cell, and possibly two of its neighbors in the plant's effort to contain the pathogen. This bright-field micrograph shows the dark staining caused by the oxidative cross-linking of phenolics in the wall that will help the effort. (From L.A.J. Mur, in K. Roberts, ed., Handbook of Plant Science, pp. 1514–1517. Chichester, UK: Wiley, 2007.)

grammed cell deaths associated with normal plant development events, such as xylem element differentiation, because it is always associated with additional local and systemic defense responses.

The key signal for HR is salicylic acid, and that in turn is dependent on the generation of reactive oxygen species as an early response to danger signals (**Figure 8.35**). The full picture involves complex interactions of SA with other regulators, but ROS and SA are essential. Experiments in which SA levels are reduced result in slow or nonexistent HR, allowing the pathogen to escape the threat. Plants that lack a superoxide dismutase, the enzyme that breaks down superoxide radicals, show HR lesions on their leaves in the absence of any pathogen. The HR response is also linked to a systemic defense response that involves both local and distant production of toxins, phytoalexins, and PR proteins. Although HR, as described here, is a host's response to a particular pathogen attack, the same sort of response can also be triggered by a variety of abiotic stresses, including strong sunlight (photooxidative stress), wounding, and ozone. All of these generate reactive oxygen species, which appear to be the common danger signal. Different possible outcomes of an encounter between pathogen and plant are summarized schematically in **Figure 8.36**.

It is also possible to induce a response similar to HR, involving programmed cell death, by perturbing cell surface AGPs. Treatment of cells in culture with Yariv reagent, an artificial lectin that binds to AGPs (see Concept 3E6), causes cell death, and mutants lacking certain AGPs show a typical SA/wound response, but whether AGPs are normally involved in pathogen-induced HR is not known. One mechanism for triggering death in adjacent cells in response to adapted bacterial pathogens is the fusion of the vacuole membrane, the tonoplast, with the plasma membrane. This process releases the vacuolar contents, which include antimicrobials and death-inducing factors, into the wall and intercellular spaces. It is not yet known whether this novel defense strategy is also induced by other danger signals.

7.   Cell wall fragments act as danger signals.[ref14]

For a pathogen to survive it must obtain sugars as nutrients from the wall of its host, and pectin is a prime target. The relatively mobile pectic polysaccharides are some of the more accessible polymers in the wall, some of the most rapidly digested; and their removal is required in order to gain access to the other wall polysaccharides, cellulose and hemicelluloses, which are slower to digest. It is not surprising, then, that many pathogens, bacteria, oomycetes, and fungi all secrete enzymes capable of degrading pectic polysaccharides. These enzymes include a wide range of polygalacturonases, lyases, and methylesterases, and most pathogens show remarkable redundancy in the number of different enzymes they secrete for this one purpose. One of the intermediate products of this enzyme attack on pectin is a range of oligogalacturonides of various lengths, and it was discovered that some of these could act as "endogenous elicitors," which trigger the

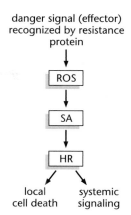

Figure 8.35 Reactive oxygen species and the hypersensitive response (HR). In this oversimplified pathway, a danger signal, in this case an effector, is recognized by a resistance protein and initiates a rapid production of reactive oxygen species (ROS), which leads to the production of salicylic acid (SA) and a hypersensitive response (HR) that includes systemic signaling (SAR) and local programmed cell death. The full picture is much more complicated, additionally involving interactions of SA with other regulators including jasmonic acid, nitric oxide, ethylene, abscisic acid, and the redox state of the cell.

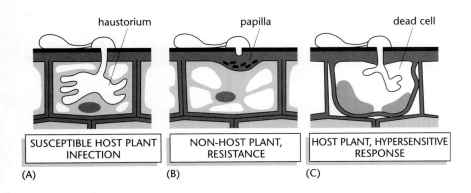

(A)   (B)   (C)

Figure 8.36 Possible interactions between a fungus and a plant. (A) The fungus has penetrated the epidermal cell of a susceptible host plant, forming a feeding haustorium and establishing an infection. (B) In a nonhost plant, the formation of a papilla and other defense responses prevent penetration. (C) In a nonhost plant, the fungus has overcome the preinvasion defenses but has triggered a rapid hypersensitive response, and the cell has committed suicide to isolate the fungus and prevent its spread.

basal defense machinery in almost all plants. We described earlier (p. 321) the concept of elicitors, danger signals that can originate from the microbe (MAMPs and effector molecules) or from the plant (DAMPs), and two of these, the heptaglucoside and chitin fragments, were described in some detail in relation to their pattern-recognition receptors and their signaling pathways. The case of the oligogalacturonides, which are DAMPs, is similar, and several lines of evidence suggest their receptors are WAK family members (see Figure 7.14). All three of these elicitors are oligosaccharides, and since this seems to be a general principle, that certain wall polysaccharide fragments can act as defense signals, such molecules are called *oligosaccharins*, regardless of their origin.

Different pectic enzymes produce different products, but it is only unesterified oligogalacturonides with a degree of polymerization (DP) of between about 10 and 15 that are active oligosaccharins, capable of eliciting the basal defense responses. The enzymes that generate oligogalacturonides, the fungal polygalacturonases (PGs), can be divided into three families. One family is composed of the classical *endo*-PGs that cleave their polymeric substrate, homogalacturonan, in a random fashion. A second family is composed of *exo*-PGs that attack their substrate in an *exo* fashion, removing one galacturonic acid at a time from the nonreducing end of homogalacturonan. The third, less studied family is composed of *endo-/exo*-PGs that have a mixed mode of action.

endo-PGs attach to their homogalacturonan substrate more or less equally anywhere along a chain that lacks methylesters. *endo*-PGs cleave one glycosidic bond at the site of attachment and then release their substrate before randomly attaching to another site, for example, on another molecule of homogalacturonan to catalyze the next hydrolytic cleavage. The oligogalacturonide products of *endo*-PGs include physiologically significant amounts of active oligogalacturonides (DP 10–15) (**Figure 8.37**).

Prolonged hydrolysis by *endo*-PG converts polygalactosyluronic acid into a mixture of mono-, di-, and trigalactosyluronic acids. The di- and trigalactosyluronic acids are converted into galacturonic acid by *exo*-galactosyluronic acid hydrolase, which is also likely to assist *endo*-PG in hydrolyzing slightly larger oligogalacturonides. *endo-/exo*-PGs, like *endo*-PGs, randomly attach to a homogalacturonan substrate before cleaving a glycosidic bond, but in contrast to *endo*-PGs, they do not release their substrate after cleaving a glycosidic bond. Instead they continue cleaving the same homogalacturonan molecule in a progressive manner, releasing one galacturonic acid residue after another from the nonreducing end of the chain. The *endo-/exo*-PGs do not finally release their substrate until they come either to the end of the chain or to an obstruction such as a methylesterified galactosyluronic acid residue. *endo-/exo*-PGs therefore digest their substrate without accumu-

Figure 8.37 Fungal polygalacturonases. Three different classes of fungal polygalacturonase (PG) can digest polygalacturonic acid, and the products can be separated and compared quantitatively by anion exchange chromatography. (A) *exo*-PG produces free galacturonic acid monomers. (B) *endo-/exo*-PG produces free galacturonic acid together with some low-DP oligosaccharides. (C) *endo*-PG produces a series of oligosaccharides, from free galacturonic acid to DP 14. Those in the higher range are DAMPs, biologically active oligosaccharides that elicit plant defense responses. (Adapted from B.J. Cook et al., *Mol. Plant Microbe Interac.* 12:703–711, 1999.)

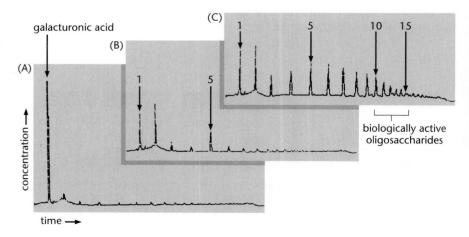

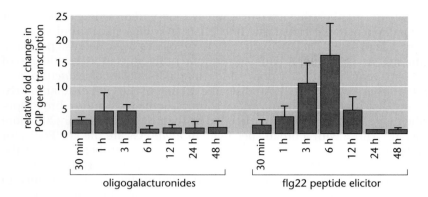

Figure 8.38 Gene expression induced by two danger signals. The genes induced in *Arabidopsis* by two elicitors, the MAMP flg22 and the DAMP oligogalacturonide, were analyzed by microarrays. In this selected example, the fold-increase in transcription of the polygalacturonase inhibitor protein (PGIP) gene is shown as a function of time. This protein is made rapidly by the plant as a way of limiting the action of the fungal polygalacturonase attacking its wall. Both elicitors induce transcription of the gene, but the response to flg22 is both stronger and longer-lasting. (Adapted from C. Denoux et al., *Mol. Plant* 1:423–445, 2008.)

lating significant levels of bioactive oligogalacturonides. Thus, biotrophic fungi that secrete *endo-/exo*-PGs can effectively use homogalacturonan as a food source without generating signal molecules from them that are detectable by their host.

Oligogalacturonides are currently the only plant-derived oligosaccharide DAMPs with known elicitor activity. Even at low concentrations, 100 nM, and mediated by the WAK family receptors, they trigger the same set of basal defense responses elicited by other danger signals, including membrane depolarization, calcium influx, generation of ROS, phytoalexin biosynthesis, and callose deposition. How important this is in effective defense has been difficult to determine genetically, partly because of redundancy in the set of pectin-degrading enzymes produced by pathogens. Removing one of the *endo*-PGs from the fungal pathogen, *Botrytis cinerea,* reduced the full virulence usually found on three different hosts, but other similar approaches have been less clear cut. Adding pectin-degrading enzymes directly to a plant is sufficient to induce soft rot symptoms. The signaling pathway from oligogalacturonides, via the WAK pattern recognition receptors, requires the calcium influx, reactive oxygen species, and MAP kinase cascade described earlier, suggesting that oligosaccharins act through exactly the same defense pathway as other danger signals.

Confirmation of this idea comes from a comparison of the transcriptional changes in response to two elicitors, flg22 (a MAMP) and oligogalacturonide (a DAMP). This showed that the same group, including hundreds of basal defense genes, were activated in each case. Early genes, related to the salicylic acid, jasmonic acid, and ethylene-dependent responses, largely against necrotrophs, were followed by slower responses, related to SAR and the hypersensitive response, largely against biotrophs. There were subtle differences in both the strength and the nature of the two responses, however, in that the flg22 responses were consistently stronger and more prolonged (**Figure 8.38**), and many of the later, SA-dependent responses, including HR, were not triggered by the oligogalacturonides. It may be that the oligogalacturonides have a much shorter half-life, but it seems they may also require other signal inputs to reach the threshold that is required to trigger the full immune response triggered by flg22.

8. Plants and pathogens battle to control the release of oligogalacturonides.[ref15]

The secretion of polygalacturonases by pathogens, and the consequent release of oligogalacturonide elicitors, provides an interesting example of the ways in which plants and their pathogens have developed attack and counterattack strategies, as each vies for the upper hand. It also emphasizes the importance of the cell wall as the front line in their two surveillance systems, as we now discuss. The plant has a vested interest in limiting the

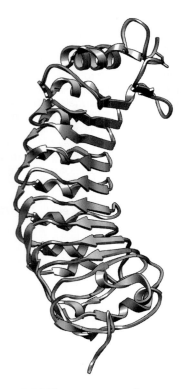

Figure 8.39 The structure of a polygalacturonase-inhibitor protein (PGIP). This schematic representation of a PGIP from the bean *Phaseolus vulgaris* reveals the repetitive LRR structure, a set of α-helices linking two parallel β-sheets. The concave front face, formed by 10 parallel β-strands, forms the binding surface for the PG.

activity of *endo*-PGs that are damaging its walls and providing nutrients for the pathogen. At the same time *endo*-PGs release oligogalacturonides of the right size to act as DAMPs, which elicit the plant's basic defense program. One way in which plants can deal with the problem is to make a lineup of proteins that inhibit PG action. Almost all plants make *polygalacturonase-inhibitor proteins*, or *PGIPs*, and secrete them into the extracellular matrix, where they inhibit PGs from various sources. Some have quite a broad specificity, while other members, encoded by their multigene family, are more specific. Those with a broad specificity to *endo*-PGs are the key PGIPs for pathogen resistance. For example, all four PGIPs from beans inhibit the *endo*-PGs from the necrotrophic fungus *Botrytis cinerea*. Plants in which the levels of PGIPs have been increased show an increased resistance to *Botrytis*, and plants in which the levels of PGIPs have been reduced show an increase in susceptibility to *Botrytis*. This suggests that such inhibitor proteins are important players in the ongoing battle in the wall.

But PGIP inhibition of the attacking PG is a subtler strategy than complete inhibition. This, after all, would remove the oligogalacturonides that trigger the plant's defense response. PGIPs are among the earliest genes to be induced as a response to elicitors, including oligogalacturonides. Structurally, PGIPs are members of a large family of extracellular proteins made largely of leucine-rich repeats (LRRs), the very same structural motif that we saw earlier in the context of pattern-recognition receptors at the cell surface (see Figure 8.5). This gently curved structure has a flattish concave surface, made from a long parallel β-sheet, and it is this surface that forms an adaptable and versatile structure for binding other proteins, in this case *endo*-PG (**Figure 8.39**). Very few amino acid changes are required to make the surface of an LRR domain adapt to new variants of the bound protein, and this surface is under strong positive selection pressures to cope with new variants of PG that the potential pathogen comes up with. A key feature of the interaction between an *endo*-PG and the corresponding PGIP is that in many cases the active site of the enzyme is only partly obscured, reducing enzyme activity but leaving just enough to mean that oligogalacturonides are still produced. Since these can no longer be so quickly converted to smaller fragments by the *endo*-PG, the half-life of active oligogalacturonides present is increased about 100-fold, enough to signal to the plant to mount a basic defense response.

There still remain many unanswered questions about PGIPs. One is that, in addition to the binding surface for PG, they also have, on the second β-sheet, a smaller, positively charged binding site for homogalacturonan, the substrate of PG. The bound pectin is displaced on binding to PG, but the exact significance of this is unclear. Another is that plant-encoded PGs and PGIPs are present in walls during various stages of normal plant development, irrespective of the presence of pathogens. How the plant avoids the potential elicitor-releasing activity of the endogenous PGs is unclear, and PGIPs are also often produced as a response to wounding. There is even one example where a PG may be recognized directly as a protein MAMP, in the absence of any enzymatic activity. We still have much to learn about the intriguing coevolution of PG/PGIP interactions.

9.   PR proteins attack the walls of fungi and bacteria.[ref16]

Earlier in this chapter we discussed the plant defense strategy of secreting chitinases. These enzymes attack the cell wall of fungi and release chitin and chitosan fragments that act as potent elicitors, or MAMPs, which are recognized by a pattern-recognition receptor and trigger the plant's basal defense response (see Figure 8.7 and Figure 8.10). Although there may be low constitutive levels of secreted chitinases, the large-scale induction of chitinases is one of the typical features of the strong local and systemic salicylic acid–dependent SAR response (see Concept 8B6), often associ-

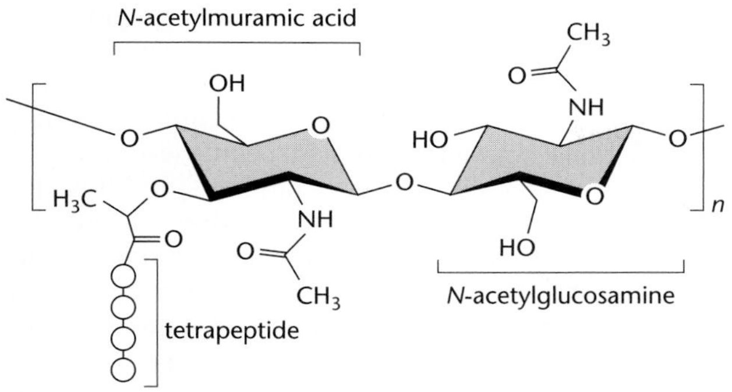

Figure 8.40 Bacterial peptidoglycan is a source of potent MAMPs. The peptidoglycan layer in both gram-positive and gram-negative bacteria is composed of a repeating disaccharide unit, shown here, with a variable tetrapeptide attached to the *N*-acetylmuramic acid residue. Oligosaccharide fragments, enzymically released from this polymer by a chitinase with lysozymal activity, are strong elicitors of the plant basal defense response.

ated with HR, cell death, and resistance protein–mediated defense. But chitinase is only one of a very large collection of proteins that are induced in this particular response, and collectively they have been called *pathogenesis-related proteins*, or PR proteins. They have been grouped into different families, and we have already come across some of them earlier, such as the defensins (in the PR-12 family) and the β-1,3-glucanase that releases the heptaglucoside MAMP from fungal walls (in the PR-2 family). Other families include peroxidases and proteinases.

Almost all of these PR proteins show some level of antifungal activity, and plants in which the chitinase and β-1,3-glucanase genes have been upregulated show enhanced resistance to fungal pathogens, although not to some other classes of pathogen. PR proteins are switched on only in a strong defense response, both locally and systemically, and this is dependent on salicylic acid (SA) signaling. Plants that cannot make SA cannot mount a SAR response and do not make PR proteins. Although some of the PR proteins are secreted to the apoplast, many are located in the vacuole. In order to reach their target molecules and prospective pathogens, they must somehow be released, and this can happen in two distinct ways. The first is during the kind of cell death caused by a necrotrophic pathogen, when the cell essentially "spills its guts" into the apoplast. The other way is a remarkable, deliberate defense strategy, induced by MAMPs and DAMPs, in which the tonoplast membrane fuses directly with the plasma membrane, in a process that in some way is dependent on proteasomes. This releases the vacuolar contents, and the PR proteins, directly into the intercellular spaces of the leaf.

The chitinases are a very complex set of enzymes, and they fall into more than one family of PR proteins (the PR-3, PR-4, PR-8, and PR-11 families). One aspect of this complexity is that while most chitinases act on chitin or chitosan, some of them have a related lysozyme activity, meaning that they can hydrolyze the sugar backbone of the peptidoglycan layer of bacteria (**Figure 8.40**). The oligosaccharides released, muropeptides, are in turn potent MAMPs, eliciting the plant's defense responses, including the production of some PR proteins. This is an interesting example of convergent evolution, since animals also recognize muropeptides as potent MAMPs, which trigger their own innate immune responses.

10. Carbohydrate-binding modules (CBMs) help enzymes attach to the wall.ref17

When pathogens secrete wall-degrading enzymes, either with the aim of penetrating the wall, or using the released sugars as nutrients, or both, it is an advantage for the enzymes involved to be anchored close to their substrate. For saprophytic microbes this is important to ensure priority access to the released nutrients, and they have developed a clever way of ensuring that their enzymes stay close by. In addition to their catalytic domain, most wall-degrading enzymes secreted by saprophytes have an additional

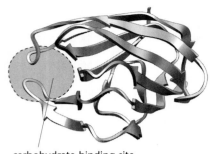

carbohydrate-binding site

Figure 8.41 The structure of a carbohydrate-binding module, or CBM. This CBM, called CsCBM6-3, is from a xylanase secreted by *Clostridium stercorarium*, a bacterium widely used in the breakdown of lignocellulosic biomass. It shows a typical β-sandwich protein fold, which can bind to either xylan or cellulose. This particular CBM is a family 6 member. (Adapted from A.B. Boraston et al., *J. Mol. Biol.* 327:659–669, 2003.)

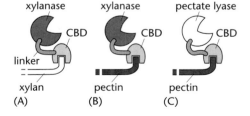

xylanase   xylanase   pectate lyase

CBD   CBD   CBD

linker

xylan   pectin   pectin
(A)   (B)   (C)

Figure 8.42 Complexity in the carbohydrate-binding domains of wall-degrading enzymes. Shown in schematic form are three wall-degrading enzymes that all have CBMs. (A, B) A xylanase with an associated CBM that can bind to either pectin or a component of xylan. (B, C) A similar CBM, which binds to pectin, attached to either a xylanase or a pectate lyase. These complexities presumably confer some advantage to the saprophytic microbe in accessing nutrients from the wall.

domain that is attached to them by a flexible linker region and is called a *carbohydrate-binding module*, or *CBM*. This is a discrete protein fold that has carbohydrate-binding activity but no catalytic activity. These domains, or modules, were previously classified as cellulose-binding domains because the first ones characterized had cellulose-binding activity, But now large numbers of CBMs have been examined, and it is clear that they can bind to almost all known cell wall carbohydrates of plants, fungi, and oomycetes. Although they show wide sequence divergence, many of them have a similar protein fold, called a β-sandwich (**Figure 8.41**). CBMs have been classified into 55 different families, and one general conclusion is that the binding affinity of the CBM usually matches the enzyme activity of the attached wall-degrading enzyme. Thus, cellulases usually have cellulose-binding CBMs and xylanases usually have xylan-binding CBMs. However, there are exceptions, and sometimes enzymes can have CBMs that can bind to different polymers, or one CBM may bind to several different substrates (**Figure 8.42**).

Because they are such ubiquitous components of microbial enzymes that threaten the health of the plant, it is not surprising to find that plants have learned to recognize some CBMs as MAMPs, as danger signals that can trigger the plant's defense responses. Although there are only a few described cases so far, it is likely that this will be a general feature of such enzymes. Indeed, it has been found that the wall-degrading enzymes of pathogenic fungi have far fewer CBMs than do the corresponding enzymes of saprophytic fungi, perhaps an indication that in the ongoing battle in the apoplast, pathogenic fungi have compromised on CBMs to reduce the risk of their acting as MAMPs. One intriguing example is a protein called CBEL, or cellulose-binding elicitor lectin, that is effectively just two CBMs coupled together, without an associated enzyme. It is produced by the oomycete pathogen *Phytophthora parasitica*, which causes a disease on tobacco. Each CBM can bind to a cellulose molecule, and each on its own can trigger the plant's basal defense response. However, when both CBMs are coupled together they elicit a much stronger response, including HR and related cell death, and SAR. Since CBEL has no effect on protoplasts, it seems clear that its detection depends not on a plasma membrane–located receptor, but on its interaction with the wall itself. How this might be accomplished is not yet known, but it is possible, since CBMs interfere with the structure of the wall in some way, that the cell can detect this perturbation and respond. Other evidence is now pointing in this direction, as we now discuss.

11. Cell wall integrity is sensed by the host.[ref18]

Many of the pathogenesis-related events we have discussed so far involve some kind of perturbation to the cell wall. These include physical damage from wall-degrading enzymes, host-derived DAMPs such as oligogalacturonides, the attempted mechanical penetration of the wall by fungi, callose deposition, lignification, and wall cross-linking. It is now clear that plants, just like fungi, in fact, have complex and robust ways of detecting such perturbations, of sensing whether or not the wall is maintaining its general integrity, and of responding accordingly. This process of *wall-integrity signaling* is ancient and operates during normal plant growth and development to maintain a competent wall of appropriate thickness and composition. But since pathogens, wounding, and other abiotic stress all affect wall dynamics, it is not surprising that the signaling systems in place have adapted to respond to them all. So plants detect danger in two different ways, directly through MAMPs, DAMPs, and effectors, and indirectly through wall-integrity sensing, and both sets of signaling inputs activate many of the same basal defense responses. Biotrophic pathogens, since they require a living host, try to avoid damaging wall integrity and, as we have seen, use stealth in their attack to avoid triggering plant defenses. Key responses to wall-

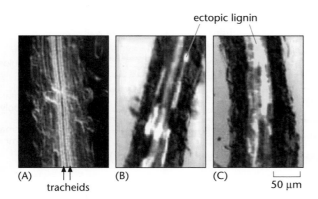

ectopic lignin

(A)   tracheids   (B)   (C)   50 μm

Figure 8.43 Perturbation of cellulose synthesis alerts the plant to danger. Loss of a component of the cellulose synthase subunit CESA3 causes reduced growth in roots of *Arabidopsis*. This is accompanied by ectopic lignification of nonvascular cells. (A) Wild-type root, showing the two central tracheids, normally the only cells in the primary root that are lignified. (B) Ectopic lignification in the *cesA3* mutant root. In this reverse-contrast image, the stained lignin appears white. (C) A similar result is seen in a wild-type root treated with the cellulose synthesis inhibitor isoxaben. In both (B) and (C) plant defense responses are also activated. (Adapted from A. Caño-Delgado et al., *Plant J.* 34:351–362, 2003.)

integrity signaling include effecting repairs and reinforcing the wall as well as defense responses, often dependent on SA, jasmonate, and ethylene, that help to restrict the source of the damage, whether it is caused by pathogen, herbivore, or wounding. In this context, callose might be seen as a temporary, amorphous filler material to be rapidly deposited at the first sign of danger to the wall's integrity. The wall itself is therefore a key sensory component in the plant's ability to detect danger, in whatever form.

In principle there are several kinds of wall perturbations that might be sensed or detected by the plant. These include damage to polysaccharides, the production of bioactive oligosaccharides, changes in wall polymer biosynthesis or assembly, and mechanical stresses within the wall network itself. One of the first indications of wall-integrity signaling came when a mutant was isolated on the basis that its jasmonate- and ethylene-responsive genes were constitutively activated. The mutated gene turned out to encode one of the subunits of the primary wall cellulose synthase, CESA3. The mutant plants have less cellulose and are reduced in size and, in addition to having activated defense responses, show ectopic lignin deposition in nonvascular cells, a common wall-reinforcing defense response (**Figure 8.43**). Mutations in other cellulose synthase genes give similar phenotypes, but, more important, wild-type plants treated with cellulose synthesis inhibitors show the same phenotype. Integrity signaling in this case appears to depend on a perturbation to the usual membrane context of the cellulose synthase. Earlier we saw that CBMs, such as CBEL, can also elicit defense responses and in this case the perturbation appears to be the physical integrity, or texture, of the cellulose network.

Cellulose, however, is only one of three main wall network components, and the hemicelluloses and pectins are also vital to wall structural integrity. And indeed, similar results have been found for perturbations of either component. It has been known for some time that altering the levels of any single wall component causes balancing adjustments in the other components, but it is now clear that these changes are also accompanied by defense responses. On the hemicellulose side, a mutant called *mur3*, defective in a xyloglucan galactosyltransferase, shows enhanced basal defenses and shows increased pathogen resistance, at least to downy mildew. On the pectin side, a screen for powdery mildew–resistant *Arabidopsis* mutants led to the isolation of two mutants, *pmr5* and *pmr6*, both with similar altered pectin composition. PMR5 is a membrane protein of unknown function, while PMR6 is a pectate lyase. Both are required for the normal pectin composition of the wall, but in their absence a robust defense response is mounted. In all the above cases a defense response is elicited by the presence of altered wall structures rather than the presence of a pathogen.

In the case of attempted penetration by a fungal pathogen, we saw earlier how physical pressure on the wall exerted by a glass needle could also elicit some of the same defense responses. Such mechanical signals transmitted

Figure 8.44 Signaling from the wall to the cell. This deliberately schematic diagram shows how three kinds of wall perturbation events might be detected by the cell and their signaling outputs converge on triggering basal defense responses. Many of the components shown here are conjectural, such as the proposed "perturbation sensor," while others have more experimental backing, such as the receptor-like kinase that monitors cellulose synthesis integrity.

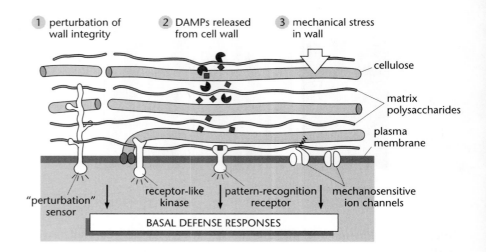

by the wall may well be registered at the plasma membrane, possibly by stretch-activated calcium channels or mechanosensitive ion channels activated by forces generated either by the curvature of the membrane or within the plane of the membrane. The idea that an interaction between plasma membrane and wall is involved is supported by some observations on the *cesA3* mutant described above. One aspect of the mutant phenotype is that cells in the root are badly swollen. If the internal osmotic pressure, which is driving expansion by pushing on the weakened wall, is compensated for by an external osmoticum, the cell swelling is reduced and the ectopic lignin and the defense responses normally associated with the phenotype are also reduced.

The molecular mechanisms for sensing wall integrity are not yet well understood, but some candidate molecules are known. Members of the large family of receptor-like kinases are likely to recognize many of these signals. Members of the WAK receptor kinase family (see Figure 7.14) recognize pectin and oligogalacturonides, and may be involved in integrity signaling. A serine-threonine receptor kinase called THESEUS (THE) has been identified that appears to monitor perturbations in cellulose synthase and/or the integrity of the membrane microdomain in which it operates. When activated, it triggers ROS production, growth suppression, and downstream defense responses. There are also many lectin-containing receptor-like kinases that could act as potential integrity sensors; and GPI-anchored AGPs, which have lectin-like activity, might also be involved. Some of the general ways in which the wall may signal to the cell are shown schematically in **Figure 8.44**. This is an active area of cell wall research, and the overall picture is far from clear, but it does seem likely that the study of defense elicitors, MAMPs and DAMPs, may well throw much needed light on this basic plant problem.

12. Responses to wounding and pathogens overlap.[ref19]

We have seen several cases now where mechanical signals—wounding or abiotic stresses—often end up triggering the same set of basal defense responses that are elicited by MAMPs, DAMPs, and effectors. Since we have argued that what the plant is really concerned with is sensing danger, it is easy to see that traumas like wounding or chewing insects (**Figure 8.45**) are just as much a danger as a fungal pathogen, and so it is no surprise that many of the signaling pathways involved are shared or converge downstream, but with at least some specialization in the final responses. We described earlier the general features of the wound response, in connection with systemic defense signaling (see Figure 8.16), but it is worth reexamining this in the context of the cell wall.

Figure 8.45 Wounding by herbivores. Caterpillars attack a nasturtium leaf.

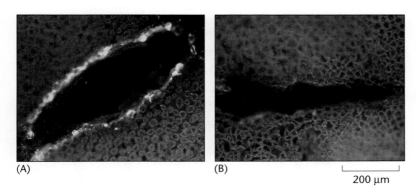

(A)   (B)

200 µm

Figure 8.46 Callose deposition as a wound response. (A) Callose (white signal) is deposited around the margins of a razor-blade cut in an *Arabidopsis* leaf. (B) No callose is deposited around the cut in a leaf from a plant that is not expressing *GSL5*, a gene encoding a callose synthase involved in callose production both in papilla formation and in the wound response. (From A.K. Jacobs et al., *Plant Cell* 15:2503–2513, 2003.)

The local result of either a wound or herbivory is three sets of cells, each affected differently. Some cells will be ruptured and killed and will release their vacuolar contents, containing several defense-related molecules. Around these will be cells that are damaged but still alive, and around these will be the remaining fully healthy cells. All of these will be signaling to each other in different ways. For example, the plasmodesmata that connect healthy cells with more damaged ones need to be promptly sealed, a job for the rapid deposition of callose. The wound needs to be sealed, not only to prevent water and solute loss but also to prevent opportunistic pathogens gaining access to the plant. Living cells lay down barriers of callose, suberin, polyphenols, and lignin in their walls to seal off the wound. The enzyme activated to make callose during the wound response is the same enzyme elicited to lay down callose in papillae, produced in response to attempted fungal penetration (**Figure 8.46**). The immediate signaling pathways invoked by wounding are the same as those described for the basal defense responses (see Figure 8.8), a rise in apoplastic pH, a calcium influx, and the generation of ROS. Downstream from their signaling pathways, jasmonic acid then plays a central role in the wound response, one refined by the effects of ethylene.

There are subtle differences between the genes activated in response to herbivores and to mechanical damage. Part of this can be put down to the continuous nature of the signal from a caterpillar compared with that from a razor blade. And some signals, in ovipositor fluids or oral secretions, for example, may arise as digestion products of host molecules, essentially acting as DAMPs. This is not the whole story, however, and it is emerging that herbivores, like microbes, produce molecules that are recognized by host pattern-recognition receptors to activate defense responses. Since these danger signals are not from microbes, the herbivore-associated molecular patterns have been called HAMPs! Examples of such elicitor molecules include several peptides and also a fatty acid–glutamine conjugate called volicitin. The latter, found in caterpillars, is recognized specifically by a PRR in the plant and triggers the production of volatile organic compounds that act as antiherbivores. The signaling complexity involved in the plant's responses to wounding, DAMPs, MAMPs, effectors, and biotic and abiotic stresses is not a set of linear pathways. Rather it is a complex network of pathways that intersect at crucial nodes and that are modulated by other inputs involving species and cultivar differences, environmental effects, and developmental effects. There is, not surprisingly, significant interplay between pathways mediating the wound response and those involving responses to pathogens, and the net result is a plant's optimized, economical, and integrated response to the dangers it encounters from a variety of sources.

13. Wall-degrading enzymes and their inhibitors coevolve.[ref20]

Pathogens can secrete a very wide range of wall-degrading enzymes, aimed at the plant cell wall, and this has initiated an arms race in the wall, in which

Figure 8.47 Schematic diagram illustrating three possible strategies that a fungal pathogen might use to evade plant surveillance.

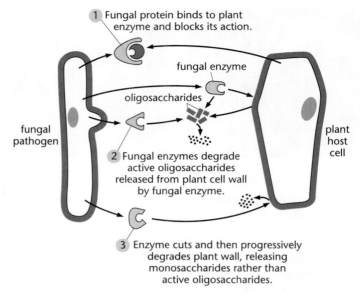

① Fungal protein binds to plant enzyme and blocks its action.

fungal enzyme

oligosaccharides

fungal pathogen

plant host cell

② Fungal enzymes degrade active oligosaccharides released from plant cell wall by fungal enzyme.

③ Enzyme cuts and then progressively degrades plant wall, releasing monosaccharides rather than active oligosaccharides.

measures are met with countermeasures and those with counter-countermeasures. For example, there are three ways, summarized in **Figure 8.47**, in which the plant's overall wall defense strategy can be minimized by the pathogen. It can make enzymes that degrade the active oligosaccharides released by other enzymes; it can make additional *exo*-enzymes so that monosaccharides rather than bioactive oligosaccharides are released; and it can make proteins that inhibit the action of plant enzymes designed to attack the pathogen's wall. Conversely, the plant can defend itself against the pathogen's array of wall-degrading enzymes by making adaptable, promiscuous inhibitor proteins that bind to and inactivate them. Using the example of *endo*-PG, we have seen that the plant has come up with a strategy for minimizing the damage while maximizing the likelihood that bioactive oligogalacturonides are produced as danger signals (see Concept 8B7). That strategy is to develop PGIPs, inhibitor proteins (Figure 8.39) that bind to the enzyme in question and inhibit its action. It is an emerging theme that almost all the classes of wall-degrading enzyme are turning out to have corresponding *inhibitor proteins* to counteract them.

Pathogens of grasses secrete xylanases. *Fusarium*, for example, makes over 30 different xylanases. These enzymes degrade xylan, the major hemicellulose of grass cell walls. But grasses in their turn have developed a formidable set of xylanase inhibitor proteins, which fall into several different structural classes active against the different classes of xylanase. For example, there is the TAXI class, evolved from a pepsin-like precursor, which inhibits family GH11 xylanases; and the XIP class, evolved from a class III chitinase precursor, which inhibits family GH10 and GH11 xylanases (carbohydrate-active enzymes have been classified into families that are described in detail in the curated CAZy database at www.cazy.org). Remarkably, none of them have any effect on the plant's own xylanases.

plant XIP1 xylanase-inhibitor protein

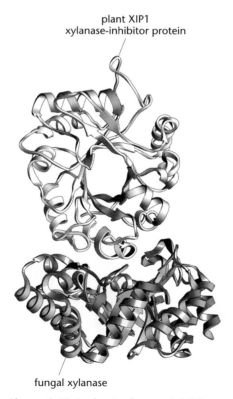

fungal xylanase

Figure 8.48 A plant xylanase-inhibitor protein, XIP1. The protein structure of the wheat XIP1 is shown in association with the xylanase that it inhibits from the fungus *Aspergillus nidulans*. XIP1 has a structure that is related to class III chitinases.

One instance is XIP1, a xylanase inhibitor protein from wheat, whose structure is shown in **Figure 8.48**. XIP1, which can also inhibit another family xylanase but using a different face of the same protein, is expressed at low constitutive levels but is also upregulated in response to elicitors. The predominant hemicellulose in dicots is xyloglucan, a target for fungal xyloglucan-specific *endo*-glucanases. Not unexpectedly, plants have developed a family of inhibitor proteins, called XEGIPs, to counter this threat. XEGIPs are structurally related to the TAXI class of xylanase inhibitors, and they are strongly induced by both wounding and elicitors. The production of all these inhibitor proteins is a regular part of the basal defense responses that we have discussed before.

The arms race in the wall suggests that pathogens in turn will develop proteins to inhibit enzymes secreted by plants, and indeed they have. As an example we can take *endo*-β-1,3-glucanaseA, one of several *endo*-glucanases secreted by soybeans. This is the constitutive enzyme (family GH17) that releases, from the fungal cell wall, the potent heptaglucoside elicitor discussed before (see Figure 7.48) and is distinct from the *endo*-glucanases that are PR proteins. The pathogen *Phytophora sojae* makes an inhibitor protein, GIP1, which has evolved from a chymotrypsin-like precursor protein that has lost its catalytic triad along the way (**Figure 8.49**).

This illustrates a general point, which is that almost all the inhibitor proteins described to date have evolved from preexisting, stable, apoplastic enzymes, in which the structural platform of the protein has been modified to lose enzyme activity but maintain binding capacity. Indeed, it seems that because of this recent explosion to create inhibitors, many proteins are not annotated correctly in databases. For example, many so-called chitinases are in fact xylanase-inhibitor proteins! Most inhibitors are members of multigene families, as are the enzymes that they inhibit, and many of them coevolve along with their target enzyme. So the existence of many isoforms probably facilitates coping with rapid small changes in the structure of either partner. As enzymes are selected that are structurally modified in various ways to avoid the inhibitor protein, either the inhibitor protein adapts in turn or new inhibitors are selected for (**Figure 8.50**). Both enzyme and inhibitor often show strong diversifying selection around the contact site residues between the two proteins.

It is also intriguing that the inhibitors known to date are all against *endo*-enzymes. Whether this is an attempt to protect the integrity of oligosaccharide products so that they can act as elicitors, as in the case of PGIP, is not yet clear. The molecular struggle to make wall-degrading enzymes that evade inhibitor proteins, and in turn to make new inhibitors that are adapted to them, is an ongoing part of the arms race, a struggle between pathogen and plant in which the cell wall is the crucial front line (**Figure 8.51**).

C. Recycling of Cell Walls

Since dead plants do not pile up around our ears, it is clear that some organisms must be capable of recycling all the cell wall polymers and finally returning the trapped carbon back into the carbon cycle. We have seen that many plant pathogens begin that process, killing cells and feeding on the sugars released from their walls, but these processes are continued after the plant is dead by groups of organisms, mainly fungi and bacteria, that are saprophytic, that is, living on the remains of plants. In Chapter 7 we looked at how plants are able to dismantle their own walls and remodel them, but we turn now to the elaborate suites of enzymes and structures that enable saprophytes to reduce even giant trees to simple sugars and other carbon skeletons. Obligate saprophytes possess weapons of mass destruction, vast batteries of enzymes, many of them organized into macromolecular

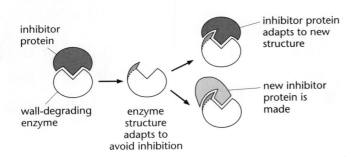

fungal endoglucanase-inhibitor protein (GIP1)

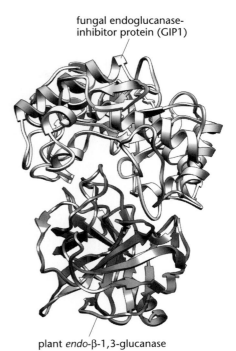

plant *endo*-β-1,3-glucanase

Figure 8.49 A fungal glucanase-inhibitor protein. The molecular structure of the soybean *endo*-β-1,3-glucanaseA is shown, docked onto the structure of GIP1, the corresponding glucanase-inhibitor protein from *Phytophthora sojae*. (Coordinates of the docked structure kindly supplied by J.K.C. Rose.)

inhibitor protein

wall-degrading enzyme

enzyme structure adapts to avoid inhibition

inhibitor protein adapts to new structure

new inhibitor protein is made

Figure 8.50 Enzyme-inhibitor protein coevolution. Schematic diagram showing changes and counterchanges that might occur during the coevolution of wall-degrading enzymes and their corresponding inhibitor proteins. The adaptive change in the enzyme structure might be a simple change in the charge of a single amino acid on the solvent-exposed residues at interacting surfaces or may be a more pronounced structural rearrangement. There is strong diversifying selection at work on the protein-protein interaction surfaces. (Adapted from J.C. Misas-Villamil and R.A.L. van der Hoorn, *Curr. Opin. Plant Biol.* 11:380–388, 2008.)

Figure 8.51 The battle in the apoplast. The rapidly evolving conflict between fungal enzymes that attack plant walls and plant enzymes that attack fungal walls, and what they do to minimize the release of MAMP and DAMP elicitors.

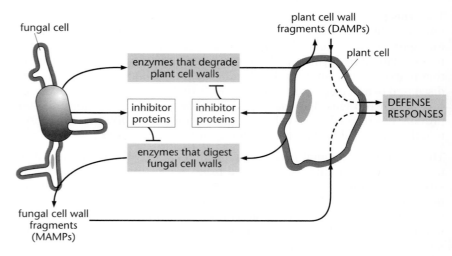

machines, whose sole function is to reduce cell walls to simple usable sugars. Collectively they reduce 10^{11} tons of plant material to recycled carbon skeletons a year. This may be done rapidly—soft fruit doesn't last long—but large hardwood trees may take decades to finally sink back into the soil. This partly depends on the plant's degree of lignification, partly on its internal anatomy, and partly on the environmental conditions. In very dry conditions, wood can last indefinitely; in wet, all rotting is accelerated (**Figure 8.52**). All wall components, including suberin, cutin, waxes, and lignin, are eventually recycled, but in this section we shall briefly examine what resources saprophytes deploy to digest the three main classes of wall polysaccharides—pectin, hemicellulose, and cellulose.

1. Removing pectin is a key early step in dismantling the wall.[ref21]

Pectin is a relatively mobile component of the wall, and its removal facilitates easier access for other wall-degrading enzymes. It is therefore a priority for both pathogens and saprophytes to begin pectin digestion as soon as possible. When fungi are grown on isolated primary cell walls as their sole carbon source, polygalacturonases (PGs) are among the first enzymes to be secreted. PG cleaves partially methylesterified homogalacturonan, a polysaccharide component of primary cell walls. At the same time, PG solubilizes almost half of the cell wall's rhamnogalacturonan l (RG-I) and most

Figure 8.52 Some plant substrates take a long time to break down. (A) The stump of an old beech tree is being slowly broken down by the mycelia of the wood-rotting basidiomycete *Meripilus giganteus*. This fungus, whose fruiting bodies are shown here, initially attacked the heartwood, but along with other saprophytic fungi and bacteria, it will eventually digest the remains of the stump, even if it does take several decades. (B) How plants are broken down by saprophytes is affected by internal plant anatomy, with cells of different shapes, sizes, and lignification, as shown here in a piece of dried maize stem (corn stover). In one case this might be fed to cattle, where it will be digested by rumen bacteria; in another it might be plowed back into the soil; and in another it might form the feedstock for producing sugars for biofuel production.

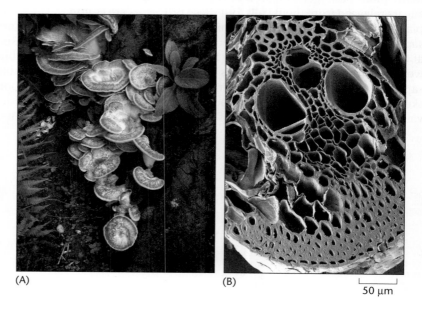

(A)

(B)

50 µm

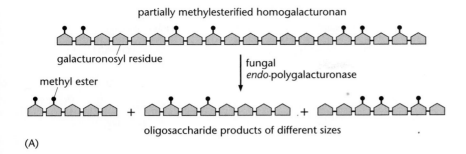

Figure 8.53 Polygalacturonase.
(A) Partially methylated homogalacturonan is cleaved by *endo*-polygalacturonase into different-size products. (B) Unesterified homogalacturonan will eventually be broken down to free galacturonic acid.

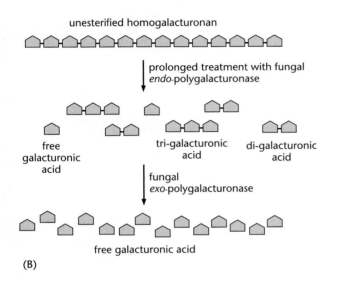

or all of the RG-II, even though neither is a substrate of PG. Treatment of isolated cell walls with PG thus makes the remaining wall polymers more accessible to the action of other *endo*-glycanases.

Quantitatively, *endo*-PGs are usually the dominant PG activity secreted by fungi grown in culture, and they play an important role during saprophytic growth. PGs appear to require several (perhaps as many as four) consecutive unesterified galactosyluronic acid residues in order to cleave homogalacturonan. The extent to which PG fragments homogalacturonan depends on its degree of methylesterification, and also on the precise arrangement of the methylester groups along the homogalacturonan chain. The cleavage products generated by an *endo*-PG will include partially methylesterified oligogalacturonides with a range of degrees of polymerization (**Figure 8.53A**). A combination of an *endo*-PG and an *exo*-PG can fully hydrolyze homogalacturonan to free galacturonic acid (Figure 8.53B).

For fungal *endo*-PGs and pectate lyases to extensively fragment the homogalacturonan backbone, many of the methylesters must be removed. Saprophytic microbes secrete one or more pectin methylesterases that hydrolyze the methylesters, releasing methanol and leaving unesterified galactosyluronic acid residues (**Figure 8.54**). Acetylesterases may also be required.

2. Lyases are important pectic enzymes for both necrotrophic pathogens and saprophytic microbes.[ref22]

Lyases are the only class of enzymes, other than the hydrolytic glycanases, known to cleave the glycosidic bonds of polysaccharides, but they are much more limited than glycanases in the range of glycosidic bonds they can cleave. Lyases in general can cleave only glycosidic bonds that are attached

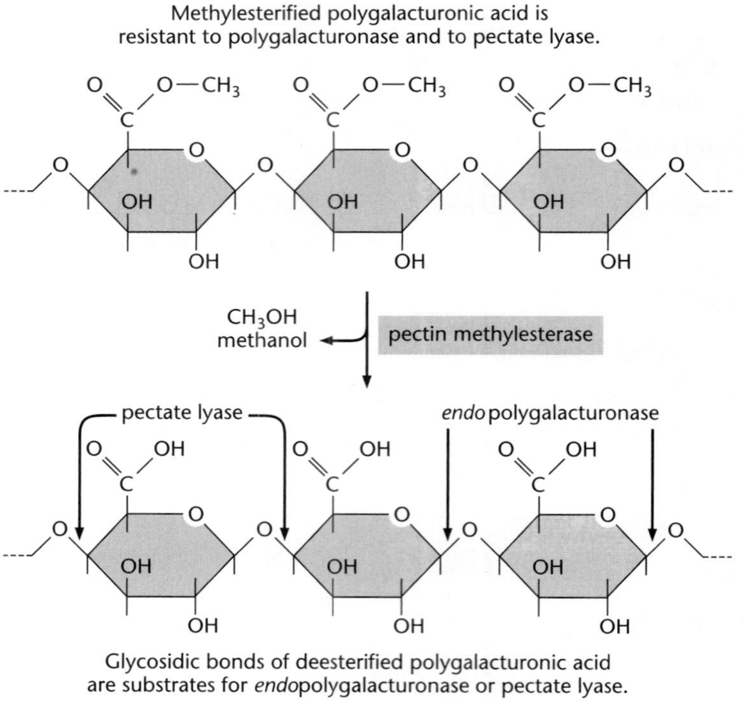

Figure 8.54 Pectin methylesterase. The action of pectin methylesterase creates the optimal substrate for both *endo*-polygalacturonase and pectate lyase.

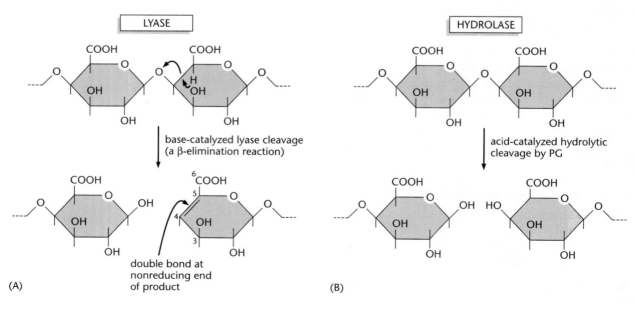

Methylesterified polygalacturonic acid is resistant to polygalacturonase and to pectate lyase.

CH₃OH methanol ← pectin methylesterase

pectate lyase

endo polygalacturonase

Glycosidic bonds of deesterified polygalacturonic acid are substrates for *endo* polygalacturonase or pectate lyase.

to O-4 of uronic acid residues (**Figure 8.55**), which for plant polymers means they can cleave only glycosidic linkages attached to O-4 of D-galactosyluronic acid residues. Only two plant polysaccharides contain such linkages: the bonds between the α-1,4-linked D-galactosyluronic acid residues of homogalacturonan, and those between the α-1,4-linked L-rhamnosyl-D-galactosyluronic acid repeating unit of the RG-I backbone. The RG-II backbone also contains α-1,4-linked D-galactosyluronic acid residues, but these are sterically inaccessible to both lyases and polygalacturonases.

Pectin lyases cleave only those glycosidic bonds attached to O-4 of methylesterified galactosyluronic acid residues, while *pectate lyases* cleave only

LYASE

HYDROLASE

base-catalyzed lyase cleavage (a β-elimination reaction)

acid-catalyzed hydrolytic cleavage by PG

double bond at nonreducing end of product

(A)

(B)

Figure 8.55 Lyases and hydrolases have different modes of action. (A) Pectate lyase and (B) *endo*-polygalacturonase are shown here cleaving the same glycosidic linkage in homogalacturonan.

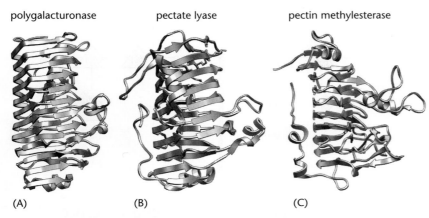

polygalacturonase   pectate lyase   pectin methylesterase

(A)   (B)   (C)

Figure 8.56 Pectin-active enzymes share an unusual protein fold. The same parallel β-helix structure can be seen in (A) an *endo*-polygalacturonase from the phytopathogenic fungus *Colletotrichum lupini*, (B) a pectate lyase II from *Xanthomonas campestris*, and (C) a pectin methylesterase from *Erwinia chrysanthemi*.

those glycosidic bonds attached to O-4 of unesterified galactosyluronic acid residues. Hydrolases cleave glycosidic bonds (Figure 8.55) via a mechanism that entails acid-catalyzed protonation of the glycosidic oxygen, followed by elimination of the aglycone (the "alcohol" to which the glycosidic bond was attached). Lyases, on the other hand, rely on a base-catalyzed removal of the proton attached to C-5 of the galactosyluronic acid residue. The proton on C-5 is made more acidic, that is, it is activated by the electron-withdrawing power of the carbonyl function attached to C-6. Following removal of the acidic proton from C-5, the glycosyl anion attached to C-4 is eliminated, resulting in the formation of a double bond between C-4 and C-5. This reaction is catalyzed in part by the fact that glycosyl anions are good leaving groups. This unsaturated residue strongly absorbs ultraviolet light at 235 nm, a useful property that is used to assay lyase-catalyzed reactions. Rather surprisingly, in view of the very different chemistries involved, pectate lyase and polygalacturonase, and indeed other pectin enzymes, have closely related structures, all being variations on an unusual parallel β-helix protein fold (**Figure 8.56**).

There are two types of glycosidic linkages that form the disaccharide-repeating unit of the backbone of RG-I. One is cleaved by rhamnogalacturonan hydrolase, which hydrolyzes the glycosidic linkages between D-galactosyluronic acid and α-1,2-L-rhamnosyl residues; and the other by rhamnogalacturonan lyase, which β-eliminates the glycosidic linkage between L-rhamnosyl and α-1,4-D-galactosyluronic acid residues (**Figure 8.57**). The rhamnogalacturonan backbone of RG-I is, to a high degree, sterically protected from these enzymes by side chains attached to O-4 of the rhamnosyl residues and by *O*-acetyl groups and perhaps glycosyl side chains attached to O-2 or O-3 of some of the galactosyluronic acid residues. These need to be partially or fully removed before rhamnogalacturonan hydrolases and rhamnogalacturonan lyases can operate.

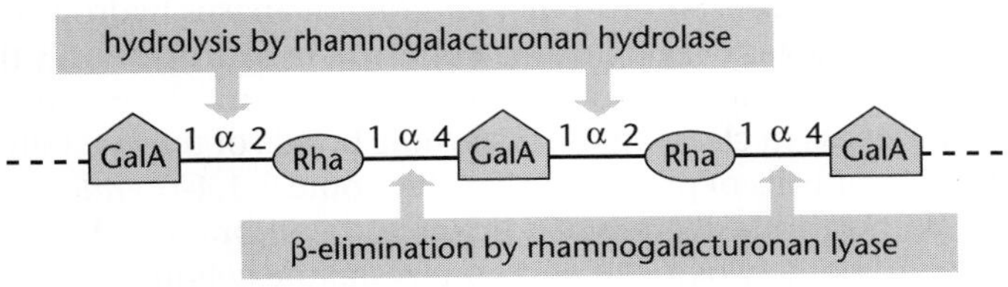

hydrolysis by rhamnogalacturonan hydrolase

- - - GalA — 1 α 2 — Rha — 1 α 4 — GalA — 1 α 2 — Rha — 1 α 4 — GalA - - -

β-elimination by rhamnogalacturonan lyase

Figure 8.57 Two enzymes are required to cleave the two types of glycosidic linkages present in the disaccharide-repeating unit of the backbone of RG-I.

Figure 8.58 Schematic diagram showing how the concerted action of a suite of fungal enzymes can degrade glucuronoarabinoxylan, the major grass hemicellulose.

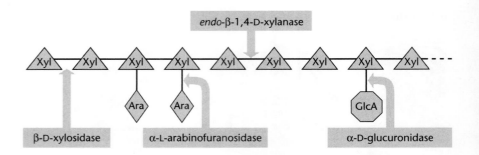

3.   To fully digest hemicelluloses, enzymes act in concert.[ref23]

Several *endo*-glycanases and glycosidases, all working in unison, are generally required to convert primary cell wall polysaccharides into usable mono- or disaccharides. The hemicellulose component of the wall is generally the next to be attacked by saprophytes after the pectins, allowing enzymes access finally to the cellulosic framework of the wall. Breakdown of the three main hemicellulose polymers will be discussed here; they all illustrate the same general principles. The hydrolytic enzymes are made in large multigene families, they are generally induceable by their substrate, they act cooperatively, they often show catabolite repression, and their activity is often aided by having attached CBMs.

Glucuronoarabinoxylan, the main hemicellulose in grasses, for example, minimally requires an *endo*-β-1,4-D-xylanase, α-L-arabinofuranosidase, β-D-xylosidase, α-D-glucuronidase, ferulic acid esterase, and perhaps a α-D-4-*O*-methylglucuronidase to convert glucuronoarabinoxylan into its monosaccharide constituents. *endo*-β-1,4-D-Xylanases and α-L-arabinofuranosidase are essential for the depolymerization (**Figure 8.58**). *endo*-Xylanases fragment the β-1,4-linked xylan backbone, while arabinofuranosidase removes the frequent arabinofuranosyl side chains that sterically block the ability of *endo*-xylanase to cleave the backbone. It is important that the two enzymes act in concert, for *endo*-xylanase has little ability to degrade heavily arabinosylated xylan without prior removal of some side chains. Arabinoxylan accounts for about 40% of the primary cell walls of grasses but for only about 5% of the walls of other plants. *Magnaporthe grisea*, the fungus that causes rice blast disease, is able to synthesize multiple members of each of three different β-1,4-D-*endo*-xylanase families. At least four are secreted into the medium of the fungus grown on arabinoxylan as the carbon source, and the fungus secretes at least one additional *endo*-xylanase when infecting rice leaves.

A suite of enzymes acting in concert is needed to degrade the second hemicellulose, galactoglucomannan. Its structure dictates that it requires at least an α-galactosidase, an *endo*-β-1,4-mannanase, an *O*-acetyl esterase, an α-mannosidase, and an *endo*-glucanase. In some cases epimerases may be important, for example, to convert mannose to glucose, which may be more easily metabolized. Unlike the first two hemicelluloses, xyloglucan has a β-1,4-D-glucan backbone that has the same structure as cellulose. Most saprophytes secrete cellulases, but they also secrete a variety of other *endo*-β-1,4-D-glucanases that specifically depolymerize xyloglucan, rather than attacking cellulose. Some of these *endo*-glucanases hydrolyze only the β-1,4-glucosyl linkages of xyloglucan, while others hydrolyze only the β-1,4-glucosyl linkages of cellulose, but none of the *endo*-glucanases, on their own, have much success in cleaving the glycosidic bonds of the crystalline cellulose in the microfibrils of primary cell walls. Some β-1,4-D-*endo*-glucanases hydrolyze, with various degrees of efficiency, the unbranched β-1,4-D-glucosyl linkages of both xyloglucan and noncrystalline cellulose.

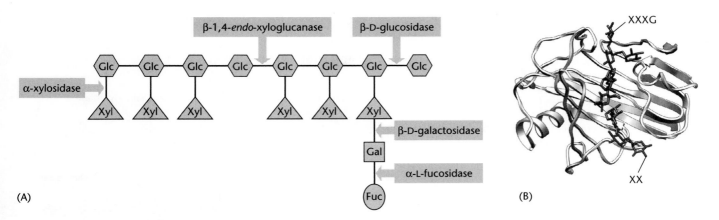

(A)

(B)

Figure 8.59 Xyloglucan breakdown. (A) The enzymes required to hydrolyze xyloglucan. The conversion of an *endo*-glucanase-generated xyloglucan oligosaccharide into catabolizable monosaccharides requires an array of glycosidases. Some esterases may also be needed. It should be noted that although the bonds cut by each enzyme are indicated, the glycosidases have to act sequentially and can cleave terminal sugars only at nonreducing ends. For example, the β-D-galactosidase cannot act until the terminal fucose has been removed. Once the galactose has been removed, the α-xylosidase can act, and so on. (B) The protein structure of an *endo*-xyloglucanase, from *Bacillus licheniformis*, with its xyloglucan substrate, shown here as XX and XXXG fragments, in the active site cleft. Remarkably, families of *endo*-xyloglucanases with at least five different protein folds have been found.

The products of the digestion of xyloglucan by β-1,4-*endo*-glucanases are for the most part heptasaccharides or larger. Conversion of these *endo*-glucanase–generated xyloglucan oligosaccharides into more usable mono- or disaccharides requires a suite of other glycosidases that include α-L-fucosidase, β-D-galactosidase, α- and β-D-xylosidase, α-L-arabinofuranosidase, and β-D-glucosidase, as well as one or more acetylesterases (**Figure 8.59**). Many of the enzymes involved in hydrolyzing hemicellulose, and indeed other polymers, have associated CBMs attached. These show different specificities for different polysaccharides, and not always for their target substrate. A mannanase, for example, may have a CBM that binds it to a cellulose microfibril. CBMs increase the local enzyme concentration, and they can often disrupt the structure of the polymer, making it more susceptible to attack. Unlike fungal enzymes, some bacterial enzymes can have more than one CBM, for example, one binding it to cellulose and the other attaching the enzyme to its target polysaccharide (**Figure 8.60**).

4. Cellulose is tough, and its disassembly usually requires special machinery.[ref24]

Crystalline cellulose is the final and most difficult of the wall polysaccharides to depolymerize. Several *endo*- and *exo*-glucanases are required to work together and with other enzymes to convert cellulose into glucose (**Figure 8.61**). Many microbes, particularly anaerobic bacteria and fungi, have evolved an organized, macromolecular enzyme complex that can successfully depolymerize crystalline cellulose, albeit still at a relatively slow rate. This complex is called a *cellulosome*. Among other components, cellulosomes are built on a nonenzymatic scaffold protein that includes a cellulose-binding module. This CBM binds to the surface of the cellulose microfibril and helps perturb its crystalline structure. In bacterial cellulosomes the scaffold also has a domain that attaches the complex to the bacterium, ensuring that it remains close to its substrate.

Along the scaffold are a number of binding sites, called cohesins, for a suite of wall-degrading enzymes that include *endo*-β-1,4-D-glucanases, cello-dextrinases (including cellobiohydrolase), *endo*-β-1,4-D-xylanases, and

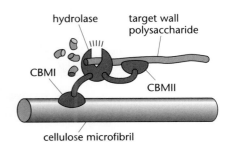

Figure 8.60 Unlike fungal wall-degrading enzymes, many bacterial hydrolases can have more than one CBM. Shown here is a schematic illustration of an enzyme with one CBM attaching it to cellulose and another anchoring it to its target polysaccharide.

Figure 8.61 Complete breakdown of cellulose requires three separate enzyme activities. An *endo*-glucanase cuts internal glycosidic bonds. *Exo*-cellobiohydrolases progressively release the disaccharide cellobiose from the ends of the chain, and cellobiose is then split by a glucosidase to release glucose. Some organisms produce *endo*-cellobiohydrolases that act both on the reducing end and on the nonreducing end, while others produce only one or the other.

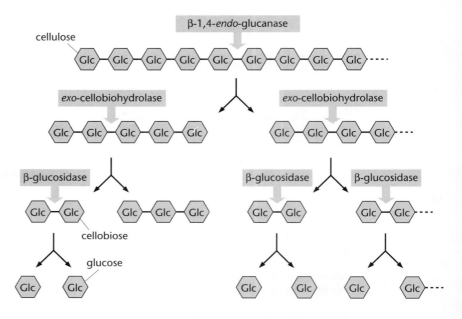

endo-β-1,4-mannanases (**Figure 8.62**). These enzymes all possess an additional dockerin domain, which binds them to the appropriate cohesins on the scaffold. Cellulosomes are macromolecular machines that are thought to orient the hydrolytic enzymes with respect to crystalline cellulose so that the enzymes work synergistically as they depolymerize the glucan chains (**Figure 8.63**). Many *endo*-glucanases, *endo*-xylanases, and other glycanases that are not associated with cellulosomes also have their own CBMs separated from their catalytic domain by a flexible linker region.

To tackle woody tissue, there is an additional major problem for saprophytes, which is lignin. Not only is this polymer very recalcitrant to digestion but it also hinders access to enzymes that can degrade the embedded wall polysaccharides. There are only a few specialized and adapted organisms that can finally break down this insoluble and very resistant polymer. Chief among these are the white rot fungi, a group of basidiomycetes that can degrade both cellulose and lignin in wood leaving only a whitish residue (see Panel 9.1). They degrade lignin in hydrogen peroxide–mediated oxidation reactions that require laccases, lignin peroxidases (also called ligninases), and manganese-dependent peroxidases, but the reactions are not well characterized. So-called brown rot fungi cannot digest lignin and therefore leave rotted wood that is colored brown from the residual lignin.

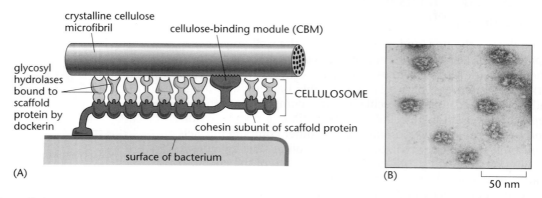

Figure 8.62 The cellulosome is a macromolecular enzyme complex produced by bacteria and fungi to degrade crystalline cellulose microfibrils. (A) The cellulosome consists of a scaffold protein that is built up from a set of repeating domains each of which binds to a different cellulase or related glycosyl hydrolase. The scaffold protein also has different domains that bind to the bacterium and to its substrate, the cellulose microfibril. (B) Free cellulosomes can be isolated from cultures of bacteria, in this case *Clostridium thermocellum*, and examined, when negatively stained, in the electron microscope. (B, courtesy of Edward Bayer.)

Figure 8.63 Dissolution of crystalline cellulose by purified cellulosomes. The tough crystalline cellulose ribbons produced by *Acetobacter xylinum* are incubated here with a preparation of cellulosomes from the anaerobic bacterium *Clostridium thermocellum*. A combination of *exo-* and *endo-*cellulases progressively break down the ribbons. (A) The starting cellulosic substrate. (B) The same material after 3 h incubation is already 60% degraded. (C) After 8 h the ribbons are 85% degraded. (Adapted from Bayer et al., in M.E. Himmel, ed., Biomass Recalcitrance: Deconstructing the Plant Cell Wall for Bioenergy p. 407-426. Chichester, UK: Blackwell, 2008.)

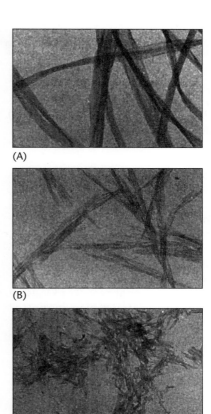

(A)

(B)

(C)

0.5 µm

In very wet timber, soft rot fungi, ascomycetes, can operate, but these again cannot digest lignin and leave a wet spongy wood as their product. Degraded lignin products are a major contributor to humus in the soil. Some enzymes may assist in the catabolism of lignin and cellulose. The fungal cellobiose oxidoreductases (cellobiose oxidase and cellobiose:quinone reductase) interconnect the catabolism of lignin with that of cellulose by concomitantly reducing depolymerization products of lignin and oxidizing depolymerization products of cellulose. Cellulose may be the most abundant biopolymer on earth, but a wide range of microbes makes sure it is efficiently broken down and slowly but surely recycled as a key part of the planet's carbon economy (**Figure 8.64A**).

It has emerged from the plant genome projects that plants themselves, perhaps unexpectedly, have the genes to encode all the enzymes required to disassemble all the polysaccharide components of their own walls. The *Arabidopsis* genome, for example, contains over 400 glycosyl hydrolases, of which probably 170 or so are concerned with the degradation of pectin, but whether their roles are in plant defense, growth and development, wall remodeling, or some other function is largely unknown territory. Bioinformatics approaches have also been revealing for microorganisms. For example, the genome sequence of the aerobic saprophytic bacterium *Cellvibrio japonicus* reveals that it has all the enzymes needed to fully degrade all the polysaccharides found in plant cell walls. These include 123 glycosyl hydrolases, 14 esterases and lyases, but no cellulosomes. Nearly a third of the enzymes have associated CBMs to assist their activity. Comparisons with related marine bacteria show that they too have a similar suite of enzymes, suggesting that plants in the marine environment are broken down in the same way as those on land.

(A)

(B)

Figure 8.64 The beginning of the end. All plant matter is eventually recycled, some rapidly, some more slowly. (A) Dead tree parts are rapidly colonized by fungal and bacterial colonies that progressively rot the wood. (B) The surfaces of fresh fruit are rapidly colonized by fungal colonies. This orange is already showing the presence of the fungus *Penicillium digitatum,* which rots citrus fruit. The white mycelial mat quickly turns green in the older central parts of the colony.

Key Terms

adaptive immune system	nonhost resistance
appressorium	oligosaccharin
biotroph	oxidative burst
carbohydrate-binding module (CBM)	papilla
cellulosome	pathogen
damage-associated molecular pattern (DAMP)	pathogenesis-related protein (PR protein)
danger signal	pattern-triggered immunity (PTI)
defensin	pattern-recognition receptor (PRR)
effector	pectate lyase
effector-triggered immunity (ETI)	pectin lyase
general elicitor	phytoalexin
hemibiotroph	plant immunity
hypersensitive response (HR)	polygalacturonase-inhibitor protein (PGIP)
inhibitor protein	reactive oxygen species (ROS)
innate immune system	resistance gene (R-gene)
leucine-rich repeat receptor kinase (LRR-RK)	saprophyte
microbe-associated molecular pattern (MAMP)	systemic acquired resistance (SAR)
molecular pattern	systemic wound response
necrotroph	wall-integrity signaling

References

[1] Most plants are immune to most pathogens.

Agrios GN (2005) Plant Pathology, 5th ed. Amsterdam: Elsevier/Academic Press.

Hückelhoven R (2007) Cell wall–associated mechanisms of disease resistance and susceptibility. *Annu. Rev. Phytopathol.* 45, 101–127.

Parker J (ed) (2009) Molecular Aspects of Plant Disease Resistance. Vol. 34 of Annual Plant Reviews. Chichester, UK: Wiley-Blackwell.

Rietz S & Parker JE (2007) Plant disease and defense. In Handbook of Plant Science (K Roberts ed), pp 1460–1467. Chichester, UK: Wiley.

Smith AM, Coupland G, Dolan L et al. (2010) Interactions with other organisms. Ch. 8 in Plant Biology. New York: Garland Science pp 499–572.

[2] Signals from both pathogen and host can elicit a common defense response.

Boller T & Felix G (2009) A renaissance of elicitors: perception of microbe-associated molecular patterns and danger signals by pattern-recognition receptors. *Annu. Rev. Plant Biol.* 60, 379–406.

Darvill AG & Albersheim P (1984) Phytoalexins and their elicitors: a defense against microbial infection in plants. *Annu. Rev. Plant Physiol. Plant Mol. Biol.* 35, 243–275.

Staal J & Dixelius C (2007) Tracing the ancient origins of plant innate immunity. *Trends Plant Sci.* 12, 334–342.

Zipfel C (2009) Early molecular events in PAMP-triggered immunity. *Curr. Opin. Plant Biol.* 12, 414–420.

[3] Cell surface receptors recognize common molecular patterns.

Aslam SN, Erbs G, Morrissey KL et al. (2009) Microbe-associated molecular pattern (MAMP) signatures, synergy, size and charge: influences on perception or mobility and host defence responses. *Mol. Plant Pathol.* 10, 375–387.

Chinchilla D, Zipfel C, Robatzek S et al. (2007) A flagellin-induced complex of the receptor FLS2 and BAK1 initiates plant defence. *Nature* 448, 497–500.

Enkerli J, Felix G & Boller T (1999) The enzymatic activity of fungal xylanase is not necessary for its elicitor activity. *Plant Physiol.* 121, 391–398.

Miya A, Albert P, Shinya T et al. (2007) CERK1, a LysM receptor kinase, is essential for chitin elicitor signaling in *Arabidopsis*. *Proc. Natl. Acad. Sci. USA* 104, 19613–19618.

Ron M & Avni A (2004) The receptor for the fungal elicitor ethylene-inducing xylanase is a member of a resistance-like gene family in tomato. *Plant Cell* 16, 1604–1615.

Zipfel C (2008) Pattern-recognition receptors in plant innate immunity. *Curr. Opin. Immunol.* 20, 10–16.

[4] The plant's responses to danger involve a common sequence of events.

Asai T, Tena G, Plotnikova J et al. (2002) MAP kinase signalling cascade in *Arabidopsis* innate immunity. *Nature* 415, 977–983.

Daxberger A, Nemak A & Mithöfer A (2007) Activation of members of a MAPK module in beta-glucan elicitor-mediated non-host resistance of soybean. *Planta* 225, 1559–1571.

Jabs T, Tschöpe M, Colling C et al. (1997) Elicitor-stimulated ion fluxes and O_2^- from the oxidative burst are essential components in triggering defense gene activation and phytoalexin synthesis in parsley. *Proc. Natl. Acad. Sci. USA* 94, 4800–4805.

Panstruga R, Parker JE & Schulze-Lefert P (2009) Snapshot: plant immune response pathways. *Cell* 136, 978.e1–978.e3.

Pitzschke A, Schikora A & Hirt H (2009) MAPK cascade signaling networks in plant defense. *Curr. Opin. Plant Biol.* 12, 421–426.

Ryan CA, Huffaker A & Yamaguchi Y (2007) New insights into innate immunity in *Arabidopsis. Cell. Microbiol.* 9, 1902–1908.

[5]Some specialist pathogens deliver effectors to suppress the basal defense response.

Abramovich RB, Anderson JC & Martin GB (2006) Bacterial elicitation and evasion of plant innate immunity. *Nat. Rev. Mol. Cell Biol.* 7, 601–611.

Boller T & Felix G (2009) A renaissance of elicitors: perception of microbe-associated molecular patterns and danger signals by pattern-recognition receptors. *Annu. Rev. Plant Biol.* 60, 379–406.

de Wit PJGM, Mehrabi R, van den Berg HA & Stergiopoulos I (2009) Fungal effector proteins: past, present and future. *Mol. Plant Pathol.* 10, 735–747.

Inglwe RA, Carstens M & Denby KJ (2006) PAMP recognition and the plant-pathogen arms race. *BioEssays* 28, 880–889.

Jones JDG & Dangl J (2006) The plant immune system. *Nature* 444, 323–329.

Ma W & Guttman DS (2008) Evolution of prokaryotic and eukaryotic virulence effectors. *Curr. Opin. Plant Biol.* 11, 412–419.

[6]Some defenses are preformed.

Aerts AM, François IE, Cammue BP & Thevissen K (2008) The mode of antifungal action of plant, insect and human defensins. *Cell Mol. Life Sci.* 65, 2069–2079.

Bednarek B & Schulze-Lefert P (2009) Role of plant secondary metabolites at the host-pathogen interface. In Molecular Aspects of Plant Disease Resistance (J Parker ed), vol 34 of Annual Plant Reviews, pp 220–260. Chichester, UK: Wiley-Blackwell.

Lay FT & Anderson MA (2005) Defensins: components of the innate immune system in plants. *Curr. Protein Pept. Sci.* 6, 85–101.

Morrisey JP & Osbourn AE (1999) Fungal resistance to plant antibiotics as a mechanism of pathogenesis. *Microbiol. Mol. Biol. Rev.* 63, 708–724.

Thomma BPHJ & Cammue BPA (2002) Plant defensins. *Planta* 216, 193–202.

Walton JD (2007) Secondary metabolites: killing pathogens. In Handbook of Plant Science (K Roberts ed), pp 1066–1071. Chichester, UK: Wiley.

[7]Defense can also be mounted at a distance.

De Bruxelles GL & Roberts MR (2001) Signals regulating multiple responses to wounding and herbivores. *Crit. Rev. Plant Sci.* 20, 487–521.

van der Ent S, Koornneef A, Ton J & Pieterse CMJ (2009) Induced resistance: orchestrating defence mechanisms through crosstalk and priming. In Molecular Aspects of Plant Disease Resistance (J Parker ed), vol 34 of Annual Plant Reviews, pp 334–370. Chichester, UK: Wiley-Blackwell.

Weigel RW (2007) Salicylic acid. In Handbook of Plant Science (K Roberts ed), pp 1401–1408. Chichester, UK: Wiley.

Yang Y (2007) Systemic signaling. In Handbook of Plant Science (K Roberts ed), pp 1338–1342. Chichester, UK: Wiley.

[8]Pathogens enter either by force, through wounds, or through natural openings.

Hückelhoven R (2007) Cell wall–associated mechanisms of disease resistance and susceptibility. *Annu. Rev. Phytopathol.* 45, 101–127.

Isaacson T, Kosma DK, Matas AJ et al. (2009) Cutin deficiency in the tomato fruit cuticle consistently affects resistance to microbial infection and biomechanical properties, but not transpirational water loss. *Plant J.* 60, 363–377.

Kankanala P, Mosquera G, Khang CH et al. (2009) Cellular and molecular analyses of biotrophic invasion by the rice blast fungus. In Advances in Genetics, Genomics and Control of Rice Blast Disease (GL Wang, B Valent eds), pp 83–91. New York: Springer Science and Business Media.

Skamnioti P & Gurr SJ (2007) *Magnaporthe grisea* cutinase2 mediates appressorium differentiation and host penetration and is required for full virulence. *Plant Cell* 19, 2674–2689.

[9]Plasmodesmata provide a route for pathogen spread.

Boevink P & Oparka KJ (2005) Virus-host interactions during movement processes. *Plant Physiol.* 138,1815–1821.

Kankanala P, Czymmek K & Valent B (2007) Roles for rice membrane dynamics and plasmodesmata during biotrophic invasion by the blast fungus. *Plant Cell* 19, 706–724.

Simpson C, Thomas C, Findlay K et al. (2009) An *Arabidopsis* GPI-anchor plasmodesmal neck protein with callose binding activity and potential to regulate cell-to-cell trafficking. *Plant Cell* 21, 581–594.

Thomas CL, Bayer EM, Ritzenthaler C et al. (2008) Specific targeting of a plasmodesmal protein affecting cell-to-cell communication. *PLoS Biol.* 6(1), e7. doi: 10.1371/journal.pbio.0060007.

[10]Callose deposition is a local response to both pathogens and wounding.

Jacobs AK, Lipka V, Burton RA et al. (2003) An *Arabidopsis* callose synthase, GSL5, is required for wound and papillary callose formation. *Plant Cell* 15, 2503–2513.

Lagaert S, Beliën T & Volckaert G (2009) Plant cell walls: protecting the barrier from degradation by microbial enzymes. *Semin. Cell Dev. Biol.* 20, 1064–1073.

Vorwerk S, Somerville S & Somerville C (2004) The role of plant cell wall polysaccharide composition in disease resistance. *Trends Plant Sci.* 9, 203–209.

[11]Wall strengthening helps contain the pathogen.

Bradley DJ, Kjellbom P & Lamb CJ (1992) Elicitor- and wound-induced oxidative cross-linking of a proline-rich plant cell wall protein: a novel, rapid defense response. *Cell* 70, 21–30.

Brisson LF, Tenhaken R & Lamb C (1994) Function of oxidative cross-linking of cell wall structural proteins in plant disease resistance. *Plant Cell* 6, 1703–1712.

Hückelhoven R (2007) Cell wall–associated mechanisms of disease resistance and susceptibility. *Annu. Rev. Phytopathol.* 45, 101–127.

Lamb C & Dixon RA (1997) The oxidative burst in plant disease resistance. *Annu. Rev. Plant Physiol. Plant Mol. Biol.* 48, 251–275.

Torres MA, Jones JDG & Dangl JL (2006) Reactive oxygen species signaling in response to pathogens. *Plant Physiol.* 141, 373–378.

[12]Host cell polarization is a common response to a pathogen.

Frei dit Frey N & Robatzek S (2009) Trafficking vesicles: pro or contra pathogens. *Curr. Opin. Plant Biol.* 12, 437–443.

Hardham AR, Jones DA & Takemoto D (2007) Cytoskeleton and cell wall function in penetration resistance. *Curr. Opin. Plant Biol.* 10, 342–348.

Lipka U, Fuchs R, Kuhns C et al. (2010) Live and let die: *Arabidopsis* nonhost resistance to powdery mildew. *Eur. J. Cell Biol.* 89, 194-199.

Meyer D, Pajonk S, Micali C et al. (2009) Extracellular transport and integration of plant secretory proteins into pathogen-induced cell wall compartments. *Plant J.* 57, 986–999.

Palma K, Wiermer M & Li X (2009) Marshalling the troops: intracellular dynamics in host pathogen defense. In Molecular Aspects of Plant Disease Resistance (J Parker ed), vol 34 of Annual Plant Reviews, pp 177–219. Chichester, UK: Wiley-Blackwell.

[13]Localized cell death restricts pathogen spread.

Heath MC (2000) Hypersensitive response-related death. *Plant Mol. Biol.* 44, 321–324.

Mur LAJ (2007) Hypersensitive response. In Handbook of Plant Science (K Roberts ed), pp 1514–1517. Chichester, UK: Wiley.

Mur LAJ, Kenton P, Lloyd AJ et al. (2008) The hypersensitive response: the centenary is upon us but how much do we know? *J. Exp. Bot.* 59, 501–520.

Pennell RI & Lamb C (1997) Programmed cell death in plants. *Plant Cell* 9, 1157–1168.

[14]Cell wall fragments act as danger signals.

Ayers AR, Ebel J, Valent B & Albersheim P (1976) Host-pathogen interactions X: fractionation and biological activity of an elicitor isolated from the mycelial walls of *Phytophthora megasperma* var. *sojae*. *Plant Physiol.* 57, 760–765.

Davis KR, Darvill AG, Albersheim P & Dell A (1986) Host-pathogen interactions XXIX: oligogalacturonides released from sodium polypectate by endopolygalacturonic acid lyase are elicitors of phytoalexins in soybean. *Plant Physiol.* 80, 568–577.

Hahn MG, Darvill AG & Albersheim P (1981) Host-pathogen interactions XIX: the endogenous elicitor, a fragment of a plant cell wall polysaccharide that elicits phytoalexin accumulation in soybeans. *Plant Physiol.* 68, 1161–1169.

Ridley BL, O'Neill MA & Mohnen D (2001) Pectins: structure, biosynthesis, and oligogalacturonide-related signalling. *Phytochem.* 57, 929–967.

Walton JD (1994) Deconstructing the cell wall. *Plant Physiol.* 104, 1113–1118.

[15]Plants and pathogens battle to control the release of oligogalacturonides.

Albersheim P & Anderson AJ (1971) Proteins from plant cell walls inhibit polygalacturonases secreted by plant pathogens. *Proc. Natl. Acad. Sci. USA* 39, 1815–1819.

Bergmann CW, Ito Y, Singer D et al. (1994) Polygalacturonase-inhibiting protein accumulates in *Phaseolus vulgaris* L. in response to wounding, elicitors, and fungal infection. *Plant J.* 5, 625–634.

Di Matteo A, Federici L, Mattei B et al. (2003) The crystal structure of polygalacturonase-inhibiting protein (PGIP), a leucine-rich repeat protein involved in plant defense. *Proc. Natl. Acad. Sci. USA* 100, 10124–10128.

Ferrari S, Galletti R, Valro D et al. (2006) Antisense expression of the *Arabidopsis thaliana AtPGIP1* gene reduces polygalacturonase-inhibiting protein accumulation and enhances susceptibility to *Botrytis cinerea*. *Mol. Plant Microbe Interact.* 19, 931–936.

Lagaert S, Beliën T & Volckaert G (2009) Plant cell walls: protecting the barrier from degradation by microbial enzymes. *Semin. Cell Dev. Biol.* 20, 1064–1073.

Luca F, Di Matteo A, Fernandez-Recio J et al. (2006) Polygalacturonase inhibiting proteins: players in plant innate immunity? *Trends Plant Sci.* 11, 65–70.

[16]PR proteins attack the walls of fungi and bacteria.

Erbs G, Silipo A, Aslam S et al. (2008) Peptidoglycan and muropeptides from pathogens *Agrobacterium* and *Xanthomonas* elicit plant innate immunity: structure and activity. *Chem. Biol.* 15, 438–448.

Gust AA, Biswas R, Lenz HD et al. (2007) Bacteria-derived peptidoglycans constitute pathogen-associated molecular patterns triggering innate immunity in *Arabidopsis*. *J. Biol. Chem.* 282, 32338–32348.

van Loon LC, Rep M & Pieterse CMJ (2006) Significance of inducible defense-related proteins in infected plants. *Annu. Rev. Phytopathol.* 44, 135–162.

van Loon LC & van Strien EA (1999) The families of pathogenesis-related proteins, their activities, and comparative analysis of PR-1 type proteins. *Physiol. Mol. Plant Pathol.* 55, 85–97.

Yang Y (2007) Systemic signaling. In Handbook of Plant Science (K Roberts ed), pp 1338–1342. Chichester, UK: Wiley.

[17]Carbohydrate-binding modules (CBMs) help enzymes attach to the wall.

Boraston AB, Bolam DN, Gilbert HJ & Davies GJ (2004). Carbohydrate-binding modules: fine tuning polysaccharide recognition. *Biochem. J.* 382, 769–781.

Gaulin E, Dramé N, Lafitte C et al. (2006) Cellulose binding domains of a *Phytophthora* cell wall protein are novel pathogen-associated molecular patterns. *Plant Cell* 18, 1766–1777.

Montanier C, van Bueren AL, Dumon C et al. (2009) Evidence that family 35 carbohydrate binding modules display conserved specificity but divergent function. *Proc. Natl. Acad. Sci. USA* 106, 3065–3070.

The different families of carbohydrate-binding module are defined and described on the CAZy website at www.cazy.org/fam/acc_CBM.html.

[18]Cell wall integrity is sensed by the host.

Caño-Delgado A, Penfield S, Smith C et al. (2003) Reduced cellulose synthesis invokes lignification and defense responses in *Arabidopsis thaliana*. *Plant J.* 34, 351–362.

Ellis C, Karafyllidis I, Wasternack C & Turner JG (2002) The *Arabidopsis* mutant cev1 links cell wall signalling to jasmonate and ethylene responses. *Plant Cell.* 14, 1557–1566.

Hématy K, Cherk C & Somerville S (2009) Host-pathogen warfare at the plant cell wall. *Curr. Opin. Plant Biol.* 12, 406–413.

Hématy K & Hofte H (2008) Novel receptor kinases involved in growth regulation. *Curr. Opin. Plant Biol.* 11, 321–328.

[19]Responses to wounding and pathogens overlap.

Bergmann C, Ito Y, Singer D et al. (1994) Polygalacturonase-inhibiting protein accumulates in *Phaseolus vulgaris* L. in response to wounding, elicitors, and fungal infection. *Plant J.* 5: 625–634.

De Bruxelles GL & Roberts MR (2001) Signals regulating multiple responses to wounding and herbivores. *Crit. Rev. Plant Sci.* 20, 487–521.

Mithöfer A & Boland W (2008) Recognition of herbivory-associated molecular patterns. *Plant Physiol.* 146, 825–831.

Schilmiller AL & Howe GA (2005) Systemic signalling in the wound response. *Curr. Opin. Plant Biol.* 8, 369–377.

[20]Wall-degrading enzymes and their inhibitors coevolve.

Bergmann C, Ito Y, Singer D et al. (1994) Polygalacturonase-inhibiting protein accumulates in *Phaseolus vulgaris* L. in response to wounding, elicitors, and fungal infection. *Plant J.* 5: 625–634.

Bishop JG, Ripoll DR, Bashir S et al. (2005) Selection on *Glycine* beta-1,3-endoglucanase genes differentially inhibited by a *Phytophthora* glucanase inhibitor protein. *Genetics* 169, 1009–1019.

Juge N (2006) Plant protein inhibitors of cell wall degrading enzymes. *Trends Plant Sci.* 11, 359–367.

Lagaert S, Beliën T & Volckaert G (2009) Plant cell walls: protecting the barrier from degradation by microbial enzymes. *Semin. Cell Dev. Biol.* 20, 1064–1073.

Misas-Villamil JC & van der Hoorn RAL (2008) Enzyme-inhibitor interactions at the plant-pathogen interface. *Curr. Opin. Plant Biol.* 11, 380–388.

Rose JKC, Ham K-S, Darvill AG & Albersheim P (2002) Molecular cloning and characterization of glucanase inhibitor proteins: coevolution of a counterdefense mechanism by plant pathogens. *Plant Cell* 14, 1329–1345.

[21]Removing pectin is a key early step in dismantling the wall.

Azadi P, O'Neill MA, Bergmann C, Darvill AG & Albersheim P (1995) The backbone of the pectic polysaccharide rhamnogalacturonan I is cleaved by an *endo*hydrolase and an *endo*lyase. *Glycobiology* 5, 783–789.

Bauer S, Vasu P, Persson S et al. (2006) Development and application of a suite of polysaccharide-degrading enzymes for analyzing plant cell walls. *Proc. Natl. Acad. Sci. USA* 103, 11417–11422.

Daas PJH, Meyer-Hansen K, Schols HA, De Ruiter GA & Voragen AGJ (1999) Investigation of the non-esterified galacturonic acid distribution in pectin with endopolygalacturonase. *Carbohydr. Res.* 318, 135–145.

de Vries RP & Visser J (2001) Aspergillus enzymes involved in degradation of plant cell wall polysaccharides. *Microbiol. Mol. Biol. Rev.* 65, 497–522.

Lang C and Dornenburg H (2000) Perspectives in the biological function and the technological application of polygalacturonases. *Appl. Microbiol. Biotechnol.* 53, 366–375.

[22]Lyases are important pectic enzymes for both necrotrophic pathogens and saprophytic microbes.

Davis KR, Lyon GD, Darvill AG & Albersheim P (1984) Host-pathogen interactions XXV: endopolygalacturonic acid lyase from *Erwinia carotovora* elicits phytoalexin accumulation by releasing plant cell wall fragments. *Plant Physiol.* 74, 52–60.

Mutter M, Colquhoun IJ, Beldman G et al. (1998) Characterization of recombinant rhamnogalacturonan α-L-rhamnopyranosyl-(1,4)-α-D-galactopyranosyluronide lyase from *Aspergillus aculeatus*: an enzyme that fragments rhamnogalacturonan I regions of pectin. *Plant Physiol.* 117, 141–152.

Olsson L, Christensen TMIE, Hansen KP & Palmqvist EA (2003) Influence of the carbon source on production of cellulases, hemicellulases and pectinases by *Trichoderma reesei* Rut C-30. *Enzyme Microb. Technol.* 33, 612–619.

[23]To fully digest hemicelluloses, enzymes act in concert.

Gilbert HJ, Stalbrand H & Brumer H (2008) How the walls came crumbling down: recent structural biochemistry of plant polysaccharide degradation. *Curr. Opin. Plant Biol.* 11, 338–348.

Pauly M, Andersen LN, Kauppinen S et al. (1999) A xyloglucan-specific *endo*-β-1,4-glucanase from *Aspergillus aculeatus*: expression cloning in yeast, purification and characterization of the recombinant enzyme. *Glycobiology* 9, 93–100.

[24]Cellulose is tough, and its disassembly usually requires special machinery.

Bayer EA, Belaich J-P, Shoham Y & Lamed R (2004) The cellulosomes: multienzyme machines for degradation of plant cell wall polysaccharides. *Annu. Rev. Microbiol.* 58, 521–554.

DeBoy RT, Mongodin EF, Fouts DE et al. (2008) Insights into plant cell wall degradation from the genomic sequence of the soil bacterium *Cellvibrio japonicus*. *J. Bacteriol.* 190, 5455–5463.

Himmel ME (ed) (2008) Biomass Recalcitrance: Deconstructing the Plant Cell Wall for Bioenergy. Chichester, UK: Blackwell.

Lopez-Casado G, Urbanowicz BR, Damasceno CMB & Rose JKC (2008) Plant glycosyl hydrolases and biofuels: a natural marriage. *Curr. Opin. Plant Biol.* 11, 329–337.

Plant Cell Walls: A Renewable Material Resource

9

Humans depended on plant cell walls for food, shelter, tools, clothing, fuel, and medicine long before the invention of agriculture. This long evolutionary relationship is still evident in our teeth and digestive tracts, which are adapted to process plant cell walls as a major component of our diet. Early in our history, we learned to use wood for construction and tools, and to use plant fibers for making textiles, netting, and ropes. The oldest man made fibers discovered to date were produced from twisted wild flax fibers by cave dwellers in the Caucasus mountains ~30,000 years ago. Furthermore, these fibers were dyed with colors derived from local plants. By about 10,000 BC, Peruvian hunter-gatherers had mastered the art of spinning and twining wild cotton. Neolithic farmers in the Near East used flax fibers for textiles by about 8000 BC, and the oldest known woven flax products made by lake dwellers in Central Europe were produced around 7000 BC. Egyptian mummies were wrapped in ramie cloth between 5000 and 3000 BC, and in linen (flax) cloths thereafter. The oldest woven cotton fabrics from Peruvian grave-sites, and from the ancient city of Mohenjo Daro in India, have been dated to around 3000 BC.

The advent of the Bronze Age in Anatolia, around 6500 BC, ushered in the use of wood as fuel to smelt ores and to shape metal products. Since then, the rise of many great civilizations, beginning with Mesopotamia over 5000 years ago, can be attributed to the availability of plentiful supplies of wood for fuel to produce metal products—and their subsequent decline to the depletion of these wood supplies. Similarly, the rise and fall of naval empires has been traced to the availability of wood for shipbuilding. Even today, wood remains the principal source of fuel for cooking, heating, and small industry in a large number of developing countries (**Figure 9.1**).

Industrialized nations continue to depend heavily on plant cell walls, because plant cell walls are the material basis for making paper, building supplies, and many other products that depend on renewable resources. The promised computer-based paperless society has yet to materialize, although electronic messaging and media have a growing impact. Computers have resulted in a greatly increased demand for high-quality paper. Utilization of and demand for plant materials will most likely continue to increase as the standard of living and the level of technology of the less developed countries of the world advance.

Biotechnology will contribute to an increased and more efficient use of plant products derived from cell walls. The engineering of plants for improved growth and yield, and improved products, will include significant modifications to the plant cell walls. Production, utilization, and reutilization of plant cell wall materials will provide the food, fiber, and energy for the future populations of our planet. How wisely this work is done will

Figure 9.1 Fifteenth-century woodcut depicting three wood gatherers carrying wood for cooking and heating their homes.

determine the fate of our farmland, our forests, our global environment, and our prosperity.

In this chapter, we discuss examples of the role of cell walls in food, medicine, textiles, building materials, paper products, and energy. Cell walls are important determinants of the nutritional value and texture of foods for humans and animals. Gums, many of which are cell wall–associated glycoproteins, and pectins are used widely in food production. Pectic polysaccharides are the bioactive compounds of a number of medicinal herbs. Plant cell wall–derived fibers such as cotton, linen (flax), and jute are used for producing textiles. Wood is one of our most important renewable industrial raw materials. Plant cell walls also constitute the most important source of renewable biomass for energy production. With the growing interest in producing liquid fuels from lignocellulosic biomass, plant cell walls are destined to play a key role in the development of sustainable human societies. Genetic engineering provides challenges and opportunities to tailor the physical and chemical properties of plant cell walls to improve their uses for human society.

A. Effects of Cell Walls on the Nutritional Quality and Texture of Foods and Forage Crops

1. Digestibility of forage crops depends upon the properties of plant cell walls and is mainly restricted by the lignin content.[ref1]

Forage (grasses and legumes consumed by grazing animals) is composed of about 40% cellulose, a roughly equivalent amount of hemicelluloses, lower but variable amounts of lignin, and small amounts of pectin, as well as proteins, fats, soluble sugars, and minerals. Fermentation of plant cell walls by microorganisms provides energy and nutrients for the metabolism and growth of many herbivores, including the *ruminants*. Ruminants chew and swallow food that is partially digested in the rumen, a modified part of the esophagus that comprises one of the specialized chambers of the stomach; the food is re-chewed periodically to reduce the size of food particles in order to increase digestibility (**Figure 9.2**). Forage grasses are a primary food for ruminants, and high live-weight gain or high milk production requires a high food intake and high digestibility of the cell walls and cytoplasm of the forage crops.

Plant cell wall structure determines how easily forage is broken down into digestible particles through chewing and rumination, and the rate of passage of particles from the rumen. Degradation depends upon cell wall toughness, microbial adhesion to the fibers, interaction of different microbial species in the rumen, and the digestive enzymes produced to degrade the plant cell wall material. In addition, secondary plant metabolites may significantly lower digestibility. The yield of energy from digestible dry matter depends upon the hydrolases produced by the rumen microorganisms and the accessibility of the polysaccharides for hydrolysis. The accessibility of polysaccharides depends upon the physical and chemical interactions between the polysaccharides and other polymers, whereas the efficiency of hydrolysis depends upon the combined action of different hydrolases with complementing activities, such as ferulate esterases, xylanases, and cellulases. Many *cellulases* have a tripartite structure that is important for efficient cellulose digestibility. They typically have a specific cellulose-binding domain, a hydrolytic domain, and a flexible linker (see also Panel 9.2). In the absence of a cellulose-binding domain, soluble cellulose is hydrolyzed but crystalline cellulose is not.

Substantial quantities of protozoa, fungi, and bacteria reside in the rumen and help with the digestion of the ingested plant cell walls. Cellulolytic rumen bacteria, such as *Ruminococcus* and *Bacteroides*, contain complexes of cellulolytic enzymes—cellulosomes—on their surface to digest the plant cell wall where the bacteria adhere (see Panel 9.2). The enzyme complexes may include *endo*-β-glucanases, *exo*-β-glucanases, and β-glucosidases (see Figure 8.61). Rumen bacteria are also able to degrade hemicelluloses and pectins. The products of fermentation in the anaerobic digestive tract are CO_2, methane, volatile fatty acids, and microbial cells themselves. Herbivores have adapted to digest, absorb, and metabolize some of these end products. The microbial cells that proliferate through fermentation are also digested and absorbed by the intestine. Making forage crops more digestible may be possible by increasing the ability of the rumen microorganisms to access and degrade the forage polysaccharides or by manipulating the structure or composition of the plant cell walls to enhance enzyme accessibility.

The relationship between the composition of the dry matter of the plant cell wall and its digestibility has been studied extensively both *in vivo* and *in vitro*. Lignin content is the most important factor to affect digestibility, because it limits accessibility of hydrolases to the cell wall carbohydrates. Lignin-carbohydrate complexes and other phenolic carbohydrate complexes, such as esters of ferulic acid or *p*-coumaric acid, are also thought to limit digestibility of forage grasses. Lignin is highly resistant to microbial degradation, and wood-degrading fungi (white rot, see Panel 9.1) do so slowly and only under aerobic conditions. The anaerobic environment of the rumen prevents the growth of these organisms, and therefore lignin cannot be degraded there.

Mutations in maize, sorghum, and pearl millet have been identified that affect digestibility by modifying the properties or quantity of lignin. These mutations show a *brown midrib phenotype* (**Figure 9.3**), a reduction in total lignin, higher ratios of ferulate to *p*-coumarate, and a more condensed and cross-linked kind of lignin. *In vitro* digestibility is generally increased in the brown midrib mutants, in spite of increased cross-linking, presumably due to the reduced quantity and modified composition of the lignin. All brown

Figure 9.2 One grazing and one ruminating cow. Cows partially digest forage fiber in the rumen.

Figure 9.3 Leaves of (A) normal maize and (B) a brown midrib corn mutant. The marked, brown midrib possesses a striking red-brown color. (Courtesy of Wilfred Vermerris.)

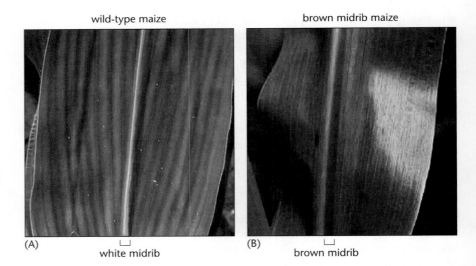

wild-type maize

brown midrib maize

(A)   white midrib

(B)   brown midrib

midrib mutant plants have been found to be more digestible by ruminants and to improve growth of the animals. Genetic modification of lignin content and composition in forage grasses, therefore, has been a subject of commercial interest (see Concept 9E6).

2. The texture of our fruits and vegetables depends to a large extent on the properties of their cell walls.[ref2]

The texture of our fruits and vegetables depends largely on three factors: crispness, mealiness, and toughness, all properties dependent on the structure and composition of their cell walls. The result of mechanical force on plant tissues is either fracture of the cells, fracture of the middle lamella between cells, or resistance to fracture.

Crispness in our foods results when fracture of the cells themselves releases liquid from the cells, providing a crisp texture and juice from the cell contents. The ability to fracture depends upon a relatively strong bonding between the cell walls and the middle lamellae throughout a tissue. Mealiness comes from tissues with less tightly adhering cells, so that the middle lamella fractures rather than the cells. Less of the cell contents is released as liquid, and a relatively dry, mealy texture is perceived. When the plant tissue is not disrupted, due to a high fiber content or the collapse of the cell wall without fracture, the result is perceived as toughness, which is an effective deterrent to herbivores.

Toughness of plants depends upon the long strands of cellulose combined with other polysaccharides, proteins, and lignin to produce stiff, strong, and flexible cell walls. As the wall comprises an increasing proportion of the cell volume, toughness and cellular density increases. Isodiametric parenchyma cells have relatively low toughness compared with elongated lignified cells. Cell breakage is favored when strong adhesion occurs in the middle lamella due to pectic polysaccharides or lignification. Many fruits and vegetables are composed mostly of parenchyma cells, which have only primary cell walls with little or no secondary cell wall thickening. Cooking (**Figure 9.4**) typically softens plant tissue through loss of turgor, depolymerization of pectic polysaccharides, cell wall swelling, and cell separation. Calcium can cross-link pectins in the middle lamella, and therefore chelating agents can separate cells from ripening fruit because of the mechanical importance of such cross-links.

Asparagus shoots become tougher during maturation because of the formation of pectin-xylan phenolic complexes, increased lignin in the middle lamella of the sclerenchyma, and because of an increased ratio of fibers and xylem rich in secondary cell walls. Beans harden during storage due to increased cell adhesion; extensive cooking counteracts cell adhesion and softens the beans. In freshly harvested beans, cooked briefly, fracture occurs along the middle lamella and separates the cells. In dried beans, briefly cooked, fracture disrupts the cell walls. Following cooking under pressure, softening occurs, and fracture also occurs along the middle lamella separating cells. Water chestnuts fail to soften with cooking. They remain very crisp even after extensive heating. Their walls are rich in phenolic derivatives, presumably cross-linked to polysaccharides (Figure 6.43). Diferulic acids have been implicated in the cross-linking that determines thermal stability, similar to the ferulic acid esters in grasses. When exposed to a UV light, these phenolics produce a yellow-green autofluorescence at the three-way junctions of adjacent cells.

3. Gums, water-soluble cell wall–associated polymers, are used to stabilize emulsions and to modify the texture of processed foods and other industrial products.[ref3]

Many food products contain one or more plant *gums* as food additives. For the most part gums are added to modify the physical properties of food. Gums can provide different levels of viscosity, form gels, stabilize solutions, or act as emulsifiers. These properties, together with their lack of toxicity, explain why gums are widely used in foods and the pharmaceutical industry.

The principal components of gum arabic are polysaccharides (mol. wt. ~0.25×10^6), arabinogalactan oligosaccharides, and large (mol. wt. ~2.5×10^6) hydroxyproline-rich AGPs with a simple repetitive polypeptide backbone containing an almost symmetrical 19-residue consensus motif. Based on structural predictions for hydroxyproline-rich glycoproteins, it has been proposed that the contiguous hydroxyprolines are attached to *oligo*-α-1,3-L-arabinofurans, and the noncontiguous hydroxyprolines to galactose residues of *oligo*-arabinogalactans. The individual molecules possess a flexible but compact conformation.

Of the *exudate gums* (see Concepts 3D7 and 3E6), those most widely used in food products are *gum arabic, karaya, ghatti*, and *tragacanth*. Gums from other sources such as seeds, microbes, and seaweeds are also utilized by the food industry, as are pectins and starches, which may be chemically modified prior to use.

Gum arabic is derived from wound-response exudates of *Acacia senegal* or *A. seyal* trees growing in the arid grassland savannahs in sub-Saharan Africa. Sudan, the largest supplier, produces over 50% of all traded gum, with most of the rest coming from Chad and Nigeria. Sudan's dominant position in the gum arabic trade is due to optimal soil and climate conditions, and the long experience of the Sudanese in producing, collecting, and sorting the gum into consistent quality grades. Historically, the gum arabic trade in Sudan can be traced to the twelfth century BC. Gum collected in Nubia was exported to Egypt for use in inks, watercolors, and dyes, as well as for embalming.

For commercial production, short, transverse cuts are made in the bark of 4- to 18-year-old trees at the beginning of the dry season starting in October or November to facilitate the flow of the gummy sap. The teardrop-shaped gum granules are harvested about one month later, and multiple collections from the same tree are possible (average yield per tree 500–5000 g/year). Secretion is most abundant during dry and very hot weather. The top-quality gum is known as Kordofan gum, named after a location in Sudan. Kordofan gum tears are hazelnut sized, nearly colorless, and translucent, and their fracture faces are shiny and vitreous (**Figure 9.5**). They are hard

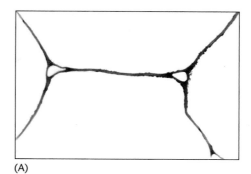

(A)

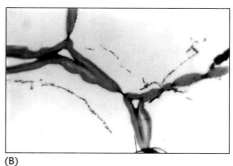

(B)

Figure 9.4 Effects of cooking on the structure of onion parenchyma cell walls. Cross-sectional light microscope images of (A) fresh tissue and (B) cooked tissue. (Courtesy of Mary L. Parker.)

Figure 9.5 Top-quality gum arabic granules (tears). These hazelnut-sized granules are nearly colorless and slightly opaque.

Figure 9.6 Kordofan-type gum arabic powder. Kordofan is the name of the top-quality gum arabic sold commercially. It has an off-white color.

Figure 9.7 Beer with gum-stabilized foam. The lacelike pattern left on the side of a freshly drained beer glass is also created by the foam-stabilizing gum.

but pulverable, are odorless, and possess a faintly sweetish and mucilaginous taste. The second-quality tears are very pale, yellow-white, yellow-red, or brown and tend to be larger. Final processing consists of dissolving the ground tears in water, and spray-drying the mixture into powder (**Figure 9.6**).

The food industry uses gum arabic as a thickener and to add smoothness to creamy products while adding essentially no calories or nutritional value. Its ability to bind and stabilize flavor molecules such as citrus oils is exploited in soft drinks. Gum arabic stabilizes emulsions in salad dressings. In soft candies such as caramels and toffees, it prevents sucrose crystallization and maintains the uniform distribution of fats. Beer brewers and marshmallow manufacturers depend on gum arabic to produce stable foams (**Figure 9.7**). The cosmetic industry uses gum arabic as a binding agent or to stabilize creams and lotions.

Many of the nonfood uses of gum arabic depend upon the presence of both hydrophilic carbohydrate and hydrophobic protein groups. One of its early documented uses was as an additive to writing ink to improve ink flow and the binding of ink to leather and paper. Today, due to its ability to produce highly uniform, thin liquid films, it is widely used in offset lithography, a common form of printing, where it is mixed with nitric acid to etch images on aluminum or zinc plates. In the textile industry, it is employed to attach pigments to fabrics as well as to increase the strength of yarn. Gum arabic is also a key ingredient for producing microencapsulated perfume and flavor molecules such as those used in detergents, in baking mixtures, and in scratch-and-sniff perfume advertisements (**Figure 9.8**). In paints, it binds pigments, increases opacity, and helps create stable elastic films. It is also the principal ingredient of the lickable adhesive of postage stamps. Other types of gums are used as rheological facilitators of concrete and as drilling agents for oil and gas wells.

4. Pectins are used as thickeners, texturizers, stabilizers, and emulsifiers in the food and pharmaceutical industries.[ref4]

Pectin is a polysaccharide with very widespread use in food, paper, textile, pharmaceutical, and other industries as a biodegradable thickener, texturizer, stabilizer, and emulsifier. It is used as a sizing (surface bonding) agent for paper and textiles, and as a film-forming agent when combined with other materials such as starch. Pectins constitute a $2 billion industry.

Pectin is primarily used for gel formation. Commercial pectin is typically derived from either *apple pomace*, the pulp residue of apples after crushing or pressing, or from orange peel because of the superior gelling properties of the pectic polysaccharides of these fruits. Molecular size and degree of methyl esterification are primary determinants of pectin gelling ability. Pectin gelation is affected by pH, solutes, the number and arrangement of side chains, and charge density. In pectins with a low degree of methyl esterification, gelation results from ionic linkages by way of calcium bridges between carboxyl groups of different pectin chains (see Concept 6A5). In highly methyl esterified pectins, hydrogen bonds and hydrophobic interactions are more important, in addition to coordinate bonding with calcium ions. High-methoxyl pectin provides improved gelling, whereas low-methoxyl pectin improves "mouth feel" and body. During cooking, depolymerization of pectic polysaccharides is enhanced by the presence of methylester groups, which aid in the solubilization of the middle lamella, and thereby soften many foods. The molecular weight and chemical properties of pectin are readily modified by industrial processing. There is a considerable potential for genetic modification of pectin properties in many plant species.

Pectin is graded by the number of equivalent parts of sugar that one part of pectin will gel under standard conditions. About 500 grades of pectin are commercially available. As a food additive, pectin has virtually no regulated limit on daily intake. Jams, jellies, and preserves are the major foods to which pectin is added for gelling purposes. Pectin is also a fat or sugar replacement in low-calorie foods. Pectin is a major ingredient in confectionery products to make flavored candies and edible coatings. Pectin reduces ice crystal size and improves texture in frozen foods. In puddings, yogurt, and mustard (**Figure 9.9**) it provides viscosity and superior flavor release properties.

B. Medicinal and Physiological Properties of Cell Wall Molecules

1. Dietary fiber in the human diet is a residual component of plant cell walls.[ref5]

The term *dietary fiber* was introduced by E. H. Hipsley in 1953 to define plant cell wall constituents that are not digested in the human digestive tract and have effects on human physiology and health. Over 20 classes of chemicals are included in a recent list of dietary fiber components in *Cereal Foods World* (**Table 9.1**). This list includes most of the cell wall polymers, with the exception of cell wall proteins and glycoproteins. Carbohydrate analogs such as maltodextrins and polydextrose—molecules produced during food processing, by chemical and/or physical processes affecting the digestibility of starches, or by purposeful synthesis—are also included in this list.

Dietary fiber has two major fractions, water-insoluble and water-soluble. Insoluble fiber consists of cell wall components such as cellulose, lignin, and hemicelluloses, whereas soluble fiber includes pectins, gums, and mucilages. The insoluble fraction is typically about 75% of the fiber content of foods. Fruits, oats, wheat, barley, and legumes are good sources of soluble fiber.

Despite the diversity of compounds, this compilation does not reflect the immense molecular diversity within the classes of the compounds and polymers produced during different stages of plant growth, under different growth conditions, and in different tissues of the thousands of members of the plant kingdom used as food. The chemical complexity highlights the problems faced by food chemists in their quest to purify and characterize plant polymers, and then to describe dietary fiber on food labels. Labeling data is often limited by inadequate knowledge and methodology.

The interest in the dietary fiber contents of foods (**Table 9.2**) is based on the dietary fiber hypothesis, which postulates that a high intake of fiber-containing foods lowers the incidence of diseases common to a Western lifestyle (for example, chronic bowel disease, diabetes, coronary heart disease, and colon cancer). A typical Western diet contains less than 20 g of dietary fiber per day (in the UK, it is typically 12 g/day) versus more than 100 g per day in the diet of our preagricultural ancestors, who consumed large amounts of uncultivated fruits and vegetables. A known risk of high fiber diets is the reduced uptake of vitamins and minerals. The beneficial effects of high fiber can be enhanced or decreased by nondietary factors such as genetic predisposition, physical activity, and stress. The current USDA recommendation for dietary fiber intake is 20 to 35 g per day.

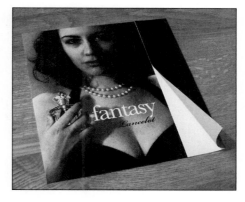

Figure 9.8 Scratch-and-sniff perfume advertisement contains gum arabic with microencapsulated perfume droplets.

Figure 9.9 Prepared mustards often contain fruit pectin to provide viscosity and superior flavor release.

TABLE 9.1 CONSTITUENTS OF DIETARY FIBER
NONSTARCH POLYSACCHARIDES AND RESISTANT OLIGOSACCHARIDES
Cellulose
Hemicelluloses: arabinoxylans and arabinogalactans
Mixed linked glucans
Pectins
Polyfructoses: inulin and oligofructans
Galactooligosaccharides
Gums
Mucilages
OTHER CARBOHYDRATES WITH DIETARY FIBER PROPERTIES
Indigestible dextrins (acid or thermally treated starches)
Resistant maltodextrins (for example, from corn)
Resistant potato dextrins
Synthesized carbohydrate compounds
Polydextrose
Methyl cellulose
Hydroxypropylmethyl cellulose
Indigestible ("resistant") starches
LIGNIN
OTHER SUBSTANCES
Waxes, cutin, and suberin
Tannins
Phytate
Saponins

Adapted from *Cereal Foods World* 46:112–126, 2001.

2. The consumption of dietary fiber has been associated with benefits in human health.[ref5]

The most detailed studies of the beneficial health effects of dietary fiber have been reported on laxation (fecal bulking and softening; increased frequency and/or regularity). Dietary fiber also produces a positive adjustment in levels of serum cholesterol, reduction in the peak level of serum glucose after eating, and the inhibition of uptake of cancer-causing chemicals.

Laxation is a physiological effect that imparts increased body comfort. Increased dietary fiber typically lowers colonic pH, increases intestinal microflora, changes microflora composition, and reduces diverticular disease and hemorrhoids. Water-holding capacity is one of the physical characteristics of dietary fiber important for laxation. Partially degraded bran fiber has about twice the water-holding capacity of typical vegetable fiber derived from primary cell walls. About 15% of Australians and Americans suffer from constipation. The yearly costs to society are billions

TABLE 9.2 DIETARY FIBER CONTENT OF FOODS

FOOD	SERVING SIZE	GRAMS OF FIBER
Whole wheat bread	2 slices	3
Spaghetti, regular (cooked)	1 cup	1
Spaghetti, whole wheat (cooked)	½ cup	4
All bran cereal	½ cup	10
Oatmeal (cooked)	¾ cup	2
Navy beans (cooked)	½ cup	6
Lentils (cooked)	½ cup	4
Tomato	1 medium	2
Lettuce	1 cup	1
Sauerkraut	½ cup	4
Peas (cooked)	½ cup	4
Carrots (cooked)	½ cup	2
Apple	1 medium	4
Strawberries	1 cup	3

of dollars. Partially hydrolyzed guar gum, a water-soluble fiber also used in the food industry (see Concept 9A3), can increase defecation frequency in constipated individuals, while decreasing the number of cases of diarrhea in patients on supplemental enteral nutrition.

One of the earliest findings in dietary fiber research was that dietary fiber reduces the risk of fatal coronary heart disease. In a 12-year California study, a 6-g increment in daily fiber intake resulted in a 25% reduction in mortality due to ischemic heart disease (restriction in blood supply to the heart). This decrease in heart disease has been correlated with a dietary fiber–related decrease in serum cholesterol, which has been traced to reduced uptake of cholesterol in the intestine. Pectin, guar, oat fiber, and legumes consistently lowered serum cholesterol levels, whereas cellulose and wheat fiber did not. Beta-glucans lower serum cholesterol levels by increasing bile acid secretion. The higher molecular weight forms were more active than the lower molecular weight forms.

Another benefit of increased consumption of soluble dietary fiber is blood glucose attenuation. Diabetes affected an estimated 30 million people worldwide in 1985, and is expected to afflict 220 million people by 2010. A high-fiber diet can both reduce the risk of developing diabetes and improve the wellbeing of people suffering from the disease. Both type 1 and type 2 diabetics show positive responses to increased consumption of soluble dietary fibers, including improved glucose tolerance, reduced insulin requirements, decreased serum cholesterol, decreased serum triglycerides, better weight control, and lower blood pressure.

The ability of dietary fiber to reduce the risk for colorectal cancer may be due to the increased absorption of carcinogens found in the diet, thereby reducing exposure of the colonic mucosa to these carcinogens. In tests with model carcinogenic compounds (heterocyclic amines), cell walls containing more lignin and suberin bound the carcinogens more strongly. Cell walls rich in phenolics were also shown to reduce the frequency of visible precursors of colon cancer in rats. *Wheat bran*, a popular source of dietary

Figure 9.10 Wheat bran–containing cereal with a high dietary fiber content.

fiber (**Figure 9.10**), is a good source of lignin, which is present in the outer layers of the grain and aleurone. Potato skins are a good source of suberin-rich cell walls.

Several epidemiologic studies have shown that dietary fiber uptake is inversely associated with risk of breast cancer in women. One hypothesis suggests that dietary fiber lowers the risk of breast cancer by modulating estrogen metabolism, for example by inhibiting the intestinal reabsorption of estrogen excreted by the biliary system. However, there are other studies that have failed to confirm this association between dietary fiber and breast cancer risk. These inconsistent results may be explained by breast cancer heterogeneity, the types of dietary fibers consumed, and/or differences in diets of different countries. More research is needed to clarify how these different factors contribute to breast cancer risk.

3. Gum arabic and pectins soothe irritated or inflamed mucosal tissues.[ref6]

The first description of the use of gum arabic (see Concept 9A3) in medicine is the ninth-century book *Ten Treatises on the Eye* by the Arabic physician Hunayn ibn Ishaq who suggested its use in eye compresses. Ancient Egyptians also employed gum arabic as a treatment for loose teeth, taking advantage of its thick, mucilaginous properties to support the tooth, and of its astringent properties to tighten the gum around a loose tooth. In Africa today, gum arabic is sold for a number of ailments, and many commercial drugs contain gum arabic as a tablet-binding agent, or to keep oral antibiotics dispersed in solution, and as a medically active tissue-coating ingredient.

The most important medical use of gum arabic is to provide a soothing and protective coating over irritated or inflamed mucosal tissues of eyes, the respiratory tract (cough and sore throat), the alimentary tract (stomach and intestinal disorders), and the urinary tract. Gum arabic forms thin protective films to treat burns and scalds. Mixed with alum, it is used to check bleeding from small cuts and wounds.

Pectins are also used to shorten coagulation time to control bleeding, and sulfated pectins are used to control clotting. In the intestine, pectins immobilize food particles and bacteria, absorb toxic cations and other molecules, and inhibit contact between food and the intestinal walls. The rate of gastric emptying can be increased significantly if a meal is fortified with pectin.

4. Pectic polysaccharides from medicinal plants stimulate the immune system via Peyer's patch cells in the intestine.[ref6]

Humans have used *medicinal herbs* since the dawn of civilization, and today, approximately 50% of synthetic medicines are derived from, or patterned after, phytochemicals. The healing agents of many herbal medicines are poorly understood, because the medicinal properties of their metabolites and their complex polysaccharides have yet to be characterized. Effects of some medicinal herbs have been traced to *bioactive pectins*. Bioactive pectins can stimulate cells of the immune system and help defend against infectious microbes. The primary target of ingested polysaccharides is the mucosal immune system, which occupies the crypt regions between the villi in the gastrointestinal tract called *Peyer's patches*. Peyer's patches contain organized clusters of three classes of *immune system cells*—lymphocytes, antigen-presenting cells, and effector cells—as well as M cells, a specialized type of intestinal epithelial cell. M cells transport endocytosed macromolecules in vesicles across the epithelium (the process is known as transcytosis) to the underlying immune system cells. The immune system cells become activated when partially digested molecules are recognized by the antigen-presenting cells. The function of these antigen-presenting cells is to induce helper T cells to stimulate the proliferation of other types of T cells, antibody-producing B cells, macrophages, and other leukocytes. Pectic polysaccharides can also stimulate the proliferation of splenic lymphocytes. The immunological detection of specific pectic polysaccharide fragments in liver cells demonstrates that some fragments are capable of passing from the intestinal lumen into the blood.

5. Bioactive pectins from some medicinal herbs promote the proliferation of immune system cells and their activities.[ref7]

Aqueous extracts of roots of *Bupleurum falcatum* (Chinese thoroughwax) are used in Japanese herbal medicine to treat autoimmune diseases, chronic hepatitis, and nephritic syndrome. Two compounds isolated from these extracts, bupleuran 2IIb and 2IIc, are pectins that activate complement (a group of serum proteins that can destroy pathogens by causing cell lysis or inducing phagocytosis). These proteins can upregulate the synthesis of macrophage Fc receptors that bind to antibodies attached to pathogens for clearance of complexes, and also stimulate the proliferation of cells in the spleen and in Peyer's patches.

Digestion of the 63,000-kD polysaccharide bupleuran 2IIc with *endo*-α(1-4)-polygalacturonase yields fragments of galacturonans, branched RG-I type molecules with neutral sugar side chains, and an RG-II type of polysaccharide. Only the branched RG-I type molecules possess the cell proliferation activity. A structure (**Figure 9.11**) has been proposed for this bioactive sub-

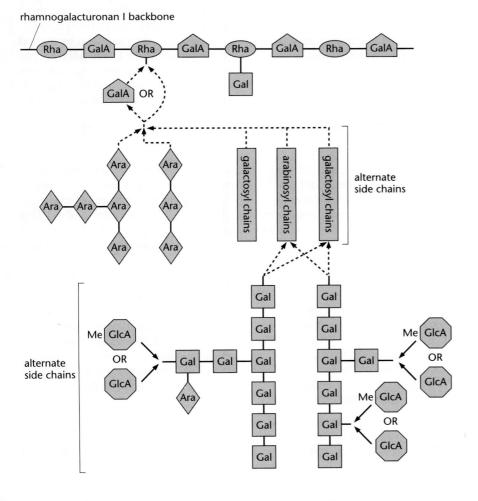

Figure 9.11 Proposed structure of the "ramified" region of RG-I of bupleuran 2IIc, a pectic polysaccharide fragment that activates B cell proliferation. (Adapted from H. Yamada, H. Kiyohara and T. Matsumoto, in F. Voragen et al., eds., Advances in Pectin and Pectinase Research, pp. 481–490. Dordrecht, Netherlands: Kluwer Academic Publishers, 2003.)

domain of bupleuran 2IIc, which appears to trigger the response by binding to specific B cell receptors and inducing them to cap. Receptor binding stimulates cell proliferation via activation of tyrosine kinase, PI phospholipase C, and protein kinase C, which, in turn, activate a multitude of cell-cycle regulating proteins including the cyclins A, B, D, and E and CDK4 and 6. Bupleuran 2IIc also induces immature B cells to mature into activated B cells, and stimulates the synthesis of interleukin-6 (IL-6) and the secretion of IgM antibodies.

Immune system stimulating cell-wall polysaccharides have also been identified in other medicinal plants. For example, the rhizomes of *Atractylodes lancea* contain three pectic polysaccharides capable of stimulating the intestinal immune system, and extracts of the leaves of *Panax ginseng* (ginseng) contain bioactive dimeric RG-II molecules that stimulate immune complex clearance and IL-6 production by macrophages. Dimerized RG-II molecules derived from bupleuran 2IIb exhibit no such activity, suggesting that the pharmacological properties of certain medicinal herbs may depend on plant-specific forms of RG-II.

Although *Aloe vera* is native to Africa, it was already recognized as a medicinal herb by the seventh century in China. When cut, the fleshy leaves of *Aloe* exude a mucilaginous gel that is used to treat burns and cuts. This gel stimulates leukocyte infiltration into wounds. Wound recovery is stimulated by the polyuronic acids in the gel. Purified acetylated mannans, which constitute the major fraction of the *Aloe* mucilage polysaccharides, stimulate leukocytes and lymphocytes and trigger the release of IL-1 and IL-6, as well as the tumor necrosis factor TNF-α.

Another African medicinal herb that contains bioactive pectins is *Veronica kotschyana*. In Mali, roots of this herb are prescribed for the treatment of gastritis and gastroduodenal ulcers, and for wound healing. Characterization of 50°C and 100°C water extracts has identified pectic polysaccharides consisting of a rhamnogalacturonan core with branched arabinose and branched galactose side chains as the principal active compounds. These polysaccharides are capable of inducing chemotaxis of human macrophages (T cells and NK cells), of triggering a T cell–independent induction of B cell proliferation, and of stimulating potent, dose-dependent complement fixation.

6. Bioactive pectins inhibit tumor growth by inducing cancer cell apoptosis.[ref8]

Thirty to forty percent of cancer cases worldwide may be preventable by modification of diet. Insoluble fiber can reduce the risk of cancer by binding cancer-causing chemicals and thereby preventing them from affecting the body. Pectic polysaccharides can also directly inhibit metastasis of existing cancers as well as cause cancer lesions to shrink. Exposure of malignant colon cells to citrus pectins increases *cancer cell apoptosis*, and reduces tumor growth.

Whether the specific bioactive pectin(s) can be identified, purified, and developed into drugs suitable for *cancer therapy* remains to be determined. Two potential advantages of such pectin-based drugs would be their low toxicity and their potential for becoming multifunctional therapeutic agents due to the inherent structural diversity of the originating pectic polysaccharides. Identifying the active compounds in commercial pectin preparations is made difficult not only by the structural complexity of the starting material, but also by the modifications introduced by the extraction and fragmentation processes.

One recent study has focused on characterization of bioactive pectins capable of inducing apoptosis of androgen-independent human prostate cancer cells, that is, cancers that do not respond to changes in hormone levels and for which treatment options are limited. The starting material for these studies was a product known as Fractionated Pectin Powder, which induces a 40-fold increase in apoptosis of both androgen-responsive and androgen-independent human prostate cancer cells. Chemical deesterification of the active starting material destroyed most of the apoptotic activity, whereas specific cleavage of methylesters by pectin methyl esterase had no effect, suggesting that an ester-based cross-link (not methylester-based) is important for the apoptotic activity. The specific molecular linkage required for the pharmacological response has yet to be identified. Also unknown is the apoptosis-inducing mechanism of these pectins.

Studies of the inhibitory effects of pectin on metastatic lesions in the lung have demonstrated that the cellular response to those pectins involves binding of the pectin molecules to galectin proteins (galactoside-binding proteins) that mediate cell-cell interactions between cancer cells. This binding mechanism does not seem to apply to prostate cancer cells, because the androgen-responsive prostate cancer cells do not express galectins on their surface, yet they are stimulated to undergo apoptosis by the bioactive pectins in Fractionated Pectin Powder.

7. Coating of medical devices with cell wall polysaccharides alters their bioactive properties.[ref8]

The interaction of cells with medical devices is controlled to a large extent by the surface properties of the materials. Whereas strong cell-material interactions are desired in some instances, for example, where attachment of a device to surrounding tissues is required, minimal interactions are essential in other devices such as stents and heart valves. For example, heparin and dextran have long been used as "biopassive" surface coating materials due to their high degree of hydration and low protein absorption properties. The physical and physiological properties of alginate, a polysaccharide produced by seaweeds as well as by *Pseudomonas* and *Azotobacter* bacteria, and hyaluronan, a nonsulfated glycosaminoglycan of the extracellular matrix of mammals, have led to their use in wound healing and postsurgery applications. Until recently, it was assumed that the advantages of coating dressing materials with alginate were due to a simple "moist healing" mechanism. However, part of the success is due to the bioactive effects of alginate on macrophages. This insight has stimulated a search for other types of bioactive polysaccharides, including pectins.

Pectins have potential uses as surface modifiers of materials used in medical devices due to their structural diversity, their long safe history of human consumption, their commercial availability, and the increasing number of bioactive domains that have been identified. Furthermore, the availability of increasing numbers of enzymes capable of cleaving specific sugar linkages (**Figure 9.12**) opens up the possibility of producing a variety of

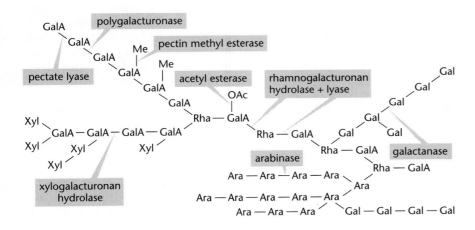

Figure 9.12 Sites of attack of enzymes capable of modifying the homogalacturonan and rhamnogalacturonan structural domains of pectins. (Adapted from M. Morra et al., *Biomacromolecules* 5:2094–2104, 2004.)

Figure 9.13 Light micrographs of fibroblast cells grown for 24 hours on plastic surfaces coated with two different pectin polymers derived from enzymatically treated apple pomace. (A) The PSMHR B-type pectin treated surfaces cause cell elongation and cell spreading, whereas in (B) the PSMHR A-type pectin treated surfaces induce cell rounding and cell aggregation. (Courtesy of Marie-Danielle Nagel.)

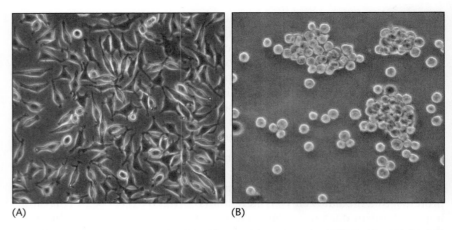

(A)   (B)

molecules with diverse chemical and biological properties from a single, well-defined source of starting material. Coupling different enzymatically modified rhamnogalacturonans from apple pectin to the surface of Petri dishes affects the growth of mammalian cells in culture (**Figure 9.13**). One type of coating stimulates cell spreading, whereas another produces an opposite effect. Pectin-coated medical devices are not yet ready for use, but the potential is evident.

C. Cell Wall Fibers Used in Textiles

1. Commercially important plant fibers are derived from many types of plants and possess physical properties that are exploited in textile products.[ref9]

Many plant species yield fibers that are used for making threads, fabrics, cordage, nets, carpets, filters, paper, brushes, fillings, and reinforced composites. This diversity of fiber origins and uses led to varying definitions of a *plant fiber*. Classical botanists use the term fiber to describe elongated sclerenchyma cells of vascular plants whose thick secondary walls contain cellulose and lignin. These cells are typically organized into tightly adhering bundles and have tapered ends with simple pits. The commercial definition of a plant fiber is much broader and includes any cell type that is long, thick-walled, and durable and has high tensile strength. This definition encompasses the classical botanical fibers, the long seed hairs of cotton that contain only cellulose, and the pulp derived from softwoods, which technically lack fiber cells. In this chapter, the term plant fiber is employed in the latter, more general commercial sense.

The plant fibers used in textiles today are often divided into four groups based on the tissue of origin (**Table 9.3**): *seed hair fibers, fruit fibers, leaf fibers*, and *bast fibers* (fibers derived from the bark of stems). Their lengths can vary from 1 to 200 cm. Cellulosic plant fibers tend to be dense; hence, fabrics produced from such yarns feel comparatively heavy. Their relatively low elasticity and low resilience explains their tendency to wrinkle easily, whereas their ability to absorb moisture slows their drying. Two highly valued properties of cellulose-rich fibers are their heat- and electricity-conducting capabilities. In hot climates cellulose fiber garments are appreciated for their ability to carry away body heat, whereas in dry climates their electrical conductivity prevents the buildup of static electricity.

2. Cotton is the most widely used plant textile fiber.[ref9]

Wild *cotton* (*Gossypium*) species are native to Eurasia, Africa, and the Americas. Domestication occurred independently in Southeast Asia and northern Peru. *Gossypium hirsutum* and *G. barbadense*, both New World

CLASS AND COMMON NAME OF FIBER	NAME OF PLANT
TABLE 9.3 COMMERCIALLY EXPLOITED NATURAL FIBERS	
Seed hair fibers	
Cotton	*Gossypium hirsutum* and *G. barbadense*
Kapok	*Ceiba pentandra*
Fruit fibers	
Coir	*Cocos nucifera*
Bast (soft) fibers	
Jute	*Corchorus capsularis*
Flax	*Linum usitatissimum*
Ramie	*Boehmeria nivea*
Hemp	*Cannabis sativa*
Leaf (hard) fibers	
Manila hemp/abaca	*Musa textilis*
Sisal	*Agave sisalana*
Henequen	*Agave fourcroydes*

cultivars, account for the bulk of the current world's crop. *Gossypium hirsutum* has greater resistance to boll weevil attack, and *G. barbadense* has a long fiber length. Longer fibers are easier to process and yield a smoother thread. Cotton still accounts for about 50% of total world textile fiber production despite the growing importance of synthetic fibers. Between 1989 and 1993 the largest cotton producers were China (33%), the United States (22%), India (16%), Pakistan (11%), Brazil (6%) and Turkey (5%).

Cotton owes its commercial success to the large amount of fiber produced per plant, as well as to the low cost of harvesting the fibers, processing the raw cotton, and the manufacturing of cotton products compared with other plant fibers (**Figure 9.14**). Before the invention of the cotton gin in 1793 by Eli Whitney, cotton was an expensive textile fiber due to the difficulty of separating the cotton fibers from the seeds. Prior to this invention, one pound of cotton thread required approximately 13 days of labor by one person, whereas one pound of wool took one person only 7 days of work, silk about 6 days, and linen 5 days.

Although the seed fibers of *kapok* (**Figure 9.15**), which is also known as Java cotton, resemble cotton in many respects, they are not used for textiles, for two reasons: their smooth, slippery surface prevents them from being spun into stable threads, and the fibers are more brittle than cotton. In contrast to cotton, kapok fibers contain lignin, which might explain their greater resilience and their resistance to shrinkage and warping on drying, as well as their slippery surface. Their exceptional resiliency and resistance to wetting make them useful as stuffing for pillows, mattresses, and upholstery, and as insulation. Their high buoyancy (30 times their weight) was exploited in life preservers prior to the use of synthetic foams.

Figure 9.14 Cotton boll that has opened to expose tightly packed cotton fibers (trichomes). Each commercial cotton boll produces 24 to 32 seeds and about 500,000 fibers.

Figure 9.15 Kapok fruit that has split, allowing the seeds, which are covered with large, silky hairs, to escape. Mature kapok trees produce hundreds of these fruits, which are up to 15 cm in length.

3. Cotton fibers are epidermal trichomes that develop on the surface of ovules.[ref10]

Cotton fibers are produced by the elongation of epidermal cells on the surface of the ovules of *Gossypium* species (**Figure 9.16A**). Each ovule gives rise to 14,000 to 18,000 fiber cells (Figure 9.16B), 16 to 20 μm in diameter and 1 to 6 cm long, depending on the cotton variety. Isolated ovules, both fertilized and unfertilized, can also be cultured on defined media and induced to form fibers. Such cultures have yielded insights into the hormonal and other physiological requirements for fiber formation and the molecular biology of fiber growth.

Cotton fiber growth starts on the day of anthesis (flower opening) by spherical expansion of a subset of epidermal cells. These cells can be identified by the presence of an outer pectin sheath that is absent from the cell walls of young cells. Spherical expansion is followed by cell elongation, which lasts for 15 to 30 days depending on growth conditions. During expansion the fibers exhibit tapered tips and grow into the air space (locule) between the ovule and the carpels, which ultimately becomes tightly packed with fibers (see Figure 9.14). The primary walls of the elongating fibers are thin (0.2 to 0.4 μm) and contain about 30% cellulose, 30 to 50% GalA-rich pectic polysaccharides and hemicelluloses, and arabinogalactans. The outer surface of the cotton fiber is covered by a hydrophobic cuticle of layered wax molecules (**Figure 9.17**).

Secondary cell wall formation begins when cell elongation is completed. During secondary cell wall formation, each cell produces copious amounts of cellulose to yield fibers comprised of 90% cellulose and containing about 130 ng of cellulose per millimeter, in contrast to the 1 ng of cellulose per millimeter for fibers possessing only primary walls. In addition, the degree of polymerization (DP) of the cellulose molecules increases from 2000 to 3000 during primary wall growth to 12,000 to 15,000 at the onset of secondary wall formation. The approximately 4-nm-thick cellulose fibrils are produced by rosette-type cellulose synthase complexes (see Figure 4.27).

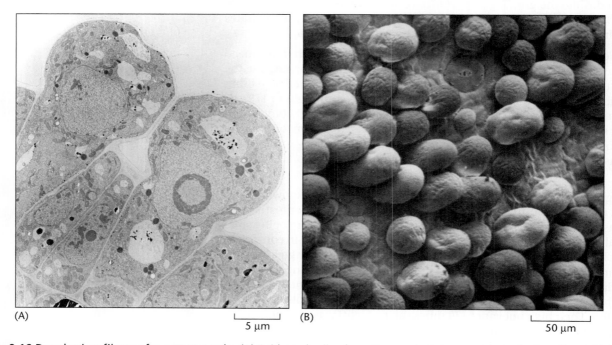

(A)   5 μm

(B)   50 μm

Figure 9.16 Developing fibers of a cotton ovule. (A) Epidermal cells of a cotton ovule 4 days postanthesis. Two fiber initials have begun to extend beyond the surface of the non-fiber-forming epidermal cells. The cells are surrounded by a thin primary wall and their vacuoles are still small. (B) Scanning electron micrograph of the surface of a cotton ovule 4 days postanthesis. The close packing of the fiber initials is clearly seen. (Courtesy of Ulrich Ryser.)

Between 45 and 60 days postanthesis, the fruit capsules burst open and the cylindrical fibers collapse into irregularly twisted ribbonlike structures during drying (**Figure 9.18**). The surface texture created by the collapse and twisting of the fibers is important for textiles in that it prevents the fibers in the cotton threads from slipping past each other and thereby weakening and unraveling the material. Mature fibers readily detach from the seeds by breakage of the fiber wall slightly above the epidermal surface. After removal of these so-called lint fibers, the seeds remain covered by a dense layer of very short *fuzz fibers*, or *linters*, whose elongation starts several days postanthesis. Fuzz fibers, too short for textiles, are used for stuffing and as substrates for making rayons, acetates, and paper. Cottonseed oil and fertilizer are produced from the seeds.

In cross sections most cotton fibers display a very uniform texture across their secondary walls, even though the cellulose fibrils are organized into layers with alternating fibril orientations. However, when *Gossypium* plants are subjected to nightly temperatures below 18°C, growth rings can be observed in walls treated with cuproammonium hydroxide. This layering has been traced to the low-temperature cellulose fibrils being less crystalline and more accessible to swelling reagents. Layered walls are also observed in the naturally colored "green lint" mutant of cotton that deposits alternating suberin and wax layers between the cellulose layers (see Concept 3G3).

4. Coconut-derived coir fibers are used extensively in tropical countries.[ref9]

Coir is a fiber obtained from the outer husk, the fibrous mesocarp, of the *coconut* (**Figure 9.19**). Coir fibers are composed of bundles of cells that are 10 to 25 cm long. They are separated from the nonfiber cells by a process known as *retting*. Retting, which is also used to release the "soft" bast fibers from stems of flax, jute, and ramie (see Concept 9C5), takes advantage of the fact that fiber cells have thicker cell walls than the cells in the surrounding tissues. This makes them comparatively resistant to the breakdown by cell wall–degrading enzymes secreted by bacteria and fungi. In the case of coir, the retting process in brackish water takes about eight months. The most useful properties of the brown-colored coir fibers are resistance to rot and salt water, lightness, elasticity, and resistance to abrasion, hence their use in ropes, floor mats, netting, textiles used in landscaping, carpet matting, and brushes.

5. The physical properties of different bast fibers make them suitable for use in specialized types of textiles.[ref9]

Bast, or "soft," fibers are derived from multicellular strands of thick-walled and highly elongated fiber cells. These supporting cells are located within or adjacent to the phloem in the stems of many dicotyledonous plants (**Figure 9.20**). Individual bast fiber bundles can be over 2 m long and are separated from the stems by retting followed by a bending or pounding process to break and remove the brittle woody parts. Linen, which comes from flax, is the best-known bast fiber. Jute, ramie, and hemp are other important commercial bast fibers.

Flax. Flax, *Linum usitatissimum* (translated from Latin as "most useful linen"), most likely originated in the eastern Mediterranean. The flax varieties planted today for fiber production are 60- to 120-cm-tall, sparsely

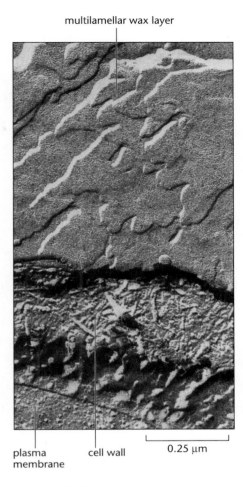

multilamellar wax layer

plasma membrane   cell wall   0.25 μm

Figure 9.17 Freeze-fracture micrograph of a growing cotton fiber 10 days postanthesis. The wax-covered cuticle appears as a multilamellar layer overlying the primary wall and the plasma membrane. (Courtesy of Werner Herth.)

100 μm

Figure 9.18 Mature cotton fiber exhibiting the typical collapsed and twisted morphology observed after boll opening and drying of the fiber. (Courtesy of Thomas Pesacreta.)

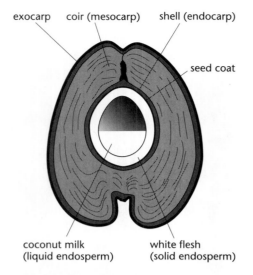

exocarp   coir (mesocarp)   shell (endocarp)

seed coat

coconut milk
(liquid endosperm)

white flesh
(solid endosperm)

Figure 9.19 Cross-sectional view of a coconut. The coconut fiber, coir, is derived from the fibrous fruit coat that surrounds the seed, which in turn contains the white, nutritious flesh (solid endosperm) and the coconut milk (liquid endosperm).

branched annuals that produce few seeds (**Figure 9.21**). Cultivars grown for linseed oil used in the production of dyes, paints, and cosmetics have a high seed yield.

Flax thrives in cooler climates not suitable for cotton. Nearly half of the world's flax is grown in Russia. It is sown in the spring and harvested after 80 to 100 days. Harvesting involves pulling the plants out of the ground, which maximizes stem length and prevents discoloration of fibers through wicking from the cut ends during retting. The retting process involves decomposition of the stems by bacteria to release the bast fibers. In drier climates, this involves immersing the stems in tanks or small ponds. In wetter areas like Ireland, stems are left in the field to become soaked by dew. Both processes require 6 to 20 days. Attempts at using chemical or enzymatic retting procedures have not progressed beyond small-scale laboratory experiments.

Groups of flax fibers are both longer (30–50 cm) and two to three times stronger than cotton fibers. Linen fabrics were used to cover the wings and bodies of early aircraft. More recently, manufacturers have added flax fibers to isocyanate-glued particleboards to increase their bending strength. The characteristic luster of linen household textiles and wearing apparel is due to the naturally smooth surface and the straightness of the flax fibers. Until it was supplanted by cotton in the nineteenth century, linen was the leading plant textile fiber in Europe. Due to the time-consuming retting process, mechanization of linen fiber production is more difficult and expensive than for cotton. Linen production has stabilized at about 2% of the world fiber market. Like cotton materials, linen fabrics wrinkle easily. However, because linen produces less lint than cotton, it is still preferred for special uses such as drying glassware.

Jute. Jute plants grow best in a fertile soil and a hot, moist climate. The principal producing countries are India, Bangladesh, China, and Thailand. Jute is the world's foremost bast fiber and second only to cotton in total tonnage. In India alone, 1.6 million tons of jute goods were produced in 2002. The fast-growing, reedlike annuals grow up to 5 m tall (**Figure 9.22**), and two crops can be harvested in a growing season. The species appears to be native to tropical Asia (for example, Malaysia and Sri Lanka).

After retting, jute fibers are light to dark brown in color. Large fiber bundles are up to 2 m long, but the individual fiber cells are much shorter. Together with their limited pliability, the shortness of individual cells makes them difficult to spin. Jute is used mostly as a packaging material in the form of burlap bags. Other products are carpet backing, canvas, and twine. In recent years, polypropylene fibers have become a major competitor for jute, which has caused the producing countries to look for novel applications. This search has led to jute being employed as reinforcement fibers in resin composites for tea and fruit shipping containers. Jute-based *geotextiles* are

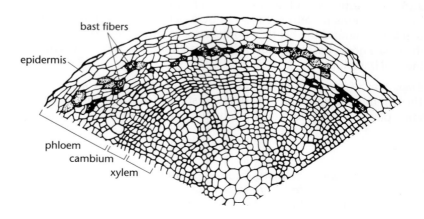

bast fibers

epidermis

phloem

cambium

xylem

Figure 9.20 Cross section of a hemp stem showing the location of the bast fibers (*dark patches*) that are released by the retting process.

being used increasingly for erosion control in European ski areas, because they are environmentally friendly, can retain five times their weight in moisture, and are cheaper than synthetic textiles.

Ramie. Ramie is a hardy perennial native to China. Today, most ramie is produced in China, Brazil, the Philippines, India, South Korea, and Thailand. Ramie plants are propagated by rhizome cuttings, and the roughly 2-m-long stems can be harvested in the third year. Three to six crops can be produced per year thereafter. Release of the fibers involves a combination of peeling, beating, scraping, and natural or chemical retting.

Although ramie produces a fine, white, easily dyed, exceptionally strong, highly lustrous, and long (individual fiber cells up to 15 cm) fiber, its commercial use has been limited by the high cost of removing the resins that bind the fibers together. However, the introduction of improved chemical retting methods has reduced the cost of production. Most often, ramie fibers are blended with polyester, cotton, linen, and acrylics for use in wearing apparel such as sweaters and suits and in fabrics for home furnishings. Due to its ability to bond with rubber, PVC, and other plastics, it is also used to strengthen panels of European cars, in protective gloves, and in geotextiles. The use of ramie fibers for making fishing nets can be traced to their resistance to bacteria and mildew, and their increase in strength when wet.

6. Leaf fibers are used for ropes, tea bags, currency paper, and filter-tipped cigarettes.[ref9]

Leaf fibers, also known as "hard" fibers, are whole vascular bundles, which are commercially produced from two plant genera, *Agave* (sisal and henequen; **Figure 9.23**) and *Musa* (abaca, or Manila hemp). The term "hard" reflects their stiffness, which makes them unsuitable for use in modern clothing. However, they make better ropes than most bast fibers, and they have found uses in a number of specialized products.

Abaca (Manila hemp) is a relative of the banana, which is native to the Philippines. Entire plants (5 to 7 m tall) are harvested when they are four years old. The fibers are extracted from the outer peripheries of the leaf bases by feeding the leaves into a peeling (decorticating) machine. Products that have been made from abaca fibers include currency paper, "Manila" envelopes, tea bags, Italian salami casings, and cigarette filters. However, the primary use of Manila hemp is in the production of Manila ropes for naval applications.

Sisal and henequen are two *Agave* species that are native to Central America. They thrive in arid regions and are grown commercially in Mexico, Brazil, and central and southern Africa. The plants are perennials that pro-

Figure 9.21 Young flax plant that already exhibits the tall growth phenotype of mature plants.

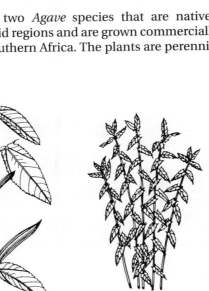

Figure 9.22 Jute plants.

Figure 9.23 *Agave* plant from which the lower leaves have been removed for sisal production.

duce long, fiber-rich, fleshy leaves for 5 to 40 years. The outermost leaves of the huge rosettes are cut at their bases. Crushing and scraping methods are employed to separate the fibers from the succulent pulp and the tough, cutinous epidermis. The creamy white fibers are washed and dried and can be used directly. The Mayans and Aztecs wove sisal fibers into garments, but by today's standards such clothes would be considered too rough and stiff. For the most part, sisal and henequen fibers are made into ropes.

D. Wood: an Essential Product for Construction and Paper Production

1. Wood is one of the world's most abundant industrial raw materials.[ref11]

Wood is an extraordinary material composed almost completely of plant cell walls. The total volume of wood in the world is estimated at 0.4×10^{12} cubic meters (www.fao.org). In the United States, wood accounts for about 25% of the value of all major industrial materials. On a weight basis, wood exceeds use of all other structural materials combined; it is also the principal source of industrial fiber. Wood has a very high strength-to-mass ratio, and a stiffness that rivals steel of equivalent weight. A major disadvantage of wood is that it is subject to microbial decay, and it is more difficult to work than steel. On the other hand, wood is renewable and can provide not only structural materials, but also a wide variety of chemicals and paper products. Wood provides much of the energy for its own processing. About two-thirds of the wood harvested in the United States is used for *pulp* and paper. Roughly half the biomass of a tree is recovered as wood, and the yield of wood converted into paper is roughly one-half also.

Wood is widely used for construction (**Figure 9.24**), furniture, paper, industrial pulp, and many other familiar products. Wood is far less demanding of energy than other major structural materials, particularly steel, aluminum, plastics, and concrete. It is an excellent thermal insulator due to the high void volume of the cells, and it is a strong electrical insulator due to its low metal content and low conductivity. Absolutely dry wood offers nearly complete electrical resistance. One of the most specialized uses of wood is in the construction of musical instruments, which exploit its unique acoustic properties. A piece of wood, fixed so that it can vibrate freely, will emit sound when struck. The frequency (or pitch) of the sound depends upon the dimensions and elasticity of the material.

Figure 9.24 The great Todaiji Temple in Nara, Japan, the world's largest wooden building. Note the size of the people near the entrance. (Courtesy of Tetsuko Noguchi.)

Figure 9.25 The structure of wood. (A) Drawing of a young pine tree showing the origin of a cross section of wood from the trunk. (B) Segment of a cross section from the stem showing bark, growth rings, and lateral rays in the wood. Dark rings represent latewood (summerwood), and light areas represent earlywood (springwood). Box shows area enlarged in (C). (C) Scanning electron micrograph of part of a polished cube of softwood illustrating three views of the vertically oriented tracheids (see Figures 7C3.1 and 3.2). The top plane depicts a cross section of a set of tightly packed tracheids as shown schematically in (B). The lower right surface shows longitudinal views of the tracheids, which appear as hollow tubes. The lower left surface, in turn, depicts both longitudinally sectioned tracheids and longitudinal views of a cluster of ray cells that run at a right angle to the tracheids and radially within the stem. The arrows point to ray cells in different planes. (A, and B, adapted from C.J. Biermann, Essentials of Pulping and Papermaking. San Diego: Academic Press, p. 14, 1993. C, courtesy of the N.C. Brown Center for Ultrastructural Studies, SUNY College of Environmental Science and Forestry.)

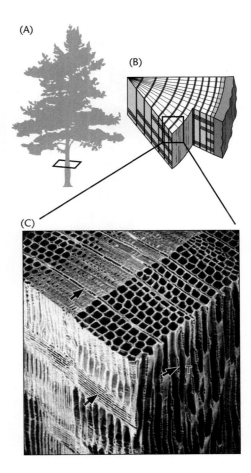

At present, only a small component of the forest products industry makes specialty chemicals from wood. However, much of the *chemical feedstock* that is now derived from fossil fuels can also be derived from wood (see Concept 9E9). As the cost of fossil fuels increases and availability decreases, the utilization of wood for chemicals becomes more economically attractive.

2. Wood and paper properties are derived from the composition and morphology of the xylem cell walls.[ref12]

Wood is a product of the terminal differentiation of the xylem tissues of woody plants and is almost entirely composed of cell walls (**Figure 9.25**). The properties of wood are important for its uses as a structural material, and as a source of pulp and paper. The major factors affecting the variation within a segment of wood and in different types of wood are the location, abundance, and distribution of different cell types. The most common cells in *softwood* are the longitudinal tracheids—long, thin cells responsible for support and water transport. They vary from 1.0 to 5.0 mm in length depending on the species, and about 0.01 to 0.02 mm in diameter. Small, isodiametric *ray parenchyma* cells (cells with similar diameters, distributed radially) are the next most predominant cell type. In *hardwoods*, vessel elements up to 0.3 mm in diameter create large pores, which give rise to the characteristic grain of many hardwoods.

Wood is highly anisotropic, that is, its physical properties differ along different axes. The strength of wood and the permeability and conductivity of water through wood are greatest along the grain due to the longitudinal orientation of the longitudinal tracheids or fibers and the structure of the their cell walls. Tracheid or fiber cell walls are composed of several layers. The predominant middle layer of the secondary wall, the *S2 layer*, may comprise up to 90% of the thickness of the wall (**Figure 9.26**). The cellulose microfibrils are highly oriented, and the angle of the fibrils contributes greatly to the tensile strength and the stiffness of the fiber and the wood. Tensile strength increases as the angle of the fibers relative to the long orientation of the cell (and the grain) decreases.

Wall thickness and cell length have important effects on wood properties and the paper derived from the wood. Thick-walled cells with small lumens have high wood density and greater strength. After removal of lignin for pulp or paper production, the length of the cells and the thickness of the walls determine many of the important properties of paper. Long cells have increased strength in the formation of paper. Thick cell walls increase coarseness, and thin cell walls, which are more easily flattened, have better bonding properties.

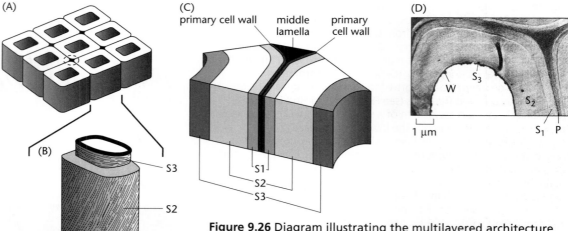

Figure 9.26 Diagram illustrating the multilayered architecture of tracheid cell walls. (A) Schematic drawing of cross-sectioned tracheids in a wood section cut at right angles to the longitudinal axis of the stem. The dotted circle marks a cell corner region where four cells meet. A segment of such a corner is shown in (C) to illustrate the layered structure of the cell walls.(B) Three-dimensional diagram of the layers of the wall of a softwood tracheid in which the different orientations of the cellulose microfibrils in the primary, S1, S2, and S3 cell wall layers are shown. The primary wall (P) microfibrils are randomly oriented, the parallel S1 microfibrils have a low pitch, those in the S2 layer a high pitch, and the S3 microfibrils again a low pitch. No cellulose microfibrils are present in the middle lamella (ML). (C) Diagram of a cell wall at a corner between adjacent tracheids, illustrating the multilayered structure of woody walls. A seven-layered composite cell wall structure is formed by the adjacent cell walls of two tracheids. (D) Thin-section electron micrograph of a composite type of cell wall as illustrated in (B) and (C). The warty layer on the inner surface of the tracheids consists of cytoplasmic remnants of the dead cells. (A and B, adapted from A.W. Côté Jr., Wood Ultrastructure: An Atlas of Electron Micrographs. Syracuse, NY: University of Washington Press, 1967. C, adapted from A.P. Schneiwind and H. Berndt, in M. Lewin and I.S. Goldstein, eds., Wood Structure and Composition. New York: Marcel Dekker, pp. 435–476, 1991. D, from H. Harada and A.W. Côté Jr., in T. Higuchi, ed., Biosynthesis and Biodegradation of Wood Components: Structure of Wood, pp 1–42. Orlando: Academic Press, 1985.)

The typical length of the fiber cells or tracheids differs greatly in hardwoods and softwoods. Typical softwoods have fibers about 3 mm long, whereas hardwoods have fibers 1 mm or less. Strong paper requires fibers longer than 2 mm. Fine writing papers require less strength, but need thin-walled fibers that flatten well to produce a good printing surface. Wood pulp from some *Eucalyptus* species has exceptional softness and absorptive properties and is therefore desirable for soft facial tissues and other specialty paper products.

The physical and chemical composition of wood determines its strength and constrains the methods employed for industrial processing. The strength of the cellulose fibers in wood and in paper is due to the crystalline structure of cellulose microfibrils, and the hydrogen bonding between cellulose and hemicelluloses is of critical importance for cell wall strength. The content and composition of the lignin affects the efficiency of pulping. Lignin content may vary between 16% and 45% and determines the choice of the pulping methods needed to make pulp or paper. The degree of methoxylation of the lignin subunits affects the extent of condensation of the lignin and the ease of pulping. Lignin with increased methoxylation is less condensed and easier to pulp.

3. The physical strength of wood depends on the multilayered structure of the wood cell walls and cell morphology.[ref13]

The morphology of the wood cell wall is a major determinant of the strength and physical properties of wood. The structure of a wood cell wall is that of a multilayered composite. The xylem tracheid cell wall has four layers: the primary cell wall, formed first in the differentiation of xylem from a cambial initial; followed by the formation of three layers of the secondary cell wall internal to the primary wall (see Figure 9.26). The middle and thickest layer of the secondary wall, the S2 layer, has the greatest structural influence on the properties of the wall. At the end of xylem differentiation, all layers of the wall are impregnated with lignin and adjacent cells are joined by the lignin-rich middle lamella. The structure of wood, therefore, can be considered to be a nine-layer structure, composed of the walls of two adjacent cells, each with a primary wall and three secondary wall layers, and sharing a middle lamella. The structure, a layered composite, is analogous to reinforced concrete, with cellulose as a fibrous strength element, surrounded by a polymeric network of lignin. The cellulose confers tensile strength, whereas the lignin confers stiffness and resistance to compression. The thickness of the cell walls and the orientation of the cellulose microfibrils determine the directional (anisotropic) differences in tensile and compressive strength of wood.

Density is one of the most important and most general determinants of wood strength (**Figure 9.27**). The density of the wall material itself is not highly variable across the layers of the wall; it is the molecular structure and architecture of the walls that confers the differences in wood density. Most cell walls have a density of about 1.5 g/cm^3, even though lignin content var-

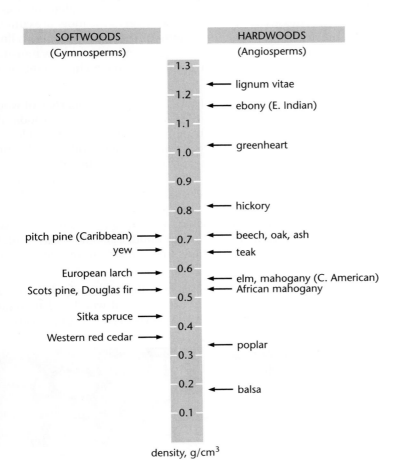

density, g/cm^3

Figure 9.27 Density values at 12% moisture content for some common "hardwoods" and "softwoods." Typically, the higher the density of a wood, the harder the wood. Softwoods are derived from gymnosperms and are, *on average*, softer than the *average* hardwood, which is derived from angiosperms. (Adapted from H.E. Desch and J.M. Dinwoodie, Timber. Forest Grove, OR: Timber Press, 1981.)

ies considerably. The density of wood, however, varies from 0.16 g/cm^3 in balsa wood to 1.25 g/cm^3 for wood of *Lignum vitae*. The variation in wood density is due to differences in morphology (cell length and diameter, and wall thickness) of the major cell types that constitute the elements in the wood (**Figure 9.28**). Balsa wood has many large vessels and thin walls compared with *Lignum vitae*, which has smaller, densely packed cells with very thick walls in the wood fibers.

Although softwoods are on average lower in density than hardwoods, many hardwoods overlap the density range typical of softwoods. Most softwoods are below densities of 0.7 g/cm^3, whereas many hardwoods are higher, ranging up to densities of 1.2 g/cm^3. Pitch pine is relatively dense (0.7 g/cm^3), roughly equivalent to beech and oak, but more dense than hardwoods such as poplar, mahogany, and some species of elm, which are significantly lower in density. These wood density values are averages and are not constant within a species or even within a tree. They are affected by age, growth rate, environmental conditions, and genetic variation within the species. Density varies greatly between earlywood and latewood, in juvenile wood and mature wood, and in *reaction wood* that is formed in response to mechanical stress. Other aspects of the physical properties of wood are defined by ability to retain shape after deformation (elasticity) and the ability to resist bending (stiffness).

Density correlates well with the strength required for a beam (resistance to static bending) and the strength required for a post (resistance to compression parallel to the grain, also called compressive strength parallel to the grain). Density also correlates well with hardness. The angle of the microfibrils of the S2 layer has a great influence on the tensile strength of wood. As the angle increases, tensile strength and stiffness both decrease. Strength is greater when the S2 microfibril angle is very low and parallel to the grain. Tracheid length is an important factor, because overlap of adjacent cells is needed to transfer stresses. Wood becomes stronger as the moisture content is decreased, presumably due to the strengthening of hydrogen bonds linking cellulose microfibrils. Drying of wood and maintenance of low moisture content during the lifetime of wood structures become an important factor in construction and architecture.

The relationship of cell wall properties to the physical properties of wood is further illustrated by a comparison of reaction woods in hardwoods and softwoods. In the *compression wood* response of gymnosperms (including conifers), which occurs on the underside of a branch bending or leaning downward, the resulting wood is stronger in resistance to longitudinal compression, higher in density, and higher in lignin content and has shorter fibers. The microfibrillar angle is increased. In contrast, in hardwoods, the wood formed on the upper side of a branch bending or leaning downward, called *tension wood*, is stronger in tension, weak in compression, increased in flexibility, and lower in density. In tension wood, a gelatinous layer forms on the innermost surface of the tracheids that is nearly pure cellulose (see also Concept 7C8).

Virtually all of the features of the cell walls of the terminally differentiating xylem have effects on the physical properties of wood and therefore affect the behavior of wood from different species in its many applications.

Figure 9.28 Cross-sectional views of woods with different densities: poplar (0.35 g/cm^3), teak (0.65 g/cm^3), greenheart (1.02 g/cm^3). Note the correlation between wood density and cell wall thickness. (Adapted from H.E. Desch and J.M. Dinwoodie, Timber. Forest Grove, OR: Timber Press, 1981.)

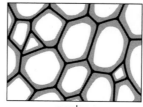
poplar

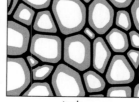
teak

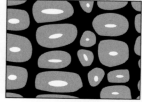
greenheart

A greater understanding of the mechanisms of wood formation that determine the morphology and composition of the cell walls has great practical value for the improvement of wood properties.

4. Cell wall–degrading enzymes are increasingly used for pulp and paper processing for environmental and economic reasons.[ref14]

The processing of wood to produce pulp and paper products usually involves chemical degradation under conditions of high temperature and pressure and high pH (for example, pH 12). Wood chips (**Figure 9.29**) are usually digested under extreme conditions to remove lignin from the middle lamella to release fibers (xylem cell walls). These fibers can be further processed into pulp or paper by bleaching, coating, and rolling. Cell wall–degrading enzymes isolated from microbes have increasing utility in the pulp and paper industry for improving pulping and bleaching, and for de-inking of recycled paper. There is environmental and economic value to this approach because of the large scale of the industry and the considerable benefits to be obtained from reducing both energy use and effluents, or producing effluents that are more environmentally acceptable.

Enzymes, particularly from fungi, provide new ways to process pulps and fibers. Xylanases can reduce the amount of chemicals required for bleaching pulp, and cellulases can produce smoother fibers and enhance the removal of water from pulp. Compared with chemical processing, enzyme-based processes can produce cleaner products with substantial environmental benefits. Chlorine and chlorine derivatives are no longer used in the bleaching processes of many mills in the United States, Scandinavia, and other places around the world, because trace components of effluents (for example, chlorolignins) from chlorine bleaching are considered harmful. New *chlorine-free bleaching* methods employ enzymes combined with oxygen for delignification, followed by ozone treatment and peroxide to produce hardwood pulps of equal brightness and strength compared with conventional chlorine-bleached pulps.

Another specific use of enzymatic processing is *enzymatic de-inking* of recycled paper products. As increasing amounts of newsprint and office paper are being recycled, enzymatic de-inking has been developed to improve efficiency and reduce the environmental impact of the processing that removes the ink from the paper. Laser-printed and xerographic papers are difficult to de-ink. Enzymatic de-inking improves removal of large ink particles, such as highly cross-linked soybean oil–based ink, by increasing the separation of fibers of the paper that are covered with ink. *Endo*-glucanases, lipases, and cellulases have been used for de-inking to

Figure 9.29 Wood chips are used for pulp and paper manufacturing as well as for biomass based heating (Concept 9E4), and for lignocellulose-based ethanol production (Concept 9E8).

wood

cork bark

Figure 9.30 Cross-sectional view of the trunk of a cork oak tree showing the thick cork bark layer used for making cork stoppers.

improve fiber yield and reduce use of chemicals. De-inking with cellulase depends on the cleavage of cellulose polymers, which in turn depends upon the accessibility of the enzyme to the fibers that bind the ink. This process can result in some reduction in fiber length and strength of bonding of the recycled fibers. Similar approaches have been used to remove printing ink from recycled cotton fabric.

Treatment of wood chips with fungi that produce lignin-degrading and other cell wall polymer–degrading enzymes, a process known as *biopulping*, can provide increased pulping efficiency, reduce energy and chemical consumption, and reduce wastes at considerable environmental benefit (**Panel 9.1**). Fungal pretreatment prior to mechanical pulping (the disruption of wood into pulp by mechanical shear) reduces electrical energy requirements during refining, increases mill throughput, and can improve paper strength. Fungal pretreatment also reduces the *pitch* content (pitch is a resinous material produced in trees as a response to pathogens, pests, or wounding), thus reducing pulp cooking time and the formation of chemical by-products that adversely affect the environment.

5. Bark is a source of cork, tannins, waxes, and drugs.[ref15]

Bark is a cell wall–derived tissue that is the protective outermost layer on stems and branches of woody plants (**Figure 9.30**). Bark protects plant stems and branches from pests, pathogens, fire, and mechanical injury. Bark is rich in secondary metabolites, with extraordinary chemical complexity, having high levels of volatiles, phenolics, and other complex macromolecules. The complexity of the chemicals may be due to the need for many chemicals in plant defense against insects and other herbivores, and pathogenic fungi. Bark is formed by the differentiation of vascular cambium to phloem, which in turn forms a second meristem in the cork cambium. This cork cambium gives rise to the inner bark and its derivative, outer bark. Outer bark is the residual cell wall material from the terminal differentiation of the cork meristem. The texture, color, and composition of bark are distinctive for many species and may change dramatically with age as trees go from juvenile to mature stages of development.

The residual cell walls in bark are, in part, chemically similar to wood fibers, and contain cellulose, hemicelluloses, and lignin. In addition, bark contains suberin, cutin, and waxes, plus a large number of other phenolic derivatives, all of which contribute to its water-repelling properties. The chemical complexity of bark reduces its commercial value, because many products of value are present in low concentrations and difficult to purify. Nevertheless, a number of commercial chemical products are derived from bark, including tannins and waxes. The bark of the Western yew has been an important source of taxol, an antitumor agent for treating ovarian cancer. Most bark is currently used for energy, because it is an inexpensive waste product when trees are debarked before processing for saw wood or for pulp and paper-making, and because of its high fuel value.

The most familiar use of bark is in cork as a stopper in wine bottles (**Figure 9.31**). The source of this cork is the bark of the cork oak, *Quercus suber* (Figures 9.30 and **Figure 9.32**). Cork remains one of Portugal's major exports and is harvested nondestructively from the trees as the bark dries and splits from the tree. The cork is rich in suberin, a polymer of aliphatic acids and phenolics. Some trees such as *Eucalyptus grandis* shed their outermost bark layers on a continuing basis (**Figure 9.33**).

Figure 9.31 Cork stopper from a traditional wine bottle.

Figure 9.32 Cork harvest. Harvested cork in front of cork oaks from which the cork bark layer has been removed. Cork harvesting usually starts after the trees reach an age of over 34 years; thereafter a new cork bark layer can be harvested nondestructively roughly every 9 years.

Figure 9.33 Bark peeling off the trunk of a *Eucalyptus grandis*, the remaining dark bark forms a "sock" at the base. The tree shown is a well-known inbred eucalypt, BRASUZ1. The genome of this tree is currently being sequenced. (Courtesy of Dario Grattapaglia.)

E. Plant Cell Walls: the Most Important Renewable Source of Biofuels and Chemical Feedstocks

1. Humankind's survival may depend on the wise use of renewable natural resources.[ref16]

The recent increase in the exponential growth rate of the human population (**Figure 9.34**) is due to the Industrial Revolution, which began roughly 200 years ago. Economies based on manual labor were supplanted by economies built on an "invisible workforce," fossil fuels, which enabled the large-scale use of machinery capable of replacing human labor (**Table 9.4**). This transition has dramatically reduced the number of farmers needed to provide food for the rest of the population and has enabled us to produce synthetic

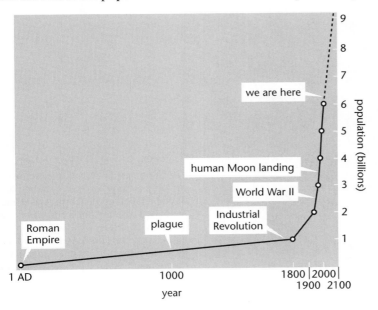

Figure 9.34 Growth of the human population between AD 1 and the present. (Adapted from J. Lewis, *Sci. Am.* 283:30–32, 2000.)

DELIGNIFICATION

The principal barrier to the conversion of lignocellulosic biomass into biofuels and chemical feedstocks is the insoluble lignin network that surrounds and shields the cellulose microfibrils from degrading enzymes. Harsh chemical and thermal treatments are typically used to remove lignin from wood chips to increase their digestibility by cellulose-degrading enzymes. However, the high energy and environmental costs of these treatments constitute a major obstacle to commercial biofuel production.

Delignification of wood chips can also be achieved by means of lignin-degrading enzymes produced by *white rot fungi* such as *Phanerochaete chrysosporium*, a basidiomycete fungus that is used for biopulping and biobleaching in the paper industry. This fungus secretes enzymes capable of breaking down lignin to enable its cellulases to efficiently degrade the cellulose microfibrils, its major source of energy (see also Concept 8C4).

WOOD ROTS

The white color of white rot wood is caused by oxidative bleaching and degradation of the brown-colored lignin.

In contrast, *brown rot fungi* digest only the cellulose microfibrils, leaving brown, oxidized lignin, and wood that shrinks and fragments into cubes upon drying.

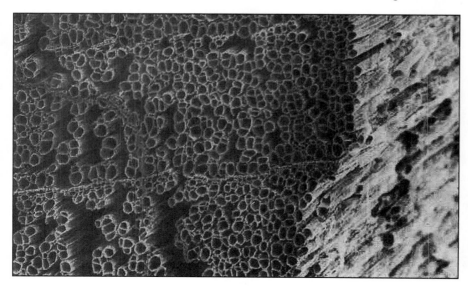

The stringiness of wood left by some white rot fungi (stringy rot) is due to the slower rate of degradation of the long vessel elements compared with the fibers and parenchyma and cells. (Courtesy of K.-EL Eriksson, R.A. Blanchette, and P. Ander, Microbial and enzymatic degradation of wood and wood components. Berlin: Springer-Verlag, 1990.)

WHITE ROT FUNGI

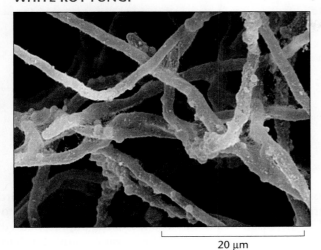

20 μm

A key member of the white rot fungi is *Phanerochaete chrysosporium*, whose mycelial hyphae are shown in this scanning electron micrograph. The 30 million base-pair genome of this important basidiomycete fungus has been sequenced (*Nature Biotechnology* 22, 695–700 (2004)), and its genome reveals an impressive array of genes encoding secreted oxidases, peroxidases and hydrolytic enzymes that cooperate in wood decay. (Photograph courtesy of Fred Michel).

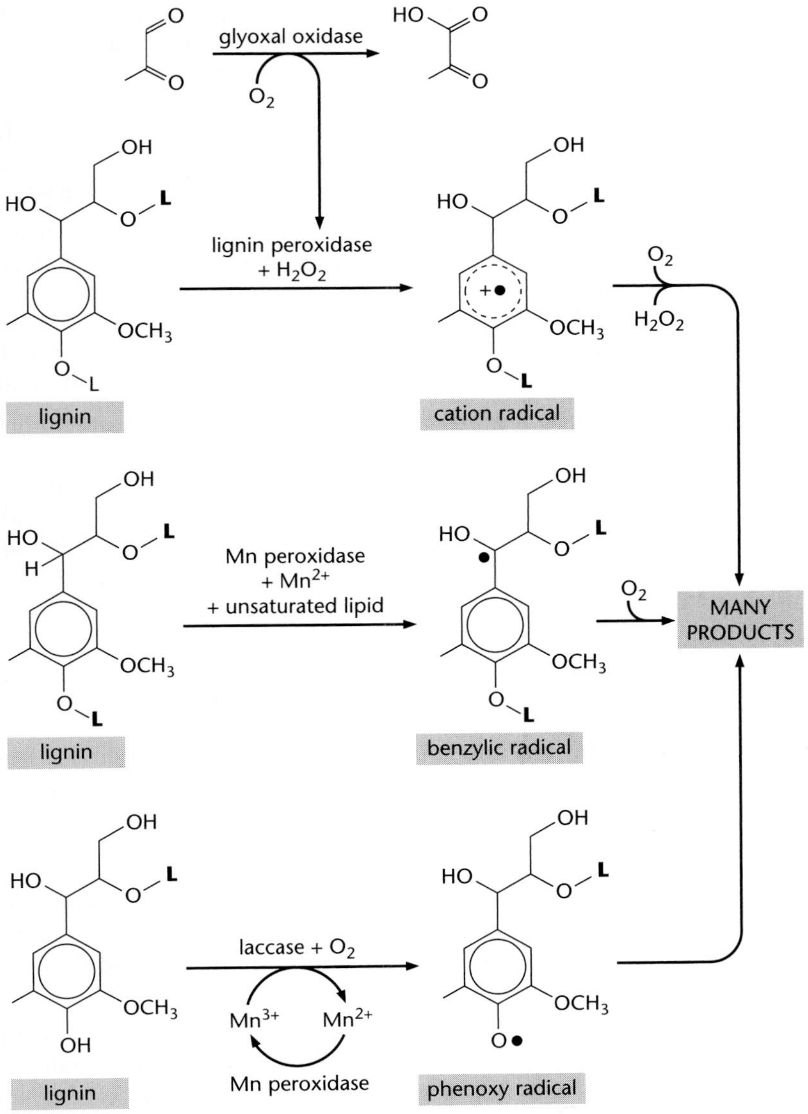

glyoxal oxidase

O_2

lignin peroxidase + H_2O_2

O_2

H_2O_2

lignin

cation radical

Mn peroxidase + Mn^{2+} + unsaturated lipid

O_2

MANY PRODUCTS

lignin

benzylic radical

laccase + O_2

Mn^{3+} Mn^{2+}

Mn peroxidase

lignin

phenoxy radical

White rot fungi produce three kinds of lignin-degrading phenoloxidases that attack lignin in a nonselective manner: lignin peroxidase (LiP), a heme-containing enzyme with a high redox potential; manganese-dependent peroxidase (MnP), another heme-containing enzyme; and laccase (Lac), a blue multi-copper oxidase. Both LiP and MnP require H_2O_2, and Lac requires O_2 to generate free radicals in the lignin subunits, and thereby induce spontaneous cleavage of the lignin polymers. White rot fungi are also used for *bioremediation* of persistent environmental pollutants such as polychlorinated biphenyls (PCBs) and dioxin, because of the strong oxidative activity and low substrate specificity of their lignolytic enzymes.

TABLE 9.4 RELIANCE ON NONRENEWABLE FOSSIL FUELS AS ENERGY SOURCES (2005)	
ENERGY SOURCE	**SHARE OF WORLD OUTPUT**
Oil	35%
Gas	21%
Coal	23%
Nuclear	7%
Other (renewable)*	14%

*Approximately 11% of which is biomass, mainly wood.

fertilizers, to increase the base of manufacturing and services, to transport people and goods around the world at high speed and low cost, and even to wage global wars. The increased wealth has also led to the unprecedented size of the human population, to faster rates of innovation, and to urban sprawl—and, above all, to vast increases in the consumption of energy. To sustain these changes, humans have commandeered ever-increasing amounts of the earth's natural resources. The rate of consumption of many of these natural resources, particularly fossil fuels, is unsustainable. Many of the critical natural resources that we have taken for granted for the past 100 years are projected to become scarce and expensive. Increasing numbers of economists, engineers, and scientists now believe that humans have about 50 years to make the transition to renewable energy–based societies to avoid a catastrophic breakdown in living conditions around the world.

The term *peak oil* has become a focal point in the emerging debate about the finite nature of natural resources on planet Earth and the need to develop renewable energy resources, especially for transportation. The peak oil theory predicts that conventional oil production will peak in the near future, and that new sources of fuels for transportation will have to be developed, for example, from unconventional fossil resources and biofuels.

Plant cell walls constitute the single largest source of renewable biomass suitable for conversion into *biofuels* and industrial chemicals. They are destined to play an increasingly important role in our energy and industrial future. Not only will plant cell walls provide starting material for making significant quantities of biofuels at competitive prices, but equally important, the substitution of biofuels for fossil fuels will reduce CO_2 emissions and thereby ameliorate global climate change.

2. Plant cell walls and their derivatives comprise a major fraction of the terrestrial biomass and carbon content.[ref17]

Plant cell walls are the major source of *biomass* in plants, particularly woody plants, and the organic material remaining in litter and humic soils. Humus is plant-derived biomass which still retains some biological organization. The upper meter of soil on Earth, where the organic material is largely undecomposed plant material, represents about 1500 Pg (**Table 9.5**), and typically turns over in less than a century. Lignin, cutin, and suberin are major cell wall components resistant to decay. Some plant materials (for example, hemicelluloses) are degraded more rapidly, and some derived materials, such as coal, persist for hundreds of millions of years.

The world's forests can be an important sink for storage of carbon, when biomass gain by tree growth exceed losses due to destruction or degradation. Forests cover about 30% of the terrestrial surface, comprising about 4 billion hectares (a hectare is 10,000 square meters, or 2.47 acres). There has been a recent increase in atmospheric carbon of anthropogenic origin, based on use of fossil fuels and burning of tropical forests. However, carbon storage in vegetation has increased considerably since the last glacial maximum (about 20,000 years ago) due at least in part to an increase in temperate and boreal forests. A large portion of the net ecosystem production (annual NEP) of carbon comes from boreal and tropical forests. Half of the carbon of our forests is in soils and litter, and about half is in living trees. The total carbon found in forest ecosystems is roughly equivalent to the carbon content of the atmosphere.

Much of the world's carbon is found in fossil fuels, particularly *oil* and *coal*. Oil (liquid petroleum) comes largely from marine algae; it is rich in hydrogen, and is fluid and mobile. Coal is the metamorphic remains of terrestrial vegetation and of humus in soil. Decay-resistant plant material, including lignin, forms peat, which is the incipient stage of coal formation. Coal is typically formed when peat formed in swamps subsides and is compressed by water and sediment, followed by heating and removal of water. Compressed peat gives rise to the relatively soft brown coal called lignite. Higher ranks of coal have increased hardness and lower concentrations of organic volatiles. In some cases, about 10 feet of peat compressed over hundreds of years can form about one foot of soft coal. Further compression of soft coals over much longer time periods gives rise to hard coals. Some coals are derived in considerable part from suberin (suberain), some from wood (xylain); and others, like anthracite, a hard coal, come primarily from lignin. Coal is also an important source of ancient fossils of plant material. Plant fossils in coal are often derived from a wide variety of plant parts including leaves, stems, roots, bark, and some reproductive structures.

Coal has been formed in many locations in the world since the Carboniferous era (354 to 290 million years ago) when the vegetation from vast swamps produced rich deposits of coal. Coal deposits in the Eastern USA are typically from the Carboniferous. Major coal deposits have also been formed in many regions of the world during the subsequent major epochs. In the Western United States most of the coal was formed during the Cretaceous era (140 to 65 million years ago). Some coal deposits are only a million years old.

3. Photosynthesis-driven biomass production by plants constitutes a major and sustainable energy source.[ref18]

Trees are very efficient at fixing carbon. About 3×10^{10} tons of wood are produced each year from fixed carbon. A pine forest can fix 3 kg of dry matter

TABLE 9.5 COMPONENTS OF GLOBAL CARBON	
COMPONENT	AMOUNT (Pg*)
Total world carbon (primarily in carbonate rock)	1 million
Atmospheric carbon (primarily in CO_2)	730
Terrestrial carbon in all types of soils and vegetation	2150
Forest biomass carbon in vegetation	321
Forest ecosystem carbon	638
Ocean total	40,000
Ocean organic carbon	2500

*1 petagram (Pg, 10^{15} g) is equal to one billion metric tons (1 Gt).

per square meter per year (30 tons/hectare). Only certain grasses that use C_4 photosynthesis, such as sugarcane, switchgrass, and *Miscanthus*, are significantly more efficient, and it is these grasses together with fast-growing trees such as poplar and *Eucalyptus* that are being developed as crops for *biofuel production*.

The Department of Energy and the United States Department of Agriculture recently concluded that the United States has sufficient land to produce over 1 billion dry tons of biomass annually on a sustainable basis, enough to displace about 30% of the current consumption of liquid transportation fuels. Of these, about 368 million dry tons would come from forests (33% of land area), and about 998 million tons from agriculture (46% of land area) providing a total of 1366 million tons. Of the estimated 368 million dry tons of forest biomass, about 140 million tons are currently being exploited for bioenergy and bioproducts. Perennial crops reduce the energy inputs needed for biomass production. The goal set for biofuel researchers in the United States is to produce 10 tons of biomass per acre (24.7 tons/hectare) with an ethanol yield of 80 to 100 gallons of ethanol per ton (360 to 450 liters/ton), and to replace 30% of the fossil liquid transport fuels by 2030. Experimental fields of switchgrass and existing pine forests have already reached the biomass production goal, but many more advances are needed to make the conversion of biomass to ethanol an economic alternative to fossil fuels in the United States. The emerging biofuels industry also has the potential for producing new employment opportunities for significant numbers of people. Current estimates suggest the creation of 10,000 to 20,000 jobs for every billion gallons (4.5×10^9 liters) of ethanol production.

4. Half of the wood harvested around the world is burnt as fuel.[ref18]

Over half of the world's consumption of wood is as fuel, and the predominant use is in simple stoves or ovens and open fires. About one-third of the world population continues to depend on the energy from plant cell walls for most of its energy needs. Most commonly, wood is collected and utilized directly for cooking and heating in individual dwellings. In some developed countries such as Sweden, plantations of fast-growing clones of *willow* (*Salix* spp.), with a cycling time of 4 years, produce wood exclusively for use in central heating systems to generate steam or to generate electricity in community energy projects. Yields of 15 tons of dry matter per hectare per year have been reported. In Brazil, China, and South Africa, *eucalyptus* is the preferred heating fuel crop (Figure 9.37C). In Brazil, large volumes of *Eucalyptus globulus* wood are used to produce charcoal for the steel industry.

The heat value of wood depends upon its density, chemical composition, and dryness. Lignin has a higher intrinsic Btu content as fuel than cellulose, and future breeding programs might strive to increase lignin content in addition to wood density in trees used for heating purposes. Resin content is a significant factor, because its fuel value is twice that of ordinary wood, even higher than lignin. However, both lignin content and wood density are often negatively correlated with growth rate.

5. Producing bioethanol from lignocellulosic biomass at competitive prices will require significant investments in research.[ref19]

The main biofuel efforts now are focused on converting biomass to *ethanol* and other liquid fuels. Most of the ethanol used as a fuel additive still comes from the bioconversion of sugars and starch, two plant products that are also central to our food supply. Whereas the use of *sugarcane* (**Figure 9.35**)

(A)   (B)

in tropical countries such as Brazil can be economically and energetically justified, the starch-to-ethanol conversion industry in the United States cannot, because corn starch is a critical component of human food and animal feed, besides having a minimal net energy gain and high cost of production. For this reason, the focus has shifted in the United States toward making ethanol and other liquid fuels from plant cell walls (wood or grass stalks), which are referred to as *lignocellulosic* or *cellulosic biomass*.

The envisaged biomass production program in the United States will have to rely on many different plant species because of the different environments in which the plants will be grown (**Figure 9.36**). In the Western states bordering the Pacific and in Hawaii, hybrid *poplars* and eucalyptus, respectively, have been identified as the most promising biomass crops, whereas in the Eastern half of the country different combinations of *switchgrass* (*Panicum virgatum*), *Miscanthus* (*Miscanthus x giganteus*), hybrid poplars, and willows as well as other trees and grasses have the greatest potential (**Figure 9.37** and **Figure 9.38**). Systematic breeding of these crops has begun only recently. Some progress has also been made in modifying the growth and the energy content of poplars and aspen by genetic means. Mixing of cultivars appears to enhance both the yield and the sustainability of biomass from prairie grasses such as switchgrass.

To reach the goal of producing 30% of transportation fuels from biomass will require major progress in the following fields of research and technology development: (1) breeding of biomass crops capable of fast growth on non-

Figure 9.35 Harvesting of sugarcane in (A) the nineteenth and (B) the late twentieth century. (Courtesy of the Agricultural Research Service, USDA.)

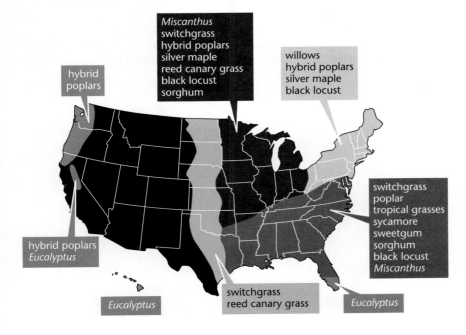

Figure 9.36 Geographic distribution of potential biomass crops in the United States. Due to differences in soil, precipitation, and climate, different biomass crops will have to be grown in different geographic regions to maximize production. (Adapted from Oak Ridge National Laboratory Biomass Program.)

(A)   (B)   (C)

Figure 9.37 Three lignocellulosic biomass crops being developed by the liquid biofuels industry in the United States. (A) Switchgrass, a prairie grass native to North America. (B) Short-rotation hybrid poplar trees (6–7 years old) at the time of harvest. A harvester known as a feller buncher cuts trees and lays them down in bunches. The felled trees are de-barked and chipped on the spot prior to transport to the factory for chemical processing. (C) Eucalyptus trees (4 years old) growing on the Big Island of Hawaii. Hybrid poplars and eucalypts are also feedstocks for the paper industry. (Courtesy of the Department of Energy National Renewable Energy Research Laboratory.)

irrigated land and of maximizing production of biomass, while minimizing input of energy and fertilizer; (2) development of biomass crops with cell walls that can be readily accessed by liquid fuel–producing enzymes and microorganisms; (3) development of low-cost and efficient enzyme systems for breaking down the diverse types of cell wall molecules and for producing ethanol; and (4) simplification and optimization of the biomass processing technologies for liquid fuel production.

6. Overcoming the lignin barrier is essential to the success of the lignocellulosic biomass–based liquid biofuel industry.[ref20]

The major hurdle in using lignocellulosic biomass such as *corn stover* (the residual aboveground plant material after harvesting of the ears, see also Figure 8.52), grasses, or wood for ethanol production is the *lignin barrier*. This barrier consists of highly cross-linked lignin polymers (see Concept 5E12) that completely encompass the cellulose fibrils and hemicelluloses and thereby limit the access of chemicals or cellulose-degrading enzymes to the cellulose fibrils and cellulose molecules. Making the lignin barrier more permeable to the cellulose-degrading enzymes is one of the key problems that have to be solved before a viable liquid biofuel industry based on lignocellulosic biomass can be developed.

Figure 9.38 Switchgrass bale. The harvesting of switchgrass can be done with conventional farm equipment, which makes it easier for farmers to convert to switchgrass farming for liquid biofuel production. (Courtesy of the Department of Energy National Renewable Energy Laboratory.)

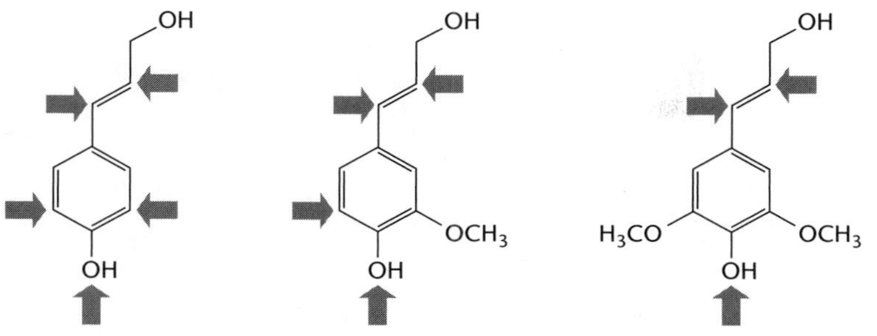

p-coumaryl
alcohol

coniferyl
alcohol

sinapyl
alcohol

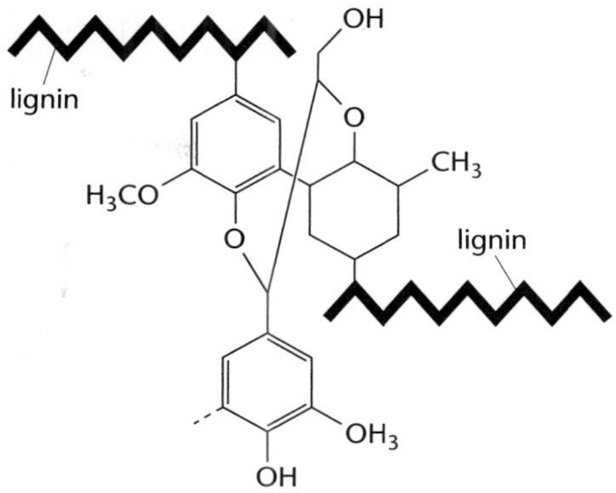

dibenzodioxocin, a cyclic ether unit
with 5-5 and β-O-4 linkages

Two types of solutions for overcoming the lignin barrier are being investigated: reducing the lignin content of the cell walls, and altering the cross-linking pattern between the lignin subunits. Because both the pulp and paper industry and animal feed researchers have been pursuing these same goals for many years, considerable progress has already been made. Thus, cell walls containing lesser amounts of lignin and/or an altered lignin composition have been found to increase the pulping efficiency of wood and to improve the digestibility of forage crops. Thus, lignin can be made more readily hydrolyzable during Kraft and other alkaline pulping by increasing the content of the more highly methoxylated sinapyl alcohol lignin precursors, which produces a less condensed polymer, that is, lignin containing fewer C-C linkages (**Figure 9.39**). The extent to which such lignin changes might make plants carrying these traits more susceptible to mechanical damage, and/or more susceptible to disease, is a concern, but the great variation in lignin content and composition among plants suggests that substantial changes may be tolerable (see Concept 5E11).

Many genetic tools are now available to enhance or suppress gene expression of individual enzymes or groups of enzymes involved in lignin biosynthesis and thereby manipulate the lignin content and composition. However, any given genetic manipulation has the potential for producing multiple changes in cell wall composition and structure. In *Arabidopsis*, a mutation in a gene encoding coniferaldehyde-5-hydroxylase (CAld5H, also called ferulate-5-hydroxylase or F5H) dramatically changes the composition of the lignin to resemble that of gymnosperms (see Figure 5E2.1 for an overview of the lignin biosynthesis pathway). Gymnosperm lignin is rich in guaiacyl units (G units). By blocking conversion of coniferaldehyde to 5-hydroxyconiferaldehyde through mutation of the 5-hydroxylase, the level of sinapyl alcohol subunits leading to syringyl subunits (S units) in lignin can be greatly reduced. In contrast, an increase in the activity of the CAld5H enzyme produces the opposite effect, that of an increase in S units. These results have been applied to alter the wood of aspen. An increase in S units in lignin is correlated with increased wood pulping efficiency, which should result in improved yield and reduced environmental impact.

In tobacco, modification of the lignin biosynthesis pathway has produced both altered lignins and changes in other metabolic products. For example, a reduction in the enzyme activity for the first step in the phenylpropanoid pathway, phenylalanine ammonia-lyase (PAL), reduced lignin content, increased the S/G ratio, and changed other nonlignin phenylpropanoid products. In poplar, suppression of caffeate-*O*-methyltransferase (COMT) not only decreased the S/G ratio, but also altered wood color (pale rose), whereas reduction of cinnamyl alcohol dehydrogenase (CAD) produced a red-colored wood without affecting the S/G ratio. When researchers suppressed the expression of 4-coumarate CoA ligase in aspen, a dramatic

Figure 9.39 Comparison of the predominant sites of linkages (*arrows*) of the monolignols *p*-coumaryl alcohol, coniferyl alcohol, and sinapyl alcohol in natural lignins. Note that *p*-coumaryl alcohol possesses more potential cross-linking sites than coniferyl alcohol and sinapyl alcohol. This is why lignin polymers rich in sinapyl alcohol–derived S subunits are less condensed than lignin polymers rich in *p*-coumaryl alcohol–derived G subunits. The cyclic ether dibenzodioxocin creates a branched structure between lignin subunits that may reduce the rate of enzymatic degradation.

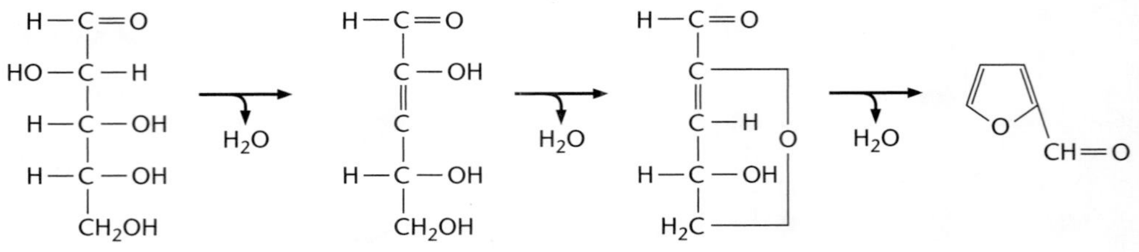

Figure 9.40 Conversion in acid of D-xylose to furfural (2-furaldehyde) by loss of 3 H_2O. (Adapted from W.A. Bonner and M.R. Roth, *J. Am. Chem. Soc.* 81:5454–5456, 1959.)

xylose   furfural

reduction in lignin content was observed, often associated with a significant increase in growth. Reduction of expression of cinnamoyl CoA reductasae (CCR) in alfalfa resulted in reduced lignin and as much as a 60% improvement in saccharification efficiency. These results emphasize the potential for genetic engineering to modify plant cell walls in useful ways, while also highlighting the many challenges that will have to be overcome to realize the full benefit of these new technologies.

7. Major improvements in fermenting enzyme systems are needed to improve the efficiency of ethanol production.[ref18]

Hemicelluloses have a deleterious effect on fermentation efficiency of cellulose. This effect is a significant problem in liquid fuel production because hemicelluloses represent about 20% of the lignocellulosic biomass. Current fermentation technologies are based on the conversion of 6-carbon sugars such as glucose to ethanol. Hemicelluloses such as xylans (see Concept 3B6) are composed of xylose (a C_5 sugar) that can form furfural, which is toxic to yeast (**Figure 9.40**). One strategy for overcoming this problem involves the development of yeast or other ethanol-producing organisms that are less sensitive to toxic fermentation intermediates and which can utilize C_5 sugars. Another strategy is to develop plants in which xylans are replaced by hemicelluloses that contain lesser amounts of C_5 sugars. Several genes controlling the biosynthesis of hemicelluloses have been identified, and the retailoring of hemicellulose biosynthesis, while challenging, appears to be a possible task.

Major progress has been achieved in the production of enzymes, for example, cellulases, required for the production of ethanol from lignocellulosic biomass. Many highly efficient enzymes that can function at elevated temperatures have been identified in a variety of thermophilic microorganisms from around the world. The genes encoding these enzymes have been transferred to yeast and bacteria to produce the enzymes in industrial quantities. The discovery and characterization of multienzyme wall-degrading complexes called cellulosomes in bacteria such as *Clostridium thermocellum* (see also Figure 8.62) has opened up new opportunities for the design of novel enzyme complexes optimized for the degradation of specific types of cell walls (see **Panel 9.2**, The Cellulosome, Figure 1).

8. New types of biorefineries are needed to produce ethanol from lignocellulosic biomass at increased efficiencies and lower cost.[ref21]

To replace one-third of the liquid fuels in the United States with ethanol will also require major improvements in the efficiency of the lignocellulosic biomass conversion process. The two main goals of the designers of the next-generation biorefineries are to reduce the number of processing steps and to reduce the energy input. The envisioned processing steps are diagrammed in the flowchart of **Figure 9.41**.

CELLULOSOMES AND CELLULASE

Plant cell walls contain a multitude of cell wall molecules that are cross-linked into three-dimensional interwoven networks to provide mechanical strength to the tissues and to resist degradation by pathogenic organisms. However, due to their high content of polysaccharides, plant cell walls also offer microorganisms an extraordinary source of carbon and energy. To gain access to this energy source, several wood-degrading microorganisms have evolved specialized multienzyme complexes, termed *cellulosomes*, which are remarkably efficient at degrading plant cell walls. Scientists interested in liquid biofuel production are now working on the design of improved cellulosomes capable of degrading biomass from different sources into fermentable sugars and ethanol.

CELLULOSOME STRUCTURE

The structural elements of a cellulosome include a scaffold, with paired docking and enzyme-binding subunits (see figure below). The scaffoldin protein contains a single cellulose-binding domain and nine repeat domains that serve as binding sites for the enzymes. These binding sites are called cohesins, and each cohesin domain can bind one enzyme via a dockerin-type subdomain. Because all of the cohesin and all of the dockerin domains have the same binding specificity, binding of the enzymes to the scaffoldin backbone is random.

NATIVE CELLULOSOME

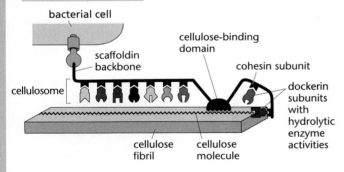

A second line of research is designed to individualize the "generic" binding domains on the cohesin and dockerin repeats. By doing so, it will become possible to place individual enzymes in defined positions along the scaffoldin backbone and thereby determine the optimal spatial relationship between pairs or groups of enzymes. By lengthening or shortening the scaffoldin backbone, the number of enzymes combined into a given cellulosome-type complex, and thus the size of the complex, can also be altered. Finally, efforts are under way to identify the best biological systems (yeast, bacteria) for producing the designer cellulosomes in industrial quantities.

References: E.A. Bayer, J.P. Belaich, Y. Shoham, and R. Lamed, *Annu. Rev. Microbiol.* 58:521–554, 2004; and H-P. Fierobe et al., *J. Biol. Chem.* 280:16326–16334, 2005.

Cellulosomes of the anaerobic thermophilic cellulolytic bacterium *Clostridium thermocellum* consist of a backbone-like scaffold protein, termed scaffoldin, with multiple binding sites for complementary cell wall–degrading enzymes including different cellulases, hemicellulases, and carbohydrate esterases, as well as a cellulose-binding domain. The cellulose-binding domain serves two functions: to pull individual cellulose molecules out of the surface of crystalline cellulose fibrils to render them accessible to endoglucanases, and to anchor the enzyme complex to the cellulose substrate during cell wall degradation (see figure below). In turn, the coupling of multiple hydrolytic enzymes into one large complex provides a means for the enzymes to work synergistically by increasing local concentrations of substrate as they degrade the different types of cell wall molecules and their intermediate breakdown products.

Cellulase enzyme with a cellulose-binding domain.

DESIGNER CELLULOSOMES

To produce the next generation of cellulosomes for industrial uses, scientists are modifying the individual parts of the complex with the goal of increasing the general enzymatic efficiency of the complex, and of producing designer-type cellulosome complexes optimized for different feedstocks. Much work has gone into identifying novel and more efficient wood-degrading enzyme systems to increase the general efficiency of the complexes. These proteins are then fused to dockerin-type binding domains so that they can be bound to scaffoldin via the cohesin domains.

DESIGNER CELLULOSOMES

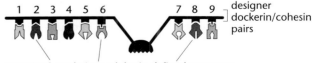

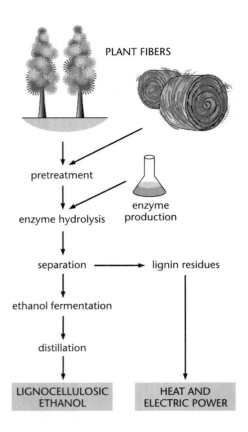

PLANT FIBERS

pretreatment

enzyme hydrolysis

enzyme production

separation ——→ lignin residues

ethanol fermentation

distillation

LIGNOCELLULOSIC ETHANOL

HEAT AND ELECTRIC POWER

Figure 9.41 Outline of a next-generation biomass-to-ethanol conversion scheme. It is hoped that by reducing the number of processing steps and by increasing efficiency, the unit cost of ethanol production can be substantially reduced.

The biomass conversion barrier, often referred to as *recalcitrance* of the lignocellulosic feedstock, constitutes one of the major obstacles to making the biomass conversion process more economical. Currently, the most effective method for rendering the wood chips permeable to cellulase enzymes is to heat the chips in acid (2% sulfuric acid to 150°C for 20 minutes). During this heating, the residual monomeric and small oligomeric lignin molecules appear to melt or vaporize, and the molten lignin oozes to the surface of the chip, where it forms discrete droplets (**Figure 9.42A**). The cavities left behind by the displaced lignin are sufficiently large to allow the celluloytic enzymes to diffuse through all of the cell wall layers (that is, through the S3 to the S1 and S2 layers) of the wood chips (see Figure 9.26 for the spatial organization of the secondary cell wall layers) and thereby gain greater access to all of the cellulose fibrils (**Figure 9.42B and C**). This is important because the S2 layer contains the largest amount of cellulose. Provided that the breeding and propagation of hardy, fast-growing plants with modified lignin in their walls can be achieved, it may be possible to eliminate the pretreatment and increase the efficiency of the wood chip degradation process and the yield of ethanol. Similarly, novel cellulosome-type multienzyme complexes with more efficient enzymes could further advance the process. The results showing that reduced lignin could allow for the elimination of an acid pretreatment step are also encouraging.

9. Plant cell walls are a renewable source of industrial chemicals.[ref22]

The chemical industry generates a small number of "platform" chemicals such as ethylene, propylene, and benzene that are used to manufacture a large variety of chemical products, including polypropylene, acrylic fibers, rubber elastomers, polyesters, and polyurethane. For more than a century, these carbon-based feedstocks have been derived from fossilized forms

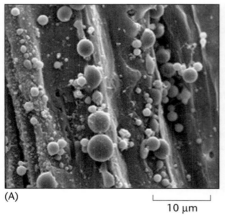

(A)

|← 10 μm →|

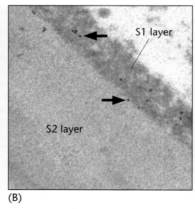

S1 layer

S2 layer

(B)

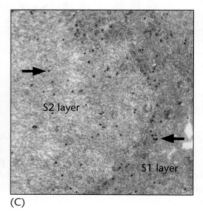

S2 layer

S1 layer

(C)

Figure 9.42 Effects of heating of wood in dilute acid on lignin distribution and enzyme permeability of secondary cell walls. A) Scanning electron micrograph of internal cell wall surfaces of a fractured corn stover stem heated in 2% sulfuric acid to 150°C for 20 minutes. The round droplets are lignin polymers that melted during the heat treatment and were driven to the free surface of the S3 cell wall layer (see Figure 9D2.2/9.26 for secondary cell wall nomenclature) by the pressure produced by the vapors of volatile cell wall components. As demonstrated in the electron micrographs of the immunolabeled thin sections (B, C), the void spaces in the cell walls created by the displacement of the lignin provide channels for the entry of the celluloytic enzymes (labeled with antibodies carrying gold labels seen as black dots) into the interior of the cell walls. Note that in the control sample (B), which was heated to only 120°C, the enzymes (*arrows*) were able to penetrate the S1 layer but not the S2 layer, whereas in the sample heated to 150°C (C), enzymes are seen throughout the S1 and S2 layers. (Courtesy of Todd Vinzant and Bryon Donohoe, Department of Energy National Renewable Energy Laboratory.)

of carbon. This strategy is currently being challenged by the high cost of oil. It is also inherently nonsustainable. Thus, new, reliable, and renewable sources of these chemicals will have to be found to enable the industry to continue to produce many of the critical products on which modern societies depend. Lignocellulosic biomass will likely be the renewable source of carbon-based chemical feedstocks in the future. It is relatively inexpensive, grows quickly, and appears to be available on a sufficiently large scale to support the needs of the chemical industries.

Before the twentieth century, a large fraction of industrial chemicals were derived from wood. Wood extractives produce turpentine and rosin. Thermal treatment produces gases, tars, and charcoal, while sugars are obtained by hydrolysis. The extraction of tannins and turpentine from wood constituted major American industries in the nineteenth century, and fermentation used to be the method of producing ethanol and metabolites. The production of industrial chemicals from wood and other lignocellulosic materials declined due to the predominance of cheaper petroleum, but could easily become significant again. Some familiar chemical compounds are derived from wood. Vanillin, the primary flavor component of vanilla bean extract, is made from wood, as a derivative of guaiacyl lignin units.

To build a biomass-based chemical feedstock industry to meet future needs will require a sustained research effort and major investments in process development. At present there is no comprehensive framework for industrial production of chemicals from biomass, but recent efforts in this field have yielded positive results. For example, it is now possible to use biological conversion methods to produce some chemical feedstocks from biomass at competitive prices, but significant infrastructure investments are still needed to make industrial quantities of these products by newly developed biomass-based processes.

Many potential biobased chemicals have relatively high oxygen to carbon ratios; therefore significant removal of oxygen (chemical reduction) is required for conversion to industrial products. In general, removal of oxygen is energy-intensive and therefore is relatively expensive. However, microbial fermentation methods have the potential to make many of the processes commercially viable. Currently, a few chemicals derived from biomass are in large-scale production, most notably glycerol, lactic acid, and 1,3-propanediol.

One promising approach for reducing the cost of biomass-based chemical feedstock production is to couple the production of sugar-derived chemicals in a biorefinery with the extraction of other types of chemicals and materials. Corn mills and pulp and paper mills are examples of biorefineries that can produce combinations of food, chemicals, power and other industrial and consumer products. One class of chemicals that could help drive the development or expansion of biorefineries (**Table 9.6**) is sugar-derived chemicals. In some instances, such chemicals can be derived by simple modifications of sugars. Examples include the production of xylitol and arabinol, two 5-carbon alcohols, from xylose and arabinose, or of sorbitol (glucitol) from glucose. Other feedstock chemicals that can be produced by fermentation are well-known metabolites such as succinic acid (**Figure 9.43**), glutamic acid, and aspartic acid.

Another goal is to develop biobased, alternative chemical feedstocks that could replace some of the current petroleum-based chemicals over time. Progress to date suggests that glucaric acid could replace adipic acid as a monomer for nylon production, and 2,5-furan dicarboxylic acid could be used for the synthesis of polyester. Such changes in feedstock composition may also lead to polymers with new or improved performance characteristics. On the downside, the introduction of a new chemical feedstock into an existing industry is expensive, typically on the order of $1 billion.

TABLE 9.6 POTENTIAL PLATFORM CHEMICALS FROM BIOMASS

C UNITS	ORIGIN	CHEMICAL	PROJECTED OR KNOWN USE
C_3	Oils, transesterification	Glycerol	Diverse products, solvents, polymers
	Fermentation	3-Hydroxy propionic acid	Fibers, polymers, acrylates
C_4	Fermentation	Succinic acid	Fibers, solvents
	Fermentation	Fumaric acid	Fibers, solvents
	Fermentation	Malic acid	Fibers, solvents
	Fermentation	Aspartic acid	Sweeteners, novel polymers
	Starch oxidation	3-Hydroxy-butyrolacetone	Pharmaceuticals, solvents, fibers
C_5	Fermentation	Itaconic acid	Copolymers, nitrile latex
	Xylose	Xylitol	Sweetener, resins, polymers
	Arabinose	Arabinitol	Sweetener, resins, polymers
	5- or 6-C sugars	Levulinic acid	Oxygenates, solvents, polymers
	Fermentation	Glutamic acid	Polyesters, polyamides
C_6	Glucose	Sorbitol	Commodity chemicals, sweetener
	Starch oxidation	Glucaric acid	Solvents, nylons
	Chemical C_6 sugars	2,5-Furan dicarboxylic acid	Plastics, polyesters, nylons

After T. Werpy and G. Peterson (eds.), Top Value Added Chemicals from Biomass, vol. 1: Results of Screening for Potential Candidates from Sugars and Synthesis Gas. Washington, DC: US Department of Energy, 2004.

10. Large-scale biofuel and chemical feedstock production is likely to affect food prices and biodiversity.[ref23]

Production (and use) of liquid transportation fuels and chemical feedstocks derived from biofuel crops has the potential to significantly raise the sustainability of worldwide energy consumption and therby reduce the amount of greenhouse gases currently released by human activities. However, producing liquid fuels and chemicals from trees and perennial grasses is advantageous only when these plants are grown on marginal lands and therefore do not compete directly for land currently used for food. Even the conversion of marginal agricultural land to biofuel production could have significant long-term effects on global food production and food prices. History has shown that when food supplies are limited, even small changes in the availability of specific foods can lead to major changes in food prices. Considering the growth of the human population, demands for more meat and nutritional foods, climate change, and the loss of water supplies, biofuel crops are likely to compete for land resources needed for future food production. We cannot predict how the resulting land use conflicts will be resolved, but one can foresee the need for establishing broad criteria and guidelines for deciding how to allocate finite land resources between different types of enterprises that produce food, fiber, building materials, paper, biofuels and chemical feedstocks on a sustainable basis.

The ecological effects of the conversion of large tracts of public and private nonagricultural land to biofuel crop production have yet to be realized and evaluated. The production of liquid transportation fuels from lignocellulosic biomass holds great promise for the future, but the scale of production that is required is unlikely to be achieved without substantial effects on the world economy, the world ecology, *biodiversity*, and the natural resources

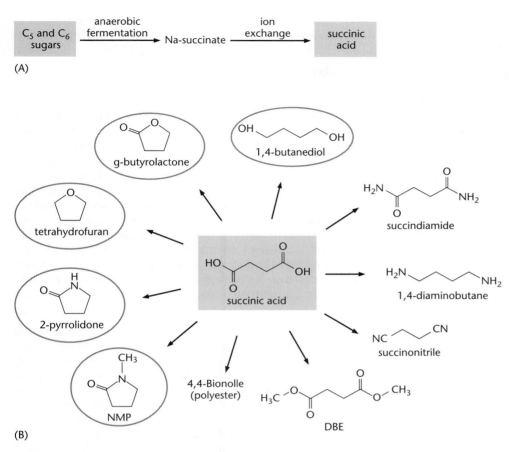

Figure 9.43 Potential uses of succinic acid, a chemical feedstock produced from C5 and C6 sugars. (A) Industrial processes involved in the production of succinic acid from C5 and C6 sugars. (B) Chemical derivatives (circled compounds) of succinic acid that are used at commercial commodity levels today. NMP is N-methyl-2-pyrrolidone, and DBE is dibasic acid ester, a mixture whose main ingredient is dimethyl succinate (shown). The diacids may be directly made into polymers, for example, Lycra, or modified to produce derivatives, including specialty chemicals. (Adapted from T. Werpy and G. Peterson, eds., Top Value Added Chemicals from Biomass, vol. 1: Results of Screening for Potential Candidates from Sugars and Synthesis Gas. Washington, DC: US Department of Energy, 2004.)

of the planet. Although the biofuels initiatives in the United States and in many other countries are driven by the necessity to develop alternative sources of liquid fuels, they may constitute global-scale experiments with an uncertain outcome.

11. Cell walls offer unique challenges and opportunities for genetic engineering.[ref24]

The human uses of plant cell walls began when undomesticated plants were collected for food and materials. Subsequent domestication involved the genetic selection and cultivation of plants for improved properties affecting cell wall structure and composition related to seed or growth characteristics and crop management. Modern plant breeding, applying quantitative genetics to domestic crops, has resulted in great improvements in yield and quality. However, breeding was limited to our ability to exploit natural genetic variation, and by the length and number of breeding cycles required to produce a new plant variety. *Genetic engineering* modifies plants through the introduction of specific genes using methods that are faster and more precise and can extend beyond species barriers. Much is being gained by integrating genetic engineering and traditional breeding to provide dependable crops with novel properties. The potential benefits are currently limited by our lack of fundamental information about the genes and proteins that direct the synthesis and assembly of plant cell walls, and by the dearth of knowledge about the diverse types of cell walls found in the biosphere.

There is great potential for manipulation of cell wall properties, because cell wall composition and morphology are determined by a specific set of gene products acting during plant growth and development (**Figure 9.44**).

In principle, any gene affecting cell wall properties may be modified to increase or decrease expression, or to alter the products. Potentially, genes from animals, fungi, algae, and bacteria may also be harnessed. Synthetic genes for which there are currently no known natural counterparts could also improve cell wall or other plant characteristics.

The use of genetic engineering technologies to modify the properties of plants has been controversial, with many people viewing this technology as a potential threat to human health and to the environment. The number of farmers planting genetically engineered crops has increased to over 13 million in 2009 (12 million of which were small farmers in developing countries) and the accumulated global biotech crop area reached approximately 1 billion hectares (2.5 billion acres, which is greater than the land area of the contiguous United States). Remarkably, none of the dire predictions of the detractors of genetically engineered crops have materialized. To the contrary, biotech crops have benefited not only farmers (better yields at lower costs), but also the environment (less use of pesticides to fight pests and pathogens; less loss of topsoil) and consumers (healthier, higher-quality foods). With this track record, we can expect the number of such crops to rapidly expand in the future, with each new crop being evaluated on its own merits.

We can also expect genetic engineering to play an increasingly important role in the development of non-food plants such as those used for biofuel production, for environmental cleanup, and for other purposes. Some modified plants are able to remove mercury and other heavy metals as well as ameliorate toxicity and salinity of soils. Fast-growing trees can also sequester more carbon, and the planting of large numbers of such trees could contribute to an offset of anthropogenic factors affecting global climate. At the same time, we should not underestimate the effects of such large plantations on biodiversity. But here too, genetic technologies might help, since such technologies can also be used to monitor and maintain biodiversity, and even to rescue plant species on the brink of extinction.

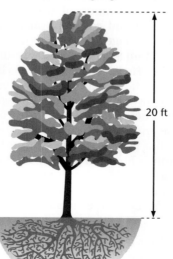

PHOTOSYNTHESIS AND CARBON ALLOCATION

- increased photosynthesis
- optimized photoperiod response
- optimized crown and leaf architecture
- greater carbon allocation to stem diameter vs. height growth

BIOMASS

- controlled and readily processable cellulose, hemicellulose, and lignin
- tailored biomass composition with value-added chemicals
- enhanced biomass production per acre by manipulation of photomorphogenic responses

TOLERANCE AND SUSTAINABILITY

- pest and disease resistance
- drought and cold tolerance
- floral sterility
- regulated dormancy
- delayed leaf senescence
- optimal nutrient acquisition and use
- rhizosphere and microbial community health

20 ft

Figure 9.44 Traits of poplar that are being targeted to improve its usefulness as a biomass feedstock for bioethanol production. (Adapted from A.J. Ragauskas et al., *Science* 311:484–489, 2006.)

Plant cell walls constitute one of the primary renewable resources on planet Earth. Future generations will have to rely more heavily on renewable resources, including plant cell walls, for producing the sustainable economies of the future. We hope that this book will foster a better understanding of the physical, chemical, and structural properties of plant cell walls, as well as of the many ways in which they have contributed to human societies in the past, and how they might serve us in the future.

Key Terms

agave
Aloe vera
apple pomace
bast fiber
bioactive pectin
biodiversity
biofuel production
biomass
biopulping
bioremediation
brown midrib phenotype
cancer cell apoptosis
cancer therapy
cellulase
cellulosic biomass
chemical feedstock
chlorine-free bleaching
coal
coconut
coir
compression wood
corn stover
cotton
dietary fiber

enzymatic de-inking
ethanol
eucalyptus
exudate gum
flax
forage
fruit fiber
fuzz fiber
genetic engineering
geotextile
ghatti
gum arabic
gum
hardwood
immune system cell
jute
kapok
karaya
leaf fiber
lignin barrier
lignocellulosic biomass
linen
linter
medicinal herb

Miscanthus
Musa
oil
peak oil
Peyer's patch
pitch
plant fiber
poplar
pulp
ramie
ray parenchyma
reaction wood
recalcitrance
retting
ruminant
S2 layer
seed hair fiber
softwood
sugarcane
switchgrass
tension wood
tragacanth
wheat bran
willow

References

[1]Digestibility of forage crops depends upon the properties of plant cell walls and is mainly restricted by the lignin content.

Jung HG, Buxton DR, Hatfield RD & Ralph J (eds) (1993) Forage Cell Wall Structure and Digestibility. American Society of Agronomy (Madison WI), Crop Science Society of America (Madison WI), and Soil Science Society of America (Madison WI).

Reddy MSS, Chen F, Shadle G, et al. (2005) Targeted down-regulation of cytochrome P450 enzymes for forage quality improvement in alfalfa (Medicago sativa L.). *Proc. Natl. Acad. Sci. USA* 102, 16573–16578.

[2]The texture of our fruits and vegetables depends to a large extent on the properties of their cell walls.

Stephen AM (1995) Food Polysaccharides and Their Applications. New York: Marcel Dekker.

Waldron KW, Parker ML & Smith AC (2003) Plant cell walls and food quality. *Comp. Rev. Food Sci. Food Safety* 2, 101–119.

[3]Gums, water-soluble cell wall–associated polymers, are used to stabilize emulsions and to modify the texture of processed foods and other industrial products.

Cecil CO (2005) Gum arabic. *Saudi Aramco World* 56, 36–39.

Showalter AM (2001) Arabinogalactan proteins: structure, expression and function. *Cell Mol. Life Sci.* 58, 1399–1417.

Whistler RL (1993) Introduction to industrial gums. In Industrial Gums: Polysaccharides and their Derivatives (RL Whistler, JN Miller eds), pp 1–19. San Diego: Academic Press.

[4]Pectins are used as thickeners, texturizers, stabilizers, and emulsifiers in the food and pharmaceutical industries.

Brett CT & Waldron KW (1999) Physiology and Biochemistry of Plant Cell Walls, 2nd ed. London: Chapman and Hall.

Thakur BR, Singh RK & Handa AK (1997) Chemistry and uses of pectin. *CRC Crit. Rev. Food Sci. Nutr.* 37, 47–73.

5The consumption of dietary fiber has been linked to benefits in human health.

American Association of Cereal Chemists (AACC) International (2001) Report of the dietary fiber definition committee. The definition of dietary fiber. Cereal Foods World 46:112–126.

Choo SS & Dreher ML (2001) Handbook of Dietary Fiber. New York: Marcel Dekker.

Gallaher DD, & Schneeman BO (1996) Dietary Fiber. In Present Knowledge in Nutrition, 7th ed. (EE Ziegler and LJ Filer Jr. eds), pp. 87– 97. Washington, DC: International Life Sciences Group ILSI Press.

6Pectic polysaccharides from medicinal plants stimulate the immune system via Peyer's patch cells in the intestine.

Nergard CS, Matsumoto T, Inngjerdingen K et al. (2005) Structural and immunological studies of a pectin and a pectin arabinogalactan from *Veronica kotschyana* Sch. Bip. ex Walp. (Asteraceae). *Carbohydr. Res.* 340, 115–130.

Tan BKH & Vanitha J (2004) Immunomodulatory and antimicrobial effects of some traditional Chinese medicinal herbs: a review. *Curr. Med. Chem.* 11, 1423–1430.

Yamada H, Kiyohara H & Matsumoto T (2003) Recent studies on possible functions of bioactive pectins and pectic polysaccharides from medicinal herbs on health care. In Advances in Pectin and Pectinase Research (F Voragen, H Schols & R Visser eds), pp 481–490. Dordrecht, Netherlands: Kluwer Academic Publishers.

7Bioactive pectins from some medicinal herbs promote the proliferation of immune system cells and their activities.

Yamada H, Kiyohara H & Matsumoto T (2003) Recent studies on possible functions of bioactive pectins and pectic polysaccharides from medicinal herbs on health care. In Advances in Pectin and Pectinase Research (F Voragen, H Schols & R Visser eds), pp 481–490. Dordrecht, Netherlands: Kluwer Academic Publishers.

Jackson CL, Dreaden TM, Theobald LK et al. (2007) Pectin induces apoptosis in human prostate cancer cells: correlation of apoptotic function with pectin structure. *Glycobiology* 17, 805–819.

8Bioactive pectins inhibit tumor growth by inducing cancer cell apoptosis.

Morra M, Cassinelli C, Cascardo G et al. (2004) Effects of interfacial properties and cell adhesion of surface modification by pectic hairy regions. *Biomacromolecules* 5, 2094–2104.

Kokkonen HE, Ilversaro JM, Morra M, et al. (2007) Effect of modified pectin molecules on the growth of bone cells. Biomacromolecules 8, 509–515.

9Commercially important plant fibers are derived from many types of plants and possess physical properties that are exploited in textile products.

Tortora PG & Collier BJ (1997) Natural cellulosic fibers. In Understanding Textiles, 5th ed., pp 65-92. Upper Saddle River, NJ: Prentice-Hall.

Simpson BB & Conner-Ogorzaly M (1986) Fibers, dyes and tannins. In Economic Botany: Plants in Our World, pp 472–507. New York: McGraw-Hill.

10Cotton fibers are epidermal trichomes that develop on the surface of ovules.

Bowman DT, Van Esbroek GA, Van't Hof J & Jividen GM (2001) Ovule fiber cell numbers in modern upland cottons. *J. Cotton Sci.* 5, 81–83.

Ryser U (1985) Cell wall biosynthesis in differentiating cotton fibers. *Eur. J. Cell Biol.* 39, 236–256.

Tokumoto H, Wakabayashi K, Kamisaka S & Hoson T (2002) Changes in sugar composition and molecular mass distribution of matrix polysaccharides during cotton fiber development. *Plant Cell Physiol.* 43, 411–418.

Vaughn KC & Torley RB (1999) The primary walls of cotton fibers contain an ensheathing pectin layer. *Protoplasma* 209, 226–237.

11Wood is one of the world's most abundant industrial raw materials.

Perrin J (1989) A Forest Journey: The Role of Wood in the Development of Civilization. New York: Norton.

12Wood and paper properties are derived from the composition and morphology of the xylem cell walls.

Desch HE & Dinwoodie JM (1981) Timber: Its Structure, Properties and Utilization, 6th ed. Forest Grove, OR: Timber Press.

Lewin M & Goldstein IS (1991) Wood Structure and Composition. New York: Marcel Dekker.

Zobel BJ & van Buijtenen JP (1984) Wood Variation. Berlin: Springer.

13The physical strength of wood depends on the multilayered structure of the wood cell walls and cell morphology.

Bruce A & Palfreyman JW (eds) (1998) Forest Products Biotechnology. London: Taylor and Francis.

Desch HE & Dinwoodie JM (1981) Timber: Its Structure, Properties and Utilization, 6th ed. Forest Grove, OR: Timber Press.

Panshin AM & deZeeuw CH (1980) Textbook of Wood Technology, 4th ed. New York: McGraw-Hill.

14Cell wall–degrading enzymes are increasingly used for pulp and paper processing for environmental and economic reasons.

Bayer EA, Belaich J-P, Shoham Y & Lamed R (2004) The cellulosomes: multienzyme machines for degradation of plant cell wall polysaccharides. *Annu. Rev. Microbiol.* 58, 521–554.

Divne C, Ståhlberg J, Teeri TT & Jones TA (1998) High resolution crystal structures reveal how a cellulose chain is bound to the 50Å long tunnel of cellobiohydrolase I from *Trichoderma reesei*. *J. Mol. Biol.* 275, 309–325.

Eriksson K-EL & Adolphson RB (1997) Pulp bleaching and deinking pilot plants use a chlorine-free process. *TAPPI J.* 80, 80–81.

Eriksson K-EL, Blanchette RA & Ander P (1990) Microbial and Enzymatic Degradation of Wood and Wood Components. Berlin: Springer.

Fischer K, Akhtar M, Blanchette RA et al. (1994) Reduction of resin content in wood chips during experimental biological pulping processes. *Holzforschung* 48, 285–290.

Scott GM, Akhtar M, Lentz MJ et al. (1998) New technology for papermaking: commercializing biopulping. *TAPPI J.* 81, 220–225.

[15]Bark is a source of cork, tannins, waxes, and drugs.

Borger GA (1973) Development and shedding of bark. In Shedding of Plant Parts (TT Kozlowski ed), pp 205–236. New York: Academic Press.

Pereira H (2007) Cork: Biology, Production and Uses. Amsterdam: Elsevier.

Srivastava LM (1964) Anatomy, chemistry, and physiology of bark. *Int. Rev. For. Res.* 1, 204–277.

[16]Humankind's survival may depend on the wise use of renewable natural resources.

Diamond J (2005) Collapse: How scoieties choose to fail or succeed, p 575. New York: Viking Books.

[17]Plant cell walls and their derivatives comprise a major fraction of the terrestrial biomass and carbon content.

Global Forest Resources Assessment (2005) Progress towards sustainable forest management. Rome: Food and Agricultural Organization of the United Nations. (http://www.fao.org/DOCREP/008/a0400e/a0400e00.htm)

Thomas L (2002) Coal geology. Chichester, UK: Wiley.

Zimov SA, Schuur EAG & Chapin FS III (2006) Permafrost and the global carbon budget. *Science* 312, 1612–1613.

[18]Photosynthesis-driven biomass production by plants constitutes a major and sustainable energy source.

Perlack RD, Wright LL, Turhollow AF, et al. (2005) Biomass as feedstock for a bioenergy and bioproducts industry: The technical feasibility of a billion-ton annual supply. US Department of Agriculture and Department of Energy. (http://www.eere.energy.gov/biomass/publications.html)

[19]Producing bioethanol from lignocellulosic biomass at competitive prices will require significant investments in research.

Tao L & Aden A (2009) The economics of current and future biofuels. *In Vitro Cell. Dev. Biol. Plant* 45, 199–217.

[20]Overcoming the lignin barrier is essential to the success of the lignocellulosic biomass–based liquid biofuel industry.

Hatfield R & Vermerris W (2001) Lignin formation in plants: the dilemma of linkage specificity. *Plant Physiol.* 126, 1351–1357.

Ralph J, Brunow G, Harris P et al. (2008) Lignification: are lignins biosynthesized via simple combinatorial chemistry or

via proteinaceous control and template replication? *Recent Adv. Polyphenols Res.* 1, 36–66.

Ralph J, Lundquist K, Brunow G et al. (2004) Lignins: natural polymers from oxidative coupling of 4-hydroxyphenylpropanoids. *Phytochem. Rev.* 3, 29–60.

Chen F, & Dixon RA (2007) Lignin modification improves fermentable sugar yields for biofuel production. Nature Biotechnology 25, 759–761.

[21]New types of biorefineries are needed to produce ethanol from lignocellulosic biomass at a competitive price.

Donohoe BS, Decker SR, Tucker MP et al. (2008) Visualizing lignin coalescence and migration through maize cell walls following thermochemical pretreatment. *Biotech. Bioeng.* 101, 913–925.

Perlack RD, Wright L, Turhollow AF et al. (2005) Biomass as a Feedstock for a Bioenergy and Bioproducts Industry: The Technical Feasibility of a Billion-Ton Annual Supply. Washington, DC: DOE/USDA publication. (http://www.osti.gov/bridge)

Scheer T (2007) Energy Autonomy: The Economic, Social and Technological Case for Renewable Energy. London: Earthscan.

Yuan JS, Tiller KH, Al-Ahmad H et al. (2008) Plant to power: bioenergy to fuel the future. *Trends Plant Sci.* 13, 421–429.

[22]Plant cell walls are a renewable source of industrial chemicals.

Nikolau BJ, Perera MADN, Brachova L & Shanks B (2008) Platform biochemicals for a renewable chemical industry. *Plant J.* 54, 536–545.

Sjostrom E (1993) Wood chemistry. New York: Academic Press.

Werpy T & Peterson G (eds) (2004) Top Value Added Chemicals from Biomass, vol 1: Results from Screening for Potential Candidates from Sugars and Synthesis Gas. Washington, DC: US Department of Energy; NREL Report No. TP-510-35523; DOE/GO-102004-1992. (http://www1.eere.energy.gov/biomass/pdfs/35523.pdf)

[23]Large scale biofuel and chemical feedstock production is likely to affect food prices and biodiversity.

Fargione J, Hill J, Tilman D, et al. (2008) Land clearing and the biofuel carbon debt. *Science* 319, 1235–1238.

[24]Cell walls offer unique challenges and opportunities for genetic engineering

Bushman F (2002) Lateral DNA Transfer: Mechanisms and Consequences. Cold Spring Harbor, NY: Cold Spring Harbor Press.

Genetic Engineering of Plants: Agricultural Research Opportunities and Policy Concerns (2000) Washington, DC: National Academy Press, National Research Council.

Genetically Modified Pest-Protected Plants: Science and Regulation (2000) Washington, DC: National Academy Press, National Research Council.

Index

Entries followed by 'F' refer to figures and those followed by 'T' and 'P' refer to tables and panels, respectively.

A

Abaca, 383
ABC (ATP binding cassette) transporters
   defense responses, 339, 339F
   lignin biosynthesis, 209–210
   wax biosynthesis, 220
Abiotic stress
   callose deposits, 142, 143F, 144–145
   growth factor signaling, 307
   hypersensitive response, 341
Abscisic acid, 23
Abscission, 299–300, 300F
Abscission zone, 299–300, 300F, 301F
Acacia gums, 85
   see also Gum arabic
Acacia senegal, 91, 369
Acacia seyal, 369
Aceric acid, 46P, 79
Acetobacter xylinum see Gluconacetobacter xylinus
Acid growth hypothesis, 294
Acidic sugars, 45, 46P
Acriflavin, 218F
Actin
   free zone, cell cortex, 9–10, 11F, 149
   tracks, Golgi stack movement, 125
Actin filaments
   cell division plane positioning, 9–10, 11F
   pathogen-induced earrangements, 338, 338F
   transcellular coordination of alignment, 279–280, 280F, 281F
Acyl carrier protein (ACP), 219–220, 219F
Adaptive immune system, 321
Adenosine, 163
Adhesion
   intercellular *see* Cell–cell adhesion
   plasma membrane–cell wall, 281–282, 282F
ADP-glucose (ADP-Glc), 164
Aerenchyma, 259–260, 260F
Agave, 383–384, 384F
Air spaces, intercellular, 25, 25F, 258

   continuum, 258–260, 259F, 260F
   within leaves, 34–35, 36F, 258, 259F
   under stomata, 22, 22F
Alditol acetates
   compositional analysis using, 55P
   methylated, 57, 59
Aldoses, 50P
Alfalfa, 126F, 400
Alginate, 377
Aloe vera, 376
Amyloids, 84
Anatomy, plant, 12–14, 14–15P
Aniline blue, 142
Anisotropic cell expansion, 287–289, 287F
   see also Cell elongation
   cell-autonomous nature, 287, 288F
   wall architecture and, 287–289, 289F
Anisotropic properties, wood, 385
Annexins, 201
Anthocyanins, 204F, 207
Antibodies, 255P
Antimicrobial compounds, 328–329, 328F, 329F
D-Apiose, 46P
Apiosyl residues, 79, 81F
Apobiose, 83
Apogalacturonan, 83, 83F
Apoplast, 11–12, 12F
   barrier to water and solutes, 25, 26F
Apoptosis, cancer cell, 376–377
Apple pomace, 370
Appressorium, 332–333, 332F, 333F
Aquaporins, 283–284
Aqueous phase, primary cell walls, 49
Arabidopsis thaliana
   callose synthase, 202
   cell differentiation, 17F
   cell division patterns, 6, 6F, 7F
   cell expansion, 285, 286
   cell plate formation, 153–155, 153F, 154F
   cell separation, 258
   cellulose synthesis, 138, 141, 195–196, 195T, 198
   cell wall-degrading enzymes, 359

   cell wall heterogeneity, 253F
   chitinases, 311–312, 312F
   defense responses, 323, 330, 336F, 338F, 340F, 343F
   defensins, 328–329
   glycoprotein processing, 179
   glycosyltransferases, 170S, 171T, 185, 192
   Golgi stacks, 125, 131, 145
   lateral root formation, 27F, 28F
   leaf hair, 18, 18F
   lignin, 97T, 207
   microtubules, 245F, 274F, 275, 275F, 276F
   nucleotide sugar transporters, 167
   plasmodesmata, 33F
   pollen grain wall, 37F
   primary cell wall, 237F
   spiral growth patterns, 288
   structural proteins, 89–90, 89F, 91
   sugar nucleotide synthesis, 164
   tissue systems, 13F
   WAK proteins, 281–282
   waxes, 111, 219F, 220, 220F, 221
   xylem vessels, 28F, 29F
Arabinans, 77, 78F
Arabinitol, 404T
α-L-Arabinofuranosidase, 356, 356F
Arabinofuranosyl precursors, 192
Arabinofuranosyl residues, 73
Arabinogalactan proteins (AGPs), 49, 87T, 90–93
   biological function, 90–91
   carbohydrate side chains, 91, 93F
   cell death and, 341
   cell wall models, 267P
   classical, 91, 91F, 92–93, 93F
   developmental regulation of expression, 260–261, 260F, 261F
   evolutionary origin, 95
   glycomodules, 87–88
   GPI-anchors, 91–92, 93F, 281
      structure, 94F
      synthesis and addition, 181, 182F
   localization methods, 255P
   non-classical, 91F, 92
   O-linked glycosylation, 179–181

outer epidermal wall, 267P
signal transduction, 93, 94F
structures, 91, 91F
turnover, 295, 295F
Arabinogalactans, branched, 85
Arabinogalactan side chains, 91, 93F
Arabinol, 403
Arabinopyranosyl residues, 79
Arabinose, turnover, 295
L-Arabinose, 46P
Arabinosylation, O-linked, 181
Arabinosyl residues, 74, 76, 77–78
α-L-Arabinosyl residues, 69, 74
Arabinosyltransferases, 192
Arabinoxylans, 47
    cross-links between, 232, 232F
    hydrogen bonding to cellulose
        microfibrils, 228
    oligosaccharides, 73–74, 73F
    structure, 72–74
        conformation analysis, 62
        methods of analysis, 73–74
    wall architecture, 238, 239F
ARAD1, 189
β-Araf residues, 69
Architecture, cell wall, 235–268
    anisotropic cell expansion and,
        287–289, 289F
    control of thickness, 242–243,
        243F
    developmental variations, 260–262
    dynamic remodeling during cell
        expansion, 290–292, 291F
    environmental influences,
        261–262
    helicoidal, 248–249, 248F, 249F
    hierarchical nature, 249–250, 249F
    lamellae, 236–237, 237F, 238F
    material mechanical properties
        and, 262–264
    models, 264–268, 266–267P
    oriented cellulose microfibril
        deposition, 244–246, 246F
    pectin network, 240–242, 241F
    principles, 235–250
    reorientation in growing cells,
        246–248, 247F, 248F
    role of proteins, 243–244, 244F
    spacing of cellulose microfibrils,
        237–240
    two-network model, 235–236
    variations, 250–268
Arylpropane-1,3-diol, 215, 215F
Asparagus, 97T, 295, 295T, 295F,
    368–369
Aspartic acid, 403, 404T
Aspen, 97T, 175, 208, 306F, 399–400
Assembly, cell wall
    hierarchical nature, 249–250, 249F
    membrane systems involved,
        119–157
        cytokinesis, 145–157
        interphase cells, 120–145
    polymer biosynthesis, 161–221
        general mechanisms, 161–175
        glycoproteins and
            proteoglycans, 175–181
        polysaccharides, 182–202

within the wall, 203–221
principles, 227–268
    architectural principles,
        235–250
    architectural variations,
        250–268
    cross-links between polymers,
        227–235
    self-ordered, 248–249
    template-based, 249–250
AtCTL1, 312
AtEP3 gene, 312, 312F
AtEXT3
    cell plate formation, 151
    networks, 90, 91F
AtPRP3, 89F
Atractylodes lancea, 376
Atrichoblasts, 253F
Auxin
    antagonism by
        oligogalacturonides, 308,
        308F
    antagonism by XXFG, 309, 309F
    control of cell expansion, 287, 294
    control of endoreduplication, 283
    cortical microtubule
        reorientation, 276
    inhibition of abscission, 300
    lateral root formation, 27
    reaction wood formation, 305
Avenacin, 328, 328F
Avr4, 326, 327F

B
Bacteria
    actions of PR proteins, 344–345
    effectors, 326
    entry into plant host, 331, 331F
    papilla formation, 336F
    rumen, 367
Bacteroides, 367
BAK1, 323, 323F
Balsa wood, 387F, 388
Bamboo, 97T
Bark, 390, 391F
    commercial uses, 390, 390F
    resistance to degradation, 304
    suberin content, 107
Barley
    β-1,3/1,4 glucans, 81, 82
    epicuticular waxes, 111
    powdery mildew fungus, 336F, 340
Bast fibers, 378, 379T, 381–383
BCSA/ACSA gene, 193
BCSB/ACSB gene, 193
BCSC/ACSC gene, 193
BCSD/ACSD gene, 193
Beans, cooking, 369
Beer foam, 370, 370F
γ-p-Benzoate conjugates, 216
Benzyl esters, 103, 105F, 234F
Benzyl ethers, 103, 105F
β-sandwich protein fold, 346, 346F
Bioactive pectins, 374–377
Biochemistry, cell wall molecules,
    67–113
Biodiversity, 404–405, 406

Bioethanol production see Ethanol
    production
Biofuels, 391–402
    biorefinery development, 400–402,
        402F
    crop development, 396–398, 397F,
        398F, 406F
    efficiency of conversion process,
        400–402
    fermenting enzyme systems, 400
    food price/biodiversity issues,
        404–405
    lignin barrier, 398–400
Biomass
    based chemical feedstocks, 403,
        404T
    cellulosic, 397
    composition of terrestrial, 394–395
    lignocellulosic see Lignocellulosic
        biomass
    photosynthesis-driven
        production, 395–396
Biopulping, 390
Biorefineries
    chemical feedstock production,
        403
    ethanol production, 400–402, 402F
Bioremediation, 393P, 406
Biotic stress
    callose deposits, 142, 143F,
        144–145
    growth factor signaling, 307
Biotrophs, 320, 320T, 346
    entry into plant host, 331–332,
        331F
    hypersensitive response, 340
Biphenyl, 101F
Birch, white, 97T
Bleaching, wood pulp, 389
Block polymer construction,
    xyloglucan, 169, 185
"Bloom," 22
Blumeria graminis, 336F, 340
Borate diester cross-links, 231–232,
    232F
    regulation of wall porosity, 241
Botrytis cinerea, 343, 344
Bran, wheat, 373–374, 374F
Brassinosteroids, 288
Breakdown, cell wall see Degradation,
    cell wall
Breast cancer, 374
Brefeldin A, 147, 285
Browning reaction, 337
Brown midrib phenotype, 367–368,
    368F
Brown rot fungi, 358–359, 392P
Bupleuran, 375–376, 375F
Bupleurum falcatum, 375
Butanedione monoxime, 125
Buttercup, 35F

C
^{13}C-nuclear magnetic resonance, 60P
C_{16} fatty acids, 108F, 219–220, 219F
C_{18} fatty acids, 108F, 219–220, 219F
C_{18} omega (ω) hydroxy acids/
    alcohols, 108–109, 108F

Caffeic acid methyltransferase (COMT), 206T, 208, 215, 399
Caffeoyl CoA, 208, 208F
Caffeoyl CoA *O*-methyltransferase (CCoAOMT), 206T, 208
Calcium (Ca²⁺)
    callose synthase requirements, 200–201
    cell plate assembly, 152
    defense responses, 325, 325F
    exocystosis, 134
    gel formation, 75, 76F, 231, 231F
Calcium-chelating agents, 240, 256
Calcofluor white, 139, 140F, 141, 254P
Callases, 142, 157, 298
Callose, 4, 5F
    biosynthesis, 142–145
        defense response, 338F, 339
        growth and development, 143–144
        inducers, 200
        primer for initiation, 169, 200, 200F
        regulation, 142, 143F
        sites, 120, 136–137, 137F, 142, 143F
        stress-induced, 144–145
    deposits, 142–145, 143F
        cell plate, 143, 143F, 154F, 155–157, 156F
        histochemical staining, 142
        hypersensitive response, 340, 340F
        pollen formation, 143–144, 144F, 157
        pollen tube growth, 144, 144F
        response to pathogen attack, 335–336, 336F
        sieve plate closure, 30F, 31, 144–145
        stomatal guard cells, 23
        wound-induced, 30F, 31, 144–145, 335–336, 349, 349F
    plasmodesmata
        gating size exclusion limits, 298, 333, 333F
        sleeve formation, 33, 143F, 144
    structure, 82
    turnover, 295
    virus-mediated hydrolysis, 334
Callose synthase, 142, 200–202
    Ca²⁺ dependent, 200–201
    Ca²⁺ independent, 200, 201
    regulation, 143F, 202
    structure, 201–202, 201F
    sucrose synthase association, 199–200
Callus, 20
Calmodulin, 200
Calreticulin, 33
CALS genes *see GSL* genes
Camalexin, 329F
Cambium
    cell division plane alignment, 10, 11F
    secondary xylem production, 29

Cancer
    cell apoptosis, 376–377
    dietary fiber consumption and, 373–374
    therapy, 376
Carbohydrate Active Enzyme (CAZy) database, 168, 170S, 171T, 188–189
Carbohydrate-binding modules (CBMs), 345–346, 346F
    hemicellulose-degrading enzymes, 357, 357F
    molecular localization using, 254P
Carbon
    based chemical feedstocks, 402–403
    fixation, 395–396
    in fossil fuels, 395
    global, components, 394–395, 395T
CarboSource, 190
Carnauba wax, 22
Carotenoids, 37
Carrot
    arabinogalactan proteins, 260F, 261
    endo-chitinase, 312
    extensin, 86F, 88, 89, 90F
Casparian band, 25–26, 26F
    apoplastic restriction, 12
    endodermal net, 26, 27F
    plasma membrane attachment, 25, 26F
    suberin, 25, 106
Castanospermine, 178
Cauliflower mosaic virus, 334, 334F
CAZy database, 168, 170S, 171T, 188–189
CBEL, 346
CDTA, 240, 241F, 256
Celery, 34
Cell(s)
    derivation, 1–3
    differentiation *see* Differentiation, cell
    fate, determination, 17–18
    morphogenesis, 18
    polarization, pathogen-induced, 338–340, 338F, 339F
    size *see* Cell volume
Cell–cell adhesion, 1, 1F, 252–256
    see also Cell separation
    at three-way junctions, 257, 258F
Cell cultures, secreted polysaccharides, 53, 53F
Cell death
    hypersensitive, 340, 340F
    necrotrophic pathogen-induced, 340
    programmed, 340–341
Cell division
    see also Cytokinesis; Mitosis
    asymmetric, 8, 8F
    cell wall assembly, 4, 5F, 6F, 147–157
    generation of new daughter cells, 10, 11F
    origin of cell walls, 3–4, 3F, 4F

    planes, 4–10
        cytoskeletal elements predicting, 8–9, 8F
        patterns during development, 6–7, 6F, 7F
        preference for three-way junctions, 9, 10F
        role of actin filaments, 9–10, 11F
Cell elongation
    see also Anisotropic cell expansion
    cellulose microfibril reorientation, 247, 247F
    cell wall thinning during, 286, 286F
    remodeling of wall architecture, 291–292
Cell expansion, 16–17, 282–294
    see also Cell growth
    anisotropic, 287–289, 287F
    auxin-mediated wall acidification, 294
    cell wall thinning, 286, 286F
    cessation, 292
    deposition of new wall material, 284–287, 285F
    diffuse growth, 286–287, 287F
    distinction from cell growth, 283
    precise control, 4–7
    proteins enhancing wall expansion, 292–293, 293F
    remodeling of wall architecture, 290–292, 290F, 291F
    tip growth, 286–287, 287F
    underlying plant growth, 282–284, 284F
    vacuolar enlargement, 16, 16F, 283
Cell growth, 273–313
    see also Cell expansion
    cellulose microfibril reorientation, 246–248, 247F, 274
    cortical microtubule reorientation, 273–276
    cytoskeleton–wall interactions, 273–282
    distinction from cell expansion, 283
    Golgi stack duplication, 145
Cellobiohydrolase, 357–358, 358F
Cellobiose, 173, 358F
Cellobiose oxidoreductases, 359
Cellobiosyl-1,3-β-D-glucose, 82
Cellodextrinases, 357–358
Cellotriosyl-1,3-β-D-glucose, 82
Cell plate, 4, 4F, 5F
    assembly, 147–152, 147F
        association with ER, 152
        callose deposition, 143, 143F
        dumbbell-shaped vesicles, 149, 151, 152F
        gelling of polysaccharides, 150–151, 151F
        Golgi stack redistribution and, 146–147
        phases of somatic-type, 147–150, 148F
        by phragmoplasts, 4, 6F, 147

xyloglucan deposition, 238, 239F
fusion with mother cell wall, 10, 11F, 148F, 149
origins of plasmodesmata, 10–11, 152
peripheral growth zone (PGZ), 148F, 149
planar fenestrated sheet (PFS) stage, 148F, 149
syncytial-type, 153–157
tubulovesicular network (TVN) stage, 148F, 149, 152F
Cell plate assembly matrix (CPAM) formation, 130, 130F
somatic-type cell plate formation, 147–149, 148F
Cell plate-forming vesicles, Golgi-derived, 155
large and dumbbell-shaped vesicle formation, 150–151, 150F, 151F
scaffold protein transport, 130, 130F
somatic-type cytokinesis, 147, 148F, 149
Cell separation
see also Air spaces, intercellular
stomatal pore formation, 22F, 258
at three-way junctions, 257–258, 258F
Cell types, 13, 21–38
functionally specialized walls, 20–37, 20F
Golgi stack specialization, 132–133, 133F
lacking walls, 37–38
redifferentiation, 19–20, 19F
Cellular heterogeneity, 251–252, 251F, 253F
Cellularization
endosperm, 153–155, 154F
meiocytes, 155–157, 156F
Cellulases, 357–358
see also Endo-β-1,4-glucanases
bioethanol production, 400, 401P
carbohydrate-binding module, 346
cellulose digestion by ruminants, 367
cellulosomes, 357–358, 359F, 401P
enzyme activities, 357, 358F
pulp and paper processing, 389–390
Cellulose, 44–45
biosynthesis, 193–200
3D backbone structure and, 172–173, 173F
CESA proteins *see* Cellulose synthase
hierarchical process, 250
KORRIGAN function, 198–199
perturbation, activating defense response, 347, 347F
primer for initiation, 169, 198, 198F, 199F
regulation, 175

site, 120, 136–142
source of UDP-glucose, 199–200, 200F
studies in bacteria, 138, 139F, 193
termination of polymerization, 169
two-site glycosyltransferase model, 173
conformation, 62
cotton fibers, 380
degrees of polymerization (DPs), 44–45, 141, 142
deposition on cell plate, 4, 10, 11F
enzymatic digestion
pathogens and saprophytes, 357–359, 358F, 359F
rumen microorganisms, 367
fermentation, 400
mechanical properties, 263–264, 263F
primary walls, 44–45, 52T
rich plant fibers, 378
secondary walls, 44–45, 49, 52T
specialized walls, 44
structure, 67–68, 68F
X-ray diffraction analysis, 63, 63F
Cellulose I, 67–68
Cellulose II, 68
Cellulose-binding elicitor lectin (CBEL), 346
Cellulose-hemicellulose network, 69, 235
determination of architecture, 238–240, 239F
in vitro studies, 229F, 239
mechanical properties, 263–264, 263F
model, 266P
organization into lamellae, 236
outer epidermal wall, 267P
porosity, 241, 241F
Cellulose microfibrils
anisotropic growth and, 289, 289F
cell plate maturation, 149–150
cellulose synthase complex association, 137, 137F, 138F
cross-linking via hemicelluloses, 229, 229F, 238–239, 239F
elementary, 137
forces generating by growing, 137–138, 139F
formation from cellulose, 67, 68F
glucan chain orientation within, 67–68
hydrogen bonding to hemicelluloses, 68, 228, 228F
lamella formation, 236, 236F
lateral aggregation, 139, 140F, 141F
length
determination, 141
primary versus secondary cell walls, 142
lignin-mediated protection, 303–304, 304F
orientation, 236, 236F, 244–248
change during growth, 246–248, 274

control during deposition, 244–246, 246F
tracheids, 385, 386F
remodeling during cell expansion, 290, 291–292, 291F
self-assembly into complex structures, 248
spacing within lamellae, 237–240
stomatal guard cells, 24F
structural studies, 63
wall construction, 235
Cellulose synthase (CESA proteins), 194–200
class-specific region (CSR), 196–197, 196F
dimerization, 197, 197F
fluorescent labeling, 138
genes *see CESA* genes
identification, 193
KORRIGAN association, 198
proof of functionality, 195–196
roles of different isoforms, 198, 198F
structure and domains, 196–197, 196F
subunits in rosette complexes, 137, 138F
sucrose synthase association, 199–200, 200F
superfamily, 182–183, 182F
zinc finger domains, 196F, 197
Cellulose synthase complexes
distribution in plasma membrane, 139–141
heteromeric model, 198, 198F
linear-type, 137, 137F
movement within plasma membrane, 137–139, 139F
guidance by microtubules, 245–246, 246F, 274, 289, 289F
phylogenetics, 137, 138F
rosette type *see* Rosette complexes
Cellulose synthase-like genes, 182–184, 182F
see also CSL genes
Cellulose synthase superfamily, 182–183, 182F
Cellulosic biomass, 397
see also Lignocellulosic biomass
Cellulosomes, 357–358, 401P
designer, biofuel production, 400, 401P
dissolution of cellulose, 358, 359F
rumen bacteria, 367
structure, 357–358, 358F, 401P
Cellvibrio japonicus, 359
Cell volume, 16, 16T, 282, 283T
increase in *see* Cell expansion
nuclear ploidy and, 283, 285F
Cell wall-degrading enzymes
battle between fungal and plant, 351, 352F
bioethanol production, 400
complexes *see* Cellulosomes
pathogen-derived, 341–342, 345–346
coevolution with their inhibitors, 349–351, 351F

pulp and paper processing, 389–390
rumen bacteria, 367
structural studies, 53–54
Cell wall-integrity signaling, 346–348, 347F, 348F
CER4 protein, 220
CER6 protein, 220
cer mutants, 221
CERT5, 220
CESA1 gene
    identification, 193–194
    mutations, 141, 195, 195T
CESA3 (cesA3) mutants, 347, 347F, 348
CESA6, 289, 289F
CESA6 gene, 175, 195, 195T
CESA genes, 193–194, 195T
    Arabidopsis mutants, 195–196, 195T, 198
    cyanobacterial origin, 194
    identification, 193–194, 194F
CESA proteins see Cellulose synthase
Cf-4, 327F
CFL genes see GSL genes
CGA 325'615, 141
Chalazal endosperm (CZE) domain, 153F
Chemical feedstocks, 385, 402–405
    biomass-based production, 402–403
    food price/biodiversity issues, 404–405
    potential biomass-based, 403, 404T, 405F
Chemical ionization (CI) mass spectrometry, 59, 59F
Chemical shift, 60P
Chinese thoroughwax, 375
Chitin, 323
    receptors, 323, 324F
Chitinases
    acidic, 311–312
    antimicrobial activity, 345, 345F
    defense response, 311, 329–330, 344–345
    lysozymal activity, 345, 345F
    plant development, 311–313
    potential plant substrates, 312–313
    protection of pathogens from, 326, 327F
    yieldin homology, 293, 311
Chitin synthase, two-site model, 173–174
Chlorine-free bleaching, wood pulp, 389
Cholesterol, serum, 373
Chorismic acid, 204, 204F
Cinnamate-4-hydroxylase (C4H), 206T, 207
Cinnamic acid (cinnamate)
    biosynthesis, 206
    monolignol biosynthesis, 205F, 206–208
Cinnamoyl CoA reductase (CCR), 206T, 207, 400
Cinnamyl alcohol dehydrogenase (CAD), 206T, 207, 208, 215, 399

Cinnamyl alcohols, 203–205
cis Golgi cisternae, 123
    pectic polysaccharide synthesis, 189, 190F
    progression/maturation model, 128, 128F
Cisternal progression/maturation model, intra-Golgi transport, 126–127, 127F
cis-trans polarity, Golgi stacks, 122–123, 122F, 123F
Cladosporium fulvum, 326, 327, 327F
Class-specific region (CSR), 196–197, 196F
Clathrin-coated vesicles (CCV), 122F
    formation, 129–130, 129F, 130F, 150
    plasma membrane recycling, 135, 136F
    site of origin, 124, 124F
    syncytial-type cell plate formation, 155
clg mutation, 179
Climacteric fruits, 300, 301F
Clostridium stercorarium, 346F
Clostridium thermocellum, 358F, 359F, 400, 401P
Clover, 189
Coal, 395
Coconut, coir fibers, 381, 382F
Coevolution
    enzyme–inhibitor protein, 349–351, 351F
    host–pathogen, 327F, 328
Cohesins, 357–358, 358F, 401P
Coir fibers, 381, 382F
Colchicine, 140, 276
Coleus, 19F, 279F
Collectotrichum lindemuthianum, 88
Collenchyma, 14P, 34, 34F
    control of wall thickness, 242, 243
    helicoidal wall architecture, 248, 248F
    regional wall heterogeneity, 251, 252F
    wall thickenings, 34, 34F
Colletia cruciata, 22F
Colorectal cancer, 373
Combinatorial model, lignin biosynthesis, 213
Companion cells, 15P, 30, 30F
Complex polysaccharides, 120
Complex-type glycans, 178–179, 179F, 180F
Composites, artificial wall-like, 263–264, 263F
Compression wood, 106, 305, 305F
    cell wall remodeling, 305, 306F
    strength, 388
Compressive strength, wood, 388
Computational methods, conformation analysis, 62–63
Conformations, investigation methods, 62–63
Coniferaldehyde, 205F, 208, 215
Coniferaldehyde-5-hydroxylase (CAld5H), 206T, 208, 399
Coniferin, 209, 209F

Coniferin β-glucosidase, 209, 210F
Coniferyl alcohol, 98F, 101F, 204, 205F
    biosynthesis, 204–208, 205F, 206T
    C-C linkages, 399F
    free radicals, 211F
    glucosylation, 209, 209F
    lignan biosynthesis, 214, 214F
    lignin biosynthesis, 212, 212F
Coniferyl alcohol oxidase, 210, 211T
Constipation, 372–373
Constitutive secretion, 129
Construction industry, 384, 384F
Cooking
    effects on plant tissues, 368–369, 369F
    use of wood for, 396
COPIa vesicles, 122F, 123, 124F
COPIb vesicles, 122F, 123–124, 124F
COPII scaffold, 126–127, 126F, 127F
COPII vesicles, 123, 124F
    transfer from ER to Golgi stacks, 125–127, 127F
Cork, 106, 106F
    cell wall architecture, 107, 107F
    commercial use, 390, 390F, 391F
    suberin isolation, 108
Cork oak, 390, 390F
Corn see Maize
Coronopus didymus, 153F
Corn stover, 352F, 398
Coronary heart disease, 373
Cortical microtubule array, 245, 245F, 274, 275F
Cosmetic industry, 370
Cotton
    cellulose biosynthesis, 193–194, 197
    suberin, 107
Cotton fibers, 378–381, 379F
    degree of polymerization of cellulose, 169, 380
    development, 193, 380–381, 380F, 381F
    mature, morphology, 381, 381F
p-Coumarate, 207, 216
p-Coumarate-3-hydroxylase (C3H), 206F, 207
4-Coumarate CoA ligase (4CL), 206T, 207, 399–400
Coumaric acid, 72, 78
p-Coumaric acid, 232
Coumarins, 204F, 207
p-Coumaroyl CoA, 205F, 207
p-Coumaroyl CoA-quinate, 207, 208F
p-Coumaroyl CoA-shikimate, 207, 208F
p-Coumaryl alcohol, 98F, 204, 205F
    biosynthesis, 204–207, 205F, 206T
    C-C linkages, 399F
    glucosylation, 209
Covalent bonds, cross-linking wall polymers, 227–228
Cowpea mosaic virus, 334, 334F
CRDS protein, 202
Crispness, food, 368
Cross-linking glycans, 238
Cross-links between wall polymers, 227–235, 256
CSLD proteins, 183

zinc finger domains, 197
CSL genes, 182–184, 182F
  cyanobacterial origin, 194
  hydrophobic cluster analysis, 194F
  mannan synthase, 303
CSL proteins, 183–184, 183F
  xyloglucan biosynthesis, 184, 184F
CtMANS gene, 191
Cucumber, 112, 292
Cuticle, 21–22, 21F, 110
  assembly, 121, 218–221
  waxy, 218
Cutin, 110, 218
  biosynthesis, 121, 219–220, 219F
  chemistry, 108F, 110, 110F
  layer, 21, 21F
  networks, 110F, 218
  outer epidermal wall, 267P
Cutinases, fungal, 332–333
Cyanobacteria, 194
Cyclic-di-GMP, 193, 193F
Cycloheximide, 141
Cysteine domain (CD), 234
Cytochalasin D, 125
Cytokinesis, 4, 4F
  see also Cell division; Mitosis
  cell wall assembly, 4, 6F, 145–157
    somatic type, 147–152, 148F
    syncytial type, 153–157, 154F,
      156F
  cessation of cytoplasmic
    streaming, 145–146
  redistribution of Golgi stacks,
    146–147
Cytoplasmic streaming
  cessation during mitosis, 146–147
  Golgi components, 129, 130
Cytoskeleton
  see also Actin filaments;
    Microtubules
  control of plane of cell division,
    8–9
  pathogen-induced
    rearrangements, 338, 338F
  transcellular coordination,
    278–280, 279F, 280F, 281F
  wall interactions during growth
    and development, 273–282

D

D, D, D, Q/RXXRW motif, 182F, 183,
  191, 196, 196F
2,4-D (2,4-dichlorophenoxyacetic
  acid), 309, 309F
Damage-associated molecular
  patterns (DAMPs), 322
  cell surface receptors, 324
  eliciting defense response, 322,
    322F, 343
  generation by pathogens, 341–343,
    342F
Danger signals, 322, 328, 341–343
Datura stramonium, 285F
DCB *see* 2,6-Dichlorobenzonitrile
Default pathway, 122
Defense proteins, 85–86, 88, 328–329
Defense responses, 320–331
  basal (immunity), 320–328

elicitors, 321–322, 322F
  pathogen-mediated
    suppression, 326–328
  pattern-recognition receptors,
    322–324, 323F, 324F
  sequence of events, 324–326,
    325, 325F
callose deposition, 145, 335–336,
  336F
cell polarization, 338–340, 338F,
  339F
chitinases, 311, 329–330
convergent evolution, 321, 322F,
  345
hydroxyproline-rich
  glycoproteins, 88, 337
localized cell death, 340–341, 340F
oligogalacturonides eliciting, 308,
  324
phytoalexins, 329
polygalacturonase-inhibitor
  proteins, 344, 344F
preformed defenses, 320, 328–329
proline-rich proteins, 88
PR proteins, 344–345, 345F
strategies for evading, 350, 350F
systemic, 329–331, 330F
wall-integrity signaling, 346–348,
  348F
wall reinforcement, 336–337, 337F
to wounding, 348–349
Defensins, 328–329, 328F
Degradation, cell wall, 294–305
  see also Cell wall-degrading
    enzymes
  rumen enzymes and
    microorganisms, 367
Degrees of polymerization (DP), 54S
Dehydrogenation polymers (DHPs),
  211–212
De-inking of paper, enzymatic,
  389–390
Delignification, 392–393P, 402, 402F
Density, wood, 387–388, 387F, 388F
3-Deoxy-D-lyxo-2-heptulosaric acid
  (DHA), 46P, 79, 163T
3-Deoxy-D-manno-2-octulosonic acid
  (KDO), 46P, 79, 163T
Deoxyhexoses, 45, 46P
1-Deoxynojirimycin, 178
Dermal tissue system, 12, 14P, 15P
  development, 13F
Desmotubule, 11, 32, 32F
Development, 273–313
  arabinogalactan protein
    expression, 260–261, 260F,
    261F
  cell wall changes, 261–262
  cell wall-derived signals, 306–313
  microtubule-defined wall
    domains, 277–278, 278F,
    279F
  organ, 4–7
  transient callose deposits, 143–144
DHA *see* 3-Deoxy-D-lyxo-2-
  heptulosaric acid
Diabetes, 373
Diaryl propane, 101F

Dibenzodioxocin, 101F, 102, 212, 399F
2,6-Dichlorobenzonitrile (DCB),
  140–141, 240, 240F, 243F
Dietary fiber, 371–374
  constituents, 371, 372T
  content of foods, 373T
  human health benefits, 371–374
Diferulate esters, 104, 105F
Diferulic acid, 232, 232F, 369
Differentiation, cell, 17–18, 17F
  cell wall changes, 18–19, 262
  Golgi stack changes, 132–133,
    133F
  reprogramming, 19–20, 19F
  terminal, 95
Diffuse growth, 286–287, 287F
Digitonin, 200
Dihydrocinnamyl alcohols, 215F
Dihydroconiferyl alcohol, 215–216,
  215F
Dihydroxyacetone, 50P
Di-isodityrosine cross-links, 89, 90F,
  234, 234F
Dilignols, 212
Diphenyl ether, 101F
Diplocarpon rosae, 321F
Dipolar coupling, 61P
Directed reorientation, 247–248, 248F
Dirigent (guide) proteins, 210, 211T,
  214–215, 214F
Disaccharides, 51P
Diseases, plant, 320–321, 321F
Disulfide bonds, 234
DNA content, nuclear *see* Ploidy
Dock, pluck and go model, ER-to-
  Golgi trafficking, 126–127,
  127F
Dockerin domains, 358, 358F, 401P
Dolichols, 122, 175–177
  N-glycan assembly, 175–177, 176F
  N-glycan removal from, 177, 177F
  structure, 175, 176F
Douglas fir, 97T, 306F
Dumbbell-shaped vesicles, 150–151
  cell plate formation, 149, 151,
    152F
  expulsion of water, 150–151, 151F
  formation, 150, 150F, 151F
Dynamin *see* Dynamin-GTPases
Dynamin-GTPases
  somatic-type cell plate formation,
    150–151, 150F, 151F
  syncytial-type cell plate
    formation, 155

E

Early TGN cisternae, 128, 129, 129F,
  130F
Eceriferum (*cer*) mutants, 221
Ectopic lignification, 95, 347, 347F
Effectors, 326–328
Effector-triggered immunity (ETI),
  327
EFR, 323
EF-Tu, 323
Egg cells, 38
Electron impact ionization (EI) mass
  spectrometry, 58–59

Electron microscopy, 255P
Electrospray ionization mass
        spectrometry (ESI-MS), 59
Elementary microfibrils, 137
Elicitors, 321–322, 322F
        cell surface receptors, 322–323,
                322F, 323F, 324F
        endogenous, 322, 341–342
        fungal heptaglucoside, 307F
        general, 321
        herbivore-derived, 349
        responses to *see* Defense
                responses
        wall reinforcement response, 337
Elongation, biopolymer, 168
Embryo development, plant
        control of cell divisions, 6–7, 6F, 7F
        establishment of tissue systems,
                13, 13F
        origin of cells, 1–3, 2F
Emulsions, stabilization of, 370
Endo-chitinases, 312
Endocytic pathway, 119
Endocytosis, 134
Endodermis, root, 25–26, 26F
        casparian bands, 26, 27F
        regional wall heterogeneity, 251,
                252F
Endo-β-1,3-glucanaseA, 351, 351F
Endo-(1→4)-β-D-glucanase, 302
Endo-glucanases, 54
        cellulose breakdown, 357–358,
                358F
        inhibitor proteins, 350–351
        wall remodeling during growth,
                290
        xyloglucan oligosaccharides
                released, 69–72, 71F, 72F
Endo-β-1,4-D-glucanases, 356–357
Endo-β-1,4-glucanases
        KORRIGAN, 198–199, 199F
        structural characterization, 47
        xyloglucan oligosaccharides
                released, 69–72, 71T
Endo-glycanases, 173, 356
Endo-β-1,4-mannanases, 358
Endomembrane system, 119
        polysaccharide assembly, 182–192
Endoplasmic reticulum (ER), 120F,
        121–122
        arabinogalactan protein assembly,
                181, 182F
        cell plate association, 152
        chloroplast attachment domains,
                121F
        dock, pluck and go model of Golgi
                trafficking, 126–127, 127F
        export sites, 121F, 123
                Golgi stack interactions, 125,
                        126–127, 127F
        functional domains, 121–122,
                121F
        glycoprotein synthesis, 175–178,
                176F, 177F
        lipid-recycling, 121F, 135, 136F
        nucleotide sugar transport into,
                166–167, 166F
        plasmodesmata formation, 152,
                296, 297F

polymer assembly, 120
rough, 121–122, 121F
smooth, 121F, 122
transfer of COPII vesicles to Golgi,
        125–127, 127F
vesicle origins/trafficking, 124F
wax biosynthesis, 219–220, 219F
α-1,4-*Endo*-polygalacturonase
        pectic polysaccharide studies, 47,
                47F, 52
        RG-I solubilization, 75–76
        RG-II solubilization, 79, 231
Endo-polygalacturonases (*endo*-PGs)
        abscission, 300
        cleavage of glycosidic bonds, 230,
                230F
        pathogens, 342–343, 342F
        plant-derived inhibitors, 344,
                344F, 350
        saprophytes, 353, 353F
        structure, 355F
Endoreduplication, 283
Endosome-mediated secretion
                pathway, defense response,
                338–339, 339F
Endosperm
        cellularization, 153–155, 154F
        domains, 153F
        mixed-linked glucans, 81
        nuclear cytoplasmic domains, 153,
                153F, 154F
        storage polysaccharides, 84, 84F
Endo-xylanases, 54
        cellulose breakdown, 357–358
        hemicellulose digestion, 356, 356F
        localization studies, 254P
        oligosaccharides released, 73, 73F
Endo-xyloglucanases, 356–357, 357F
Energy sources
        nonrenewable, 394T
        renewable, 391–407
Environmental changes, adaptive
                responses to, 261–262
Enzymatic de-inking, recycled paper,
                389–390
Enzymes
        cell wall, 48–49, 85–86
        molecular localization studies,
                254P
        wall degrading *see* Cell
                wall-degrading enzymes
Epicuticular waxes *see* Waxes,
                epicuticular
Epidermal cells
        see also Leaf epidermal cells
        anisotropic growth, 288
        cortical microtubules, 274, 275,
                276F, 277F
        cotton, fiber development,
                380–381, 380F
        functionally specialized walls, 20,
                20F
        microtubule guidance of cellulose
                synthase, 289, 289F
        petal, 20, 20F, 21
        waterproof cuticle, 21–22, 21F
        wax biosynthesis, 219–220, 220F
Epidermis, 15P, 21
        outer wall, 267P

penetration by fungi, 332–333,
        332F, 333F
Equisetum (horsetail)
        β-1,3/1,4 glucans, 81, 82, 82F
        silica, 111, 112, 112F
ER *see* Endoplasmic reticulum
ER export sites, 121F, 123
        Golgi stack interactions, 125,
                126–127, 127F
Ethanol production, 396–402
        biorefinery development, 400–402,
                402F
        crop development, 396–397, 397F,
                398F, 406F
        fermenting enzyme systems, 400
        lignin barrier, 398–400
Ethylene, 276, 277F, 307
        abscission, 300, 300F
        defense responses, 330, 330F
Ethylene-inducing xylanase (EIX),
        323, 323F, 324F
Eucalyptus, biofuel production, 396,
        397, 398F
Eucalyptus botryoides, 97T
Eucalyptus globulus, 396
Eucalyptus goniocalyx, 106
Eucalyptus grandis, 390, 391F
Exine, 37, 37F
Exocytosis, 133–134
Exo-galactosyluronic acid hydrolase,
        342
Exo-polygalacturonases (*exo*-PGs),
        342–343, 342F
Exosomes, 339
Expansins, 292–293, 293F
        fruit ripening, 302
Extensins
        cell plate formation, 151
        isodityrosine cross-links, 233–234,
                234F
        LRR, 281
        networks, 90, 91F, 267P
        O-linked arabinosylation, 181
        peroxide-mediated cross-linking,
                89, 90F
        spatial distribution, 86F, 88
        structure, 89–90, 90F
Exudate gums, 84–85, 369

F

Fatty acid elongases, 220
Fatty acid reductase, 220
Fatty acid synthase complex, 219,
        219F
Fenugreek, 191, 295–296, 296F, 303F
Fermenting enzyme systems
        biofuel production, 400
        chemical feedstock production,
                403
Ferulate-5-hydroxylase (F5H) *see*
                Coniferaldehyde-5-
                hydroxylase
Ferulate esters
        lignin-carbohydrate complexes,
                103–104, 105F
        polysaccharide cross-linking, 232
Ferulic acid
        arabinoxylan, 72

cutin, 108F, 110
  derivatives in lignin, 216
  rhamnogalacturonan I, 78
  suberin, 108F, 109
Feruloyl-CoA, 205F, 208
Feruloyl esters, suberin, 109F
FG, 70P
Fiber, dietary *see* Dietary fiber
Fiber cells, sclerenchyma, 34, 35F
Fiber composite materials, 235, 235F, 266P
Fibers, plant
  definitions, 378
  used in textiles, 378–384, 379T
Flagellin, bacterial, 322–323, 322F
FLAGELLIN-SENSING 2 (FLS2), 323, 323F
Flavonoids, 204F, 205F, 207
Flax, 295, 365, 381–382, 383F
flg22, 322–323
  genes activated by, 343, 343F
  receptor, 323, 323F
  structure, 322F
Fluorescent immunocytochemistry, 255P
Foods, 366–371
  availability and prices, 404–405
  dietary fiber content, 373T
  digestibility, 366–368
  processing, 369–371
  texture, 368–369
Forage crops, digestibility, 366–368
Forests, 394–396
Fossil fuels, 391
  see also Oil
  carbon in, 395
fra8 mutants, 192
Fractionated Pectin Powder, 376–377
Fructose, 50P
Fruit
  abscission, 299–300
  climacteric, 300, 301F
  degradation, 359F
  nonclimacteric, 300, 301F
  ripening and softening, 264, 300–302
  texture, 368–369
Fruit fibers, 378, 379T
L-Fucose, 46P
α-1,3-Fucose, 179
α-L-Fucosyl-(1→2)-β-D-galactosyl residues, 69
Fucosyl residues
  rhamnogalacturonan I, 76
  xyloglucan biosynthesis, 185–186, 185F
Fucosyltransferases, 185–186, 185F
  identification methods, 172
Fuels
  see also Biofuels; Fossil fuels
  renewable sources, 391–407
  use of wood, 365, 366F, 396
Fumaric acid, 404T
Fungal infections, 321F
  callose synthesis, 145
  HRGP and PRP synthesis, 88
  hypersensitive response, 340F
  papilla formation, 336, 336F, 337F

protective role of silica, 112
Fungi
  actions of PR proteins, 344–345
  biopulping, 390
  cell wall-degrading enzymes, 342–343, 342F
  effectors, 326
  entry into plant host, 331–333, 331F, 332F, 333F
  interactions with plants, 341F
  spread via pit fields, 335, 335F
2,5-Furan dicarboxylic acid, 403, 404T
Furfural, 400, 400F
β-Furfuryl-β-glucoside, 200, 200F
Fusarium, 350
Fusicoccin, 294
Fuzz fibers, 381

G

Galactans, 77–78
Galactoglucomannans, 52, 74
  degrading enzymes, 356–357
  structure, 83–84
Galactomannan galactosyltransferase, 303
Galactomannans
  biosynthesis, 191
  seed cell wall, 84
  as storage materials, 303, 303F
  turnover, 295–296, 296F
Galactose, 46P, 50P
Galactosyl residues
  fruit ripening, 301
  galactoglucomannans, 74, 84
  galactomannan biosynthesis, 191
  rhamnogalacturonan I, 76, 77–78
  xyloglucan, 69, 185–186
Galactosyluronic acid residues
  backbone synthesis, 172, 173
  homogalacturonan, 75
  rhamnogalacturonan I, 76, 77, 78F, 85
  rhamnogalacturonan II, 79, 80
  transesterification reactions, 233, 233F
Galacturonans, substituted, 83, 83F
Galacturonic acid, 50P
  fungal enzymes releasing, 342–343, 342F, 353, 353F
  glucuronoxylans, 74
D-Galacturonic acid, 46P
GalA-transferase, 190–191
Galectins, 377
Gal-transferase, 191
Gametes, 38
Gametophytic cells, cell plate assembly, 153–157
Gaseous exchange, within leaves, 34–35
Gas-liquid chromatography (GLC), 57
Gas-liquid chromatography–mass spectrometry (GLC-MS), 57, 58–59
GATL enzymes, 189
GAUT1, 186–189
GAUT7, 188–189
GDP-fucose (GDP-Fuc), 165F
GDP-glucose (GDP-Glc), 164

GDP-mannose (GDP-Man), 165F, 191
Gel, 230
Gelatinous layer (G layer), 106, 106F
  tension wood, 305, 306F
Gel electrophoresis, polysaccharide synthases, 172
Gel formation
  cell plate polysaccharides, 151
  homogalacturonans, 75, 76F, 230–231, 231F
  industrial applications, 370–371
General elicitors, 321
Genetic engineering, 405–406
  biofuel crops, 399–400, 406F
Geotextiles, 382–383
Germination, seed, 302–303
Ghatti, 85, 369
Gibberellic acid (GA3), 276, 277, 277F
Ginseng, 376
GIP1, 351, 351F
G layer, 106, 106F
  tension wood, 305, 306F
GlcNAc transferase I, 168, 178, 179, 180F
Glossy (*gl*) mutants, 221
Glucanase inhibitor protein (GIP1), 351, 351F
β-1,3-Glucanases, 298, 329–330, 333–334
  see also Callases
β-1,3-Glucan hydrolase, 144F
β-D-Glucanohydrolase, 82
β-1,3/1,4-Glucans (mixed-linked glucans), 81–82, 82F
  biosynthesis, 202
    in vitro studies, 202
    site, 120, 136–137, 202
  structural characterization, 82
β-1,3-D-Glucans, 82
  see also Callose
β-1,4-Glucans, 44–45, 47, 69
  see also Cellulose; Xyloglucan
  cellulose microfibril formation, 67–68
β-1,3/1,4-(mixed-link) Glucan synthase, 202
β-1,3-Glucan synthase, 200–201
Glucaric acid, 403, 404T
Glucomannans, 52, 74
  effect on cellulose mechanics, 264
  seed cell wall, 84
Gluconacetobacter xylinus
    (*Acetobacter xylinum*)
  artificial wall-like composites, 263–264, 263F
  cellulose synthase genes, 193
  cellulose synthesis, 138, 139, 139F
Glucose, 50P
  linear polymers, 120, 136–137
  turnover, 295
D-Glucose, 46P, 51P
L-Glucose, 51P
Glucose-1-phosphate (Glc-1-P), 163, 165F
Glucosidases, 177F, 178, 358F
  monolignol glucosides, 209, 210F
Glucosinolate metabolites, 339, 339F
Glucosyl residues

backbone synthesis, 172–173,
    184–185, 185F
β-1,3/1,4 glucans, 82
galactoglucomannans, 74
glucomannan, 74
xyloglucan, 69
Glucosyluronic acid residues
    arabinoxylans, 72, 73–74
    mannoglucuronans, 85
    rhamnogalacturonan I, 76
β-D-Glucosyl-Yariv reagent, 91, 92F
D-Glucuronic acid, 46P
Glucuronoarabinoxylans, 52
    biosynthesis, 172–173, 192
    degrading enzymes, 356, 356F
    solubilization and purification, 53
    structural characterization, 58
    structure, 74
Glucuronoxylans, 52
    biosynthesis, 192
    reducing end oligosaccharide, 74
    structure, 74
Glutamic acid, 403, 404T
Glycans, cross-linking, 238
Glycan synthases
    CSL-type, 183–184, 183F
    xyloglucan biosynthesis, 184, 184F
Glyceraldehyde, 50P
Glycerol
    industrial production, 404T
    suberin, 108F, 109
Glyceryl esters, suberin, 109F
Glycine-rich proteins (GRPs), 48, 87T,
    94
    evolutionary origin, 95
    integration into wall architecture,
        244, 244F
Glycogen, biosynthesis, 169
Glycomodules, 87–88, 88T
Glycoproteins, 48
    assembly, 175–181
        membrane-bound enzymes,
            167–168
        N-linked glycosylation,
            177–178, 177F
        O-linked glycosylation,
            179–181
        role of dolichols, 175–177, 176F
        sites, 120, 122
        source of glycosyl residues,
            162–165
    biochemistry, 85–95
    building blocks, 161–162
    misfolded, 178
    processing in Golgi, 178–179, 180F
Glycosidases
    common topology, 168, 168F
    mechanisms of retention in Golgi,
        168
    N-linked glycan processing,
        178–179, 180F
    numbers of genes, 161, 161F
Glycosidic linkages (bonds)
    enzymatic cleavage, 353–355, 354F
    glycopeptides, 87F
    oligosaccharides and
        polysaccharides, 51P
    pectic polysaccharides, 229–230,
        230F

Glycosylation, 87–88
    N-linked, 177–178, 177F
    O-linked, 179–181
    sites, 120
Glycosyl linkage composition
    analysis, 56P, 57
Glycosylphosphatidyl inositol (GPI)-
    anchored proteins, 91,
    92–93, 94F, 281
Glycosylphosphatidyl inositol (GPI)
    anchors
    signal sequence, 91, 93F
    structure, 94F
    synthesis and attachment, 181,
        182F
Glycosyl residue composition, 54
    analysis, 54–57, 55P
Glycosyl sequencing, 57–58
Glycosyltransferases (GTs), 167–173
    common topology, 168, 168F
    families, 168, 170S
    galactomannan biosynthesis, 191
    hydrophobic cluster analysis, 194
    identification methods, 172
    inverting reaction mechanism,
        170F, 170S, 171T
    low abundance, 167
    mechanisms of retention in Golgi,
        168
    membrane bound, 167
    N-linked glycan processing,
        178–179, 180F
    numbers of genes, 161, 161F
    pectic polysaccharide synthesis,
        186–189, 187T, 188T
    processive, 183, 183F, 196
    regulation, 175
    retaining reaction mechanism,
        170F, 170S, 171T
    two active sites, 172–174, 174F
    xylan biosynthesis, 192
    xyloglucan biosynthesis, 184, 184F
Glycosyluronic acid residues, 57, 233
Golgi apparatus, 120, 122–133
    N-linked glycan processing,
        178–179, 180F
Golgi belt, 146, 146F
Golgi cisternae
    acidification, 131–132
    classes, 123–124
    glycosidases and
        glycosyltransferases, 168
    nucleotide sugar synthesis, 164,
        166F
    nucleotide sugar transport into,
        166–167, 166F
    pectic polysaccharide synthesis,
        189, 190F
    progression/maturation model,
        127–128, 128F
Golgi-derived vesicles, 123–124, 124F
    cell plate forming see Cell plate-
        forming vesicles, Golgi-
        derived
    defense responses, 338, 339F
Golgi matrix see Golgi scaffold/matrix
Golgi scaffold/matrix, 126–127, 126F,
    127F

Golgi stacks, 120F, 122–128, 122F
    cell-type specialization, 132–133,
        133F
    cisternal progression/maturation
        model, 127–128, 128F
    cis-trans polarity, 122–123, 122F,
        123F
    dispersion in cytoplasm, 124–125,
        125F
    dock, pluck and go model of ER
        trafficking, 126–127, 127F
    duplication, 145, 145F
    ER export site interactions, 125,
        126–127, 127F
    movements, 125, 125F
    pH gradients, 131–132
    redistribution in dividing cells,
        146–147, 146F
    transfer of COPII vesicles to,
        125–127, 127F
Gossypium species see Cotton
GPI-anchored AGPs, 281
    see also under Arabinogalactan
        proteins
GPI anchors see Glycosylphosphatidyl
    inositol (GPI) anchors
Grapevine fanleaf virus, 334F
Grasses
    β-1,3/1,4 glucans, 81, 82
    carbon fixation, 396
    cell wall-degrading enzymes, 356,
        356F
    forage, 367
    hemicelluloses, 72
    lignin-carbohydrate complexes,
        103–104
    lignin composition, 97T, 99, 104
    lignin content, 96, 97T
    polymer cross-links, 228, 232
    silica, 111, 111F, 112
    xylanase inhibitor proteins, 350,
        350F
Gravity sensing, 304–305
Green fluorescent protein (GFP), 125,
    125F, 254P
Ground tissue system, 12–13, 14P
    development, 13F
    parenchymal cell shapes, 24–25,
        24F, 25F
Growth, plant, 273–313
    see also Cell growth
    cell wall-derived signals, 306–313,
        307F
    mechanisms, 282–284, 284F
    transient callose deposits, 143–144
Growth anisotropy, 287F
Growth factors, plant, 276, 307
GRPs see Glycine-rich proteins
GSL5 gene, 202, 349F
GSL5 protein, 335, 338F
GSL (CALS; CFL) genes, 201, 202
GSL proteins, 201–202
    see also Callose synthase
Guaiacyl glycerol-β-aryl ether, 101F
Guaiacyl units (G units), 98F, 99
    content of different plants, 97F,
        104

different wall regions and cell types, 217, 218F
intermolecular linkages, 100–102, 101F
manipulation in biofuels, 399
monolignol precursor, 204, 205F
Guanine, 163
Guanosine diphosphate *see* GDP
Guar, 191
Guar gum, 303F, 372–373
Guide proteins *see* Dirigent proteins
Gum arabic, 91, 369–370
industrial uses, 85, 370, 370F, 371F
medicinal uses, 374
production, 369–370, 369F, 370F
Gums
exudate, 84–85, 369
as food additives, 369–370
G units *see* Guaiacyl units

H

¹H-nuclear magnetic resonance, 59, 60P
Hairs *see* Trichomes
Hard fibers *see* Leaf fibers
Hardwoods
density, 387F, 388
length of fiber cells, 386
lignin-carbohydrate complexes, 104
lignin composition, 96–98, 97T, 104, 105
lignin content, 97T
lignin linkages, 101–102
reaction wood, 105–106, 106F, 305
structure, 385, 385F
H⁺-ATPase, vacuolar (V-ATPase), 131–132
Haustoria, fungal, 331–332, 331F, 332F
encasement, 336, 337F
Headward growth, polysaccharides, 169, 169F
Heating, 396
Hechtian threads/strands, 281, 282F
Helical growth patterns, 288
Helicoidal architecture, 248–249, 248F, 249F
Hemibiotrophs, 320, 320T
entry into plant host, 331–332
Hemicelluloses
see also Cellulose-hemicellulose network
biofuel production and, 400
cellulose microfibril cross-links, 229, 229F, 238–239, 239F
conformation analysis, 62
effect on cellulose mechanics, 263–264, 263F
enzymatic digestion
pathogens and saprophytes, 356–357, 356F
rumen microorganisms, 367
hydrogen bonding to cellulose microfibrils, 68, 228, 228F
lignin cross-linking, 103, 234–235
perturbations activating defense responses, 347
primary cell walls, 45, 47, 52T, 68–69

architectural role, 235, 238–240, 239F
methods of studying, 47
secondary cell walls, 49–52, 52T, 74
self-assembly into complex structures, 248
solubilization and purification, 53
as storage materials, 303
structure, 68–74
Hemp stem, 382F
Henequen, 383–384
Heptaglucoside elicitor, fungal, 307F
Herbal medicines, 374–376
Herbicide 2,6-dichlorobenzonitrile *see* 2,6-Dichlorobenzonitrile
Herbivore-associated molecular patterns (HAMPs), 349
Herbivores, 320
digestion of forage crops, 366–367
responses to, 349
wounding by, 348, 348F
Heterogeneity, cell wall, 250–252, 251F
Hexadecanoic acid (C16) derivatives, 108F, 110, 219–220, 219F
Hexose monophosphates, 163–164, 165F
Hexoses, 45, 46P, 50P
HG *see* Homogalacturonan
HG-methyltransferase, 191
Hierarchical construction, 250
High-mannose-type *N*-linked glycans, 178, 179F, 180F
High-performance liquid chromatography (HPLC), 54
Homogalacturonan (HG), 45–47, 75
biosynthesis, 172, 173, 173F
enzymes, 187T, 188T
in vitro studies, 190–191
pathway, 189, 190F
enzymatic digestion, 342–343, 342F, 352–353, 353F, 354
ester cross-links, 233, 233F
gel formation, 75, 76F, 230–231, 231F
glycosidic cross-links, 229–230, 230F
localization studies, 231, 231F
methyl esterification, 75
partially methylesterified, gel formation, 75, 230–231, 231F
structural analysis, 47, 47F, 62–63
three-way junctions, 231, 231F, 257F
Homogalacturonan (HG)-methyltransferase, 191
Horsetail *see* Equisetum
HRGPs *see* Hydroxyproline-rich glycoproteins
Humus, 394
H units *see* Hydroxyphenyl units
Hyaluronan, 377
Hybrid-type *N*-linked glycans, 178, 179, 179F, 180F
Hydrogen bonds

cellulose microfibril formation, 67, 68F
between hemicelluloses and cellulose microfibrils, 228, 228F, 229F
Hydrogen peroxide (H2O2), 197
cross-link formation, 232, 234
defense response, 325, 336–337
insolubilization of HRGPs, 244
lignin polymerization, 211, 211F
Hydrolases, 354F
Hydrophobic cluster analysis, 193–194, 194F
4-Hydroxybenzaldehyde, 96, 98F, 99
Hydroxybenzaldehydes, incorporation into lignin, 215, 215F
3-Hydroxy-butyrolacetone, 404T
Hydroxycinnamaldehydes, incorporation into lignin, 215, 215F
p-Hydroxycinnamoyl-CoA:shikimate/quinate *p*-hydroxycinnamoyl transferase (HCT), 206T, 207, 208F
p-Hydroxycinnamyl alcohol, 209
Hydroxycinnamyl alcohols, 215F
5-Hydroxyconiferaldehyde, 208
5-Hydroxyconiferyl alcohol, 205F, 208
incorporation into lignin, 215, 215F
Hydroxyl radical, 291
Hydroxyphenyl units (H units), 98F, 99
content of different plants, 97F, 104
intermolecular linkages, 100–102, 101F
monolignol precursor, 204, 205F
Hydroxyproline-rich glycoproteins (HRGPs), 48, 87T
control of expression, 88–89
defense responses, 88, 337
evolutionary origin, 95
extensin-type *see* Extensins
glycomodules, 87–88, 88T
glycosylation, 179–181
integration into wall architecture, 244
repeated sequence motifs, 87
three-way junctions, 257, 257F
3-Hydroxy propionic acid, 404T
Hypersensitive response (HR), 340–341
cell death, 340, 340F
signals inducing, 341, 341F
Hypocotyl
anisotropic growth, 288
cell wall thinning during elongation, 286, 286F
cortical microtubules, 245F, 275, 276F
helicoidal wall architecture, 248, 249F
Hypoxia, adaptation to, 259–260

I

Immune system
adaptive, 321

cells, effects of medicinal herbs,
374–376
innate, 321
Immunity, plant, 320
see also Defense responses
effector-triggered (ETI), 327
pattern-triggered (PTI), 322
Immunocytochemistry, 255P
Immunofluorescence, 255P
Immunogold labeling, 255P
Industrial Revolution, 391–394
Inhibitor proteins, enzyme, 350–351,
351F
Initials (stem cells), 2–3, 13
Initiation, 168–169
Innate immune system, 321
Insects, 320
chewing damage, 330, 331F, 348,
348F
responses to, 349
In-situ hybridization, 254P
Intercellular spaces *see* Air spaces,
intercellular
Intercisternal elements, 122F, 124
Interphase cells
cortical array of microtubules,
245, 245F
sites of cell wall polymer assembly,
120–145
Intine, 37, 37F
Intracuticular wax, 218
Ion flux, defense response, 325, 325F
Ionophores, 131
irx1 mutants, 195, 195T
irx3 mutants, 195, 195T
irx5 mutants, 195T
irx8 mutants, 192
irx9 mutants, 192
Isodityrosine cross-links, 233–234,
234F
Isotropic cell expansion, 287F
Isoxaben, 196, 240, 286, 347F
Itaconic acid, 404T
ixr1 mutants, 195T, 196
ixr2 mutants, 195T

J

Jasmonic acid (JA), 330, 330F
Java cotton *see* Kapok
Jojoba, 220
Jute, 382–383, 383F

K

KAM1/MUR3, 186
Kapok, 379, 380F
Karaya, 85, 369
KDO *see* 3-Deoxy-D-manno-2-
octulosonic acid
Kenaf, 97T, 99, 216
β-Ketoacyl-CoA synthases (KCS), 220
Ketoses, 50P
Klason lignin, 96
Kordofan gum, 369–370, 370F
KORRIGAN (KOR), 198–199, 199F

L

Laccase, 211, 211F, 211T
white rot fungi, 358, 393P

Lamellae, 236, 236F
see also Middle lamella
arrangement, 236, 236F, 237F
cellulose microfibril reorientation,
246–248, 247F, 248F
cell wall models, 266P
helicoid architecture, 248–249,
248F, 249F
oriented cellulose microfibril
deposition, 244–246, 246F
primary walls, 236–237, 237F, 238F
spacing of cellulose microfibrils,
237–240, 239F
Lateral meristems, 2
Late TGN cisternae, 129, 129F, 130F
Latewood, 385F
Laxation, 372–373
Leaf
abscission, 299–300
air spaces, 34–35, 36F, 258, 259F
development, role of expansins,
293, 293F
lobe formation, 280, 281F
mesophyll cells *see* Mesophyll cells
Leaf epidermal cells
lobe formation, 280, 281F
preprophase band, 9F
silica deposits, 111–112, 111F, 112F
wall heterogeneity, 251, 252F
waterproof cuticle, 21, 21F
wounding, 279, 280F
Leaf fibers, 378, 379T, 383–384
Lectins
see also Yariv reagent
cellulose-binding elicitor (CBEL),
346
chitin-binding (Avr4), 326, 327F
solanaceous, 86, 87T, 179–180
wall-integrity sensing, 348
wheat germ agglutinin (WGA),
312–313
Lecythis chartacea, 22F
LeEix1, 323, 324F
LeEix2, 323, 324F
Lemna minor, 83
Leucine-rich repeat (LRR)-extensin
proteins, 281
Leucine-rich repeat (LRR) proteins,
327, 344, 344F
Leucine-rich repeat receptor kinases
(LRR-RKs), 323
Leucoplasts (colorless plastids)
export of fatty acids, 220, 220F
fatty acid biosynthesis, 219, 219F,
220F
Levulinic acid, 404T
Lignans, biosynthesis, 203, 204F, 214,
214F
Lignification, 95
ectopic, 95, 347, 347F
stages, 216, 216F, 217F
Lignin, 95–106, 95F
barrier, 398
biofuel industry, 398–400
wall turnover and degradation,
303–304, 304F
biological functions, 96, 96F
biosynthesis, 203–217
combinatorial model, 213

manipulation in biofuel crops,
399–400
monolignol polymerization,
210–216
monolignol production,
204–210, 205F, 208F
regulation in time and place,
216, 216F, 217F
related metabolic pathways,
203, 204F
sites, 121, 203
chemical degradation, 392P, 402,
402F
composition
analysis, 96–99, 98F
changes over time, 262
different plants, 97T, 104–105
different wall domains and cell
types, 217, 218F
manipulation in biofuel crops,
399–400
reaction wood, 105–106, 106F
content
determination, 96
different plants, 96, 97T
forage crop digestibility and,
367–368
industrial uses and, 386
manipulation in biofuel crops,
399–400
reaction wood, 105–106
defense responses, 337
enzymatic degradation, 358–359,
390, 392–393P
hemicellulose complexes,
103–104, 234–235
histochemical staining, 99–100,
100F, 254P
intermolecular linkages, 100–102,
101F
Klason, 96
models for higher-order structure,
102, 103F, 104F
monomer endgroups, 102, 102F
nitrobenzene oxidation, 96–98,
98F, 99
polymerization, 210–216
coupling of two oligomers, 212
enzymatic oxidation of
monolignols, 210–212,
211F, 211T
monolignol coupling to
polymer, 212–213, 212F
nonenzymatic model, 214–215,
214F
oxidative carriers/radical
mediators, 213, 213F
using nontraditional
monomers, 215–216, 215F
polymer subunits, 98F, 99
see also Guaiacyl units;
Hydroxyphenyl units;
Syringyl units
polysaccharide cross-linking,
103–104, 105F
precursors *see* Monolignols
protoxylem elements, 19, 28
purification from wood, 96
secondary walls, 49, 52T

stability, 95
thioacidolysis, 98, 98F, 99
Lignin-carbohydrate complexes
(LCCs), 103–104, 105F,
234–235, 234F
Lignin peroxidases (ligninases), 358,
393P
Lignocellulosic biomass, 397–402
chemical feedstock production,
403
crop development, 397–398, 398F,
406F
ethanol production, 400–402, 402F
lignin barrier, 398–400
recalcitrance, 402
Lignum vitae, 387F, 388
Linear glucose polymers, 120, 136–137
Linen, 381–382
Linters, 381
Lipid rafts, 336, 338
Lipids
recycling of membrane, 135, 136F
transfer between organelles, 220,
220F
Lipid transfer proteins (LTPs), 49,
85–86
wall-loosening activity, 293, 293F
wax biosynthesis, 219F, 220
Lipo-oligosaccharides, 310–311, 310F,
312
Load-bearing components, 292
Loblolly pine, 215F, 261, 261F
Localization methods, molecular, 252,
254–255P
LRR *see* Leucine-rich repeat
LRR-extensin proteins, 281
Lyases, 353–355, 354F
Lysigeny, 260
LysM repeats, 324F

M

Magnaporthe oryzae, 332–333, 332F,
333F, 335, 335F, 356
Magnesium ions (Mg2+),
polysaccharide
biosynthesis, 196, 202
Mahogany, Southern, 97T
Maiden grass, 97T
Maize (corn) (*Zea mays*)
brown midrib mutants, 367–368,
368F
callose, 82
lignin composition, 97T, 104
plasmodesmata, 32F
root meristem, 2F, 5F
rosette complexes, 142F
silica, 112
stover, 352F, 398
tangled mutants, 6–7, 7F
waxes, 111, 221
Malic acid, 404T
MAMPs *see* Microbe-associated
molecular patterns
Manganese-dependent peroxidase
(MnP), 358, 393P
Manganese(II) ions (Mn^{2+}),
polysaccharide synthesis,
191, 202

Manganese(III) ions (Mn^{3+}), lignin
biosynthesis, 213, 213F
Manila hemp, 383
Mannans
biosynthesis, 191
seed cell wall, 84
as storage materials, 303
Mannan synthase, 191, 303
Mannoglucuronans, 85
Mannose, 46P, 50P
Mannose-1-phosphate (Man-1-P),
163, 165F
Mannosidase I
green fluorescent protein fusion,
125, 125F, 254P
localization, 123, 124F
processing of *N*-linked glycans,
178, 180F
Mannosidase II, 178, 180F
Mannosyl residues, 74, 85, 191
Man synthase, *see* mannan synthase,
191, 303
MAP kinases, 325, 325F
Mass spectrometry (MS), 58–59, 59F
Material mechanical properties,
263–264
Mäule reaction, 99–100, 100F
Mealiness, food, 368
Mechanical properties
cellulose, 263–264, 263F
plants and plant products,
262–264, 263F
Mechanical stress
cell wall remodeling, 304–305
determining cell division plane, 9
effects on lignin, 105–106, 106F
Mechanical support, role of lignin,
96, 96F
Medial Golgi cisternae, 123, 124F
pectic polysaccharide synthesis,
189, 190F, 191
Medicago truncatula, 310F
Medical devices, coating of, 377–378,
378F
Medicinal herbs, 374–376
Medicinal properties, 371–378
Meiocytes, cell plate assembly, 153,
155–157, 156F
Membrane proteins, type II, 164, 168,
168f
Membrane recycling, 134, 135, 136F
Membrane systems, cell wall
assembly, 119–157
during cytokinesis, 145–157
interphase cells, 120–145
Meripilus giganteus, 319F, 352F
Meristematic cells, 2–3, 3F
see also Initials
volumes, 16, 16T, 282, 283T
Meristems, 2–3, 2F
cell division patterns, 6, 7F
Mesophyll cells, 34–35, 36F
air space creation between, 35,
36F
role of microtubules in shaping,
277–278, 279F
spongy, 35, 36F, 258, 259F
Methylation analysis, 56P, 57

β-Methyl cellobioside, 60P
2-*O*-Methyl fucosyl residues, 79
4-*O*-Methyl glucosyluronic acid
residues, 72, 74
Methylsalicylic acid (MSA), 330, 330F
2-*O*-Methyl xylosyl residues, 79
Microasterias denticulata, 139, 140F,
141F, 142F
Microbe-associated molecular
patterns (MAMPs),
321–322, 322F
bacterial flagellin, 322–323, 322F
carbohydrate-binding modules,
346
mechanisms of release, 326, 327F,
345
receptors recognizing, 322–323,
323F, 324, 324F
Microbes
see also Bacteria; Fungi; Pathogens
defense responses *see* Defense
responses
interactions with plants, 319–360,
319F
recognition by plants, 321–324
Micropylar endosperm domain
(MCE), 153F, 154F
Microtubule-associated proteins
(MAP), 274, 275F, 277
Microtubule end-binding protein
(EB1), 274F, 275, 276F
Microtubules (MTs)
anisotropic growth and, 288–289,
289F
bundles
defining specific wall domains,
277–278, 278F, 279F
organization into, 274, 275F
cell plate formation, 147
cellulose microfibril orientation,
245–246, 245F, 246F
control of alignment, 289
cortical array, 245, 245F, 274, 275F
dynamics in live cells, 274–275,
274F, 276F
preprophase band, 8–9, 9F, 149
reorientation in growing cells,
273–277
factors affecting, 276–277, 277F
guiding cellulose microfibrils,
274
reorientation responses, 277
somatic-type cytokinesis, 147,
148F, 149
syncytial-type cell plate
formation, 153, 153F, 154F,
155, 156F
transcellular coordination,
278–279, 279F, 280, 281F
Middle lamella, 1F, 252–256, 253F
barrier function, 238F
cell wall models, 266P
control of cell separation, 257–258
dividing cells, 10, 11F
homogalacturonan localization,
231, 231F
lignification, 216, 217F
lignin composition, 217, 218F

Mini-phragmoplasts, 153, 153F, 154F
Miscanthus, 396, 397
Mitosis
    see also Cytokinesis
    cell plate assembly, 147–152, 148F
    cessation of cytoplasmic
        streaming, 145–146
    redistribution of Golgi stacks,
        146–147
Mixed-linked glucans *see* β-1,3/1,4-
    Glucans
MMNO (4-methyl-morpholine-*N*-
    oxide-hydrate), 228
Models, cell wall architecture,
    264–268, 266–267P
Molecular patterns
    cell surface receptors recognizing,
        322–324, 322F, 323F, 324F
    eliciting defense responses,
        321–322, 322F
Molecules, wall, 43–52
    biochemistry, 67–113
    localization methods, 252,
        254–255P
Monensin, 131, 141
Monoamine oxidase, 86F
Monoclonal antibodies, 255P
Monolignol conjugates, 216
Monolignol glucosides, 209–210, 209F,
    210F
Monolignols, 95, 98F, 203–207, 205F
    see also Coniferyl alcohol;
        p-Coumaryl alcohol;
        Sinapyl alcohol
    biosynthesis, 204–210, 205F, 208F
        coniferaldehyde branch point,
            208
        enzymes, 206T
        phenylalanine modifications,
            206–207
        ring hydroxylation and
            O-methylation, 207–208,
            208F
    C-C linkages, 399, 399F
    defense responses, 337
    free radicals, 210
    glucosylation, storage and
        transport, 208–210, 209F,
        210F
    lignin polymerization from,
        210–216
        enzymatic oxidation, 210–212,
            211F, 211T
        monolignol coupling to
            polymer, 212–213, 212F
        nonenzymatic model, 214–215,
            214F
        oxidative carriers/radical
            mediators, 213, 213F
        using nontraditional
            monomers, 215–216, 215F
    related metabolic pathways, 203,
        204F
Monosaccharides
    α and β configuration, 50P
    d and l forms, 51P
    derivatives, 50P
    glycosyl residue composition
        analysis, 54–57

isomers, 50P
nucleotide sugars, 162–163, 164
primary cell walls, 45, 46P
ring formation, 50P
salvage pathways, 164
structural features, 50–51P
Morphactin, 305
Mosaic wall, 250–268
Mucilage
    root cap, 94
    synthesis, Golgi stack changes,
        132, 133F
Multivesicular bodies (MVBs), 130,
    339
Mung bean hypocotyls, 248, 249F
mur1 mutation, 164
mur2 mutation, 185
mur3 mutants, 347
Muropeptides, 345
Musa, 383
Musical instruments, 384
Myosin, Golgi stack movement, 125
Myrsine laetevirens, 84F

N

NAC domain factors, 299
N-acetylglucosamine
    oligosaccharides, modified,
        310–311, 310F, 312–313
NaD1, 328F
NADPH oxidase, 325, 325F
Nasturtium, 84, 295–296, 303, 331F,
    348F
Necrotrophs, 320, 320T
    cell death, 340
    pectic enzymes, 353–355
New cell walls, origin of, 3–4, 3F, 4F, 5F
Nicotiana see Tobacco
Nigericin, 131, 131F, 141
Nitrobenzene oxidation, 96, 98F, 99
N-linked glycans
    assembly, 120, 122, 175–177, 176F
    complex-type, 178–179, 179F, 180F
    high-mannose-type, 178, 179F,
        180F
    hybrid-type, 178, 179, 179F, 180F
    paucimannosidic-type, 178, 179,
        179F, 180F
    processing in Golgi, 178–179, 180F
    structure, 175F
    transfer to nascent polypeptides,
        177–178, 177F
N-linked glycoproteins, sites of
    assembly, 120, 122
Nocodazole, 125
Nod factors, 310–311, 310F, 312
nod genes, 311
Noncarbohydrate primers *see*
    Primers, noncarbohydrate
Nonclimacteric fruits, 300, 301F
Noncovalent bonds, cross-linking wall
    polymers, 227–228, 228F
Nonhost resistance, 320
Nonreducing end, polysaccharides,
    51P
Norway spruce, 97T
Nuclear magnetic resonance (NMR)
    spectroscopy, 60–61P

conformation analysis, 62–63
    solid state, 61P
    solution, 61P
    structural characterization, 59
Nuclear Overhauser effect (NOE), 61P
Nuclei, syncytial endosperm, 153
Nucleocytoplasmic domains, 153,
    153F, 154F
Nucleoside diphosphate sugars (NDP-
    sugars; nucleotide sugars),
    162–166, 163T
    biosynthesis, 163–164, 164F, 165F,
        166F
    interconversions, 164, 164F, 165F
    pectic polysaccharide synthesis,
        189–190
    polysaccharide biosynthesis from,
        170S
    regulation of levels, 175
    structure, 162–163, 162F
    transport into ER and Golgi,
        166–167, 166F
Nucleotide sugars *see* Nucleoside
    diphosphate sugars
Nucleotide sugar transporters (NSTs),
    166–167, 166F

O

Octadecanoic acid (C18) derivatives,
    218–220, 219F
    cutin, 108F, 110
    suberin, 108–109, 108F
Oil (liquid petroleum), 395
    chemical feedstock production,
        402–403
    peak, 394
Oilseed rape, 10F, 38F
Oligogalacturonides (OGAs)
    control of release, 343–344
    eliciting defense response, 308,
        324, 343
    genes activated by, 343, 343F
    modulation of development, 308,
        308F
    production by pathogens,
        341–343, 342F
    systemic wound response, 330,
        330F
Oligosaccharides, 51P
    conformations, study methods,
        62–63
    degrees of polymerization (DP),
        54S
    eliciting defense responses, 324
    glycosyl linkage composition
        analysis, 56P, 57
    glycosyl residue composition
        analysis, 54–57, 55P
    glycosyl sequencing, 57–58
    mass spectrometry, 58–59
    modified *N*-acetylglucosamine,
        310–311, 310F, 312–313
    NMR spectroscopy, 59
    nomenclature, 47S
    reducing and nonreducing ends,
        51P
    structural characterization, 54, 58
    xyloglucan-derived *see under*
        Xyloglucan

Oligosaccharins, 307
    effects on elongation growth, 308–309, 309F
    generation by pathogens, 342–343
Oligosaccharyltransferase, 177, 177F
O-linked glycans, sites of assembly, 122
O-linked glycoproteins
    biosynthesis, 179–181
    sites of assembly, 120, 122
Onion, 147F, 169, 241F, 369F
Oocystis, 137, 137F, 236F
Oomycetes, 326
Opal, biogenic, 112
Organs, plant, 12, 14–15P
    abscission, 299–300
    development, 4–7
    mechanisms of growth, 283–284, 284F
Outer epidermal wall, 267P
Oxidative burst, 325, 325F
    see also Reactive oxygen species
    wall reinforcement, 336–337, 337F
Oxidative polymerization, 203, 210–212
Oxygen (O$_2$)
    adaptation to conditions of low, 259–260
    lignin polymerization, 211, 211F

P

p115 protein, 126
Paints, 370
Palisade cells, 35, 36F
PAMPs (pathogen-associated molecular patterns), 322
Panax ginseng, 376
Panicum virgatum see Switchgrass
Paper, 365, 384
    enzymatic de-inking, 389–390
    processing, 99, 389–390
    properties, factors affecting, 385–386
Papillae, 335–336, 336F, 337F
    actin rearrangements, 338
    polarized secretions at, 338–340, 339F
Parallel configuration, 68
Paramural bodies, 336, 338
Parenchyma, 14P, 24
Parenchymal cells, 24–25
    effects of cooking, 368, 369F
    polyhedral, 24, 24F
    shapes, 24–25, 24F
Passive reorientation, 247
Pathogen-associated molecular patterns (PAMPs), 322
Pathogenesis-related proteins (PR proteins)
    see also Chitinases; Defensins; β-1,3-Glucanases
    antimicrobial activity, 344–345
    induction by salicylic acid, 330, 345
Pathogens, 319–351
    cell wall-degrading enzymes, 341–343, 342F
        coevolution with inhibitors, 349–351

coevolution with host, 327F, 328
defense responses see Defense responses
effectors, 326–328
entry into plant host, 331–333, 331F, 332F
immunity to, 320
interactions with plants, 341F
lifestyles, 320, 320T
recognition by plants, 321–324
spread via plasmodesmata, 333–335, 334F
strategies for evading plant surveillance, 350, 350F
Pattern-recognition receptors (PRRs), 322–324, 323F, 324F
    effects of activation, 325, 325F
Pattern-triggered immunity (PTI), 322
Paucimannosidic-type N-linked glycans, 178, 179, 179F, 180F
PDCB proteins, 333, 333F
PDLP1 protein, 33F
PDLP proteins, 334, 334F
Pea
    casparian bands, 26F, 27F
    cell separation, 258, 258F
    fucosyltransferase, 185, 185F
    growth effects of XXFG, 308–309, 309F
    microtubules, 274, 277, 277F
    monoamine oxidase, 86F
Peak oil, 394
Pear, 34, 35F, 94F
Peat, 395
Pectate lyase, 301, 302, 354–355, 354F
    carbohydrate-binding module, 346F
    structure, 355F
Pectic polysaccharides (pectins), 352F
    bioactive, 374–377
    biosynthesis, 186–191
        enzymes, 186–189, 187T, 188T
        in vitro studies, 189–191
        localization, 189
        pathway, 189, 190F
    changes over time, 262
    effect on cellulose mechanics, 263F, 264
    enzymatic digestion
        pathogens, 341–342
        rumen microorganisms, 367
        saprophytes, 352–353, 354–355
    fruit ripening, 300
    gel formation, 75, 76F, 151, 230–231, 231F
    glycosidic cross-links, 229–230, 230F
    heterogeneous wall deposition, 252, 253F
    industrial uses, 370–371, 371F
    medical uses, 374–378, 377F, 378F
    methylesterification, 262, 301
    middle lamella, 252–256, 253F
    outer epidermal wall, 267P
    perturbations activating defense responses, 347
    primary cell walls, 45–47, 52T
        see also Pectin network

secondary walls, 49, 52T
solubilization and purification, 47, 47F, 52–53
structure, 75–80
three-way junctions, 257, 257F
Pectin, 47
Pectin esterase, 302
Pectin lyases, 354–355
Pectin methylesterase (PME), 231, 292
    fruit ripening, 301–302
    saprophytes, 353, 354F
    structure, 355F
Pectin network, 235, 240–243
    control of wall thickness, 242–243, 243F
    effects on wall porosity, 241–242, 241F
    in vitro studies, 240, 240F
    model, 266P
    remodeling during cell expansion, 291
    self-assembly, 249
Pectins see Pectic polysaccharides
Pedicel, elongation, 286
PEN2, 339, 339F
Pentoses, 45, 46P, 50P
Peptidoglycan, bacterial, 345, 345F
Perforation plates, 298–299, 298F
    complex, 299, 299F
    simple, 299, 299F
Perfumes, microencapsulated, 370, 371F
Pericycle, 26–27, 27F
Periderm, wound, 106
Periodic Acid-Schiff stain, 254P
Peripheral endosperm (PEN) domain, 153F
Peroxidases
    cross-link formation, 232, 234
    lignin polymerization, 211, 211F, 211T
Peroxisomes, 339, 339F
Pests, 320, 321F
Petal epidermal cells, 20, 20F, 21
Petroleum, liquid see Oil
Peyer's patches, 374–375
Phanerochaete chrysosporium, 393P
Pharmaceutical industry, 374
Phaseolin, 329F
Phase separation, 248
PhD1, 328F
Phenolic acids, 72, 78, 110
Phenolics, wall reinforcement, 337
Phenylalanine
    monolignol biosynthesis, 204–207, 205F, 206T
    origins, 204, 204F, 206
Phenylalanine ammonia-lyase (PAL), 205F, 206, 206T, 399
Phenyl coumarin, 101F
Phenylglucoside, 103, 105F
Phenylpropanoids, 203–204
    general metabolic pathway, 203, 204F
    lignin, 99
    suberin, 109
pH gradients, 131
Phloem, 15P, 29–31
    cell types, 13

companion cells, 15P, 30, 30F
    sieve tube elements, 17F, 29–31,
        30F
    in vascular bundles, 29, 29F
Phloroglucinol, 99, 100F, 215, 254P
Photosynthesis-driven biomass
        production, 395–396
Phragmoplast initials, 147, 148F
Phragmoplasts, 4
    cell plate assembly, 4, 6F, 147
    mini-, 153, 153F, 154F
    polysaccharide synthesis by
        isolated, 4, 5F
    ring, 148F, 149
    solid-type, 148F, 149
    transitional, 148F, 149
Phytoalexins, 329, 329F
Phytoanticipins, 328
Phytohormones, 307
Phytoliths, 112
Phytophthora parasitica, 346
Phytophthora sojae, 351, 351F
Pilus, 326
Pine wood, 29F, 97T, 385F
Pinoresinol, 101F, 214, 214F
Pinus radiata, 218F
Pitch, 390
Pit fields, 32F
    fungal spread via, 335, 335F
    primary, 297–298, 297F
Pits, xylem vessel elements, 299, 299F
Plant fibers see Fibers, plant
Plant immunity see Immunity, plant
Plant–microbe interactions, 319–360,
        319F
Plasma membrane, 120F
    assembly of cell wall polymers,
        120, 121
    association proteins, 281–282,
        282F
    attachment to casparian band,
        25, 26F
    callose synthesis, 136–137, 137F,
        142–145
    cellulose microfibril synthesis,
        136–142
    cell wall adhesion, 281–282, 282F
    continuity, 11, 12F
    defense responses, 325, 325F
    fusion with tonoplast, 341, 345
    mixed-linked glucan synthesis,
        136–137
    phloem sieve elements, 30
    plasmodesmata formation, 296,
        297F
    polysaccharide assembly, 193–202
    preprophase band "memory," 9,
        149
    recycling, 134, 135, 136F
    role in solute transport, 12
    sensing of wall integrity, 348
    underlying papillae, 336, 338
Plasmodesmatal-located proteins
        also plasmodesma-located
        proteins (PDLPs), 334, 334F
Plasmodesmata, 32–33, 32F
    callose sleeve, 33, 143F, 144
    changes over time, 262

endoplasmic reticulum, 121F, 122
    gating by callose, 298, 333, 333F
    paired, 296–297, 297F
    pathogen spread via, 333–335,
        334F
    primary, 296
        assembly, 152
    protein targeting, 33, 33F
    secondary, de novo insertion,
        296–298, 297F
    separating apoplast and symplast,
        11, 12F
    symplastic domains, 33, 33F
    wound-induced closure, 145
Plasmolysis, 281, 282F
Plastids, nonphotosynthetic see
        Leucoplasts
Ploidy
    cell size and, 283, 285F
    trichome size and, 285, 286F
pmr4 mutants, 145
PMR5, 347
PMR6, 347
Poaceae
    β-1,3/1,4 glucans, 81, 82
    hemicelluloses, 69, 73, 74
Pollen
    development
        role of callose, 143–144, 144F
        syncytial-type cell plate
            formation, 155–157, 156F
        tapetal cells, 37, 38F
    grains
        cell wall layers, 37, 37F
        germination, 38, 38F
        sculptural patterns, 37, 37F
    tube
        callose synthase, 201
        role of callose, 144, 144F
Polygalacturonase-inhibitor proteins
        (PGIPs), 49, 344, 344F, 350
Polygalacturonases (PGs)
    see also Endo-polygalacturonases;
        Exo-polygalacturonases
    cell wall degradation
        pathogens, 342–343, 342F
        saprophytes, 352–353, 353F
    fruit softening, 301–302
Polymerization, polysaccharides, 169,
        169F
Polymers
    biosynthesis, 161–221
        general mechanisms, 161–175
        glycoproteins and
            proteoglycans, 175–181
        membrane systems involved,
            119–157
        polysaccharides, 182–202
        within the wall, 203–221
    building blocks, 161–162
    cross-links between, 227–235, 256
    linear glucose, 120, 136–137
    localization methods, 252,
        254–255P
    primary walls, 43–47, 52T
    secondary walls, 49–52, 52T
    uneven distribution within wall,
        250–252

wall construction, 235–250
Polyphenolics
    hypersensitive cell death, 340,
        340F
    wall reinforcement, 337
Polysaccharides, 43–63
    biochemistry, 67–85
    biosynthesis
        backbones, 168–173, 173F,
            174F, 182–185
        control, 175
        elongation, 168, 169, 169F
        endomembrane system,
            182–192
        initiation, 168–169
        membrane-bound enzymes,
            167–168
        plasma membrane, 193–202
        sites, 120
        source of glycosyl residues,
            162–165
        termination, 168, 169
    complex, 120
    degrees of polymerization (DP),
        54S
    diferulic acid cross-links, 232,
        232F
    enzymatic digestion
        pathogens and saprophytes,
            341–343, 352–359
        rumen microorganisms, 367
    ester cross-links, 233, 233F
    extracellular secreted, 53, 53F
    exudate gums, 85
    lignin cross-linking to, 103–104,
        105F
    localization methods, 254–255P
    nomenclature, 47S
    nonreducing end, 51P
    primary cell walls, 44–47, 52T
        cellulose microfibril spacing
            and, 237–240
        construction principles,
            235–236, 235F
    primary structure determination,
        47–48, 52–59
    reducing end, 51P
    secondary cell walls, 49–52, 52T
    storage, seed cell wall, 84, 84F,
        295–296, 296F, 302–303
    structural characterization, 47–48,
        52–63
        conformational analysis, 62–63
        enzymatic and chemical
            cleavage, 53–54
        glycosyl linkage composition
            analysis, 56P, 57
        glycosyl residue composition
            analysis, 54–57, 55P
        glycosyl sequencing, 57–58
        mass spectrometry, 58–59, 59F
        NMR spectroscopy, 58–59,
            60–61P
        solubilization and purification,
            52–53
    structural features, 50–51P
    turnover, 294–296

Polysaccharide synthases
  membrane-bound, 167
  methods of identification, 172
Poplar
  biofuel production, 396, 397, 398F, 406F
  lignin composition, 104
  tension wood, 306F
  xyloglucan *endo*-transglycosylase, 290F
Populus nigra, 218F
Porosity, wall
  lignin polymerization and, 203
  role of pectin network, 241–242, 241F
Positional information, determining cell fate, 18
Potato
  defense responses, 330, 331
  suberin, 107, 109
  three-way junctions, 259F
P-proteins, 30F, 31
prc1 (Procuste1) mutants, 195, 195T
Precession, 60P
Preprophase band, 8, 8F
  actin-free zone replacing, 9–10, 11F, 149
  control of plane of cell division, 8–9, 8F
  "memory" on plasma membrane, 9, 149
  microtubules, 8–9, 9F, 149
Primary cell walls, 18–19
  aqueous phase, 49
  architecture, 235–244
    control of thickness, 242–243, 243F
    lamellae, 236–237, 237F, 238F
    microtubule-mediated control, 245–246, 245F, 246F
    models, 266–267P
    pectin network, 240–242, 241F
    role of proteins, 243–244, 244F
    spacing of cellulose microfibrils, 237–240
    two-network model, 235–236
  cellulose, 44–45
  cellulose microfibril length, 142
  composition, 18T, 43–49, 52T
  hemicelluloses *see under* Hemicelluloses
  methods of preparation, 43–44, 44F
  monosaccharide components, 45, 46P
  pectic polysaccharides, 45–47
  pores, 203, 241–242, 241F
  tissues rich in, 43S
Primary pit fields, 297–298, 297F
Primary plasmodesmata, 152, 296
Primary structure, determination, 47–48, 52–59
Primers, noncarbohydrate, 169
  callose synthesis, 200, 200F
  cellulose synthesis, 198, 198F, 199F
Printing industry, 370
Processive glycosyltransferases, 183, 183F, 196

Procuste1 (prc1) mutant, 195, 195T
Product entrapment, 172, 193
Proline hydroxylation, 87
Proline-rich proteins (PRPs), 48, 87T
  control of expression, 88–89, 89F
  evolutionary origin, 95
  integration into wall architecture, 244, 244F
Prolylhydroxylase, 180–181, 181F
Proso millet, 295
Prostate cancer, 376–377
Proteinase inhibitor proteins (PINs), 330, 330F
Proteins
  cell wall, 48–49
    architectural integration, 243–244, 244F
    architectural models, 267P
    biochemistry, 85–95
    cross-linkage to lignin, 104
    enhancing wall expansion, 292–293, 293F
    insolubilization by covalent cross-links, 233–234, 234F
    localization methods, 254–255P
    sites of synthesis, 122
    spatial distribution, 86, 86F
    structural *see* Structural cell wall proteins
  plasma membrane-associated, 281–282, 282F
  retention sequences, 122
  retrieval sequences, 122
  vacuolar targeting sequences, 122
Proteoglycans, 49, 90–93
  assembly and processing, 175–181
Proton pumps, 131
Protoxylem elements, 17F
  see also Xylem vessel elements
  lignin deposition, 19, 28
  wall thickening, 27–29, 28F
PRP *see* Proline-rich proteins
PR proteins *see* Pathogenesis-related proteins
PRRs *see* Pattern-recognition receptors
Puccinia graminis, 340
Pulp, 384
  processing, 389–390, 389F
  properties, factors affecting, 385–386

Q

Q/RXXRW motif, 194F, 196, 196F
  see also D, D, D, Q/RXXRW motif
Quercus suber (cork oak), 390, 390F

R

Radial heterogeneity, 251, 251F
Radiolabeling, 254P
Ramie, 383
Ray parenchyma cells, 385, 385F
Reaction wood, 105–106, 106F
  cell wall remodeling, 304–305, 305F, 306F
  density, 388
  lignin composition, 218F

Reactive oxygen species (ROS), 325, 325F
  see also Hydrogen peroxide
  hypersensitive response, 341, 341F
  wall reinforcement, 336–337, 337F
Recalcitrance, lignocellulosic biomass, 402
Receptors
  recognizing common molecular patterns, 322–324, 322F, 323F, 324F
  signal molecules, 307
Recycling, cell wall, 351–359
Redifferentiation, cell, 19–20, 19F
Redox shuttle model, lignin biosynthesis, 213, 213F
Reducing end, polysaccharides, 51P
Red wine, 79
ref8 mutant, 207
Regional heterogeneity, 251, 251F, 252F
Remodeling, cell wall, 294–305
  de novo insertion of plasmodesmata, 296–298, 297F
  facilitating cell expansion, 290–292, 290F, 291F
Renewable natural resources, 391–407
Reorientation (of cellulose microfibrils)
  directed, 247–248, 248F
  passive, 247
Resistance
  see also Defense responses
  genes, 326–327, 327F
  nonhost, 320
  systemic, 329–331, 330F
  systemic acquired (SAR), 329–330, 330F
Resonance frequency, 60P
Resveratrol, 329F
Retention sequences, 122
Retrieval sequences, 122
Retting, 381, 382
Reversibly glycosylated protein, 167
R-genes, 326–327, 327F
RG-I *see* Rhamnogalacturonan I
RG-II *see* Rhamnogalacturonan II
Rhamnogalacturonan I (RG-I), 45–47
  bioactive molecules, 375–376, 375F
  biosynthesis, 172, 173, 173F
    enzymes, 187T, 189
    pathway, 189, 190F
  enzymatic digestion, 354, 355, 355F
  glycosidic cross-links, 229–230, 230F
  heterogeneous wall deposition, 252F, 253F
  regulation of wall porosity, 241, 242
  related exudate gum, 85
  structure, 75–78, 78F
    composition of side chains, 77–78
    disaccharide backbone, 76–77

methods of analysis, 47, 47F, 52, 58, 75–76
Rhamnogalacturonan II (RG-II), 45–47
  bioactive molecules, 376
  biosynthesis, 172
    enzymes, 188T, 189
    termination of polymerization, 169
  dimers (d-RG-II), 79, 79F
    borate ester cross-links, 231–232, 232F
    self-assembly, 249
  glycosidic cross-links, 229–230, 230F
  regulation of wall porosity, 241
  structure, 78–80, 81F
    methods of analysis, 47, 47F, 52–53, 58, 79, 80
    oligogalacturonide backbone, 80
    side chains, 80, 81F
Rhamnogalacturonan hydrolase, 355, 355F
Rhamnogalacturonan lyase, 355, 355F
L-Rhamnose, 46P
Rhamnosyl-galactosyluronic acid residues, 172
Rhamnosyl residues, 76, 77, 78F, 79, 85
Rhizobial bacteria, 310–311
Ribose, 50P
Ribosomes, endoplasmic reticulum-bound, 121–122
Ribulose, 50P
Rice
  glycosyltransferases, 171T, 183–184
  lignin composition, 97T
  silica, 111F, 112
Rice blast disease, 332–333, 332F, 333F, 335, 335F
RING-finger domains, 197
Ring phragmoplast, 148F, 149
Ripening, fruit, 264, 300–302
RNA localization, 254P
Root cap
  development, Golgi changes, 132, 133F
  mucilage, 94
Root hairs
  cell wall heterogeneity, 251, 251F
  Nod factor-induced deformation, 310F, 311
  tip growth, 287, 287F
Root meristem, 2, 2F
  trans Golgi network cisternae, 129, 129F
Roots
  anisotropic growth, 288
  arabinoglycan proteins, 260F, 261
  cell differentiation, 17–18, 17F
  cell wall heterogeneity, 251, 252, 253F
  endodermis *see* Endodermis, root
  lateral, formation, 26–27, 27F, 28F
Rosette complexes, 137, 138F, 141
  see also Cellulose synthase complexes

chemicals affecting, 140–141
distribution patterns, 139–141, 141F, 142F, 277, 278F
half-life, 141
heteromeric model, 198, 198F
laterally aggregating cellulose microfibrils, 139, 140F, 141F
movement within plasma membrane, 138, 139F
  microtubule-mediated guidance, 245–246, 246F, 274, 289, 289F
transport to plasma membrane, 141, 142F
Rosin, 403
Rs-AFP1, 328F
rsw1 mutant, 195, 195T
Ruminants, 366–367, 367F
Ruminococcus, 367
Rye, 8F
Ryegrass, 112

S

S1 layer
  cellulose microfibril orientation, 247–248, 248F
  formation, 216
  reaction wood, 305, 306F
S2 layer
  cellulose microfibril orientation, 247–248, 248F
  formation, 216
  reaction wood, 305, 306F
  strength of wood and, 388
  tracheids, 385, 386F, 387
S3 layer
  cellulose microfibril orientation, 247–248, 248F
  formation, 216
  reaction wood, 305, 306F
Salicylic acid (SA), 307, 329–330, 330F
  hypersensitive response, 341, 341F
  induction of PR proteins, 330, 345
Salvage pathways, monosaccharide, 164
Saprophytes, 320, 351–359
  breakdown of plant substrates, 351–352, 352F
  wall-degrading enzymes, 345–346, 352–359
Scaffoldin, 401P
Scalar coupling, 61P
Schizogony, 260, 260F
Sclereids, 34, 35F
Sclerenchyma, 14P, 34, 35F
Scots pine, 97T
Secondary cell walls, 19, 19F
  cellulose, 44–45, 49
  cellulose microfibril length, 142
  composition, 49–52, 52T
  hemicelluloses, 49–52, 52T, 74
  layers *see* S1 layer; S2 layer; S3 layer
  lignification, 216, 216F, 217F
  lignin, 95–96
  lignin–hemicellulose complexes, 234–235, 234F
  lignin polymerization, 203, 217F

specialized architecture, 20–37, 20F
turnover and degradation, 303–304
Secondary plasmodesmata, 296–298, 297F
Secondary thickening, 19, 286
Secretion
  constitutive, 129
  vesicle-mediated, 134–135, 135F
Secretory pathway, 119, 119F
Secretory vesicles (SV), 122F, 129
  cell plate-forming, 130F, 147
  delivery to cell surface/plate, 133–134
  discharge of contents, 134–135, 135F
  formation, 129–130, 129F
  mechanism of exocytosis, 134
  scaffold acquisition, 130, 130F
  site of origin, 124, 124F
  transport of rosette complexes, 141, 142F
Seed cell wall, storage polysaccharides, 84, 84F, 295–296, 296F, 302–303
Seed hair fibers, 378–381, 379T
Selaginella moellendorffii, 97T, 104
Self-ordering properties, polylamellate walls, 248–249, 248F, 249F
Shikimate, 204, 204F
Shoot apical meristem, 2, 2F
Sieve elements *see* Sieve tube elements
Sieve plates, 15P, 30F, 31
  pores, 144
  wound-induced closure, 30F, 31, 144–145
Sieve tube elements (sieve elements), 17F, 29–31, 30F
Sieve tubes, 15P, 30
Signal molecules, 306–313, 307F
  binding to receptors, 307
  cell wall-derived, 85–86, 307–313
  eliciting defense responses, 321–322
  structural specificity, 307, 307F
Signal transduction pathways, 307
Silica, 111–112
  bodies, 112, 112F
  transporters, 112
Sinapaldehyde, 208, 215
Sinapyl alcohol, 98F, 204, 205F
  biosynthesis, 204–207, 205F, 206T
  C-C linkages, 399, 399F
  coupling to lignin polymer, 212
  glucosylation, 209, 209F
Sinapyl alcohol dehydrogenase (SAD), 206T, 208
Sisal, 383–384, 384F
Sitosterol-β-glucoside, 198, 198F, 199F
Small interfering RNAs (siRNAs), 175
Snapdragon, 20F
Soft fibers *see* Bast fibers
Softwoods
  density, 387F, 388
  length of fiber cells, 386

lignin-carbohydrate complexes, 104
lignin composition, 96, 97T, 104, 105
lignin content, 97T
lignin linkages, 101
reaction wood, 105–106, 305, 306F
structure, 385, 385F
Solanaceae, 69, 83–84
Solanaceous lectins, 86, 87T, 179–180
Solid phragmoplast, 148F, 149
Solute transport, 12
apoplastic barrier, 25, 26F
phloem, 29–31
transfer cells, 31, 31F
Solvolysis, 98
Sorbitol, 403, 404T
Sorghum, 112, 367
Soybeans, 88, 112, 244F
Spaces, intercellular *see* Air spaces, intercellular
Specialized cell walls, functionally, 20–37, 20F
cellulose content, 44
Sperm cells, 38, 38F
Spilocea oleaginia, 321F
Spiral growth patterns, 288
Spongy mesophyll cells, 35, 36F, 258, 259F
Sporophytic self-incompatibility, 144
Sporopollenin, 37
Springwood, 385F
Stains, histochemical, 254P
Starch, 169, 302–303
Stem cells (initials), 2–3, 13
Stomata, 15P, 22–24
entry of pathogens via, 331–332, 331F
opening and closing, 23, 24F
Stomatal complex, 22–24, 22F
development, 8F, 9F, 22–23
different types, 23F
Stomatal guard cells, 22–23, 22F
asymmetrically thickened wall, 23, 24F
symplastic restriction, 12
Storage polysaccharides
seed cell wall, 84, 84F, 302–303
turnover, 295–296, 296F
Stress
see also Abiotic stress; Biotic stress; Mechanical stress
callose synthesis, 144–145
gum secretion, 84–85
insolubilization of HRGPs, 244
signal tranduction pathways, 307
Structural cell wall proteins, 48, 85–95
biological functions, 86
classes, 86, 87T
evolutionary origin, 95
integration into wall architecture, 243–244, 244F
intra- and intermolecular cross links, 88
posttranslational modifications, 87–88, 87F
repeated sequence motifs, 86–87
Suberin, 106–109, 218
biosynthesis, 121, 219–220, 219F

casparian bands, 25, 106
defense response, 330–331
isolation and purification, 108
lamellar structure, 107–108, 107F
monomeric building blocks, 108–109, 108F
oligomeric degradation products, 109, 109F
resistance to degradation, 304
Substituted galacturonans, 83, 83F
Succinic acid, 403, 404T, 405F
Sucrose, 51P
translocation, 29–31
UDP-glucose synthesis from, 199–200
Sucrose synthase (SuSy), 199–200, 199F, 200F
Sugar beet, 22F
Sugarcane, 9F, 396–397, 397F
Sugar-ester cross-links, 233, 233F
Sugar nucleotides *see* Nucleoside diphosphate sugars
Sugars, 50–51P, 162
chemical feedstock production, 403, 404T
nucleotide sugar synthesis, 162–164, 164F
polysaccharide backbone synthesis, 168–169
Suicide, plant cell, 327, 340–341, 340F
S units *see* Syringyl units
Superoxide (O_2^-), 325, 337
Superoxide dismutase, 341
Switchgrass, 396, 397, 398F
Sycamore
Golgi compartments, 131, 179
pectic polysaccharides, 76, 77, 189, 190F, 230
xyloglucans, 69, 71–72
Symplast, 11–12, 12F
Symplastic domains, 33, 33F
Syncytial-type cell plate, 153–157
Syringaldehyde, 98, 98F, 99
Syringin, 209, 209F
Syringyl units (S units), 98F, 99
content of different plants, 97F, 104
different wall regions and cell types, 217, 218F
intermolecular linkages, 100–102, 101F
manipulation in biofuels, 399
monolignol precursor, 204, 205F
Systemic acquired resistance (SAR), 329–330, 330F
Systemic resistance, 329–331, 330F
Systemic wound response, 329, 330–331, 330F, 331F
Systemin, 330, 330F

T

Tailward growth, polysaccharides, 169, 169F
Tamarind, 84, 172, 185–186, 303
Tangled mutant, 6–7, 7F
Tannins, 403
Tapetal cells, 37–38, 38F
TAXI xylanase inhibitors, 350
Taxol, 288

Template-based assembly, 249–250
Tensile strength, wood, 385, 388
Tension wood, 106, 106F, 305, 305F
cell wall remodeling, 305, 306F
lignin composition, 218F
strength, 388
Teosinte, 112
Terminal differentiation, 95
Termination, biopolymer synthesis, 168, 169
Textiles, 365, 378–384
plant fibers, 378, 379T
use of gum arabic, 370
TGN *see trans* Golgi network
THESEUS (THE) kinase, 348
Thickness, wall
properties of wood and, 385
regulation, 242–243, 243F
things fall apart (tfa) mutants, 258
Thioacidolysis
lignin, 98, 98F, 99
suberin, 109
Thioglucosidase, 339, 339F
Thionins, 49
Three-way junctions, 256–257, 256F
3D network, 258–259, 259F
adhesion at, 257, 258F
cell separation at, 257–258, 258F
generation, 10, 11F
molecules localized to, 231, 231F, 257, 257F
preference for, 9, 10F
Thymidine, 163
Tinopal LPW, 141
Tip growth, 286–287, 287F
Tissue print, 254P
Tissues, plant, 13
Tissue systems, plant, 12–16, 14–15P
early development, 13, 13F
Tobacco (*Nicotiana*)
callose, 82
expansins, 293, 293F
galactoglucomannans, 83–84
genetic engineering, 399
leaf architecture, 36F
membrane systems, 120F, 125
oligogalacturonides, 308, 308F
phragmoplasts, 5F
plasmolysis, 282F
polysaccharides, 53F
preprophase band, 8F
prolylhydroxylase, 180–181
stomatal complex, 22F
vascular bundles, 29F
wall proteins, 88, 244F
xyloglucans, 69
Tobacco BY-2 cells
callose assembly, 137F
cell plate assembly, 146–147
Golgi apparatus, 125, 125F, 129, 146, 146F
Tobacco mosaic virus, 334
Tomato, 21F
abscission zone, 301F
cell wall architecture, 237, 238F, 240, 240F, 243F
delayed fruit deterioration cultivar, 302
leaf mold, 326, 327F

middle lamella, 253F, 256
   ripening, 264, 301–302, 302F
   xyloglucans, 69
Tonoplast, 283–284
   fusion with plasma membrane, 341, 345
Toughness, food plants, 368
Tracheids, 28F, 29
   cellulose orientation, 247–248, 248F
   developing, polar end wall removal, 298–299, 298F
   length, 386
   microtubule-defined domains, 277, 278F
   multilayered wall architecture, 385, 386F, 387
   reaction wood, 305, 306F
   regional wall heterogeneity, 251, 252F
   wood structure, 385, 385F
Tradescantia, 280F
Tragacanth, 369
Transesterification, 233, 233F
Transfer cells, 14P, 31, 31F
   regional wall heterogeneity, 251, 252F
trans Golgi cisternae, 123–124
   pectic polysaccharide synthesis, 189, 190F
   progression/maturation model, 128, 128F
   transformation into TGN cisternae, 128, 129F
   xyloglucan biosynthesis, 184–185, 184F, 186
trans Golgi network (TGN), 122F
   functions, 128–129
   scaffold, 126F, 130, 130F
   vesicle origins/trafficking, 124F
trans Golgi network (TGN) cisternae, 123F, 124, 128–130
   acidification, 131–132
   early, 128, 129, 129F, 130F
   formation, 128, 129F
   late, 129, 129F, 130F
   vesicle release, 129–130
   xyloglucan biosynthesis, 186
Transitional phragmoplast, 148F, 149
Trichoblasts, 253F
Trichoderma viride, 323
Trichomes (hairs), 15P, 18F
   cotton, development, 380–381, 380F
   diffuse growth, 287
   effect of ploidy on size, 285, 286F
   morphogenesis, 18
Trimethylsilyl (TMS) methyl glycosides, 55P
Trioses, 50P
Tubule-forming viruses, 334, 334F
Tubulin
   antibodies, 8F, 9F, 153F
   fluorescently labeled, 274
   fluorescent protein fusion, 245F, 246, 274
α-Tubulin, detyrosination, 277
Tufted hair grass, 112
Tunicamycin, 176–177

Turgor pressure, 134–136
   mechanical consequences, 264
   membrane recycling and, 135, 136F
   vesicle-mediated secretion and, 134–135, 135F
   wall microscopy and, 264, 265F
   wall tension generation, 237–238
Turnover, cell wall, 294–305
Turpentine, 403
Two-network model, cell wall architecture, 235–236
Type II membrane proteins, 164, 168, 168F
Type III secretion, 326
Tyrosine- and lysine-rich proteins (TLRP), 234

U
UDP-apiose (UDP-Api), 165F
UDP-arabinose (UDP-Ara), 164, 165F
UDP-arabinose mutase, 167
UDP-galactose (UDP-Gal), 162, 165F, 191
UDP-galacturonic acid (UDP-GalA), 164, 165F, 189–190
UDP-glucose (UDP-Glc), 162, 165F
   biosynthesis from sucrose, 199–200, 200F
   callose synthesis, 201
   cellulose synthase binding, 193, 196
   interconversion reactions, 164, 165F
   mixed-linked glucan synthesis, 202
UDP-glucose coniferyl alcohol glucosyltransferase, 209
UDP-glucose (UDP-Glc):glycoprotein glucosyltransferase, 178
UDP-glucose (UDP-Glc) transferase, 201–202
UDP-glucuronic acid (UDP-GlcA), 164, 165F
UDP-glucuronic acid (UDP-GlcA) 4-epimerase, 164, 166F
UDP-rhamnose (UDP-Rha) synthase, 164, 165F
UDP-xylose (UDP-Xyl), 164, 165F
UDP-xylose (UDP-Xyl) 4-epimerase, 164, 166F
UDP-xylose (UDP-Xyl) synthase, 164, 166F
Ultraviolet (UV) absorbance, 54
Uridine, 163
Uridine diphosphate see UDP

V
Vacuolar H⁺-ATPase (V-ATPase), 131–132
Vacuolar H⁺-pyrophosphatase (V-PPase), 131
Vacuolar targeting sequences, 122
Vacuolation
   cell expansion by, 16, 16F, 283
   regulation, 283–284
Vanillin, 96–98, 98F, 99, 403
Vascular bundles, 15P, 29, 29F
   glycine-rich proteins, 94

Vascular tissue system, 13, 14P, 15P
   development, 13F
V-ATPase, 131–132
Vegetables, texture, 368–369
Veronica kotschyana, 376
Very-long-chain fatty acids (VLCFAs), 219–220, 219F
Vesicles, 123–124, 124F
   see also Clathrin-coated vesicles; Golgi-derived vesicles; Secretory vesicles
Vessel elements, xylem see Xylem vessel elements
Vessels, 27
Viruses
   entry into plant host, 333
   spread via plasmodesmata, 333–334, 334F
   tubule-forming, 334, 334F
Volicitin, 349

W
WAK proteins (wall-associated kinases), 281–282, 282F, 324, 348
Wall-integrity signaling, 346–348, 347F, 348F
Walnut, black, 97T
Water chestnuts, 257, 258F, 369
Water content, primary cell walls, 49
Water transport, 12
   apoplastic barrier, 25, 26F
   role of lignin, 96, 96F
   xylem vessels, 27
Waxes, 218–221
   biosynthesis, 219–220, 219F, 221
   epicuticular, 111, 218
      chemistry, 111, 111T
      regulation of synthesis, 221
   intracuticular, 218
   outer epidermal wall, 267P
   transporters, 121, 220
WAX INDUCER 1 (WIN1), 221
Wax layer, 21–22, 21F
   cotton fibers, 380, 381F
   crystalline morphology, 22F
wax mutants, 221
Wax synthases, 219F, 220
Waxy cuticle, 218
WBC1, 220
Weisner reaction, 99, 100F
Wetland plants, aerenchyma, 259–260, 260F
Wheat
   bran, 373–374, 374F
   leaf wax layer, 22F
Wheat germ agglutinin (WGA), 312–313
White rot fungi, 319F, 358, 392–393P
Whole mounts, 255P
Willow, 396
Wood, 28F, 29, 29F, 384–390
   breakdown, 352, 352F, 358–359, 359F
   chemical feedstock production, 403
   compression see Compression wood

delignification, 392–393P, 402, 402F
density, 387–388, 387F, 388F
exploitation by humans, 365, 366F, 384–385
heat value, 396
lignin-carbohydrate complexes, 104
lignin content, 96, 97T
lignin extraction from, 96–98
multilayered cell wall structure, 385, 386F, 387
properties, factors affecting, 385–389
reaction see Reaction wood
resistance to degradation, 304
rots, 352, 358–359, 392P
silica content, 112
strength, determinants, 387–389
structure, 385, 385F, 386F
tension see Tension wood
use as fuel, 365, 366F, 396
Wood chips, 389F
    see also Lignocellulosic biomass
    delignification, 392P, 402, 402F
    pulp and paper manufacture, 389, 390
Wound-induced callose, 30F, 31, 144–145, 335–336, 349, 349F
Wound periderm, 106
Wound response
    arabinogalactan proteins, 91
    callose deposition, 30F, 31, 144–145, 335–336, 349, 349F
    cell redifferentiation, 19–20, 19F
    cell wall changes, 262
    cytoskeletal changes, 278–279, 279F, 280F
    extensins, 89, 90F, 267P
    gum secretion, 84–85, 91
    local, 349
    overlap with response to pathogens, 348–349
    suberin production, 106
    systemic, 329, 330–331, 330F, 331F
    wall reinforcement, 337
WRKY transcription factors, 325

X

XEGIPs, 350
XEH see Xyloglucan endohydrolase
XET see Xyloglucan endotransglycosylase
XFFG, 71, 309, 309F
XG see Xyloglucan
XG-Fuc transferases, 185–186, 185F
XG-Gal transferases, 185–186, 186
XG-Glc synthase, 184
XIP1, 350
XLFG, 70P, 71
XLLG, 70P
XLXG, 70P
X-ray diffraction, 63, 63F
XTH see Xyloglucan transglucosylase-hydrolases
XXFG, 62, 69–71, 70P, 71F, 308–309, 309F

XXFGAXXG, 70P
XXG, 69–71, 70P, 71F
XXLG, 70P, 309
XXXG, 62, 69–71, 71F, 309
XXXGBXFG, 71, 72F
XXXGol, 70P
XXXGXXXG, 70P
Xylanase inhibitor proteins (XIPs), 350, 350F
Xylanases
    see also Endo-xylanases
    carbohydrate-binding modules, 346, 346F
    ethylene-inducing (EIX), 323, 323F, 324F
    pulp and paper processing, 389
Xylans
    see also Arabinoxylans;
        Glucuronoxylans
    biosynthesis, 192
    lignin cross-links, 234F
    localization methods, 254P
    secondary walls, 74
Xylem, 15P, 27
    glycine-rich proteins, 94
    primary, 28F
    secondary, 28F, 29, 29F
    vascular bundles, 29, 29F
    wood, 385–386, 385F
Xylem vessel and tracheid differentiation, 277
Xylem vessel elements, 15P, 27–29, 28F
    arabinoglycan proteins, 261, 261F
    cell redifferentiation into, 19–20, 19F
    cellulose orientation, 247–248, 248F
    microtubule-defined domains, 277, 278F
    perforation plates, 298–299, 298F, 299F
    pits in lateral walls, 299, 299F
    polar end wall removal, 298–299, 298F
    regional wall heterogeneity, 251, 252F
    rosette complexes, 140, 142F
    wall thickenings, 27–29, 28F
    wound-induced differentiation, 278–279, 279F
Xylem vessels, 27
Xylitol, 403, 404T
Xylogalacturonan, 83, 83F
Xyloglucan (XG), 47
    biosynthesis, 184–186
        3D backbone structure and, 172–173, 173F
        block polymer construction, 169, 185
        fucosyl and galactosyl residue addition, 185–186
        glucosyl and xylosyl residue assembly, 184–185, 185F
        localization, 186
    cellulose microfibril cross-links, 229, 229F
    degrading enzymes, 356–357, 357F
    deposition in cell plate, 238, 239F

effect on cellulose mechanics, 263–264, 263F
hydrogen bonding to cellulose microfibrils, 228, 228F, 229F
localization methods, 255P
oligosaccharides, 69–72, 71F
    17 to 21 glycosyl residue, 71–72, 71T, 72F
    conformational models, 62, 63F
    effects on growth, 308–309, 309F
    enzymatic release, 47, 69–71
    nomenclature, 70P
remodeling during cell expansion, 290–291, 290F
secreted by cultured cells, 53
storage, 84, 303
structure, 68–72
    methods of analysis, 47, 53, 58, 69–71
    methylation analysis, 56P
wall architecture, 238–239
Xyloglucan-endohydrolase (XEH), 290
Xyloglucan-endo-transglycosylase (XET), 290, 290F, 309
Xyloglucan-fucosyl (XG-Fuc) transferases, 185–186, 185F
Xyloglucan-galactosyl (XG-Gal) transferases, 185–186
Xyloglucan-glucosyl (XG-Glc) synthase, 184
Xyloglucan transglucosylase-hydrolases (XTH), 290, 302
Xyloglucan-xylosyl (XG-Xyl) transferases, 184
D-Xylose, 46P
    conversion to furfural, 400, 400F
β-1,2-Xylose, 179
Xylosyl residues
    assembly, backbone synthesis, 173, 184–185, 185F
    glucuronoxylans, 192
    xylan backbones, 72, 74
    xyloglucan, 69

Y

Yariv reagent, 91, 92F, 341
Yeast, fermentation of cellulose, 400
Yieldins, 293, 311

Z

Zea mays see Maize
Zinc (Zn) finger domains, 196F, 197, 197F
Zinnia
    cellulose synthase complexes, 138F
    leaf architecture, 36F
    lignin biosynthesis, 212
    sucrose synthase, 199F
    transfer cells, 31F
    xylem vessel elements, 28F, 278F, 299F
ZmGRP4, 94
Zygote, 38